Plant Diversity in Biocultural Landscapes

Siva Ramamoorthy • Inocencio E Buot, Jr •
Rajasekaran Chandrasekaran
Editors

Plant Diversity in Biocultural Landscapes

Springer

Editors
Siva Ramamoorthy
Department of Biotechnology, School of Bio Sciences and Technology
Vellore Institute of Technology
Vellore, Tamil Nadu, India

Inocencio E Buot, Jr
Institute of Biological Sciences
College of Arts and Sciences
University of the Philippines Los Baños
Laguna, Philippines

Rajasekaran Chandrasekaran
Department of Biotechnology, School of Bio Sciences and Technology
Vellore Institute of Technology
Vellore, Tamil Nadu, India

ISBN 978-981-19-8648-2 ISBN 978-981-19-8649-9 (eBook)
https://doi.org/10.1007/978-981-19-8649-9

This Springer imprint is published by the registered company Springer Nature Singapore Pte Ltd.
The registered company address is: 152 Beach Road, #21-01/04 Gateway East, Singapore 189721, Singapore

Contents

About the Editors

Siva Ramamoorthy is Professor and Dean at School of Bio Sciences and Technology, Vellore Institute of Technology, India. He obtained his undergraduate and postgraduate degrees in Plant Science from Madurai Kamaraj University. He did his PhD in *Plant Genetic Diversity* from Bharathidasan University and postdoctoral research at Ben Gurion University and Gyeongsang National University. His research interests are plant biotechnology and pigments biology, publishing more than 165 manuscripts in peer-reviewed journals with high impact factor and received many recognitions. In the past 3 years, consequently Dr. Siva has been recognized as one among the Top 2% scientists of the world per Stanford University and Mendeley survey. He is also the associate editor of many reputed journals.

Inocencio E Buot, Jr is professor of botany, curator of the Plant Biology Division Herbarium and head of Plant Systematics Laboratory, Institute of Biological Sciences, a Center of Excellence for Biology since 1984, University of the Philippines Los Baños. He obtained his PhD (Ecology) from Chiba University and his postdoctoral certificate in Landscape Ecology from the Ministry of Forestry in Kyoto under the auspices of the Japan Society for the Promotion of Science. His laboratory is now focusing on biodiversity conservation and plant systematic studies. He collaboratively published more than 200 articles in indexed journals and received numerous academic awards.

Rajasekaran Chandrasekaran is a professor in the Department of Biotechnology at Vellore Institute of Technology. He graduated in Botany at the University of Madras and Annamalai University. He specialized in plant ecophysiology for his doctoral degree at H.N. B. Grahwal University. For 27 years, his laboratory has been focusing on ecophysiology, conservation biodiversity, bioprospecting of medicinal and aromatic plants, phycoremediation, and climate change indicators. He has more than 140 publications in the form of journal articles and book Chapters. For his impactful works, Dr. Rajasekaran has been recognized as Fellow of the Royal Soc of Biology, Linnean Society of London, Indian Society for Plant Physiology, Academy of Sciences Chennai, National Academy of Biological Sciences, among others. He is a commission member of many theme groups of IUCN and also a recipient of many awards.

Part I

Biocultural Landscapes During the Anthropocene

Understanding Plant Diversity Dynamics in Biocultural Landscapes During the Anthropocene

1

Inocencio E Buot, Jr, Rajasekaran Chandrasekaran, and Siva Ramamoorthy

Abstract

Plant diversity in biocultural landscapes faces a critical challenge at the onset of the Anthropocene with its numerous challenges aggravated by the climate change issues. Two opposing perspectives or philosophies, i.e., *anthropocentrism* and *ecocentrism*, prevail. Hence, understanding plant diversity dynamics is imperative in order to formulate appropriate and strategic solutions beneficial to all organisms on earth. The chapter aims to look into the following objectives: (1) describe plant diversity in biocultural landscapes, (2) investigate reports on plant diversity responses during the Anthropocene, and (3) propose a framework on how to manage plant diversity in biocultural landscape during the Anthropocene. Results indicated that in the olden times, plant diversity has been protected and enhanced by the local people through direct mentoring, education, and conversation with elders. During the Anthropocene, triggered with population explosion, there have been more land use type conversion, leading to reduction of plant population size and even species extinction. Species distribution has been altered especially along elevational gradients of many tropical mountains. Lower dominants ascend to higher elevations when disturbances destroyed upper forest zones. Succession at various seral stages seems unconsummated destabilizing ecosystem structure and processes, remarkably diminishing ecosystem services. A framework to harmonize anthropocentrism and ecocentrism is proposed using systems thinking approach, to

I. E. Buot, Jr (✉)
Institute of Biological Sciences, College of Arts and Sciences, University of the Philippines Los Baños, Laguna, Philippines
e-mail: iebuot@up.edu.ph

R. Chandrasekaran · S. Ramamoorthy
Department of Biotechnology, School of Bio Sciences and Technology, Vellore Institute of Technology, Vellore, Tamil Nadu, India

S. Ramamoorthy et al. (eds.), *Plant Diversity in Biocultural Landscapes*,
https://doi.org/10.1007/978-981-19-8649-9_1

sustainably manage plant diversity in biocultural landscapes amidst the escalating challenges brought about by the Anthropocene epoch. The framework which appears idealistic can evolve to a more detailed and pragmatic one incorporating philosophical, sociopsychological, and economic theories and realities.

Keywords

Anthropocentrism · Ecocentrism · Bioethics · Biodiversity · Elevational gradients · Human population · Systems thinking

1.1 Introduction

Plant diversity refers to the variety of plant species in an ecosystem landscape. In nature, species are always occurring in association with other species which they have evolved with through the years. They had been experiencing the same set of ecosystem processes and had been interacting and adapting to some changes of the surroundings to maintain homoeostasis and be able to perpetuate life. At certain times, there are rough and tumultuous events and plants need to overcome seemingly insurmountable environmental disturbances together as a plant association and as a community.

In some landscapes, plant diversity is believed to be harmoniously related with the village people. And hence, plant diversity has always been sustained by the people. This is particularly true in biocultural landscapes or cultural landscapes to some (Birnbaum 1994), where humans and nature, as has been said, make adjustments and have the so called give and take relationship for the purpose of securing shelter, food, and/or for pleasure (Ciftcioglu et al. 2016). Humans take only what they need, seeing to it that nature and its ecosystem services are never compromised with whatever activity has been planned and done for the purpose of just getting what the family or the village need. People also protect what they deemed essential for their daily existence. These essential plants could be for food (for humans, farm animals, and pets), medicine, shade plants, construction materials, home for birds or other wild organisms, or simply, plants for occasional rituals like dances and village festivities for the Gods and the deities which they consider important and obligatory.

However, during the Anthropocene where disturbances and all forms of ecosystem landscape degradation abound, apprehensions can arise, whether the harmony between humans and nature is sustained (Radcliffe 2000; Buot 2008a, b; Buot and Osumi 2004; Buot and Osumi 2011; Buot and Buhay 2022; Ellis 2018). This is especially fueled by the aggravating and alarming climate change crises that beset the world and its populace. Several studies had been conducted to determine the current condition of plant diversity in biocultural landscapes (Buot and Okitsu 1998, 1999; Buot 2007, 2008a, b, c; Banaticla and Buot 2004, 2005; Buot and Osumi 2004, 2011; Malaki and Buot 2011; delos Angeles and Buot 2015, 2019; Chanthavong and Buot 2017, 2019a, b; Obena and Buot 2019; Vallena and Buot

2018; Martinez and Buot 2018a, b; Sulistiyowati and Buot 2013, 2016; Cadiz and Buot 2009; Sopsop and Buot 2009, 2011a, b; Doydee and Buot 2011; Santiago and Buot 2018a, b; Caringal et al. 2019, 2020, 2021; Villanueva and Buot 2020; Obemio and Buot 2021; Lillo et al. 2020, 2021; Villanueva et al. 2021a, b). In Mount Pulag, the highest mountain peak on Luzon Island, Philippines, and a revered biocultural landscape of the local tribe in the Cordilleras, shifting cultivation has been observed in steep slopes of up to 2300 masl (Buot and Okitsu 1998, 1999). The elevational zones of woody vegetation have been altered contrary to what is normal in the tropics (Buot 2001). The same is true with Mount Akiki, a high peak near Mount Pulag. Vegetable farms are frequently seen on steep slopes (Buot 2007; Buot and Buhay 2022). And again, the elevational zones of woody vegetation are quite anomalous. Pines ascended in higher altitudes following cutting of oaks for farms. This is very similar to what is happening in Mount Pulag (Buot 2014) and perhaps in many mountain peaks in the Cordillera mountain range on Luzon Island, Philippines. Incidentally, similar observations were recorded for Mount Makiling (Buot 2001; Buot and Osumi 2011; Lasco et al. 2001; Luna et al. 1999; Payawal 1983; Payawal and Markgraf 1981), Mount Mayon (Buot 2008c), Mount Tabunan (Cadiz and Buot 2009), and in Aborlan Guba System in Palawan (Sopsop and Buot 2009, 2011a, b). This means that even biocultural landscapes where plant diversity is supposedly protected by the local peoples have never been spared from the ongoing massive ecosystem landscape degradation brought about by the search for more food sources and spaces for food production.

Indeed, there is a need to investigate plant responses to extensive human disturbances in many ecosystem landscapes. This will give us insights and various leads on how to craft appropriate landscape approaches and management strategies to sustain plant diversity amidst the escalating challenges brought about by the Anthropocene.

The chapter aims to look into the following objectives: (1) describe plant diversity in biocultural landscapes, (2) investigate reports on plant diversity responses during the Anthropocene, and (3) propose framework on how to manage plant diversity in biocultural landscape during the Anthropocene.

1.2 Monitoring Plant Diversity in Biocultural Landscapes

Plant diversity is a unique property of a community which needs to be assessed, monitored, and evaluated periodically (Buot 2008a; Morin 1999). Monitoring plant diversity especially in biocultural landscapes is crucial since provision of adequate ecosystem services would all depend on the status of plant diversity, faunal diversity, and microbial diversity of any ecosystem landscapes (Ciftcioglu et al. 2016). Checking plant diversity in biocultural landscape is of interest. It is given that plant diversity is only attained when there is a good interaction of the biophysical, sociocultural, and politico-economic components of the environment. When diversity is good, then ecosystem services (food, medicine, unpolluted freshwater and fresh air, etc.) will be abundant, resulting to a high degree of community well-being

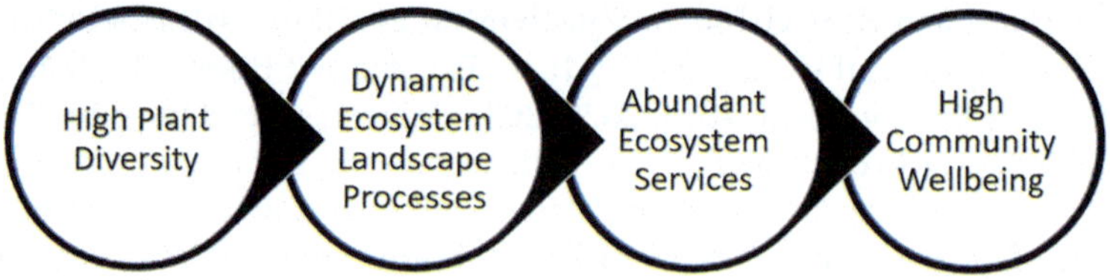

Fig. 1.1 Impact of high plant diversity to ecosystem/landscape processes, ecosystem services, and community well-being. Note that this is the ideal case scenario which in reality, considering the Anthropocene, does not exist. See discussions for suggested activities to attain this

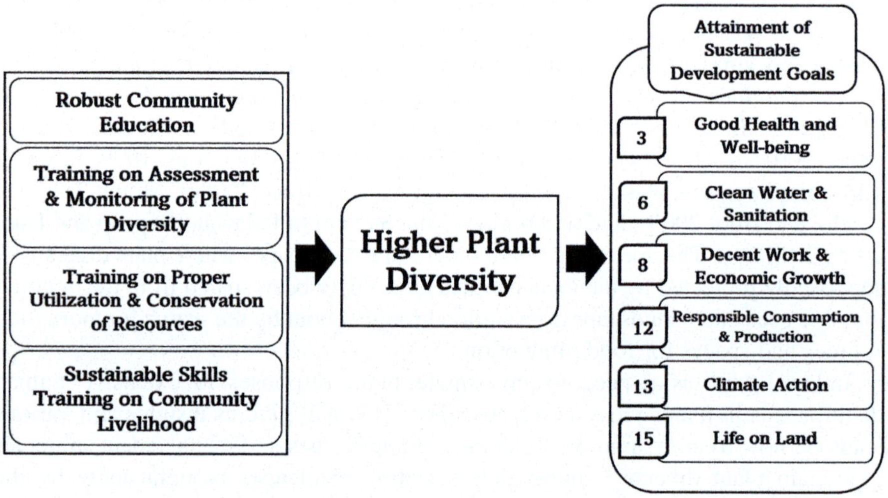

Fig. 1.2 A framework showing what are urgently needed to have a higher plant diversity that would result in achievement of the UN Sustainable Development Goals

(Buot and Buot 2022; Buot et al. 2020; Buot and Dulce 2019; Buot and Cardenas 2018; Buot 2017; Buot et al. 2017). This is illustrated in Fig. 1.1. In these papers, Buot and colleagues were using the community capital framework (Flora and Flora 2013) in evaluating the well-being of the community, comprising natural, sociocultural, financial, political, and built capitals. They have been emphatic in stressing the significance of a good and diverse forest cover (natural capital) to incredibly increase community well-being index (cwbi). However, this is the ideal case scenario. In reality, the ideal does not exist. We have to take into account and understand the impact of several intervening processes and disturbances, natural or man-made.

It is therefore imperative for everyone across age groups, gender classes, sectors and beliefs to actively participate in aggressively sustaining a high plant diversity in biocultural landscapes. Some activities contributing towards this end include robust community education and awareness program, persuasive local training on biodiversity assessment, monitoring and proper utilization and conservation, and sustained skills training on community livelihood, among others (Fig. 1.2). These activities are the components of many sustainable biodiversity conservation and management frameworks (Buot 2008a, b; Fowler 2009; Flora and Flora 2013; Hong 2007, 2011, 2014). Doing these will indeed contribute to the achievement of the UN Sustainable Development Goals 3 (good health and well-being), 6 (clean water and sanitation), 8 (decent work and economic growth), 12 (responsible consumption and production), 13 (climate action), and 15 (life on land).

1.3 Plant Diversity Responses During the Anthropocene

Plant diversity responds to whatever stimulus in the surrounding environment (May and McLean 2007). When the environment is well and good, generally, plant diversity is well and good too, assuming that management implements sustainable practices. Corollarily, there is better ecosystem structure and processes resulting to abundant ecosystem services and excellent community well-being (Fig. 1.3). However, when the adjacent surroundings are exposed to destructive activities, such as logging, shifting cultivation, mining, flooding, etc., plant diversity is destroyed and endangered of degradation and even extinction. This is followed by disruption of dynamic biogeochemical cycling (Morin 1999), resulting to low productivity of food, medicine and other necessities; air and water pollution; vulnerability to global warming and drastic climatic changes, prevalent during the Anthropocene; and decreasing community well-being of the locals. This is critical, especially in biocultural landscapes, where there are human populations largely dependent on the natural resources and the biodiversity. Hence, investigating the plant diversity dynamics in biocultural landscapes, during the Anthropocene, is very important.

Plant diversity responses can be very erratic (Fig. 1.3), with the aggravating climate change issues during the Anthropocene (Ellis 2018). Several studies revealed varying responses of plant species diversity to varying sets of disturbances. The most common is the stunting of growth (Stewart and Globig 2021), remarkable

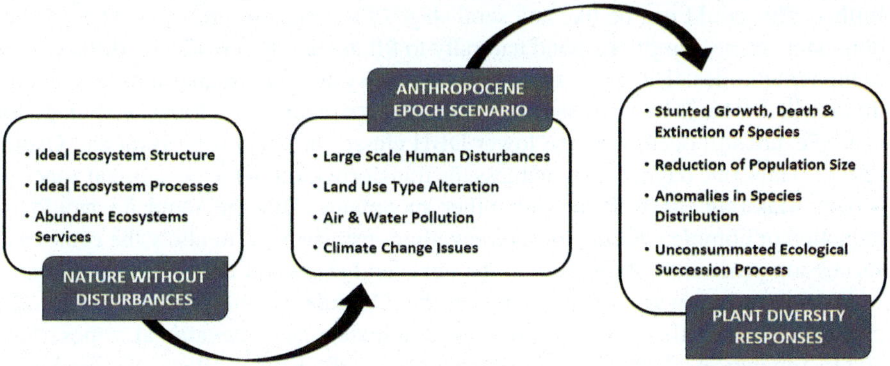

Fig. 1.3 A diagram illustrating some plant diversity responses to disturbances brought about by the Anthropocene

reduction of population size (May and McLean 2007), decreasing diversity (Buot and Osumi 2011), and in extreme cases, death and extinction (May and McLean 2007; Buot and Buhay 2022; Caringal et al. 2021). In terms of elevational zonation of vegetation, studies reported anomalies in species distribution. Lower dominants ascend to higher elevation and occupy the niche of the higher elevation dominants destroyed with shifting cultivation, deforestation, mining, or other deliberate change of land use (Buot 2007, 2008c; Buot and Okitsu 1998; Buot and Osumi 2004; Banaticla and Buot 2005; Cadiz and Buot 2009; Sopsop and Buot 2011a, b). A drastic change in microclimatic conditions occurs with changes in land use types in upper elevations favorable for the lower dominants to colonize and proliferate.

In Mount Makiling, *Diplodiscus* climbed up to the niche of the dipterocarps when the latter was logged over in the 1940s (Payawal 1983; Payawal and Markgraf 1981; Luna et al. 1999; Lasco et al. 2001; Buot and Osumi 2011; Castillo et al. 2018). *Pinus* of Mount Pulag, Mount Akiki, and other peaks of the Cordillera mountain range, ascended up to more than 2400 masl, replacing the oaks after cutting and high altitude farming (Buot 2001; Buot and Okitsu 1998, 1999; Tamayo et al. 2021). In Mayoyao, Ifugao, a UNESCO World Cultural Heritage Site, several invasives had been reported in the lower elevation of the *pinuchu* or *muyong*, a managed forest believed to sustain soil fertility and water on the age-old culture-laden rice terraces paddies (Santiago and Buot 2018a, b), a similar scenario in Mount Ilong of the Halcon Range, in Mindoro Island (Villanueva and Buot 2018). In southern Luzon (Buot 2008c, 2014), there is ascent of *Astronia* populations from lower (500–800 masl) to higher altitudes (900–1500 masl). This is a very similar pattern as that of the Cordillera mountains of the Philippines. In Cebu Island, Cadiz and Buot (2009) reported the same trend for *Artocarpus* from Zone 2 (ca 580 masl) to Zone 4 (ca 700 masl) of Mount Tabunan, a remnant of a primary forest but now has been subjected to some sort of human disturbances as well. In Palawan island, Sopsop and Buot (2013) reported *Artocarpus* to dominate at the lower elevation (150 masl) since the mountain (Aborlan Guba System) has been sustainably managed by the Tagbanua tribe, one of the indigenous people in Palawan.

This alteration of the elevational vegetation zonation pattern should be studied further. This could just be like any seral stage of a succession process. These lower dominant species simply respond naturally to fill in the gaps created by disturbances in higher altitudes. This means that once the ecosystem landscape stabilizes, everything will be back to normal as well. The dipterocarps of Mount Makiling, for example, are still occupying the lower DBH classes in a report by Buot and Osumi (2011). Thus, one day, we may witness the transformation of the elevational zones of Mount Makiling, as well as with other mountains. But the requisite would be reduction or complete halting of the prevailing disturbances, to allow the ecological succession process to progress onto the next seral stage and up to the climax state.

We are more wary of the fate of the Cordillera mountains, knowing the characteristics of the pine, which now dominate many elevational zones (Paw 2022; Tamayo et al. 2021; Buot and Okitsu 1999; Kowal 1966). As we know, pines are aggressive successional species, but can be topo-edaphic climax as well, as has been observed in the case of *Pinus taiwanensis* in Taiwan (Hsieh et al. 1994).

Buot and Okitsu (1998, 1999) argued that the proliferation and dominance of *Pinus* along the steep slopes of the Cordillera is a demonstration that it is a topo-edaphic climax. The topography is very steep and the original oaks have difficulty in reestablishing in this stressful environs. Kowal (1966) maintains that pines are simply successional species in the Cordillera, citing the conclusion that pines are simply result of human disturbance and fire (Bogaert et al. 2014; Hong 1998, 2001; Hong et al. 1995; Denevan 1961) in areas with shallow, poor soils, and low temperature. Kowal (1966) argued that if fire and human disturbances will be eliminated in the Cordillera mountains, the direction would be towards the development of a montane forest mostly dominated by oaks. Much work has to be done to better understand the pines and the ecology of the Cordillera mountains. Robust studies that will elucidate the plant diversity responses to huge disturbances brought about by the Anthropocene in biocultural landscapes are urgent.

1.4 Framework in Managing Plant Diversity in Biocultural Landscapes During the Anthropocene

Currently, two worldviews or perspectives prevail on how humans view our ecosystem landscapes, including, biocultural landscapes (Fig. 1.4). Some scholars refer to these as the philosophies of anthropocentrism (Kopnina et al. 2018; Simkins 2014; Boddice 2011; Grey 1993; Routley and Routley 1980) and ecocentrism (Milstein and Castro-Sotomayor 2020; Smith 2019; Lupinacci 2017; Frim 2017;

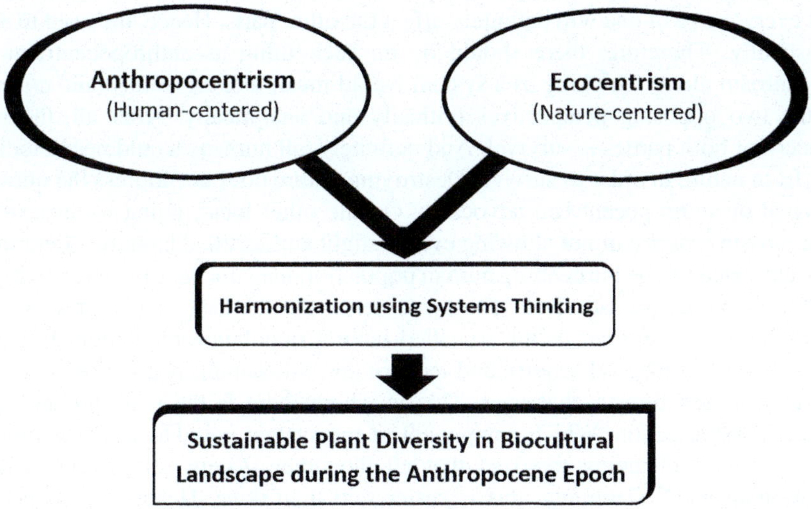

Fig. 1.4 A framework harmonizing the philosophies of anthropocentrism and ecocentrism using systems thinking to achieve sustainable plant diversity and all forms of life in biocultural landscapes during the critical times of the Anthropocene. See discussion for lacking details in the framework, like interdisciplinary and transdisciplinary brainstorming activities and negotiations

Shoreman-Ouimet and Kopnina 2016; Rowe 1994). Anthropocentrism believes that humans are the main element of the planet (Kopnina et al. 2018) and argues on the right of humans to live at all costs. Ecocentrism, on the other hand, is nature-centered (Milstein and Castro-Sotomayor 2020; Smith 2019; Shoreman-Ouimet and Kopnina 2016) and believes that there is no separation between nature and humans with the other beings. It recognizes that all species, including human race, are results of long history of coevolution with nature, and hence, focuses on the living community and works to sustain ecosystem structure and processes which would help sustain humans and all life on earth. As a philosophy, ecocentrism is ideal. But in practice, sometimes there could be the tendency to go to the extremes. Anthropocentrism and ecocentrism are two opposing perspectives causing problems in sustainable management of plant diversity in biocultural landscapes, particularly, during the critical times under the Anthropocene. Consequently, policy makers might get confused with which perspective has to be considered in the crafting of relevant bills and laws, like allocating alienable and disposable land to be used for human habitation, for example, from mangrove forests or from portions of mountain landscapes, or the construction of road network across mountain landscapes with rare native species.

For the planet earth and all the biotic flora, fauna, microbes, including every one of us to survive, something urgent has to be done. As Fig. 1.4 illustrates, there is an immediate need to harmonize anthropocentrism and ecocentrism views. Perhaps, there has to be a sort of compromise. The systems thinking approach (Government Office for Science 2012; Ramage and Shipp 2009; Fowler 2009) can be the answer to sustain plant diversity and all life in biocultural landscapes during the critical times of the Anthropocene epoch. Systems thinking approach will see the whole as a system with interconnected and deeply interrelated component parts. This means that every move of one will definitely affect the other parts. Hence, the need to think holistically. Therefore, there should be no such thing as anthropocentrism and ecocentrism alone. Thinking as a system would mean looking at the valid concerns of the two opposing perspectives critically and analytically. After all, the main concern of both parties is survival. And definitely, all humans would need biodiversity from nature in order to survive. Destroying nature does not address the quest for survival of anthropocentrism advocates. On the other hand, going to the extreme ecocentrism practice of not allowing even a slight and justified land use alteration to allocate space for the burgeoning human population may not contribute favorably, as well to human race survival. The two views should have a compromise. For example, in nature, there is the so-called natural population regulation. But, with the advances in medical science and technology, this natural population regulation mechanism has been a challenge. This is where there is the need for an urgent systemic examination. Which component of the system now has to act in order to address this challenging trend so that all the other system components will be functioning well? Definitely, this is easier said than done. Hence, the framework (Fig. 1.4) as presented appears impractical. The many interdisciplinary and transdisciplinary brainstorming activities and negotiations are not shown, but are understood as part of the framework. The state of the social, cultural, politico-economic, and natural environments of the biocultural landscape affects the quality of the conversation and discussions, as well as the ultimate decision making process. But at the very least, the framework guides us in showing the broad strokes.

Once these two opposing perspectives are harmonized, the planet earth will be an ideal ecosystem with diverse structure and dynamic processes, enhancing rich plant biodiversity and abundant ecosystem services for everyone, despite the challenges brought about by the Anthropocene.

1.5 Summary and Conclusion

Plant biodiversity in biocultural landscapes is important for the proper functioning of the ecosystem processes, resulting in abundant ecosystem services enhancing community well-being. Knowing the numerous challenges associated with the Anthropocene, scientific reports on plant diversity responses to stresses even in biocultural landscapes were investigated. Notable responses include stunted growth, reduction of population size, and anomalous species distribution and unconsummated ecological succession. A framework using the systems thinking approach is proposed to harmonize anthropocentrism and ecocentrism, the two prevailing perspectives or philosophies affecting the management of plant biodiversity in biocultural landscape during the Anthropocene. Although the framework presented is more ideal than practical, the discussions point out the realities that may hinder attainment of the main goal, and therefore should be taken into consideration. In the future, the framework can be revised to have a more pragmatic approach incorporating philosophical, sociopsychological, and economic theories and realities.

Acknowledgments The authors are thankful to the late Dr. Anacleto Caringal of Batangas State University for a fruitful discussion about plant diversity, anthropocentrism, ecocentrism, and systems thinking. Dr. Emilio Padoa-Schioppa of the University of Milan, once Visiting Professor of the University of the Philippines, has introduced the senior author to the interesting studies on the impact of the Anthropocene on biodiversity. The assistance of the University of the Philippines Los Banos' Conserve-Kaigangan Research Staff, Forester Marne G. Origenes and Ms. Ren Divien R. Obena, in illustrating the frameworks and in ordering the references, respectively, is much appreciated.

References

Banaticla MCN, Buot IE Jr (2004) Fern patch structure and species diversity along the altitudinal gradient on Mount Banahaw de Lucban, Luzon Island, Philippines. Philipp Agric Sci 87:49–60

Banaticla MCN, Buot IE Jr (2005) Altitudinal zonation of pteridophytes on Mount Banahaw de Lucban, Luzon Island, Philippines. Plant Ecol 180:135–151

Birnbaum CA (ed) (1994) Preservation forum. "Focus on landscape preservation", vol 7, no 3. National Trust for Historic Preservation, Washington, DC. https://www.nps.gov/tps/how-to-preserve/briefs/36-cultural-landscapes.htm. Accessed 23 Aug 2022

Boddice R (ed) (2011) Anthropocentrism: humans, animals, environments. Brill, Leiden, 348p

Bogaert JB, Vranken I, Andre M (2014) Anthropogenic effects in landscapes: historical context and spatial pattern. In: Hong SK, Bogaert K, Min Q (eds) Biocultural landscapes: diversity, functions and values. Springer, Berlin

Buot IE Jr (2014) Conservation and sustainability of socio-ecological production landscapes and seascapes. Int J Ecol Conserv 10:93–102

Buot MM (2017) Benchmarking concepts of community wellbeing in an area vulnerable to disaster due to volcanic eruption. Int J Environ Rural Dev 8(1):58–62

Buot MM, Buot IE Jr (2022) Volcanic seascapes: disaster recovery for community wellbeing and habitat protection. In: Pungetti G (ed) Routledge handbook of seascapes. Routledge, New York, pp 385–398

Buot MM, Cardenas VR (2018) Community Wellbeing Index (CWBi) in the area exposed to Mayon Volcanic eruption in Camalig, Albay, Philippines. Sylvatrop 28(2):27–46

Buot MM, Dulce MZ (2019) An index to determine community wellbeing along coastal community in Leyte, Philippines. Environ Asia 12(1):56–67

Buot IE Jr (2001) Latitudinal comparison of the vertical vegetation zonation on Mt. Pulog (Philippines) with other southeast and east Asian mountains. Philipp Sci 38:9–25

Buot IE Jr (2007) Vegetation zonation of Mt. Akiki, Cordillera Mountain range, Benguet Province, Philippines. Asia Life Sci 16(1):53–70

Buot IE Jr (2008a) A new way of looking at environmental health: focused on man and his environment. Asia Life Sci Suppl 2:1–5

Buot IE Jr (2008b) Sustaining environmental health in Philippine Satoyama landscapes. Asia Life Sci Suppl 2:129–138

Buot IE Jr (2008c) Vertical distribution and zonation pattern of woody vegetation on the north-western slope of Mt. Mayon, Philippines. Asia Life Sci 17:189–205

Buot IE Jr, Buhay AFV (2022) Types of socioecological production landscapes of the Philippines based on dominant biodiversity: status, problems and future directions. Biodiversitas 23(7): 3755–3770

Buot IE Jr, Okitsu S (1998) Vertical distribution and structure of the tree vegetation in the montane forest of Mt. Pulag, Cordillera Mountain range, the highest mountain in Luzon Is., Philippines. Veg Sci 15:19–32

Buot IE Jr, Okitsu S (1999) Leaf size zonation pattern of woody species along the altitudinal gradients of Mt. Pulag, Philippines. Plant Ecol 145:197–208

Buot IE Jr, Osumi K (2004) Satoyama landscapes in the Philippines. J Nat Stud 3:1–11

Buot IE Jr, Osumi K (2011) Land use type pattern and woody species composition near human disturbed landscapes on Mount Makiling, Luzon Island. Am J Environ Sci 7(4):306–315

Buot MM, Cardenas VR, Nelson GLM, Quimbo MAT, Dizon JT (2017) Developing Community Wellbeing Index (CWBi) in disaster-prone areas of the Philippines. J Nat Stud 16(1):63–75

Buot MM, Cardenas VR, Dizon JT, Quimbo AT, Nelson GLM (2020) Multisectoral participation in the development of an index for community wellbeing. In: Buot IE (ed) Methodologies supportive of sustainable development in agriculture and natural resources management: selected cases in Southeast Asia. Southeast Asian Regional Center for Graduate Study and Research in Agriculture (SEARCA) and the University of the Philippines Los Baños (UPLB), Laguna, pp 243–260

Cadiz GO, Buot IE Jr (2009) An enumeration of the woody plants of Cantipla forest fragments, Cebu Island, Philippines. Philipp J Syst Biol 3:1–7

Caringal AM, Buot IE Jr, Villanueva ELC (2019) Woody plant communities in the Philippine teak forest landscape along Verde Island Passage, Batangas, Luzon, Philippines. Biodiversitas 20(11):3189–3198

Caringal AM, Buot IE Jr, Villanueva ELC (2020) Analysis of human and Philippine teak forest interaction in the lasang-Baybay landscape along Verde Island Passage Marine Corridor, Batangas Province, Philippines. J Mar Island Cult 9(1):1–19

Caringal AM, Buot IE Jr, Villanueva ELC (2021) Endemic Philippine teak (*Tectona philippinensis* Benth. & Hook. F.) and associated flora in the coastal landscapes of Verde Island Passage, Luzon Island, Philippines. Curr Sci 120(6):1057–1065

Castillo ML, Castillo LA, Canceran MS, Gonzalvo KJP, Barua LD, Alegre AC, Barredo-Parducho VO, Gestiada EC, Breva RV, Bantayan NC (2018) Distribution, diversity and spatial analysis of tree species in a long-term ecological research plot in Molawin-Dampalit Watershed, Mount Makiling Forest Reserve. Asian J Biodivers 9:12–36

Chanthavong S, Buot IE Jr (2017) Plant diversity in Dong Na Tard provincial protected area, Lao People's Democratic Republic (Lao PDR): species structure and Forest zonation. Environ Asia 10:52–62

Chanthavong S, Buot IE Jr (2019a) Priority areas for conservation planning in Dong Na Tard provincial protected area, Lao People's Democratic Republic (Lao PDR). Environ Asia 12(2): 116–125

Chanthavong S, Buot IE Jr (2019b) Conservation status of plant diversity at Dong Na Tard provincial protected area, Lao, PDR. Int J Conserv Sci 10:393–402

Ciftcioglu GC, Uzun O, Nemutlu FE (2016) Evaluation of biocultural landscapes and associated ecosystem services in the region of Suğla Lake in Turkey. Landsc Res 41(5):538–554

delos Angeles MD, Buot IE Jr (2015) Diversity and distribution of pteridophytes altitudinal gradient of a secondary forest in Mt Makiling, Philippines. Int J Ecol Conserv 16:25–46

delos Angeles MD, Buot IE Jr (2019) A checklist of Fern species in disturbed areas of mount Makiling, Philippines. Thailand Nat Hist Mus J 13(2):163–172

Denevan WM (1961) The upland pine forests of Nicaragua, a study in cultural plant geography. Geography 12:251–320

Doydee P, Buot IE Jr (2011) Mangrove vegetation zones in Ranong Coastal Wetland Ecosystem, Thailand. KU Fish Res Bull 35:14–28

Ellis EC (2018) Anthropocene, a very short introduction. Oxford University Press, Oxford, 183p

Flora CB, Flora JL (2013) Rural communities: legacy and change, 4th edn. Westview Press, Boulder

Fowler CW (2009) Systemic management-sustainable human interactions with ecosystems and the biosphere. Oxford University Press, New York, 295p

Frim L (2017) Humanism, biocentrism, and the problem of justification. Ethics Policy Environ 20(3):243–246

Government Office for Science (2012) Introduction to systems thinking. Report of GSE and GORS seminar. Civil Service Live, 3 July 2012

Grey W (1993) Anthropocentrism and deep ecology. Australas J Philos 71(4):463–475. https://doi.org/10.1080/00048409312345442. Archived from the original on 2008-07-26. Accessed 2006-11-12

Hong SK (1998) Changes landscape patterns and vegetation process in the far-eastern cultural landscapes: human activity, vol 28, p 45

Hong SK (2001) Factors affecting landscape changes in Central Korea: cultural disturbance on the forested landscape systems. In: Zonneveld IS, van der Zee D (eds) Landscape ecology applied in land evaluation, development and conservation. ITC Publisher, Enschede

Hong SK (2007) Linking man and nature landscape systems: landscaping blue-green network. In: Hong SK et al (eds) Landscape ecological applications in man-influenced areas: linking man and nature systems. Springer, Dordrecht, pp 505–523

Hong SK (2011) Biocultural diversity and traditional ecological knowledge in island regions of Southwestern Korea. J Ecol Field Biol 34(2):137–147

Hong SK (2014) Philosophy and background of biocultural landscapes. In: Hong SK, Bogaert K, Min Q (eds) Biocultural landscapes: diversity, functions and values. Springer, Berlin

Hong SK, Nakagoshi N, Kamada M (1995) Human impacts on pine-dominated vegetation in rural landscapes in Korea and western Japan. Vegetatio 116:161–172

Hsieh C-F, Shen C-F, Yang K-C (1994) Introduction to the flora of Taiwan, 3: floristics, phytogeography and vegetation. Flora Taiwan 1:7–18

Kopnina H, Washington H, Taylor B, Piccolo JJ (2018) Anthropocentrism: more than just a misunderstood problem. J Agric Environ Ethics 31:109–127

Kowal NE (1966) Shifting cultivation, fire and pine forest in the Cordillera Central, Luzon, Philippines. Ecol Monogr 36:383–419

Lasco RD, Visco RG, Pulhin JM (2001) Secondary forests in the Philippines: formation and transformation in the 20th century. J Trop For Sci 13:652–670

Lillo EP, Malaki AB, Alcazar SMT, Rosales R, Redoblado BR, Diaz JLB, Pinote JP, Rosales R, Buot IE Jr (2020) Native trees in Nug-as forest key biodiversity area, Cebu Philippines. Biodiversitas 21(9):4162–4167

Lillo EP, Malaki AB, Alcazar SMT, Rosales R, Redoblado BR, Diaz JLB, Pantinople EM, Buot IE Jr (2021) Inventory of native and mother trees in key biodiversity areas of Cebu Island, Philippines for species selection in local reforestation programs. Biodiversitas 22:4740–4749

Luna AC, Osumi K, Gascon AF, Lasco RD, Palijon AM (1999) The community structure of a logged-over tropical rain forest in Mt. Makiling Forest Reserve, Philippines. J Trop For Sci 11: 446–458

Lupinacci JJ (2017) Addressing 21st century challenges in education: an Ecocritical conceptual framework toward an ecotistical leadership in education. J Transform Prof Pract 2(1):20–27

Malaki ABB, Buot IE Jr (2011) Conservation status of indigenous trees in Argao River Watershed Reserve, Cebu Island, Philippines. Asia Life Sci Suppl 6:45–59

Martinez MR, Buot IE Jr (2018a) Mangrove assessment in Manamoc Island for coastal retreat mitigation. J Mar Island Cult 7(1):65–83. https://doi.org/10.21463/jmic.2018.07.1.05

Martinez MR, Buot IE Jr (2018b) Altitudinal distribution of some Hoya species in the Philippines. Thailand Nat Hist Mus J 12(1):9–18

May RM, McLean AR (2007) Theoretical ecology. Oxford University Press, Oxford, 257p

Milstein T, Castro-Sotomayor J (2020) Routledge handbook of ecocultural identity. Routledge, London. https://doi.org/10.4324/9781351068840

Morin PJ (1999) Community ecology. Wiley-Blackwell Press, Hoboken

Obemio CD, Buot IE Jr (2021) Checklist of pteridophytes in Mount Matutum protected landscape, Southern Mindanao, Philippines. Sylvatrop 31(2):87–124

Obena RDR, Buot IE Jr (2019) Enumeration of Hoya species in Mindanao Island, Philippines: conservation concerns. Biodiversitas 20(6):1699–1707

Paw JN (2022) Spread of winged seeds: the biogeography of pine trees in the Philippines. Gorilla Trading, Diliman Quezon City, 133p

Payawal PC (1983) Vegetation and modern pollen rain of Mt. Makiling, Philippines. II. Pollen analysis of surface soil. Kalikasan 12:317–327

Payawal PC, Markgraf V (1981) Vegetation and modern pollen rain of Mount Makiling, Philippines. 1. Vegetation analysis of the NE slope. Kalikasan 10:255–267

Radcliffe J (2000) Humanity and nature: the problem of alienation. In: Green politics. Palgrave Macmillan, London

Ramage M, Shipp K (2009) Systems thinkers. Open University/Springer, London

Routley R, Routley V (1980) Human chauvinism and environmental ethics. In: Mannison DS, McRobbie M, Routley R (eds) Environmental philosophy. ANU Research School of Social Sciences, Canberra, pp 96–189

Rowe SJ (1994) Ecocentrism: the chord that harmonizes humans and earth. Trumpeter 11(2): 106–107. http://www.ecospherics.net/pages/RoweEcocentrism.html

Santiago JO, Buot IE Jr (2018a) Assessing the status of pinuchu as indicator of socio-ecological resilience of Chaya Socio-Ecological Production Landscape, Ifugao, Philippines. Biodiversitas 19(6):2010–2019

Santiago JO, Buot IE Jr (2018b) Conceptualizing the socio-ecological resilience of the Chaya Rice terraces, a socio-ecological production landscape in Mayoyao, Ifugao, Luzon Island, Philippines. J Mar Island Cult 7(1):107–126

Shoreman-Ouimet E, Kopnina H (2016) Culture and conservation: beyond anthropocentrism. Routledge, Oxon, pp 8–9. ISBN 9781315858630

Simkins R (2014) The Bible and anthropocentrism: putting humans in their place. Dialect Anthropol 38(4):397–413. https://doi.org/10.1007/s10624-014-9348-z. JSTOR 43895115. S2 CID 145211787. Retrieved 15 September 2020

Smith W (2019) The role of environment clubs in promoting ecocentrism in secondary schools: student identity and relationship to the earth. J Environ Educ 50(1):52–71

Sopsop LB, Buot IE Jr (2009) The endangered plants of Palawan Island, Philippines. Asia Life Sci 18(2):251–279

Sopsop LB, Buot IE Jr (2011a) The importance of non-wood Forest products in the household economy of the direct users of Aborlan Guba System, Palawan Island, Philippines. J Environ Sci Manag 14(2):50–59

Sopsop LB, Buot IE Jr (2011b) Human forests interaction in Aborlan-Guba System, Palawan Island, Philippines: implications for conservation and management. Asia Life Sci 20(1): 153–171

Sopsop LB, Buot IE Jr (2013) The forest types in Aborlan Guba System, Palawan Island, Philippines. Int J Ecol Conserv 7:88–104

Stewart P, Globig S (2021) Plant physiology. Apple Academic Press, Florida, 298p

Sulistiyowati H, Buot IE Jr (2013) Integrated biodiversity valuation framework: ecological approach. J Wetlands Biodivers 3:7–16

Sulistiyowati H, Buot IE Jr (2016) Ecological valuation tools to appraise biomass and soil organic matter in a natural forest ecosystem. J Wetlands Biodivers 6:97–108

Tamayo MN, Agcaoili JMM, Tandang DN, Baoanan ZG (2021) New insights on the landscapes and flowering plants of Mount Pulag National Park. PABIDACO, Cordillera Studies Center, UP Baguio, Baguio City, 138p

Vallena MG, Buot IE Jr (2018) Checklist and conservation status of Hoya species in Bicol Peninsula, Luzon Island, Philippines. Thailand Nat Hist Mus J 12(2):91–102

Villanueva ELC, Buot IE Jr (2020) Useful plants of the Alangan Mangyan of Halcon Range, Mindoro Island, Philippines. J Mar Island Cult 9(1):76–102

Villanueva ELC, Fernandez DAP, delos Angeles M, Tolentino PJS, Obeña RDR, Buot IE Jr (2021a) Checklist of the flora and fauna of the Karst Forests in Basey, Samar, Philippines. Thailand Nat Hist Mus J 15(2):147–160

Villanueva ELC, Fernandez DAP, delos Angeles M, Tolentino PJS, Obeña RDR, Buot IE Jr (2021b) Biodiversity in forests over limestone in Paranas, Samar Island Natural Park (SINP), a UNESCO World Natural Heritage Site Nominee. Trop Nat Hist 21(1):119–145

Building Sustainability in Community-Managed Mangrove Forest Using Biocultural Approach to Conservation

2

Mario A. Cudiamat and Blesshe L. Querijero

Abstract

Building sustainable development in mangrove conservation requires an integrated understanding of the different factors affecting socio-ecological system such as the case of community-managed mangrove forest. In the context of this study, biocultural approach to conservation links local people and the mangrove environment. This study aims to assess various factors influencing biocultural approach to mangrove conservation such as socioeconomics, conservation practices, local ecological knowledge, gender roles, and management. It employed mixed method research design to describe these factors. The results revealed that the local community has sufficient awareness on mangrove forest in terms of ecosystem services and local knowledge which significantly affect how they utilize and manage their resources. Conservation and management of mangrove using community-based conservation are evident in the stakeholders' practices and activities. The use of biocultural approach to conservation shows potential to build sustainability.

Keywords

Mangrove · Sustainability · Biocultural · Conservation · Local community

---

M. A. Cudiamat (✉) · B. L. Querijero
College of Arts and Sciences, University of the Philippines Los Baños, Los Baños, Laguna, Philippines
e-mail: macudiamat@up.edu.ph

S. Ramamoorthy et al. (eds.), *Plant Diversity in Biocultural Landscapes*,
https://doi.org/10.1007/978-981-19-8649-9_2

2.1 Introduction

Biocultural approach refers to conservation actions made in the service of sustaining the biophysical and sociocultural components of social-ecological systems (Gavin et al. 2015). It is a progressive approach to conservation that recognizes and honors the intrinsic relationships between nature and humanity (National Tropical Botanic Garden 2021).

Integrated conservation is a biocultural approach to conservation initiatives that attempt to ensure the conservation of biological diversity by reconciling the management of protected areas (PAs) with the social and economic needs of local people. Integrated community-based conservation can improve environmental integrity, economic efficiency of actions, and equity for local community in sustainable benefit distribution (Nandy and Islam 2010).

Community-led local resource protection can draw participation and collective actions to deal with climate change impacts and build resilience of socially disadvantaged communities (Allen 2006). Management of mangrove resources and sustainable livelihoods of vulnerable social groups are indeed great challenges to promote adaptation and build resilience.

Application of the biocultural approach in community-based management is challenging. The voluntary role of coastal communities in guarding mangrove forests is the ultimate outcome of integrated approach and has been referred to as a positive transformation.

Local communities and stakeholders also develop adaptive capability through their management capacity, supported by effective legislation that enables mangrove protection from non-climate stressors. Effective sustainable management promotes mangrove resilience, a concept linked with adaptive capacity as the ability to absorb and recover from the effects of disturbance (Turner et al. 2000; Intergovernmental Panel on Climate Change (IPCC) 2014).

On the other hand, the lack of community participation is one of the causes of mangrove degradation in many parts of the world (Kairo 2007). Lack of stakeholders' participation and ownership in the conservation project is an issue and leads to different problems in project implementation (Macintosh et al. 2002) and failure to meet the management of a natural resource (Crona 2006; Kairo 2007; Creencia and Querijero 2018).

The long-term goal of coastal adaptation is to increase overall social and ecological resilience. While the need to improve the current state of coastal marine ecosystems and fisheries remains to be a core resource management objective, there is now an urgent need to also focus on how to improve coastal communities' ability to cope and eventually adapt to climate change (Quibilan 2014).

The management of natural resources focuses on the dynamics between natural resources and people. The use of natural resources, and decision-making regarding natural resources, is not shaped purely by biophysical characteristics of the resources, but in large part by who can use and manage them. Unequal relations—usually defined by social differences of gender, ethnicity, socioeconomic class, age, and so on—translate into inequalities among people in their access to and control over resources and thereby influence the benefits and adverse effects that different people and nature experience.

There is growing recognition that natural resource management (NRM) requires a diversity of stakeholders in the decision-making process to meet sustainability objectives. Programs that employ participatory approaches have risen in popularity in response to this. However, contrary to its aim, participatory NRM may build on and reinforce social inequalities unless there is a conscious effort to address them. There is therefore a need for analyses of the power dynamics and sociocultural norms that give rise to social inequalities.

The socio-demographic-economic dimensions are important considerations towards mangrove conservation and utilization (Fortes and Salmo III 2017). A need to look at mangrove management and conservation requires systems approach (Fortes 2010) and holistic or global perspectives to help reduce threats and pressures that lead to vulnerability of mangrove ecosystem as socio-ecological system. Most of the vulnerability assessments proposed in the current literature have focused on the ecosystem itself of the mangroves, its socio-ecological characteristics, and also the economic dimension. Based on the report of the Municipality of Calatagan Coastal Resources Management (2015), the local people's organizations were mainly composed of male fishers whose focus of management is on coral reefs and seaweeds.

Many rehabilitation efforts on mangroves have been done in the Philippines (Primavera and Esteban 2008), but very few of these efforts integrate sustainability and resilient-based approaches. Some of the issues that hinder that integration of sustainability paradigm include insufficient knowledge base at all levels of planning, implementation, monitoring and evaluation, limited focus and scale, lack of consensus among stakeholders on objectives, roles, and responsibilities, inadequate stakeholder participation, insufficient focus on institutional and capacity-building of all key groups involved in conservation, and weak monitoring and evaluation.

As illustrated by Tobin (1999), sustainable and resilient communities are those that are more prepared to minimize the effects of, and recover quickly from, disaster. Thus, to increase local resiliency, there must be careful planning and organization for both immediate and long-term biophysical impacts, stemming from strong social networks within and between segments of society (Tobin 1999).

While the need to improve the current state of coastal marine ecosystems and fisheries remains to be a core resource management objective, there is now an urgent need to also focus on how to improve coastal communities' ability to cope and eventually adapt to climate change using community-based conservation. Thus, assessing the impacts of management of mangroves, integrated community-based conservation is important towards sustainability. This research attempted to bridge this gap through mainstreaming of sustainability criteria and indicators for mangrove forest.

To achieve this, the specific objectives are as follows:

1. To document the socioeconomic profile of the stakeholders of the people's organization;

2. To assess the conservation practices, local ecological knowledge (LEK), and gender roles in context with the criteria and indicators for sustainable mangrove forest; and
3. To evaluate the governance of the community-based mangrove forest using the harmonized Marine Protected Area Management Effectiveness Assessment Tool (MPA MEAT).

2.1.1 Research Design

The present study employed a mixed methods research (MMR) design that uses both quantitative and qualitative inquiries (Tashakkori and Teddlie 2010) to investigate the sociocultural, economic, and political dimensions of a community-managed mangrove forest in Calatagan, Batangas. It specifically utilized a case study approach and descriptive-quantitative design. The concurrent design type of MMR was employed where both qualitative and quantitative data are collected in a single phase.

Because the general aim of this approach was to better understand the sociocultural, economic, and political dimensions of community-based conservation of mangrove ecosystem from the biocultural lens, the data were collected from the community stakeholders of the people's organization (PO). The goal was to obtain different but complementary data that cross-validate the overall results, and so, concurrent triangulation or transformational design should be conducted for corroboration purposes.

There are basically two (2) approaches in studying environmental issues: qualitative and quantitative methodologies. Qualitative research methodologies are used to explore why or how a phenomenon occurs, to develop a theory, or describe the nature of an individual or group's experience, while quantitative methodologies address questions about causality, generalizability, or magnitude of effect (Fetters et al. 2013). Mixed methods research, frequently referred to as the 'third methodological orientation' (Tashakkori and Teddlie 2008), draws on the strengths of both qualitative and quantitative research.

According to Greene (2008), MMR is a type of thinking and orientation toward social inquiry that actively invites the researchers in dialogue about multiple ways of seeing and hearing, multi ways of making sense of the social world, and multiple standpoints on what is important and to be valued and cherished.

Triangulation allows one to identify aspects of a phenomenon more accurately by approaching it from different lenses or vantage points using different methods and techniques. In context with this study, triangulation was incorporated through interviews, panel or group discussions, document analyses, and survey research instruments. Semi-structured interviews (IDIs), in situ observations, and focus group discussions (FGDs) with the members of the PO as regard to their conservation practices and document analysis of the secondary sources of data (e.g., management plan, mangrove ecotourism business plan, constitution and by-laws, SEC registration, etc.) comprise the qualitative methods of data gathering procedures.

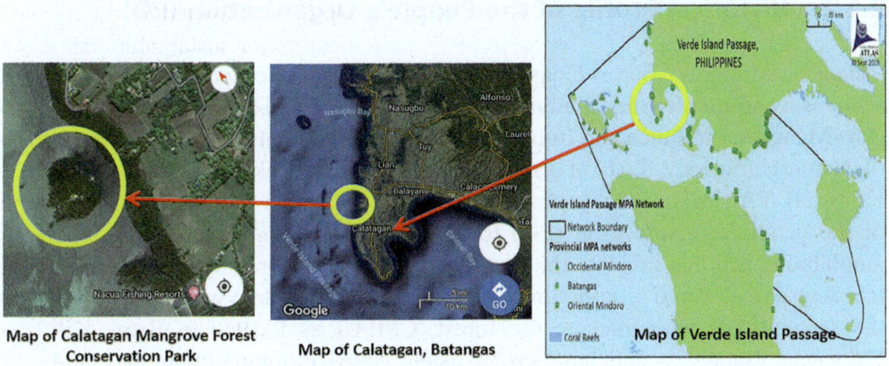

Fig. 2.1 Map showing the community-managed mangrove forest in Calatagan, Batangas, Philippines

Fig. 2.2 PALITAKAN (Pro-mangrove Alliance and Implementing Team and Arm as Kilitisan's Advocates of Nature), a community-based organization composed of various stakeholders managing the mangroves

On the other hand, survey questionnaire on the effectiveness of the community-based management of the coastal community was used to collect quantitative data of the biophysical, socioeconomic, and institutional details of the PO and the mangrove systems (Figs. 2.1 and 2.2).

2.1.2 Historical Profile of the People's Organization (PO)

Calatagan Mangrove Forest Conservation Park (CMFCP) or locally known as "Ang Pulo" is a community-managed mangrove forest governed locally by PALITAKAN (Pro-Mangrove Alliance and Implementing Team and Arm as Kilitisan's Advocates of Nature), formerly TALIMUSAK (Tagapaglinis ng Likas na Yamang-dagat sa Kilitisan). TALIMUSAK was changed to PALITAKAN in 2010 as the official name of the people's organization (PO) to symbolize the *palitakan* (mud skipper), an amphibious fish that inhabits the mangrove pneumatophores with 360° vision. The fish is also considered biomonitor of the mangrove ecosystem which is analogous to the stakeholders who manage the forest. CMFCP is located in Brgy. Kilitisan, Calatagan, Batangas, and has two (2) islets called Pulong Buhangin 1 and 2. It follows a restricted zone scheme in contrast with a no-take zone to give benefits for the local people to extract other aquatic resources.

The conceptualization of the forest park started from a barangay resolution of *Sangguniang Kabataan* (SK), a youth organization, headed by Ms. Hannah Esguerra. Afterwards, it is declared a municipal nature park and a community-managed marine protected area (MPA) through the *Sangguniang Bayan* (SB) Resolution No. 76 on September 8, 2009. The CMFCP was initially supported by the United States Agency for International Development (USAID) through the Conservational International Philippines and the Local Government of Calatagan in 2009. Further socioeconomic developments have taken place in terms of the introduction of the community-based sustainable tourism (CBST) through the Coral Triangle Support Project (CTSP). The PO of this forest park is composed of various stakeholders such as women, fishers, farmers, youth, academe, LGU, and other internal stakeholders who participate in the conservation and management of the mangrove ecosystem. There are 40 individual members of the PO who reside near the mangrove area. They served as the respondents of the study.

2.1.3 Participatory Rural Appraisal (PRA)

Participatory rural appraisal (PRA) is a growing combination of approaches and methods that enable rural people to share, enhance, and analyze their knowledge of life and conditions, to plan and act and to monitor and evaluate (Chamber 1997; Macandog et al. 2005). A PRA is a set of tools that include the engagement of the community. It is an enabling mechanism that allows them to participate and be full partners in their own development. The Food and Agriculture Organization (FAO) defined PRA as a research methodology in which a local community studies an issue that concerns the population, prioritizes problem, evaluates options for solving the problem(s), and comes up with an action plan to address the concerns (Macandog et al. 2014).

Various approaches of PRA were employed to explore the sociodemographic, economic, and political drivers of community-based mangrove restoration in Calatagan, Batangas, using appreciative inquiry. The following were the general activities involved in the rapid rural appraisal conducted in PALITAKAN (Pacardo et al. 2000): preparatory activities, focus group discussion (FGD), key informant interview (KII), reconnaissance survey, field data collection, and data analysis.

2.1.4 Socioeconomic Profiling and Gender Roles in Mangrove Conservation

The sociodemographic profile of the 40 respondents was gathered from the stakeholders using a survey questionnaire. An informant consent agreement form was presented to the respondents before the data collection based on the ethics clearance approved by the DLSU-D University Research Office. The survey questionnaire is composed of two (2) parts: (1) sociodemographic profile of the stakeholders such as location and household information including income, family size, position, sex, age, civil status, educational attainment, and occupation; (2) gender roles in mangrove conservation (Bagsit and Jimenez 2013) which is composed of three indicators namely: mangrove nursery development and maintenance; mangrove planting; and mangrove management, protection, and maintenance.

It was validated in terms of content validity using the Aiken's content validity test by three (3) experts in environmental science and coastal resources management. These experts came from the following institutions: LGU-MENRO, Batangas State University, and Calatagan Senior High School-Ecology Laboratory. A coefficient of 0.91 was obtained which indicated very good validity. On the other hand, it was piloted to ten (10) mangrove stakeholders in Calatagan who were not part of the study to check the reliability of the tool. The Cronbach alpha statistics was used for this purpose. A coefficient of 0.88 was obtained which signified that there is internal consistency of the items in the revised tool, thus, there is good reliability.

2.1.5 Assessment of Ecological Knowledge (EK) and Conservation Practices on Mangroves

EK was assessed in terms of the different thematic aspects of ecosystem services such as provisioning services, regulating services, supporting services, and cultural services. Focus group discussion was conducted to determine the benefits derived from mangrove ecosystem by the stakeholders. It was conducted through in-depth interview of the different groups of stakeholders. A survey tool was also used to supplement the data needed. Conservation practices were assessed using social indicators such as stakeholders' engagement, extent of government and external organizations' participation, and Mangrove Awareness and Uses (MAU) index. The MAU tool is divided into two subcategories namely: ecological knowledge and socioeconomic knowledge.

2.1.6 Marine Protected Area Management Effectiveness Assessment Tool (MPA MEAT)

The MPA MEAT aims to assess governance in terms of enforcement, implementation, and maintenance (see Appendix G). However, MPA management in the context of governance approaches in the Philippines is not limited to the physical management of the MPA only, but also includes direct and indirect uses, threats, people, and the systemic interaction between the people and the resource.

Management effectiveness is defined, in the context of the MPA MEAT, according to four different levels: (1) established, (2) strengthened, (3) sustained, and (4) institutionalized. MPA effectiveness, on the other hand, is based on several criteria and/or governance indicators in combination with the biophysical and socioeconomic impact indicators and socioeconomic impact indicators. MPA MEAT is a management tool to help measure MPA effectiveness using simplified tools allowing an objective evaluation of MPAs. It can be applied to locally managed MPAs and marine areas declared under the National Integrated Protected Area System Act (RA 7586). It can be implemented through an assisted self-evaluation or key informant interviews. Documents provide proof of completion of targets. Assessment will be conducted through the PRA approaches with the PALITAKAN.

2.2 Data Gathering and Statistical Analysis

The demographic profile of the respondents, livelihood sources, perception on their socioeconomic and ecological knowledge, gender participation, and conservation practices were gathered from the PO. Mean, standard deviation, frequency, and percentage were used for the descriptive statistics of the sociodemographic and economic profile and the MPA management effectiveness assessment tool.

Nonparametric test such as Mann-Whitney test (U), Kruskal-Wallis test (H), and Spearman (r_s) rank coefficient of correlation were used in analyzing data in the local ecological knowledge and gendered perspectives on mangrove conservation. This was due to the problem of the normality assumption of the data or normal distribution in spite of several attempts to data transformation.

All statistical analyses were carried out at $p < 0.05$ level of significance. Statistical Package for the Social Sciences (SPSS) software version 25 was used for the descriptive and inferential tests. Paleontological Statistics (PAST) software was also used in the multivariate analyses.

2.3 Results and Discussion

2.3.1 Socioeconomic Profile of the PO of Calatagan Mangrove Forest Conservation Park

The socioeconomic profile of the people's organization was assessed in terms of the different demographic variables such as sex, age, civil status, educational attainment, monthly income, and occupation (Table 2.1). Based on the results of the socioeconomic profiling, majority of the stakeholders were female (n = 22). In addition, majority of the stakeholders are in age range of 21–30, while the least number of the stakeholders lie in ages 11–20. In terms of civil status, 24 of the stakeholders are married which composed its majority, while widow (n = 2) and live-in (n = 2) are the least numbers of stakeholders. In terms of educational attainment, majority of the stakeholders are high school graduate (n = 14), while the least number are composed of both elementary level and high school level with three participants each. Majority of the stakeholders (n = 26) have a monthly income of Php1–5000.00, while the least (n = 4) have monthly income of Php10,001–15,000.00. In terms of occupation, boat operators hold most of the stakeholders (n = 7), while MPA Patroller had the least number of stakeholders (n = 2).

2.3.2 Ecosystem Services, Diversification of Livelihood, and Ecotourism in CMFCP

The results showed that majority of the stakeholders observed that the mangrove forest provides ecosystem services under cultural category (n = 122) with four ecosystem services (Table 2.2). Ecotourism holds the majority (n = 40) under this category, while spiritual/recreation had the least frequency (n = 21). In terms of regulating fresh air and protection from strong wind, waves and storms had the highest frequency (n = 36) while carbon sequestration had the least frequency (n = 10). Under supporting/habitat category, majority of the stakeholders perceived that mangrove forest provides habitat of invertebrates (n = 36), while nursery of juvenile fishes had the least frequency (n = 33). In terms of provisioning, fishes had the highest frequency (n = 32), while fuel wood had the least (n = 14).

The stakeholders perceived that the mangroves provide cultural services such as education, research, spiritual/recreation aesthetic value/happiness, and ecotourism. This is because the conservation site is a frequent venue for mangrove study tour of the academe and a recreational place. Ecotourism plays a very significant role in the livelihood of the community stakeholder. The PO members provide various services such as boat/raft and table/hut rental, tour assistance, and income derived from entrance and conservation fees.

Regulating services serve as the second dominant ecosystem services perceived important by the respondents, particularly on providing fresh air and protection from strong waves and storm. Few (25%) recognized the role of mangroves to sequester carbon dioxide, which is related to providing fresh air. Education on how this is done

Table 2.1 Socioeconomic profile of the stakeholders of Calatagan Mangrove Forest Conservation Park (CMFCP)

Demographic profile	Frequency (*n*)	Percentage (%)
Sex		
Male	22	55
Female	18	45
Total	*40*	*100*
Age (in years)		
11–20	3	7.5
21–30	13	32.50
31–40	6	15
41–50	11	27.5
51–60	7	17.5
Total	*40*	*100*
Civil status		
Single	12	30
Married	24	60
Widow	2	5
Live-in	2	5
Total	*40*	*100*
Educational attainment		
Elementary level	3	7.5
Elementary graduate	4	10
High school level	3	7.5
High school graduate	14	35
College level	10	25
College graduate	6	15
Total	*40*	*100*
Monthly income (in peso, Php)		
1–5000	26	65
5001-10,000	10	25
10,001–15,000	4	10
Total	*40*	*100*
Occupation		
Boat operator	7	17.5
Farmer	4	10
Fisher	6	15
Mollusks gleaner	4	10
MPA patroller	2	5
Tour guide	6	15
Vendor	3	7.5
LGU worker	8	20
Total	*40*	*100*

Table 2.2 Ecosystem services provided by the mangrove forest as perceived by the stakeholders

Category	Ecosystem services	Frequency	Percentage (%)
Cultural	Educational/research area	32	80
	Spiritual/recreation	21	52.5
	Aesthetic value/happiness	29	72.5
	Ecotourism	40	100
	Total	*122*	
Regulating	Fresh air	36	90
	Prevention of floods and soil erosion	16	40
	Water quality	19	47.5
	Carbon sequestration	10	25
	Protection from strong wind, waves, and storms	36	90
	Total	*117*	*58.5*
Supporting/ habitat	Nursery of juvenile fishes	33	82.5
	Habitat of invertebrates	36	90
	Total	*69*	
Provisioning	Fishes	32	80
	Crustaceans	28	70
	Mollusks	29	72.5
	Fuel wood	14	35
	Total	103	

by mangrove trees can be an area for knowledge enhancement for the members. Most of the respondents who were earlier trained by an international conservation organization had ecological knowledge of the different roles of mangroves as an ecosystem and their knowledge about the ecosystem services of the mangrove forest was evident in the results of the survey.

The stakeholders are also aware of the supporting services of mangroves as habitat for invertebrates and nursery ground for fishes. They also recognized that mangroves provide them fish, crustaceans, mollusks, and other invertebrates, the last two are usually gleaned during low tide and are used as food for the family. It is interesting to note that only a third recognized mangrove forest as a source of firewood indicating their awareness on the prohibition to cut mangrove trees.

Table 2.3 presents the different livelihood and mangrove-related ecotourism activities engaged in by the stakeholders. The findings showed that there is diversification of livelihood and the stakeholder. Based on the KII, they do not depend solely on ecotourism because there are lean seasons when there are limited guests or tourists availing the ecotourism services of the park.

Table 2.3 Diversity of livelihood and mangrove-related ecotourism activities of the stakeholders based on the focus group discussion (FGD)

Additional livelihood	Mangrove-related ecotourism activities
Mollusk gleaning	Mangrove study tour
Fishing	Adopt-a-mangrove project
Sugar cane farming/crop growing	Bird watching
Sea cucumber collecting	Balsa riding (rental basis)
Selling in sari-sari store	Kayaking
Fruit vending	Motor boating
Swine production	Kayaking
Poultry raising	Snorkeling
Cattle/carabao/goat raising	Eco-camp
Construction works	Seminar-workshop
Carpentry	
Sewing	

Table 2.4 Sociopolitical indicators of conservation in terms of stakeholders' engagement

Social indicators		Mean ± S. D.	Verbal interpretation
1.	Volunteer in planning and managing MPA	4.33 ± 0.89	Always
2.	Participate in mangrove-*related* trainings and meetings	4.05 ± 1.01	Usually
3.	Participate in mangrove reforestation	4.38 ± 0.87	Always
4.	Participate in mangrove monitoring and the adjacent ecosystem*s*	4.10 ± 0.93	Usually
5.	Follow strictly the policies and laws on mangroves	4.45 ± 0.68	Always
	Grand mean ± S.D.	4.26 ± 0.68	Always

Extent of stakeholders' participation in conservation is presented using mean ± standard deviation (s.d.). (Threshold legend: Never = 1.00–1.80, Seldom = 1.81–2.60, Sometimes = 2.61–3.40, Usually = 3.41–4.20, and Always = 4.21–5.00)

2.3.3 Conservation Practices and Mangrove Awareness and Uses (MAU) Index of the Stakeholders

Sociopolitical dimensions of mangrove conservation practices revealed that the stakeholders are engaged very much in conservation-related activities such as planning, managing, participation in trainings, meetings, reforestation, afforestation, mangrove monitoring, and following policies and laws on mangroves (Table 2.4). According to the stakeholders, their motivation to participate in the community-based project includes giving back to nature or sense of gratitude, new learnings and insights, social interaction with people from all walks of life and of course for the purpose of sustainability that will benefit the younger generation.

Although most of them are volunteers, they only get a very few amounts from the revenue. A local guide would earn only 25% of the daily income. Say, for instance, there are ten guests who paid Php150.00 each and there are five local guides who

Table 2.5 Sociopolitical indicators of conservation in terms of the extent of government and external stakeholders' support in the community-based program on mangroves

Political indicators		Mean ± S.D.	Verbal interpretation
1.	Provide logistics and training and technical assistance	3.65 ± 0.77	Usually
2.	Provide financial support for the sustainability of the mangrove project	3.33 ± 1.27	Sometimes
3.	Lead in involving people's organization in government's development projects	3.93 ± 0.94	Usually
4.	Assist in the legislation and enforcement of laws related to mangroves	4.20 ± 0.91	Usually
5.	Encourage the community to participate in the mangrove projects	4.48 ± 0.75	Always
	Grand mean ± S.D.	3.92 ± 0.59	Usually

Threshold legend: Never = 1.00–1.80, Seldom = 1.81–2.60, Sometimes = 2.61–3.40, Usually = 3.41–4.20 and Always = 4.21–5.00

assisted in terms of guiding and operating the boat raft. Unfortunately, they would only receive Php75.00 each.

In terms of the extent of support of the LGU and other external organization such as NGOs, the stakeholders assessed that there is usual support mechanism (Table 2.5). Based on the FGD, the LGU particularly the Municipal Environment and Natural Resources Office (MENRO) of Calatagan has been providing capacity-building among the members of the PO. Empowering activities focused mainly on the systematics and biology of mangroves and livelihood diversification.

However, financial support remains to be a challenge for community-based programs to finance sustainably the different management and conservation-related activities and operations due to budget constraint. With the small revenues from the user fees/guests, they found it very difficult to finance the rehabilitation of the facilities such as board walk made from bamboo and the *balsa* (boat raft) and other maintenance and operating expenses. And of course, much more of the benefit sharing/honorarium for the services was rendered by the stakeholders.

2.3.3.1 Mangrove Awareness and Uses (MAU) Index of the Stakeholders

The stakeholders' knowledge in the context of mangrove conservation was divided into knowledge indicators namely: ecological knowledge and socioeconomic knowledge. These were assessed using the developed MAU index. The results showed that the stakeholders have high awareness/knowledge on most of the ecological roles of mangroves in the marine ecosystem (Table 2.6). However, while the average results showed high awareness, the head tour guide of the PO still felt the need for some stakeholders to undergo capacity-building on science-based ecological restoration for sustainable mangrove planting/growing.

As to socioeconomic knowledge on mangrove (Table 2.7), MAU indices showed high scores on the economic importance of mangroves such as source of food, livelihood, and ecotourism. Stakeholders were also highly aware about the social

Table 2.6 Mangrove awareness and uses (MAU) index of ecological knowledge

Knowledge indicators	Mean ± S. D.	Verbal interpretation
1. Mangroves are plants in the marine ecosystem	3.70 ± 0.65	HA
2. Mangrove can sequester carbon dioxide from the atmosphere	3.30 ± 0.61	HA
3. Mangroves serve as habitats and nurseries to marine animals	3.38 ± 0.52	HA
4. Mangroves protect coastal communities from storm surge and strong winds	3.85 ± 0.53	HA
5. There are *different* species of mangroves in Ang Pulo	3.48 ± 0.82	HA
6. Mangroves help in nutrient recycling	3.35 ± 0.74	HA
7. Mangroves prevent erosion and siltation	3.40 ± 0.87	HA
8. Sudden Sea level rise in coastal area can be reduced by mangrove conservation	3.08 ± 0.89	MA
9. There are vulnerable or threatened species in Ang Pulo	2.88 ± 0.97	MA
Grand mean ± S.D.	3.38 ± 0.29	HA

HA high awareness, *MA* moderate awareness, *LA* low awareness, and *NA* no awareness

Table 2.7 Mangrove Awareness and Uses (MAU) Index of socioeconomic knowledge

Knowledge indicators	Mean ± S. D.	Verbal interpretation
1. Mangroves indirectly provide food and livelihood	3.65 ± 0.66	HA
2. Destruction of mangrove forests results to low fishery production	3.60 ± 0.67	HA
3. Mangroves can be used as firewood and for construction materials	2.73 ± 1.06	MA
4. Mangrove can be a source of tourism	3.68 ± 0.76	HA
5. Mangroves provide livelihood and income to the community	3.70 ± 0.61	HA
6. Mangrove areas can be used for aquaculture activities.	3.25 ± 0.98	MA
7. Mangroves can be source of pharmaceuticals and medicine	2.20 ± 1.04	LA
8. Conserving mangroves builds camaraderie and volunteerism	3.78 ± 0.62	HA
9. Mangrove conservation can increase community participation	3.78 ± 0.62	HA
Grand mean ± S.D.	3.37 ± 0.56	HA

HA high awareness, *MA* moderate awareness, *LA* low awareness, and *NA* no awareness

benefits of mangroves as evidenced by community participation, camaraderie, and volunteerism. They also assessed themselves to have moderate awareness on mangrove uses as firewood and for construction materials and aquaculture activities in the mangrove areas and low awareness on their use as medicine. Moderate and low awareness indicates that the stakeholders are not using the mangrove for these purposes either as a prohibition of the law (cutting of trees and conversion to aquaculture ponds) or plain lack of knowledge on how to use them as medicine.

Still, there were reports especially in the past of occasional mangrove logging and unsustainable aquaculture activities according to some stakeholders.

2.3.4 Analysis of the Social Dimensions of Integrated Community-Based Conservation

Summary statistics of the social dimensions of ICBC are presented in Table 2.8. Stakeholders' engagement and participation and perceived extent of support by the external stakeholders are evident. They had higher awareness on the ecological and socioeconomic aspects of mangroves. However, their knowledge on the ecological aspect of conservation needs to be augmented with science-based restoration approaches as there are areas in the mangrove forest that did not follow zonation pattern in the case of monospecific plantation of *Rhizophora* spp. This is supported by the findings of Creencia and Querijero (2018) that knowledge awareness is a key driver of community-based conservation.

Statistical analysis of the different dimensions showed that stakeholders' engagement is significantly correlated with ecological knowledge and socioeconomic knowledge (Table 2.9). Awareness plays an important role in mangrove

Table 2.8 Summary of the descriptive statistics of the dimensions of conservation-related practices

	N	Minimum	Maximum	Mean	S.D.
Stakeholders' engagement (SE)	40	2.60	5.00	3.38	0.29
Perceived extent of support (EE)	40	2.60	5.00	3.37	0.56
Ecological knowledge (EK)	40	1.33	3.67	3.92	0.59
Socioeconomic knowledge (SEK)	40	1.33	3.56	4.26	0.68

Table 2.9 Spearman rank coefficient of correlation of the conservation practices and stakeholders' knowledge on mangroves

		SE	EE	EK	SEK
Stakeholders' engagement (SE)	Correlation coefficient	1.000	0.096	0.601*	0.384**
	Sig. (2-tailed)		0.558	0.000	0.015
Perceived extent of support (EE)	Correlation coefficient	0.096	1.000	−0.131	0.051
	Sig. (2-tailed)	0.558		0.420	0.756
Ecological knowledge (EK)	Correlation coefficient	0.601*	−0.131	1.000	0.486*
	Sig. (2-tailed)	0.000	0.420		0.001
Socioeconomic knowledge (SEK)	Correlation coefficient	0.384**	0.051	0.486*	1.000
	Sig. (2-tailed)	0.015	0.756	0.001	

*Correlation is significant at the 0.01 level (2-tailed)
**Correlation is significant at the 0.05 level (2-tailed)

Table 2.10 Gender-differentiated perception on conservation practices and stakeholders' knowledge using Mann-Whitney *U* test

Gender	Variable	Mean rank	*U*	*p*-value*
Male	Stakeholders' engagement	10.61	171.5	0.470
Female		9.89		
Male	Perceived extent of support	11.29	196.5	0.978
Female		9.26		
Male	Ecological knowledge	9.45	125	0.047*
Female		11.05		
Male	Socioeconomic knowledge	8.13	72	0.001**
Female		12.38		

*Significant at $p < 0.05$
**Significant at $p < 0.01$

conservation towards sustainability (Creencia and Querijero 2018). There was also significant relationship between ecological knowledge and socioeconomic knowledge. However, there was no marked significant relationship between stakeholders' engagement and the perceived extent of support. This further implies that community engagement may be driven by intrinsic motivation.

The statistical analyses of the gender-differentiated perception between men and women (Table 2.10) revealed that there is no significant difference on the stakeholders' participation ($p = 0.470$) in mangrove conservation and the extent of support of the external stakeholders ($p = 0.978$). This result further means that members of the PO participated in the community-based conservation regardless of their gender. They also agreed that there is support mechanism from external stakeholders.

However, there were significant differences between the ($p = 0.047$) ecological knowledge of men and women. The findings showed that women had higher ecological knowledge than men as evidenced by the mean rank of 11.05 for women compared to men with mean rank of 9.45. Similarly, women also possess higher awareness than men in terms of socioeconomic knowledge on mangroves as shown by the mean rank of 12.38 for women and a mean rank of 8.13 for men.

2.3.5 Gender-Based Analysis of the Impacts of Integrated Community-Based Conservation

Gender analysis using the framework of Bagsit and Jimenez (2013) was used to assess the gender roles in mangrove conservations of the people's organization based on mangrove nursery development and maintenance, mangrove planting, and mangrove management, protection, and maintenance (Table 2.11). Based on the findings, both men and women are engaged in majority of the mangrove conservation-related tasks and activities in the PO, although there are still tasks that are more gendered based on the perception of the stakeholders.

Table 2.11 Gender roles in mangrove conservation as perceived by the stakeholders. Frequency and percentage (enclosed in parenthesis) are shown

	Activity	Female	Male	Both
A. Mangrove Nursery Development and Maintenance (MNDM)				
1.	Preparing planting materials (MNDM1)	2 (5)	0	38 (95)
2.	Preparing nursery (MNDM2)	3 (7.5)	5 (12.5)	32 (80)
3.	Potting the seeds (MNDM3)	3 (7.5)	1 (2.5)	36 (90)
4.	Cutting bamboo stakes (for the fence) (MNDM4)	0	27 (67.5)	13 (32.5)
5.	Fencing (MNDM5)	0	28 (70)	12 (30)
6.	Purchasing polybags and other materials for the nursery (MNDM6)	11 (27.5)	2 (5)	27 (67.5)
B. Mangrove planting (MP)				
1.	Transferring of mangrove seedlings/propagules to the planting area (MP1)	5 (12.5)	1 (2.5)	34 (85)
2.	Digging holes prior to planting (MP2)	1 (2.5)	6 (15)	33 (82.5)
3.	Planting mangroves (MP3)	0	2 (5)	38 (95)
C. Mangrove management, protection, and maintenance (MMPM)				
1.	Attending meetings (MMPM1)	2 (5)	0	38 (95)
2.	Planning (MMPM2)	2 (5)	0	38 (95)
3.	Decision-making (MMPM3)	0	1 (2.5)	39 (97.5)
4.	Coordinating with other agencies/organizations (MMPM4)	5 (12.5)	0	35 (87.5)
5.	Sales (MMPM5)	3 (7.5)	1 (2.5)	36 (90)
6.	Monitoring (MMPM6)	0	22 (55)	18 (45)
7.	Clean-up of mangrove area (MMPM7)	0	0	40 (100)
8.	Accounting of income and expenses (MMPM8)	19 (47.5)	1 (2.5)	20 (50)
9.	Safekeeping of funds (MMPM9)	26 (65)	0	14 (35)
10.	Installing sign boards (MMPM10)	1 (2.5)	27 (67.5)	12 (30)

Data on the factor analysis of the different activities showed that men and women are engaged in differentiated tasks (Table 2.12). Scores were shown for each of the identified factors generated from the three major mangrove conservation-related tasks. Gender-specific activities were categorized based on these scores in each factor. High scores greater than 0.50 were considered and were in boldfaced. The PCA biplot of the three axes showing factors was also shown to cluster the different gender-specific tasks (Fig. 2.3). Conservation activities that require muscular strength such as cutting bamboo stakes (MNDM4), putting up fences (MNDM5), monitoring and patrolling (MMPM6), and installation of sign boards (MMPM10) were identified as activities performed by majority of the male members (Factor 2).

Table 2.12 Factor analysis of the gender roles in mangrove conservation related activities performed by the stakeholders

	Factor 1	Factor 2	Factor 3
MNDM1	**0.954**	0.206	−0.220
MNDM2	**0.900**	0.351	−0.257
MNDM3	**0.940**	0.231	−0.249
MNDM4	0.227	**0.971**	−0.079
MNDM5	0.184	**0.980**	−0.072
MNDM6	0.814	0.243	**−0.527**
MP1	**0.923**	0.228	−0.310
MP2	**0.905**	0.378	−0.196
MP3	**0.951**	0.259	−0.168
MMPM1	**0.954**	0.206	−0.220
MMPM2	**0.954**	0.206	−0.220
MMPM3	**0.958**	0.233	−0.168
MMPM4	**0.931**	0.200	−0.306
MMPM5	**0.940**	0.231	−0.249
MMPM6	0.446	**0.888**	−0.112
MMPM7	**0.964**	0.208	−0.168
MMPM8	**0.578**	0.157	**−0.800**
MMPM9	0.314	0.062	**−0.948**
MMPM10	0.192	**0.976**	−0.108

Bold values show significant factors affecting mangrove conservation ($p < 0.05$)

On the other hand, traditionally assigned roles for women such as purchasing polybags for mangrove seedlings (MNDM6), accounting of income and expenses (MMPM8), and safekeeping of funds (MMPM9) were also performed by the majority of the women in this study (Factor 3). However, MNDM1, MNDM2, MNDM3, MP1, MP2, MP3, MMPM1, MMPM2, MMPM3, MMPM4, MMPM5, MMPM7, and MMPM8 were mangrove-related tasks performed by both men and women. This implies that the community-based organization has a big potential towards gender-equality. This is supported by the practices of the PO where women were also part of the mangrove governance and planning. These results lead to a greater opportunity for environmental initiatives to achieve multiple benefits from the mangroves, amplify results, and increase effectiveness, as gender equality and women's empowerment lead to more successful, efficient, and equitable environment and conservation outcomes (USAID 2019).

2.3.6 Effectiveness of Governance of the Marine Protected Area (MPA)

MPA MEAT of the governance of Calatagan Mangrove Forest Conservation Park revealed that the present MPA is at Level 3 based on the scores obtained in the assessment (Table 2.13). This is due to the fact that community participation was evident among the stakeholders. There was also legal identity and site development

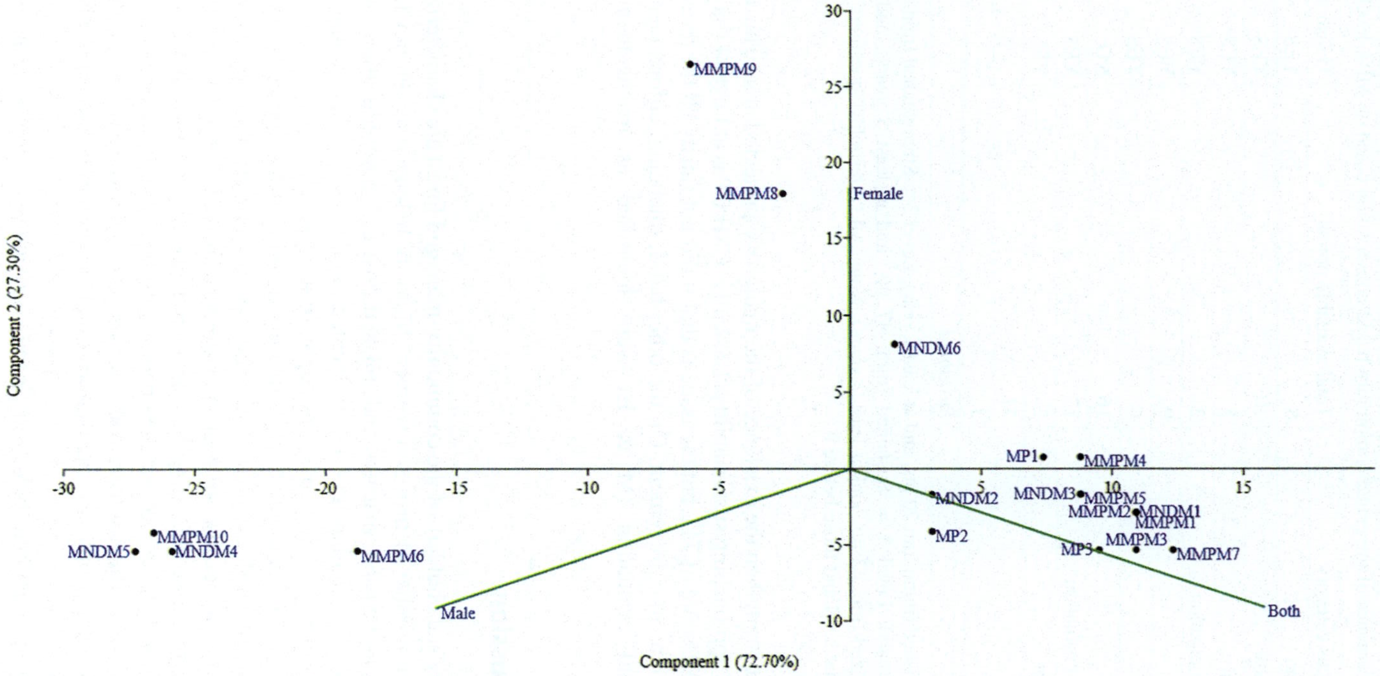

Fig. 2.3 PCA ordination biplot of the different mangrove conservation-related activities performed by the stakeholders based on gender. Gender-specific activities are shown using the given legend: *MNDM* mangrove nursery development and maintenance, *MP* mangrove planting, and *MMPM* mangrove management, protection and maintenance. *MMPM* mangrove management, protection, and maintenance

Table 2.13 Marine-Protected Area Management Effectiveness Assessment Tool (MPA MEAT) results of the governance of integrated community-based conservation based on MPA level

Management focus category	Total available points	Threshold met
Management plan	8	YES
Management body	10	YES
Legal instrument	5	YES
Community participation	2	YES
Financing	9	YES
IEC	6	YES
Enforcement	16	**NO**
Monitoring and evaluation	13	YES
Site development	2	YES
Total	*71*	
Level	*3*	
Remarks	*Sustained*	

Level 1 = Established (minimum 20 pts. and at least 1 year), Level 2 = Strengthened (min. 31 pts. and at least 2 years), Level 3 = Sustained (min. 47 pts. and at least 5 years), and Level 4 = Institutionalized (min 63 pts. and at least 7 years). Note: All "thresholds" questions should be satisfied for that level and those before it

plan that might strengthen the existence of the mangrove protected area. This means that the park is approaching sustainability. Basically, CMFCP is at Level 4 (based on the MPA MEAT score of 71), but the issue is that it did not satisfy all the thresholds in Level 4 such as in enforcement. On the other hand, financing, enforcement, and monitoring and evaluation (M & E) were identified as management foci (Table 2.14).

2.4 Discussion

The community stakeholders of the community-managed forest are characterized by an almost equal number of men and women. Majority belonged to age brackets of 31–40, 41–50, and 51–60, married, and finished postsecondary education level. A study conducted by Paglinawan (2016) revealed similar findings in a community-managed mangrove forest in Papaya, Nasugbu, Batangas.

Majority of the PO members have a monthly income (in Philippine pesos, Php) within the range of Php1–5000.00. The relatively low average monthly income of the members may be because 40% of the members are still young and are beyond 51 years old age group that may be considered not in active work force. It is interesting to note that 75% of the members are high school graduates, some even have college degrees. The results also showed that there is diverse occupation across the members of the people's organization which is predominantly coastal-based source of income.

Majority of the members belonged to the community where the mangrove is proximate. The other members are working at the barangay and municipal levels of the local government unit in Calatagan. The overall findings show similarity with the

Table 2.14 Marine-Protected Area Management Effectiveness Assessment Tool (MPA MEAT) results of the governance of integrated community-based conservation based on management focus category

Management focus category	2009	2010	2011	2012	2013	2014	2015	2016	2017	2018	2019	Rank
Management plan	0	56	56	78	78	78	89	89	89	89	89	5
Management body	36	73	82	82	91	91	91	91	91	91	91	4
Legal instrument	100	100	100	100	100	100	100	100	100	100	100	2
Community participation	100	100	100	100	100	100	100	100	100	100	100	2
Financing	0	33	50	58	67	67	67	75	75	75	75	9
IEC	14	71	71	71	71	71	86	86	86	86	86	6
Enforcement	15	50	55	60	75	75	75	75	80	80	80	8
Monitoring and evaluation	6	50	63	75	75	75	81	81	81	81	81	7
Site development	0	50	50	50	100	100	100	100	100	100	100	2

findings of Carandang et al. (2013) whose study also focused on community-based management of mangroves in Palawan and Bohol.

There are four (4) identified categories of ecosystem services provided by the mangrove forest as perceived by the stakeholders. These ecosystem services were categorized as to cultural, regulating, supporting/habitat, and provisioning services, indicating that the stakeholders have deep understanding on the ecological and cultural services of mangrove forest and the personal benefits they derived from protecting it.

The livelihood of the community stakeholders can be categorized into agricultural, small-scale business, and mangrove ecotourism-related income generating activities. Knowledge on mangroves is high in terms of socioeconomic and ecological importance. Conservation practices of the stakeholders are evident. Gender equality is a powerful lever for change. Women are vital to conservation and resilience-building efforts, and gender equality is intrinsically linked to achieving sustainable development goals (USAID 2019). Analysis of the stakeholders' participation shows gender mainstreaming because men and women are both involved in the conservation-related activities of the organization.

Community participation was evident among the stakeholders. There was also legal identity and site development plan that might strengthen the existence of the mangrove-protected area. This means that the park is approaching sustainability. Basically, CMFCP is at Level 4 (based on the MPA MEAT score of 71), but the issue is that it did not satisfy all the thresholds in Level 4 such as in enforcement. On the other hand, financing, enforcement, and monitoring and evaluation (M & E) were identified as management foci.

2.5 Conclusions and Recommendations

The socioeconomic profile of the stakeholders of the community-managed mangrove forest is characterized by various demographic variables such as sex, age, civil status, educational attainment, monthly income, and occupation. There is equal number of men and women stakeholders who also participated in the study. Majority are mid-adult, married, and finished postsecondary education. There is diversity of livelihood among the members of the people's organization, but their monthly income is at the low range.

The stakeholders had high awareness on the four categories of ecosystem services namely: cultural, regulating, supporting/habitat, and provisioning services. There is a number of additional livelihoods aside from mangrove-related ecotourism activities engaged in by the stakeholders. Diversification of livelihood helps the community to reduce pressures on mangrove ecosystem such as timber harvesting. Sociopolitical indicators of conservation reveal that stakeholders' engagement and external stakeholders' support are evident.

The ecological knowledge and socioeconomic knowledge on mangroves of the stakeholders showed that men and women are knowledgeable based on the MAU index. In addition, gender roles in mangrove conservation indicate that men and women are both performing in most mangrove-related conservation activities such as mangrove nursery development and maintenance, mangrove planting and mangrove management, protection, and maintenance, except in specific activities that require physical strength such as gathering of bamboos and fencing wherein men are often the ones involved. On the other hand, women usually handled marketing, accounting, and finance-related activities. This implies that gender equality in terms of gender roles in the community-based conservation is possible. Integrated community-based conservation from 2009 to 2019 shows satisfactory level of conservation and management based on MPA MEAT. The assessment reveals that the management is at Level 3 which means that it is approaching sustainability. Community participation has been a best practice of the organization, while financial self-sufficiency remains a management challenge in the organization.

The use of biocultural approach to conservation shows potential to build sustainability. To scale up the community-based conservation, provision of alternative livelihood and diversification of the sources of income of the local people are recommended. Continuous research, capacity-building, and empowerment programs may be done to strengthen the participation and sustain the motivation of the community in managing the mangroves.

Acknowledgments The authors would like to extend their sincerest gratitude to the local people of Brgy. Kilitisan, Calatagan, Batangas for sharing their time, experiences, and knowledge during the conduct of this study; to the Local Government Unit of Calatagan for the permit to conduct the study, and to Ms. Lucena E. Duman, the Chairperson of the people's organization, for her support to this undertaking and sharing her experiences as a mangrove conservationist.

References

[IPCC] Intergovernmental Panel on Climate Change (2014) Impacts, adaptation and vulnerability. Part B: regional aspects contribution of Working Group II to the fifth assessment report of the Intergovernmental Panel on Climate Change

[USAID] United States Agency for International Development (2019) Biodiversity conservation and forestry programs: FY 2017 results and funding. www.usaid.gov. Accessed 4 June 2019

Allen CR (2006) Multi-model inference and the understanding of complexity, discontinuity and nomadism. Ecosystems 9:694–699

Bagsit FU, Jimenez CN (2013) Gender roles in the mangroves reforestation programmes in Barangay Talokgangan, Banate, Iloilo, the Philippines: a case study where woman have sustained efforts. SPC Women Fish Inf Bull 23:40–44

Carandang AP, Camacho LD, Gevaña DT, Dizon JTD, Camacho SC, De Luna CC, Pulhin FB, Combalicer EA, Paras FD, Peras RJ, Rebugio LL (2013) Economic valuation for sustainable mangrove ecosystems management in Bohol and Palawan, Philippines. For Sci Technol 9(3): 118–125. https://doi.org/10.1080/21580103.2013.801149

Chamber R (1997) Responsible well being – a personal agenda for development. World Dev 25(11):1743–1754

Creencia GBA, Querijero BVL (2018) Community-based management of Calatagan mangrove forest conservation park in Batangas, Philippines: a case study on environmental sustainability. Asia Pac J Acad Res Soc Sci 3:21–27

Crona B (2006) Of mangroves and middleman: a study of social and ecological linkages in a coastal community. Dissertation, Department of Systems Ecology Stockholm University, Stockholm

Fetters MD, Curry LA, Creswell JW (2013) Achieving integration in mixed methods designs-principles and practices. Health Serv Res 48:2134–2156

Fortes MD (2010) The seagrass-mangrove connection as a climate change mitigation and adaptation factor in East Asian coasts. Paper presented at international symposium on integrated coastal management for marine biodiversity in Asia, Kyoto, Japan

Fortes MD, Salmo S III (2017) Status of mangrove research and management in the Philippines: challenges and opportunities. State of the Mangrove Summit, pp 50–60

Gavin MC, McCarter J, Mead A, Berkes F, Stepp JR, Peterson D, Tang R (2015) Defining biocultural approach to conservation. Trends Ecol Evol 30(3):140–145

Greene JC (2008) Is mixed methods social inquiry a distinctive methodology? J Mix Methods Res 2:7–22

Kairo JG (2007) Sustainable management of mangrove resources through a participatory approach. Examples of successful experiences in coastal community development sharing innovative experiences. United Nations Office for South-South Cooperation in the United Nations Development Programme (UNDP) and The International Ocean Institute

Macandog DM, Mallion FK, Ani PAB (2005) Role of agroforestry in the food security of upland farmers in the degraded soils of Claveria, Northern Mindanao, Philippines. University of the Philippines

Macandog DM, de la Cruz CP, Edrial JD, Reblora MA, Pabico JP, Salvacion AR, Marquez TL Jr et al (2014) Eliciting local ecological knowledge and community perception on fishkill in Taal Lake through participatory approaches. J Environ Sci Manag 17(2):1–16

Macintosh DJ, Ashton EC, Hawanon S (2002) Mangrove rehabilitation and intertidal biodiversity: a study in Ranong mangrove ecosystem, Thailand. Estuar Coast Shelf Sci 55(3):331–345

Nandy P, Islam MA (2010) Climate resilient coastal zone development in Bangladesh: participatory governance for common resources management. In: Management and sustainable development of coastal zone environments. Springer. 978-90-481-3067-2

National Tropical Botanic Garden (2021). https://ntbg.org/science/conservation/biocultural

Pacardo EP, Briones ND, Bout IE, Castillo LV, Rebancos CM (2000) Environmental research methods. UP Open University 4031, Laguna

Paglinawan RN (2016) Comparative study of stand structure and carbon sequestration of mangrove forest in Verde Island Corridor: basis for a sustainable environmental action plan. [Master's Thesis]. Dr. Francisco L. Calingasan Memorial Colleges Foundation, Inc., Nasugbu, Batangas

Primavera JH, Esteban JMA (2008) A review of mangrove rehabilitation in the Philippines: successes, failures and future prospects. Wetl Ecol Manag 16(5):345–358

Quibilan C (2014) Coastal vulnerability assessments: tools, training, and networking. Agric Dev Notes 5(3):1–4

Tashakkori A, Teddlie C (2008) Quality of inferences in mixed methods research: calling for an integrative framework. In: Advances in methods research: theories and applications. Sage Publications, London, pp 101–119

Tashakkori A, Teddlie C (2010) SAGE handbooks of mixed methods in social and behavioral research, 2nd edn. Sage, Thousand Oaks

Tobin B (1999) Information technology effects on firm performance as measured by Tobin's Q. Manag Sci 45(7):1008–1024

Turner RK, Jeroen CJM, Iore S, Aat B, Jan S, Maltby E, Ekko C (2000) The values of wetlands. Landscape and institutional perspectives ecological-economic analysis of wetlands: scientific integration for management and policy. Ecol Econ 35(2000):7–23

Plant Diversity in Biocultural Landscapes During Anthropocene: The Need for Conservation, Challenges, and Future Prospects in Today's World

3

Janani Manochkumar, Rajasekaran Chandrasekaran, Inocencio E Buot, Jr, C. George Priya Doss, R. Seenivasan, S. Usha, and Siva Ramamoorthy

Abstract

The proposal on the Anthropocene suggests that the Earth has entered a contemporary ecological era, in which climate changes and human activities massively impact the environment universally. This new anthropogenic epoch brings exceptional challenges to the plant diversity in biocultural landscapes, thus preventing their benefits to human well-being. Additionally, the loss of biodiversity is exacerbating along with the inexorable variations in the Anthropocene, and biodiversity conservation could not cope with the elevating anthropogenic disturbances. Despite the enduring biodiversity loss, it is asserted that the effective amalgamation of several conservational strategies could contribute to the efficient conservation of endangered plants. The strategies should be specified with proper operative guidelines for monitoring the effect once it gets implemented. Therefore, the current conservational practices should be reconsidered, reexamined, and updated accordingly so that a more consistent, logical, and integrated universal strategy could be devised which will alleviate the

J. Manochkumar · C. G. P. Doss
School of Bio Sciences and Technology, Vellore Institute of Technology, Vellore, India

R. Chandrasekaran · S. Ramamoorthy (✉)
Department of Biotechnology, School of Bio Sciences and Technology, Vellore Institute of Technology, Vellore, Tamil Nadu, India

I. E. Buot, Jr
Institute of Biological Sciences, College of Arts and Sciences, University of the Philippines Los Baños, Laguna, Philippines

R. Seenivasan
Centre for Nanobiotechnology, Vellore Institute of Technology, Vellore, India

S. Usha
School of Social Sciences and Languages, Vellore Institute of Technology, Vellore, India

S. Ramamoorthy et al. (eds.), *Plant Diversity in Biocultural Landscapes*,
https://doi.org/10.1007/978-981-19-8649-9_3

limitations of the existing policies and proposed protocols. This book chapter addresses the major issues that address our failure to meet current conservational strategies and also discusses the essential changes and recommendations to existing conservational practices.

Keywords

Anthropogenic disturbance · Biodiversity · Climate change · Conservation · Human activities

3.1 Introduction

Down the ages, humans are known to have a vivid influence on altering the planet's biosphere right from elevating the level of acidity in the ocean to shattering of the land sites and change of climate (Meng et al. 2021; Otto 2018). It has been hypothesized that the earth might have entered a new ecological era, termed the "Anthropocene", where humans influence the environmental atmosphere worldwide (Lewis and Maslin 2015; Zalasiewicz et al. 2017). Along with human activity, hasty climatic change has inextricably and persistently restructured the biological diversity of the environment. In the hypercompetitive world, the knitted impact of climatic changes and human activities is proposed to have an intense consequence on eco-diverse neighborhoods and the prioritization of human-acquainted conservation plans. Additionally, recent anthropogenic alterations have overtaken the former changes in the environment as well as the history of humans (Meng et al. 2021; Ordonez et al. 2016).

It is evident that biodiversity indispensably contributes to the welfare of human beings as well as augments several profits of natural metabolites to society (Meng et al. 2021). The consumption, utilization, conservation of biodiversity, and allocation of its profits from the use of genetic resources were proposed to be guaranteed by the United Nations Convention on Biological Diversity (CBD) (Locke et al. 2019). Nevertheless, there is a gradual decrease of biodiversity throughout antiquity and it began to tremendously decline, thus contributing to the sixth mass annihilation of the earth (Isbell et al. 2017; Johnson et al. 2017).

Over the past few decades, countless intercontinental contracts, nationwide guidelines, and scientific organizations have been crucially focusing on conserving biodiversity as it has become an impending disaster recently (Meng et al. 2021). However, unprecedented delay in the response of biodiversity to anthropogenic commotion leads to the advent of novel assemblies and reformation of threatened areas. Biodiversity has also provoked incredible interest in researchers, legislative bodies, landlords, and public to understand the roots and topological patterns of biodiversity (Vellend et al. 2017). Therefore, it is the sole responsibility of students, researchers, educational institutes, and legal authorities to renovate the conservation policies of biodiversity according to the altering anthropogenic effects.

It has been proposed that anthropogenic effects have altered about 95% of the terrestrial area to some extent globally and around 75% of the land has been drastically altered by the influence of human activities (Kennedy et al. 2019; Meng et al. 2021). Furthermore, the predicted range of temperature rise elevates with high carbon utilization, monetary progress, and population (Raftery et al. 2017). The impact of humans on climatic changes is evident from the recent reports by the Intergovernmental Science-Policy Platform on Biodiversity and Ecosystem Services (IPBES) and the Intergovernmental Panel on Climate Change (IPCC) which empathized the impressions of climate change and hazards due to indefensible consumption of natural sources (IPBES Report 2019; IPCC 2019). Climate change also contributes to the rapid loss of biodiversity (Arneth et al. 2020). Hence, anthropogenic disturbances inexorably affect the biological diversity worldwide. Therefore, novel strategies for conservation of biological sources at the local or universal level are to be implemented based on the anthropogenic disturbances and influence of climatic changes. Hence, anthropogenic disturbances inexorably affect the biological diversity worldwide. The major factors driving the extinction of plant species in the Anthropocene epoch are pictorially represented in Fig. 3.1.

In this chapter, we examine the influence of climate change and human activities against the loss of biodiversity. Additionally, the major issues concerned with today's conservational policies and the possible ways to tackle it were discussed in brief.

3.2 Vulnerable Biodiversity and Anthropogenic Disturbance

The notion of plant diversity in biocultural landscapes is predicted to conserve the delicate biodiversity in today's changing world. The areas where buffering of anthropogenic disturbances shelters the vulnerable biological diversity are termed "Biodiversity arks in the Anthropocene". This notion is vital to seal the gaps amid conservational strategies and management beneath climatic changes in the present as well as in the future. Human activities like financial evolution, development, infrastructure, and consumption of natural resources as well as climatic changes have perpetually contributed to the loss of biodiversity (Meng et al. 2019a, b). Additionally, the influence of climatic alterations and anthropogenic effects on biodiversity is ostensible in key issues like loss of tropical forests, extinction of plant species, the profusion of marine fishes, invertebrates, tamed lands, external temperature, increase in population, emission of CO_2, and destruction of the environment. Therefore, it is hypothesized that areas of biodiverse arks are the regions where brittle biological plant diversity can be protected from antagonistic effects in the Anthropocene. But these areas are usually omitted in the conservational strategies for many years (Meng et al. 2021).

3.2.1 Influence of Human Activities

Along with the societal growth of anthropological influences, the escalating request for natural resources has markedly wedged the biological diversity of plant species

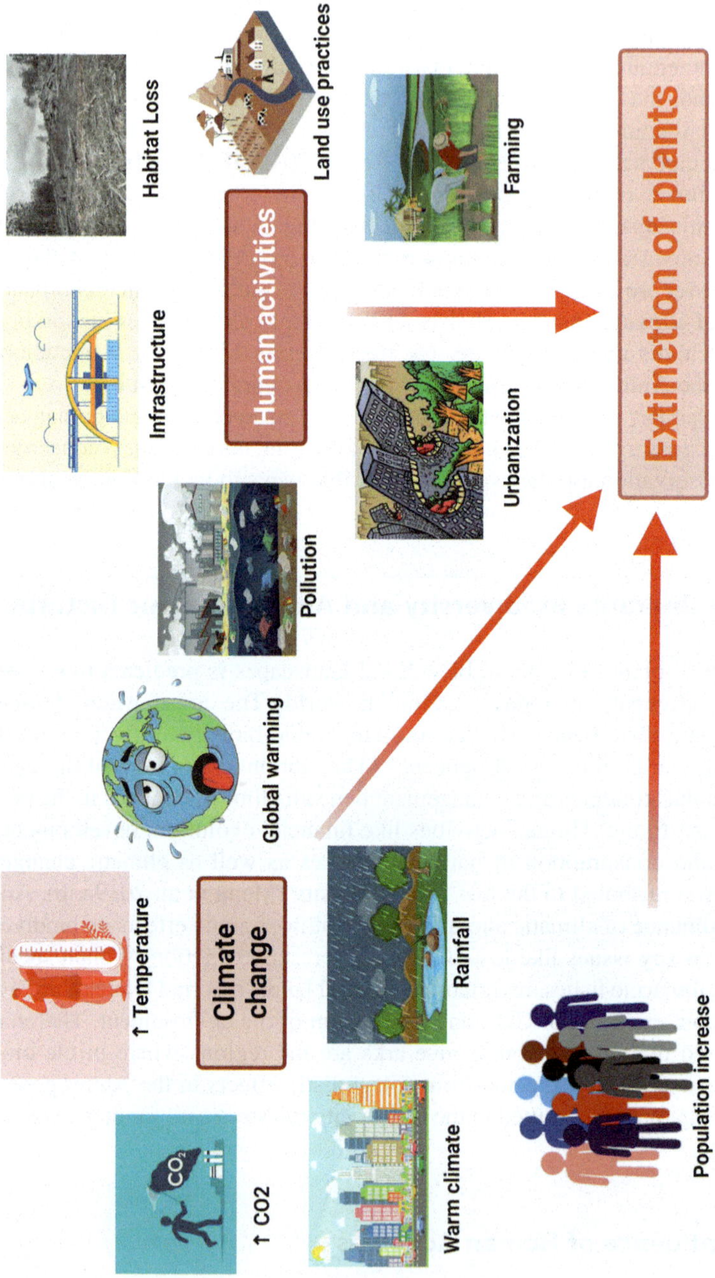

Fig. 3.1 Major factors contributing to plant extinction in the Anthropocene era

worldwide. Anthropogenic activities hasten the human-caused extinctions of plant species, thus ultimately leading to tremendous loss of biodiversity (Braje and Erlandson 2013; Meng et al. 2021). Fortunately, anthropogenic influences on plant diversity have gained enormous attention which is evident from the proposals of appropriate regulations and the establishment of national nature reserves and national parks.

Fortunately, the influence of human activity on biodiversity has received considerable attention. Particularly, many nature reserves have been established, and relevant legislation has been proposed. These areas, where biodiversity is sheltered from human activity, can be considered as parts of biodiversity arks; e.g., national nature reserves and national parks have been set up to protect biodiversity in comprehensive ecological ranges. The existence of such "safe sites" is a basic requirement for biodiversity conservation. Yet, few vicinities are unexploited and must be included in the conserved areas. For instance, the Fengshui forests have a huge cultural as well as economic value and are considered holy by localists, hence the diversity of plant species in those areas has recurrently eluded from anthropogenic disturbances (Meng et al. 2021; Tang et al. 2013). These areas play a significant role in conserving the diversity of plant species, but are simply deemed as enclaves in former conservational policies and are also omitted from the protected sites. Biodiversity arks can therefore seal the gaps between existing conservational patterns and the imminent edge of biodiversity conservation (Meng et al. 2021).

Even though the ex-situ conservation sites are considered "protected areas" for threatened or sporadic plant species, they could not shelter several plant species inside an entire flora and fauna. Presently, the sites enriched with exorbitant biodiversity are given higher priority for conservation. These sites are termed "Hotspots of biodiversity" (Meng et al. 2021; Myers et al. 2000). Yet, those hotspots were inexorably exposed to anthropogenic activities. For instance, financial evolution and refining lifestyle change involuntarily affect the conservation of plant species worldwide, especially in developing nations (Meng et al. 2019a, b). Therefore, we could conclude that the hotspots of biodiversity are slowly extinguished due to the exploitation of most of the natural sources in order to lessen the crisis of poverty. Here, it leads to the arousal of a question, on "Whether biodiversity conservation strategies could also be altered according to the changes in biodiversity hotspots?"

In addition to the impact of humans on biodiversity-related factors, the global diversity of plant species worldwide is also greatly affected by environmental and evolutionary alterations at both inter- and intraspecies levels (Meng et al. 2021). Therefore, more consideration should be given to the conservation of biodiversity by lessening the impact of humans and protecting the plant species in the anthropogenic environment in addition to the already protected areas.

3.2.2 Influence of Climate Change

For ages, humans were considered the ideal cause of biodiversity loss in conservation biology. But only a few studies reported climatic change to be the major factor influencing the significant modification of plant diversity. The reports showed positive, neutral as well as negative correlations between climate change and biodiversity loss (Vellend et al. 2013; Meng et al. 2021). Arneth et al. (2020) stated that climate change must be included as a target for the conservation of biodiversity in future evaluations. Factors like elevated temperature, warm weather, and intense rain have an effect on plant diversity. Hence, it is crucial to understand the association between plant diversity and climate change in an anthropogenic environment to evaluate and alleviate the ecological impact on biodiversity. It is assumed that global warming is increasing due to the emission of greenhouse gases. The release of greenhouse gas, especially CO_2, increases the earth's temperature by trapping the atmospheric heat. Similarly, an increase in global warming inexorably affects biodiversity. Regardless of the elevating distress about the increased risk of plant species eradication and global temperature increase, it is still tough to corroborate the interlink between the extinction of species and climate change (Panetta et al. 2018). Even though it is hard to perceive the climate-induced deterioration, change in climate exaggerates the extinction of plants as well as animals on a larger scale (Dirzo et al. 2014; Panetta et al. 2018; Harrison 2020).

Biodiversity arks emphasize the development of protecting sectors in order to safeguard the plant diversity from adverse effects of climate changes in the Anthropocene. Elevated temperature particularly is known to have a negative impact on the topographical dispersal of plant species found in high altitudes and cold temperatures (Meng et al. 2019b). The impact of climate change on the diversity of plants has also been demonstrated.

Due to an increase in temperature, it is expected that there will be a drastic decrease in the turnover of plant species in mere future which is evident from limitless distribution and shift in the species range (Alexander et al. 2018). Over the last few decades, environmental patterns elucidated the response of species diversity to the ecological deviations in the Holocene epoch. This led to the identification of prevailing refugia that have conserved existing species origin. Elevation and reduction in habitat diversity along with shifts in latitude and altitude have facilitated the plant species to thrive in unfavorable climatic shifts in the Anthropogenic epoch. Accordingly, the areas that are shielded from temperature rise in the Anthropocene could lead to the conservation of habitat diversity (Meng et al. 2021). On the other hand, alterations in climate had no impact on the thermophilic species. Surprisingly, in some instances, it has even augmented the tropical and subtropical species diversity. Nevertheless, alterations in climate have a huge impact on the plants growing in high altitudes of tropical forests as well as plants adapted to extremely cold conditions (Meng et al. 2019b). Generally, biodiversity arks emphasize species-enriched environments. Therefore, it is mandatory to give more consideration to plant species that respond slowly to ecological alterations due to global warming when compared to climatic shifts. It is necessary to include endangered plants existing in high altitudinal and latitudinal regions which are adapted to cold temperatures in conservation strategies. It is estimated that the plants adapted to cold climates will illustrate diversified levels of susceptibility, thus resulting in assembly and segmentation of habitat diversity. Therefore, it is well-known that alterations in climate have a negative impact on the physiological responses of plant species. It is evident from earlier reports that there is a tremendous downfall in species diversity of plants and also extinction of plants is higher than convergence due to global warming (Gao et al. 2020; Harrison 2020). The geological discrepancies occurring in some micro-refugia sites show experimental data on the reduction of eradication hazards owing to climatic buffering (Suggitt et al. 2018).

3.3 The Scrutiny of CBD

The Convention on Biological Diversity has been implemented around 33 years ago which is a crucial worldwide tool casing the viable progress in the conservation of plant diversity. It has been approved by around 196 countries in the world as of 2017. This evidenced the watershed on the legal responsiveness to create awareness among people on the issues in the conservation of biodiversity and the responsibilities of the treaties to pledge to understand the necessities of the Convention law and succeeding verdicts consented by the parties (Heywood 2017). The contract was documented in the international regulations that biodiversity conservation was considered as a "communal fear of mankind". The treaty cherished that the nations recollect control over their own biodiversity and regulate the approach towards their hereditary resources. It is therefore obliged that the countries are accountable for the inventory, monitoring, conservation as well as access to their own biological diversity. Yet the convention makes it vibrant that conservation depends on the possibility and suitability (Hagerman and Pelai 2016).

Now, this gives rise to a valid question on the effectiveness of CBD in attaining its major targets as signed on behalf of international authorities (Harrop and Pritchard 2011; Heywood 2017). It was formerly termed the "hard law" with the probability to impose enactment via officially required etiquettes (Abbott and Snidal 2000). Over time, it demanded the submission of countrywide reports and biodiversity policies and acted by the signed parties. Yet only two strategies have reached an agreement, namely, Cartagena and Nagoya, which refer to biosafety and access to hereditary resources, respectively. The major disadvantage of global law is the nonexistence of rebellious procedures. The convention forms a constitution that offers the pliability to the signed parties on how the endowments are instigated (Heywood 2017). The sequential upgrade of CBD started back in 2000 when new approaches were developed aiming at a more articulate and efficient execution of conservation. Then an obligation, "2010 Biodiversity Target", was implemented which is a drastic failure. Thus, to meet these targets, a proposal was made termed as "2020 Aichi Targets". The goals were systematized in a way to attain the targets of the "2010 Biodiversity strategy". In 2012, owing to the national circumstances, the treaties came to an agreement on setting up their individual goals which led to the 2020 Aichi aspiring targets as an aftermath effect. But several nations were unsuccessful in setting up their own targets which resulted in the omission of some strategies from the 2010 biodiversity targets. For instance, well-established conservation strategies and wealthy countries like Canada also lacked to meet the targets (Heywood 2017; MacKinnon et al. 2015).

As the political influence on the accomplishment of necessities of CBD was high in most of the signed countries, there rises a massive competition for resources between conservation of biodiversity and other equally important preferences like industrialization, robustness, and educational development (Heywood 2017). These issues were boldly commented on by Laikre et al. (2008) as most of the allegations of the CBD were influenced by political members rather than researchers. On whole, these deliberations are applied not only to developing countries but to all signed countries. But the developing nations usually fail to meet the targets due to poor infrastructure, resource deficit, and an unequal load of biodiversity they are enforced to tackle.

3.4 Key Issues to Be Addressed

The imminent analysis of triggering elements that have restricted improving the conservation of plant species as well as the lack of achieving the goals signed under the Convention of Biological Diversity (CBD) and other parties discloses several disputes. Roughly, few issues deal with worldwide social as well as commercial factors, some with environmental modification, some with the concern of saving plants, and some were methodological and research-oriented conservation problems.

3.4.1 Data Gaps in Terms of Plant Diversity

Generally, the absence of basic reference data makes the conservation of plants very difficult. The human acquaintance with the diversity of plants was very much limited at all points and classes with regard to species identity, dispersal, evolution, anthropology, biological activity, and socioeconomic benefits. There exist severe gaps in the perception of humans on the subjects of evolution, nomenclature, demographics, analysis of threatened or endangered species, and prominence of conservation (Havens et al. 2014; Heywood 2017). The lack of sufficient data contributes to the human inability to engender info and gain familiarity with the subject area. Hence, building maps and modeling frameworks are impossible which restrict us to achieve the aims, goals, and targets of biodiversity. It is vital to "Know what we don't know" to achieve the targets on time (Heywood 2017). After data retrieval, they have to be deposited in plant-related databases, digital servers, or any other information storing software established in resident, national, district-wise, and international levels. Even though many authorized information tools are available aiding the conservation of plants, lack of funds, inadequate maintenance, and financial support were essential. The funding information is vital to upgrade the International Union for Conservation of Nature (IUCN) red catalog of endangered/threatened and conserved plants and the World Database of Key Biodiversity Areas (KBA) and worldwide expertise scientists retrieved secondary information from the available primary datasets (Juffe-Bignoli et al. 2016). The upsurge of high volumes of datasets and advanced cloud-based computing technologies and increased funding agencies provides greater scope for plant species conservation and renewable utilization of diverse plants. In contrast, there exists an equal number of challenges like formulation of appropriate data analyzing tools so that huge complicated data could be processed accordingly. In addition to this, poorly organized monitoring agendas with fewer environmental concerns make the administrative intrusions more strenuous (Heywood 2010, 2017). Generally, less importance was inclined to implementation of monitoring programs as it is crucial, costly, and insufficient funding was only available. Therefore, young scientists could make effective alterations in monitoring programs which could also be aided by the members of plant species conservation.

3.4.2 Threatened Plant Species

Over a decade, the most obsessive topic with relevance to the conservation of plants is the estimation of the approximate numeral or proportion of species becoming extinct or threatened. Mostly, humans contemplate that plant biodiversity could be conserved by protecting endangered or vulnerable species. This specifically elevates the risk of losing the existing biodiversity (Heywood 2017; Wilcove 2010). The reported estimation of vulnerable species varies in the range of 10–60%, but then they are usually organized in a way that all species ranging from less endangered to desperately vulnerable species clump together which in turn results in forthcoming extinction (Kew 2016). The red list provided by IUCN gives a worldwide summary of threatened species, but only a limited number of samples have been evaluated. For effective plant conservation, each country should take appropriate measures to safeguard the prominence of their own natural flora and fauna with more significance given to endemic plant species, identify the possible risks, and protect them from becoming extinct. As a result, novel conservation programs could be implemented from the baseline. This is evidently seen in the tropical regions with varied flowering plants.

The IUCN along with Royal Botanic Gardens and Natural History Museum globally analyzed the extinction risk for plants grown worldwide (http://threatenedplants.myspecies.info/). This data suggested that one out of five plants is at the risk of becoming extinct. The analysis by “State of the World’s Plants 2016” also provided similar data (Kew 2016).

The next question that arises will be whether it is necessary to know the accurate number of plants that are at high risk of extinction worldwide.

The status, significance, and consequences of these conditions are well-known by the members of the conservation committee. Mostly the governments are bothered about the issues happening within their own nation rather than having a universal depiction or illustration. People should also be equally aware of the species loss at an unusual rate. Some proofs indicate that stimulating the communal people could bring about effective changes in plant conservation rather than pestering the government authorities for financial support. The government will generally give less priority to conservation programs rather than education, physical well-being, and mitigation of poverty. A workshop conducted in March 2017, "Vatican workshop on Biological Extinction" revealed that about fifty percent of the plant species will become extinct by the end of this era (Heywood 2017).

But these conclusions are not given much significance usually as they are stated at frequent breaks and show no visible aftermath effects. When the consequences slightly begin to show in the environment worldwide in addition to climate change, it is evident that most of the plant species shift to a higher risk of extinction.

3.4.3 Effectiveness of Legal Policies in Conserving Global Biodiversity

It is incredible that despite the fact that the ideation of conservation of biological diversity has been extensively recognized locally, regionally, nationally, and internationally, very less consideration was given to the abundance of essential organizations and skilled professionals for the execution of the objectives of the convention and altering the verdicts of the meetings.

It is our responsibility to make conservation the major concern. The areas that are already protected were usually excluded, but it is pathetic that no efforts were taken to conserve the ex-situ sites and for setting up the essential organizational framework worldwide. In contrast, when the agricultural and forests face the consequences of loss of crop diversity at the genetic level, administrations like Food and Agricultural Organization (FAO), Consultative Group for International Agricultural Research (CGIAR), and International Board for Plant Genetic Resources (IBPGR) created a gene bank. They also specified etiquettes to be followed while collecting, storing, and accessing the seeds. In addition to this, several gene banks have been installed statewide as well as countrywide. In terms of conservation of ex-situ plant species, the government has left it to the botanical gardens with the allotment of funds or skilled professionals (Heywood 2010, 2017). Spain is the only country to identify this emerging consequence and some gene banks for seeds have been set up in botanical gardens which are aided by some independent governmental agencies. The condition becomes even worse when in situ conservation of aimed plant species is considered owing to the unavailability of academic measures.

The major goal of CBD is to include institutional activities for attaining its proposed targets. In 2015, Global Environmental Facility (GEF) funded over $12 billion to the developing nations for conserving the overall plant diversity and has included more than 3300 conserved areas (GEF 2015). In the past 25 years of commencement of CBD, the High-level Panel reported that the 2010–2020 targets of Aichi could not be achieved without the correct academic frameworks, resource accessibility, and legal assistance (CBD High-Level Panel 2012, 2014).

3.4.4 Too Many Strategies: A Contradicting Factor

Over the last few decades, one noteworthy achievement is that organizations like United Nations societal charities, Intergovernmental Organizations (IGO), several Non-Governmental Organizations (NGO) and private sectors and various conservational treaties and agreements reported the propagation of evaluations, publications, analyses, policies, strategies, and schemes. Besides the exceptional price of printing these documentary papers with many schemes and illustrations, it is unclear for whom these documents are made and what measures are taken for reaching these documents to the audience at the specified time (Heywood 2017). But the reality is that these documents are preserved in shelves rucked with dust particles or remain unsold in cache/warehouse and very few people only will be aware of these documents. It is hearth wrecking that extensive research remains unnoticed which results in repetition of efforts at a relevantly higher cost. It may seem that much concern is not given to making the process economical or cost-efficient. It is unanticipated how these documents are processed and come to one's table.

Sometimes, the abundance of excessive info is now witnessed in most fields, but recently it has been depicted even in the field of environmental conservation of biodiversity owing to its broad scope, interconnection, and high intricacy. It includes not only the scientific and custom evaluation of conservation, analysis, and improvements at genetic and computational levels, but also the economic, socio-cultural, and legal frameworks. It is hard to be updated on the progress happening in this field due to our limited capability to recite and precise about only a part of these evaluations. Several statistical analyses suggest that we should read more information and process it, but that in turn could always lead to a superficial understanding of the consequences.

3.5 Crucial Issues with Today's Conservation Policies

The conservation strategies followed today are mostly similar to those that were implemented a centennial ago. The major elements were the development of organization of threatened sites, imposition of action plans to protect extinct species, implementation of genebanks for storage, and ex-situ conservation of endangered plants as an insurance plan. But the consequence is that these steps failed to attempt the reintroduction of the threatened plant species to sites from where they have become extinct with the aim of reestablishing the vanished biodiversity (Heywood 2019). It has been specified that the conservation policies followed today are built on the narrative history and superficial myths rather than an organized assessment of the research findings. Nevertheless, we now have enormous information from literature, knowledge, understanding, and capacity along with the high-tech facilities and a substantial amount of gathered skills and practices (Heywood 2019; Sutherland et al. 2004). According to Adams, "The issues concerned with conservation at the beginning of this century were almost the same issues that have been identified and handled by environmentalists about 100 years ago." The same goes with the solution of conservation as well which remains to be more alarming. Therefore, the solutions were also advancing, complicated, sophisticated, and acclimatizing to altering surroundings. But the irony is no solution guarantees a positive swing in the never-ending chess play of species loss (Adams 2004). The perspective of today's conservation practice is drastically diverse from the initial time as now we exist in a phase where human activities have a huge impact on the resource availability and surrounding environment (Seddon et al. 2016; Steffen et al. 2016; Johnson et al. 2017). It is now understood that there is an uprising demand for a novel conservational approach even if no solidarity exists on how to progress further.

The various conservation approaches were interlinked, even if it is not echoed in routine. Over the past decades, the demand for ideal plant conservation deliberates the application of different conservational practices in combination according to the prevailing circumstances, but practically it is hard to come into practice as the way in which conservation is systematized like species and area-based practices were generally established by various expertise of the field (Heywood 2019). The lack of synchronization between two strategies, namely, management of species and conserved sites, occurs as they both epitomize diverse communities in the ecosystem and conservation of biodiversity (Heywood 2017). Therefore, this depicts as a major consequence depending on how the environmental science is practiced and is therefore significant in the split-up of in situ and ex-situ in CBD and the separation of area-specific approaches from species-specific approaches in the country-wise conservation policies and Global Strategy for Plant Conservation (GPSC). It is heart wrecking that not even a single conservational strategy is completely effective, particularly for plants (Havens et al. 2014; Heywood 2017).

The crucial issues with today's conservational policies are listed below (Heywood 2019):

- A widespread universal organization of conserved areas which is floundering in the conservation of biodiversity,
- A hesitant and undeveloped implementation of "other efficient area-specific conservation approaches",
- A rising thrust for the appliance of public-based conservational targets,
- Significant attempts to recognize Key Biodiversity Areas (KBA) (IUCN 2016) and Imported plant areas (IPA) for plants and their relevant association (Darbyshire et al. 2017) and a combination of all these measures at the species level along with
  - Considerable financing for Red Listed threatened species even when such plans are not included in conservation action plans.
  - Conservation and restoration of species and enactment of the same on a partial scale in many countries (Heywood 2015).
  - Restoration, reinstatement of species and efficient recovery are aided by:
  - Enormous ex-situ collection of native plant species from botanical gardens and germplasm and seed banks, environmental organizations, and the worldwide network of agri-based genome banks (particularly for wild crops and aromatic and medicinal plants).

The typical attribute of conservational plans today is that most natural resources are frequently utilized for planning rather than for execution. For instance, plans targeting species conservation and recovery strategies are already prepared, processed, and even permitted but not executed (Dorey and Walker 2018). Enormous effort has been put into training and preparing the national report on "The IUCN Red List of Threatened species", but no substantial action on conservation was implemented (Juffe-Bignoli et al. 2016). A huge amount of funds and money was spent each year for the organization and service of the international NGO, IG, and UN assemblies and aiding the presence of several thousand agencies and summation of conclusions of the proceedings and future recommendations, most of which are either elapsed or forgotten, the infinite number of small-scale pilot schemes that do not have any follow-up, and many more.

3.6 Implementation of Better Strategies in the Future

The continual increase in humanoid population, financial growth, consumption, and overexploitation of resources has led to indefensible biodiversity loss, which further worsened the battle between anthropogenic activities and conservation of biodiversity (Meng et al. 2019a, b). Therefore, future perspectives should be suggested based on a clear understanding of the fundamental values of biodiversity, augmentation of conventional conservation sites, and highlighting the beneficial features of nature to humans. Recent environmentalists are supporters of conserved sites only when they show beneficial effects to mankind. The conservational plans today encourage association with financial agencies and resource utilizers (Meng et al. 2021). A framework should be formulated in a way it gets executed in conservational plans to seal the gaps between human understanding and profitable conservation in the Anthropogenic epoch. The intrusions of conservation should guarantee the sustainability of biodiversity arks for a longer period and improvement of self-monitoring diverse environment. Mostly, this institutes a novel conservational strategy to evaluate and cope with the influences of human activities and climate change on recent and future plans of dispersal of biodiversity. This unravels the impacts of numerous chauffeurs of anthropogenic disturbance, contributing to an improved perception of altering effects of the ecosystem. It also highlights the possibilities of conservational plans and their effective administration in the evolving world. Therefore, future scientists working on the conservation of biodiversity and earth science must illuminate the smooth proceedings and dispersal patterns that safeguard biodiversity against anthropogenic notions.

Even though the attempts for conserving biodiversity have given heartening results in the previous decade, only little has succeeded in preventing the loss of biodiversity due to the failure of tackling the indefensible consumption of resources and reduction of biodiversity in the anthropogenic environment. Hence, conservationists suggest to implement region-wise conservational plants (Meng et al. 2021). Initially, the foremost priority of conservational plans is to minimalize the influence of human activities in hotspots of biodiversity in developing nations.

Nevertheless, humans unintentionally affect the conservation of biodiversity worldwide, especially in developing nations with people suffering a deficit of resources and low standards of living. It is stated that poverty goes in hand with the loss of biodiversity, therefore poverty deficit and conservational issues should be handled in a collaborative form, while there is very less chance of attaining success (Adams 2004; Meng et al. 2021). Then, ecosystems with a greater number of extinct and threatened species should be transformed into protected sites. For instance, the consecrated sites flourished completely with the ecosystem (Fengshui forestry where the locals are restricted to enter) were targeted for in situ conservation of biodiversity. Human activities have devastated some areas with highly endangered species, which have to be monitored very closely (Meng et al. 2021). At last, identification and recognition of the possible climate shielding sites are crucial for alleviating the influence of climate change on biodiversity. But, the shielding zones with warm climates are to be given high consideration in terms of preserving the biodiversity to face the forthcoming climate changes.

As far as Anthropocene is concerned, the future of conservation exists in the restoration of habitats and reintroduction of plants at a wide scale, not only within, but also outside the recognized classical range of species (Volis 2016, 2017). A novel strategy "conservation-oriented restoration" has been proposed which embraces inter-situ and quasi-in-situ conservation as major constituents. This strategy is based on two proceedings:

- There exists no active plan to manage the elevation of threatened species and prevent them from becoming extinct.
- Reintroduction of plants not only within but also outside the known conventional range of species.

Mostly, the reintroduction of plant species outside conventional ranges is not encouraged. As far as the restoration of the ecosystem is concerned worldwide, it requires precise resolutions to scientific, technical, financial, and societal consequences and therefore more funds are required (Lamb 2018). Therefore, resources should be equally utilized for conservation as well as restoration of resources, taking into consideration the financial limitations.

3.7 Conclusion and Future Recommendations

The global threat of climate change is majorly due to the influence of humans and their conservational policies. Scientists are still exploring ways to manage this devastating threat. It is vital to guarantee that the conservational approaches are effective at all stages to gradually decrease the continuous loss of biodiversity so that species become adapted to the evolving ecosystem in the future.

Throughout history, humans are known to cause significant biodiversity loss despite it being beneficial for their well-being. Biodiversity loss and destruction of ecosystems were predicted to increase further in the future. It is intimidating to address the challenges of the influence of the Anthropocene and the conservation of biodiversity. Humans usually fail to focus on the importance of biodiversity at the cost of their well-being, rather they usually concentrate on resource recreation. Therefore, understanding the prevailing crisis of biodiversity is essential and the relevant authorities along with environmentalists and scientists should pledge to prevent the loss.

The failure of conservation of plant diversity to meet national and international goals could be clarified only when different strategies are synchronized right from the initial stages to execution and efficient collaboration should be formed with sponsors, landlords, and residents. Subsequently, the major concern depicting the prevailing biodiversity loss is the letdown of integration of species-specific and area-specific conservational plans. Mostly, the countries focus on protecting the in situ sites with threatened species as a major approach without any effort to eradicate the hazards to indigenous species. This ultimately results in the extinction of threatened species. Therefore, countries should focus on the conservation of in situ species and much effort has to be implemented in the retrieval of vulnerable species. Developed countries with advanced recovery plans also try to implement novel strategies for species restoration frequently (Dorey and Walker 2018).

The success of the conservation of protected areas lies in their ability to face and overcome legal and economic issues in mere future (Dudley et al. 2018). The proper maintenance of protected areas could also sometime lead to biodiversity loss when the existing species and habitats are at high risk of extinction. For species recovery, policies targeting the removal of threats to species should be implemented (Heywood et al. 2018; Monks et al. 2019). Therefore, the balance between the different action plans and their coordinated implementation could positively impact the conservation of plant diversity.

Acknowledgments The authors acknowledge the Vellore Institute of Technology management and the Dean, School of Bio Sciences and Technology, for providing the infrastructure, constant support, and encouragements.

References

Abbott KW, Snidal D (2000) Hard and soft law in international governance. Int Organ 54(3): 421–456

Adams WM (2004) Against extinction: the story of conservation. Earthscan, London

Alexander JM, Chalmandrier L, Lenoir J, Burgess TI, Essl F, Haider S et al (2018) Lags in the response of mountain plant communities to climate change. Glob Chang Biol 24(2):563–579

Arneth A, Shin YJ, Leadley P, Rondinini C, Bukvareva E, Kolb M et al (2020) Post-2020 biodiversity targets need to embrace climate change. Proc Natl Acad Sci 117(49):30882–30891

Braje TJ, Erlandson JM (2013) Human acceleration of animal and plant extinctions: a Late Pleistocene, Holocene, and Anthropocene continuum. Anthropocene 4:14–23

CBD High-Level Panel (2012) High-Level Panel on global assessment of resources for implementing the strategic plan for biodiversity 2011-2020. Resourcing the Aichi Biodiversity Targets: a first assessment of the resources required for implementing the strategic plan for

biodiversity 2011-2020. http://www.cbd.int/doc/meetings/fin/hlpgar-sp-01/official/hlpgar-sp-01-01-report-en.pdf
CBD High-Level Panel (2014) Resourcing the Aichi Biodiversity Targets: an assessment of benefits, investments and resource needs for implementing the strategic plan for biodiversity 2011-2020. Second report of the High-Level Panel on global assessment of resources for implementing the strategic plan for biodiversity 2011-2020, Montreal, Canada
Darbyshire I, Anderson S, Asatryan A, Byfield A, Cheek M, Clubbe C, Radford EA (2017) Important plant areas: revised selection criteria for a global approach to plant conservation. Biodivers Conserv 26(8):1767–1800
Dirzo R, Young HS, Galetti M, Ceballos G, Isaac NJ, Collen B (2014) Defaunation in the Anthropocene. Science 345(6195):401–406
Dorey K, Walker TR (2018) Limitations of threatened species lists in Canada: a federal and provincial perspective. Biol Conserv 217:259–268
Dudley N, Jonas H, Nelson F, Parrish J, Pyhälä A, Stolton S, Watson JE (2018) The essential role of other effective area-based conservation measures in achieving big bold conservation targets. Glob Ecol Conserv 15:e00424
Gao JG, Liu H, Wang N, Yang J, Zhang XL (2020) Plant extinction excels plant speciation in the Anthropocene. BMC Plant Biol 20(1):1–11
GEF (2015) Behind the numbers 2015. A closer look at GEF achievements. https://www.thegef.org/sites/default/files/publications/GEF_numbers2015_CRA_bl2_web_1.pdf
Hagerman SM, Pelai R (2016) "As far as possible and as appropriate": implementing the Aichi Biodiversity Targets. Conserv Lett 9(6):469–478
Harrison S (2020) Plant community diversity will decline more than increase under climatic warming. Philos Trans R Soc B 375(1794):20190106
Harrop SR, Pritchard DJ (2011) A hard instrument goes soft: the implications of the Convention on Biological Diversity's current trajectory. Glob Environ Chang 21(2):474–480
Havens K, Kramer AT, Guerrant EO Jr (2014) Getting plant conservation right (or not): the case of the United States. Int J Plant Sci 175(1):3–10. https://doi.org/10.1086/674103
Heywood VH (2010) Developing new biodiversity conservation strategies in response to global change. BMIB-Bollettino dei Musei e degli Istituti Biologici 72:95–122
Heywood VH (2015) In situ conservation of plant species–an unattainable goal? Isr J Plant Sci 63(4):211–231
Heywood VH (2017) Plant conservation in the Anthropocene–challenges and future prospects. Plant Divers 39(6):314–330
Heywood VH (2019) Conserving plants within and beyond protected areas–still problematic and future uncertain. Plant Divers 41(2):36–49
Heywood V, Shaw K, Harvey-Brown Y, Smith P (2018) BGCI and IABG's species recovery manual. Botanic Gardens Conservation International, Richmond
IPBES Report (2019) Transforming Changes Are Necessary to Restore and Protect Nature [2021-01-28]. https://www.ipbes.net/
IPCC (2019) The IPCC and the sixth assessment cycle [2021-01-28]. http://www.ipcc.ch/
Isbell F, Gonzalez A, Loreau M, Cowles J, Díaz S, Hector A et al (2017) Linking the influence and dependence of people on biodiversity across scales. Nature 546(7656):65–72
IUCN (2016) A global standard for the identification of key biodiversity areas, Version 1.0, 1st edn. IUCN, Gland
Johnson CN, Balmford A, Brook BW, Buettel JC, Galetti M, Guangchun L, Wilmshurst JM (2017) Biodiversity losses and conservation responses in the Anthropocene. Science 356(6335): 270–275
Juffe-Bignoli D, Brooks TM, Butchart SH, Jenkins RB, Boe K, Hoffmann M et al (2016) Assessing the cost of global biodiversity and conservation knowledge. PLoS One 11(8):e0160640. https://doi.org/10.1371/journal.pone.0160640

Kennedy CM, Oakleaf JR, Theobald DM, Baruch-Mordo S, Kiesecker J (2019) Managing the middle: a shift in conservation priorities based on the global human modification gradient. Glob Chang Biol 25(3):811–826

Kew (2016) Royal Botanic Gardens, Kew. State of the World's Plants 2016. https://stateoftheworldsplants.com/report/sotwp_2016.pdf

Laikre L, Jonsson BG, Ihse M, Marissink M, Gustavsson AMD, Ebenhard T et al (2008) Wanted: scientists in the CBD process. Conserv Biol 22(4):814–815

Lamb D (2018) Undertaking large-scale forest restoration to generate ecosystem services. Restor Ecol 26(4):657–666

Lewis SL, Maslin MA (2015) Defining the Anthropocene. Nature 519(7542):171–180

Locke H, Ellis EC, Venter O, Schuster R, Ma K, Shen X et al (2019) Three global conditions for biodiversity conservation and sustainable use: an implementation framework. Natl Sci Rev 6(6): 1080–1082

MacKinnon D, Lemieux CJ, Beazley K, Woodley S, Helie R, Perron J et al (2015) Canada and Aichi Biodiversity Target 11: understanding 'other effective area-based conservation measures' in the context of the broader target. Biodivers Conserv 24(14):3559–3581

Meng HH, Zhou SS, Li L, Tan YH, Li JW, Li J (2019a) Conflict between biodiversity conservation and economic growth: insight into rare plants in tropical China. Biodivers Conserv 28(2): 523–537

Meng HH, Zhou SS, Jiang XL, Gugger PF, Li L, Tan YH, Li J (2019b) Are mountaintops climate refugia for plants under global warming? A lesson from high-mountain oaks in tropical rainforest. Alp Bot 129(2):175–183

Meng H, Gao X, Song Y, Cao G, Li J (2021) Biodiversity arks in the Anthropocene. Reg Sustain 2(2):109–115

Monks L, Barrett S, Beecham B, Byrne M, Chant A, Coates D et al (2019) Recovery of threatened plant species and their habitats in the biodiversity hotspot of the Southwest Australian Floristic Region. Plant Divers 41(2):59–74

Myers N, Mittermeier RA, Mittermeier CG, Da Fonseca GA, Kent J (2000) Biodiversity hotspots for conservation priorities. Nature 403(6772):853–858

Ordonez A, Williams JW, Svenning JC (2016) Mapping climatic mechanisms likely to favour the emergence of novel communities. Nat Clim Chang 6(12):1104–1109

Otto SP (2018) Adaptation, speciation and extinction in the Anthropocene. Proc R Soc B 285(1891):20182047

Panetta AM, Stanton ML, Harte J (2018) Climate warming drives local extinction: evidence from observation and experimentation. Sci Adv 4(2):eaaq1819

Raftery AE, Zimmer A, Frierson DM, Startz R, Liu P (2017) Less than 2 C warming by 2100 unlikely. Nat Clim Chang 7(9):637–641

Seddon N, Mace GM, Naeem S, Tobias JA, Pigot AL, Cavanagh R et al (2016) Biodiversity in the Anthropocene: prospects and policy. Proc R Soc B Biol Sci 283(1844):20162094

Steffen W, Leinfelder R, Zalasiewicz J, Waters CN, Williams M, Summerhayes C et al (2016) Stratigraphic and earth system approaches to defining the Anthropocene. Earth's Future 4(8): 324–345

Suggitt AJ, Wilson RJ, Isaac NJ, Beale CM, Auffret AG, August T et al (2018) Extinction risk from climate change is reduced by microclimatic buffering. Nat Clim Chang 8(8):713–717

Sutherland WJ, Pullin AS, Dolman PM, Knight TM (2004) The need for evidence-based conservation. Trends Ecol Evol 19(6):305–308

Tang CQ, Yang Y, Ohsawa M, Momohara A, Mu J, Robertson K (2013) Survival of a tertiary relict species, Liriodendron chinense (Magnoliaceae), in southern China, with special reference to village Fengshui forests. Am J Bot 100(10):2112–2119

Vellend M, Baeten L, Myers-Smith IH, Elmendorf SC, Beauséjour R, Brown CD et al (2013) Global meta-analysis reveals no net change in local-scale plant biodiversity over time. Proc Natl Acad Sci 110(48):19456–19459

Vellend M, Baeten L, Becker-Scarpitta A, Boucher-Lalonde V, McCune JL, Messier J et al (2017) Plant biodiversity change across scales during the Anthropocene. Annu Rev Plant Biol 68:563–586

Volis S (2016) Conservation meets restoration–rescuing threatened plant species by restoring their environments and restoring environments using threatened plant species. Isr J Plant Sci 63(4): 262–275

Volis S (2017) Complementarities of two existing intermediate conservation approaches. Plant Divers 39(6):379–382

Wilcove DS (2010) Endangered species management: the US experience. In: Conservation biology for all, pp 220–235

Zalasiewicz J, Waters CN, Wolfe AP, Barnosky AD, Cearreta A, Edgeworth M, Williams M (2017) Making the case for a formal Anthropocene Epoch: an analysis of ongoing critiques. Newsl Stratigr 50(2):205–226

Extinction of Medicinal Plants in Anthropocene Epoch: Special Reference to *Rauwolfia serpentina*

4

Ressin Varghese, K. M. Gothandam, Inocencio E Buot, Jr, Rajasekaran Chandrasekaran, and Siva Ramamoorthy

Abstract

Rauwolfia serpentina is a valuable medicinal plant belonging to *Apocynaceae* family. The plant is rich with various phytochemicals particularly indole alkaloids like reserpine. Root extracts of plant have been used from centuries for the ailment of neurological disorders. Successful clinical studies have unravelled the properties like antihypertensive, antidiabetic nature, etc. However, the plant which was available widely in southern western ghats of India is now under threat of extinction. Unrestrained human exploitation of medicinal plants in anthropocene epoch has led to the reduction of plants like *R.serpentina*. An elaborate literature survey of phytochemicals and the so far proven medicinal properties of *R. serpentina* was performed. Advent of in vitro propagational strategies and the accomplishment of *Rhizogenes*-induced roots were also presented as strategies to conserve the plant in vitro. Moreover, the other biotechnological approaches used to raise the production of secondary metabolites as well for conservation were discoursed.

Keywords

Anthropocene · Endangered plants · Medicinal plants · *Rauwolfia serpentine* · Conservation

R. Varghese · K. M. Gothandam · R. Chandrasekaran · S. Ramamoorthy (✉)
Department of Biotechnology, School of Bio Sciences and Technology, Vellore Institute Technology, Vellore, Tamil Nadu, India

I. E. Buot, Jr
Institute of Biological Sciences, College of Arts and Sciences, University of the Philippines Los Baños, Laguna, Philippines

S. Ramamoorthy et al. (eds.), *Plant Diversity in Biocultural Landscapes*,
https://doi.org/10.1007/978-981-19-8649-9_4

4.1 Introduction

Medicinal plants are gaining attention of pharmaceutical industry at global level because of their immense medicinal value and reliability. As a matter of fact, these are the driving factors causing the extensive exploitation and endangering of these treasures. *Rauwolfia serpentina* (Linn.) Benth. Ex Kurz in *Apocynaceae* family is one such plant which was exploited indiscriminately due to their unparalleled effects on nervous system. Some other plants belonging to this genus are *R. tetraphylla, R. vomitoria, R. hookeri, R. micrantha,* and *R. verticillate*. *Rauwolfia serpentina* was named in honour of Leonard Rauwolf, a German physician (Goenka 2007). The plant is popular in Sanskrit name *Sarpagandha* (Snake root) which was first mentioned in historic works of Charak (1000–800 BC). The medicinal properties were also deliberated in other Ayurvedic literatures of Sushruta, Vrindamadha, and Bhavprakasha (Chauhan et al. 2017a).There are a lot of controversies regarding this naming. But the most accepted argument behind the naming is the antivenom property of this plant. Moreover, some assumptions are regarding the resemblance of roots with snake (sarpa). Another unlikely belief is that snakes run away because of the smell of the plant. Some other vernacular names are *Chandra*, *Naakooli*, *Amalapori*, *Patalagaruda,* etc. The plant is widely observed and cultivated in India, China, Japan, Sri Lanka, and Bangladesh. Appreciable vegetative growth is observed in acidic sandy soil and at an annual rainfall of 200–250 cm. It is found to be indigenous in Himalayas, Punjab, Assam, Meghalaya, Uttarakhand, Sikkim, Andaman, and Western Ghats (Agrawal 2019; Bhanwaria et al. 2021; Dey and De 2010).

This tropical shrub is 6 in. to 2 ft. long and their thick, lengthy roots possess a characteristic lens-shaped structure. These greyish brown clustered roots with uneven cervices generally penetrate deep into the soil. Leaves in whorls are ecliptic or lanceolate. Another unique feature is the clump of pink as well as white flowers. The fleshy, tiny drupe type green fruits become black purple on maturing. The flowering period is usually between November and December (Chauhan et al. 2017a; Agrawal 2019).

The therapeutic significance of plant was dated centuries ago. Advent of use of dried root powder as '*Pagal ki dawa*' for treatment of mentally violent patients was a milestone. The beginning of twentieth century witnessed several attempts to determine phytochemical constituents of the plant (Chatterjee 1953; Monachino 1954). A revolutionary step in the saga of *Rauwolfia serpentina* was put forward by Rustom Jal Vakil. This renowned scientist in cardiology carried out tremendous clinical trials for studying the antihypertensive property of dried root powder of the plant. His unparalleled efforts led to the worldwide recognition of the plant as an antihypertensive drug and he was later entitled as 'Father of *Rauwolfia serpentina*' (Goenka 2007). Unfortunately, the overexploitation of *R. serpentina* had led to a considerable reduction in its overall number and was later categorized as an endangered plant by International Union for the Conservation of Nature and Natural Resources (IUCN) and also in CITES (Convention on International Trade in Endangered Species) (Mehrotra et al. 2015). *R. serpentina* is a victim of unconstrained human exploitation during the anthropocene epoch, characterized by escalating anthropogenic activities amidst climate anomalies, aggravating climate change impact on the environment, biodiversity, and society as a whole. Tremendous and swift measures are necessary for the safeguarding of these plants.

This chapter emphasizes on the significance of conservation of *Rauwolfia serpentina* through highlighting renown pharmacological properties and its phytochemicals rendering those activities. Besides, various in vitro propagation and biotechnological approaches utilized in the *R. serpentina* research have also been discussed.

4.2 Phytochemicals

Phytochemicals are naturally available chemical compounds which attribute beneficial properties to plants such as defence against pathogen attack and abiotic stresses. Interestingly, these phytochemicals also have therapeutic properties like antimicrobial, anticarcinogenic activity, etc. (Njerua et al. 2013). An array of diverse extraction techniques have been employed for qualitative and quantitative phytochemical profiling of extracts of *R. serpentina* (Hussain et al. 2015; Kumar et al. 2016b; Namita and Yadav 2016; Das et al. 2019). The concentration of phytochemicals estimated via various detection strategies is depicted in Table 4.1.

4.2.1 Alkaloids

Rauwolfia serpentina is rich with monoterpenoid indole alkaloids which are the key factors that attribute the antihypertensive properties (Geissler et al. 2016). For instance, reserpine, serpentine, reserpiline, ajmalicine, ajmaline, ajmalimine, aricine, rescinnamidine, rescinnamine, deserpidine, corynanthine, isoreserpiline, isoreserpine, indobinine, indobine, yohimbine, isorauhimbinic acid, yohimbinic acid, 3 hydroxysarpagine, *N*(b)-methylisoajmaline, and *N*(b)-methylajmaline. Alkaloids are basically classified as reserpine-like and yohimbine-like. They considerably vary in their chemical structure (Fig. 4.1). The alkaloid profiling of different tissue extracts of wild plant, cultivated plant, and in vitro regenerated plant has been performed (Panwar et al. 2011). Furthermore, they have been successfully isolated and crude forms were used for various experimental studies (Sagi et al. 2016; Singh et al. 2017b). On the other hand, large bunches of root are necessary for the isolation of minor alkaloids (Falkenhagen et al. 1993). Ammoniacal chloroform was elucidated to be the best solvent for alkaloid extraction (Bindu et al. 2014). The techniques employed for the alkaloid examination were thin layer chromatography (Le Xuan et al. 1980), High-performance liquid chromatography (Wachsmuth and Matusch 2002), High-performance thin layer chromatography (Klyushnichenko et al. 1995; Panwar and Guru 2011), spectrophotometry (Singh et al. 2004), and GC-MS (Hong et al. 2013; Hussain et al. 2015).The advance techniques with more productivity like liquid chromatography with quadrupole time-of-flight mass

Table 4.1 Estimation of important phytochemicals from different parts of *R.serpentina* and the respective detection methods

Phytochemical	Plant part	Concentration	Solvent used	Detection	References
Reserpine	Root	35.18 mg/L	Ethanol	UHPLC[a]	Kumar et al. (2016a, b)
		0.31%	Methanol	HPTLC	Negi et al. (2014)
		90%	Methanol	HPLC	Panwar et al. (2011)
		0.955	–	Spectrophotometry	Deshmukh et al. (2012)
		254.8 ± 4.31 μg/g	Chloroform	HPLC–ESI-QToF-MS/MS	Bindu et al. (2014)
	Wild root	26.74 mg/g	Methanol	HPLC	Panwar and Guru (2011)
	Cultivated root	25.16	Methanol	HPLC	
		11.7 mg/g	Methanol	UHPLC-QToF-MS	Sagi et al. (2016)
	Leaf	2.40 mg/L	Ethanol	UHPLC[a]	Kumar et al. (2016a, b)
		80%	Methanol	HPLC	Panwar et al. (2011)
		0.880 mg/g	–	Spectrophotometry	Deshmukh et al. (2012)
	Stem	2%	Methanol	HPLC	Panwar et al. (2011)
	Stem callus	6.81 mg/g	Methanol	HPLC	Panwar and Guru (2011)
	Leaf callus	7.34 mg/g	″	″	
	In vitro root	33 mg/g	″	″	
Ajmaline	Root	52.27 mg/L	Ethanol	UHPLC[a]	Kumar et al. (2016a, b)
		0.817 mg/g	–	Spectrophotometry	Deshmukh et al. (2012)
	Leaf	3.41 mg/g	Methanol	UHPLC-QToF-MS	Sagi et al. (2016)
		0.74 mg/L	Ethanol	UHPLC[a]	Kumar et al. (2016a, b)
		0.485 mg/L	–	Spectrophotometry	Deshmukh et al. (2012)
Ajmalicine	Root	3.14 mg/L	Ethanol	UHPLC[a]	Kumar et al. (2016a, b)
		0.440 mg/g	–	Spectrophotometry	Deshmukh et al. (2012)
	Leaf	0.48 mg/g	Methanol	UHPLC-QToF-MS	Sagi et al. (2016)
		0.22 mg/L	Ethanol	UPHLC[a]	Kumar et al. (2016a, b)
		0.753 mg/g	–	Spectrophotometry	Deshmukh et al. (2012)

Yohimbine	Leaf	5.86 mg/L	Ethanol	UPHLC[a]	
		0.537 mg/g	–	Spectrophotometry	Kumar et al. (2016a, b)
	Root	0.584 mg/g	″	″	Deshmukh et al. (2012)
		3.11 mg/L	Ethanol	UPHLC[a]	Kumar et al. (2016a, b)
		0.24 mg/g	Methanol	UHPLC-QToF-MS	Sagi et al. (2016)
Serpentine	Leaf	0.52 mg/L	Ethanol	UPHLC[a]	Kumar et al. (2016a, b)
	Root	76.38 mg/L	Ethanol	UPHLC[a]	Kumar et al. (2016a, b)
		4.46 mg/g	Methanol	UHPLC-QToF-MS	Sagi et al. (2016)
	Root	0.19 mg/g	Methanol	UHPLC-QToF-MS	Sagi et al. (2016)
Corynanthine	Root	1.76 mg/g	Methanol	UHPLC-QToF-MS	Sagi et al. (2016)
Flavonoids	Young leaves	127.7 ± 1.49 mg/g	Methanol	Aluminium chloride spectrophotometry	Keshavkant et al. (2008)
	Old leaves	123.96 ± 0.07 mg/g			
	Root	41.56 ± 0.04 mg/g			
Phenols	Young leaves	223.32 ± 3.21 mg/g	Methanol	Using Folin ciocalteu reagent	Keshavkant et al. (2008)
	Old leaves	189.97 ± 1.19 mg/g			
	Root	84.13 ± 0.7 mg/g			

UHPLC ultra-high-performance liquid chromatography, *HPTLC* high-performance thin layer chromatography, *HPLC* high-performance liquid chromatography, *HPLC–ESI-QToF-MS/MS* high-performance liquid chromatography coupled with electrospray ionization quadrapole time of flight tandem mass spectrometry

[a] With hybrid triple quadrupole linear ion trap mass spectrometry

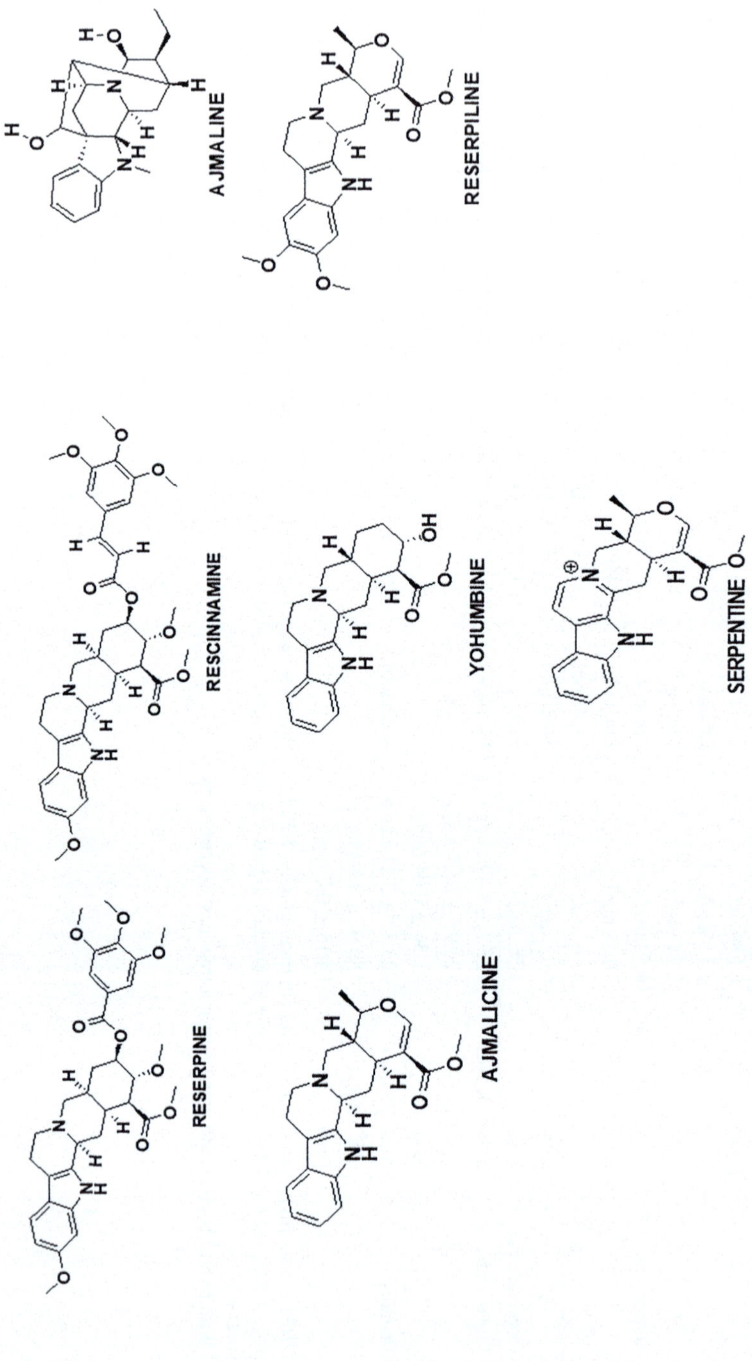

Fig. 4.1 Chemical structure of major indole alkaloids

spectrometry (Kumar et al. 2016a), UHPLC with Hybrid Triple Quadrupole Linear Ion Trap Mass Spectrometry (Kumar et al. 2016b), UPHLC-photo diode array-mass spectrometry (Sagi et al. 2016), and high-performance liquid chromatography coupled with electrospray ionization quadrapole time of flight tandem mass spectrometry (HPLC–ESI-QToF-MS/MS) (Bindu et al. 2014) were also later employed. The pharmacological properties of these alkaloids, particularly remedies on mental discomforts, have been established in the mid-1900s itself (Banerjee and Lewis 1954). Even though the concentration of alkaloids varies with plant parts, the roots are considered to be most therapeutically significant part of the plant (Lucas 1963). Later, studies were conducted to evaluate the dependency of alkaloid content on geographical conditions. Large variations were attained underlining the necessity of genetic breeding programmes for developing high yield variety (Usmani et al. 2015).

4.2.1.1 Reserpine

Reserpine is considered as the most potent and therapeutically efficient indole alkaloid derived from the root extracts of *R. serpentina.* Chemically, the compound is 11, 17 α-Dimethoxy-18 β-[(3,4,5-Trimethoxybenzoyl) Oxy]-3 β, 20 α-yohimban-16 β-carboxylic acid methyl ester ($C_{33}H_{40}N_2O_9$) (Singh et al. 2017b). It is whitish yellow powder, insoluble in water. The compound was extracted and used by Robert Wallace Wiggins (Lobay 2015). The bioavailability of the compound ranges from 50 to 70%. The ingested reserpine distributed throughout the body even in breast milk. Tissue-specific biosynthesis of reserpine in roots was spotted in many investigations (Panwar et al. 2011). In addition to this, reserpine has been successfully isolated and quantified (Negi et al. 2014). The mass production of reserpine from the plant at industrial scale later began (Zafar et al. 2020a).

Reserpine is now a commercially valuable pharmaceutical drug approved by FDA in 1953 for hypertensive patients (McQueen et al. 1954). Reserpine can act on sympathetic-parasympathetic equilibrium of central nervous system and evokes a sedative sensation. Along with this, some studies pointed out the effect of reserpine on level of glycogen, acetylcholine, and anti-diuretic hormones (Goel et al. 2009). Interestingly, it drastically reduced the blood pressure and exhibited vasodilating effect on comparison with synthetic drugs. The action on blood vessels was found to be persistent (Mcqueen and Blackman 1955; Shamon and Perez 2016). It was also perceived that frequent administration of minimum doses of reserpine induced variations in motor signals in pharmacological models of Parkinson's disease (Fernandes et al. 2012). Moreover, apoptosis inducing ability of reserpine has been noted in many studies. It inhibited DNA synthesis as well as destabilized the mitochondrial membrane potential. Recently, it had been proven that reserpine is an efficient therapeutic agent in cancer treatment. This alkaloid supressed cell proliferation and DNA repair and even induced apoptosis by regulating TGF-β signalling pathway (Ramu et al. 2020). However, certain assumptions circulated regarding the chances of genetic mutation on regular administration of reserpine. But in early 1990s itself, these controversies were proved to be wrong (von Poser et al. 1990). Various drug formulations with distinct trade names like Regroton, Demi-Regroton,

Salutensin, Hydroserpine, and Hydropres-50 are now available in market (Singh et al. 2017b).

4.2.1.2 Rescinnamine

Rescinnamine ($C_{35}H_{42}N_2O_9$), the 3,4,5-trimethoxycinnamic acid ester of methyl reserpate, is another bioactive compound with antihypertensive as well as cytotoxic property (AbdelHafez et al. 2013). It was reported that reserpine could easily dissolve and exert vasodilating effect on blood vessels (Mcqueen and Blackman 1955). The underlying factor behind this result was the ability of rescinnamine to block the conversion of angiotensin I to angiotensin II. Further studies proved that it also reduced aldosterone secretion. The earlier studies evidenced the hypotensive, bradycardic, and depressive nature of this compound (Klohs et al. 1954). Rescinnamine, its similar compounds, and the synthetic production procedure were officially patented in 1975. Commercially, now it is available in the name 'Tsuruselpi S' (Singh et al. 2017b).

4.2.1.3 Ajmaline

Ajmaline is another therapeutically consumed indole alkaloid for antiarrhythmic effect. Ajmaline ($C_{20}H_{26}N_2O_2$) is chemically a monoterpenoid alkaloid with ajmalan backbone in which hydroxyl groups are substituted at 17 and 21 positions. This alkaloid was isolated in 1931 and named ajmaline in honour of Hakim Ajmal, a legendary person in Unani medical system (Agrawal 2019).

Ajmaline is a class I antiarrhythmic agent which can alter the electrocardiogram characteristics (Roten et al. 2012). It decreases J wave and extends P-Q interval, Q-T interval, QRS complex, and R wave. It is a potent sodium channel blocker and employed in the diagnosis of Brugada syndrome and its subtypes (Rolf et al. 2003). Moreover, it is also administrated in the treatment of tolerated monomorphic ventricular tachycardias and atrial fibrillation cases in Wolff-Parkinson-White syndrome (Singh et al. 2017b). Cardenolide and bufadienolide are two important synthetic derivatives of ajmaline with antiarrhythmic and cardiotonic property. These compounds as well as the protocol for production was patented in 1975 (Makarevich et al. 1979). Now it is commercially available in brand names like Ajmalinum, Gilurytmal, Ajmalina, etc. (Singh et al. 2017b).

4.2.1.4 Yohumbine

Yohumbine ($C_{21}H_{26}N_2O_3$), the α-adrenergic blocker, is widely used in erectile dysfunction disorders as it could increase blood flow towards penis (Ernst and Pittler 1998). It specifically blocks presynaptic neuron and is experimentally validated in rabbits (Starke et al. 1975). It was reported that administration of yohumbine stimulated the fight or flight responses like anxiety, shivering, palpitations, hot, cold flashes, pupil dilation, etc. in experimental subjects (Charney et al. 1984). Yohumbine is also often considered as a sexual desire-inducing agent. It is also well-known for its hypnotic activity (Mehrotra et al. 2015).

4.2.1.5 Ajmalicine

Ajmalicine ($C_{21}H_{24}N_2O$) is an indoline alkaloid extracted from *R. serpentina* as well as *Catharanthus* species (Zenk et al. 1977). The concentration of ajmalicine was reported to be more in stem extract when compared to that of roots (Deshmukh et al. 2012). It was reported that the amount of ajmalicine in roots could be raised up by optimizing growth media conditions like utilizing sucrose, phosphate, and 6.5 pH (Bhagat et al. 2020). The commercial production has also been enhanced by use of $CdCl_2$ (Zafar et al. 2020a) Interestingly, similar to other phytochemicals, ajmalicine maintains cerebral blood pressure and also possesses psychotic effects (Agrawal 2019). Ajmalicine also possesses renal vasodilation properties and it selectively inhibits postganglionic functioning of sympathetic system (Mehrotra et al. 2015).

4.2.1.6 Other Alkaloids

Reserpiline is a yohimban alkaloid with antipsychotic property (Gupta et al. 2012). Serpentine is an anhydronium alkaloid with sedative effect. However, it also excites heart rate. Interestingly, serpentine also exhibited antihistaminase activity and was testified in guinea pigs (Sachdev et al. 1961). It also inhibits succinate dehydrogenase in liver and brain tissues. Isoajmaline, neoajmaline, and rauwolfinine are some other alkaloids with apparent actions on nervous system (Singh et al. 2017b). 3-hydroxysarpagine, yohimbinic acid, isorauhimbinic acid, N_b-methylajmaline, and N_b-methylisoajmaline are some other indole alkaloids. Remarkably, yohimbinic acid exhibited significant inhibition on topoisomerase which could be further researched for anticancer therapy (Itoh et al. 2005). It was noted that alkaloids like vomilenine and perakine were produced from cell suspension cultures in contrast to wild plant (Nikolaeva and Alterman 1993).

4.2.2 Other Phytochemicals

R. serpentina is also rich in phenols, flavonoids, and tannins. Phenols are intensely valuable because of its antioxidant, antimicrobial, anticancer, antidiabetic, anti-inflammatory, hepatoprotective, neuroprotective, and cardioprotective nature (Saibabu et al. 2015). Flavonoids are free radical scavengers and widely studied phytochemical in fields of cancer, inflammation, etc. (Panche et al. 2016). The presence of radical scavenging potential of phenols and flavonoids of *R. serpentina* was assessed and found to be higher (Keshavkant et al. 2008). Later, the use of salicylic acid (elicitor) and $AcCN/H_2O$ (solvent) was suggested with the intention of elevating the production of phenolic acids and flavonoid rutin at commercial scale (Nair et al. 2013). Quercetin, a most potent flavonoid antioxidant, was also isolated from the leaves of *R. serpentina*. It exhibits anti-inflammatory, anticancer properties and is used in the treatment of arthritis, bladder infections, asthma, eczema, and diabetes (Gupta et al. 2015). Similarly, rutin was also detected. Rutin prevents DNA alteration, lipoprotein peroxidation, and thus mutations (Gupta and Gupta 2015). Tannins are peculiar with its antiseptic property (Vieira Pereira et al. 2015). In addition to this, the presence of saponins and fat was identified in *n*-

hexane extract of roots of *R. serpentina* (Hussain et al. 2015). The presence of carbohydrates, glycosides, starch, coumarins, emodins, and phlobatanins was also later revealed. Anthraquinones were also ascertained in stem extracts of the plant (Deshmukh et al. 2012; Das et al. 2020; Vaishnav and Sahoo 2020).

The plant is also nutritionally copious with vitamins like riboflavin, ascorbic acid, thiamine, and niacin. The macronutrients like calcium, phosphorous, magnesium, and sodium were detected in the plant extract. However, zinc and iron are major microelements. (Singh et al. 2017b). All these elements are essential for the proper growth and development of our body. Unfortunately, presence of some heavy metals like arsenic, chromium, and cadmium was also detected in trace amounts via atomic absorption spectroscopy (Gupta and Gupta 2016). Recently, the presence of titanium, tin, vanadium, molybdenum, chromium, aluminium, silicon, mercury, strontium, nickel, cadmium, lead, and bismuth in limited amounts was identified in leaves and seeds by means of direct current arc optical emission spectroscopy (Bharti et al. 2020). Above all, antioxidant enzyme superoxide dismutase has been identified in the extracts (Kirillova et al. 2001).

4.3 Pharmacological Properties

Curative properties of *R. serpentina* have been efficiently utilized by tribal people in different parts of the world for the treatment of various ailments. According to literature surveys, tribal groups in Similipal Biosphere Reserve are using the root extract of *R. serpentina* for curing malaria (Panda 2014). As reported in Indian Ayurvedic literature, the root extracts of the plant were used in doctoring mental illness, sleeplessness, cardiovascular diseases, erotic violent behaviours, snake bites, diarrhoea, and severe pain (Lobay 2015; Bunkar 2017a). It was also administrated to ease uterine contraction during delivery (Singh and Singh 2009). It is now available in the market in the form of herbal drugs (Rungsung et al. 2014). When compared to the synthetic drug, extracts of *R. serpentina* have the following advantages like fewer side effects, no digestive issues, increased bioavailability, and reduced chances of drug resistance development (Bunkar 2017a). Figure 4.2 represents the phytochemical properties of *R. serpentina*.

4.3.1 Antioxidant Nature

The formation of oxidative stress in our body leads to various diseases like cancer and diabetes. Medicinal plants are chemically rich in antioxidants. In this context, extracts of the *R. serpentina* could be efficiently utilized to prevent and treat oxidative damages. Various studies have assessed the antioxidant potential of extracts of *R. serpentina* via 2, 2′-diphenyl-1-picryl Hydrazyl Radical (DPPH) assay (Fazal et al. 2011). Interestingly, methanolic extracts of roots showed higher free radical scavenging activity than leaves and buds (Dey et al. 2016). In addition to this, further studies exposed that wild varieties showed more activity than cultivated ones (Chauhan et al. 2017b).

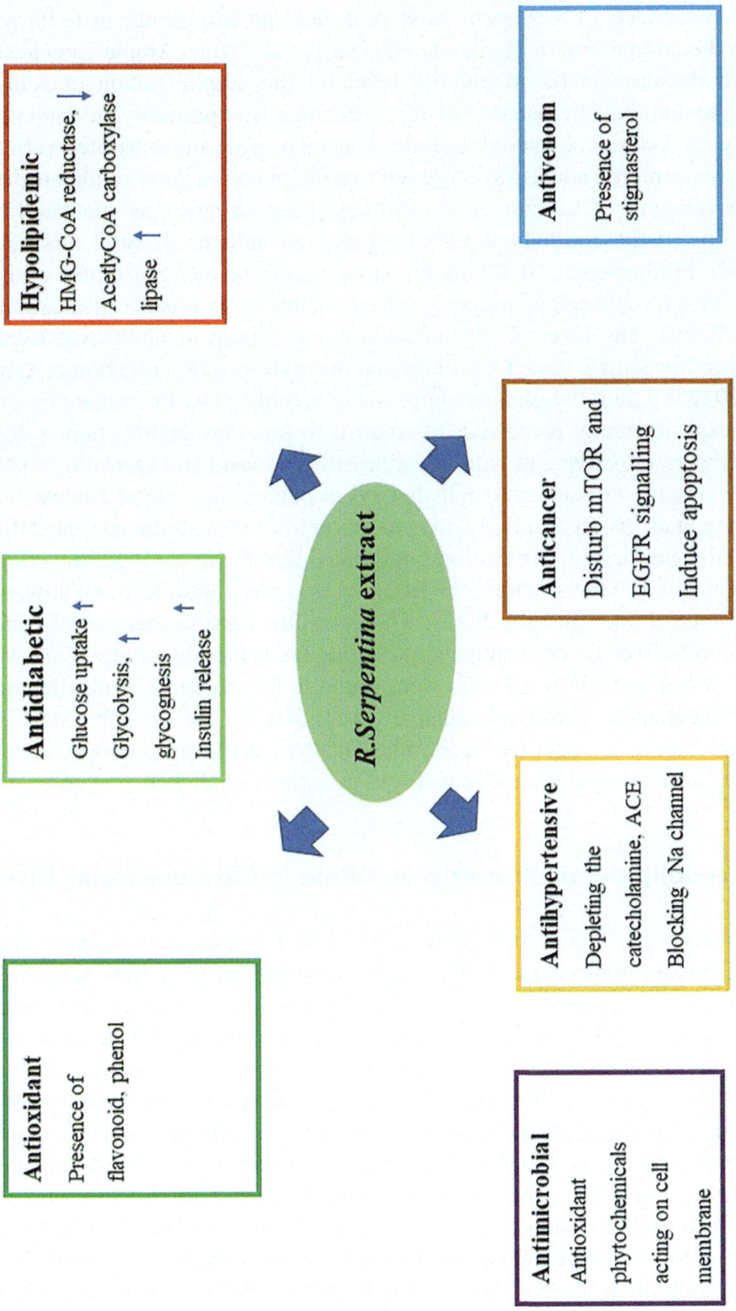

Fig. 4.2 Depicts the various pharmacological properties of *R. serpentina* and the possible reasons or ways behind the action

4.3.2 Antidiabetic Activity

The aqueous extracts of *R. serpentina* were used as an antidiabetic drug for periods by the local clan in northern Thailand (Manosroi et al. 2011). Ample clinical studies revealed a decrease in blood glucose level on the administration of Rauwolfia therapy. The antidiabetic activity of *R. serpentina* was primarily investigated via alpha-amylase assay. The hypoglycaemic efficacy of wide and cultivated varieties of the plant was explored and it was observed that the plant is a good candidate drug for diabetic treatments (Chauhan et al. 2017b). Interestingly, the methanolic root extracts showed the antidiabetic effect on alloxan-induced diabetic rats (Qureshi et al. 2009). Furthermore, 10–60 mg/kg was found to be the appropriate dose of an extract with hypoglycaemic property based on trials in Wistar mice (Azmi and Qureshi 2012b). The level of glycosylated haemoglobin in blood was used as a parameter to investigate the effects of methanolic extract of *R. serpentina* (Azmi and Qureshi 2012a). Later, the glucose suppressing activity of hydromethanolic extracts was also experimentally perceived. In contrast to previous studies, higher doses of the extract were effective and with less side effects (Azmi and Qureshi, 2014). The action of extract over pancreatic cells helped in maintaining blood glucose homeostasis (Azmi et al. 2018). Similarly, investigation in fructose-induced type 2 diabetic mice was also performed. The findings suggested that the hypoglycaemic effect was either because of reduced intake of fructose into the intestine or of low insulin resistance (Azmi and Qureshi 2013). These results were further corroborated by docking studies. The phytochemicals like ajmaline, serpentine, rauwolfine, yohimbine, and sarpagine in *R. serpentina* were found to be activators of insulin and thus mediated the diabetic effect (Ganugapati et al. 2012). The dry lab studies using aldose reductase as a target led to the identification of potent phytochemicals like indobine in attributing antidiabetic nature (Pathania et al. 2013).

4.3.3 Hypolipidemic Property and Role in Cardiovascular Diseases

The hypolipidemic effect of methanolic extracts of *R. serpentina* was experimentally proven in Wistar albino mice. Remarkably, a significant reduction in the level of total cholesterol, triglyceride, and low density lipoprotein was observed underlining the role of the herbal drug in maintaining lipid equilibrium (Azmi and Qureshi 2013). Parallel studies were performed to evaluate the antilipidemic activity of hydromethanolic extract. These actions were endorsed most likely probably because of inhibition of HMG-CoA reductase and AcetlyCoA carboxylase or by promoting lipase (Azmi et al. 2018). More studies were conducted on albino mice and alloxan-induced mice, validating the antihyperlipidemic activity with less side effects (Qureshi et al. 2009; Shah et al. 2020). Additionally, MolDock studies revealed the contribution of phytochemicals like reserpine, ajmalicine, yohimbine, and indobine in inhibiting the 3-hydroxy-3-methyl-glutaryl-CoA reductase and thus the cholesterol biosynthetic pathway (Azmi et al. 2021).

Herbal medicines are good alternatives for the expensive pharmaceuticals for healing cardiovascular diseases (Rastogi et al. 2016). The hypolipidemic property of

R. serpentina root extract is the core for its healing role in cardiovascular ailments. Clinical trials using human subjects in the 1950s proved the efficacy of Rauwolfia therapy in treating angina pectoris and coronary artery disease (Lewis et al. 1956). In vivo studies were performed by inducing oxidative stress on left anterior coronary artery ligation in dogs, followed by administration of the herbal formulation of *R. serpentina* with other medicinal plants. The plant formulation succeeded in maintaining the biochemical, haemodynamic parameters, and thus the cardiac health (Afsheen et al. 2018). The blood pressure moderating ability of the reserpine was also reported (Bunkar 2017a). Increased blood pressure leads to ailments like hypertension. The action of the reserpine on the vaso-motor centre leads to generalized vasodilation and subsequently reduced blood pressure. Several clinical trials have been carried out to appraise the antihypertensive efficacy of root extracts of *R. serpentina.* In mid of 1900s, Vakil conducted a notable work on 50 hypertensive patients. The uptake of rauwolfia resulted in a fall of blood pressure for around 80% of patients (Vakil 1949). Alseroxylon, a drug derived from Rauwolfia, was administrated to a group of 84 patients with varying intensity of hypertension. Most of them showed positive results with harmless side effects (Livesay and Moyer 1954). Correspondingly, studies on patients with hypertensive disorder for almost 11 months by giving *R. serpentina* resulted in a decrease in blood pressure (Markvotiz et al. 1955). A series of studies succeeding Vakil's findings were performed during 1950s. All of them unequivocally pointed out the impact of *R. serpentina,* specifically reserpine on hypertension. Moderate side effects like sedation and drowsiness were observed in most cases. But it was considered as blessing in disguise for patients (Lobay 2015). A review work was performed to expound the dose-dependent action of reserpine by exploring all the available clinical trials. They reported the effectiveness of reserpine similar to the available drugs, but couldn't make a conclusion regarding exact dosage of reserpine. However, it was suggested that low dosage of drug is acceptable with least side effects. (Shamon and Perez 2016) Recently, a major fall in diastolic BP, systolic BP, and level of angiotensin converting enzyme was observed as a result of experimental tests using a combination of *R. serpentina* and *Curcuma longa* on hypertension-induced dogs (Farieha et al. 2019). Besides, the cardioprotective indices were elevated and the risk of coronary attacks was reduced on the use of methanolic extracts of *R. serpentina* in alloxan-induced mice (Azmi and Qureshi 2012a).

4.3.4 Antimicrobial Activity

Medicinal plants have been extensively used for microbial infections for centuries ago. Being an important medicinal plant, the antimicrobial potential of *R. serpentina* was also investigated extensively. The bactericidal activity of the root extract was first validated via MIC and MBC against *S. typhi*. It was also observed that indole alkaloids like reserpine confer this property (Deshmukh et al. 2012). Later, methanol, chloroform, and aqueous extracts of the plant exhibited inhibition against a number of pathogenic microbes. Among them, methanol extract had shown the highest MIC against *S. aureus.* The root extract was found to be effective against bacteria like *Saccharomyces cerevisiae, Bacillus subtilis, Enterococcus faecalis, Micrococcus luteus, Escherichia coli, Klebsiella pneumoniae, Pseudomonas aeruginosa, Streptococcus pneumonia and Proteus vulgaris* as well as against fungi like *Aspergillus niger and Candida albicans* (Owk and Lagudu 2016). Interestingly, one of the bactericidal activity-bestowing factors was identified to be stigmasterol in *n*-hexane fraction of methanolic extract of the plant (Dey et al. 2016). The ethanolic extract of the plants has also showed similar repressing effect on bacteria (Bunkar 2017a, b). Analogously, the methanolic, ethanolic, and acetonic extracts of the plant were considered for investigation for fungi like *Trichoderma viride, Fusarium oxysporum, and Penicillium notatum.* The antifungal and antibacterial nature of *R. serpentina* was once again substantiated (Singh et al. 2017a, b).

4.3.5 Role in Mental Health

R. serpentina is legendary in Ayurvedic literature for its extraordinary effects on the nervous system and subsequent curing of various mental disorders. This antipsychotic herb was widely employed in the treatment of insomnia, tremor, and frenzy conditions. Studies on schizophrenic patients revealed the impact of rauwolfia in reducing aggressiveness. Further studies on neurotic patients induced dreams (Azima et al. 1957). Similarly, raudixin, a drug derived from the roots of *R. serpentina,* was proven to be a probable dream-inducing agent. But more elaborate studies are required to overcome the objections on the efficiency of the drug as well as to explain the underlying mechanism (Azima 1958). Clinical studies on 118 patients suffering from mental illness revealed the sedative potential of reserpine. Substantial reduction in nervousness and the comforting effect was produced on administration (Glynn 1955). The herbal drug was also found to be effective in reducing overanxiety in psychiatric patients. Unfortunately, no impact was observed in depression (Lowinger 1957). Interestingly, studies on autistic children with low doses showed hopeful changes like reduced hyperactivity, calmness, and better sleep (Lehman et al. 1957). Another study proved that reserpine in the form of drug serpasil had lessen the insane treatments of drug and alcohol addicts (Avol and Vogel 1955). The mitigating effect of phytochemicals like ajmaline and reserpine on Alzheimer's disease was evaluated via in vitro and in silico studies. Both of the compounds exhibited β-site amyloid cleaving enzyme (BACE-1) suppression, anticholinesterase, and monoamine oxidase-B (MAO-B) inhibition. The neuroprotective ability of these bioactive compounds promises a new treatment strategy for alleviating Alzheimer's disease (Kashyap et al. 2020).

4.3.6 Anticancer Property

Anticancer potential of reserpine, an important phytochemical extracted from roots of *R. serpentina,* was an experimental subject among researchers in the nineteenth centuries. But was not prized because of the toxic nature. At the same time, lot of rumours were circulated regarding the incidence of breast cancer on use of *R. serpentina*. Later, proper clinical studies contradicted these false assumptions (Lobay 2015). Fortunately, exceptional role of reserpine in killing resistant tumour cells was later identified and is now a hope in the anticancer studies. Resazurin reduction assay, doxorubicin uptake assay, was employed to study cytotoxicity and further confirmed by molecular docking studies (Abdelfatah and Efferth 2015). Moreover, it was evidenced that reserpine suppressed cell proliferation and also induced apoptosis in oral cancer cells via TGF-β signalling (Ramu et al. 2020). It was also found that anticancer activity of leaves extract was comparatively higher through studies on HeLa cell lines (Deshmukh et al. 2012). The anticancer property of alstonine, serpentine, was experimentally proven on groups of mice bearing either Ehrlich ascites carcinoma cells or lymphoma cells. These are bioactive compounds present in *R. serpentina* (Beljanski and Beljanski 1986). More elaborate studies are demanded to explore the anticancer potential of this wonder drug.

4.3.7 Antivenom Property

The Indian ethnic folks have been reported to use the paste of roots and leaves of *R. serpentina* as an herbal antidote (Upasani et al. 2017, 2018). It was found to be effective with combination of *Tylophora indica* paste (Ignacimuthu et al. 2006). This traditional knowledge was later proven to be true by identifying lead compounds in the plant extract against the venom of cobra. The docking studies demonstrated that phytochemicals have a suppressing effect on phospholipase A2 like potential factors in venom (Sreekumar et al. 2014). Stigmasterol is an important phytochemical attributing this antivenom nature. It was also noted that the concentration of stigmasterol in tropical plants varies considerably with the altitudinal location of growth. The factors like wind, rainfall, temperature, and humidity have influential roles in the content of stigmasterol and thus antidote nature (Dey and Pandey 2014).

4.3.8 Miscellaneous Properties

Investigation on guinea pigs with extracts of *R. serpentina* unwrapped the antihistaminase nature. Serpentine was the potent phytochemical that hindered histaminase in vivo (Sachdev et al. 1961). Fascinatingly, studies on HIV antigens uncovered the anti-HIV activity of the root extracts of the plant (Sabde et al. 2011). Aqueous and ethanolic extracts were considered for testing the antimalarial activity in mice diseased with malaria. The results indicated substantial suppression in the growth of plasmoids. Topoisomerase II inhibiting the nature of serpentine was the crucial factor that leads to the death of parasites. This promises *R. serpentina* as a candidate herbal for developing an antimalarial drug (Omoya et al. 2019). The presence of symbiotic actinomycetes has also been discovered within *R. serpentina*. These actinomycetes are efficient in the production of bioactive chemicals (Gohain et al. 2015). Correspondingly, *Cladosporium* sp., an antibactericidal fungus, has been isolated from the leaves of *R. serpentina*. Additionally, potential metabolites like naphthoquinones were also found to be synthesized from this fungi (Khan et al. 2016). A unique strain of rhizobacterium named *Delftia tsuruhatensis* was identified from *R. serpentina* with plant promoting properties. Above all, these bacterium also produced a novel antibiotic(amino (5-(4-methoxyphenyl)-2-methyl-2-(thiophen-2-yl)-2,3-dihydrofuran-3-yl)methanol) with profound fungicidal activity (Prasannakumar et al. 2015). Twenty fungal batches belonging to *Fusarium* sp., *Aspergillus* sp., *Xylaria* sp., *Phomopsis* sp., *Cladosporium* sp., *Alterneria* sp., *Gleomastix* sp., and *Colletotrichum* sp. have been isolated and four of them showed signs of noticeable antibacterial activity (Singh et al. 2016). Lately, another *fusarium* sp. with antistaphylococcal action was observed in leaves of *R. serpentina*. Sixteen more potent antibacterial fungal strains were also identified in this category. The potential of these fungal groups was not only limited to antibacterial activity, but they also mimicked the antioxidant and hypolipidemic effect. These studies also pointed out that *C. gloeosporioides* is the most colonized fungi within *R. serpentina* (Nath et al 2013) Another noteworthy property is the action of the plant against the genotoxic impact produced on the exposure of carbofuran in freshwater fishes. The antioxidant nature of *R. serpentina* attributes to this mechanism of protection (Tiwari and Trivedi 2019). Recently, the hepatoprotective and renoprotective nature of methanolic extracts of *R. serpentina* was revealed through investigations on albino mice. These studies shed new light on the pharmacological properties of *R. serpentina* and cancelled the proposed side effects of the use of these drugs (Shah et al. 2020). One more striking property of *R. serpentina* was corrosion inhibition activity on aluminum alloys. The inhibitory activity was greatly increased by intensifying the concentration of plant extracts and temperature (Chaubey et al. 2017). Likewise, the alkaloid extracts of the plant demonstrated the formation of protective layer over steel supressing the corrosion. Reserpine was also proposed as the possible candidate involved in this biological protection (Raja and Sethuraman 2010).

4.4 Growth and Conventional Propagation Methods

Generally, the plant grows in wide geographical area with an annual rainfall of 1500–4000 mm and a temperature range of 10–38 °C (Agrawal 2019). It grows well in humus-rich loamy soil. Seeds are usually used for propagation. Seeds are collected around the beginning of June and can be maintained for 6 months in proper storage conditions. Tap roots with filiform secondary roots are chosen sometimes to propagate. Soft stem cuttings could also be employed. Sproutings are seen within a month. Comparatively, low germination rates were observed on the use of these plant parts. Moreover, generated plants are susceptible to pathogen attacks and climate fluctuations. Most importantly, each generation may show genetic variability and which will subsequently affect the alkaloid content within the plant (Mukherjee et al. 2019).

4.5 Biotechnological Approaches

The various disadvantages of conventional propagation strategies, increased economic value of phytochemicals like reserpine, and the categorization of *R. serpentina* as an endangered plant attracted the biotechnologists more towards the vast realm of this medicinal plant.

4.5.1 In Vitro Propagational Strategies

Growing cells, tissues, and organs of plants under artificial synthetic media and appropriate environmental conditions is termed as tissue culture (Van Eck and Smith 1997). Tissue culture is a better option for the mass propagation of endangered plants like *R. serpentina*. Shoot tips, hairy root, leaf, callus, axillary meristems, nodal and internodal segments, and seeds could be used as explant to develop the plant. Even inflorescence was efficaciously exercised to develop callus (Kaur 2018). Young leaves have been reported as an assuring explant in many studies (Nurcahyani 2008). Nodal, internodal segments, and leaves were compared to develop callus and it was observed that leaf is a better choice for callus induction and plant regeneration (Panwar et al. 2011). Apposite sterilization is a major step ensuring the success of tissue culture. A proper sterilization pattern is followed for *R. serpentina* cultures; Bavistin (0.1–0.2%) followed by Tween-20 (10%) for 5 min, Sodium hypochlorite for 5 min, cetrimide solution (1%) for 5 min, ethanol for 30 s, and $HgCl_2$ (0.1%)for 5 min (Mukherjee et al. 2019). Several other variations in terms of concentration and exposure of chemicals have been practiced (Bahuguna et al. 2011; Mallick et al. 2012; Pan et al. 2013). Murashige and Skoog (MS) medium is generally used for organogenesis, embryogenesis, and callus induction. Half concentration of MS media (pH—5.8) was employed for root induction as well as for maintaining shoot and leaves (Benjamin et al. 1993; Goel et al. 2009). MS medium with BAP (0.5–1.5 mg/L) with KIN (0.5–1.0 mg/L) or NAA (0.5–1.0 mg/L) was reported to be best medium for callus induction. The use of other media like Gambrog's B5 medium (Pan et al. 2013), Linsmaier Skoog medium (Yamamoto and Yamada 1986), and Woody plant medium (Alatar 2015) were also successfully established. (Yamamoto and Yamada 1986) They used LS media for 13 years and successfully alleviated the reserpine production via persuading stress conditions by amending hormone concentration. One of the studies proved that woody plant medium supplemented with 5.0 μM BA and1.0 μM NAA was found to be the best choice to generate shoot artificially (Abdurahman 2012). Sucrose (3%) is the major carbon source for nourishment of growing tissues. However, MS media with1.5% sucrose was also to be effective in inducing root (Bhadra et al. 2009). Additionally, fruitful use of MS with 4% sucrose for producing roots was also reported (Pandey et al. 2007). Optimum temperature and light intensity were reported to be 20–25 ± 1–2 °C and 3000 lux, respectively. Moreover, 50–70% of relative humidity and 16 h exposure to light were enough for the growth of callus in synthetic media. Auxins, cytokinins, and gibberellic acid were the common growth regulators used by most investigators and variations in the concentration and combination ratios were also attuned to favour root and shoot initiation (Mukherjee et al. 2019). The impact of concentrations of growth regulators is depicted in Table 4.2. Various other addictives like coconut milk (Aryal and Joshi 2009), mixture of proteins, ascorbic acid, citric acid (Kataria and Shekhawat 2005), copper sulfate (Ahmad et al. 2001), and antibiotics (Benjamin et al. 1993) have also been utilized to induce and enhance the growth of in-vitro plants. Use of thidiazuron in media caused profused shooting in explants (Pandey et al. 2007). Moreover, prior treatment of thidiazuron before culturing explants in liquid MS media resulted in amplified number of shoots (Alatar 2015). Total alkaloid quantity was also elevated by adequate supply of nitrogen (Akram and Ilahi 1986). Interestingly, the shoot tips were treated with colchicine to induce autotetraploidy with the intention of ameliorating alkaloid content (Mathur et al. 1987). Direct and indirect regeneration systems were compared to study the augmentation in reserpine production. In vitro regeneration was proved to be more effective (Mukherjee et al. 2020). These direct and indirect organogeneses have been employed in many studies to explore the ways of micropropagation (Bahuguna et al. 2011). Somatic embryos were also successfully established from roots of *R. serpentina* in liquid MS media. Additionally, flow cytometry studies evidenced the stability of regenerated plants from these embryos (Zafar et al. 2019). Proper acclimatization must be followed after all these for the generation of healthy plants. Commonly used are soil, sand, and leaf manure or farmyard wastes (Mukherjee et al. 2019). Different combinations of vermiculite, soil, and sand were other good alternatives to generate plantlets (Kad et al. 2017).

4.5.1.1 Use of Elicitors

The massive production of phytochemicals like reserpine could be achieved by use of elicitors. Elicitors can be either biotic or abiotic (David Paul Raj et al. 2020). Use of elicitors like copper and zinc imparted the increase in number of shoots generated (Nurcahyani 2008; Ahmad et al. 2014). Use of other chemicals like melaphene exhibited solid impact on alkaloid content (Kozlova et al. 2005). Studies conducted on different concentrations of methyl jasmonate showed that higher concentrations caused a 7.3 times rise in total reserpine content (Harisaranraj et al. 2009). Similarly, acetyl salicylic acid (ASA), salicylic acid (SA), and methylsalicylic acid were applied directly onto the plants in form of sprays. This caused tremendous increase in the phenolic acid-derived metabolites like cichoric acid (Nair et al. 2013). The eliciting property of salicylic acid (SA), dimethyl sulphoxide (DMSO), and abscisic acid (ABA) was assessed. A positive effect was imparted by all except DMSO. An association between the action of SA and trypamine (precursor) was also uncovered. About 57.64 mg g^{-1} of reserpine was produced on 36 days of elicitation (Panwar and Guru 2013). Similarly, the use of $CdCl_2$ caused considerable increase in reserpine and ajmaline content by inducing stress (Zafar et al. 2020b). An ascent in reserpine synthesis was also reported on the use of $AlCl_3$ (0.15 mM). The antioxidant assays evidenced the stress induced by $AlCl_3$. However, it became the source of a positive effect (Zafar et al. 2017). Outstandingly, the 2.9-fold of rise in amount of ajmaline was recorded on use of mannan from *Saccharomyces cerevisiae*

Table 4.2 Depicts the influence of various growth regulators on the induction of explants

Plant parts	Media with suppliments (mg L^{-1}/μM*)	Growth conditions (temperature, photoperiod, RH)	Response (%)	Inducing	References
Nodal segments	MS with BA (1.5) + NAA (0.2)	25 ± 1 °C, 16 h, 3000 lux	90	S	Salma et al. (2008a)
	MS with IAA+ IBA (1)		80	R	
	MS with BA (10*) + NAA (0.5*)	25 ± 2 °C, 16 h, RH- 60–65%	85	S	Ahmad et al. (2014)
	MS with BAP (1.5) + IAA (0.5) + 2,4-D (1.5)	25 ± 2 °C,14 h L, 10 h D	80	C	Bhadra et al. (2009)
	MS with BAP (17.74*) + ADS (32.57*)	25 °C, 16 h (40 μmol m^{-2} s^{1})	–	S	Saravanan et al. (2011)
	MS with BAP (10*) + IAA (0.5*) + AA (50) + Arg + CA + ADS (25)	26–30 ± 2 °C, 12 h/dark for 2–3 days	–	S	Kataria and Shekhawat (2005)
	MS with BAP (5*) + IAA (0.5*)			R	
	MS with BA with BA (0.5) + NAA (0.2)	25 ± 1 °C,16 h, 3000 lux	80	C	Salma et al. (2008b)
	MS with BA (2) + NAA (0.2)			S	
	MS with IAA (1) + IBA (1)			R	
	MS with IBA (0.125) + BAP (1.0)	20–25 °C, 16/8 h L and D, 3000 lux	–	C	Bhatt et al. (2008)
	MS with BAP (2) + IAA (0.02) + NAA (0.5)	25 ± 2 °C, 16 h	98	S	Khan et al. (2018)
				R	
	MS with IBA (0.2)	25 ± 2 °C,16 h L and D, 60 μE m^{-2} s^{-1}	96 ± 2.33	S	Mallick et al. (2012)
	MS with BAP (2.5)		100	R	
	MS with NAA (0.5)				
Shoot tips	MS with 2,4-D (3) and BAP (2)	25 ± 2 °C, 16/8 h L and D, 3000 lux	–	C	Pandey et al. (2007)
	MS with BAP (2.0) + NAA (0.5)		75	S	

(continued)

Table 4.2 (continued)

Plant parts	Media with suppliments (mg L^{-1}/μM*)	Growth conditions (temperature, photoperiod, RH)	Response (%)	Inducing	References
	MS with 4% sucrose + NAA (0.5 mg/L)		85	R	
	MS with BAP (5.0) + IBA (0.5)	25 °C, 16 h, RH- 55–60%	85.6	S	Jain et al. (2003)
	MS with IBA (0.5)		75.4	R	
	MS with TDZ (0.8*)	24 ± 2 °C, 16 h (50 mmol m^2 s^{-1})	75	S	Alatar (2015)
	MS with 2,4-D (1.5)	25 ± 2 °C, 16/8 L and D, 2000 lux	80 ± 4.5	C	Kumari et al. (2015)
	½ MS with IBA + IAA (1)	25 ± 2 °C, 16 h, RH- 50–55%	100	R	
	MS with NAA (0.1) and BA (2.5)		92	S	Susila et al. (2013)
	½ MS NAA (0.4) + IBA (0.1)	25 ± 3 °C, 16/8 L and D 3000 lux	91	R	
	MS with BAP (4.0) and NAA (0.5)		85	S	Baksha et al. (2007)
	½ MS with NAA (0.5)	25 ± 2 °C, 16 h, 2000 lux	100	R	
	MS with BAP (4) and NAA (0.4)		–	S	Chikte et al. (2013)
	½ MS with IAA + IBA (1)	25 ± 1 °C, 16 h	90	R	
Nodal segments with shoot apices	MS with BAP (2.5) followed by multiple shoot proliferation via MS with BA (0.2) + NAA (0.5))	25 ± 1 °C, (16 h 50 μmol/m^2/s)	90	S	Mukherjee et al. (2020)
	MS with BA (1.0) + NAA (0.1)	26 ± 1 °C, 16 h, 2000 lux	79 ± 4.9	S	Sarker et al. (1996)
	½ MS with NAA (1)		83	R	
Auxillary bud	MS with BAA (5.0*) + NAA (1.0*)	25 ± 2 °C, 60 ± 10% RH	90	S	Abdurahma (2012)
	WPM with BAA (5.0*) + NAA (1.0*)				
	B5 with BAA (5.0*) + NAA (1.0*)				

(continued)

Table 4.2 (continued)

Plant parts	Media with suppliments (mg L^{-1}/μM*)	Growth conditions (temperature, photoperiod, RH)	Response (%)	Inducing	References
	SH with BAA (5.0*) + NAA (1.0*)				
	MS with BA (2.2) + KIN (2.32) + NAA (0.54) after regenerating roots and shoots	24 ± 2 °C, 12 h (40 μM $m^{-2}s^{-1}$)	90	F	Mondal et al. (2011)
	B5 with 6-BAP (3.0) + NAA (0.54)	25 °C, 8 h (55 μmol m^{-2} s^{-1})	–	S	Tiwari et al. (2003)
	MS+NAA (0.1) + BA (1.0)	–	–	S	Roja et al. (1987)
Shoot	MS with IAA (0.5) + BAP (0.5)	24 ± 2 °C, 2000 lux, 16 h	77.33 ± 1.2	S	Rani et al. (2014)
	MS with IBA (20)	24 ± 2 °C, 16 h, 3 months	–	R	Yahya et al. (2007)
	MS with BA (1) +2,4-D (2)	–	–	C	Roja et al. (1987)
	MS with BAP (1.5) + IAA (0.5) + 2,4-D (1.5)	25 ± 2 °C, 14 h L, 10 h D	30	C	Bhadra et al. (2009)
Leaf	MS with PABA (1) + NAA (4)	25 ± 2 °C, 16/8 h L and D, 3000 lux	97.33 ± 0.45	R	Pandey et al. (2010)
	MS with IBA (0.125) + BAP (1.0)		–	C	Bhatt et al. (2008)
	MS with 2,4-D (1.5-2)	20–25 °C, 16/8 h L and D, 3000 lux	90	C	Mukherjee et al. (2020)
	MS with NAA (0.2) + KIN (1.5)	25 ± 1 °C, 16 h (50 μmol/m^2 /s) after 72 h D	75.0 ± 7	S	
	MS with BAP (1.0) + IAA (0.5)		77.77	C	Singh et al. (2009)
	MS with BAP (2.5) +IAA (0.4)	25 ± 1 °C, after 72 h dark 16 h L, 60% RH	75	S	
			100	R	
	MS with BAP (2.5) + IAA (0.5) + NAA (0.5)	16/8 h L and D			
	MS with 2,4-D (2.5)	(80 μM per m^{-2}/ s^{-1})	89 ± 4.8	C	Mallick et al. (2012)

(continued)

Table 4.2 (continued)

Plant parts	Media with suppliments (mg L^{-1}/μM*)	Growth conditions (temperature, photoperiod, RH)	Response (%)	Inducing	References
	MS with NAA (2.5 ppm) + Kinetin (2.5 ppm)	25 °C ± 2 °C, 16 h L and D, 60 μE m^{-2} s^{-1}	–	C	Subandi et al. (2018)
		–	–	C	Roja et al. (1987)
	MS with BA (1) +2,4-D (2)	–	93.65	C	Panwar et al. (2011)
	MS with 2,4-D (2.0) + BA (1.0)	25 ± 1 °C, 16 h	95.34	S	
	MS with BAP (2.0) + NAA (0.5)		99.95	R	
	MS with NAA + IBA (0.2)		–	S	Chikte et al. (2013)
	MS with BAP (4) and NAA (0.4)	–	60	R	
	½ MS with IAA + IBA (1)		91.54	C	Sinha and Kumar (2020)
	MS with BAP (5)	–			
	MS with 2, 4-D (2.0) + BAP (0.5) followed by $AlCl_3$ treatment	24 ± 1 °C, 12 h (100 μmol/m^2/s)	90.1 ± 2	C	Zafar et al. (2017)
Root	MS with IBA (20)	24 ± 2 °C, with or without light, 3 months	–	R	Yahya et al. (2007)
	White's media with coconut milk (100), Biotin (10), IAA (10), NAA (0.8), BAP (10), ENS (10)	–	–	C	Akram and Ilahi (1986)

(100 mg/L). Meanwhile, use of sodium chloride elevated the production of ajmalicine (Srivastava et al. 2016). All the reports underline the fact that induction of stress via chemicals caused a considerable advance in phytochemical production.

4.5.1.2 Synthetic Seed Development

The artificial seed technology was a breakthrough in the field of in vitro propagation. It can be defined as encapsulated somatic embryo, any meristematic tissue like shoot tips that can be cultivated as seeds and which are capable to regenerate the whole plant (Chandrasekhara 2012). Meristematic nodal segments, shoot tips are most often used for producing synthetic seeds. The explants were encapsulated using

3% sodium alginate and calcium chloride. The efficacy of different media in generating artificial seeds was also considered in the above study. Woody plant medium supplemented with plant hormones like NAA and BA showed sproutings within 2 weeks of growth (Faisal et al. 2012). Later, studies testified the optimum concentration of Sodium alginate as 3% and of Calcium chloride as 75 mM. In contrast to the above study, comparatively higher germination rates were observed in ½ MS liquid and semisolid media (Gantait et al. 2017). Storage at 4 °C was evaluated as a means for maintenance of alginated shoot ends up to 18 weeks (Ray and Bhattacharya 2008). It was found that proper storage at 8 °C will retain the reserpine content and viability of seeds (Gantait and Kundu 2017).

4.5.2 Use of Genetic Markers

Variation in genetic stability was a common consequence of in vitro propagation studies. Therefore, fidelity of the clone produced must be assessed via various gene markers. RAPD (Random Amplified Polymorphic DNA) and ISSR (Inter Simple Sequence Repeats) markers were used to study the genetic stability of artificial seeds. Genomic DNA was extracted from young leaves using CTAB method followed by PCR. Out of 20 RAPD primers, 19 generated distinct, consistent bands. In case of ISSR, all the seven primers produced bands. All the clones produced bands monomorphic to mother, endorsing the efficacy of in vitro propagation methodology (Faisal et al. 2012). Similarly, it was reported that 10 RAPD primers produced 97 reproducible bands, while ISSR produced 68 consistent maps. This study once again cemented the reliability of synthetic seed approach for micropropagation (Gantait et al. 2017). Finger printing bands generated via RAPD markers confirmed the uniformity between mother and clones produced by direct organogenesis (Goel et al. 2009; Senapati et al. 2014). The ISSR markers were fruitfully employed to investigate the genetic diversity within *R. serpentina* collected from different geographical locations of Southern Western Ghats. Gene flow (Nm), Nei's gene diversity (h), Degree of genetic differentiation (Gst), and Shannon index of genetic diversity (I) were also considered. The high polymorphism observed signified the need of proper conservation strategies for this vital plant. Moreover, it also strengthened the reliability of ISSR markers in genetic studies (Pillai et al. 2012). They have been used to evaluate the clonality of indirect organogenesis-regenerated plants. Out of 18 primers used, one that was reported to be not monomorphic band was further characterized (Saravanan et al. 2011). By the same token, genetic diversity was assessed via RAPD markers in *R. serpentina* variants of Western ghats. Large variation in the quantitative traits within *R. serpentina* varieties was observed. Along with RAPD, researchers also analysed the horticultural and chemical profiling. Implementation of proper conservation strategies localizing each small area is mandatory for protecting this natural treasure (Nair et al. 2014). In contrast to above all, DNA barcoding of different tissues and time periods of *R. serpentina* was performed. They detected species unique indels in the rps16 intron for *R. serpentina* (Eurlings et al. 2013).

4.5.3 *Rhizogenes*-Induced Hairy Roots

It is a declared fact that roots are the major source of alkaloids in *R. serpentina.* Transformation strategies were mainly focussed to produce bulk amount of roots via *Agrobacterium rhizogenes.* Ri-induced cultures are a better conservational strategy to guard and propagate this endangered plant (Srivastava and Misra 2017). The first attempted was reported in 1983 where shoot cultures were infected with AT 15834. Remarkably, callus with hairy roots was produced and later fully developed into plants. Unfortunately, much variation in alkaloid content was not reported. However, the advent of transformation techniques gave an economic preface to the studies of *R. serpentina* (Benjamin et al. 1993). Relatedly, the strain A4 was used to produce roots in wounded leaf explants. This study also led to the discovery of a new alkaloid called 12-hydroxyajmaline (Falkenhagen et al. 1993). Analogous studies were reported in the other species of *Rauwolfia* (Sudha et al. 2003). Another, *Agrobacterium* strain LBA 9402, was used to transform shoots and leaves of *R. serpentina.* The presence of *rolA* and *rolB* genes was detected by PCR to mark the transformation. Variations in the alkaloid content were observed in transformed root lines. Physiological condition of resultant cell, copy number of T-DNA, insertion site, and expression of rol genes were most possible reasons (Ray et al. 2014a). Genetic variability in root lines was also reported on studies using strains of LBA 9402 and A4. Various cytogenic studies showed the difference in T_L and T_R DNA integration pattern and in ajmalicine content too. It was recommended to choose the variant with more alkaloid content for scale-up studies (Ray et al. 2014b). Relatively, transformation efficiency of A4 was reported to be more. Plants generated from R_i-induced roots were found to be normal and produced higher amount of secondary metabolites compared to cultivated plants. It unfolds the possibility of using Ri-induced plants for reserpine production commercially (Mehrotra et al. 2013a). Successful in vitro studies were later upgraded to industrial level by means of Erlenmeyer flasks and bioreactors. The use of 5 L bioreactor with provisions for air supply and all other necessities was reported for the scale up of root cultures of *R. serpentina.* The growth rate of cultures was considerably high in bioreactor than flasks (Mehrotra et al. 2015). A standardized protocol by the above-mentioned authors for mass production of root cultures via AT 5 strain in B 5 medium in a 5 L bench top bioreactor was later published (Mehrotra et al. 2016). Interestingly, a ANN-based combinatorial model was predicted by them standardizing the culture conditions to obtain maximum metabolite yield from root cultures of *R. serpentina* (Mehrotra et al. 2013b). In between, the idea of using table sugar instead of sucrose was auspicious which will reduce the heavy expenses of industrial manufacturing. They also reported the upturn in the ajmaline, yohimbine, and reserpine contents on a long-term culture of 6 years and also suggested that this longevity elevated the genetic stability of strain (Pandey et al. 2014). Since the inoculants are highly susceptible to contamination, a better cryopreservation storage strategy was established. Root tips were first encapsulated using sodium alginate and were priorly cultured in sucrose. Gel beads were stored in cryovials and kept in liquid N_2 for storage. It was successfully revived later (Mehrotra et al. 2015).

4.5.4 Metabolic Engineering

Increased demand for bioactives like reserpine for pharmacological needs caused a considerable upsurge in metabolic engineering approaches. Metabolic engineering is defined as the manipulation of metabolic pathway in a selected organism with the intention of producing economically valuable metabolites. The decade of 1900s witnessed early attempts of researchers to alleviate the alkaloid content by incorporating microbes as bioreactors. Interestingly, strictosidine synthase, a key catalytic agent in alkaloid biosynthesis, was isolated from *R. serpentina* and was expressed in *E. coli.* According to the report, this is the first enzyme from plant secondary metabolism to be expressed in an microorganism (Kutchan 1989). The enzyme was structurally and chemically similar to the plant-derived enzyme. Later, microsomes were artificially synthesized using cell cultures of *R. serpentina* and guinea pig liver. Vomeline, an intermediate in ajmaline biosynthetic pathway, was successfully produced via vinorine hydroxylase (Cytochrome P 450-Dependent) at 40 °C and pH 8.3 (Falkenhagen and Stöckigt 1995). Interestingly, phytoene desaturase (PDS) was silenced via biolistic-mediated VIGS in *R. serpentina.* The construct (pTRV2-RsPDS) was introduced in eight plantlets and bleaching of leaves was perceived indicating the transformation (Corbin et al. 2017). In addition to this, Catharanthus tryptophan decarboxylase (tdc) was overexpressed in roots of *R. serpentina* via AT 4 strain. tdc is elicitor responsive gene in TIA pathway. An intensification in reserpine and ajmalicine biosynthesis was observed (Mehrotra et al. 2013c). Sadly, very few metabolic engineering works are reported. Elaborate researches are recommended to study the possible ways to ameliorate the bioactives in *R. serpentina.*

4.5.5 Novel Approaches

Nanotechnology has been efficaciously employed in the research of *R. serpentina.* Green synthesized Copper oxide nanoparticles (CuO Nps) were subjected to antibacterial study. These highly porous, spongy nanoparticles were found to be active against both gram positive and gram negative bacteria. Interestingly, these particles decayed trypan blue dye on exposure to ultraviolet rays (Lingaraju et al. 2015). Ample studies resulted in the development of transcriptomic assembilies for *R. serpentina*(Góngora-Castillo et al. 2012). In silico approaches carried out in *R. serpentina* resulted in the identification of functional MicroRNAs. MicroRNAs are short non-coding RNAs with specific functional roles. Remarkably, 15 conserved miRNAs were detected and most of them are proposed to be involved in developmental metabolic roles of the plant. These studies can be considered as benchmark in recognizing the possibilties of genetic pathways involved in metabolite production. Interestingly, the investigators also performed the phylognetic study of the evoluntionary relationships between rse-miRNA precursors and stem-loop sequence found in some other plants (Prakash et al. 2016).

4.5.6 Conclusion

R. serpentina plants are well-known for their medicinal properties and were found easily in many regions of India. However, the colossal exploitation of these valuable medicinal plants during the Anthropocene epoch has led to the considerable reduction in their number. Rampant effective strategies are mandatory to sustain a considerable population needed by society. Effectual in vitro propagation and biotechnological approaches should be utilized to curtail the endangering of plant in the wild. Unfortunately, a very few metabolic engineering approaches have only been reported. More intricate and effective studies are indispensable to get insights of the unexplored possibilities of this plant. Ample computational studies will be revoluntionary in determining the genes involved in secondary metabolite production and subsequent engineering. Besides, stringent regulations are required for the efficacious protection of plants without exploiting them.

References

Abdelfatah SAA, Efferth T (2015) Cytotoxicity of the indole alkaloid reserpine from Rauwolfia serpentina against drug-resistant tumor cells. Phytomedicine 22:308–318. https://doi.org/10.1016/j.phymed.2015.01.002

AbdelHafez EMN, Diamanduros A, Negureanu L et al (2013) Computational and synthetic studies towards improving rescinnamine as an inducer of MSH2-dependent apoptosis in cancer treatment. Mol Cancer Biol 1:44. https://doi.org/10.9777/mcb.v1i1.42

Abdurahman AA (2012) High frequency shoot regeneration and plant establishment of Rauvolfia serpentina: an endangered medicinal plant. J Med Plants Res 6:3324. https://doi.org/10.5897/jmpr12.111

Afsheen N, Khalil-Ur-Rehman JN et al (2018) Cardioprotective and metabolomic profiling of selected medicinal plants against oxidative stress. Oxidative Med Cell Longev 2018:1–17. https://doi.org/10.1155/2018/9819360

Agrawal S (2019) Rauvolfia serpentina: a medicinal plant of exceptional qualities. Altern Med Chiropr Open Access J 2:1–6

Ahmad S, Amin MN, Azad MAK, Mosaddik MA (2001) Micropropagation and plant regeneration of Rauvolfia serpentina by tissue culture technique. Pak J Biol Sci 5:75–79. https://doi.org/10.3923/pjbs.2002.75.79

Ahmad N, Alatar AA, Faisal M et al (2014) Effect of copper and zinc on the in vitro regeneration of Rauvolfia serpentina. Biol Plant 59:11–17. https://doi.org/10.1007/s10535-014-0479-5

Akram M, Ilahi I (1986) Plantlet formation in root callus of Rauwolfia serpentina. Pak J Bot 18:15–19

Alatar AA (2015) Thidiazuron induced efficient in vitro multiplication and ex vitro conservation of Rauvolfia serpentina - a potent antihypertensive drug producing plant. Biotechnol Biotechnol Equip 29:489–497. https://doi.org/10.1080/13102818.2015.1017535

Aryal S, Joshi SD (2009) Callus induction and plant regeneration in Rauvolfia Serpentina (L.) Benth Ex. Kurz. J Nat Hist Mus 24:82–88. https://doi.org/10.3126/jnhm.v24i1.2245

Avol M, Vogel PJ (1955) Treatment of delirium tremens with reserpine (serpasil): a preliminary report. J Am Med Assoc 159:1516–1520. https://doi.org/10.1001/jama.1955.02960330016005

Azima H (1958) The possible dream inducing capacity of the whole root of Rauwolfia serpentina. Can Psychiatr Assoc J 3:47–51. https://doi.org/10.1177/070674375800300110

Azima H, Cramer-Azima FJ, De Verteuil R (1957) Effects of Rauwolfia derivatives on psychodynamic structure. In: 113th annual meeting of the American Psychiatric Association, Chicago

Azmi MB, Qureshi SA (2012a) Methanolic root extract of Rauwolfia serpentina benth improves the glycemic, antiatherogenic, and cardioprotective indices in alloxan-induced diabetic mice. Adv Pharmacol Sci 2012:376429. https://doi.org/10.1155/2012/376429

Azmi MB, Qureshi SA (2012b) Methanolic root extract of Rauwolfia serpentina improves the glucose tolerance in Wister mice. J Food Drug Anal 20:484–488. https://doi.org/10.6227/jfda.2012200208

Azmi MB, Qureshi SA (2013) Rauwolfia serpentina improves altered glucose and lipid homeostasis in fructose-induced type 2 diabetic mice. Pak J Pharm Sci 29:1619–1624

Azmi MB, Qureshi SA (2014) Glucose lowering potential of hydromethanolic extract of Rauwolfia. World J Pharm Sci 2:219–223

Azmi MB, Qureshi SA, Ahmed SDH et al (2018) Ameliorative effect of hydro-methanolic extract from roots of Rauwolfia serpentina on some biochemical parameters of type 1 diabetic mice. Pak J Pharm Sci 31:663–668

Azmi MB, Sultana S, Naeem S, Qureshi SA (2021) In silico investigation on alkaloids of Rauwolfia serpentina as potential inhibitors of 3-hydroxy-3-methyl-glutaryl-CoA reductase. Saudi J Biol Sci 28:731–737. https://doi.org/10.1016/j.sjbs.2020.10.066

Bahuguna RN, Joshi R, Singh G et al (2011) Micropropagation and total alkaloid extraction of Indian snake root (Rauwolfia serpentina). Indian J Agric Sci 81:1124–1129

Baksha R, Akhter A, Khatun R, Liton J (2007) In vitro rapid clonal propagation of Rauvolfia serpentina (Linn.) Benth. Bangladesh J Sci Ind Res 42:37–44

Banerjee JN, Lewis JJ (1954) A note on the pharmacology of the alkaloids of Rauwolfia Serpentina Benth. J Pharm Pharmacol 8:50–54. https://doi.org/10.1111/j.2042-7158.1956.tb12127.x

Beljanski M, Beljanski MS (1986) Three alkaloids as selective destroyers of cancer cells in mice. Synergy with classic anticancer drugs. Oncology 43:198–203. https://doi.org/10.1159/000226363

Benjamin BD, Roja G, Heble MR (1993) Agrobacterium rhizogens mediated transformation of Rauvolfia serpentina: regeneration and alkaloid synthesis. Plant Cell Tissue Organ Cult 35:253–257. https://doi.org/10.1007/BF00037278

Bhadra SK, Akhter T, Hossain MM (2009) In vitro micropropagation of Rauvolfia serpentina L. through induction of direct and indirect organogenesis. Plant Tissue Cult Biotechnol 19:169–175. https://doi.org/10.3329/ptcb.v19i2.5434

Bhagat P, Verma SK, Yadav S et al (2020) Optimization of nutritive factors in culture medium for the growth of hairy root and ajmalicine content in sarpgandha, Rauwolfia serpentina. J Environ Biol 41:1018–1025. https://doi.org/10.22438/JEB/41/5/MRN-1316

Bhanwaria R, Singh B, Gochar R (2021) Indian snake root and devil root as distinctive medicinal plant for curing human disease: biology, chemistry and cultivation practices of Rauwolfia serpentina and Rauwolfia tetraphylla. In: Plants for novel drug molecules: ethnobotany to ethnopharmacology. New India Publishing Agency, New Delhi, pp 445–465

Bharti AS, Sharma S, Uttam KN (2020) Elemental assessment of the leaf and seed of Rauwolfia serpentina (Sarpagandha) by direct current arc optical emission spectroscopy. Natl Acad Sci Lett 43:361–365. https://doi.org/10.1007/s40009-019-00872-4

Bhatt R, Arif M, Gaur AK, Rao PB (2008) Rauwolfia serpentina: protocol optimization for in vitro propagation. African J Biotechnol 7:4265–4268. https://doi.org/10.5897/AJB08.718

Bindu S, Rameshkumar KB, Kumar B et al (2014) Distribution of reserpine in rauvolfia species from India - HPTLC and LC-MS studies. Ind Crop Prod 62:430–436. https://doi.org/10.1016/j.indcrop.2014.09.018

Bunkar AR (2017a) Phytochemical and antimicrobial study of Rauwolfia serpentina. Int J Biol Res 2:13–16

Bunkar RA (2017b) Therapeutic uses of Rauwolfia serpentina. Int J Adv Sci Res 2:23–26

Chandrasekhara M (2012) Synthetic seeds: a review in agriculture and forestry. African J Biotechnol 11:14254–14275. https://doi.org/10.5897/ajb12.770

Charney DS, Heninger GR, Breier A (1984) Noradrenergic function in panic anxiety: effects of yohimbine in healthy subjects and patients with agoraphobia and panic disorder. Arch Gen Psychiatry 41:751–763. https://doi.org/10.1001/archpsyc.1984.01790190025003

Chatterjee A (1953) Rauwolfia alkaloids. In: Progress in the chemistry of organic natural products. Springer, Vienna, pp 390–470

Chaubey N, Yadav DK, Singh VK, Quraishi MA (2017) A comparative study of leaves extracts for corrosion inhibition effect on aluminium alloy in alkaline medium. Ain Shams Eng J 8:673–682. https://doi.org/10.1016/j.asej.2015.08.020

Chauhan S, Kaur A, Pareek RK (2017a) Pharmacobotanical and pharmacological evaluation of ayurvedic crude drug: Rauwolfia serpentina (Apocynaceae). Int J Green Pharm 11:S686–S693

Chauhan S, Kaur A, Vyas M, Khatik GL (2017b) Comparison of antidiabetic and antioxidant activity of wild and cultivated variety of Rauwolfia serpentina. Asian J Pharm Clin Res 10:404–406. https://doi.org/10.22159/ajpcr.2017.v10i12.21287

Chikte SS, Dhakolkar P, Chikhale NJ, Burghate SK (2013) In vitro evaluation and comparison of differentiated roots for alkaloids of Rauwolfia serpentine. Sch J Agric Sci 3:573–575

Corbin C, Lafontaine F, Sepúlveda LJ et al (2017) Virus-induced gene silencing in Rauwolfia species. Protoplasma 254:1813–1818. https://doi.org/10.1007/s00709-017-1079-y

Das S, Saraf A, Sharma D et al (2019) Qualitative screening of bioactive secondary metabolites present in Withania somnifera and Rauwolfia serpentina root and stem extracts. Int J Res Anal Rev 6:69–74

Das S, Saraf A, Sharma D et al (2020) Study of phytochemical analysis of Withania somnifera and Rauwolfia serpentina plant extract Study of phytochemical analysis of Withania somnifera and Rauwolfia serpentina plant extract 8:1–4

David Paul Raj RS, Regi R, Mathew AA, Beena Kanimozhi R (2020) In vitro production of secondary metabolites from Andrographis paniculata and Rawvolfia serpentina against snake venom. Drug Invent Today 14:179–182

Deshmukh SR, Ashrit DS, Patil BA (2012) Extraction and evaluation of indole alkaloids from Rauwolfia serpentina for their antimicrobial and antiproliferative activities. Int J Pharm Pharm Sci 4:329–334

Dey A, De JN (2010) Rauvolfia serpentina (L).Benth. ex Kurz.-a review. Asian J Plant Sci 9:285–298

Dey A, Pandey DK (2014) HPTLC detection of altitudinal variation of the potential antivenin stigmasterol in different populations of the tropical ethnic antidote Rauvolfia serpentina. Asian Pac J Trop Med 7:S540–S545. https://doi.org/10.1016/S1995-7645(14)60287-X

Dey A, Mukherjee S, De A, Pandey DK (2016) A Stigmasterol containing n-hexane fraction of Rauvolfia serpentina methanolic extract shows tissue-specific variation of biocidal and antioxidant activities. Int J Geogr Inf Syst 22:81–91. https://doi.org/10.1080/10496475.2015.1042617

Ernst E, Pittler MH (1998) Yohimbine for erectile dysfunction: a systematic review and meta-analysis of randomized clinical trials. J Urol 159:433–436. https://doi.org/10.1016/S0022-5347(01)63942-9

Eurlings MCM, Lens F, Pakusza C et al (2013) Forensic identification of Indian snakeroot (Rauvolfia serpentina Benth. ex Kurz) using DNA barcoding. J Forensic Sci 58:822–830. https://doi.org/10.1111/1556-4029.12072

Faisal M, Alatar AA, Ahmad N et al (2012) Assessment of genetic fidelity in rauvolfia serpentina plantlets grown from synthetic (encapsulated) seeds following in vitro storage at 4 °C. Molecules 17:5050–5061. https://doi.org/10.3390/molecules17055050

Falkenhagen H, Stöckigt J (1995) Enzymatic biosynthesis of vomilenine, a key intermediate of the ajmaline pathway, catalyzed by a novel cytochrome P 450-dependent enzyme from plant cell cultures of Rauwolfia serpentina. Zeitschrift fur Naturforsch Sect C J Biosci 50:45–53. https://doi.org/10.1515/znc-1995-1-208

Falkenhagen H, Stockigt J, Kuzovkina IN et al (1993) Indole alkaloids from "hairy roots" of Rauwolfia serpentina. Can J Chem 71:2201–2203. https://doi.org/10.1139/v93-276

Farieha, Jahan N, Khalil-Ur-Rahman, Ali S (2019) Effect of herbal mixture as angiotensin converting enzyme inhibitor in angiotensin-II dependent hypertension. Pak Vet J 39:25–30. https://doi.org/10.29261/pakvetj/2018.103

Fazal H, Ahmad N, Khan MA (2011) Physico-chemical, phytochemical evaluation and dpph-scavenging antioxidant potential in medicinal plants used for herbal formulation in Pakistan. Pak J Bot 43:63–67

Fernandes VS, Santos JR, Leão AHFF et al (2012) Repeated treatment with a low dose of reserpine as a progressive model of Parkinson's disease. Behav Brain Res 231:154–163. https://doi.org/10.1016/j.bbr.2012.03.008

Gantait S, Kundu S (2017) Does synthetic seed storage at higher temperature reduce reserpine content of Rauvolfia serpentina (L.) Benth. ex Kurz.? Rend Lincei 28:679–686. https://doi.org/10.1007/s12210-017-0637-8

Gantait S, Kundu S, Yeasmin L, Ali MN (2017) Impact of differential levels of sodium alginate, calcium chloride and basal media on germination frequency of genetically true artificial seeds of Rauvolfia serpentina (L.) Benth. ex Kurz. J Appl Res Med Aromat Plants 4:75–81. https://doi.org/10.1016/j.jarmap.2017.01.005

Ganugapati J, Baldwa A, Lalani S (2012) Docking studies of Rauwolfia serpentina alkaloids as insulin receptor activators. Int J Comput Appl 43:32–37. https://doi.org/10.5120/6173-8600

Geissler M, Burghard M, Volk J et al (2016) A novel cinnamyl alcohol dehydrogenase (CAD)-like reductase contributes to the structural diversity of monoterpenoid indole alkaloids in Rauvolfia. Planta 243:813–824. https://doi.org/10.1007/s00425-015-2446-6

Glynn JD (1955) Rauwolfia serpentina (serpasil) in psychiatry. J Neurol Neurosurg Psychiatry 18: 225–227. https://doi.org/10.1136/jnnp.18.3.225

Goel MK, Mehrotra S, Kukreja AK et al (2009) In vitro propagation of Rauwolfia serpentina using liquid medium, assessment of genetic fidelity of micropropagated plants, and simultaneous quantitation of reserpine, ajmaline, and ajmalicine. Methods Mol Biol 547:17–33

Goenka AH (2007) Rustom Jal Vakil and the saga of Rauwolfia serpentina. J Med Biogr 15:195–200. https://doi.org/10.1258/j.jmb.2007.06-53

Gohain A, Gogoi A, Debnath R et al (2015) Antimicrobial biosynthetic potential and genetic diversity of endophytic actinomycetes associated with medicinal plants. FEMS Microbiol Lett 362:1–10. https://doi.org/10.1093/femsle/fnv158

Góngora-Castillo E, Childs KL, Fedewa G et al (2012) Development of transcriptomic resources for interrogating the biosynthesis of monoterpene indole alkaloids in medicinal plant species. PLoS One 7:e52506. https://doi.org/10.1371/journal.pone.0052506

Gupta J, Gupta A (2015) Isolation and identification of flavonoid rutin from Rauwolfia serpentina. Int J Chem Stud 3:113–115. https://doi.org/10.13140/RG.2.2.22346.88003

Gupta J, Gupta A (2016) Determination of trace metals in the leaves of Rauwolfia serpentina by using atomic absorption spectroscopy. Int J Chem Stud 4:94–96

Gupta S, Khanna VK, Maurya A et al (2012) Bioactivity guided isolation of antipsychotic constituents from the leaves of Rauwolfia tetraphylla L. Fitoterapia 83:1092–1099. https://doi.org/10.1016/j.fitote.2012.04.029

Gupta J, Gupta A, Gupta AK (2015) Extraction and identification of flavonoid natural antioxidant in the leaves of Rauwolfia serpentina. Int J Chem Stud 3:35–37

Harisaranraj R, Suresh K, Saravana Babu S (2009) Production of reserpine in somatic embryos of Rauwolfia serpentina cultured in bioreactors by the induction of elicitor (methyl jasmonate). Glob J Biotechnol Biochem 4:143–147

Hong B, Li WJ, Song AH, Zhao CJ (2013) Determination of indole alkaloids and highly volatile compounds in Rauvolfia verticillata by HPLC-UV and GC-MS. J Chromatogr Sci 51:926–930. https://doi.org/10.1093/chromsci/bms191

Hussain A, Neupane P, Jha R (2015) Phytochemical and GC-MS analysis of n-hexane extract of Rauvolfia serpentina L. Benth. ex Kurz. Chem Sci Rev Lett 4:223–229

Ignacimuthu S, Ayyanar M, Sivaraman KS (2006) Ethnobotanical investigations among tribes in Madurai District of Tamil Nadu (India). J Ethnobiol Ethnomed 2:1–7. https://doi.org/10.1186/1746-4269-2-25

Itoh A, Kumashiro T, Yamaguchi M et al (2005) Indole alkaloids and other constituents of Rauwolfia serpentina. J Nat Prod 68:848–852. https://doi.org/10.1021/np058007n

Jain V, Singh D, Saraf S, Saraf S (2003) In-vitro micropropagation of Rauwolfia serpentina through multiple shoot generation. Anc Sci Life 23:44–49

Kad A, Pundir A, Sharma S, Sood H (2017) Ex-vitro rooting in rauwolfia serpentina for rapid and large scale multiplication. Int J Innov Res Sci Eng 3:135–143

Kashyap P, Kalaiselvan V, Kumar R, Kumar S (2020) Ajmalicine and reserpine: indole alkaloids as multi-target directed ligands towards factors implicated in Alzheimer's disease. Molecules 25: 1–26. https://doi.org/10.3390/molecules25071609

Kataria V, Shekhawat NS (2005) Cloning of Rauvolfia serpentina - an endangered medicinal plant. J Sustain For 20:53–65. https://doi.org/10.1300/J091v20n01_04

Kaur S (2018) In vitro callus induction in inflorescence segments of medicinally important endangered plant Rauwolfia serpentina (L.) Benth. ex Kurz – a step towards ex situ conservation. Ann Plant Sci 7:1986–1991. https://doi.org/10.21746/aps.2018.7.2.1

Keshavkant S, Sukhdev T, Srinivasarao C, Naithani SC (2008) Antioxidant activities, phenols and flavonoid contents of Withania somnifera and Rauwolfia serpentina. Indian J Plant Physiol 13: 394–399

Khan MIH, Sohrab MH, Rony SR et al (2016) Cytotoxic and antibacterial naphthoquinones from an endophytic fungus, Cladosporium sp. Toxicol Reports 3:861–865. https://doi.org/10.1016/j.toxrep.2016.10.005

Khan S, Banu T, Akter S et al (2018) In vitro regeneration protocol of Rauvolfia serpentina L. Bangladesh J Sci Ind Res 53:133–138. https://doi.org/10.3329/bjsir.v53i2.36674

Kirillova NV, Smirnova MG, Komov VP (2001) Sequential isolation of superoxide dismutase and ajmaline from tissue cultures of Rauwolfia serpentina Benth. Appl Biochem Microbiol 37:181–185. https://doi.org/10.1023/A:1002875714260

Klohs MW, Draper MD, Keller F (1954) Alkaloids of Rauwolfia serpentina benth. 111.1 rescinnamine, a new hypotensive and sedative principle. J Am Chem Soc 76:2843. https://doi.org/10.1021/ja01639a079

Klyushnichenko VE, Yakimov SA, Tuzova TP et al (1995) Determination of indole alkaloids from R. serpentina and R. vomitoria by high-performance liquid chromatography and high-performance thin-layer chromatography. J Chromatogr A 704:357–362. https://doi.org/10.1016/0021-9673(95)00082-X

Kozlova RY, Vinter VG, Fattakhov SG et al (2005) Melamine salt of bis(methanol)phosphinic acids (melaphene) as a regulator of Rauwolfia serpentine specialized metabolism. Dokl Akad Nauk 401:556–559

Kumar S, Singh A, Bajpai V et al (2016a) Structural characterization of monoterpene indole alkaloids in ethanolic extracts of Rauwolfia species by liquid chromatography with quadrupole time-of-flight mass spectrometry. J Pharm Anal 6:363–373. https://doi.org/10.1016/j.jpha.2016.04.008

Kumar S, Singh A, Bajpai V, Srivastava M (2016b) Simultaneous determination of bioactive monoterpene indole alkaloids in ethanolic extract of seven Rauvolfia species using UHPLC with hybrid triple quadrupole linear ion trap mass spectrometry. Phytochem Anal 27:296. https://doi.org/10.1002/pca.2631

Kumari A, Kumar S, Anam A et al (2015) In vitro cloning of an endangered medicinal plant, Rauwolfia serpentina (L.). J Plant Dev Sci 7:555–561

Kutchan TM (1989) Expression of enzymatically active cloned strictosidine synthase from the higher plant Rauvolfia serpentina in Escherichia coli. FEBS Lett 257:127–130. https://doi.org/10.1016/0014-5793(89)81802-2

Le Xuan P, Munier RL, Meunier S (1980) Two-dimensional separations and behaviour of Rauwolfia, Corynanthe and Pseudocinchona alkaloids on unmodified silica gel. Chromatographia 13: 693–697. https://doi.org/10.1007/BF02303439

Lehman E, Haber J, Stanley R (1957) The use of reserpine in autistic children. J Nerv Ment Dis 125: 351–356

Lewis BI, Wild JB, Material C (1956) Rauwolfia serpentina in the treatment of angina pectoris. In: 28th annual scientific sessions of the American Heart Association, New Orleans, pp 227–232

Lingaraju K, Raja Naika H, Manjunath K et al (2015) Rauvolfia serpentina-mediated green synthesis of CuO nanoparticles and its multidisciplinary studies. Acta Metall Sin 28:1134–1140. https://doi.org/10.1007/s40195-015-0304-y

Livesay WR, Moyer JH (1954) Treatment of hypertension with Rauwolfia serpentina alone. J Am Med Assoc 155:1027–1035

Lobay D (2015) Rauwolfia in the treatment of hypertension. Integr Med 14:40–46

Lowinger P (1957) Rauwolfia serpentina in the control of anxiety. Psychiatry Q 31:445–453. https://doi.org/10.1007/BF01568739

Lucas RA (1963) The chemistry and pharmacology of the Rauwolfia alkaloids. Prog Med Chem 3: 146–186. https://doi.org/10.1016/S0079-6468(08)70118-8

Makarevich IF, Khadzhai YI, Nikolaeva AV, Pavlova VV (1979) Cardenolide and bufadienolide derivatives of ajmaline. Chem Nat Compd 15:466–468. https://doi.org/10.1007/BF00565048

Mallick SR, Jena RC, Samal KC (2012) Rapid in vitro multiplication of an endangered medicinal plant Sarpgandha (Rauwolfia serpentina). Am J Plant Sci 03:437–442. https://doi.org/10.4236/ajps.2012.34053

Manosroi J, Moses ZZ, Manosroi W, Manosroi A (2011) Hypoglycemic activity of Thai medicinal plants selected from the Thai/Lanna Medicinal Recipe Database MANOSROI II. J Ethnopharmacol 138:92–98. https://doi.org/10.1016/j.jep.2011.08.049

Markvotiz M, Koik VJ, Hick KF, Grissom JR (1955) Hexamethonium, hydralazine and Rauwolfia serpentina therapy in hypertension. Am J Med Sci 229:486–489

Mathur A, Mathur AK, Kukreja AK et al (1987) Establishment and multiplication of colchi-autotetraploids of Rauvolfia serpentina L. Benth. ex Kurz. through tissue culture. Plant Cell Tissue Organ Cult 10:129–134. https://doi.org/10.1007/BF00035910

Mcqueen EG, Blackman JG (1955) Direct effects of reserpine, rescinnamine and canescine on blood vessels. Australas Ann Med 4:250–254. https://doi.org/10.1111/imj.1955.4.4.250

McQueen EG, Doyle AE, Smirk FH (1954) Mechanism of hypotensive action of reserpine, an alkaloid of Rauwolfia serpentina. Nature 174:1015. https://doi.org/10.1038/1741015b0

Mehrotra S, Goel MK, Rahman LU, Kukreja AK (2013a) Molecular and chemical characterization of plants regenerated from Ri-mediated hairy root cultures of Rauwolfia serpentina. Plant Cell Tissue Organ Cult 114:31–38. https://doi.org/10.1007/s11240-013-0302-6

Mehrotra S, Prakash O, Khan F, Kukreja AK (2013b) Efficiency of neural network-based combinatorial model predicting optimal culture conditions for maximum biomass yields in hairy root cultures. Plant Cell Rep 32:309–317. https://doi.org/10.1007/s00299-012-1364-3

Mehrotra S, Srivastava V, Rahman LU, Kukreja AK (2013c) Overexpression of a Catharanthus tryptophan decarboxylase (tdc) gene leads to enhanced terpenoid indole alkaloid (TIA) production in transgenic hairy root lines of Rauwolfia serpen ... Overexpression of a Catharanthus tryptophan decarboxylase (tdc). Plant Cell Tissue Organ Cult 115:377–384. https://doi.org/10.1007/s11240-013-0369-0

Mehrotra S, Goel MK, Srivastava V, Rahman LU (2015) Hairy root biotechnology of Rauwolfia serpentina: a potent approach for the production of pharmaceutically important terpenoid indole alkaloids. Biotechnol Lett 37:253–263. https://doi.org/10.1007/s10529-014-1695-y

Mehrotra S, Srivastava V, Goel MK, Kukreja AK (2016) Scale - up of Agrobacterium rhizogenes-mediated hairy root cultures of Rauwolfia serpentina: a persuasive approach for stable reserpine production. In: Protocols for in vitro cultures and secondary metabolite analysis of aromatic and medicinal plants, 2nd edn. pp 241–257

Monachino J (1954) Rauvolfia serpentina-its history, botany and medical use. Econ Bot 8:349–365. https://doi.org/10.1007/BF02908608

Mondal S, Silva J, Ghosh P (2011) In vitro flowering in Rauvolfia serpentina (L.) Benth. ex. Kurz. Int J Plant Dev Biol 5:75–77

Mukherjee E, Gantait S, Kundu S et al (2019) Biotechnological interventions on the genus Rauvolfia: recent trends and imminent prospects. Appl Microbiol Biotechnol 103:7325–7354. https://doi.org/10.1007/s00253-019-10035-6

Mukherjee E, Sarkar S, Bhattacharyya S, Gantait S (2020) Ameliorated reserpine production via in vitro direct and indirect regeneration system in Rauvolfia serpentina (L.) Benth. ex Kurz. 3 Biotech 10:294. https://doi.org/10.1007/s13205-020-02285-3

Nair VD, Panneerselvam R, Gopi R, Hong-bo S (2013) Elicitation of pharmacologically active phenolic compounds from Rauvolfia serpentina Benth. Ex. Kurtz. Ind Crops Prod 45:406–4415. https://doi.org/10.1016/j.indcrop.2013.01.008

Nair VD, Raj RPD, Panneerselvam R, Gopi R (2014) Assessment of diversity among populations of Rauvolfia serpentina Benth. Ex. Kurtz. from Southern Western Ghats of India, based on chemical profiling, horticultural traits and RAPD analysis. Fitoterapia 92:46–60. https://doi.org/10.1016/j.fitote.2013.09.017

Namita B, Yadav M (2016) Evaluation of the chemical composition of Rauwolfia serpentina and Leucas Aspera - a comparative study. World J Pharm Pharm Sci 5:914–920. https://doi.org/10.20959/wjpps201610-7812

Nath A, Chattopadhyay A, Joshi SR (2013) Biological activity of endophytic fungi of Rauwolfia serpentina Benth: an ethnomedicinal plant used in folk medicines in Northeast India. Proc Natl Acad Sci-Sect B Biol Sci 85. https://doi.org/10.1007/s40011-013-0184-8

Negi JS, Bisht VK, Bhandari AK et al (2014) Quantification of reserpine content and antibacterial activity of Rauvolfia serpentina (L.) Benth. ex Kurz. African J Microbiol Res 8:162–166. https://doi.org/10.5897/ajmr2013.5847

Nikolaeva L, Alterman I (1993) Alkaloid formation in hairy roots and cell suspensions of rauwolfm serpentina benth. Nat Prod Lett 3:107–112. https://doi.org/10.1080/10575639308043846

Njerua SN, Matasyoh J, Mwaniki CG et al (2013) A review of some phytochemicals commonly found in medicinal plants

Nurcahyani N (2008) The reserpine production and callus growth of Indian Snake root (Rauvo/vla serpentlna (L.) Benth. Ex Kurz) culture by addition of Cu2+. Biodiversitas 9:177–179. https://doi.org/10.13057/biodiv/d090305

Omoya FO, Ade- O, Falusi M, Ogundare AO (2019) In vivo antimalarial activity and toxicological effects of ethanolic and hot water extracts of Rauwolfia serpentina leaf in mice infected with chloroquine- sensitive Plasmodium berghei. J Microbiol Antimicrob Agents 3:3–12

Owk AK, Lagudu MN (2016) In-vitro antimicrobial activity of roots Rauvolfia serpentina L. Benth Kurz. Natulac Sci Biol 8:312–316. https://doi.org/10.15835/nsb.8.3.9831

Pan S, Neeraj A, Srivastava K et al (2013) Effects of growth regulators on in vitro response and multiple shoot induction in some endangered medicinal plants. OA Biotechnol 2:1–7. https://doi.org/10.13172/2052-0069-2-1-428

Panche AN, Diwan AD, Chandra SR (2016) Flavonoids: an overview. J Nutr Sci 5:e47. https://doi.org/10.1017/jns.2016.41

Panda SK (2014) Ethno-medicinal uses and screening of plants for antibacterial activity from Similipal Biosphere Reserve, Odisha, India. J Ethnopharmacol 151:158–175. https://doi.org/10.1016/j.jep.2013.10.004

Pandey VP, Kudakasseril J, Cherian E, Patani G (2007) Comparison of two methods for in vitro propagation of Rauwolfia serpentina from nodal explants. Indian Drugs 44:514–519. https://doi.org/10.5281/zenodo.3597204

Pandey VP, Cherian E, Patani G (2010) EffEct of growth regulators and culture conditions on direct root induction of Rauwolfia serpentina L. (apocynaceae) benth by leaf explants. Trop J Pharm Res 9:27–34. https://doi.org/10.4314/tjpr.v9i1.52031

Pandey P, Kaur R, Singh S et al (2014) Long-term stability in biomass and production of terpene indole alkaloids by hairy root culture of Rauvolfia serpentina and cost approximation to endorse commercial realism. Biotechnol Lett 36:1523–1528. https://doi.org/10.1007/s10529-014-1495-4

Panwar GS, Guru SK (2011) Alkaloid profiling and estimation of reserpine in rauwolfia serpentina plant by TLC, HP-TLC and HPLC. Asian J Plant Sci 10:393–400. https://doi.org/10.3923/ajps.2011.393.400

Panwar GS, Guru SK (2013) Stimulation of reserpine production in the whole plant culture of Rauwolfia serpentina L. by elicitors and precursor feeding. J Plant Biochem Biotechnol 24:49–55. https://doi.org/10.1007/s13562-013-0235-5

Panwar G, Attitalla I, Guru S (2011) An efficient in vitro clonal propagation and estimation of reserpine content in different plant parts of Rauwolfia serpentina L. Am Eurasian J Sci Res 6:217–222

Pathania S, Randhawa V, Bagler G (2013) Prospecting for novel plant-derived molecules of Rauvolfia serpentina as inhibitors of aldose reductase, a potent drug target for diabetes and its complications. PLoS One 8:1–16. https://doi.org/10.1371/journal.pone.0061327

Pillai PP, Sajan JS, Menon KM et al (2012) ISSR analysis reveals high intraspecific variation in Rauvolfia serpentina L. - a high value medicinal plant. Biochem Syst Ecol 40:192–197. https://doi.org/10.1016/j.bse.2011.10.019

von Poser G, Andrade HHR, da Silva KVCL et al (1990) Genotoxic, mutagenic and recombinogenic effects of rauwolfia alkaloids. Mutat Res 232:37–43. https://doi.org/10.1016/0027-5107(90)90107-F

Prakash P, Rajakani R, Gupta V (2016) Transcriptome-wide identification of Rauvolfia serpentina microRNAs and prediction of their potential targets. Comput Biol Chem 61:62–74. https://doi.org/10.1016/j.compbiolchem.2015.12.002

Prasannakumar SP, Gowtham HG, Hariprasad P et al (2015) Delftia tsuruhatensis WGR-UOM-BT1, a novel rhizobacterium with PGPR properties from Rauwolfia serpentina (L.) Benth. ex Kurz also suppresses fungal phytopathogens by producing a new antibiotic-AMTM. Lett Appl Microbiol 61:460–468. https://doi.org/10.1111/lam.12479

Qureshi SA, Nawaz A, Udani SK, Azmi B (2009) Hypoglycaemic and hypolipidemic activities of Rauwolfia serpentina in alloxan-induced diabetic rats. Int J Pharmacol 5:323–326. https://doi.org/10.3923/ijp.2009.323.326

Raja PB, Sethuraman MG (2010) Studies on the inhibition of mild steel corrosion by Rauvolfia serpentina in acid media. J Mater Eng Perform 19:761–766. https://doi.org/10.1007/s11665-009-9541-4

Ramu AK, Ali D, Alarifi S et al (2020) Reserpine inhibits DNA repair, cell proliferation, invasion and induces apoptosis in oral carcinogenesis via modulation of TGF-β signaling. Life Sci 264:1–10. https://doi.org/10.1016/j.lfs.2020.118730

Rani A, Kumar K, Sanjeev K (2014) Effect of growth regulators on micropropagation of Rauvolfia serpentina (L.) Benth. J Appl Nat Sci 6:507–511. https://doi.org/10.31018/jans.v6i2.490

Rastogi S, Pandey MM, Rawat AKS (2016) Traditional herbs: a remedy for cardiovascular disorders. Phytomedicine 23:1082–1089. https://doi.org/10.1016/j.phymed.2015.10.012

Ray A, Bhattacharya S (2008) Storage and plant regeneration from encapsulated shoot tips of Rauvolfia serpentina - an effective way of conservation and mass propagation. S Afr J Bot 74:776–779. https://doi.org/10.1016/j.sajb.2008.06.002

Ray S, Majumder A, Bandyopadhyay M, Jha S (2014a) Genetic transformation of sarpagandha (Rauvolfia serpentina) with Agrobacterium rhizogenes for identification of high alkaloid yielding lines. Acta Physiol Plant 36:1599–1605. https://doi.org/10.1007/s11738-014-1536-6

Ray S, Samanta T, Majumder A et al (2014b) Cytogenetic characterization of Agrobacterium rhizogenes transformed root lines of Rauvolfia serpentina. Nucleus 57:105–112. https://doi.org/10.1007/s13237-014-0112-1

Roja PC, Slpahimalani AT, Heble MR, Chadha MS (1987) Multiple shoot cultures of Rauvolfia Serpentina: growth and alkaloid production. J Nat Prod 50:872–875. https://doi.org/10.1021/np50053a016

Rolf S, Bruns HJ, Wichter T et al (2003) The ajmaline challenge in Brugada syndrome: diagnostic impact, safety, and recommended protocol. Eur Heart J 24:1104–1112. https://doi.org/10.1016/S0195-668X(03)00195-7

Roten L, Derval N, Sacher F et al (2012) Ajmaline attenuates electrocardiogram characteristics of inferolateral early repolarization. Hear Rhythm 2:232–239. https://doi.org/10.1016/j.hrthm.2011.09.013

Rungsung W, Dutta S, Mondal DN et al (2014) Pharmacognostical profiling on the root of Rauwolfia serpentina. Int J Pharmacogn Phytochem Res 6:612–616

Sabde S, Bodiwala HS, Karmase A et al (2011) Anti-HIV activity of Indian medicinal plants. J Nat Med 65:662–669. https://doi.org/10.1007/s11418-011-0513-2

Sachdev KS, Aiman R, Rajapurkar MV (1961) Antihistaminase activity of serpentine. Br J Pharmacol Chemother 16:146–152. https://doi.org/10.1111/j.1476-5381.1961.tb00307.x

Sagi S, Avula B, Wang YH, Khan IA (2016) Quantification and characterization of alkaloids from roots of Rauwolfia serpentina using ultra-high performance liquid chromatography-photo diode array-mass spectrometry. Anal Bioanal Chem 408:177–190. https://doi.org/10.1007/s00216-015-9093-4

Saibabu V, Fatima Z, Khan LA, Hameed S (2015) Therapeutic potential of dietary phenolic acids. Adv Pharmacol Sci 2015:1–10

Salma U, Rahman MS, Islam S et al (2008a) Mass propagation of Rauwolfia serpentina L.Benth. Pakistan J Biol Sci 11:1273–1277

Salma U, Rahman MS, Islam S et al (2008b) The influence of different hormone concentration and combination on callus induction and regeneration of Rauwolfia serpemtina L. Benth. Pakistan J Biol Sci Biol 11:1638–1641

Saravanan S, Sarvesan R, Vinod MS (2011) Identification of DNA elements involved in somaclonal variants of Rauvolfia Serpentina (L.) arising from indirect organogenesis as evaluated by ISSR analysis. Indian J Sci Technol 4:1241–1245. https://doi.org/10.17485/ijst/2011/v4i10/30164

Sarker KP, Islam A, Islam R et al (1996) In vitro propagation of Rauvolfia serpentina through tissue culture. Planta Med 62:358–359. https://doi.org/10.1055/s-2006-957903

Senapati SK, Lahere N, Tiwary BN (2014) Improved in vitro clonal propagation of Rauwolfia serpentina L. Benth – an endangered medicinal plant. Plant Biosyst 148:885–888. https://doi.org/10.1080/11263504.2013.845264

Shah SMA, Naqvi SAR, Munir N et al (2020) Antihypertensive and antihyperlipidemic activity of aqueous methanolic extract of Rauwolfia serpentina in albino rats. Dose Resp 18:1–7. https://doi.org/10.1177/1559325820942077

Shamon SD, Perez MI (2016) Blood pressure-lowering efficacy of reserpine for primary hypertension. Cochrane Database Syst Rev. Rev 21:CD007655

Singh A, Singh PK (2009) An ethnobotanical study of medicinal plants in Chandauli District of Uttar Pradesh, India. J Ethnopharmacol 121:324–329. https://doi.org/10.1016/j.jep.2008.10.018

Singh DK, Srivastava B, Sahu A (2004) Spectrophotometric determination of Rauwolfia alkaloids: estimation of reserpine in pharmaceuticals. Anal Sci 20:571–573. https://doi.org/10.2116/analsci.20.571

Singh P, Singh A, Shukla AK et al (2009) Somatic embryogenesis and in vitro regeneration of an endangered medicinal plant sarpgandha (Rauvolfia serpentina L.). Life Sci J 6:57–62

Singh SK, Verma M, Ranjan A, Singh RK (2016) Antibacterial activity and preliminary phytochemical screening of endophytic fugal extract of. Open Conf Proc J 7:104–113. https://doi.org/10.2174/2210289201607010104

Singh HK, Charan AA, Charan AI, Prasad SM (2017a) Antifungal and antibacterial activity of methanolic, ethanolic and acetonic leaf extracts of sarpagandha (Rauwolfia serpentina). J Pharmacogn Phytochem 6:152–156

Singh M, Kaur R, Rajput R, Mathur G (2017b) Evaluating the therapeutic efficiency and drug targeting ability of alkaloids present in Rauwolfia serpentina. Int J Green Pharm 11:132–142
Sinha NK, Kumar G (2020) Rauwolfia serpentina: protocol optimization for callus induction from leaf explants. Int J Innov Sci Res Technol 5:289–292
Sreekumar S, Nisha NC, Biju CK, Krishnan PN (2014) Identification of potential lead compounds against cobra venom in Rauvolfia serpentina (L.) Benth. Ex Kurz. through molecular docking. Int J Pharm Res Dev 6:32–43
Srivastava M, Misra P (2017) Enhancement of medicinally important bioactive compounds in hairy root cultures of Glycyrrhiza, Rauwolfia, and Solanum through in vitro stress application. In: Production of plant derived natural compounds through hairy root culture. Springer, Cham, pp 117–132. https://doi.org/10.1007/978-3-319-69769-7_6
Srivastava M, Sharma S, Misra P (2016) Elicitation based enhancement of secondary metabolites in Rauwolfia serpentina and Solanum khasianum hairy root cultures. Pharmacogn Mag 12:S315–S320. https://doi.org/10.4103/0973-1296.185726
Starke K, Borowski E, Endo T (1975) Preferential blockade of presynaptic alpha-adrenoceptors by yohimbine. Eur J Pharmacol 34:385–388
Subandi M, Dikayani D, Firmansyah E (2018) Production of reserpine of rauwolfia serpentina [L] kurz ex benth through in vitro culture enriched with plant growth regulators of NAA and kinetin. Int J Eng Technol 7:274–278. https://doi.org/10.14419/ijet.v7i2.29.13331
Sudha CG, Reddy BO, Ravishankar GA, Seeni S (2003) Production of ajmalicine and ajmaline in hairy root cultures of Rauvolfia micrantha Hook f., a rare and endemic medicinal plant. Biotechnol Lett 25:631–636. https://doi.org/10.1023/A:1023012114628
Susila T, Satyanarayana Reddy G, Jyothsna D (2013) Standardization of protocol for in vitro propagation of an endangered medicinal plant Rauwolfia serpentina Benth. J Med Plants Res 7: 2150–2153. https://doi.org/10.5897/jmpr11.066
Tiwari V, Trivedi SP (2019) Investigations on remedial role of Rauwolfia serpentina root extract against carbofuran formulation induced genotoxicity in Channa punctatus. J Exp Biol 40:1023–1028
Tiwari S, Shah P, Singh K (2003) Efficient in vitro clonal propagation of Rauwolfia serpentina L.: an important medicinal plant. Plant Cell Biotechnol Mol Biol 4:157–162
Upasani SV, Beldar VG, Tatiya AU et al (2017) Ethnomedicinal plants used for snakebite in India: a brief overview. Integr Med Res 6:114–130. https://doi.org/10.1016/j.imr.2017.03.001
Upasani MS, Upasani SV, Beldar VG et al (2018) Infrequent use of medicinal plants from India in snakebite treatment. Integr Med Res 7:9–26. https://doi.org/10.1016/j.imr.2017.10.003
Usmani G, Chawhaan PH, Mishra Y et al (2015) Geographical variation of total alkaloid and reserpine content in Rauwolfia serpentina (L.) Benth. ex. Kurz. Euphytica 202:427–434. https://doi.org/10.1007/s10681-014-1311-1
Vaishnav V, Sahoo D (2020) Comparative phytochemical analysis of stem and root extracts from Rauwolfia serpentina. Int J Adv Sci Res Manag 5:5–7. https://doi.org/10.36282/ijasrm/5.1.2020.1681
Vakil RJ (1949) A clinical trial of Rauwolfia serpentina in essential hypertension. Br Heart J 11: 350–355. https://doi.org/10.1136/hrt.11.4.350
Van Eck JM, Smith FD (1997) Tissue culture. In: Environmentally safe approaches to crop disease control, 1st edn, pp 317–331
Vieira Pereira A, Santana GM, Góis MB, Sant'ana DMG (2015) Tannins obtained from medicinal plants extracts against pathogens: antimicrobial potential. In: The battle against microbial pathogens: basic science, technological advances and educational programs, pp 228–235
Wachsmuth O, Matusch R (2002) Anhydronium bases from Rauvolfia serpentina. Phytochemistry 61:705–709
Yahya AF, Hyun J-O, Lee J-H, Jung M-S (2007) Effect of explant types, auxin concentration and light condition on root production and alkaloid content of (L.) Benth. ex Kurz. J Korean For Soc 96:178–182

Yamamoto O, Yamada Y (1986) Production of reserpine and its optimization in cultured Rauwolfia serpentina Benth. cells. Plant Cell Rep 5:50–53. https://doi.org/10.1007/BF00269717

Zafar N, Mujib A, Ali M et al (2017) Aluminum chloride elicitation (amendment) improves callus biomass growth and reserpine yield in Rauvolfia serpentina leaf callus. Plant Cell Tissue Organ Cult 130:357–368. https://doi.org/10.1007/s11240-017-1230-7

Zafar N, Mujib A, Ali M et al (2019) Genome size analysis of field grown and tissue culture regenerated Rauvolfia serpentina (L) by flow cytometry: histology and scanning electron microscopic study for in vitro morphogenesis. Ind Crop Prod 128:545–555. https://doi.org/10.1016/j.indcrop.2018.11.049

Zafar N, Mujib A, Ali M et al (2020a) Cadmium chloride (CdCl2) elicitation improves reserpine and ajmalicine yield in Rauvolfia serpentina as revealed by high-performance thin-layer chromatography (HPTLC). 3 Biotech 10:1–14. https://doi.org/10.1007/s13205-020-02339-6

Zafar N, Muzamil A, Mujib A et al (2020b) Cadmium chloride - (CdCl2) elicitation improves reserpine and ajmalicine yield in Rauvolfia serpentina as revealed by high-performance thin-layer chromatography (HPTLC). 3 Biotech 10:344. https://doi.org/10.1007/s13205-020-02339-6

Zenk MH, El-Shagi H, Arens H et al (1977) Formation of the indole alkaloids serpentine and ajmalicine in cell suspension cultures of Catharanthus roseus, pp 27–43

Part II

Keystone and Other Significant Species in Bicultural Landscapes

Ethnobotany of Yams (Dioscoreaceae) Used by Local Communities in the Northwest of Luzon Island

5

Menisa A. Antonio and Inocencio E Buot, Jr

Abstract

Much of our underutilized and neglected species can be leveraged for food security, and dietary and nutrient diversity. In this chapter, we documented species of Dioscoreaceae used by local communities in the Province of Ilocos Norte, Philippines, as we gather informed decisions toward mainstreaming their sustainable conservation and utilization.

There are ten taxa belonging to nine species and two genera of Dioscoreaceae documented. This is an update on the digital checklist of family Dioscoreaceae in the Philippines, which listed only two species growing in Ilocos Norte. Eight taxa are edible, two of which [*Dioscorea hispida* Dennst. and *Tacca leontopetaloides* (L.) Kuntze] are reported toxic but could be consumed after processed and detoxified. The top species are *Dioscorea alata* L., *T. leontopetaloides, Dioscorea esculenta* (Lour.) Burkill var. *spinosa* (Roxb. ex Prain & Burkill) R. Knuth, and *D. hispida* based on cultural importance. The traditional uses include: as food, boiled feedstuff for hogs, laundry starch to stiffen fabrics and linens, medicinal remedy for skin problems, and live amulet or ritual. Majority of the informants source the plants from the wild, process, and sell them, earning additional income for their families. Ethnobotanical evidences such as ethnotaxonomic, ethnolinguistic, tradition of continued utilization, and traditional

M. A. Antonio (✉)
Mariano Marcos State University, Batac, Ilocos Norte, Philippines

Institute of Biological Sciences, College of Arts and Sciences, University of the Philippines Los Baños, Laguna, Philippines
e-mail: maantonio@mmsu.edu.ph

I. E. Buot, Jr
Institute of Biological Sciences, College of Arts and Sciences, University of the Philippines Los Baños, Laguna, Philippines

S. Ramamoorthy et al. (eds.), *Plant Diversity in Biocultural Landscapes*,
https://doi.org/10.1007/978-981-19-8649-9_5

knowledge and practices, among others, confirm that these species occupy an important position in the lives and subsistence as well as culture of the locals. Giving additional taste and flavor in local dining tables, these species reflect the simplicity of the Ilocano people. Wild-sourcing them likewise reflects the frugality and hard work that also shine their resilience as people.

Keywords

Ethnobotany · Dioscoreaceae · *Dioscorea* · *Tacca* · Traditional uses · Cultural importance

5.1 Introduction

Species of Dioscoreaceae, locally called yams, are economically important plants for their tuberous starchy roots, which are staple foods in many tropical regions and sources of raw materials, i.e., steroids in the pharmaceutical industry. In the Province of Ilocos Norte, situated in Northwestern Luzon, Philippines, the traditional use of species of *Dioscorea* and *Tacca,* two of the four genera comprising the family Dioscoreaceae (APG II 2003), is common in many remote and upland communities. They are subsistence food, sometimes alternative to rice during the lean months or hunger season that precedes crop harvest. Despite the traditional utilization, the species remain underutilized. Thus, as a foundation activity towards harnessing the species' potential, documentation of the plant utilization for food and nonfood needs to be done.

There are limited studies done on the wild, semi-wild, and cultivated species of Dioscoreaceae in Ilocos Norte. Earlier studies have documented only five species which include *Dioscorea alata* L. (greater yam or *ube*), *Dioscorea esculenta* (Lour.) Burkill var. *spinosa* (Prain) R.Knuth (wild lesser yam or *buga*), *Dioscorea hispida* Dennst. (asiatic bitter yam, nami or *karot*), *Dioscorea luzonensis* Schauer (*kamangeg*), and *Tacca leontopetaloides* (L.) Kuntze (*pannarien*) (Agbigay et al. 2020; Antonio 2008; Antonio et al. 2011; Legaspi 2004; Legaspi et al. 2018; Tomas et al. 2020). Reconnaissance survey suggests more species growing in the province and there is no ethnobotanical documentation done on them yet. A more comprehensive documentation is sought for all Ilocandia species in the rural communities of the Province where these are greatly depended on for subsistence and income. We need to document them in order to make an informed decision on prioritization both in conservation planning and leveraging their potential for intended applications.

This chapter therefore presents the ethnobotanical study conducted on yams in Ilocos Norte to identify the species known and used by the locals, document the uses and associated traditional knowledge of the people, and assess related evidences confirming that the species have rooted in the lives and food culture of the people in the province.

5.2 The Survey Process

5.2.1 The Survey Area

The Province of Ilocos Norte is situated in the northwestern tip of the Luzon Island, Philippines, at latitude 17° 48′ and 18° 29′ North and longitude 120° 25′ and 120° 58′ East. The province consists of 21 municipalities and two component cities and is subdivided into four agroecological zones: central lowland, northern coastal, southern coastal, and eastern interior (Provincial Planning and Development Office-Ilocos Norte (PPDO-IN) 2002). Of these, about 25% are upland areas. The ethnobotanical survey was conducted in these upland or mountainous communities in the Municipalities of Badoc, Vintar, Banna, Nueva Era and Pasuquin, and City of Batac. These municipalities were selected with the help of Municipal Agriculture Offices (MAO) based on agroecological zone, species abundance, and prevalence of utilization of the target species. The Municipality of Badoc represents the southern coastal zone; the Municipality of Vintar and City of Batac, the central lowland; the Municipalities of Banna and Nueva Era, the eastern interior; while the Municipality of Pasuquin, the northern coastal zone (Fig. 5.1). Prior Informed Consent (PIC) and required permits were secured from concerned local government units, community leaders, and private land owners.

5.2.2 Data Gathering and Analysis

The survey employed a key informant interview using a modified structured interview schedule. Information gathered include: (1) demographic data on the key informants, (2) species known or used, (3) species' growing areas and seasonal abundance, (4) informants' perceptions on the plants' socioeconomic importance, (5) status of utilization and conservation; (6) indigenous knowledge, e.g., method of preparation and recipe prepared; and (7) other experiences or knowledge of informants on the target plants. The interview was aided with electronic images of the species listed in Pelser et al. 2011 onwards). A follow-up interview using free-listing method was also made to generate more comprehensive data.

The Key Informants. A total of 60 key informants (Table 5.1) were purposively selected to give sound data on the target species. They were selected with the help of community leaders based on credibility, familiarity or knowledge, and experiences on the target plants found in the area. The informants were between the ages of 19 and 90 with an average of 45. Mostly are married, female, residing in the area since birth (21–90 years), and farming is their main source of livelihood (Table 5.2).

Nature Walk. A nature walk or visual inspection of the plants cited by the informants in each area was done to validate the species' identities. Voucher specimens were prepared and deposited at the Plant Biology Division Herbarium of the Institute of Biological Sciences, University of the Philippines Los Baños,

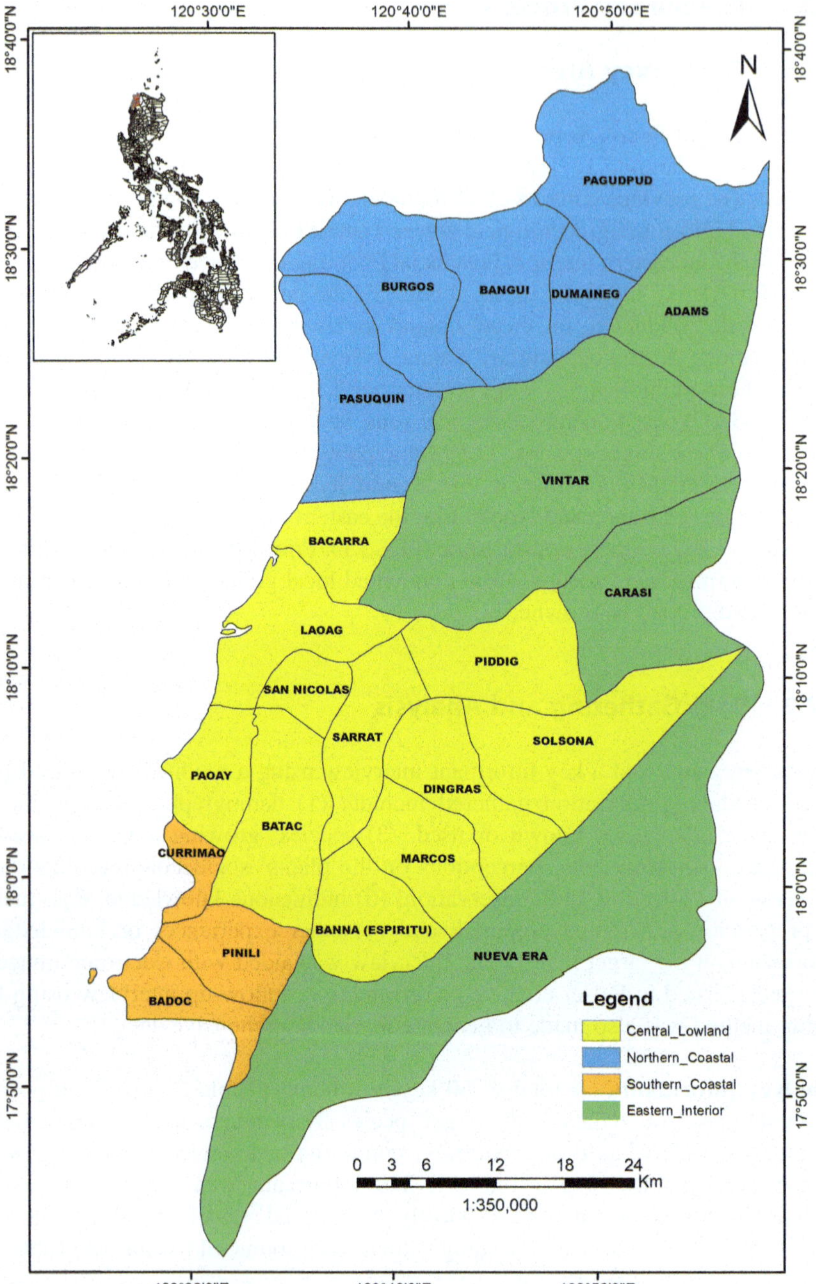

Fig. 5.1 Map of the Province of Ilocos Norte showing the survey areas and the agroecological zones. (Map by Engr. Rodel T. Utrera, Mariano Marcos State University)

Table 5.1 Number of key informants in the survey areas

Agroecological zone	Municipality	Number	Percentage
Southern Coastal	Badoc	10	16.7
Northern Coastal	Pasuquin	15	25
Central Lowland	Batac	10	16.7
	Vintar	15	25
Eastern Interior	Nueva Era	5	8.3
	Banna	5	8.3
Total		60	100

Table 5.2 Demographic profile of the key informants

Parameter	State	Number	Percentage
Age (years)	19–30	12	20
	31–60	43	72
	>61	5	8
Civil status	Married	55	92
	Single	5	8
Duration of residency in the area (year)	Since birth (21–90 years)	57	95
	≤20 years	3	5
Educational attainment	High School level/graduate	49	82
	Elementary level/graduate	11	18
Source of income[a]	Farming	50	83
	Market/ambulant vending	7	12
	Employment in government agency	6	10
	Others (i.e., construction)	4	7

[a]Multiple responses

College, Laguna. Taxonomic keys, online botanical herbarium database, online plant databases, flora, plant dictionaries, and taxonomic references were used in the identification. We subscribed to the APG classification, in which Dioscoreaceae is composed of genera *Dioscorea, Tacca, Stenomeris,* and *Trichopus* (APG II 2003) in this study. Biophysical characteristics of the growing areas which could affect species distribution and diversity were not investigated in the study.

Data Analysis. Data were analyzed using means, weighted means, frequencies, percentages, rank, relative frequency of citation (RFC), and cultural importance (CI). Species' relative frequency of citation (RFC) was calculated adopting the formula of Maghirang et al. (2018), as follows:

$$\mathrm{RFC} = \frac{\text{No. of informants citing a species}}{\text{Total no. of informants}}$$

The species cultural importance index (CI) was also computed, expressed through the formula adopting Barcelo (2014):

$$\mathrm{CI} = \sum_{i}^{\mathrm{NU}} \frac{\mathrm{UR}i}{N}$$

where

i—varies from only one use to the total number of uses (NU) mentioned for a species, and
N—total number of informants in the survey.

For example, *D. hispida* (karot) was cited as used in food by 60 informants, for medicinal by nine, and for ritual by five. The total number of informants is 60. Hence, the CI of *D. hispida* is 1.23.

5.3 Species of Dioscoreaceae Identified in Northwestern Luzon

5.3.1 Taxonomic Diversity, Distribution, and Growth

Taxonomic Diversity. There are ten taxa in nine species under two genera of Dioscoreaceae identified in Ilocos Norte: eight taxa belonging to seven species of *Dioscorea,* and two species of *Tacca* (Plate 5.1). All species except two were cited by the informants. The most cited species are *D. alata, Dioscorea esculenta* var. *fasciculata, D. hispida, D. luzonensis,* and *T. leontopetaloides* (Table 5.3). The non-cited species were *Dioscorea pentaphylla* L. and *Tacca palmata* Blume, which were spotted during the nature walk in Batac, and Badoc and Nueva Era, respectively.

Local Distribution. The local distribution of the species varies across study areas. Seven species which include *D. alata, Dioscorea bulbifera* L., *D. esculenta* var. *spinosa, D. esculenta* var. *fasciculata, D. hispida, D. luzonensis,* and *T. leontopetaloides* are cited and utilized in all the survey areas (Table 5.4).

D. pentaphylla, and *D. divaricata* and *T. palmata* were known and/or spotted in only one or two sites, respectively. *D. pentaphylla* was spotted in Batac during the forest walk, but not cited during the interview in Batac including the rest of the municipalities. *D. divaricata* is both cited and used by the informants in Badoc; while same species were not cited nor used by the informants in Nueva Era. Meanwhile, *T. palmata* was not cited in the interview, but found growing wild within bolo (*Gigantochloa levis* (Blanco) Merr.) stand in Badoc, and in thickets in Nueva Era. It is extant, but the residents are not aware of its existence and have not given considerable attention until recently during the pandemic when interest on ornamentals surfaced.

Plate 5.1 The identified species of Dioscoreaceae in Ilocos Norte. (**a**) *D. alata*, (**b**) *D. bulbifera*, (**c**) *D. divaricata*, (**d**) *D. esculenta* var. *fasciculata*, (**e**) *D. esculenta* var. *spinosa*, (**f**) *D. hispida*, (**g**) *D. luzonensis*, (**h**) *D. pentaphylla*, (**i**) *T. leontopetaloides*, (**j**) *T. palmata*. (Photos by M.A. Antonio)

Based on the survey and nature walk, species' richness per municipality is in the order: Badoc (10 species) > Nueva Era (9 species) > Batac (8 species) > Banna, Pasuquin, and Vintar (7 species each) (Table 5.4).

Genetic Status. Based on informants' responses and the researcher's observations, majority of the species are wild-growing. These include *D. divaricata, D. hispida, D. esculent* var. *spinosa, D. luzonensis, D. pentaphylla, T. leontopetaloides,* and *T. palmata*. These grow in thickets, low to medium-elevation or foot of secondary

Table 5.3 The relative frequency of citation (RFC) of the species

Species	Common, local, or vernacular name	RFC	Rank
Dioscorea			
Dioscorea alata L.	Greater yam	1	3
	Ubi		
Dioscorea bulbifera L.	Aerial yam	0.12	7
	Ubi-bunga		
Dioscorea divaricata Blanco	Dolian, dorian	0.08	8
Dioscorea esculenta (Lour.) Burkill var. *fasciculata* (Roxb.) R. Knuth	Lesser yam	1	3
	Tugui, Lukto		
Dioscorea esculenta (Lour.) Burkill var. *spinosa* (Roxb. ex Prain & Burkill) R.Knuth	Wild lesser yam, buga	0.88	6
Dioscorea hispida Dennst.	Asiatic bitter yam, nami, karot	1	3
Dioscorea luzonensis Schauer	Kamangeg	1	3
Dioscorea pentaphylla L.	Lima-lima	0	9.5
Tacca			
Tacca leontopetaloides (L.) Kuntze	Pannarien	1	3
Tacca palmata Blume	–	0	9.5

Table 5.4 Local distribution of the species of Dioscoreaceae

	Municipalities					
Species	Badoc	Banna	Batac	Nueva Era	Pasuquin	Vintar
D. alata	✓	✓	✓	✓	✓	✓
D. bulbifera	✓	✓	✓	✓	✓	✓
D. divaricata	✓	×	×	✓	×	×
D. esculenta var. *fasciculata*	✓	✓	✓	✓	✓	✓
D. esculenta var. *spinosa*	✓	✓	✓	✓	✓	✓
D. hispida	✓	✓	✓	✓	✓	✓
D. luzonensis	✓	✓	✓	✓	✓	✓
D. pentaphylla	×	×	✓	×	×	×
T. leontopetaloides	✓	✓	✓	✓	✓	✓
T. palmata	✓	×	×	✓	×	×
Species Richness	9	7	8	9	7	7

✓—present; ×—absent

forests, hills, and roadsides (Table 5.5). Some *D. luzonensis* and *D. hispida* plants are now in semi-wild or domesticated state through the efforts of some local residents to cultivate them. Recognizing the importance and uses of *D. luzonensis* and *D. hispida*, some residents propagate them for easier access and assured supply of tubers for utilization later. There were two observed domestication practices in Pasuquin and Badoc.

Table 5.5 Genetic status and seasonal abundance of the species of Dioscoreaceae observed by the informants

Species	Frequency[a] (%)			Growing area	Growing season	Harvest month
	Cultivated	Semi-wild/ domesticated	Wild			
D. alata	20	–	100	Upland farms, marginal areas, hill, thickets, home garden	May to Oct	Oct to Dec
D. bulbifera	11.7	–	–	Home garden, marginal fields, fallow/idle lands	May/Jun to Oct	Oct to Dec
D. divaricata	–	–	8.3	Hill (in Badoc), low to medium-elevation secondary forest (in Nueva Era)	May/Jun to Dec	*In Badoc:* Dec–Jan of next year
D. esculenta var. *fasciculata*	100	–	–	Rainfed farms, marginal areas	Jan/Feb to Nov	Sep (early)
						Nov to Dec (for market)
						Jan to Feb (for planting material)
D. esculenta var. *spinosa*	–	–	88.3	Hill, foot of secondary forest, roadsides, thickets	May to Oct/Nov	Oct to Nov
D. hispida	–	16.7	100	Hill, foot of secondary forest, roadsides, thickets	May to Oct/Nov	*In Pasuquin:* Aug to Sep
						In Badoc, Batac, Solsona: Aug to Dec
D. luzonensis	–	15	100	Hill, foot of the mountain, roadside, thickets, roadside	May to Nov	*In all towns, esp. Pasuquin:*
						Aug to Sep (early)
						Oct to Nov (regular)
						In Badoc:
						Sep to Oct (first harvest)
						Nov to Dec (second harvest)
D. pentaphylla	–	–	[b]	Hill, roadside	May/Jun to Dec	Not harvested

(continued)

Table 5.5 (continued)

Species	Frequency[a] (%)			Growing area	Growing season	Harvest month
	Cultivated	Semi-wild/ domesticated	Wild			
T. leontopetaloides	–	–	60	Hill, foot of the mountain, roadside, thickets	May/Jun to Dec	Nov to Dec
T. palmata	–	–	[b]	Hills, foot of the mountain	May/Jun to Dec	Not harvested (no known utilization)

[a] Multiple responses
[b] Wild-growing (as observed during the nature walk)

First, they follow more of the *rubbish heap model*, i.e., when wildlings are seen growing in their areas, they tend to them by cultivating around the hills, clearing other wayward plants and weeds, and erecting trellis for support. Second, they employ an informal seed system by setting aside the small tubers of *D. hispida* for sowing in succeeding season. In both practices, the informants indicated that they propagate in the usual natural growing areas and have not tried growing them in prime cultivated areas, e.g., upland farms (*bangkag*) and rainfed lowland, nor applying chemical inputs. This indicates that, even if we promote the cultivation of these wild species, it would not compete with the 'big-ticket' crops on the cultivated areas. Likewise, the application of chemical inputs such as fertilizer to increase the species' productivity is a future researchable area.

D. alata have both cultivated and wild forms in the province (Table 5.5). The much-preferred varieties such as those with purple, yellowish, and globose tubers are usually cultivated in upland farms, marginal areas, and home gardens. The less-preferred and those used as boiled feedstuff for swine are usually wild-growing in upland farms, idle and fallow areas, thickets, and low-elevation secondary forests. Compared to the adjacent Provinces of La Union, Ilocos Sur and Benguet, *D. alata* culture in the survey areas in Ilocos Norte is not as popular as in the former provinces where the greater yam (*ubi*) is commercially produced and processed into *halaya*, jams, ice cream, cakes, and other pastries.

D. bulbifera and *D. esculenta* var. *fasciculata* are both cultivated crops (Table 5.5). *D. bulbifera* is grown in marginal fields, fallow areas, and home gardens, while that of *D. esculenta var. fasciculata* is grown commercially in rainfed lowland and upland farms (*bangkag*).

Seasonal Abundance. Based on the Corona's system of classification, the Province of Ilocos Norte has Type I climate (Lantican 2001), which is characterized by two pronounced seasons: wet and dry. The dry season (DS) is from November to April, while the wet season (WS) is during the remaining months of the year. The active growing season of yam in natural vegetation coincides with the wet season (WS) (May to October) in Ilocos Norte (Table 5.5). Sprouting of yams in natural vegetation is triggered by the first rains in the month of May. The plants generally mature, characterized by the start of yellowing of leaves, by the month of October, but senescence extends until October or November (sometime December) depending on rainfall frequency during the last months of the year. The schedule of harvesting tuber varies among species (Table 5.5).

D. alata, D. bulbifera, D. esculenta var. spinosa, and *T. leontopetaloides.* These are harvested between October to December. Since these species except *T. leontopetaloides* are usually intended for home consumption and boiled feedstuff for swine, harvest schedule is dictated by the food need of the household and farm and off-farm activities. To the informants, the tubers are usually used as supplement to rice and for viand, hence they only dig and harvest when the household needs it. Harvesting is also scheduled earlier or later so as not to coincide with the peak of farm and off-farm activities like rice harvesting, planting, etc.

D. divaricata. The informants in Badoc usually harvest it the last in December or January of succeeding year or during the lean months. To them, it is their reserve food, and they leave it un-dugged (*ilebben Ilk.*) until needed.

D. esculenta var. fasciculata. This cultivated species is usually grown from January or February to November and harvested at different months of the year depending on the purpose. Early harvesting can be made in September, usually for home consumption, during which tuber flesh is still loose and friable. Regular harvesting is between November and December, usually intended for market, during which tuber flesh has turned chewy and firm. Late harvesting of tubers intended for planting materials is done in January to February.

D. hispida. Tubers of *D. hispida* are harvested earlier in August to September in Pasuquin to catch up available running fresh water (river water) for washing. This species needs detoxification prior to consumption due to its toxic dioscorine component, and they need ample running water like the river for repeated washing during detoxification process. The August to September harvest months are also beneficial to the processors as they could gain more income during these times. Balled *D. hispida* tubers are sold at PhP10 each, and the turnover of the commodity is faster during these early months. Meanwhile, in Badoc, Batac, and Solsona, harvesting is done between August and December.

D. luzonensis. In most of the survey areas especially Pasuquin, harvesting is done as early as September when tubers are sold at P100 per kilogram despite the loose and friable texture of the tubers. More matured tubers are harvested in October to November during which tuber flesh has turned chewy and firm. These are sold at PhP40 to PhP50 per kilogram. Harvest practices differ in Badoc where many informants follow two-time harvesting in a single hill. The first harvest is in September to October, then they get back to the same hill for a second round of harvest in November, December, and January. The tuber yield at the second harvest is a regrowth of the upper sett, usually smaller and amorphous which is in contrast to the first harvest which is an elongated cylindric tuber.

D. pentaphylla and T. palmata. These species are not harvested and there is no documented utilization for food or feed. *T. palmata*, however, was observed to be used as ornamental by a few plant hobbyists in Nueva Era, Ilocos Norte, for its penta-foliate leaves.

5.3.2 Cultural and Socioeconomic Importance of the Species

Traditional Uses. All of the documented species except *D. pentaphylla* and *T. palmata* are consumed by the local communities in the Province of Ilocos Norte. Consumption of the tubers as food is the foremost use of the species (Table 5.6). They are consumed as supplement to rice, boiled and/or broiled snack, processed products like jams, *halaya,* and *guinatan* (a local snack cooked with coconut milk), or mixed with other vegetables in viand or in pork and beef stews. Detoxified tubers of *T. leontopetaloides*, in particular, are dried and powdered into flour to enhance the taste and aroma of fried rice (*sinangag*). Detoxified tubers

Table 5.6 Traditional uses and cultural importance (CI) of the species of Dioscoreaceae in Ilocos Norte

Species	Traditional uses	Cultural importance	Rank
D. alata	Food	1.33	1
	Boiled feedstuff for swine		
D. bulbifera	Food	0.12	7
D. divaricata	Food	0.08	8
D. esculenta var. *fasciculata*	Food	1	5.5
D. esculenta var. *spinosa*	Food	1.22	3
	Boiled feedstuff for swine		
D. hispida	Food	1.08	4
	Folkloric medicine		
	Used for ritual		
D. luzonensis	Food	1	5.5
D. pentaphylla	–	0	9.5
T. leontopetaloides	Food	1.25	2
	Laundry starch (stiffen fabric)		
T. palmata	Ornamental[a]	0	9.5

[a] Not cited by informants but observed by the researcher

of *D. hispida* are mixed as topping to fried rice or cooked with a little oil or butter for breakfast or snack.

Other documented traditional uses include feed, folkloric medicinal, and for ritual. Tubers of lesser-known and lesser-preferred *D. alata* and *D. esculenta var. spinosa* are used as boiled feedstuff for swine. *D. hispida* is a folkloric topical cure for cuts, bruises and skin itchiness, and as live amulet to ward off bad spirits. *T. leontopetaloides* flour, after a quick boil, is used as laundry starch to stiffen linen and fabrics.

The statement that "people cannot use what they don't know" holds true to *D. pentaphylla* and *T. palmata*. While these species are present in one or two study areas, people are not aware of their presence and identity. Hence, there was no known or documented use. *T. palmata*, however, was observed by the researcher as a potted ornamental by a few hobbyists during the COVID pandemic.

Importance Level. The level of importance of the Dioscoreaceae plants in general was based on the informants' perception. The informants strongly agree that the yams provide more food to their households, thus enhancing dietary diversity (Table 5.7). They moderately agree that wild-harvesting or cultivation supplements family income. They only agree that these enhance nutrition and have other uses such as medicinal, animal feedstuff, and laundry starch. They moderately disagree that the plants have esthetic and ritual values (Table 5.7), despite a very few responses on *D. hispida* being used as live amulet, and observation on *T. palmata* being used as ornamental.

Table 5.7 Key informants' perception of the importance of the species of Dioscoreaceae

Importance	Importance level
Provide more food to the family	Strongly agree
Provide additional household income	Moderately agree
Enhance nutrition	Agree
Has medicinal uses	Agree
Use as animal feedstuff	Agree
Has ornamental/esthetic value	Moderately disagree
Use in ritual (live amulet)	Moderately disagree
Has other uses (i.e., laundry starch or stiffen fabric)	Agree

Table 5.8 Utilization of Dioscoreaceae tubers by informants

Parameter	Frequency	Percentage
Are yams utilized		
Yes	60	100
No	0	0
Where are they taken[a]		
Gathered from the wild	54	90
Own harvest	17	28
Bought	6	10
Reason for using[a]		
Readily available in the area	42	70
Abundant	42	70
Love/used to the taste	42	70
Provides income	37	62
Nutritious	30	50
Emergency food during the pandemic	27	45

[a] Multiple responses

All the informants confirmed they utilize the yam tubers for family consumption or market. Majority (72%) of them gather tubers from the wild or naturally growing Dioscoreaceae plants. The other informants derive raw tubers from their own harvest (28%) or buy raw as well as processed tubers from neighbor-gatherers/processors or market (10%) (Table 5.8). Among the reasons the informants use the species include: readily available in the area, abundant, and taste is loved or used to and thus have become integral part of their diet at particular months of the year.

Tubers of some species are known to be toxic and exude bitter taste while others result to health inconveniences. Thus, four of the species are used in moderation, with precaution or some limitations (Table 5.9). *D. hispida* and *T. leontopetaloides* are believed to be poisonous (with bitter taste), hence the tubers need to be detoxified prior to cooking. Together with *D. luzonensis* and *D. divaricata*, extra care is needed during tuber preparation as these can cause itchy skin. Other reasons given by the informants include crop management-related concerns like tedious digging and

Table 5.9 Informants' reasons for moderated use of some species

Reason	Species	Frequency	Percentage
Plant-related			
Believed to be poisonous	*D. hispida*	9	15
Cause health inconveniences, i.e., diarrhea, itchy skin	*D. hispida*	3	5
	T. leontopetaloides *D. luzonensis*		
No longer found in the area	*D. divaricata*	3	5
Difficult to harvest	*D. luzonensis*	7	12
	D. divaricata		
Tedious processing	*D. hispida*	11	18
	T. leontopetaloides		
Human-related			
Health/age restrictions (i.e., old age)	–	11	18

detoxification, and scarce supply like in the case of *D. divaricata*; and gatherer/processor' health- or age-related restrictions.

Cultural Importance. Based on cultural importance (CI) value, the top species, in descending order, are *D. alata, T. leontopetaloides, D. esculenta* var. *spinosa, D. hispida* and *D. luzonensis,* and *D. esculenta* var. *fasciculata* (Table 5.6). CI measures the frequency of citation of the species taking into account its multiple uses. Comparing it to RFC, which has a maximum value of 1.0 (Table 5.3), RFC $\neq$ CI as the later takes into consideration the multiple uses of the species. Hence, the higher values of above species are due to more frequent citation and multiple uses. In the case of *D. esculenta var. fasciculata* (a cultivated crop) and *D. luzonensis* (a much-loved yam among Ilocanos), they are both commercial crops but obtained lower CI than the others due to single use (as food).

Lower CI is estimated for the lesser-known and lesser-used *D. bulbifera* and *D. divaricata*, while zero CI for *T. palmata* and *D. pentaphylla*.

Socioeconomic Importance. In addition to the traditional uses mentioned above, the species of Dioscoreaceae have also played important role in income supplementation in the study areas. Almost all (90%) of the informants sell tubers they harvested or gathered from the wild. They sell in village market (83%), neighborhood (83%), downtown or poblacion market (48%), and at farmgate through assemblers and private orders (28%) (Table 5.10). About 80% of the selling informants disclosed that cultivation and/or wild-harvesting and processing of yams is a good source of income. The mainly marketed tubers are *D. luzonensis, D. hispida, D. esculenta* var. *fasciculata,* and *T. leontopetaloides*, especially in Pasuquin, Badoc, Banna, and Batac. The two major processor-sellers in Pasuquin earn as much as PhP32,000 gross income per season (4 months) from *D. hispida*. The three major gatherers of *D. luzonensis* in Brgy. Turod, Badoc earn an estimated

Table 5.10 Marketing of Dioscoreaceae tubers in Ilocos Norte

Parameter	Frequency	Percentage
Selling?		
Yes	54	90
No	6	10
Selling place[a] ($n = 54$)		
Village market	45	83
Downtown or supermarket	26	48
Neighborhood	45	83
Others (farmgate with assembler, orders)	15	28
Is it a good source of income ($n = 54$)		
Yes	45	83
No	9	17
Not selling, why? ($n - 6$)		
Just enough for own household	6	100
Bartered/traded with other goods	2	33
Give away to neighbor	2	33
Not saleable	0	0
Not known/popular among consumers	0	0

[a] Multiple responses

income of PhP9,000 to PhP15,000 per season. This venture gives them a handsome amount of additional income; however, such income is seasonal and short-duration (about 4 months) only. Some informants reported that the scope of work and income derived from yams is better than a fixed daily wage-earner or farm laborer who earns P400 a day. However, others added that it is not a stable income source due to seasonality of the plants.

The social impact of wild-gathering yams has developed adaptive values among rural families in the study areas, making them more resilient especially during the lean months and crises, e.g., COVID pandemic. Additionally, wild-gathering has perpetuated the values of frugality and hard work among the locals.

5.4 Ethnobotanical Evidences on the Utilization of Dioscoreaceae in Northwestern Luzon

Yams have rooted in the subsistence and survival, food culture, and tradition of Ilocanos and other cultural groups living in Ilocos Norte. Four major evidences were documented pointing to the significant position of the species in the daily living of the people in this part of the country.

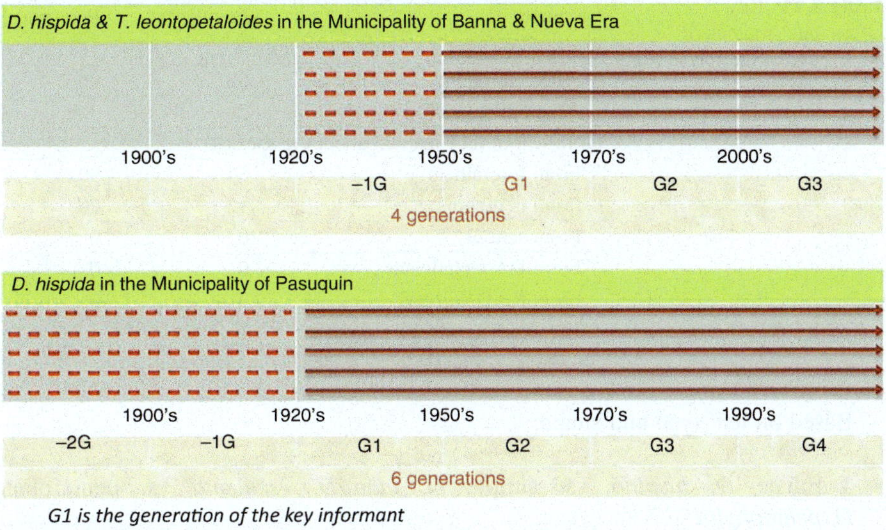

Fig. 5.2 The use of Dioscoreaceae in Ilocos Norte can be traced from four to six generations as suggested by the key informants. G1 is the generation of the key informant

5.4.1 Long History of Continued Utilization

The long history of continued utilization of yams dates back to the olden days of the informants' forefathers (Fig. 5.2), thus suggesting the plants have become a tradition across the communities in the province. The eldest (90 years old) informant in the Municipality of Pasuquin can trace the use of *D. hispida* for six generations (two generations before her, and four generations from her time to her great grandchildren). Meanwhile, the use of *D. hispida* and *T. leontopetaloides* can be traced for only four generations in the Municipalities of Banna and Nueva Era. But, this does not imply that these are used only during these generations. A limitation on older informants has restricted the short duration documented.

Majority of the informants learned about the plants since their childhood (21–90 years) (Table 5.11). They learned about the plants mainly from their forefathers (90%), and a few (22%) learned from neighbors and personal experience. Knowledge on plants which are socioeconomically important, such as yams, are oftentimes transmitted by oral tradition from generations to generations.

5.4.2 Ethnotaxonomic Knowledge and Practice

The people's familiarity of the plants enables them to discriminate one species from the other and identify toxic from nontoxic yam. Leaf type and shape; vine spines, shape, and strength; phyllotaxy; and tuber habit are important discriminatory characters. Using them, the people can discern the following:

Table 5.11 Key informants' sources of knowledge of the species of Dioscoreaceae

Parameter	Frequency	Percentage
When they learned about the species		
Since childhood (21–90 years)	58	96.7
About 20 years ago	2	3.3
From whom they learned about the species[a]		
From forefathers	54	90
From neighbors	12	20
From school	0	0
From DA technicians	0	0
From R&D workers	0	0
Through experience	1	1.7

[a] Multiple responses

Based on leaf type and shape:

- Trifoliate *D. hispida* vs. simple *D. alata/D. esculenta* vs. pentafoliate *D. pentaphylla*
- Cordate *D. alata* vs. sagittate *D. luzonensis*

Based on vine spines and shape:

- Spiny *D. hispida* vs. non-spiny *D. luzonensis*
- Round *D. esculenta* vs. winged *D. alata*

Based on phyllotaxy

- Alternate *D. luzonensis* vs. opposite *D. divaricata*

Based on tuber habit:

- Fascicled or clustered *D. esculenta var. fasciculata* (tugui) vs. with runner *D. esculenta var. spinosa* (buga)

Yam vines exhibit directional twining, either twining to the right (clockwise) or twining to the left (counter-clockwise). The twining direction is constant in each species, and is a characteristic peculiar to particular sections within the genus *Dioscorea.* The wings present in some species such as *D. alata* support the twining habit. This character is not being used by the locals in species discretion and identification.

5.4.3 Indigenous Knowledge and Practices

Yams generally produce poisonous components in their tubers (either underground or aerial). These are secondary metabolites, alkaloids, or saponins. The presence of the bitter or toxic components provides protection to the plant, enabling them to

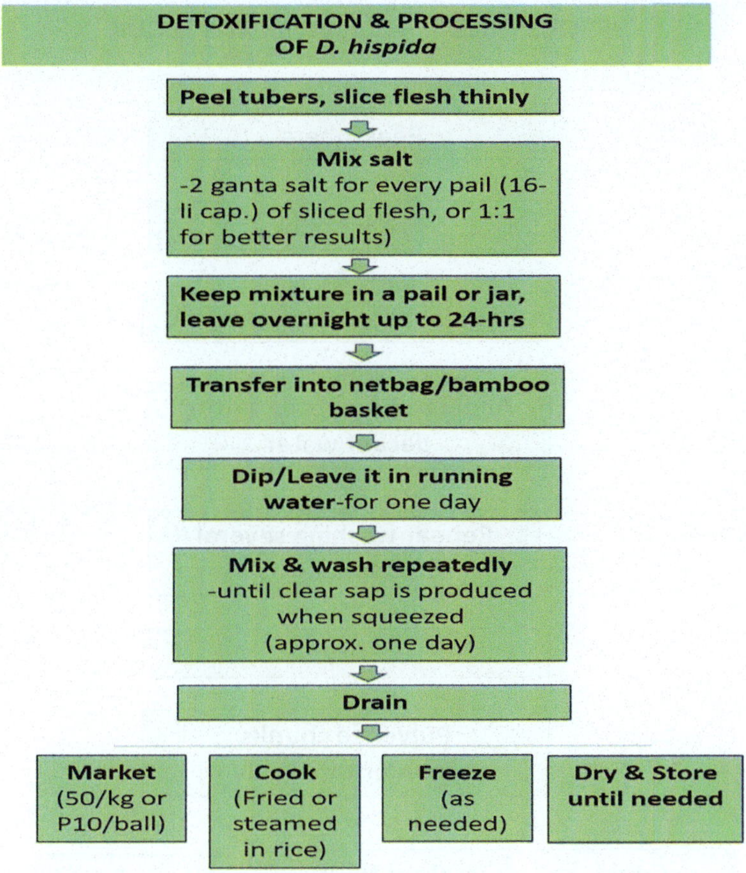

Fig. 5.3 Detoxification process for *Dioscorea hispida*

escape extinction in the wild by restricting overcollection by man or got eaten by wild herbivorous animals.

With the bitterness and perceived toxicity of the bitter tubers, locals have developed indigenous knowledge and practices which include methodologies in cooking, detoxifying, processing, and local dishes and recipes.

Among the documented species in the Province, eight of them are edible, two of which are toxic (*D. hispida* and *T. leontopetaloides*). Through experience, the locals are aware of the toxicity and bitterness in the two species. But, this observation did not restrict them from eating the tubers as they have grown fond of the unique taste and flavor of these wild foods. Instead, they have developed crude methodologies and indigenous knowledge and practices.

Figure 5.3 shows the schematic process documented from the locals in the Municipality of Pasuquin, Nueva Era, and Banna in the detoxification and processing of *D. hispida* tubers. Important steps in the process integrate brining

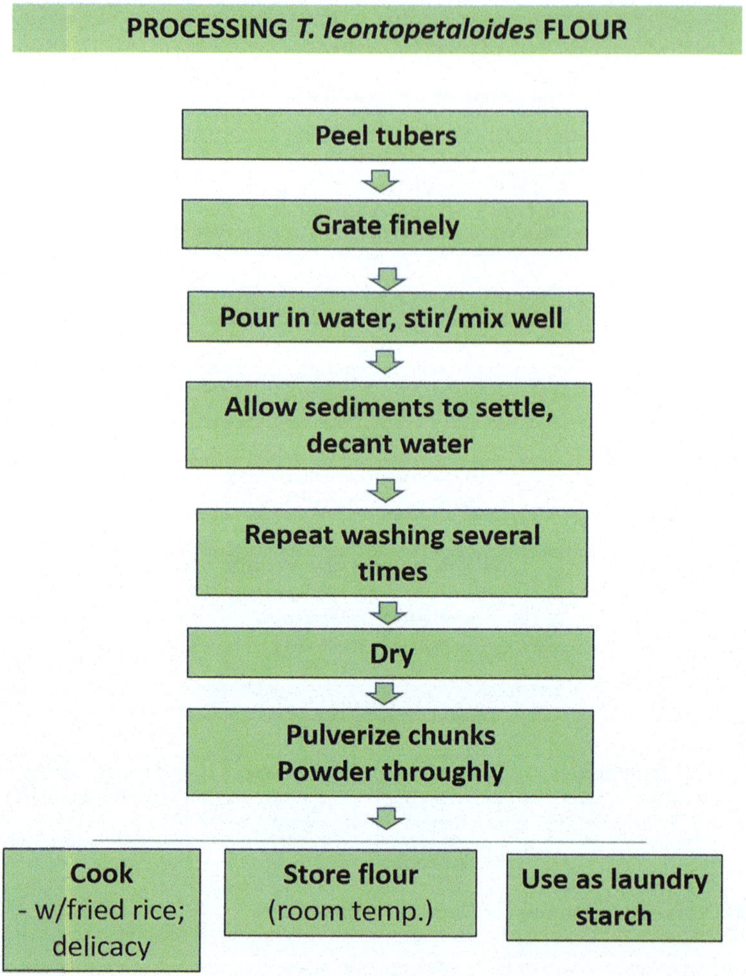

Fig. 5.4 Detoxification process for *Tacca leontopetaloides*

and repeated washing in running water. Washing in running water is recommended in order to evade itchiness when the skin gets in contact with the exudate. They also follow some indicators for properly detoxified *D. hispida* tubers, which include: (1) Pliable slices, (2) Clear sap or exudate, (3) No biting taste, (4) No karot aroma or smell, (5) Taste good when cooked, and (6) No dizziness effect when ingested. A rapid detection kit should be developed for a systematic and scientific assessment of the safety of the tuber after processing.

Detoxified *D. hispida* tuber is fried or steamed in rice for snack or breakfast meal. It can also be kept frozen, or sundried and stored in a jar under room temperature until needed. It is sold at PhP50 per kilogram or PhP10 per fist-sized ball.

T. leontopetaloides tubers are processed into flour prior to cooking. For the detoxification of the bitter content, the schematic process is shown in Fig. 5.4.

Unlike *D. hispida* tubers which are sliced thinly during processing, *T. leontopetaloides* tubers are grated finely, followed by repeated sedimentation and washing, drying, and pulverizing. Here, the process can be done in the kitchen, not needing running water as there is no documented itchiness on the skin upon exposure to the wash water.

Tacca dough is cooked into snack and delicacies such as *guinata*n, fritter, baked *Tacca* (*bibingka*), etc. The flour is also sprinkled in fried rice during cooking and gives a unique aroma or taste and a gluten texture to fried rice. Flour can be stored in a jar under room conditions. In the Municipality of Pasuquin, *Tacca* flour is boiled with plenty of water and the mixture is used to stiffen fabric or linen.

5.4.4 Ethnolinguistic Information

Yams are known to the locals in corresponding vernacular names (Table 5.3). Except for *D. pentaphylla* (locally called lima-lima), the etymology of the vernacular names cannot be easily defined. *D. pentaphylla* is a pentafoliate species, hence the vernacular name *lima-lima*.

Several local terminologies associated with yams are also coined by the locals. These include terms such as (1) *panagilebben*, which pertains to a postharvest practice or storage by burying in the soil, (2) *lappatan,* to mark a yam hill and serve as reference for digging later even when the leaves and vines have dried up, (3) *unaban*, the process of detoxifying or washing thoroughly.

Above evidences suggest that yams have indeed rooted in the diet, food culture, and plant heritage of the people in northwestern Philippines.

5.5 Enhancing the Utilization of Species of Dioscoreae

Generally, members of family Dioscoreaceae are major tuber crops and numerous wild species, which form staple foods in tropical regions and are of pharmacological interest, especially some species of *Dioscorea* which are frequently rich in steroidal sapogenins and usually accumulating chelidonic acid and lactone alkaloids.

In the local arena, much of our local plant resources, which include the members of family *Dioscoreaceae*, are being mined and harnessed for their potential in achieving food sufficiency, health and nutrition, livelihood creation, and income generation.

Three *Dioscorea* spp. (*D. esculenta* var. *spinosa*, *D. hispida,* and *D. luzonensis*) and *Tacca leontopetaloides* were earlier documented and characterized as among the indigenous food plants in Ilocos Norte (Antonio et al. 2011). These are important substitutes for rice in the upland communities and prepared into boiled snacks, delicacies, or as vegetable dishes. Three yam species were also reported by Vidad (2016) as local feed resource for native pigs in Ilocos Norte. These include *D. esculenta* var. *spinosa*, *D. luzonensis,* and *D. alata* '*ube ti baboy*' or '*ube-bantay*'.

In addition to the food and feed uses, there has been heightened exploration of the lesser-known yams for food product development. *D. luzonensis* tubers were successfully processed by Legaspi et al. (2018) into flour which is noted to be *at par* with the commercial flour and unique for being gluten-free. Thus, the produced flour is good for gluten-intolerant individuals, those with autoimmune diseases, persons with special needs, and the health and wellness enthusiasts. The produced flour is further processed into healthy snack products which are now being promoted for commercialization (Legaspi et al. 2018).

The documented use as food and feed in the province is supported by the proximate composition of the species. Vidad (2016) reported the proximates of two species in both raw and boiled forms. The raw *D. alata* contains 32.24%, 5.15%, 5.85%, 3.85%, 0.83%, 84.32%, and 3833.33 kcal/kg of air dry matter, ash, crude protein, crude fiber, crude fat, nitrogen free extract, and gross energy, respectively; boiled *D. alata* had 30.77%, 4.31%, 12.91%, 2.61%, 0.42%, 79.75%, and 3925.33 kcal/kg, respectively; raw *D. esculenta 'buga'* contains 33.65%, 4.35%, 3.84%, 3.96%, 0.04%, 87.81%, and 3806.00 kcal/kg, respectively; boiled "buga" (*Dioscorea esculenta var. spinosa*) contains 31.78%, 04.07%, 07.02%, 04.04%, 1.38%, 83.50%, and 3869.33 kcal/kg, respectively (Vidad 2016).

The data of Legaspi et al. (2018) and Agbigay et al. (2020) proved the extent how Ilocanos patronize one of the *Dioscorea* species, *D. luzonensis* or locally called 'kamangeg'. An estimate of more than 13 tons of *D. luzonensis* tubers are being gathered from the wild and subsequently sold in public markets of Ilocos Norte per season at PhP32.00 to PhP57.00 per kilogram (Agbigay et al. 2020). Among the 20 municipalities and cities in Ilocos Norte, the public markets of the City of Batac and Laoag City had the highest volume of *D. luzonensis* tubers sold (3110 kg and 3010 kg, respectively), while the municipality of Pagudpud had the lowest (20 kg). There is no record of *D. luzonensis* being sold in the public markets of Badoc, Pagudpud, Adams, Carasi, and Dumalneg; but, there were indications that gathered *D. luzonensis* are limited for home consumption and ambulant vending in the community. In the neighboring Province of Ilocos Sur, about 1–2 kg are gathered per hill of *D. luzonensis*, 3–4 kg for *D. esculenta* 'buga', and >7 kg for *D. alata* (Tomas et al. 2020). *D. alata* and *D. luzonensis* were most preferred by the people in Ilocos Sur.

Yams are known to contain bitter or toxic principles, which render them less palatable, cause health inconveniences such as in the case of histamine which triggers itching (Shim and Oh 2008), and oftentimes poisonous such as in the case of dioscorine which triggers fatal paralysis of the nervous system (Reddy 2015). In yam-eating regions, various techniques are used by the people to reduce, if not eliminate, the bitter and toxic components of yams. Boiling, steaming, and/or baking over coals after cleaning and peeling the tubers are among the most common techniques reported by Bhandari and Kabawata (2005) (as cited by Kumar et al. 2017).

Much of the known ethnobotanical uses of species of Dioscoreaceae highlight their potential for food. But, recent studies reveal various bioactivities with pharmacological potential and use in dietary supplements and cosmetics (Son et al., 2007; Black et al., 2007) (as cited by Kumar et al. 2017).

The many uses are documented and growing efforts on their exploration and promotion show the harmonious blend of the people with the abundance of nature. In the study areas, the locals live by the wild foods they gather. They earn supplemental income for their families. And recently, local processors and technology adopters are aiming at value-addition using the lesser-known yams.

To accelerate the utilization of yams, thus giving optimum benefits for human gains, a harmonious blend of science and indigenous knowledge has to be emphasized. Various ethnobotanical studies highlight the utilization of yams for food, feeds, and folkloric medicine, among others. While indigenous knowledge systems in the consumption as well as elimination of the toxic components of yams are in place, food safety remains a concern. Thus, science-based and systematic technologies eliminating the toxic components will enhance optimum utilization for food and nonfood. For rural communities which do not have ready access to laboratories, rapid detection kits should be developed for use in the mountainous and remote landscapes. Likewise, a proactive search for the nutritive and nonnutritive components of the lesser-known and lesser-documented species will also aid utilization.

5.6 Summary, Conclusions, and Recommendations

A survey was done in the Province of Ilocos Norte to document the species of Dioscoreaceae known and used by the locals, their uses, and ethnobotanical evidences implying the significant role of the species among the people of the province. There are ten taxa of yams identified in Ilocos Norte. These include *D. alata*, *D. bulbifera*, *D. divaricata*, *D. esculenta* var. *fasciculata*, *D. esculenta* var. *spinosa*, *D. hispida*, *D. luzonensis*, *D. pentaphylla*, *T. leontopetaloides*, and *T. palmata*. Majority of them are wild-growing and seasonally available.

All the species except *D. pentaphylla* and *T. palmata* have documented uses. Traditionally, these are mainly used for food; the others, at a lower extent, for feed, folkloric medicine, laundry starch, and live amulet. There is no enumerated use of *T. palmata* by the informants, but observed to be propagated by some hobbyist for ornamental purposes.

All informants utilize the tubers for home consumption and market. Additional income is derived by families in rural communities from sourcing, processing, and vending or marketing. Tubers are sold as raw tubers (i.e., *D. luzonensis*, *D. esculenta*, *D. alata*), detoxified slices (*D. hispida*), or flour (*T. leontopetaloides*).

Of the eight edible species, two species (*D. hisida* and *T. leontopetaloides*) have bitter taste. The locals have developed crude methodologies of detoxification and processing of the tubers. Detoxification of *D. hispida* entails slicing, brining, soaking, and repeated washing in running water, while that of *T. leontopetaloides* entails grating, repeated washing, sedimentation, drying, and pulverizing.

Several ethnobotanical evidences confirm the significant role of the documented species in the diet and livelihood of the people in the study areas. These include long history of continued utilization dating back since the olden days of their forefathers, people's familiarity of the species, indigenous knowledge and developed methodologies in detoxification and processing, and the presence of vernacular names and associated terminologies, among others.

With the documented uses of the species and informal efforts on domestication and culture, there is felt harmonious blend of the people with the abundance of nature in this part of the country. Likewise, a harmonious blend between science-based methodologies and indigenous knowledge has to be emphasized to ensure food safety and leverage on the potential of the species to the fullest.

Since many yam species are wild-growing, they are exposed to threats in their natural habitats. They are exposed to risks caused by changes in land use, environmental degradation or destruction; climate change; and overharvesting and exploitation. Thus, a conservation and management framework should be developed and mainstreamed. Such framework should encompass interventions such as domestication, development of crop production technologies, heightened exploration of the potential, and role assessment for various stakeholders, among others. Lastly, a strong policy support should be instituted for the implementation of the conservation and management framework.

Acknowledgments The authors would like to thank the people of Ilocos Norte for sharing their knowledge and insights on the ethnobotany of local yams; and the leaders and officials of the local communities and local government units for the permit and assistance granted during the implementation of the research.

References

Agbigay LC, Legaspi NB, Quiapo CD (2020) Accelerating promotion and adoption of less-known yam production and utilization technologies through capacity building approach (unpublished). Paper presented in the ILAARRDEC commodity review on gender and development and gender and socio-economic studies. ILAARRDEC, MMSU, City of Batac Ilocos Norte, 27 Oct 2020

Antonio MA (2008) Assessment of genetic diversity in, and identification of wild Dioscorea "Buga" through morphological and cytological characterization. MS thesis, University of the Philippines Los Banos, College, Laguna

Antonio MA, Utrera RT, Agustin EO, Jamias DL, Badar AJ, Pascua ME (2011) Survey and characterization of indigenous food plants in Ilocos Norte, Philippines. SEARCA agriculture & development discussion paper series no. 2011-2. Southeast Asian Regional Center for Graduate Study and Research in Agriculture, Los Banos, Laguna

APG II (2003) An update of the Angiosperm Phylogeny Group classification for the orders and families of flowering plants: APG II. Bot J Linn Soc 141(4):399–436. https://doi.org/10.1046/j.1095-8339.2003.t01-1-00158.x. Accessed 12 June 2020

Barcelo RTC (2014) Ethno-botanical survey of edible wild fruits in Benguet, Cordillera Administrative Region, The Philippines. Asian Pac J Trop Biomed 4:S525–S538. https://doi.org/10.12980/APJTB.4.201414B36. Accessed 10 June 2022

Kumar S, Das G, Shin HS, Patra JK (2017) *Dioscorea* spp. (a wild edible tuber): a study on its ethnopharmacological potential and traditional use by the local people of Similipal Biosphere

Reserve, India. Front Pharmacol 8:52. https://doi.org/10.3389/fphar.2017.00052. https://www.ncbi.nlm.nih.gov/pmc/articles/PMC5306286/. Accessed 10 June 2022

Lantican RM (2001) The science and practice of crop production. SEAMEO SEARCA and UPLB, College, Laguna

Legaspi NL (2004) Germplasm collection, multiplication, evaluation and maintenance of native and exotic varieties of ubi and tugui. Paper presented in the ILARRDEC Regional Symposium on R&D Highlights. PhilRice, Batac, Ilocos Norte, 8 Aug 2004

Legaspi NB, Agbigay LC, Quiapo C (2018) Techno-economic potentials of "Kamangeg" (*Dioscorea luzonensis*). Unpublished, Mariano Marcos State University, City of Batac, Ilocos Norte. Paper presented in the 30th Regional Symposium on R&D Highlights, DMMMSU, Bacnotan La Union, 27–29 Nov 2018

Maghirang RG, Oraye CD, Antonio MA, Cacal MS (2018) Ethnobotanical studies of some plants commonly used as vegetables in selected provinces of the Philippines. J Nat Stud 17(2):30–43

Pelser PB, Barcelona JF, Nickrent DL (eds) (2011 onwards) Co's Digital Flora of the Philippines. www.philippineplants.org

Provincial Planning and Development Office-Ilocos Norte (PPDO-IN) (2002) Ilocos Norte provincial physical framework plan/comprehensive Landuse plan (1993-2002). Province of Ilocos Norte, Ilocos Norte

Reddy DS (2015) Ethnobotanical studies of Dioscorea hispida Dennst. in Nallamala forest area AP, India. Rev Res 4:1–4

Shim WS, Oh U (2008) Histamine-induced itch and its relationship with pain. Mol Pain 4:29–32. https://doi.org/10.1186/1744-8069-4-29. Accessed 1 Aug 2022

Tomas P, Etrata R, Gautani A, Sanidad R (2020) Wild yam assessment: cultural management, processes and preferences. Unpublished, Ilocos Sur Polytechnic State College, Santa Maria, Ilocos Sur

Vidad SR (2016) Identification and characterization of locally available feed resources for native pigs in selected towns of Ilocos Norte. MS Thesis, College of Agriculture. Benguet State University, La Trinidad, Benguet

A Preliminary Survey of the Genus Hoya R. Br. (Apocynaceae) in Papua, Indonesia, with Notes on Hoya as Larval Food Plant of *Euploea netscheri* Snellen

6

Daawia, Inocencio E Buot, Jr, Krisantini, and Dwi Suratman

Abstract

The genus *Hoya* (Apocynaceae) has been known as an ornamental plant for 100 years in Europe, USA, and Australia, while in Indonesian it is still poorly known. There were about 350–450 species of *Hoya* in the world, and 74 species among them were distributed in Papuasia. Most of the *Hoya* species in Papuasia were recorded from Papua New Guinea, and there has been very poor record from western New Guinea (Papua Province and West Papua Province). This preliminary survey of *Hoya* species diversity was carried out in several Regencies in Papua and Papua Provinces using purposive sampling method. We conducted 12 field expeditions in Yapen, Biak, Sorong, Raja Ampat, Maybrat, Jayapura, and Keerom. From this survey, we discovered 10 species of Hoya, namely: *Hoya stenakei, Hoya megalaster, Hoya pachyphylla, Hoya patella, Hoya globulifera,* and *Hoya inconspicua.* Four other species belong to Eriostemma Section (*Hoya lauterbachii, Hoya coronaria, Hoya sussuela, Hoya* sp1). We recorded *Hoya pachyphylla* and *Hoya globulifera* as larval food plants for *Euploea netscheri*. To

Daawia (✉)
Biology Department, Faculty of Mathematics and Natural Sciences, Cenderawasih University, Jayapura, Papua, Indonesia

I. E. Buot, Jr
Institute of Biological Sciences, College of Arts and Sciences, University of the Philippines Los Baños, Laguna, Philippines

Krisantini
Department of Agronomy and Horticulture, Faculty of Agriculture, IPB University, Bogor, Indonesia

D. Suratman
Hoya Grower, Kelurahan Malawili, Distrik Aimas, Kabupaten Sorong, Papua Barat, Indonesia

S. Ramamoorthy et al. (eds.), *Plant Diversity in Biocultural Landscapes*,
https://doi.org/10.1007/978-981-19-8649-9_6

the best of our knowledge, these are new record of larval food plants for *Euploea netscheri.*

Keywords

Hoya · Apocynaceae · *Euploea* · Papuasia · Papua · Indonesia

6.1 Introduction

Hoya belongs to family Apocynaceae, subfamily Asclepiadoideae (Endress et al. 2014). Many home gardeners have enjoyed Hoya species for decades as ornamental plants. In Sweden, hoyas are called "porcelain flowers" and "wax plants" in English. The stems, leaves, and flowers are covered by a waxed layer and produce very unique looking flowers. The flowers grow in umbels and the number of flowers per umbel varies from one single flower up to 80 flowers. The collective number of flowers per umbel form a bell shape. The flowers often release unique, sweet fragrances to attract pollinators. The leaves are opposite, thick, and usually in pairs. Most of the Hoya species are epiphytic vines, while some are terrestrial vines and only a few species are woody shrubs.

Hoya R.Br is the largest genus of Apocynaceae in which Rodda (2015) estimated to have 350–450 species. *Hoya* has a wide distribution in tropical and subtropical regions; it is distributed from India to the Pacific Islands (Wanntorp et al. 2006). Hoya is also recorded widespread in perhumid tropical lowlands of the Philippines (Kleijn and Van Donkelaar 2001). There are over 172 species of *Hoya* in the Philippines (Martinez and Buot 2017).

The most species-rich regions for Hoya are in tropical and subtropical South Asia, South East Asia, and Papuasia (Juhonewe and Rodda 2017; Lamb and Rodda 2016). Papuasia is a level 2 botanical region defined in the World Geographical Scheme for Recording Plant Distribution (WGSRPD). Papuasia lies in the Southwest Pacific Ocean in the Melanesia Eco-Region of Oceania and tropical Asia. Papuasia comprises Aru Islands, West Papua (Papua Province and West Papua Province) in Eastern Indonesia, Papua New Guinea, and Solomon Islands (excluding Santa Cruz Islands).

New Guinea Island consists of two countries; in the western half it is Indonesia New Guinea (Papua Province and west Papua Province), and the eastern half is Papua New Guinea. Most of New Guinea's described species of Hoya were from Papua New Guinea, while the inventory of *Hoya* in Indonesia New Guinea is still very poor.

According to Forster (1996), New Guinea Island has 74 species of *Hoya* and less than half of them have been found in Indonesia New Guinea (Papua and West Papua provinces). While Moore (1916) published and described *Hoya* species from Indonesian New Guinea, Hemsley (1891) published *Hoya* species from Solomon Islands.

The new publications to the *Hoya* of New Guinea were made by Forster and Liddle (1991, 1992, 1993) and Forster et al. (1995). Indonesian New Guinea's *Hoya* was poorly explored compared to Papua New Guinea's *Hoya* (Forster 2006) and so far only 16 species of *Hoya* have been recorded similar for both areas (Forster 1996, 2006).

6.2 Methodology

The preliminary inventory on *Hoya* diversity in Papua and West Papua Provinces was conducted using Purposive Sampling Method. We did 12 field expeditions in 5 Regencies in Papua Province and West Papua Province which were Jayapura, Keerom, Biak, Sorong, Raja Ampat, and Yapen during 2018 and 2019 as shown in Fig. 6.1. We did one field expedition in Serui Japen 3–7 February 2019, one in Biak (2–4 October 2018), one in Sorong (Mei 2019), one in Raja Ampat (April 2019), one in Maybrath (April 2019), two field expeditions in Jayapura (August 2018 and April 2019), and five field expeditions in Keerom (April 2018, July 2018, September 2018, June 2019). All of the occurrences of *Hoya* species were recorded, and specimens were collected for identification purpose. The collected specimens were also planted in *Hoya* Garden which belongs to Yayasan Pendidikan Alam Papua (YADIKAP) in Arso 2, Keerom Regency, for further study of identification of Papua *Hoya* and for ex-situ conservation purposes. Every time they produced flowers in

Fig. 6.1 Map of Papua and West Papua Provinces and surveyed sites shown by red dots

Hoya Garden, the flowers were photographed and sent to Papuasia's Hoya Expert (Ms. N. S. Juhonewe) for further identification.

6.3 Result and Discussion

Despite our limited explorations, we believe that Papua Land (Western New Guinea) has a very rich diversity of *Hoya* species as mentioned by Forster (1996) that New Guinea has 74 species. During this preliminary inventory (2018–2019), we only encountered 10 species of *Hoya* in both Papua and West Papua Provinces. The number of these species will increase if *Hoya* inventory is carried out more intensively in the Western New Guinea. The list and the description of the species are described as follows:

1. ***Hoya stenakei*** Simonsson & Rodda
   Stems cylindrical 2–6 mm in diameter, dark green, young stem often slightly pubescent: old stem greyish-brown, glabrous with peeling bark; internode 4–18 cm long. ***Leaves*** petiolate; petiole terete, 2–3.5 cm in diameter, green to brown, glabrous (pubescent when young); lamina lanceolate, succulent, 10–17 cm long and 3–6 cm wide, dark green on adaxial surface, light green on abaxial surface, glabrous, apex acute, base cordate; pinnate. ***Inflorescences*** one per node, 2–10 flowers per umbel, ***peduncle*** terete, 2–7 cm long, 2–5 mm in diameter. ***Corolla*** spreading, pink margin with pale cream yellow at the center of inside corolla, cream color outside; lobes fleshy, triangular-acuminate, 3 cm long, 2 cm wide, inside corolla lobes fuzzy, densely pubescent along margin and on apices, outside corolla glabrous; ***Corona*** pentagonal, cream color.
2. ***Hoya megalaster*** Warb. ex K. Schum & Lauterb
   Stem cylindrical 1–3 mm in diameter, very thin new stem, dark green, wiry first and easily damaged; old stems with grey waxy coating, glabrous; internode 4–16 cm. ***Leaves*** petiolate; petiole terete 1.7–2 cm long, 3–5 mm in diameter, green to brown, glabrous; lamina thin, not succulence, glabrous, broadly ovate, cordate at the base with the basal lobes often overlapping, 8–14 cm long and 5–6 cm wide, apex acute. ***Inflorescences*** one per node, 6–10 flowers per umbel, ***peduncle*** terete, 2–7 cm long, 2–5 mm in diameter ***Corolla*** saucer-shaped (concave), red, glabrous except the center of corolla covered with fine hairs, corolla lobes are triangular in shape, 8 mm long and incurved. ***Corona*** star-shaped, lobes are relatively thick, dark red, apex round. The flowers release strong fragrance of vanilla in the evening.
3. ***Hoya pachyphylla*** K. Schum & Lauterb
   Stem cylindrical, stiff, 2–4 m long, 2–5 mm in diameter, green greenish brown, glabrous; internode 10–25 cm. **Leaves** petiolate; petiole 1–2 cm long and 3–5 mm in diameter, greyish green; lamina ovate, fleshy, wavy on the edge, 6.5–12 cm long, 3.5–7.5 wide, light green to red mauve on abaxial surface, light green on adaxial surface, apex acuminate, base rounded; venation palmate, visible vein on adaxial surface, new leaves are often red mauve in color and

turn into light green as they mature. ***Inflorescences*** one per node, positively geotropic, 30–40 flowers per umbel; peduncle terete, 5–10 cm long, 2–3 mm in diameter. ***Corolla*** light yellow-cream. ***Corona*** pale pinkish white with darker pink in center.

4. ***Hoya patella*** Schltr

   Stem wiry and thin, cylindrical, glabrous, 1–3 mm in diameter, internode 1–11 cm, young stem dark green, old stem greenish grey. ***Leaves*** petiolate; petiole terete, 5–13 mm long and 2–3 mm in diameter; lamina thin, dark green, small to medium size, very thick hairs on young leaf, linear-lanceolate, apex acute, acute at the base. Flowers are pink, large, widely and shallowly cup-shaped. ***Corolla*** widely cup-shaped with mostly fused petals. ***Corona*** ovate from above, inner angles are acute and raised, outer angles are rounded and horizontal, pink reddish.

5. ***Hoya globulifera*** Blume

   Stem cylindrical, stiff, 2–3 m long, 1.5–3 mm in diameter, greenish brown, glabrous, sparsely peeling bark, internode 5–10 cm. ***Adventitious roots*** sparsely produced along the stem. ***Leaves*** petiolate; petiole terete, 1–2 mm in diameter, 5–7 mm long, greyish green, glabrous; lamina linear-lanceolate, 1–2 mm thick, fleshy, 2.5–3 cm wide, 4–8 cm long dark green on adaxial surface and light green on abaxial surface; apex acute, base acuminate, pinnate. ***Inflorescence*** one per node, 10–30 flowers, ***peduncle*** terete, 2–7 cm long and 2–5 mm in diameter. ***Corolla*** rotate, fleshy, fuzzy, red mauve with dense whitish hairs on margin of inside corolla. Corona inner angles are acuminate and red maroon, outer angles are acute, pinkish white.

6. ***Hoya inconspicua*** *Hemsl*

   Stems cylindrical, green. ***Leaves*** petiolate, petiole 2–5 mm long and 2–3 mm in diameter; lamina small, shiny, ovate to lanceolate. 5–6.5 long and 2.5–3.5 wide, apex acute to acuminate, base acute. ***Inflorescences*** one per node, 10–20 flowers per umbel, flowers are small, orange-pink.

7. ***Hoya lauterbachii*** K. Schum (Section Eriostemma)

   Hoya lauterbachii form 1

   Stems cylindrical, soft hairs cover young stem. ***Leaves*** petiolate; petiole terete, 2–3 cm long and 3–4 mm in diameter; lamina oblong elliptic, apex acuminate, rounded at the base; venation pinnate. ***Inflorescences*** one per node, very big flower up to 7 cm in diameter, 3–6 flowers per umbel; ***peduncle*** terete up to 10 cm long; ***pedicels*** terete up to 8 cm long. ***Corolla*** bowl-shaped (campanulata), densely velvety-hairy inside and outside, inside of the corolla mauve red with cream colored at the center, light green outside corolla. ***Corona*** outer angles are rounded, yellowish green.

   Hoya lauterbachii form 2

   Stems cylindrical. ***Leaves*** petiolate; petiole terete, 1–1.5 cm long and 3–4 mm in diameter; lamina oblong elliptic, apex acuminate, rounded at the base, 4–5 wide and 8–10 cm long; venation pinnate. ***Inflorescences*** one per node, 4 flowers per umbel. ***Corolla*** bowl-shaped (campanulata), up to 5 cm in diameter, shiny, glabrous without densely velvety-hairy inside and outside, pink inside and yellowish light green outside corolla. ***Corona*** yellowish dark red.

8. ***Hoya coronaria*** Blume (Section Eriostemma)
   We noticed that there are two forms of flowers in this species, hence it is recorded as *Hoya coronaria* form 1 and form 2.
   ***Hoya coronaria* form 1**:
   Stems cylindrical, 3 mm in diameter, up to 3 m long, pubescence, green. ***Leaves*** petiolate, petiole 0.5–1 cm long thick; lamina thick, oblong to ovate, 7–8.5 cm long and 3.5–4.5 cm wide, apex acuminate, base rounded; venation pinnate. ***Inflorescence*** one per node 3–7 flowers per umbel, convex, erect peduncle. ***Corolla*** star-shaped, fleshy, shiny, thick, nearly flat or slightly campanulate, yellowish white green. C***orona*** apex rounded, light yellow.
   ***Hoya coronaria* form 2**:
   Stems cylindrical, 3–5 mm in diameter, up to 5 m long, pubescence, green. ***Leaves*** petiolate, petiole 0.5–1 cm long; lamina thick, oblong to ovate, 6–7 cm long and 3–5 cm wide, apex acuminate, base rounded; venation pinnate. ***Inflorescence*** one per node; 2–7 flowers per umbel; erect peduncle. ***Corolla*** star-shaped, fleshy, shiny, thick, nearly flat or slightly campanulate, red mauve. C***orona*** apex rounded, red mauve.
9. ***Hoya sussuela*** (Roxb.) Merr (Section Eriostemma)
   Stem cylindrical, 2 cm in diameter. ***Leaves*** petiolate, petiole 1–1.5 cm long; lamina blade, thick, fleshy, oblong to ovate, 6–9 cm long and 2.5–4.5 cm wide, apex acuminate, base rounded; venation pinnate. ***Inflorescences*** one per node; positively geotropic, 20–25 flowers, peduncle 10–30 mm long, pedicels 20–25 mm long. ***Corolla*** stiff and leathery, mauve red, about 45 mm in diameter. ***Corona*** apex blunt, maroon red.
10. ***Hoya* sp 1.** (Section Eriostemma)
   ***Hoya* sp 1 form 1**
   Stem cylindrical, green, 5–10 mm in diameter, internode 9–15 cm long. ***Leaves*** petiolate, petiole 2–5 cm long and 2–5 mm in diameter; lamina oblong to ovate, apex acuminate, base rounded; venation pinnate. ***Inflorescence*** one per node; positively geotropic, 20–25 flowers, peduncle 12–15 cm long. ***Corolla*** fleshy, hairy, 1.7–2 cm in diameter, corolla lobus broadly ovate, maroon red, apex acute to acuminate. ***Corona*** yellowish cream with maroon red on the center, lobus ovate, apex blunt.
   ***Hoya* sp 1 form 2**
   Stem cylindrical, green, 3–8 mm in diameter, internode 10–17 cm long. ***Leaves*** petiolate, petiole 2–5 mm long and 2–3 mm in diameter; lamina oblong to ovate, apex acuminate, base rounded; venation pinnate. ***Inflorescence*** one per node; positively geotropic, 20–30 flowers, peduncle 8–13 cm long. ***Corolla*** fleshy, 1.5–2 cm in diameter, corolla lobus broadly ovate, inside corolla mauve red, outside corolla pale green, apex acute. ***Corona*** yellowish cream with maroon red on the center, lobus ovate, apex blunt.

H. lauterbachii form 1, *H. lauterbachii* form 2, *H. coronaria* form 1, *H. coronaria* form 2, *H. sussuela*, *Hoya* sp1 form 1, and *Hoya* sp1 form

2 belong to Section Eriostemma. Species that belong to Section Eriostemma have certain characteristics like a free-standing hairy column that stand in a cup at the point of attachment with the corolla. Eriostemma means "wolly crown" in Greek; they are vines; have woody corona (Baltazar and Buot 2019), and can grow in alkaline environment. According to Juhonewe and Rodda (2017), members of Hoya section Eriostemma Schltr. are common and widespread throughout the Malesian region and Papuasia. Species delimitation within this section is particularly problematic. Eriostemma section is hard to divide into separate species. They tend to hybridize with each other so it looks like a colorful mix in New Guinea (private comm. Juhonewe 2019). For instance, *H. lauterbachii* form 2 could be a hybrid between *H. coronaria* and *H. lauterbachii. Hoya* sp1 form 1 and *Hoya* sp1 form 2 both have very similar form and color of corona, but totally different color of corolla as shown in Fig. 6.3.

6.3.1 Habitat and Distribution of *Hoya* Species in Papua and West Papua Provinces

Hoya species encountered during this preliminary inventory scattered in different locations and habitats as shown in Table 6.1. *Hoya* species were found in almost all types of habitats, including coastal areas, gardens, plantations, and secondary to primary forests. *Hoya* species were found in each locality visited. Although this preliminary inventory did not yet cover all localities in the Provinces of Papua and West Papua, it can be predicted that *Hoya* was widespread in both provinces.

All *Hoya* species found in this preliminary survey grew in lowlands. Juhonewe and Rodda (2017) recorded lowland rainforests and lower montane forests appear to harbor the majority of *Hoya* species in Papua New Guinea. Martinez and Buot (2018) also recorded similar result that most of *Hoya* species (33) in the Philippines were found at the lower elevation ranges 0–700 m a.s.l.

There were eight species of *Hoya* discovered in Papua Province, namely: (1) *Hoya Stenakei*; (2) *Hoya megalaster*; (3) *Hoya pachiphylla*; (4) *Hoya globulifera*; (5) *Hoya lauterbachii* form 1; (6) *Hoya coronaria* form 2; (7) *Hoya* sp1 form 1; (8) *Hoya* sp1 form 2 as shown in Table 6.1 and Fig. 6.2.

Hoya globulifera was commonly found widespread in almost all types of habitats. The plant was found climbing on shrubs and trees ranging from beaches, the primary and secondary forests. While several species of *Hoya* were observed to only grow in certain types of habitats, *H. pachyphylla, H. patella, and H. incospicua* were encountered growing on the branches of trees in the primary and secondary forests. *Hoya incospicua* was observed growing on the trees in the primary and secondary forest around the seashore in Raja Ampat. *Hoya stenakei* and *H. megalaster* were found only growing in low land primary forest in Arso Keerom.

Hoya stenakei was observed growing on the forest floor and climbing until the top of big trees in the primary forests in Arso, Keerom, Papua Province, in 2018. This

Table 6.1 Habitat and distribution of *Hoya* species in Papua, Indonesia

Hoya species	Habitat	Location	Regency/city	Province
H. stenakei	PF	Arso 4	Keerom	Papua
H. megalaster	PF, SF	Yamarub Village, Kwimi Village (Arso)	Keerom	Papua
H. pachyphylla	PF, SF	Arso	Keerom	Papua
H. patella	SF, PF	Kaili District	Sorong	West Papua
H. globulifera	PF, SF, C	Arso, Holtecamp, Raja Ampat	Keerom, Jayapura, Raja Ampat	Papua & West Papua
H. incospicua	SF, PF	Raja Ampat	Raja Ampat	West Papua
H. lauterbachii form 1	P	PIR2, Arso	Keerom	Papua
H. lauterbachii form 2	G	Aimas District	Sorong	West Papua
H. coronaria form 1	C	Misool	Raja Ampat	West Papua
H. coronaria form 2	C	Holtecamp	Jayapura City	Papua
H. sussuela	C	Maybrat	Maybrat	West Papua
H. sp. 1 form 1	C	Sarahwandori Bay	Yapen	Papua
H. sp. 1 form 2	C	Biak	Biak City	Papua

PF primary forest, *SF* secondary forest, *C* coastal, *G* garden, *P* plantation

was the second *H. stenakei* found during the survey. *H. stenakei* was a new species and was newly published in 2017; the first was found in Baiberi, Sandaun Province, PNG, near the border line between Indonesia and PNG, in 2011 (Juhonewe and Rodda 2017). Both *H. stenakei* records in Papua Province, Indonesia, and Papua New Guinea were found in logging areas. Provisional IUCN conservation assessment categorized *H. stenakei* as Critically Endangered CR B2ab (ii,iii), C1, C2(ai), D (IUCN 2016).

Hoya megalaster was encountered for the first time during field expedition in Yamarub Village, Arso, Keerom Regency, Papua Province, and it was also found at Kwimi Village, Arso District. *Hoya megalaster* was observed growing in the forest floor and climbed until the top of big tree in primary forest. *H. megalaster* was also recorded in Papua New Guinea. It was first found in Papua New Guinea in the late of nineteenth century growing in shade at the base of rainforest trees near the Ruru River at about 400 m above sea level. Other early locations are the upper reaches of the Sepik River, the forest of the Ibo Mountains at Boral, at about 700 m above sea level, and the forest at the foot of the Bismarck Range, near Saugueti Stage, at about 300 m above sea level (Forster et al. 1995).

Hoya lauterbachii was found in the Palm Oil Plantation, PIR 2, Arso, Keerom. They grow very well at the base of the plantation and climbed the palm oil trees. However, *Hoya coronaria* form 1 and form 2 grew very well along the beach in Misool and Holtecamp beach in Jayapura. They rambled through the trees in search

Fig. 6.2 *Hoya* species found in Papua Province. (1) *Hoya stenakei*; (2) *Hoya megalaster*; (3) *Hoya pachyphylla*; (4) *Hoya globulifera*; (5) *Hoya lauterbachii* form 1; (6) *Hoya coronaria* form 2; (7) *Hoya* sp1 form 1; (8) *Hoya* sp1 form 2

of light. All Eriostemmas originate and grow on the ground and seem to thrive under bright sunlight. Eriostemmas are considered terrestrial rather than epiphytics.

Meanwhile, there were five *Hoya* species encountered in West Papua Province as shown in Fig. 6.3: (1) *Hoya patella*; (2) *Hoya inconspicua*; (3) *Hoya lauterbachii* form 1; (4) *Hoya coronaria* form 1; (5) *Hoya sussuela.*

6.3.2 *Hoya* as Larval Foodplants for *Euploea netcheri*

All specimens collected from the fields during the survey were planted in Yadikap's *Hoya* Garden. From observations for 2 years, there were two *Hoya* species used by *Euploea netsheri* females to lay eggs, namely, *H. pachyphylla* and *H. globulifera.* All stages of metamorphosis development took place in both *Hoya* species as their host plants. The females of *E. netscheri* laid their eggs on young leaves and usually were put on the underside of the leaves (Fig. 6.4).

Fig. 6.3 *Hoya* species encountered in West Papua Province. (1) *Hoya patella*; (2) *Hoya inconspicua*; (3) *Hoya lauterbachii* form 1; (4) *Hoya coronaria* form 1; (5) *Hoya sussuela*

Fig. 6.4 The early stages of *Euplea netscheri* on *H. pachyphylla* and *H. globulifera.* (1) An Egg of *E. netscheri* on underside leaf *of H. pachyphylla*. (2) The fifth instar larva of *E. netsheri* feeding on *H. globulifera.* (3) The fifth instar larva of *E. netsheri* feeding on *H. pachyphylla*

Euploea species are polyphagous which utilize many species of plants that belong to Apocynaceae, Asclepiadaceae, and Moraceae as their larval food plants (Goode 2014). These findings agree with Parsons (1999) who recorded that *Euploea* species in Papua New Guinea feed on the same plant families.

6.4 Conclusion and Recommendations

This survey provided list of ten species of *Hoya* found in Papua and West Papua Provinces, Indonesia. Conservation priorities for *Hoya* in Papua should address two species, *H. stenakei* and *H. megalaster,* located at logging areas in the low land rainforests in Arso, Keerom. More intensive *Hoya* exploration should cover the whole of Papua and West Papua Provinces to discover the richness of *Hoya* species in the western of New Guinea. *Hoya pachyphylla* and *Hoya globulifera* are larval food plants for *Euploea netsheri.*

Acknowledgments The authors would like to express deepest gratitude to Ms. N.S. Juhonewe, the expert of New Guinea's Hoya, who kindly helped in the Hoya identification and provided a lot of valuable information related to New Guinea's Hoya.

References

Baltazar AMP, Buot IE Jr (2019) Controversies on *Hoya* R. Br. taxonomy. Thailand Nat Hist Mus J 13(1):59–68

Endress ME, Liede-Schumann S, Meve U (2014) An updated classification for Apocynaceae. Phytotaxa 159(3):175–194

Forster PI (1996) A checklist of the Asclepiadaceae of Papuasia. Scientiae 22(1):15–22

Forster PI (2006) The Asclepiad flora of New Guinea. In: Marshall AJ, Beehler BM (eds) The ecology of Papua, pt. 1. Periplus Editions, Hong Kong, pp 371–378

Forster PI, Liddle DJ (1991) Variation in *Hoya australis* Br. ex Traill (Asclepiadaceae). Austrobaileya 3:503–521

Forster PI, Liddle DJ (1992) Taxonomic studies on the genus *Hoya* R. Br. (Asclepiadaceae) in Papuasia, 1–5. Austrobaileya 3:627–641

Forster PI, Liddle DJ (1993) Taxonomic studies on the genus *Hoya* R. Br. (Asclepiadaceae: Marsdenieae) in Papuasia, 6. Austrobaileya 4:51–55

Forster PI, Liddle DJ, Liddle IM (1995) Taxonomic studies on the genus *Hoya* R. Br (Asclepiadacea: Marsdenieae) in Papuasia, 7. Austrobaileya 4(3):401–406

Goode M (2014) Notes on a new subspecies of *Euploea phaenareta* (Schalier, 1785) (Lepidoptera: Nymphalidae) from Papua, Indonesia. Suara Serangga Papua 8(4):93–104

Hemsley WA (1891) New Solomon Islands plants. Ann Bot 5:501–508

IUCN Standards and Petitions Subcommittee (2016) Guidelines for using the IUCN red list categories and criteria. Version 12. http://www.iucnredlist.org/documents/RedListGuidelines.pdf. Accessed 15 July 2022

Juhonewe NS, Rodda M (2017) Contribution to a revision of *Hoya* (Apocynaceae: Asclepiadoideae) of Papuasia. Part I: ten new species, one new subspecies and one new combination. Gard Bull Singapore 69(1):97–147

Kleijn D, Van Donkelaar R (2001) Notes on the taxonomy and ecology of the genus *Hoya* (Asclepiadaceae) in Central Sulawesi. Blumea 46:457–483

Lamb A, Rodda M (2016) A guide to Hoyas of Borneo. Natural History Publication, Sabah

Martinez MR, Buot IE Jr (2017) Distribution of *Hoya* species across climate types in the Philippines. Thailand Nat Hist Mus J 11(2):77–89

Martinez MR, Buot IE Jr (2018) Altitudinal distribution of some Hoya species in the Philippines. Thailand Nat Hist Mus J 12(1):9–18

Moore S (1916) Asclepiadaceae. In: Ridley, H.N. Report on the botany of the Wollaston expedition to Dutch New Guinea, 1912–1913. Transactions of the Linnean Society of London. Botany 9: 112–116

Parsons M (1999) The butterflies of Papua New Guinea: their systematics and biology. Academic Press, London

Rodda M (2015) Two new species of *Hoya* R.Br. (Apocynaceae, Asclepiadoideae) from Borneo. PhytoKeys 53:83–93

Wanntorp L, Kocyan A, Renner SS (2006) Wax plants disentangled: a phylogeny of Hoya (Marsdenieae, Apocynaceae) inferred from nuclear and chloroplast DNA sequences. Mol Phylogenet Evol 39(3):722–733

Species Diversity and Habitat Association of Ferns and Lycophytes in Mts. Palay-Palay Mataas na Gulod Protected Landscape

7

Kristiane R. De Villa and Ronaldo D. Lagat

Abstract

Ferns and lycophytes are distinct lineages of free-sporing vascular plants that are frequently used to study composition and diversity patterns in tropical forests. These plant groups are known to be sensitive to environmental changes and stresses, leading them to possible ecological threats. Not much is known about their ecology in one of the Philippines' Key biodiversity areas, the Mts. Palay-Palay Mataas na Gulod Protected Landscape situated in the Island of Luzon. Hence, this chapter aims to discuss the ferns and lycophytes species diversity found in the landscape and describe the environmental variables that affect them. Based on the data generated from the conducted study, a total of 21 species of ferns and lycophytes belonging to 7 families and 11 genera were documented from the sampling plots established along the mountain's elevational gradients. As elevation increased, an increasing trend in the Shannon-Wiener Diversity (H') was observed. Further, ordination techniques suggested that the most influential factors affecting pteridophytes' diversity and distribution were elevation and light intensity. Other factors like soil moisture, air temperature, soil temperature, and relative humidity were also considered essential. Pteridophyte species were found to be preferential in unique environmental constraints, indicative of

K. R. De Villa (✉)
Department of Mathematics and Natural Sciences, Southern Luzon State University, Lucban, Quezon, Philippines

Biological Science Department, De La Salle University-Dasmariñas, Dasmariñas City, Cavite, Philippines
e-mail: krdevilla@dlsud.edu.ph

R. D. Lagat
Biological Science Department, De La Salle University-Dasmariñas, Dasmariñas City, Cavite, Philippines

S. Ramamoorthy et al. (eds.), *Plant Diversity in Biocultural Landscapes*,
https://doi.org/10.1007/978-981-19-8649-9_7

environmental stress and forest quality. A few species were documented to have a distinct association with habitats suggestive of environmental stress making them good indicators of disturbances.

Keywords

Canonical Correspondence Analysis · Environmental variables · Habitat association · Ordination · Pteridophytes · Rarefaction curve

7.1 Introduction

Pteridophytes is a term traditionally used to describe the two separate evolutionary classes of plants, particularly lycophytes (Lycopodiopsida; fern allies) and ferns (Polypodiopsida). It is broadly interpreted as vascular plants that do not produce flowers, fruits, or seeds (Callado et al. 2015). Comparing it to the seed plants, this plant group exhibits lower endemism and shows less frequent speciation (Guo et al. 2003). Though often neglected, this has been frequently used in studying community pattern composition and diversity of tropical forests (Kessler 2001). Their diversity and populations are often at risk from increasing social pressures, loss of habitat, and climate change (Fernández et al. 2010).

A genus-based estimate revealed that there are 1338 lycophyte species and 10,578 fern species in the world (PPG 1 2016). In Asia, the pteridophytes flora are remarkably rich. In fact, in the southeast Asian Region alone, an estimated 4500 species were recorded contributing to more than one-third of the world's diversity (Moran 2008; Ebihara et al. 2012). Further, in the Philippine archipelago, the fern and lycophyte species are represented by more than 1048 species, which are distributed among 176 genera and 40 families, respectively (Pelser et al. 2011). However, the speed of deforestation in Asia is remarkably high (Sodhi and Brook 2006), and pteridophytes, the major vegetation of the forest understory, are pressured under threat (Ebihara et al. 2012). The conversion of forests to agricultural or industrial lands leads to pollution, affecting pteridophyte species richness (Amoroso et al. 2016). These gravest threats to biodiversity are linked to anthropogenic activities resulting in habitat loss, fragmentation (Jackson and Fahrig 2013), and latter change in biocultural landscapes. Hence, "megadiverse" and biodiversity "hotspots" countries like the Philippines (Myers et al. 2000; Posa et al. 2008) are considered priority areas for pteridophyte and lycophyte conservation (Brummitt et al. 2016).

Pteridophytes are known to be sensitive to changes in different environmental variables. Though direct sunlight may not be appropriate for ferns, a shady area with an average of only 23 °C would be enough for ferns to thrive (Catapang et al. 2012). Studies revealed that the diversity of pteridophytes is influenced by the availability of the requirements that enable them to grow and thrive. On account, the environmental variables that affect them are associated with edaphic factors (Tuomisto et al. 2014) such as soil nutrients (nitrogen, phosphorus, and potassium content)

(Catapang et al. 2012; Zuquim 2015; Delos Angeles and Buot 2015), variation in pH, soil moisture, and other extrinsic factors, like geographic location, elevation, temperature, and humidity (Kessler 2001; Kessler et al. 2016). However, not much is known about the association between pteridophyte species and the environmental variables affecting them. Though extensive studies about Philippine pteridophytes have just started during the turn of the twenty-first century (Barcelona 2005), most of the available data are focusing on species inventory, discovery, and conservation status. Thus, this chapter will focus on the ecological associations of the documented pteridophyte species and the important environmental variables affecting their diversity and distribution in one of the Philippines' key biodiversity areas (Ong et al. 2002; Conservation International et al. 2006), Mts. Palay-Palay Mataas na Gulod Protected Landscape.

7.2 Mts. Palay-Palay Mataas na Gulod Protected Landscape

Mts. Palay-Palay Mataas na Gulod Protective Landscape (MPPMNGPL) situated in the island of Luzon, Philippines (Fig. 7.1) is located within the Municipalities of Ternate and Maragondon, Cavite and Nasugbu, Batangas (Protected Areas in CALABARZON cited 2020 Aug 2). It was proclaimed as a national park, game refuge, and bird sanctuary by virtue of Proclamation No. 1594 in 1976. With

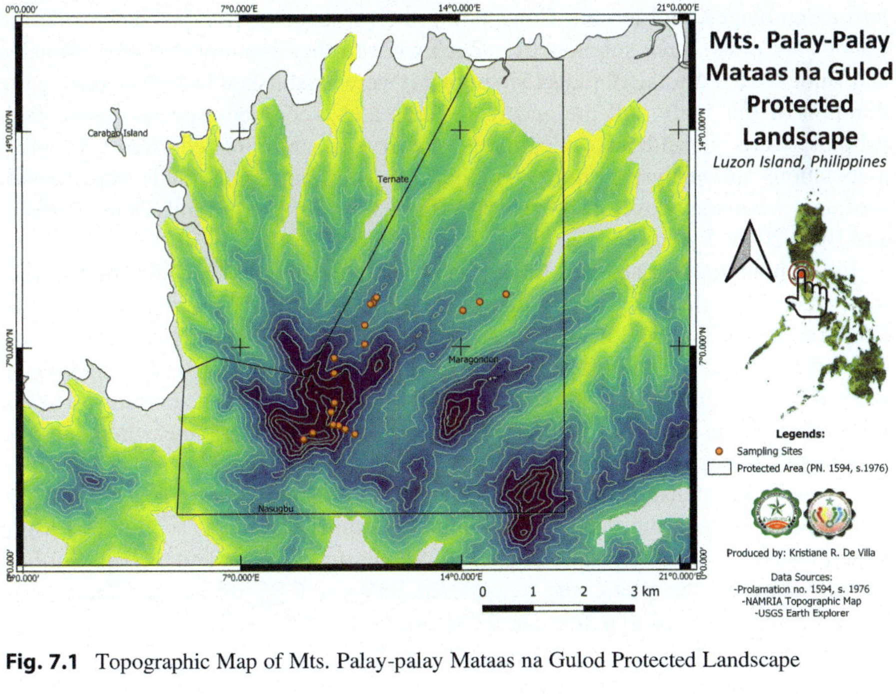

Fig. 7.1 Topographic Map of Mts. Palay-palay Mataas na Gulod Protected Landscape

coordinates of 14.233311 N 120.654502 E, it is located 60 km southeast of Manila and 60 km northwest of Mt. Makiling (Solis et al. 2015).

The landscape is composed of these major peaks which are: Mataas na Gulod (622 masl), Pico de Loro (595 masl), Palay-Palay (648 masl), and Dos Picos (442 masl) (MPPMNGPL Forest Assessment and Geospatial Report retrieved 2021). The topography of the park varies from low, moderate to steep terrains with forest floors and low altitude elevations. Vegetation in low elevations includes grasses and herbs, which gradually transition into dipterocarp trees, cogon grasses, shrubs, herbs, vines, and ferns at higher elevations (Solis et al. 2015). Though declared as a protected landscape, the area has been logged over time. The remaining vegetation is considered as a secondary growth forest with remnants of primary lowland forest (Causaren 2009). Approximately 62.5% of the area is forest, and 37.5% is non-forest vegetation. The landscape has been an ecotourism and a popular destination for hikers and mountaineers until its recent temporary moratorium in 2016 due to observed ecological distress.

7.3 Methods and Data Analysis

In compliance with Republic Act no. 9147 (Wildlife Resources Conservation and Protection Act of 2001), a Gratuitous Permit was secured to collect and sample the study site. With the granted permission, the sampling plots were laid along the elevational gradients of the landscape using the Plot Technique Method. Quadrats with a dimension of 20 × 20 m were established based on accessibility, vegetation, fern assemblages, and terrain. Considering the presence of fern patches inside the established plots, three subplots (5 × 2 m) were further assembled (Delos Angeles and Buot 2015). A total 19 plots (57 subplots) were established in the site during the duration of the study. All the occurring ferns and lycophytes species inside each subplot were considered, identified, preserved, and treated for ecological associations and analyses. Further, epiphytic and climbing individuals were considered only when they developed fronds less than 2-m from the ground (Delos Angeles and Buot 2015; Syfert et al. 2018).

Environmental variables (climatic and edaphic factors) in the plots were measured. The climatic factors that were considered and determined were: air temperature (°C), elevation (masl), relative humidity (%), and light intensity (l×/cd). Additionally, the soil factors (edaphic factors) were characterized by measuring the soil temperature (°C), slope (%), soil pH, organic matter (%), soil moisture (%), and soil texture. The relative humidity and air temperature measurement was determined using a digital psychrometer, while a digital light meter was used to determine light intensity. The percentage slope was computed based on the quotient of the rise and run of the slope of the subplots' terrain, which was acquired by using an improvised "A-frame" made of pegs, a premeasured string, and a level. Further, the elevation of the plots was determined using a GPS device. In situ, a soil thermometer was used to determine the temperature of the soil.

Other edaphic factors were determined in the laboratory after enough samples from each plot were gathered. Soil samples that are kept and secured in ziplock plastic bags were subjected to soil analysis parameters such as pH, content moisture, organic matter content, and texture. Specific techniques performed were the Gravimetric Technique (Shukla et al. 2014) and the Loss on Ignition Method (Konare et al. 2010). Ultimately, the soil texture was identified through the particle size distribution analysis through the hydrometer method (Gavlak et al. 2005).

The species diversity of pteridophytes was evaluated and analyzed using the Shannon-Wiener Index (H'), while the measure of evenness (J) was quantified using the Shannon Evenness Index. In addition, the overall diversity of the landscape was qualitatively described using the Modified Fernando (1998) Diversity scale. The acquired values were analyzed using the software EstimateS® version 9.1.0. A Species Rarefaction Curve was extrapolated using Microsoft Excel to estimate the possible number of species accumulated in the increase of sampled sites. Further, other important species data were collected (e.g., Relative frequency, relative density, relative abundance, etc.), and the Dominance Candidate Index (DC_i) of each species was computed (Avolio et al. 2019).

Ordination analyses were employed to determine the relationship between floristic and environmental data. The raw data were normalized prior to ordination analysis, a step essential to transform the data necessary for accurate comparison of statistics taken from different measurements (Weiss et al. 2017). The ordination techniques performed were the factor analysis or the Principal Component Analysis (PCA) and the Canonical Correspondence Analysis (CCA). Furthermore, assumption checks and tests for significance were carried out to support the adequacy and appropriateness of the ordination. Ultimately, an ordination biplot was generated for a holistic data visualization. The software PAST version 3.17 and Jamovi® version 2.2.2 were used in these analyses.

7.4 Species Diversity of Ferns and Lycophytes

A total of 1159 individuals belonging to 21 species, 7 families, and 11 genera were documented inside the established plots investigated along the elevational gradients of MPPMNGPL (Tables 7.1 and 7.2). The most represented fern taxa with 9 species is the family Pteridaceae. It is followed by Dryopteridaceae (4 spp.), Lygodiaceae (3 spp.), and Thelypteridaceae (2 spp.), respectively. Polypodiaceae and Tectariaceae are the fern families with the least representation having one (1) species recorded each. On the other hand, lycophytes are represented by the family Lycopodiaceae. In addition, the most represented pteridophyte genera is *Pteris* (7 spp.) followed by *Lygodium* (3 spp.) and *Bolbitis* (2 spp.). Determinately, much of the growth form of the species was terrestrial and lithophytic. This reflects the common forms of ferns as two-thirds of the extant species are terrestrial (Scheuttpelz and Pryer 2009).

Though a few numbers of species were documented in the sampling plots, the extrapolated species rarefaction curve shows that encountering new species is

Table 7.1 Checklist of the ferns and lycophytes species documented in the sampling plots established in Mts. Palay-Palay Mataas na Gulod Protected Landscape. Growth forms: A-arborescent, *E* epiphytic, *H* hemiepiphytic, *L* lithophytic, *T* terrestrial

Family	Taxa	Growth form	Exsiccata (DLSUD-Herbarium accession no.)	Status (Pelser et al. 2011))	Distribution (Pelser et al. 2011)
Dryopteridaceae	*Bolbitis rhizophylla* (Kaulf.) Hennipman	L/T	0479	Native	Philippines, Taiwan. BALABAC, LUZON, MINDORO, PALAWAN
	Bolbitis scalpturata (Fée) Ching	L	0480	Native	China, Java, Lesser Sunda Isls, Myanmar, Philippines, Sulawesi, Sumatra, Taiwan, Thailand, Vietnam. LUZON, PALAWAN
	Polystichum horizontale C. Presl	T	0491	Native	Bismarck Arch, Borneo, Moluccas, New Guinea, ? Pacific Ocean, Philippines, Solomon Isls, Sulawesi. MINDANAO: Bukidnon
	Dryopteris sp.	T	–	–	–
Lycopodiaceae	*Huperzia* sp.	L	–	–	–
Lygodiaceae	*Lygodium circinnatum* (Burm.fil.) Sw.	T	–	Native	Andaman Isls, Bangladesh, Bismarck Arch, Borneo, China, Hong Kong, ?India, Java, Laos, Lesser Sunda Isls, Malay Peninsula, Moluccas, New Guinea, Nicobar Isls, Pacific Ocean, Philippines,

(continued)

Table 7.1 (continued)

Family	Taxa	Growth form	Exsiccata (DLSUD-Herbarium accession no.)	Status (Pelser et al. 2011))	Distribution (Pelser et al. 2011)
					Singapore, Solomon Isls, Sri Lanka, Sulawesi, Sumatra, Thailand, Vietnam
	Lygodium japonicum	T	0485	Native	Bangladesh, Bhutan, Cambodia, China, India, Japan, Java, Korea, Laos, Lesser Sunda Isls, Moluccas, Nepal, New Guinea, Pakistan, Philippines, Ryukyu Isls, Sulawesi, Sumatra, Taiwan, Thailand, Tibet, Vietnam. LUZON: Ilocos Norte
	Lygodium flexuosum (L.) Sw.	T	0484	Native	Andaman Isls, Australia, Bangladesh, Bhutan, Cambodia, China, India, Laos, Lesser Sunda Isls, Malay Peninsula, Moluccas, Myanmar, Nepal, New Guinea, Philippines, Ryukyu Isls, Singapore, Sri Lanka, Sulawesi,

(continued)

Table 7.1 (continued)

Family	Taxa	Growth form	Exsiccata (DLSUD-Herbarium accession no.)	Status (Pelser et al. 2011))	Distribution (Pelser et al. 2011)
					Thailand, Vietnam. PALAWAN
Polypodiaceae	*Microsorum longissimum* J. Sm. ex Fée	L/T	–	Native	Borneo, Philippines. LUZON
Pteridaceae	*Adiantum philippense L.*	T	–	Native	Africa, Andaman Isls, Australia, Bangladesh, Bhutan, Cambodia, China, India, Indian Ocean, Laos, Lesser Sunda Isls, Malay Peninsula, Moluccas, Myanmar, Nepal, Pacific Ocean, Philippines, ? Solomon Isls, Sri Lanka, Sulawesi, Taiwan, Thailand, Vietnam. PALAWAN
	Pteris cretica L.	T	–	Native	Africa, Bangladesh, Bhutan, China, Europe, India, Indian Ocean, Japan, Korea, ? Malay Peninsula, Middle East, Myanmar, Nepal, Pacific Ocean, Pakistan, Philippines, Ryukyu Isls, Sri Lanka,

(continued)

Table 7.1 (continued)

Family	Taxa	Growth form	Exsiccata (DLSUD-Herbarium accession no.)	Status (Pelser et al. 2011))	Distribution (Pelser et al. 2011)
					Taiwan, Thailand, Tibet, Vietnam, W Asia. PALAWAN
	Pteris distans J. Sm.	T	0494	Endemic	Endemic to the Philippines. LUZON
	Pteris ensiformis Burm.	T	0503	Native	Bangladesh, Bhutan, Bismarck Arch, Borneo, Cambodia, China, India, Laos, Lesser Sunda Isls, Malay Peninsula, Moluccas, Myanmar, Nepal, New Caledonia, New Guinea, Nicobar Isls, Pacific Ocean, Philippines, Ryukyu Isls, Singapore, Solomon Isls, Sri Lanka, Sulawesi, Taiwan, Thailand, Vietnam
	Pteris heteromorpha Fée	T	0495	Native	Borneo, China, India, Laos, Myanmar, Philippines, Sulawesi, Thailand, Vietnam. LUZON
	Pteris oppositipinnata Fée	T	0496	Native	Borneo, Philippines, Sulawesi. PALAWAN

(continued)

Table 7.1 (continued)

Family	Taxa	Growth form	Exsiccata (DLSUD-Herbarium accession no.)	Status (Pelser et al. 2011))	Distribution (Pelser et al. 2011)
	Pteris pellucida C.Presl	T	0493	Native	Borneo, Java, Lesser Sunda Isls, New Guinea, Philippines
	Pteris philippinensis Fée	T	–	Native	Bhutan, Borneo, China, India, Laos, Lesser Sunda Isls, Myanmar, Nepal, Philippines, Sulawesi, Taiwan, Thailand, Vietnam. PALAWAN
	Adiantum sp.	T	–	–	–
Tectariaceae	*Tectaria hilocarpa* (Fée) M.G.Price	T	0501	Native	Borneo, Philippines, Sulawesi. LEYTE, LUZON, MINDANAO, PALAWAN, PANAY
Thelypteridaceae	*Amphineuron terminans* (J.Sm. ex Hook.) Holttum	T	–	Native	Sri Lanka to Polynesia. Throughout the Philippines in the more seasonal regions
	Christella dentata (Forssk.) Brownsey & Jermy	T	–	Native	Pantropic. MINDANAO: Bukidnon

probable (Fig. 7.2). This is presumed by the predicted trendline ($R^2 = 0.9946$), suggesting that if more sampling sites were explored, new additional species can still be encountered. This trend as explained by Jost et al. (2010) and Magurran and McGill (2011), is likely observed in large study areas with expanded sampling techniques, as the instance includes diverse habitat types that support distinct species assemblages. Further, this shows how specialized pteridophytes species are in their

Table 7.2 Species data of the ferns and lycophytes documented in MPPMNGPL

Taxa	Total frequency	Total relative frequency	Total density	Total relative density	Total relative abundance	Total Dci
Bolbitis rhizophylla (Kaulf.)	539	0.47	0.95	46.58	0.47	0.71
Microsorum longissimum J.Sm. ex Fée	175	0.56	0.31	15.12	0.15	0.64
Lygodium circinnatum (Burm.fil.) Sw.	62	0.30	0.11	5.36	0.05	0.32
Bolbitis scalpturata (Fée) Ching	61	0.11	0.11	5.27	0.05	0.13
Pteris philippinensis Fée	52	0.19	0.09	4.49	0.04	0.22
Pteris ensiformis Burm.	51	0.11	0.09	4.41	0.04	0.13
Polystichum horizontale C.Presl	36	0.11	0.06	3.11	0.03	0.12
Tectaria hilocarpa (Fée) M.G. Price	28	0.11	0.05	2.42	0.02	0.12
Pteris oppositipinnata Fée	26	0.09	0.05	2.25	0.02	0.10
Huperzia sp.	25	0.02	0.04	2.16	0.02	0.03
Pteris heteromorpha Fée	22	0.12	0.04	1.90	0.02	0.13
Pteris pellucida C. Presl	18	0.04	0.03	1.56	0.02	0.04
Lygodium flexuosum (L.) Sw.	14	0.09	0.02	1.21	0.01	0.09
Christella dentata (Forssk.)	11	0.04	0.02	0.95	0.01	0.04
Dryopteris sp.	10	0.04	0.02	0.86	0.01	0.04
Amphineuron terminans (J.Sm. ex Hook.) Holttum	9	0.05	0.02	0.78	0.01	0.06
Pteris distans J.Sm.	7	0.05	0.01	0.60	0.01	0.06
Pteris cretica L.	6	0.04	0.01	0.52	0.01	0.04
Lygodium japonicum (Thunb.) Sw.	3	0.04	0.01	0.26	0.00	0.04
Adiantum philippense L.	3	0.02	0.01	0.26	0.00	0.02
Adiantum sp.	1	0.02	0.00	0.09	0.00	0.02
TOTAL	1159	2.579	2.003	100.164	1.00	3.079

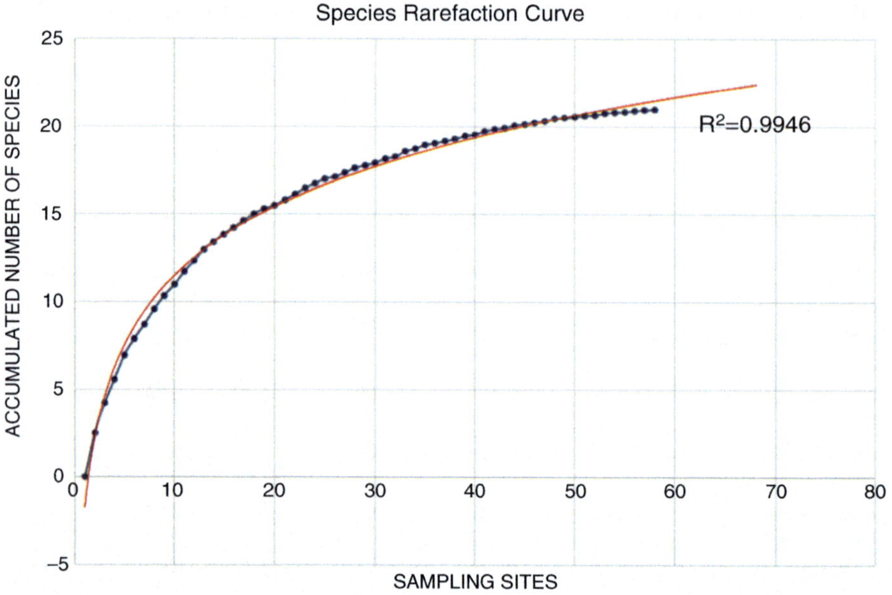

Fig. 7.2 Species rarefaction curve for the sampled sites in MPPMNGPL. Seen in red is the extrapolated predictive trendline with an R^2 of 0.9946

distribution; additional other species were further observed outside the plots and were not reported in this paper. Ultimately, the same trend was also observed in the study of Kessler (2000), Amoroso et al. (2009), Aureo et al. (2020), and Coritico et al. (2020).

The species data (frequency, relative frequency, density, relative density, relative abundance, and DCi) were determined (Table 7.2). Of the documented species, *Bolbitis rhizophylla* represents 46.58 % of the total recorded individuals encountered in the study. This lithophytic fern with free venation pattern and terminal bulbil in the fronds (Dong and Zhang 2005) was also reported as the dominant fern species in the areas of Upland Cavite (Medecilo and Lagat 2017). The sites by which this species was found are generally areas in the landscape that has a terrain characterized by a steep to a very steep slope and rocky substrates, conditions that create microenvironments which enhance unique and selective species of ferns to grow (Banaticla and Buot 2004). Aside from these edaphic conditions, this dominant species is found thriving in the areas in closed forest canopy with low light penetration; confirming its preference to grow in areas with low light intensity seen in the ordination (Fig. 7.5). Further, Raymundo et al. (1991) reported this as a potential indigenous medicinal flora due to its strong antimicrobial activity against *Micrococcus luteus*, *Staphylococcus aureus*, and *Candida utili*.

The second dominant species in the landscape was *Microsorum longissimum* constituting 15.12% of the total frequency. Like *B. rhizophylla,* patches of *M. longissimum* were documented in the closed forest canopy of the landscape, but were significantly associated and observed in much higher elevation. This observation was also confirmed by the ordination analysis (Fig. 7.5). Both dominant species were native to the Philippines (Pelser et al. 2011). Further, the only lycophyte species encountered in the sampling plots was *Huperzia* sp. This species was found in the forest's understory and accounted for 2.16 of the relative density and a relative abundance of 0.02. Species of this genera was reported to be a potential source for various pharmaceutical compounds in treating Alzheimer's Disease (Kim Thu et al. 2020). Ultimately, majority of the documented pteridophytes are considered native to the Philippines, showing the rich pteridophyte flora of the archipelagic country.

The Shannon-Weiner Diversity Indexes (H') and Evenness (J) of the established plots were determined (Table 7.3). High measures of H' were generally recorded in the elevational range of 300–646 masl wherein the most diverse plot is Site 4 ($H' = 1.290$, $J' = 0.331$), followed by site 13 ($H' = 1.143$, $J' = 0.351$) and 14 ($H' = 1.113$, $J' = 0.306$), respectively. On the other hand, the lowest diversity index was computed in site 12 with a value of $H' = 0.413$ and $J' = 0.009$. Other sites with low diversity are at the elevational range of 400–646 masl (Table 7.4).

The subplots of the most diverse site, Site 4, are situated at an elevation of 341.40 masl. It is located near a stream and characterized by a rocky substrate with enclosed canopies of trees. The area is less disturbed compared to the other sites due to its far proximity to the road and trail; a potential factor that can be attributed to its high diversity index. Though it is difficult to specify the variable/s responsible for the diversity (Whittaker et al. 2001), the other environmental variables recorded in the site such as the high relative humidity (94.26%) and the very steep slope (66.75%) were noteworthy to be considered (Table 7.3). According to Sánchez-González et al. (2016), changes in humidity are associated with ferns' diversity values. Further, the increase in the number of species can be linked to the rockiness of the substrate (Kreutz et al. 2015) and high and constant humidity, especially in areas near bodies of water—as the presence of water enhances the sexual reproduction of ferns (Gonzatti et al. 2016). In cognizance, the very steep slope of the site is held significant to diversity because it creates microenvironments with fewer disturbances leading to unique fern species to thrive (Banaticla and Buot 2004).

The least H' was calculated at site 12 (located at 477.56 masl). This site is considered an open area with less tree cover. It is situated near the trail which is indicative of possible environmental disturbance and stress. Only two species of ferns (*B. rhizophylla* and *M. longissimum*) were encountered in the area. Moreover, the environmental variable that can potentially explain the site's low diversity is its recorded high air temperature. The high-temperature reading (28.030 °C) was noted to be remarkably higher than the ideal temperature for fern growth (Catapang et al. 2012) which greatly decreases the area's relative humidity (Nassif et al. 2021). This occurrence further lessens the available moisture in the soil—important factor in fern propagation (Delos Angeles and Buot 2015), distribution, and community assemblage (Karst et al. 2005). Other sites with low fern diversity are sites 5 and 11, which also recorded a low diversity index. Climactically, the overall species diversity of the

Table 7.3 Mean summary of the ecological factors measured inside the established plots

SITE	Relative humidity (%)	Light intensity (lx)	Air temperature (°˜C)	Slope (%)	Soil temperature (° C)	Soil moisture (%)	Soil pH	Organic matter (%)
1	86.250	5526.167	27.420	31.000	23.667	30.963	6.647	3.355
2	94.263	174.167	26.820	66.750	23.333	33.917	5.963	2.127
3	84.247	3718.000	28.467	24.750	23.333	27.814	6.590	2.739
4	94.263	174.167	26.820	66.750	23.333	33.917	5.963	2.127
5	88.847	422.667	25.707	49.333	24.333	25.469	5.990	2.346
6	85.727	1108.500	26.977	48.500	24.667	18.552	5.733	3.682
7	86.707	566.167	26.810	61.333	24.667	22.954	5.967	3.271
8	83.773	801.167	27.080	60.667	24.333	22.456	6.433	2.223
9	89.777	831.000	25.563	55.667	23.333	33.408	6.353	2.644
10	92.443	1192.000	25.707	60.667	22.333	36.535	6.350	1.943
11	89.000	583.500	27.283	38.767	23.000	30.316	6.647	2.802
12	86.550	513.500	28.030	34.233	23.333	29.091	5.970	2.492
13	87.980	1194.333	25.197	34.250	22.000	41.301	6.503	2.840
14	90.320	1596.000	25.037	49.000	22.333	34.650	6.453	3.510
15	83.870	599.667	28.200	43.833	24.333	26.149	5.940	2.397
16	99.450	391.333	23.103	50.333	21.333	40.902	6.930	3.929
17	84.363	488.000	27.770	39.633	23.000	26.866	6.603	1.931
18	99.973	413.333	22.453	77.333	21.000	39.856	6.760	2.516
19	99.890	1851.500	21.997	40.667	21.000	45.813	5.417	3.059

Table 7.4 The Shannon-Weiner Index of diversity (H') and Shannon Evenness Index (J') of the Sampling Sites in MPPMNGPL

Sampling sites	Elevation (masl)	H'	J'
1	230.28	0.797	0.266
2	245.18	0.693	0.228
3	282.30	0.730	0.182
4	341.40	1.290	0.331
5	375.90	0.583	0.137
6	379.60	0.987	0.204
7	388.00	0.793	0.189
8	398.60	0.847	0.218
9	420.27	1.057	0.263
10	443.70	0.647	0.152
11	456.23	0.620	0.162
12	477.56	0.413	0.009
13	506.49	1.143	0.351
14	544.47	1.113	0.306
15	580.00	1.050	0.264
16	587.23	0.693	0.189
17	628.00	1.093	0.266
18	639.14	1.023	0.203
19	646.04	0.693	0.171
Average		0.856	4.091

landscape is $H' = 0.856$. This value is qualitatively regarded as "very low" in diversity based on the Modified Fernando Biodiversity Scale (Fernando 1998), supporting the less frequent speciation of this plant group (Guo et al. 2003).

Generally, an increasing trend of both the diversity and evenness in the landscape is observed, despite some fluctuations at some sites as elevation increases. These fluctuations may be attributed to the formation of microenvironments with unique environmental factors or abiotic conditions. Further, the same trend was also observed in Mt. Banahaw De Lucban and the secondary forest of the North-eastern Slope of Mt. Makiling (Banaticla and Buot 2004; Delos Angeles and Buot 2015), which are the proximal mountains to the landscape. Typically, the diversity of ferns is exhibited by a well-defined hump-shaped pattern seen both at the local and regional scales (Karger et al. 2011). However, since MPPMNGPL is considered a low mountain (<2000 m) (Kromer et al. 2013), this distribution pattern was not observed.

7.5 Most Important Environmental Variables Affecting Diversity and Distribution

Physicochemical parameters were associated with fern ecology. Previous studies cite that edaphic factors (Tuomisto et al. 2014), such as soil nutrients (nitrogen, phosphorus, and potassium content) (Catapang et al. 2012; Zuquim 2015; Delos Angeles

Table 7.5 Results of the principal components analysis (PCA) for the ecological variables observed in MPPMNGPL

	Principal component axis			
Ecological variables	1	2	3	4
Relative humidity	0.917			
Air temperature	−0.895			
Soil temperature	−0.893			
Soil moisture	0.874			
Elevation		0.770		
Light intensity		−0.747		
Soil organic matter			0.812	
Soil pH				0.896
Eigenvalue	3.89	1.64	1.34	1.16
% of variance	38.9	16.4	13.4	11.6
Cumulative %	39.8	55.3	68.7	80.4

Values below the factor loading of 0.71 were hidden from the table

Table 7.6 Results of the assumption checks: KMO measure of sampling adequacy, Bartlett's test of sphericity, Cronbach's reliability test

KMO measure of sampling adequacy		0.667
Bartlett's test of sphericity	χ^2	317
	df	45
	p	<0.001
Cronbach's α reliability test	sd	0.675
	α	0.801

and Buot 2015), variation in pH, soil moisture, and extrinsic factors, such as geographic location, elevation, temperature, and humidity significantly affect pteridophytes' diversity, richness, and distribution (Kessler et al. 2011). In connection, results obtained from factor analysis through PCA (Principal Component Analysis) revealed that the important environmental variables affecting pteridophyte flora in MPPMNGPL are relative humidity, air temperature, soil temperature, soil moisture, elevation, light intensity, soil organic matter, and soil pH. Set at a factor loading of 0.71, a stringent cut-off (Tabachnick and Fidell 2007), these ecological variables contributed to an accumulated variance of 80.4% and cover PC1, PC2, PC3, and PC4, respectively (see Tables 7.5 and 7.6). The assumption checks (KMO = 0.68; Bartlett = 0.001) confirm that the factor analysis is useful and acceptable for the observed data (Table 7.5).

A total eigenvalue of 8.04 was computed from the first four principal components (PC1, PC2, PC3, and PC4) that scored higher than 1. The large portion of the scores was attributed to PC1 (3.89) and PC2 (1.64), contributing to 38.9% and 16.4% of the variance. Moreover, the eigenvalues of the third and fourth components are 1.34 and 1.16, contributing to the 13.4% and 11.6% variance, respectively. Though four components scored higher than 1, the scree plot (Fig. 7.4) prescribed that the first 3 (PC1, PC2, and PC3) components (68.7% of variance) are enough for interpretation. This is due to the presence of the sharp reduction or the "elbow" (Wilcox 2012) of the eigenvalues between PC2 and PC3 (Fig. 7.3). Thus, for a more robust analysis, only the significant environmental variables in the first three components were

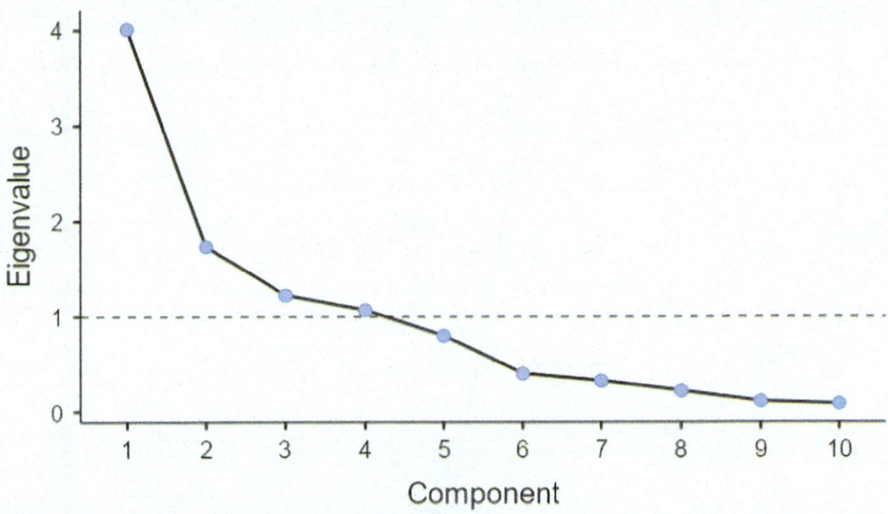

Fig. 7.3 Scree plot of the eigenvalue of the principal components

considered in the CCA; ensuring that the overfitting of the data is minimized. The environmental variables included were relative humidity, air temperature, soil temperature, soil moisture, elevation, light intensity, and soil organic matter. In support, the inclusion of these seven variables was identified with good internal consistency and is reliable for analysis (Cronbach's α index = 0.801) (Table 7.6).

7.6 Factors Affecting Fern and Lycophytes Diversity

Using the seven environmental variables determined as most important from the performed factor analysis, an ordination technique through Canonical Correspondence Analysis (CCA) was carried out. This will provide an integrated description of the species-environment relationship construing the relationship between the fern and lycophytes community and the environmental variables measured.

Results of the ordination through Canonical Correspondence Analysis (CCA) revealed that the fern and lycophytes community in MPPMNGPL are influenced by the determined environmental variables. A total variance of 1.87 was summated from the eigenvalues of all the axis generated from the analysis (Table 7.7). The first two axes of the ordination were accounted for an accumulative percent variation of 61.79% of the total inertia, a value sufficient to explain the correlation between the species data and the environmental data (David 2017). Canonical axis 1 (eigenvalue = 0.607) is correlated with elevation, while Canonical axis 2 (eigenvalue = 0.547) is correlated with light intensity. Furthermore, the percentages of the explained variance calculated from Axis 1 and Axis 2 are significant based on the

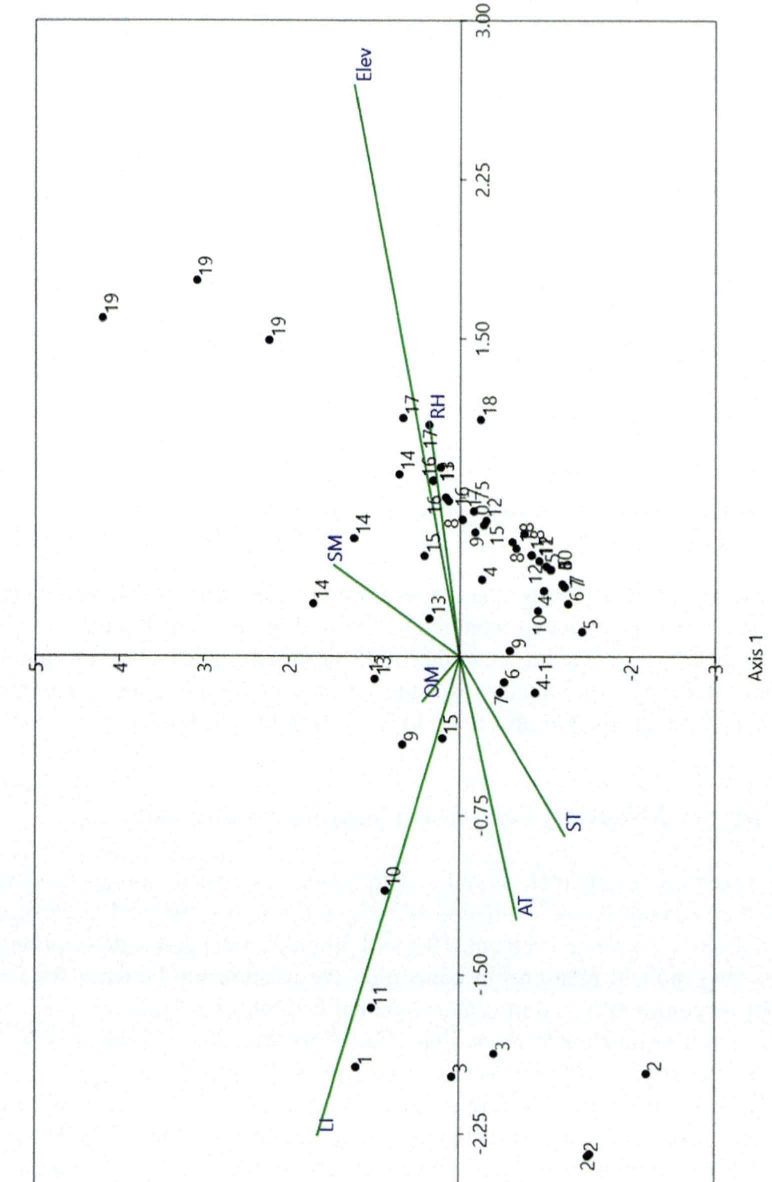

Fig. 7.4 CCA plot of the relationship of the environmental factors and the sampling sites as depicted per elevational zone. *LI* light intensity, *AT* air temperature, *ST* soil temperature, *OM* organic matter, *SM* soil moisture, *Elev* elevation, *RH* relative humidity

Table 7.7 Results of canonical correspondence analysis (CCA) for the ecological constraints and the species of ferns and lycophytes in MPPMNGPL

	Axis 1	Axis 2	Total inertia
Eigenvalue	0.60662	0.54699	1.866922
Cumulative % of the variance	32.49	61.79	
Permutation test ($N = 999$)			
p-value	0.001	0.001	

performed permutation test ($N = 999$) (Axis 1 p-value $= 0.001$, Axis 2 p-value $= 0.001$).

The environmental factors on the right-hand side of the first axis (eigenvalue = 0.60662, variance = 32.49 %) are associated with areas in high elevation (Figs. 7.4 and 7.5). This association is further supported by high relative humidity, high soil moisture, low organic matter, and low light intensity, which may implicate the forested areas of the landscape. On the other hand, the left-hand side of the axis are variables associated with the non-forested areas of the site. This is evidenced by the influence of high light intensity, high soil temperature, and high air temperature—factors that may indicate the openness of the forest and its possible disturbances. Thus, this axis is suggestive of the forest quality of the landscape as influenced by elevation. On the other hand, the second axis (eigenvalue = 0.54699) constituted 29.3% of the variance. Vectors on the upper portion of the axis are variables associated with the areas with a rich source of soil moisture. The influence of high relative humidity, low soil temperature, and low air temperature further supports this. In cognizance, the lower portion of this axis shows an association with low organic matter and low light intensity. This negative correlation is supported by the increasing air temperature and soil temperature. Thus, axis 2 is the axis of the landscape's understory quality and soil factors.

The level of influence of the standardized environmental data can be proportionally associated by the length of their vector in the plot (Sarker et al. 2014). Hence, it is suggested that the most affecting factors influencing the distribution and diversity of ferns and lycophytes in MPPMNGPL are elevation and light intensity (Fig. 7.4). Other influential factors, in order of importance, are soil moisture, air temperature, and soil temperature, respectively. In addition, the angle depicted between the vectors shows their degree of correlation. As regards a positive correlation between relative humidity, elevation, and soil moisture, air temperature and, soil temperature are detected. Elevation and relative humidity, on the other hand, are negatively correlated with air temperature and soil temperature. These relationships support that as elevation increases, the relative humidity and soil moisture increase (Banaticla and Buot 2004; Catapang et al. 2012), while the air temperature and soil temperature decrease (Heegaard 2002; Roundy and Chambers 2021).

Sites with low species diversity are generally influenced by high light intensity, high air temperature, and high soil temperature (Fig. 7.4). Further, sites with a higher index of diversity are mostly influenced by high relative humidity and elevation, low organic matter, and low light intensity. Elevation as an important environmental

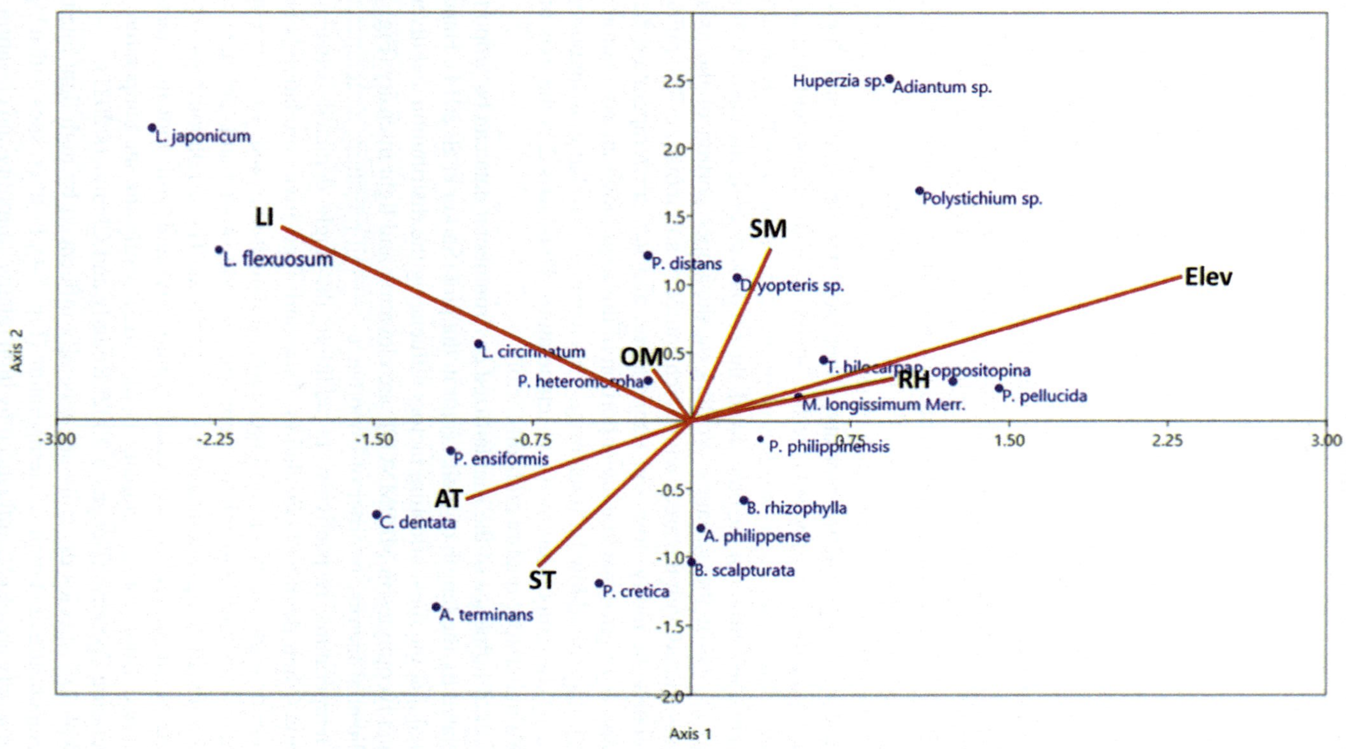

Fig. 7.5 CCA plot of the relationship between the ferns and lycophytes species in MPPMNGPL and the environmental factors that affect them. *LI* light intensity, *AT* air temperature, *ST* soil temperature, *OM* organic matter, *SM* soil moisture, *Elev* elevation, *RH* relative humidity

factor is associated strongly with fern distribution and diversity (Banaticla and Buot 2004; Catapang et al. 2012; Delos Angeles and Buot 2015). Hence, it is presumed that the changes in elevation potentially drive the distribution of pteridophytes in MPPMNGPL. In addition, low organic matter being linked with high fern diversity (Rodrıguez-Loinaz et al. 2008) was observed in the study area.

7.7 Environmental Preferences of Ferns and Lycophytes

The environmental preferences of the documented ferns and lycophytes species in MPPMNGPL show notable associations (Fig. 7.5). Most species are associated with identical environmental conditions such as high elevation, high relative humidity, low organic matter, and low light intensity. However, a few pteridophyte species show particularity in the constraints that they can thrive, deviating from most samples. This preferential nature of ferns and lycophytes in specific environmental conditions shows their adaptive nature (Rathinasabapathi 2006).

Initially are the "sun-loving" fern species of the landscape: *L. japonicum, L flexuosum*, and *L. circinnatum*. These ferns species are considered "sun-loving" because they are associated with and thrive in environments with high light intensity. Notably, this association is atypical of most ferns since direct and high light intensity is a deterrent to the germination process of their spores as spore germinates optimally in low light intensity (Catapang et al. 2012; Suo et al. 2015). Hence, it can be deduced that the aforementioned fern species are well-adapted to this condition.

Furthermore, *C. dentata*, *A. terminans*, *P. cretica,* and *P. ensiformis* are the fern species that can thrive in a harsher landscape environment. These ferns prefer areas with high air temperature and soil temperature, constraints indicative of disturbances, and a forest with an open canopy (Medecilo and Lagat 2017). In cognizance, the association *C. dentata* and *A. terminans* substantiate its previously described particularity in a specific type of environment, making it a good indicator of stress and disturbances (Alcala et al. 2019; Delos Angeles et al. 2020; Dai et al. 2020). Climactically, species that significantly flourish in high soil moisture are *Huperzia* sp., *P. horizontale,* and *Dryopteris* sp. These species require more moisture than other pteridophytes in the landscape.

Seen in the biplot are the other pteridophyte species that generally thrive in an environment with high elevation, high humidity, low organic matter, and low light intensity. The species influenced by these environmental variables are *P. pellucida*, *T. hilocarpa*, *M. longissimum* , *P, oppositipinnata* , *P. philippinensis*, *B. rhizophylla*, *A. philippense,* and *B. scalpturata*. Further, these constraints are indications of intact forest vegetation with thick canopy cover.

7.8 Threats and Biocultural Importance of Pteridophytes

Ferns and lycophytes are considered effective ecological markers; additionally, though often neglected, these plants also have various ethnobotanical usages that strengthen the innate relations between plants and people. The usage of ferns and lycophytes transcends from ornamental, handicrafts, food sources, and medicinal uses. Of the documented species in MPPMGPL, ferns and lycophytes were mostly utilized for their esthetic values (Table 7.8). *Lygodium* species such as *Lygodium circinnatum*, *Lygodium japonicum*, and *Lygodium flexuosum* were utilized as raw materials for handicrafts (PCAARD 2017) because of their durable rachis. These

Table 7.8 Harvesting risk and economic use of the ferns and lycophytes in MPPMNGPL

Family	Taxa	Harvesting risk	Economic use
Dryopteridaceae	*Bolbitis rhizophylla* (Kaulf.) Hennipman	Whole plant	Ornamental
	Bolbitis scalpturata (Fée) Ching	–	–
	Polystichum horizontale C. Presl	–	–
	Dryopteris sp.	–	–
Lycopodiaceae	*Huperzia* sp.	Whole plant	Ornamental, medicinal
Lycopodiaceae	*Lygodium circinnatum* (Burm. fil.) Sw	Fronds, stem	Handicraft, tying element, medicinal
	Lygodium japonicum	Whole plant	Food, handicraft, medicinal
	Lygodium flexuosum (L.) Sw.	Whole plant	Food, animal feed, handicraft, medicinal
Polypodiaceae	*Microsorum longissimum* J.Sm. ex Fée	–	–
Pteridaceae	*Adiantum philippense* L.	Fronds	Ornamental, medicinal
	Pteris cretica L.	Fronds	Medicinal
	Pteris distans J.Sm.	–	–
	Pteris ensiformis Burm.	Whole plant	Medicinal, ornamental, food flavoring
	Pteris heteromorpha Fée	–	–
	Pteris oppositipinnata Fée	–	–
	Pteris pellucida C.Presl	–	–
	Pteris philippinensis Fée	–	–
Tectariaceae	*Tectaria hilocarpa* (Fée) M.G. Price	Whole plant	Ornamental
Thelypteridaceae	*Amphineuron terminans* (J.Sm. ex Hook.) Holttum	Fronds	Medicinal
	Christella dentata (Forssk.) Brownsey & Jermy	Fronds	Medicinal

"–" unidentified

species, commonly referred to as “Nito” were considered a common raw material used in basket making (Pazon and Del Rio 2018), basket wares, and other handicraft products. Though the native usage of nito in weaving is linked to the practices of the indigenous people in the southernmost regions of Luzon Island (e.g., Iraya Mangyan Tribes in Mindoro Province and Palawan), the people in Cavite and nearby provinces use weaved nito food plates and food platters during festivities and gatherings in celebration of important cultural gatherings and commemoration. *Huperzia* sp., *Bolbitis rhizophylla,* and *Adiantum philippense* L. were utilized for horticultural purposes such as gardening and landscaping. At the same time, the other documented species have attributes to be utilized for medicinal purposes locally and in other countries showing their biocultural importance.

Even so, some pteridophyte species are only harvested for their fronds; several species were identified wherein the whole plant is harvested for usage. These species (*Huperzia* sp., *L japonicum, L. flexuosum, Pteris ensiformis,* and *Tectaria hilocarpa*) face high risk due to how they were harvested and their various economic uses. Though no hard quantifiable evidence was available to support the harvesting of pteridophytes in the site and the region as of the time of writing, other emerging threats in the landscape were observed during the study; threats which are not only harming the pteridophytes but also the other flora and fauna.

Despite its temporary moratorium in 2016, the following threats in the landscape were still documented: the occupation of land claimants inside the protected area, river poisoning, installation of traps and baits for birds and other animals, burning and land clearing activities, and presence of plastic wastes. If continuous, these anthropogenic threats will significantly contribute to habitat loss, diminishing of wildlife populations, habitat destruction, land conversion, and pollution. These problems will later negatively affect the landscape’s pteridophyte flora, putting some preferential species at risk, together with the local biodiversity. The impacts of these environmental problems can be addressed by the strict adherence to the existing laws governing the land and the involvement of the locals in protecting the landscape. The latter can be achieved by initiating a local-based conservation program and assessment to capacitate and engage them in protecting the landscape (Lagat and Causaren 2018).

7.9 Summary and Conclusion

The species of ferns and lycophytes documented in MPPMNGPL were all native to the Philippines. This estimate provided an update on the list of the known pteridophyte flora in the province, region, and the country as a whole. Pteridophyte diversity is seen to increase as elevation increases, though a typical hump-shaped distribution is not observed. Further, a low diversity in ferns and lycophytes was generally attributed to the open areas of the landscape.

Based on the species-habitat association generated from the ordination, the environmental variables: relative humidity, air temperature, soil temperature, soil moisture, elevation, light intensity, and soil organic matter, were considered the most important environmental variables affecting the distribution and diversity of ferns and lycophytes in MMPMGPL. Of these, the most influential factors are elevation and light intensity. Generally, sites with high diversity index tend to be influenced by elevation, while sites with low diversity index are influenced greatly by light intensity—main factors that play important roles in fern reproduction, development, and distribution. Further, a majority of fern and lycophytes species are thriving preferably in areas with high elevation, high humidity, low organic matter, and low light intensity. However, some fern species can thrive in harsher environments with specificity. These constraints are areas with high light intensity, high air temperature, and high soil temperature indicating possible habitat and other ecological disturbances. Hence, these adaptive ferns can potentially be a good indicator of environmental stresses. Furthermore, pteridophytes have known biocultural importance, providing important ecological services to man. However, anthropogenic threats put them at risk and possible harm. Using of the estimates and the scientific data described in this study can potentially strengthen the existing biodiversity and conservation programs to protect these vital plant groups and all life forms in the landscape.

Acknowledgments The authors would like to express their utmost gratitude to the following individuals and institutions who extended their help in the completion of the study successfully. The most profound appreciation to the following: Department of Mathematics and Natural Sciences (CAS-SLSU), Graduates Studies and Biological Sciences Department (CSCS-DLSUD), and the SHS Department of PAREF-Southridge School for the untiring support and encouragements; The Department of Environment and Natural Resources-Region IV-A and the Protected Area Management Board (DENR-PAMB) of Mts. Palay-Palay for the approval and accommodation to successfully complete this study.

References

Alcala A, Delos Angeles M, Buot I (2019) Fern species diversity across various land-use types of Mt. Makiling, Luzon Island, Philippines. Biodiversitas 20(9):2437–2445. https://doi.org/10.13057/biodiv/d200902

Amoroso VB, Obsioma LD, Arlalejo JB, Aspiras RA, Capili DP, Polizon JJA, Sumile EB (2009) Inventory and conservation of endangered, endemic and economically important flora of Hamiguitan Range, Southern Philippines. Blumea 54(1–3):71–76

Amoroso VB, Coritico FP, Fritsch PW (2016) Species richness and conservation status of ferns and lycophytes in Mt. Hamiguitan Range Wildlife Sanctuary, Davao Oriental, Philippines. Philipp J Sci 145:127–137

Aureo W, Reyes T, Mutia FC, Jose R, Sarnowski MB (2020) Diversity and composition of plant species in the forest over limestone of Rajah Sikatuna Protected Landscape, Bohol, Philippines. Biodivers Data J 8:e55790. https://doi.org/10.3897/BDJ.8.e55790

Avolio M, Forrestel EJ, Chang C, La Pierre K, Burghard K, Smith M (2019) Demystifying dominant species. New Phytol 223:1106–1126. https://doi.org/10.1111/nph.15789

Banaticla MCN, Buot IE (2004) Fern patch structure and species diversity along the altitudinal gradient of Mt. Banahaw de Lucban, Luzon Island, Philippines. Philipp Agric Sci 87:49–60

Barcelona JF (2005) Noteworthy fern discoveries in the Philippines at the turn of the 21st century. Fern Gaz 17:139–146

Brummitt N, Aletrari E, Syfert MM, Mulligan M (2016) Where are threatened ferns found? Global conservation priorities for pteridophytes. J Syst Evol 54:604–616. https://doi.org/10.1111/jse.12224

Callado JRC, Adji B, Suksathan P, Lestari WS, Darnaedi D (2015) Field guide to the pteridophytes of Chiang Mai, Thailand. ACB field guide series no. 2. ASEAN Centre for Biodiversity and Japan-ASEAN Integration Fund, 174p

Catapang MVL, Reyes PJD, Medecilo MP (2012) Factors influencing species diversity of ferns in Mt. Makulot, Cuenca, southern Luzon, Philippines. In: 2nd international conference on environment and industrial innovation; 2-3 June 2012; Hongkong, China. IACSIT Press, Singapore. IPCBEE, vol 35, pp 98–102

Causaren RM (2009) Preliminary report on the anurans of Mts. Palay-Palay Mataas-na-Gulod Protected Landscape, Luzon Island, Philippines. Philipp J Syst Biol 3:40–56

Conservation International, the Haribon Foundation, the Department of Environment and Natural Resources-Protected Areas and Wildlife Bureau, and the Critical Ecosystem Partnership Fund (2006) Philippines Outcomes Database

Coritico F, Amoroso V, Acma F, Cariño YL, Fritsch P (2020) Ferns and lycophytes of Mt. Tago Range, Bukidnon, Southern Philippines: species richness, distribution, and conservation status. Philipp J Sci 149(3-a):773–790

Dai X, Chen C, Li Z, Wang X (2020) Taxonomic, phylogenetic, and functional diversity of ferns at three differently disturbed sites in Longnan County, China. Diversity 12:135. https://doi.org/10.3390/d12040135

David V (2017) Multivariate approach. Data treatment in environmental sciences. Elsevier, Amsterdam

Delos Angeles M, Buot IE (2015) Diversity and distribution of pteridophytes along the altitudinal gradient of the northeastern slope of a secondary forest in Mt. Makiling, Philippines. IAMURE Int J Ecol Conser 16:25–47. https://doi.org/10.7718/ijec.v16i1.1015

Delos Angeles M, Alcala A, Buot I (2020) Ecological diversity of pteridophytes across land use types in Mt. Makiling Forest Reserve, Luzon Island, Philippines. J Mar Island Cult 9(2) 2212-6821

Dong S, Zhang X (2005) A taxonomic revision of the fern genus Bolbitis (Bolbitidaceae) from China. Zhiwu Fenlei Xuebao 43(2):97–115. https://doi.org/10.1360/aps030105

Ebihara A, Fraser-Jenkins CR, Parris BS, Zhang X-C, Yang Y-H, Chiou W-L, Chang H-M, Lindsay S, Middleton D, Kato M et al (2012) Rare and threatened pteridophytes of Asia 1. An enumeration of narrowly distributed taxa. Bull Natl Mus Nat Sci Ser B 38:93–119

Fernández H, Kumar A, Revilla MA (2010) Working with ferns: issues and applications, vol 2. Springer Science & Business Media, Berlin, pp 1–386. https://doi.org/10.1007/978-1-4419-7162-3

Fernando ES (1998) Forest formations and flora of the Philippines: handout inFBS 21. College of Forestry and Natural Resources, University of the Philippines, Los Baños

Gavlak R, Horneck D, Miller R (2005) Plant, soil and water reference methods for the Western Region. Western Regional Extension Publication (WREP) 125, WERA-103 Technical Committee

Gonzatti F, Machado L, Windisch PG (2016) Distribution patterns of ferns and lycophytes in the Coastal Region of the state of Rio Grande do Sul, Brazil. Acta Bot Brasilica 30(2):239–253. https://doi.org/10.1590/0102-33062016abb0058

Guo Q, Kato M, Ricklefs RE (2003) Life history, diversity and distribution: a study of Japanese pteridophytes. Ecography 26:129–138

Heegaard E (2002) A model of alpine species distribution in relation to snowmelt time and altitude. J Veg Sci 13(4):493–504

Jackson H, Fahrig L (2013) Habitat loss and fragmentation. Encyclopedia of biodiversity (second edition). Academic Press, New York. https://doi.org/10.1016/B978-0-12-384719-5.00399-3

Jost L, De Vries P, Walla T, Greeney H, Chao A, Ricotta C (2010) Partitioning diversity for conservation analyses. Divers Distrib 16:65–76. https://doi.org/10.1111/j.1472-4642.2009.00626.x

Karger DN, Kluge J, Kromer T, Hemp A, Lehnert M, Kessler M (2011) The effect of area on local and regional elevational patterns of species richness. J Biogeogr 38:1177–1185

Karst J, Gilbert B, Lechowicz MJ (2005) Fern community assembly: the roles of chance and the environment at local and intermediate scales. Ecology 86:2473–2486. https://doi.org/10.1890/04-1420

Kessler M (2001) Pteridophyte species richness in Andean forest in Bolivia. Biodivers Conserv 10: 1473–1495

Kessler M, Krömer T, Jimenez I (2000) Inventario de grupos selectos de plantas en el Valle de Masicurí (Santa Cruz – Bolivia). Rev Boliviana Ecol Conserv Ambiental 8:3–15

Kessler M, Moguel Velázquez AL, Sundu M, Labiak PH (2011) Alansmia, a new genus of grammitid ferns (Polypodiaceae) segregated from Terpsichore. Brittonia 63:233–244

Kessler M, Karger DN, Kluge J (2016) Elevational diversity patterns as an example for evolutionary and ecological dynamics in ferns and lycophytes. J Syst Evol 54:617–625

Kim Thu D, Vui D, Ngoc Huyen N, Duyen D, Thanh TB (2020) The use of Huperzia species for the treatment of Alzheimer's disease. J Basic Clin Physiol Pharmacol 31(3):20190159. https://doi.org/10.1515/jbcpp-2019-0159

Konare H et al (2010) Loss on ignition: measuring soil organic carbon in soils of the Sahel, West Africa. Afr J Agric Res 5(22):3088–3095

Kreutz C, de Paul Athayde Filho F, Sanchez M (2015) Spatial and seasonal variation in the species richness and abundance of ferns and lycophytes in gallery forests of Cerrado in Central Brazil. Rev Bras Bot 39:315–326. https://doi.org/10.1007/s40415-015-0236-9

Kromer T, Acebey A, Kluge J, Kessler M (2013) Effects of altitude and climate in determining elevational plant species richness patterns: a case study from Los Tuxtlas, Mexico. Flora 208: 197–210

Lagat R, Causaren R (2018) Initial terrestrial vertebrate diversity assessment in upland Cavite, Philippines. Philipp J Syst Biol. 12(2):70–91

Magurran AE, McGill BJ (2011) Biological diversity: frontiers in measurement and assessment. Oxford University Press, Oxford

Medecilo MM, Lagat M (2017) Floristic composition of the remaining forests in Upland Cavite, Luzon Island, Philippines. Philipp J Syst Biol 11(1):74–94

Moran RC (2008) Diversity, biogeography and floristics. In: Ranker TA, Haufler CH (eds) Biology and evolution of ferns and lycophytes. Cambridge University Press, Cambridge, pp 367–394

Mts. Palay-Palay Mataas Na Gulod Protected Landscape Forest Assessment and Geospatial Report. [Internet] [cited 2021 December 28]. https://www.academia.edu/26822472/Mounts_Palay_Palay_Mataas_na_Gulod_Protected_Landsacape_Forest_Assessment_and_Geospatial_Analysis_Technical_Report?source=swp_share

Myers N, Mittermeier RA, Mittermeier CG, da Fonseca GAB, Kent J (2000) Biodiversity hotspots for conservation priorities. Nature 403:853–858

Nassif WG, Jaber SH, Naif SS, Al-Taai OT (2021) Estimate of the dynamical change of air temperature, relative humidity and dew point temperature for some selected station in Iraq. Earth Environ Sci 910:012010. https://doi.org/10.1088/1755-1315/910/1/012010

Ong P, Afuang L, Ambal RG (2002) Philippine biodiversity conservation priorities: a second iteration of the National Biodiversity Strategy and Action Plan

Pazon ANR, Del Rio JMP (2018) Materials, functions, and weaving patterns of Philippine Indigenous Baskets. Asian J Multidiscip Stud 1(2)

Pelser PB, Barcelona JF, Nickrent DL (eds) 2011 onwards.Co's Digital Flora of the Philippines. www.philippineplants.org

Philippine Council for Agriculture, Aquatic and Natural Resources Research and Development (2017) Economically valuable Philippine ferns. Los Baños, Laguna. DOST-PCAARRD, 2017. 14p. (PCAARRD Information Bulletin No. 77/2017)

Posa MRC, Diesmos AC, Sodhi NS, Brooks TM (2008) Hope for threatened tropical biodiversity: lessons from the Philippines. Bioscience 58:231. https://doi.org/10.1641/B580309

PPG I - The Pteridophyte Phylogeny Group (2016) A community-derived classifcation for extant lycophytes and ferns. J Syst Evol 54:563–603

Rathinasabapathi B (2006) Ferns represent an untapped biodiversity for improving crops for environmental stress tolerance. New Phytol 172:379–381

Raymundo AK, Tan BC, Asuncion AMC (1991) Anti-microbial activities of some Philippine cryptogams. Philipp J Sci 118:59–75

Rodrıguez-Loinaz G, Onaindiaa M, Amezaga I, Mijangos I, Garbisu C (2008) Relationship between vegetation diversity and soil functional diversity in native mixed-oak forests. Soil Biol Biochem 40(2008):49–60

Roundy B, Chambers J (2021) Effects of elevation and selective disturbance on soil climate and vegetation in big sagebrush communities. Ecosphere 12(3):e03377. https://doi.org/10.1002/ecs2.3377

Sánchez-González A, Álvarez-Zúñiga E, López-Mata L (2016) Diversity and distribution patterns of ferns and lycophytes in a cloud forest in Mexico. Revista Chapingo Serie Ciencias Forestales y del Ambiente XXII(3):235–253. https://doi.org/10.5154/r.rchscfa.2015.09.042

Sarker SK, Rashid S, Sharmin M, Haque MM, Sonet SS, Nur Un Nabi M (2014) Environmental correlates of vegetation distribution in Tropical Juri Forest, Bangladesh. Trop Ecol 55:177–193

Scheuttpelz E, Pryer K (2009) Evidence for a Cenozoic radiation of ferns in an angiosperm-dominated canopy. Proc Natl Acad Sci U S A 106(27):11200–11205

Shukla A, Panchal H, Mishra M, Patel PR, Srivastava HS, Patel P, Shukla AK (2014) Soil moisture estimation using gravimetric technique and FDR probe technique: a comparative analysis. AIJRFANS 14-465:89–92

Sodhi NS, Brook BW (2006) Southeast Asian biodiversity in crisis. Cambridge University Press, Cambridge

Solis MJ, Dela Cruz TE, Schnittler M, Unterseher M (2015) The diverse community of leaf-inhabiting fungal endophytes from Philippine natural forests reflects phylogenetic patterns of their host plant species Ficus benjamina, F. elastica and F.religiosa. Mycoscience 57(2):96–106. https://doi.org/10.1016/j.myc.2015.10.002

Suo J, Chen S, Zhao Q, Shi L, Dai (2015) Fern spore germination in response to environmental factors. Front Biol 10:358–376. https://doi.org/10.1007/s11515-015-1342-6

Syfert M, Brummitt N, Coomes D, Bystriakova N, Smith M (2018) Inferring diversity patterns along an elevation gradient from stacked SDMs: a case study on Mesoamerican ferns. Global Ecol Conserv J 16:1–15

Tabachnick BG, Fidell LS (2007) Using multivariate statistics, 5th edn. Pearson Education Inc., London

Tuomisto H, Zuquim G, Cárdenas G (2014) Species richness and diversity along edaphic and climatic gradients in Amazonia. Ecography 11:1034–1046

Weiss S, Xu ZZ, Peddada S (2017) Normalization and microbial differential abundance strategies depend upon data characteristics. Microbiome 5:27

Whittaker RJ, Willis KJ, Field R (2001) Scale and Species richness: towards a general hierarchical theory of species diversity. Journal of Biogeography 28:453–470

Wilcox R (2012) Chapter 6 - Some multivariate methods. In: Wilcox R (ed) Statistical modeling and decision science, Introduction to Robust estimation and hypothesis testing, 3rd edn. Academic Press, pp 215–289. ISBN 9780123869838. https://doi.org/10.1016/B978-0-12-386983-8.00016-0. https://www.sciencedirect.com/science/article/pii/B9780123869838000160

Zuquim G (2015) Ferns and lycophytes in Amazonia: diversity patterns and usefulness as habitat indicators. University of Turku, Turku

Remaining Subpopulations of *Impatiens manillensis* Walp. 1843 (Balsaminaceae) in Mts. Palay-Palay-Mataas-Na-Gulod Protected Landscape, Luzon Island, Philippines

8

Rubie M. Causaren, Myra N. Lagat, and Ronaldo D. Lagat

Abstract

The current distribution and some ecological information of *Impatiens manillensis* are very limited. This paper reports additional information on the species' distribution, seasonality, and microhabitat requirements. The previously reported sites, where *I. manillensis* was documented and visited and habitats and microhabitats were described. Physical parameters such as light intensity, relative humidity, ambient, and soil temperature were also measured, and the soil texture was determined. Small subpopulations were still observed in the previously reported sites. The species, which is active during the rainy months, prefer shallow sandy loam on rock beds in partially exposed forest edge areas. To thrive, this herb requires these microhabitat parameter ranges: light requirement (171 × 10–193 × 100 lx), soil temperature (20–30 °C), ambient temperature (27–31 °C), and relative humidity (96–100%). Numerous flora/organisms were also associated with its habitats. Currently, the species is evaluated as data-deficient given its restricted range and distribution. It is highly recommended to be included in the national list of threatened Philippine flora.

Keywords

Cavite · *Impatiens caviteana* · MPMNGPL · Philippine endemic · Rizal

R. M. Causaren · M. N. Lagat · R. D. Lagat (✉)
Biological Sciences Department, College of Science and Computer Studies, De La Salle University-Dasmariñas, Dasmariñas, Cavite, Philippines
e-mail: rdlagat@dlsud.edu.ph

S. Ramamoorthy et al. (eds.), *Plant Diversity in Biocultural Landscapes*,
https://doi.org/10.1007/978-981-19-8649-9_8

8.1 Introduction

Plant biologists, agronomists, and conservationists are alarmed by the threat of climate change exacerbated by anthropogenic activities to plants. Expected aberrant responses of about 42% of the 4000 plant species investigated revealed the possible irreversible damage of Anthropogenic Climate Change (ACC) to plants (Parmesan and Hanley 2015). This grave threat is even compounded to a greater degree for endemic plants with a restricted distribution like *Impatiens manillensis* (Hooker 1909; Medecilo 2004; Pelser et al. 2011; Tandon 2022). A subpopulation of *I. manillensis* is present in Mts. Palay-Palay-Mataas-Na-Gulod Protected Landscape (henceforth MPMNGPL; 14.23° N, 120.22° E; Fig. 8.2) situated in the municipalities of Ternate and Maragondon in Cavite and Nasugbu in Batangas (Medecilo 2004; R.A. 11038 2018). The information on the current distribution and population estimates of *I. manillensis* is very limited, if not lacking. Available information already qualifies the species for immediate conservation and protection; however, it is not yet included in the National List of Threatened Philippine Flora (DAO No 2017–11).

Impatiens manillensis Walp., Nov. Actorum Acad. Caes. Leop.-Carol. Nat. Cur. 19(Suppl. 1): 317 (1843) IPNI 2022. The recognized synonyms for this species are *Impatiens biganensis* Hook.f., *Impatiens cleistogama* Hook.f., *Impatiens cryptogama* Hook.f., *Impatiens curranii* Hook.f., and *Impatiens caviteana* Hook.f.

8.2 Description

Impatiens manillensis is a member of the large herbaceous *Impatiens* genus under the family Balsaminaceae. Impatiens are native to Asia, Africa, and North America (Britannica 2017). It is an annual herb, 50–400 cm; stems are erect nodes with intervals of 45–70 cm, succulent, sparsely branched, and glabrous. Leaves alternate, the terminal young leaves start as whorls of three's or more (Fig. 8.1a), the lowermost occasionally pseudo-opposite, sessile, or shortly petiolate (Fig. 8.1a); petioles 1–54 mm; blades 3.5–9.9 × 1.0–2.5 cm, lanceolate to narrowly elliptic or oblanceolate, glabrous or finely pubescent above and below, the base cuneate-acuminate, the margins serrate-dentate, with pointed trichome like structure between margin dentation, the apex acute (Fig. 8.1). Inflorescences of solitary flowers or groups of 2–3 flowers borne at node of the terminal portion of the stem together with the whorled young leaves; peduncles are absent (Fig. 8.1a, b); bracts 2 × 7 mm, axillary at the base of the pedicel; pedicels 64–67 mm. The flowers are light purple and white at the center (Fig. 8.1b, c), with upper petal 15–17 × 13–20 mm, apically emarginate (Fig. 8.1c) with an appendage 2–4 mm; lateral petals 11 × 13–14 mm, also with emarginate tips (Fig. 8.1c), the upper petal of each lateral pair smaller than the lower; ovary smooth and shiny. Capsules 10–15 × 4–6 mm, pendent, pubescent, and slightly unipennate in shape. Seeds few from 9 to 14, brown, globose, 1.5–3 mm in diameter, tuberculate. Taproot superficial (Fig. 8.2a), adventitious roots present (Fig. 8.2b, c).

Fig. 8.1 *Impatiens manillensis* terminal branch with flowers. (**a**) Clustered fruits at a node with the whorled leaves in three's. (**b**) Flowers are borne at the terminal end. (**c**) Flowers are light purple in color with white at the center

Fig. 8.2 Root characteristics of *I. manillensis*. (**a**) Superficial. (**b**, **c**) Taproot type with adventitious roots growing from the node and internodes of the stem

8.3 Seasonality and Microhabitat Characterization

This species is active during the rainy months from May to November. It is dormant for the remaining months of the year as seeds in association with other plants. The reported sites (Medecilo 2004) were visited last July 9, 2022. Habitats were described, physical parameters such as light intensity, relative humidity, ambient, and soil temperature were measured, and the soil texture was determined.

The substrates preferred by this species are shallow sandy loam on rock beds occasionally flooded during rainy days. This explains the characteristic superficial root growth of the plant that occurs actively during the rainy season (Fig. 8.3b inset). The habitats are found in partially exposed forest edge areas. The light measurement obtained from a partially exposed habitat area ranges from 171 × 10 to 193 × 100 lx. Soil temperature depends on the time of the day which ranges from 20 °C to 30 °C.

Fig. 8.3 (**a**) Map of the Philippines showing MPMNGPL where subpopulations are reported. (**b**) The preferred habitat of *Impatiens manillensis*. (Inset) *I. manillensis* superficial root growth characteristic. (**c**) *I. manillensis* and associated plants

The ambient temperature ranged from 27 °C to 31 °C. Relative humidity ranges from 96% to 100%. Most of these habitats are situated near rivers and streams (Fig. 8.3b) which help maintain the high relative humidity conditions. Associated flora/organisms include begonias, lichen, moss, thallose liverworts, and different species of ferns (Fig. 8.3c). These typically inhabit areas with high relative humidity.

8.4 Existing Threats and Conservation Status

This Philippine endemic is reported to have been described as early as 1848 according to "A Review of the Known Philippine Islands Species of Impatiens by Hooker" (Hooker 1909). The distribution of *I. manillensis* is very restricted which includes forests in the provinces of Cavite and Rizal in Luzon Island (Medecilo 2004; Pelser et al. 2011). There is no information or updates if a population still exists in Rizal province. A recent ocular visit at MPMNGPL last July 9, 2022, confirmed the presence of the remaining subpopulations of this species in their preferred habitats. MPMNGPL, which is characterized by mixed forest habitats with an area of more or less 3900 ha, is possibly one of the remaining suitable habitats for this Philippine endemic (Medecilo 2004; Pelser et al. 2011).

Currently, the conservation status of *I. manillensis* is not clearly defined or is still unresolved due to data deficiency (The Plant List 2013). In addition to the restricted distribution of *I. manillensis*, its current habitat, especially those that are in areas exposed to intense anthropogenic activities like construction and maintenance of road infrastructures can negatively impact its survival. There is an increase in soil erosion related to these activities due to physical modification of the habitat that can displace dormant forms (seeds) of the plant and pollution (air and water) contributed by vehicles. Seeds displaced to unsuitable habitats may no longer germinate. In addition, the threats linked to climate change such as altered duration of dry and wet seasons and increased ambient temperature exacerbated by continuous vegetation clearing (Causaren et al. 2018) can modify specific microhabitat conditions that may result in habitat loss for this species. According to Causaren et al. (2018), serious dangers exist in the area such as illegal logging, firewood gathering, poaching, quarrying, slash-and-burn or 'kaingin', harvesting of non-timber forest products, and the use of some portions of the park as pasture for livestock.

8.5 Summary and Conclusion

Impatiens manillensis is a Philippine endemic with a very restricted distribution in Luzon particularly in the provinces of Rizal and Cavite. Recent investigations confirmed the presence of a subpopulation in MPMNGPL situated in Cavite and Batangas. It is fortunate that these subpopulations are within the protected areas; however, other possible subpopulations still present in Rizal and other areas may not be accorded with the same protection. With a data deficient species given its restricted range and distribution, the species is highly recommended to be included in the national list of threatened Philippine flora.

Acknowledgments We are grateful to Dr. Inocencio E Buot Jr. for extending this invitation to contribute to the book "*Plant Diversity in Biocultural Landscapes During the Anthropocene*". We are also indebted to DENR-PENRO Cavite for allowing us to conduct the preliminary survey for this purpose. Special mention to Mr. Roberto "Tatay Obet" L. Bautista for serving as our field guide during the site visit. Likewise, we acknowledge the continued support of the Biological Sciences Department of De La Salle University—Dasmariñas in this endeavor.

References

Britannica (2017) "Impatiens". The Editors of Encyclopedia. https://www.britannica.com/plant/Impatiens. Accessed 14 Aug 2022

Causaren RM, Lagat RD, Agoo MG (2018) Tree species diversity of the remaining forest fragments in cavite, Luzon Island, Philippines. Philippine J Sys Biol 11:56–73

DAO No (2017–11) Updated National list of threatened plants and their categories

Hooker JD (1909) A Review of the Known Philippine Islands Species of Impatiens. Bull Misc Inform 1909(7):281–289. Accessed 14 Aug 2022. https://doi.org/10.2307/4111583

IPNI (2022) International Plant Names Index. The Royal Botanic Gardens, Kew, Harvard University Herbaria & Libraries, and Australian National Botanic Gardens. Published on the Internet http://www.ipni.org. Accessed 10 Aug 2022

Medecilo M (2004) Plant diversity of the Mts. Palaypalay/Mataas na Gulod National Park, Philippines. In: Food and Agriculture Organization of the United Nations: AGRIS. University Library, University of the Philippines, Los Baños, p 80. agris.fao.org: https://agris.fao.org/agris-search/search.do?recordID=PH2005000395

Parmesan C, Hanley ME (2015) Plants and climate change: complexities and surprises. Ann Bot 116:849–864

Pelser PB, Barcelona JF, Nickrent DL (2011) Co's Digital Flora of the Philippines. Copyright © 2011, Co's Digital Flora of the Philippines. www.philippineplants.org. Accessed 17 Apr 2022

R.A. 11038 (2018) Expanded national integrated protected area systems act

Tandon A (2022) Carbon brief clear on climate. Carbon Brief.Org: https://www.carbonbrief.org/climate-change-will-hit-endemic-plants-and-animals-the-hardest-study-warn s/

The Plant List (2013) Version 1.1. Published on the Internet; http://www.theplantlist.org/. Accessed 1 Jan

Pteridophytes in Mount Matutum Protected Landscape

9

Christine Dawn Obemio, Inocencio E Buot, Jr, and Maria Celeste N. Banaticla-Hilario

Abstract

Inventory of the floristic composition is imperative to biodiversity conservation. Pteridophytes is one of the plant groups in Southern Mindanao that lack substantial data on its floristics. The composition and diversity of fern and fern allies from base to summit of the northwestern slope of Mount Matutum Protected Landscape (MMPL) are highlighted in this chapter. A total of 265 species under 94 genera and 29 families were noted, 29 of which are keystone endemics, while 27 are considered threatened. *Polypodiaceae* and *Aspleniaceae* are the families largely represented in MMPL. Moreover, *Asplenium, Pteris, Diplazium, and Elaphoglossum* were noted to be highly abundant and widely distributed genera.

Species' richness and diversity vary across elevational gradients with generally an increasing trend from 1500 to 1900 masl. A decline was observed <1500 and >2000 masl with the least recorded pteridophyte species at 2275 masl. Ground pteridophytes including tree ferns were the most abundant in the patch structure, while the epiphytic community is rich from mid-elevations (1700 masl) to the peak (2200 masl).

Keywords

Ferns · Mount Matutum · Species composition · Fern distribution

C. D. Obemio
Mindanao State University, General Santos, Philippines

I. E. Buot, Jr (✉) · M. C. N. Banaticla-Hilario
Institute of Biological Sciences, College of Arts and Sciences, University of the Philippines Los Baños, Laguna, Philippines
e-mail: inocencio.buot@upou.edu.ph

S. Ramamoorthy et al. (eds.), *Plant Diversity in Biocultural Landscapes*,
https://doi.org/10.1007/978-981-19-8649-9_9

9.1 Introduction

A significant understory flora growing in MMPL are ferns and lycophytes, collectively referred to as pteridophytes, owing to their non-seed reproductive strategy. They are widely distributed both in the tropic and subtropical regions especially at higher elevations, and they flourish in moist, shaded habitats (Delos Angeles and Buot 2012; Franco et al. 2011; Barcelona 2003a, b; Bhattarai et al. 2004; Sharpe et al. 2010; Gomez 1985; Holttum 1948; Flora Malesiana Series II n.d.). Around the world, there are about 12,000 estimated species of this plant group distributed in 51 families, around the world (PPG I 2016). In the Philippines, an estimate of about 1100 species are known belonging to 34 families (Amoroso et al. 2016).

Pteridophytes form a very delicate sub-canopy layer in the forest and constitute the largest portion of the understory community. These plants are also host to diverse faunal and other floral species. One interesting example for this can be presented by the tree ferns *Cyathea*, in which about 26 species are found solely in the country. These species have been known to house several other epiphytic plants, including the threatened ferns and fern allies. However, they are also highly priced as anchor material for many ornamentals. Several species of this group have been associated with distinct environmental characteristics (i.e., heavy metal accumulation) as well as changes that have occurred (i.e., fragmentation, clearings, climate changes) leading to significant population decline due to increasing exposure of their habitats to fragmentation and climate change (Magtoto and Austria 2017; Ibars and Estrelles 2012). Their remarkable adaptations to various disturbances suggest an important role for pteridophytes in conservation and restoration.

As economic development ensues, utilization for this understory group has also increased over the years. Several species have high economic value as food and ornamentals (Oloyede 2012), medicine (Liu et al. 2012; Mannan et al. 2008; Sen and Ghosh 2011; Upreti et al. 2009), phytoremediation agent (Sharpe et al. 2010), and important habitats of other plants as well as animals (Goswami et al. 2016; Keller and Prance 2015). Hence, the inclusion of pteridophytes in biodiversity and conservation interventions is demanded.

Mount Matutum Protected Landscape is a majestic landmark, in the southern part of Mindanao, Philippines. It was declared as a Key Biodiversity Area (KBA) (Key Biodiversity Areas: Conservation International—Philippines, Haribon Foundation and the Department of Environment and Natural Resources, year) and an Important Bird Area (IBA) (BirdLife International 2015), holding in its bosom vast forest and agricultural wealth and significant natural resources that are in danger of loss with the increasing human population in the surrounding communities. Moreover, it is a key watershed supplying 25% of the water requirements in the region (DENR–PAWB 2006). Towering at 2286 masl, this forest reserve occupies 14,000 ha, of which 3000 ha is reported to still be a primary forest (DENR–PAWB 2006).

Floristics of pteridophytes in Mindanao have been determined by several botanical surveys, but largely is confined in northern part. This is reflected in the works of Amoroso et al. since 1995. As for the southern regions, botanical surveys are highlighted in project reports (Foundation for the Philippine Environment 2003; Gonzalez 2000; Obemio and Buot 2021), mostly unpublished. In 2016, published reports on flora of Matutum presented only the seed plants and bryophytes, but not pteridophytes. Interestingly, the earliest pteridophyte exploration in this protected area dates back more than a 100 years ago (1917) by Copeland where he observed about 99 species, two of which, namely, *Gleichenia peltophora* and *Diplazium calliphyllum,* remained the only record in the Philippines based on collections published online by various herbaria worldwide. With this, an updated account on this plant group, its composition, richness, and distribution are needed to come up with a comprehensive representation of MMPL flora and that of southern Mindanao. This would be vital support to developing effective measures on protecting important habitat zones in the protected area. More so that today, conservation efforts attempt to integrate fern research, and hence, entail a great deal of commitment to more research interventions.

9.2 The Mount Matutum Protected Landscape

Mount Matutum Protected Landscape (MMPL) is a landmark of South Cotabato and Sarangani Provinces in Southern Mindanao, covering four (4) municipalities (3 in South Cotabato; 1 Sarangani Province) and 14 barangays (12 South Cotabato, 2 Sarangani Province). It rises to about 2,286 meters above sea level (masl), covering an approximate area of 14,000 hectares with 3,000 hectares considered as primary forest. It is recognized as a symmetrical stratovolcano located in southern Mindanao (6°22′N, 125°06.5′E), Northwest of Sarangani Bay, and 30 km Northwest of General Santos City.MMPL is a significant network of ecosystems—forests, watershed, wetlands, and air towering 2, 286 m above sea level (masl). Its orogenesis is the result of subduction events between Celebes Sea microplates under the greater Mindanao island, believed to have occurred more than 2000 years ago. An active stratovolcano, MMPL exhibits an andesitic geologic nature with a 320 m well-preserved dome-shaped crater, and interestingly, a 120 m summit deep vegetated area.

With reference to Yumul et al. (2008) and Philippine Institute of Volcanology and Seismology (PHIVOLCS), MMPL is recognized as a symmetrical stratovolcano located in southern Mindanao (6°22′N, 125°06.5′E), Northwest of Sarangani Bay and 30 km Northwest of General Santos City. By stratovolcano, it means that the mountain was built by multilayers of hardened alternating lava and ash. This stratovolcano belongs to the Cotabato Arc which emerged by the eastward subduction of the oceanic Celebes Sea microplate along the Cotabato trench. The trench is under the southern tip of Mindanao Island, which belongs to the Philippine Mobile Belt and Mt. Matutum is located 130 km from the trench. However, deep-seismic studies show that the subducting slab is only 75–85 km below Mt. Matutum volcano.

The climate in the northwest and southwest parts of this protected area is tropical with significant rainfall throughout the year even in the driest months. Köppen-

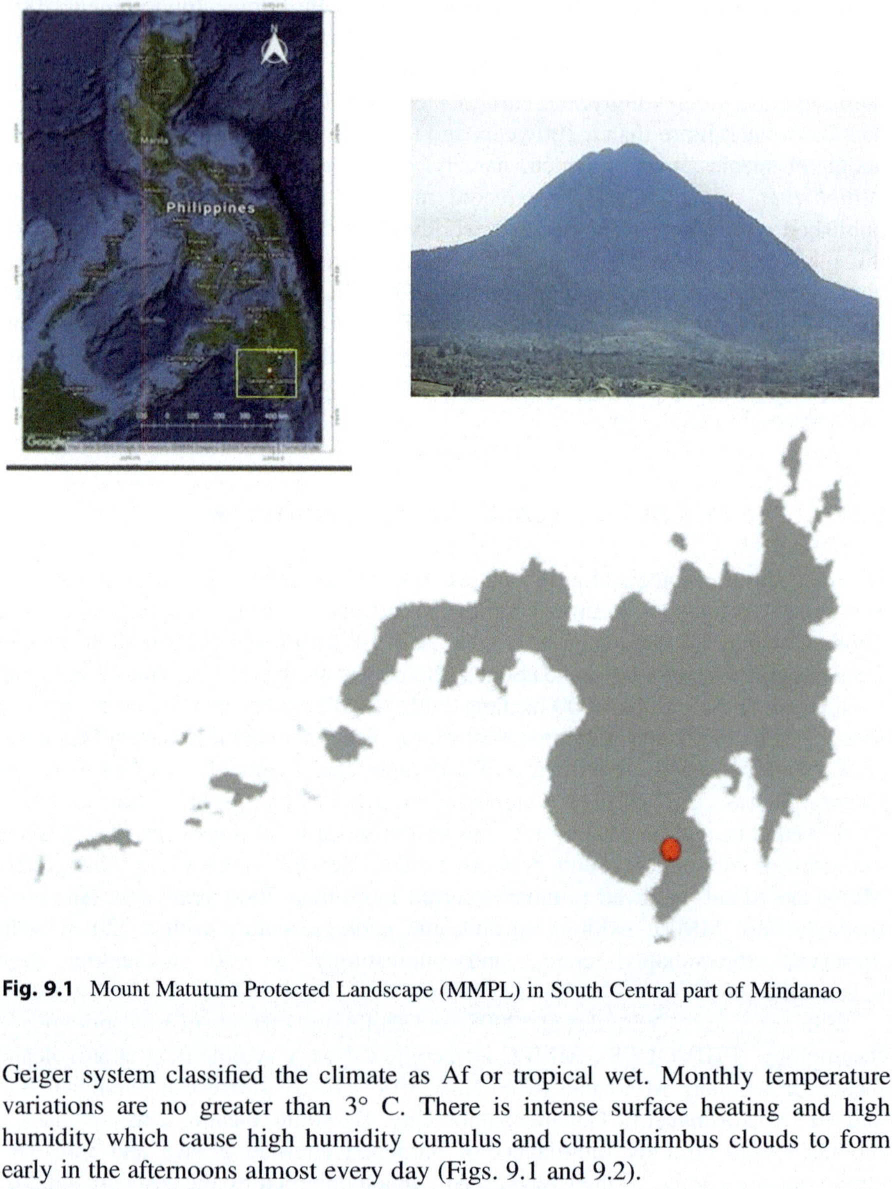

Fig. 9.1 Mount Matutum Protected Landscape (MMPL) in South Central part of Mindanao

Geiger system classified the climate as Af or tropical wet. Monthly temperature variations are no greater than 3° C. There is intense surface heating and high humidity which cause high humidity cumulus and cumulonimbus clouds to form early in the afternoons almost every day (Figs. 9.1 and 9.2).

The protected forest area boundary is at 1300–2286 masl. Site 1 (Glandang trail) has been the official monitoring area of the Department of Environment and Natural Resources. Several monitoring stations have been marked which are assessed quarterly. Also, two one-hectare plots have been permanently delineated in MMPL which are also assessed at least once a year by DENR (Fig. 9.3).

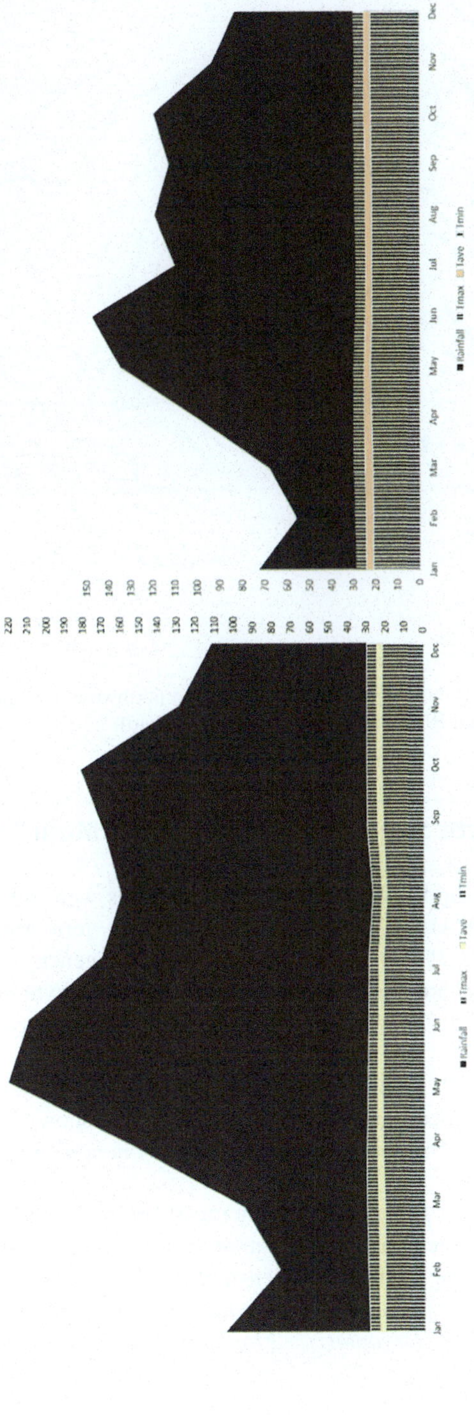

Fig. 9.2 Climate Measurements for Temperature and Precipitation in two sites (A1 Tupi) and B1 (Polomolok) and their Average Temperature (A2, B2) (© (Climate-Data.Org n.d.)/AM OP/OpenStreetMap contributors)

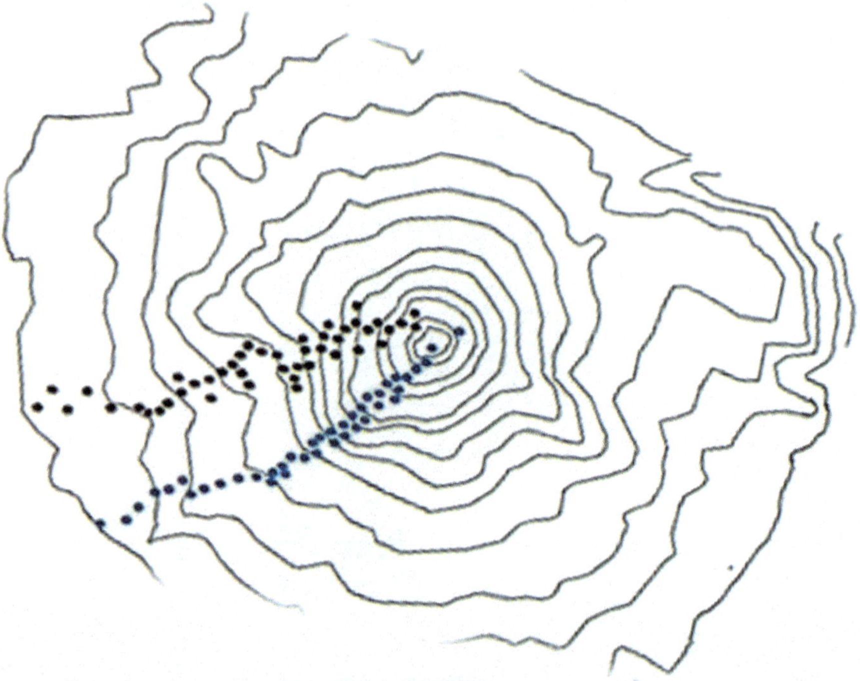

Fig. 9.3 Topographic map showing the study sites in the northwestern slope of MMPL (Black dots—Tupi, South Cotabato; Blue—Polomolok, South Cotabato)

9.3 Pteridophyte Species Composition in Mount Matutum

A total of 265 species under 97 genera and 30 families were observed in MMPL (Table 9.1 and Fig. 9.4). Of these, 21 species were fern allies belonging to *Lycopodiella*, *Huperzia*, *Phlegmariurus,* and *Selaginella* genera under 2 lycophyte families while the rest are distributed in the 26 monilophyte families. Around 217 were identified up to the species level, while 48 were named up to the genus level.

The most species-rich families were *Polypodiaceae* and *Aspleniaceae* with 52 and 30 species, respectively. Three other families showed >20 species in composition namely: *Dryopteridaceae* (18), *Hymenophyllaceae* (24), and *Pteridaceae (20)*.

Asplenium and those in *Polypodiaceae* are mostly epiphytes. They are among the families of plants identified to contain native and ecologically important species (Amoroso et al. 2011) as well as threatened flora (Fernando et al. 2008). Other genera with epiphytic species are *Abrodictyum*, *Antrophyum, Calymmodon*, *Callistopteris, Davallia*, *Dryopteris*, *Elaphoglossum*, *Haplopteris*, *Huperzia*,

Table 9.1 Pteridophyte families-species composition in Mount Matutum protected landscape showing cover and frequency values, distribution, and conservation status

Pteridophyte families and species composition	Total cover	Frequency	Distribution
I. Lycopodiopsida (Fern Allies)			
1. Lycopodiaceae			
Huperzia javanica (Sw.) Fraser-Jenk	100	0.07	Widespread
Huperzia serrata (Thunb.) Trev	30	0.019	Southeast Asia
Lycopodium clavatum L	51	0.019	Widespread
Palhinhaea cernua (L.) Vasc & Franco	162	0.07	Tropics, Mindanao
Phlegmariurus banayanicus (Herter) A.R. Field & Bostock	151	0.019	Endemic
Phlegmariurus delbrueckii A.R. Field & Bostock		–	Endemic
Phlegmariurus pinifolius (Trevis) Kiew	316	0.096	Southeast Asia
Phlegmariurus squarrosus (G. Forst.) A. Love & D. Love**	52	0.038	Australia, Southeast Asia
Phlegmariurus verticullatus (L.f) A.R. Field *Pseudodiphasium volubile* (G. Forst) Holub	15	0.019	Africa, Australia, Southeast Asia Widespread
2. Selaginellaceae			
Selaginella boninensis Baker	357	0.058	Taiwan, Vietnam, Philippines
Selaginella biformis A. Br. Ex Kuhn	35	0.019	Widespread -India, China, New Guinea, Philippines
Selaginella cupressina (Willd) Spring	28	0.019	Widespread-Moluccas, Java, Sulawesi, Borneo, Philippines
Selaginella cf. *delicatula* (Desv.) Alston	16	0.019	WS India-China, New Guinea, Philippines
Selaginella cf. *denticulata* (L.) Spring	82	0.038	India-China, New Guinea, Philippines
Selaginella elmeri Hieron	533	0.109	Endemic (Luzon, Mindanao)
Selaginella engleri Hieron	50	0.038	Widespread—Borneo, Philippines
Selaginella gastrophylla Warb	115	0.122	Borneo, Philippines
Selaginella involvens (Sw.)Spring	104	0.077	Widespread in Aisa
Selaginella magnifica Hieron**	48	0.038	Endemic
Selaginella plana (Desv. Ex Poir) Hieron	120	0.077	Widespread
Selaginella remotifolia Spring	31	0.049	Widespread
Selaginella cf. *ornata* (Hook&Grev) Spring	–		Asia
II. Polypodiopsida(Ferns)			
1. Aspleniaceae			
Asplenium affine Sw	–		Widespread
Asplenium apoense Copel	48	0.019	Endemic

(continued)

Table 9.1 (continued)

Pteridophyte families and species composition	Total cover	Frequency	Distribution
Asplenium callipteris (Fee) Baker	149	0.122	Endemic
Asplenium cf *bulbiferum* G. Forst	139	0.122	Asia
Asplenium cf. *pacificum* Holttum	20	0.024	Polynesia
Asplenium cf. *rhizophyllum* L	15	0.024	America
Asplenium cuneatum Lam	18	0.024	Widespread
Asplenium cymbifolium Christ	309	0.13	Borneo, Philippines
Asplenium elmeri Christ	289	0.163	Borneo, Philippines
Asplenium cf. *haenkei* Copel	–		Moluccas, Philippines
Asplenium horridum Kaulf	663		Moluccas, New Guinea, Pacific, Solomon Islands, Philippines
Asplenium laserpitiifolium Lam	–		
Asplenium lepturus J. Sm	–		China, Laos, Sulawesi, Vietnam, Philippines
Asplenium lobulatum Mett	1178	0.272	New Guinea, Pacific, Solomon Isls., Sulawesi, Philippines
Asplenium longgisimum Blume	579	0.122	Southeast Asia
Asplenium nidus L	1847	0.288	Widespread
Asplenium nigrescens Blume	3189	0.38	Southeast Asia
Asplenium normale D. Don	204	0.077	Widespread
Asplenium pellucidum Lam	181	0.077	Australia, Borneo, South Asia, Madagascar, Micronesia, Philippines
Asplenium persicifolium Bedd	417	0.096	Borneo, Sulawesi, Philippines
Asplenium phyllitidis D. Don	32	0.019	Southeast Asia
Asplenium polyodon G. Forst	49	0.077	Australasia, Micronesia, New Caledonia, New Guinea, Philippines
Asplenium tenerum G. Forst	32	0.077	Widespread
Asplenium tenuifolium D. Don	–		Southeast Asia
Asplenium thunbergii Kunze	2131	0.348	Asia, Mindanao,Philippines
Asplenium vittaeforme Cav*	261	0.043	Asia
Hymenasplenium excisum (C. Presl) S. Linds	2853	0.38	Australia, Asia
Hymenasplenium subnormale Copel	75	0.038	Borneo, China, Moluccas
Asplenium unilaterale (Lam.) Hayata	578	0.058	Asia, Africa
Asplenium sp. *1*	348	0.019	
Asplenium sp. *2*	–		
2. Athyriaceae			
Athyrium drepanopterum (Kunze) A.Br.ex Milde	707	0.163	Southeast Asia
Athyrium elmeri Copel	42	0.096	Endemic
Athyrium nakanoi Makino**	60	0.058	Asia, Mindanao
Athyrium puncticaule (Blume) T. Moore	45	0.038	Asia

(continued)

Table 9.1 (continued)

Pteridophyte families and species composition	Total cover	Frequency	Distribution
Cornopteris banaohensis (C. Chr.) K. Iwats & M.G.Price	–		South Asia
Cornopteris decurrenti-alata (Hook) Nakai	49	0.049	Asia
Cornopteris opaca (D. Don) Tagawa	–		Asia
Cornopteris philippinensis M. Kato	80	0.073	Moluccas, Philippines
Diplazium calliphyllum M.G. Price	0		Moluccas, Philippines
Diplazium dilatatum Blume	4204	0.25	Asia, Australia
Diplazium esculentum (Retz.) Sw	158	0.022	Asia, Australia
Diplazium forbesii C. Chr	–		Borneo, Java, Sumatra, Philippines
Diplazium geophilum Alderw	340	0.065	Moluccas, Sulawesi, Philippines
Diplazium pallidum T. Moore	72	0.077	Australia, Asia
Diplazium pseudocyatheifolium Rosenst	53	0.038	Endemic
Diplazium sibuyanense Alderw	202	0.109	Endemic
Diplazium sorzogonense (C. Presl.) C. Presl	52	0.058	Southeast Asia
Deparia sp.	18	0.024	
Diplazium sp.	–		
Diplazium sp. 1	–		
3. Blechnaceae			
Austroblechnum melanocaulon (Brack.) Gasper & V.A.O Dittrich	2710	0.239	Asia, Philippines (Mindanao)
Blechnopsis orientalis L. C. Presl			Asia, Philippines (Mindanao)
Cranfillia vulcanica (Blume) Gasper & V.A.O Dittrich	37	0.019	Southeast Asia
Parablechnum vestitum (Blume) Gasper and Salino	189	0.043	Asia, Philippines (Mindanao)
Stenochlaena palustris (Burm.f) Bedd	107.1	0.038	Asia
Brainea cf. *insignis*		–	
Blechnopsis sp.		–	
4. Cibotaceae			
Cibotium cf. *barometz* (L.) J. Sm**	49	0.019	China, West Malesia, Philippines
5. Cyatheaceae			
Alsophila apoensis (Copel.) R.M. Tryon**	60.96	0.073	Endemic
Alsophila fuliginosa Christ*	2725	0.239	Endemic
Alsophila hermannii R.M. Tryon**	2445	0.272	Endemic

(continued)

Table 9.1 (continued)

Pteridophyte families and species composition	Total cover	Frequency	Distribution
Alsophila heterochlamydea (Copel.) R.M.Tryon*	2929	0.272	Endemic
Sohaeropteris glauca (Wall ex Hook) Copel**	3262	0.293	New Guinea to India, Philippines
Sphaeropteris elmeri (Copel.) R. M.Tryon*	83	0.058	Moluccas, Sulawesi, Philippines
Sphaeropteris lepifera (Hook.) R. M.Tryon**	22.24	0.049	East Asia, Philippines
Sphaeropteris tripinnata (Copel.) R.M.Tryon	1254	0.217	Borneo, Java, Moluccas, Malaysia, Sulawesi, Philippines (widespread)
Alsophila sp.	–		
6. Davalliaceae			
Davallia repens (L.f.) Kuhn	273.24	0.038	Asia, Australia, Philippines (Mindanao)
Davallia wagneriana Copel	212	0.022	Moluccas, Sulawesi, Sumatra, Philippines (Mindanao)
Davallia denticulata (Burm.f.) Mett. ex Kuhn	483	0.065	Australia, Asia, Philippines (Mindanao)
Davallia solida (L.f.) Kuhn	48.12	0.038	Australia, Asia, Philippines (Mindanao)
Davallia trichomanoides Blume	–		Asia
7. Dennstaedtiaceae			
Dennstaedtia cf. *hooveri* Christ	61	0.019	Moluccas, Philippines (Mindanao)
Dennstaedtia scandens (Blume) T.Moore	–		Asia, Philippines (Mindanao)
Histiopteris incisa (Thunb.) J.Sm	544	0.154	Widespread, Philippines (Mindanao)
Microlepia strigosa (Thunb.) C. Presl	45	0.038	Asia
Monachosorum henryi Christ	3346	0.348	Asia
Monachosorum subdigitatum (Blume) Kuhn	–		Asia
Pteridium aquilinum (L.) Kuhn***	216	0.038	Widespread
8. Dicksoniaceae			
Dicksonia amorosoana Lehnert & Coritico	67	0.073	Endemic
Dicksonia mollis Holttum**	660	0.109	Borneo, Philippines
9. Diplaziopsidaceae			
Diplaziopsis javanica (Blume) C. Chr	1718	0.217	Asia
10. Dipteridaceae			
Dipteris conjugata Reinw	200	0.022	Widespread
11. Dryopteridaceae			
Arachniodes amabilis (Blume) Tindale	559	0.098	Asia, Philippines (Mindanao)

(continued)

Table 9.1 (continued)

Pteridophyte families and species composition	Total cover	Frequency	Distribution
Bolbitis heteroclita (C. Presl) Ching	278.7	0.073	Asia, Philippines
Bolbitis rhizophylla (Kaulf.) Hennipman	30	0.038	Taiwan, Philippines
Bolbitis quoyana (Gaudich) Ching	200	0.054	Australia, Asia
Dryopteris hendersonii (Bedd.) C. Chr	140	0.038	Asia, Philippines
Dryopteris pseudocaenopteris (Kunze) Li Bing	393	0.038	Asia
Dryopteris hirtipes (Blume) Kuntze	744	0.038	Asia
Dryopteris sparsa (D. Don) Kuntze	386	0.065	Widespread
Dryopteris subarborea (Baker) C. Chr	174	0.033	Asia, Melasia
Dryopteris nodosa (C. Presl) Li Bing Zhang	118	0.043	Asia, Melasia
Dryopteris permagna M. Price** *Dryopteris purpurascens* (Blume) Christ	–		Endemic
Elaphglossum blumeanum (Fee) J. Sm	1878	0.283	Asia, Melasia
Elaphoglossum caliifolium (Blume) T. Moore	1242	0.239	Asia, Melasia, Philippines (Mindanao)
Polystichum moluccense T. Moore**	1702	0.141	Asia, Melasia
Polystichum elmeri Copel	51	0.022	Endemic (Mindanao)
Polystichum biaristatum (Blume) T. Moore	296	0.054	Southeast Asia
Teratophyllum aculeatum (Blume) Mett ex Kuhn	–		Southeast Asia
Arachniodes sp. 2	384	0.122	
Arachnoides sp. 3	128	0.073	
Bolbitis sp.	–		
Lommogramma sp. *Polystichum* sp. 2	– 670	0.043	
12. Gleicheniaceae			
Dicranopteris linearis (Burm) Underw *Dicranopteris linearis var. rigida* (Blume) Holttum *Gleichenia peltophora* Copel	716 31	0.049 0.024	Widespread Borneo, Sulawesi, Philippines Borneo, Sulawesi, Philippines
13. Hymenophyllaceae			
Abrodictyum pluma (Hook.) Ebihara & K.Iwats	620	0.098	Asia, Melasia

(continued)

Table 9.1 (continued)

Pteridophyte families and species composition	Total cover	Frequency	Distribution
Abrodictyum cumingii C. Presl	–		Moluccas, New Guinea, Solomon Isls., Sulawesi, Taiwan, Philippines
Abdrodictyum obscurum (Blume) Ebihara & K. Iwats	61	0.054	Australia, Asia
Callistopteris apiifolia (C. Presl) Copel	654	0.109	Asia, Melasia
Crepidomanes bipunctatum (Poir.) Copel	24	0.024	Australia, Asia
Crepidomanes brevipes (C.Presl) Copel	18	0.024	Asia, Micronesia
Crepidomanes minutum (Blume) K. Iwats	–		Australia, Asia
Crepidomanes grande (Copel.) Ebihara & K. Iwats	36	0.024	Bornea, Micronesia, Melasia, Southeast Asia
Hymenophyllum cf. *acanthoides* (Bosch) Rosent	78	0.024	Asia
Hymenophyllum cf. *badium* Hook. & Grev	144	0.033	Asia, Micronesia
Hymenophyllum cf. *blandum* Racib	–		Asia
Hymenophyllum cf. *ramosii* Copel	–		Endemic
Hymenophyllum denticulatum Sw	55	0.019	*Asia*
Hymenophyllum emarginatum Sw	21	0.019	Asia
Hymenophyllum fimbriatum J. Sm	464	0.163	Asia
Hymenophyllum cf. *holochilum* (Bosch) C. Chr	–		Australia, Asia
Hymenophyllum imbricatum Blume	788	0.152	Asia
Hymenophyllum cf. *nitidulum* (Bosch) Ebihara & K. Iwats	32	0.024	Asia
Hymenophyllum cf. *pallidium* (Blume) Ebihara & K. Iwats	476	0.163	Australia, Asia, Micronesia
Hymenophyllum cf. *serrulatum* (C. Presl) C. Chr	90	0.033	Asia, Micronesia
Hymenophyllum cf.*thiudium* Harrington	70	0.022	Borneo, New Guinea, Philippines
Hymenophyllum cf. *polyanthos*(Sw.) Sw	–		Widespread
Vandenboschia auriculata (Blume) Copel	36	0.019	Asia
Vandenboschia maxima (Blume) Copel	–		Micronesia, Asia
14. Hypodematiaceae			
Leucostegia truncata (D. Don) Fraser-Jenk	386	0.154	Asia

(continued)

Table 9.1 (continued)

Pteridophyte families and species composition	Total cover	Frequency	Distribution
15. Lindsaceae			
Lindsaea apoensis Copel	577	0.122	Endemic
Lindsaea pulchella (J. Sm.) Mett. Ex Kuhn	40	0.024	Southeast Asia, Philippines (Mindanao)
Odontosoria chinensis (L.) J. Sm	62	0.073	Asia
Odontosoria retusa (Cav.) J. Sm	11	0.024	Asia, Micronesia
Osmolindsaea sp.	–		
Osmolindsaee odorata (Roxb.) Lehtonen & Lehtonen	15	0.024	Southeast Asia
Tapeinidium gracile (Blume)Alderw	26	0.024	Southeast Asia
Tapeinidium pinnatum (Cav) C. Chr	58	0.049	Asia
Tapeinidium sp.	–		
Lindsaea sp. 1	15	0.024	
Lindsaea sp. 2	–		
16. Lygodiaceae			
Lygodium japonicum (Thunb.) Sw***	15	0.024	Widespread
17. Mariatiaceae			
Angiopteris evecta Sw	1390	0.272	Mindanao
Angiopteris anguistifolia C. Presl	245	0.058	Southeast Asia
Ptisana pellucida (C. Presl) Murdock	1554	0.228	Sulawesi, Taiwan, Vietnam, Philippines
Ptisana sylvatica (Blume) Murdock	490	0.163	South Asia
18. Nephrolepidaceae			
Nephrolepis cordifolia (L.) C. Presl	562	0.087	Australia, Southeast Asia
Nephrolepis falcata (Cav.) C. Chr	431	0.058	Southeast Asia
Nephrolepis flexuosa Colenso	-		Polynesia, Australia
Nephrolepis hirsutula (G. Forst.) C Pres	84	0.038	Asia, Micronesia, Melasia
19. Oleandraceae			
Oleandra pistillaris (Sw.) C. Chr	741	0.043	Southeast Asia
Oleandra sibbaldi Grev	25	0.024	Polynesia, Southeast Asia
20. Osmundaceae			
Plenasium banksiifolium (C. Presl) C. Presl	20	0.024	Asia
21. Ophioglossaceae			
Ophioderma pendulum L.	289	0.087	Old World Tropics (Madagascar to Tropical Asia to Polynesia)
Botrychium daucifolium Wall ex Hook & Grev/	610	0.13	Asia, Micronesia

(continued)

Table 9.1 (continued)

Pteridophyte families and species composition	Total cover	Frequency	Distribution
22. Plagiogyriaceae			
Plagiogyria glauca (Blume) Mett	1391	0.239	South Asia
Plagiogyria pcnophylla (Kunze) Mett	728	0.12	Southeast Asia
Plagiogyra euphlebia (Kunze) Mett	530	0.152	Southeast Asia
23. Polypodiaceae			
Aglaomorpha sagitta (Christ) Hovenkamp & S. Linds*	101	0.038	Endemic
Calymmodon gracilis (Fee) Copel	273	0.195	Asia, Australia, Micronesia
Calymmodon gracillimus (Copel) Nakai ex H. Ito	15	0.0192	Malesia (except Peninsular Malaysia and Lesser Sunda islands)
Chrysogrammitis glandulosa (J. Sm.) Parris	54	0.024	Borneo, Malaysia, Sri Lanka, Sumatra, Taiwan, Philippines
Chrysogrammitis sp.	5	0.024	
Dasygrammitis malaccana (Baker) Parris	48	0.077	Southeast Asia
Drynaria aglaomorpha Christenh*	134	0.038	Sulawesi, Philippines
Drynaria cornucopia (Copel.) Alderw*	15	0.038	Endemic
Drynaria descensa Copel	75	0.019	Endemic
Goniophlebium persicifolium (Desv.) Bedd	1245	0.174	Southeast Asia, Solomon Islands
Goniophlebium pseudoconnatum (Copel) Copel	1114	0.25	Southeast Asia, Philippines—Luzon, Mindanao
Goniophlebium subauriculatum (Blume) C. Presl	–		Southeast Asia, Solomon Islands, Australia, New Caledonia
Lepisorus accedens (Blume) Hosok	–		Southeast Asia, Australia, Micronesia
Lepisorus mucronatus (Fee) Li Wang	40	0.019	Southeast Asia, Micronesia, Philippines - Palawan, Mindanao
Lepisorus sp.	737	0.141	Southeast Asia
Lepisorus spicatus (L.f.) Li Wang	163	0.022	Africa, Southeast Asia, Melanesia, New Caledonia
Lepisorus validinervis (Kunze) Li Wang	–		Southeast Asia
Leptochilus insignis (Blume) Fraser-Jenk	–		Southeast Asia
Loxogramme avenia (Blume) C. Presl	11	0.011	Southeast Asia, Micronesia, Philippines—Palawan, Mindanao
Loxogramme cf. *salicifolia* (Makino) Makino	158	0.087	East and Southeast Asia

(continued)

Table 9.1 (continued)

Pteridophyte families and species composition	Total cover	Frequency	Distribution
Loxogramme paralella Copel	13	0.011	Southeast Asia, Philippines -Luzon, Mindanao
Oregrammitis beddomeana (Alderw) T.C. Hsu	75	0.033	Southeast Asia, Philippines - Luzon, Mindanao
Oreogrammitis jagoriana (Mett ex Kuhn) Parris & Sundue	16	0.024	Borneo, Solomon Isls, Sulawesi, Thailand, Vietnam, Philippines
Oreogrammitis reinwardtii (Blume) Parris	15	0.024	Australia, Malesia, Melanesia, Polynesia
Oreogrammitis setigera (Blume) T.C. Hsu	674	0.065	Southeast Asia, Philippines
Oreogrammitis sumatrana (Baker) Parris	58	0.024	Southeast Asia
Oreogrammitis tenuisecta (Blume) T.C. Hsu	–		Southeast Asia
Oreogrammitis torricelliana (Brause) Parris	18	0.024	Philippines, New Guinea, Solomon Islands
Prosaptia celebica (Blume) Tagawa & K.Iwats			Southeast Asia
Prosaptia contigua (G. Forst.) C. Presl	77	0.019	Southeast Asia, Australia, New Caledonia, Solomon Islands
Prosaptia multicaudata (Copel.) Parris			Malesia, Australia, Polynesia
Prosaptia obliquata (Blume) Mett	849	0.13	Southeast Asia
Prosaptia venulosa (Blume) M.G. Price	41	0.024	Southeast Asia
Scleroglossum pusillum (Blume) Alderw	18	0.019	Southeast Asia
Tomophyllum macrum (Copel.) Parris	28	0.024	Endemic
Tomophyllum millefolium (Blume) Parris	15	0.024	Southeast Asia
Phlebodium sp. 1	60	0.022	
24. Pteridaceae			
Adiantum hosei Baker	966	0.13	Borneo, Sulawesi, Philippines
Adiantum capillus-veneris L *Adiantum diaphanum* Blume	15 –	0.024	Widespread Australia, Southeast Asia, Melanesia
Adiantum aff. mindanaoense Copel**	–		Moluccas, New Guinea, Philippines
Antrophyum latifolium Blume	89	0.033	Southeast Asia
Antrophyum parvulum Blume	26	0.024	Africa, Southeast Asia
Antrophyum reticulatum (G. Forst) Kaulf	134	0.065	Southeast Asia
Antrophyum sessifolium (Cav.) Spreng	300	0.13	Borneo, Sulawesi, Taiwan, Philippines

(continued)

Table 9.1 (continued)

Pteridophyte families and species composition	Total cover	Frequency	Distribution
Antrophyum sp.	86	0.058	
Haplopteris ensiformis (Sw.) E.H. Crane	37	0.019	Southeast Asia, Melanesia
Haplopteris alternans (Copel.) S. Linds. & C.W. Chen	22	0.024	Sulawesi, Mindanao
Haplopteris scolopendrina (Bory) C. Presl	153	0.058	Southeast Asia
Haplopteris aff. mindanaoensis S. Linds & C.W. Chen	–		Endemic
Pityrogramma calomelanos (L.) Link	–		America, Mindanao
Pteris longipinnula Wall ex. J. Agardh	–		Southeast Asia
Pteris oppositipinnata Fee	1212	0.217	Borneo, Sulawesi, Philippines
Pteris biaurita L.	–		Widespread
Pteris schlecteri Brause	54	0.024	Moluccas, New Guinea, Sulawesi, Philippines
Pteris tripartita Sw	477	0.135	
Pteris whitfordii Copel	389	0.076	Endemic
Vaginularia junghunii Mett	221	0.038	Southeast Asia, Mindanao
25. Psilotaceae			
Tmesipteris zamorae Gruezo & Amoroso**	521	0.154	Endemic
26. Thelypteridaceae			
Chingia ferox (Blume) Holttum	–		Southeast Asia
Macrothelypteris polypodioides (Hook) Holttum	62	0.038	Australia, Asia
Phegopteris cf. *paludosa*	–		Southeast Asia
Pronephrium nitidum Holttum	–		Widespread
Sphaerostephanos unitus (L.) Holttum	34	0.019	Widespread
Sphaerostephanos urdanetensis (Copel.) Holttum	55	0.019	Endemic
Christella acuminata (Hout) Holttum	550	0.154	Southeast Asia
Christella dentata (Forssk.) Brownsey & Jermy	48	0.019	Widespread
Coryphopteris squamipes (Copel) Holttum	–		
Coryphopteris pubirachis (Baker) Holttum	–		
Pneumatopteris laevis (Mett.) Holttum	551	0.13	Endemic
Pneumatopteris nitidula (C. Presl) Holttum	–		Endemic

(continued)

Table 9.1 (continued)

Pteridophyte families and species composition	Total cover	Frequency	Distribution
27. Tectariaceae			
Tectaria melanocaulos (Blume) Copel	–		Southeast Asia
Tectaria dissecta (G. Forst) Lellinger			Southeast Asia

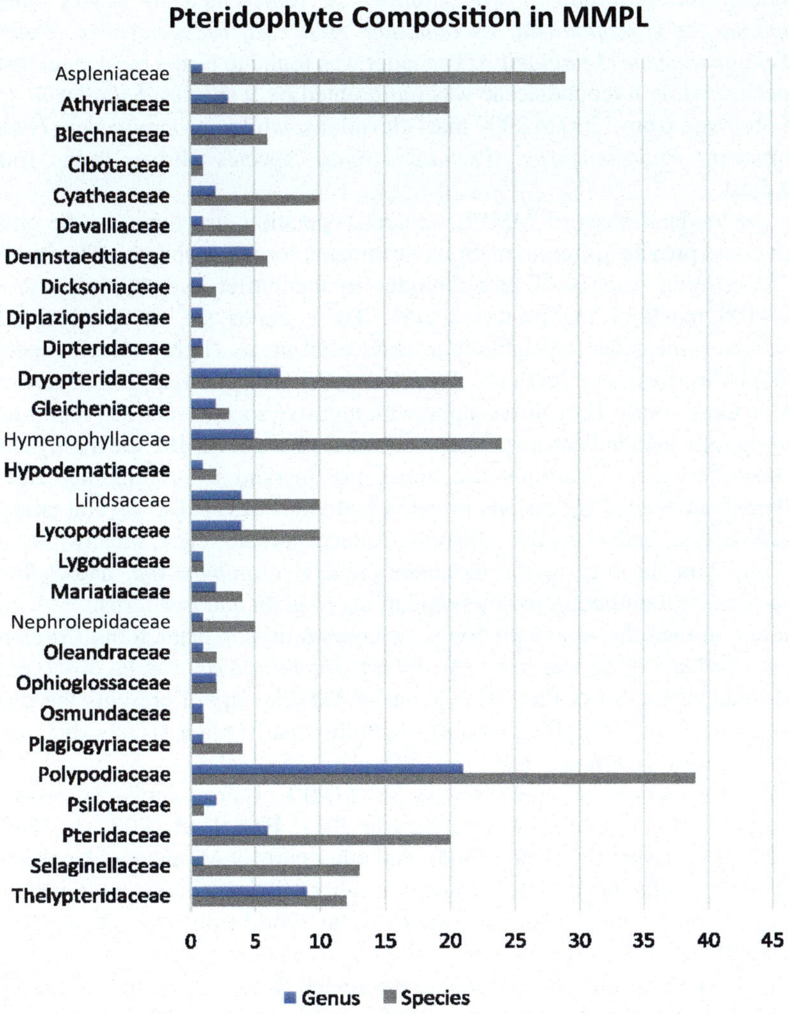

Fig. 9.4 Family-Genus-Species Composition of Pteridophytes in MMPL

Lindsaea, *Lycopodiella*, *Nephrolepis*, *Oleandra*, *Ophioglossum*, *Osmolindsaea*, *Phlegmariurus*, *Polystichum*, *Teratophyllum*, and *Tmesipteris*. Among these, *Tmesipteris zamorae* and *Polystichum moluccense* were found to exclusively prefer tree ferns as their habitat. Meanwhile, the lithophytic community is composed of species in the following genera: *Adiantum, Antrophyum, Austroblechnum, Bolbitis, Hymenophyllum, Nephrolepis, Pyrossia, Selaginella,* and *Thylacopteris*. Two genera, *Oleandra* and *Davallia,* are found to be hemi-epiphytes, exhibiting both epiphytic and terrestrial habits. Tree ferns under Alsophila (4 spp.), Sphaeropteris (4 spp.), and Dicksonia (2 spp.) and giant ferns under Angiopteris were the tallest terrestrial ferns exhibiting tree-like habit.

Among the fern allies, two families are represented in MMPL, namely, *Lycopodiaceae* (*Lycopodiella, Lycopodium, Huperzia, Phlegmariurus, Palhinlea)* and *Selaginellaceae* (*Selaginella*). The latter was found to be the most abundant with 13 species, while Lycopodiaceae was represented by 9 species. *Selaginella* species were observed from 1200 to 2000 masl elevations, while *Lycopodiaceae (Huperzia, Palhinhaea, Phlegmariurus, Pseudophasium)* species thrive from 1600 to 1900 masl.

In the lowland areas of MMPL, several vegetation structures can be observed which could provide different microenvironments for pteridophytes. This study was able to observe patches of pteridophytes in the buffer and multiple use zones (900–1100 masl) of the protected area. These zones are occupied greatly by plantations (pineapple, vegetable) and residential areas. Highly tolerant species of pteridophytes such as *Christella, Pteridium, Nephrolepis, and Diplazium* can be found in these zones. They thrive along with invasive species such as *Piper aduncum* (buyo-buyo) which indicates that these sites at the base of MMPL are highly exposed and disturbed areas. Multiple use zones are marked by community and farm dwellings surrounded by various private plantations devoted to several crops such as corn, coffee, and vegetables (pepper, lettuce, tomato, etc.). Meanwhile, across steep cliffs on the way to the mountain, several *Osmunda spp.* and *Selaginella* species can be found bordered by populations of giant and tree ferns.

Going further the mountain forest, a community of giant ferns (*Angiopteris, Ptisana*) and emerging tree ferns (*Alsophila, Sphaeropteris*) can be observed. This is now part of the Strict Protection Zone of MMPL. Uphill presents the different pteridophyte flora forming community with the rest of plant groups such as trees, palms, and flowering understory.

The composition of pteridophytes in MMPL shares similarity with other landscapes in Luzon (Magtoto and Austria 2017; Barcelona 2003a, b; Barcelona et al. 2006), Visayas (Barcelona 2004), and other parts of Mindanao (Amoroso et al. 2011, 2016; Coritico et al. 2020). Moreover, these richness counts only validate the significant contribution of the Philippines in the world's diversity of pteridophytes (Table 9.2).

About 1100 species of pteridophytes under 144 genera and 3 families of lycophytes and 34 families of ferns are known to thrive in the Philippines (Amoroso et al. 2011, 2016; Delos Angeles and Buot 2012; Pelser et al. 2011; Smith et al. 2008). From this, it can be recognized that MMPL houses an interesting and

Table 9.2 Pteridophyte richness in Mount Matutum and other mountain ecosystems in the Philippines

Mountains	Number of taxon			Reference
	Species	Genera	Family	
Mt. Makiling, Laguna	305	103	31	Price (1975)
Mt. Pangasugan, Leyte	93	55	24	Belonias and Banoc (1994)
Mt. Santo Tomas, Benguet	100	59	21	Tolentino (1988)
Marilog Forest	183	78	28	Amoroso et al. (1996)
Mt. Apulang, Bukidnon	249	85	28	Amoroso et al. (1996)
Mt. Bali-it, Kalinga	167	67	30	Barcelona et al. (2003a, b)
Mt. Iraya, Batanes	89	47	25	Barcelona et al. (2003a, b)
Mt. Banahaw, Quezon	93	47	24	Banaticla and Buot (2005)
Rajah Sikatuna Protected Landscape, Bohol	169	63	27	Barcelona et al. (2006)
Mt. Kitanglad, Bukidnon	439	100	25	Amoroso et al. (2011)
Mt. Makulot, Batangas	40	25	13	Catapang et al. (2012)
Mt. Hamiguitan, Davao Oriental	155	72	27	Amoroso et al. (2009)
Adams,Cordillera Mountain Range	47	34	21	Magtoto and Austria (2017)
Mt. Tago Range, Bukidnon	203	89	29	Coritico et al. (2020)
Mt. Matutum, Socsargen	265	86	28	This study

significant diversity of pteridophytes, constituting 24% and 43% of the total species composition in the country and Mindanao Island, respectively.

Species composition of pteridophytes in MMPL presents the floral multiplicity in the protected area, comparable with other terrestrial landscapes in the country. Moreover, the influence of topographic, atmospheric, and soil factors on the diversity of ferns and fern allies results in a variety of microenvironments that favor the growth of pteridophytes and create various niches for them which explains the preferential or restricted ranges of the majority of these species.

9.4 Species Diversity of Pteridophytes in Mount Matutum

Species diversity using richness and Shannon-Wiener (H') diversity index showed varying levels from base to summit (Figs. 9.5 and 9.6). The relationship of pteridophyte richness and altitude presented general fluctuation in richness across elevation; however, the highest number has been recorded with the 1500–2100 masl (Fig. 9.5). It can be observed that for both sites, there is the rise and fall of peaks, but generally the montane forest (1500–1900) harbors the most number of species.

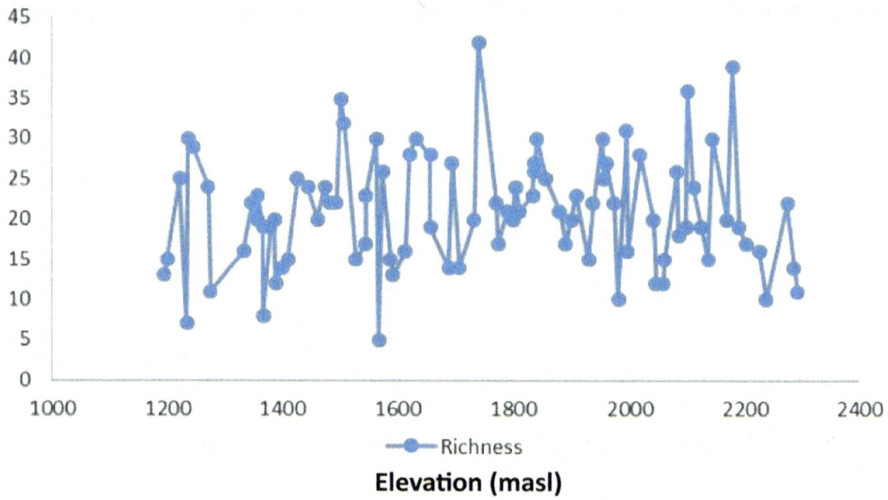

Fig. 9.5 Species richness in the NW slope of MMPL (1195–2286 masl)

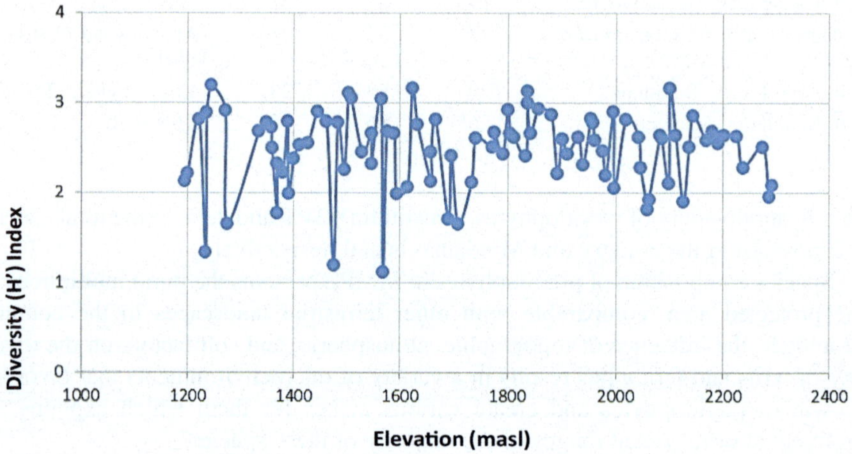

Fig. 9.6 Diversity (H′) in the NW slope of MMPL (1195–2286 masl)

Shannon-Wiener diversity index (Shannon and Weaver 1949) and evenness showed a range 1.199–3.194 and 0.3337–0.9194, respectively. It was further observed that most of the sites that have the lowest diversity were from base zone and the summit zone, while evenness was variable across elevation.

It can be gleaned that diversity peaks at certain points from 1200 to 1900 (Fig. 9.6). There are, however, sharp declines in some sites and the pattern generated is much like the ones presented for species richness. Moreover, moderate (>2.5–3.0)

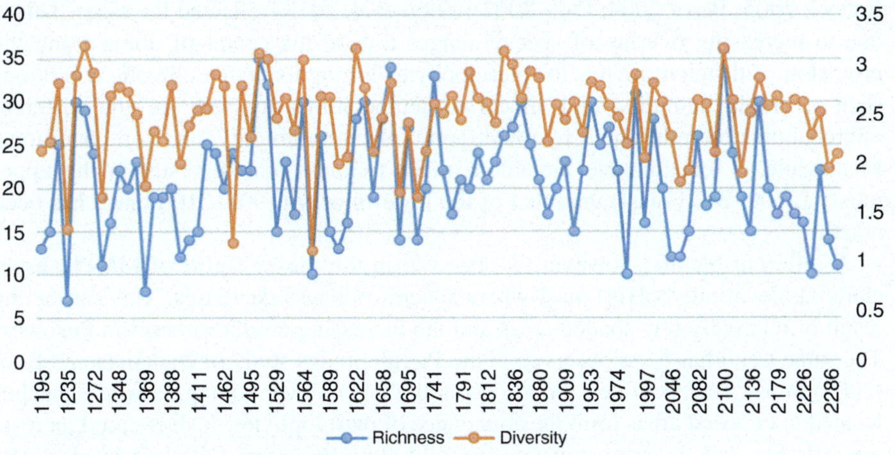

Fig. 9.7 Relationship of Species Richness and Diversity (H′)

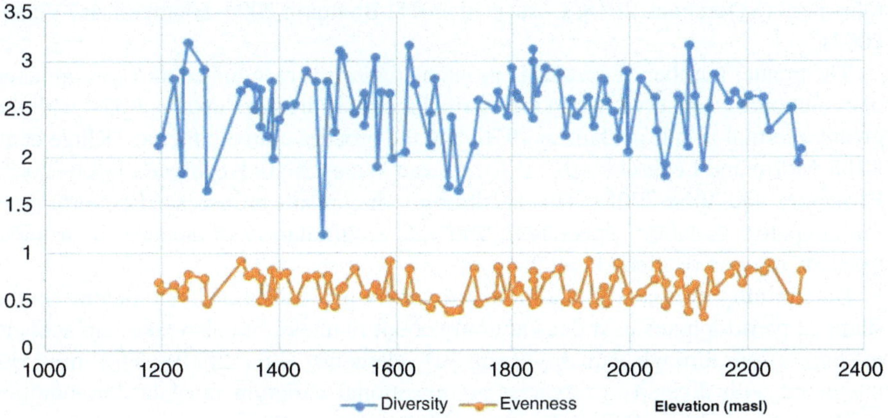

Fig. 9.8 Comparison of diversity, and evenness across elevation gradients

to high (3.0–3.5) diversity of species was observed within 1500–1900 followed thereafter with a decreasing trend in the upper mossy forest (1900–2286 masl).

A direct relationship can be observed between diversity and richness where peaks of richness often meet that of diversity (Fig. 9.7). The same with evenness which shows congruence with diversity (Fig. 9.8).

The peak of species richness and diversity index of ferns and fern allies at middle elevations gives a hump-shaped pattern (Colwell & Lees 2000; Lou 2010; Othman et al. 2015; Zhao et al. 2005). Several studies (Catapang et al. 2012; Kessler et al. 2011; Jones et al. 2011; Tu et al. 2010; Kluge et al. 2006; Amoroso et al. 2011; Bhattarai et al. 2004; Hawkins et al. 2003; Kessler et al. 2011; McCain 2005;

Rahbek 2005; Perez 2004; Page 2002); Zhou et al. 2019) reflected the same. This is due to increasing overlap of species ranges toward the center of domain and the provision of different niches for pteridophytes through suitable climatic conditions such as high humidity and moderate temperatures as well as the different microhabitats brought about by topography, plant community, and soil conditions. Meanwhile, a recognizable reduction in the richness and diversity in the upper mossy (2100–2200 masl) than that of the lower mossy (1900–2100 masl) has been noted.

A different picture, however, is observed in the mossy forest of MMPL, especially at elevations >2100 masl where species richness decreases. This can be the result of relatively few shaded areas and the increasing abiotic stresses in this zone. The same trend has been observed in the Pteridophytes study of Banaticla and Buot (2005) in Mt. Banahaw and (Amoroso et al. 2011) in Mt. Kitanglad where patches located in exposed areas limit the abundance of pteridophytes. It showcased increasing richness and diversity towards the mid-altitude occurs followed by declines. Interestingly, different peaks of richness and diversity occur at mid-altitude range. This unique pattern is the result of the landscape's topography and other environmental gradients, a result of their different ecophysiological processes and climatic tolerances (Syfert et al. 2018; Costa et al. 2018; Bhattarai 2003; Bhattarai and Vetaas 2003).

The greater number of species in montane regions is recognized as a consequence of evolutionary conservatism in the thermal niches of tropical taxa, coupled with the greater thermal zonation (Janzen 1976) and the broader climatic factors (Kluge et al. 2017; Morin and Lechowicz 2011; Kozak and Wiens 2010; Kozak and Wiens 2007; Fernández and Vrba 2005). This accordingly may result in increased opportunities for allopatric isolation, speciation, and the accumulation of species in tropical montane regions (Cadena et al. 2012).

Determining evenness in addition to diversity provides a concrete picture on the status of pteridophytes as it does not only count richness, but also take into account its proportion through equality/inequality elements (Lou 2010). This must be integrated with diversity to express compositional variation in plant communities (Gillespie et al. 2008; Hill 1973). Jost (2010) explains the importance of evenness measures much like the distribution of wealth in the community: that if species were households and abundance was wealth, then the distribution of wealth among these households (evenness) could provide a picture of how equal or unequal the proportions are with respect to the entire community. Hence, to understand biological diversity, the number of different species in a community and number of individuals within each of these species must be assessed (Leibold et al. 2004).

In MMPL, evenness measures have lows and highs which are variable across the landscape. This supplements the Shannon index by reiterating that highly diverse areas do not mean that cover or abundance was equitable. There are sites with high diversity, but only few species hold a large percentage of the entire cover. Factors that account for the low evenness in several sites were low slopes (<25), low temperatures, and soil pH.

Overall, diversity in most areas in MMPL have moderate (2.5–3.0) to high (3.0–3.5) (Fernando et al. 2008) H′ diversity index. This range is also observed in other ecological landscapes (Ifo et al. 2016; Amoroso et al. 2006, 2012; Belonias and Banoc 1994). High diversity in an ecosystem enhances its ability to withstand disturbances such as pollution, maintain soil fertility and microclimates, cleanse water, and provide other invaluable services (Goeltenboth et al. 1999). However, with the observed local threats in the area including small scale slash and burn activities, the occurrence of invasive species, and conversion of forest lands for residential areas, such high diversity may in the long run be put at risk if no sustainable measures for long-term conservation are undertaken (Gao et al. 2009).

9.5 Factors Influencing Species Diversity in Mount Matutum

Different microenvironments and abiotic factors may influence the fluctuations observed in species richness and diversity in MMPL across altitude. In fact, ferns and fern allies are known to be affected consistently by environmental factors primarily topographic and edaphic ones.

In Mount Matutum, diversity and richness patterns were influenced largely by slope, while evenness was attributed to pH and PO_4 levels. The strength of the relationship, however, needs to be drawn and further analysis is warranted (Table 9.3).

Several studies (Carvajal-Hernandez et al. 2017; Nettesheim et al. 2014; Yuan et al. 2014; Kessler et al. 2011; Banaticla and Buot 2005) also support the impact of slope in the composition and abundance of fern and fern allies. It has been observed that as the slope increases towards less steep gradients, richness and diversity increases. Variation in slope can affect the degree of moisture conservation through exposure to varying incidence angle of sunlight (Banaticla and Buot 2005). As such, increase in slope creates steeper incline which provide a unique environment for the growth of specialist species which may not be the case for flatter areas. Hence, diversity along this site can be high.

Soil pH showed direct relationship with evenness with evenly abundant areas having pH range of 5–6. pH influences the bioavailability of soil nutrients, soil microbial activity and crop growth and development in soil (Wang et al. 2019;

Table 9.3 Regression coefficients showing the relationship of dependent variables—diversity, richness, and evenness and significant independent variables—pH, Slope, NO_2, and PO_4

Dependent variables	Factors	Coeff	Std. err	t	p	R
Diversity	Constant	−2.532	6.083	−0.416	0.678	
	Slope	0.385	0.154	2.498	0.015	0.27019
Richness	Constant	52.931	93.956	−0.563	0.575	
Evenness	Constant	−0.621	2.241	−0.277	0.782	
	pH	−0.664	0.208	−3.189	0.002	0.25298
	PO_4	−0.101	0.034	−2.989	0.004	0.16125

Zhang et al. 2019; Kim-Bui et al. 2013; Fageria et al. 2011, 2014). It is also recognized as the greatest factor in the metal hyperaccumulation ability observed in some ferns which generally occurs in acidic soils (Kim-Bui et al. 2013).

pH was further associated with climatic factors where low pH is realized along low temperature and high humidity in higher areas. This along with the increasing environmental stresses like wind pressure and steep slopes substrates becomes like a filter on fern and lycophyte community. On the other hand, the upper mossy forest (2100–2200) in MMPL was observed to have many unique species found nowhere else in sites below. The same has been observed in the mountains of Southern Mindanao where species becomes very distinct and higher in number (Amoroso et al. 2011) as the altitude increases.

Phosphates (PO4) are essential for building the cell membranes and activate carbohydrates during respiration. It is also one of the mineral nutrients for soil fertility which is bound to organic matter in soil (Guignard et al. 2017). As such, their bioavailability is also dependent on factors that also affect levels of organic matter such as microbial degradation, temperature, soil pH, and relative humidity (Fig. 9.9).

9.6 Patch Structure of Pteridophytes

Community structure based on height and habit of pteridophytes was first presented by Banaticla and Buot (2005) in Mount Banahaw de Lucban, Quezon Province, Philippines. Based on this study, the patch community structure of pteridophytes in MMPL was analyzed. There are four general classes of patch, namely; Class 1 Terrestrial <100 cm tall (Selaginella spp.); Class 2 Terrestrial >100 cm tall (*Alsophila and Sphaeropteris* spp., *Angiopteris* spp.); Class 3 $<$ 100 cm Lithophytic and Epiphytic (Hymenophyllum spp.); Class 4 $>$ 100 cm Epiphytic on relatively high trees (Fig. 9.10).

Majority (64%) of the northwestern portion are represented by all the height and habit classes (Class 1–4). Transects lacking at least one class have generally lower diversity compared to those completely represented. In addition, sites where low diversity showed low percentage of class 2 height and habit forms and uneven percent share of classes 1–4. The lowest diversity was found at 1246, 1475 masl, and 2275 masl, while the highest diversity was recorded at 2100 masl where all classes are present.

There is a huge lithophytic and epiphytic community at higher elevations starting at 1700 masl towards the peak (2286). On the other hand, the lowland elevations <1700 are marked by the dominance of class 1 height and habit forms. These are the patches predominated by aggressive and highly tolerant species such as *Diplazium dilatatum, Pteris spp., Nephrolepis spp., Ptisana spp.*, and giant fern *Angiopteris evecta*. Class 2 demonstrates height and habit forms of large and tall ferns found to occupy wide range of elevation, but has second to the last percent share. It is especially profound along 1195–1462 masl marked mostly by *Alsophila and Sphaeropteris spp. Ptisana spp.* and *Angiopteris evecta*. At the peak, class 2 forms

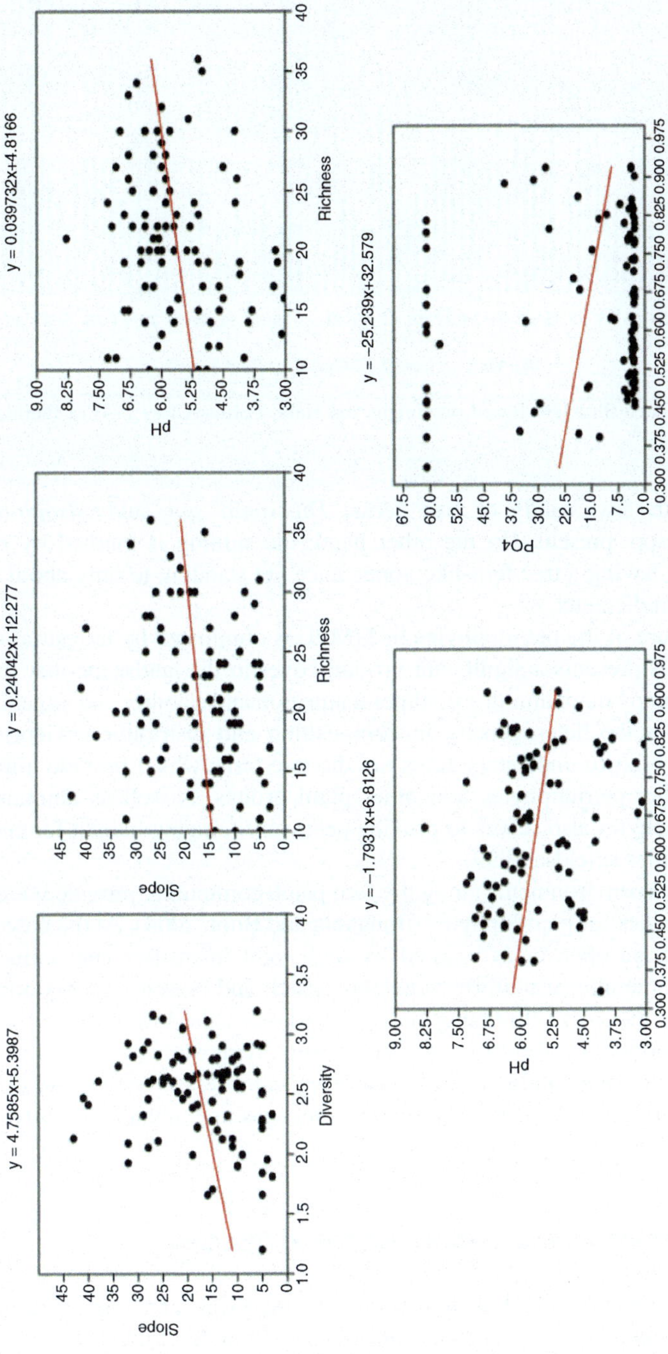

Fig. 9.9 Generalized linear model presenting extent of linear relationship among factors found with significant coefficients to possibly explain diversity, richness, and evenness across altitudinal gradient in MMPL

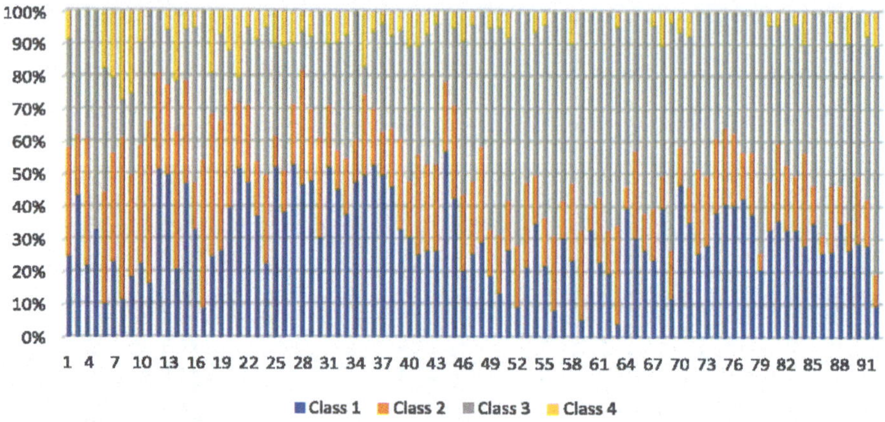

Fig. 9.10 Fern Patch Structure Based on Height and Habit Forms (Class 1–4) of Pteridophytes in MMPL

are sparse with few stands of tree ferns, *Dicksonia spp*. and *Alsophila* and *Sphaeropteris* spp. present. On the other hand, the summit is marked by minute *Coryphopteris*, having a tree fern-like appearance yet standing to only about 2 ft in height and 7 cm diameter.

Patch structure of the pteridophytes in MMPL is dominated by terrestrial understory ferns. This presents a significant ground cover in the landscape that forms a delicate sub-canopy maintaining soil mineral nutrition and stability and as ecological filter that shapes the heterogeneity in composition and distribution (George and Bazzaz 1999). Next to understory ferns are the tree ferns which provide important habitats for other pteridophytes and other plant groups as well as animals. The presence of epiphytes also serves to provide necessary foraging habitat for endemic avian communities (Fjeldsa 1999).

Different microenvironments along the fern patch community provide wide array of ecological niches for pteridophytes (Banaticla and Buot. 2004). At the base of the mountain are large open areas marked by high light intensities and temperature which explains the almost uniform vegetation patch and presence of opportunistic ferns especially those of scrambling habit. This is also the low-diversity areas as portions of pteridophytes were remarkably uneven. Meanwhile, the montane forest zones with lower temperatures and lower canopy openness influence the different patch community of ferns marked by greater richness and abundance (Silva et al. 2018).

9.7 Preferential and Widely Adapted Species

Pteridophytes with frequency less than or equal to 0.20 were considered preferential while those that exceed such value are referred to as widely adapted (Banaticla and Buot 2004). Few (8%) pteridophytes in Mount Matutum are widely adapted species, occurring in more than 18 designated sites. This indicates the narrow range distribution of most ferns and fern allies as presented by over 90% of the pteridophyte community (Table 9.4).

Most frequently encountered ferns are *Monachosorum henryii* ($r = 0.40$) and *Asplenium nigrescens* ($r = 0.38$). The former is terrestrial, while the latter is an epiphyte. Both were observed from 1600 to 2200 masl. Following them are *Hymenoasplenium excisum* and *Asplenium thunbergii,* having >0.30–0.40 frequency.

Widely adapted species in ferns only form a small proportion in the landscape which was the same observation in Mount Banahaw (Banaticla and Buot 2004). However, there are widely adapted species that have been tagged as problem ferns and their presence in the protected area in huge covers poses a threat to other patch structures especially when more areas become cleared by slash and burn practices or natural phenomenon (i.e., landslides). Example of these that can be found in MMPL are *Lygodium japonicum, Pityrogramma calomelanos, and Pteridium aquilinum* (Robinson et al. 2010).

Preferential or restricted-range species can represent the microenvironment specific to areas where they thrive and hence are sensitive to alterations in their environment. (Silva et al. 2018). Interestingly, they could serve as indicators in the

Table 9.4 Widely adapted Pteridophyte species in MMPL showing cover and frequency values

Species	Cover	Frequency
Monachosorum henryi Christ	3346	0.402
Asplenium nigrescens Blume	3189	0.380
Hymenasplenium excisum (C. Presl) S. Linds.	2853	0.348
Asplenium thunbergii Kunze	2131	0.326
Sphaeropteris glauca (Blume) R.M. Tyron	3262	0.293
Asplenium nidus L.	1847	0.288
Elaphglossum blumeanum (Fee) J.Sm.	1878	0.283
Asplenium lobulatum Mett.	1178	0.272
Alsophila hermannii. R.M. Tyron	2445	0.272
Alsophila heterochlamydea (Copel) R.M. Tyron	2929	0.272
Angiopteris evecta Sw.	1390	0.272
Diplazium dilatatum Blume	4204	0.293
Goniophlebium pseudoconnatum (Copel) Copel.	1114	0.250
Austroblechnum melanocaulon (Brack.) Gasper & V.A.O Dittrich	2710	0.239
Alsophila fuliginosa Christ.	2725	0.239
Elaphoglossum caliifolium (Blume) T. Moore	1242	0.239
Plagiogyria glauca (Blume) Mett.	1391	0.239
Ptisana pellucida (C. Presl) Murdock	1554	0.228
Sphaeropteris tripinnata (Copel) R.M. Tyron	1254	0.217
Diplaziopsis javanica (Blume) C. Chr.	1718	0.217
Pteris oppositipinnata Fee	1212	0.217

monitoring of biodiversity and ecosystem health of the landscape (Banaticla and Buot 2005; Delos Angeles and Buot 2015).

Distribution and range as well as endemism of pteridophyte species have been attributed to terrain, altitude, climate variation, soil composition as well as landscape heterogeneity (Bickford and Laffan 2006; Banaticla and Buot 2005; Lehmann et al. 2002; Duncan et al. 2001; Lillo et al. 2020). The greater proportion of restricted-range pteridophytes found mostly along montane to mossy forest in MMPL would imply the necessity for these forest zones to be closely monitored to have an accurate picture of quality of environment in the protected area.

9.8 Endemic and Threatened Species of Pteridophytes

The conservation status of pteridophytes in Mount Matutum was sought with reference to Department of Environment and Natural Resources (DENR) Administrative Order (DAO) (DAO 2017–11). This is the recent national list for threatened plants in the Philippines based on the criteria for threatened plants by DENR.

DAO (2017–11) is an administrative order from the Department of Environment and Natural Resources (DENR) which presents different categories from International Union for the Conservation of Nature (IUCN) Global Redlist except for CR and EN. This is a good supplement as IUCN only presents 47% of the plants that can be found in the Philippines. It also included two (2) other categories as Other Threatened Species (OTS) and Other Wildlife Species (OWS). OWS may be compared to the LC of IUCN, while OTS may be similar to NT of IUCN. Classification was based on studies on rate of decline, population size, area of geographic distribution, and degree of population distribution and fragmentation (Tables 9.5 and 9.6).

There are about 27 species considered to be threatened, while about 29 Philippine species are endemic (Table 9.7) with reference to Amoroso et al. (2016) and Pelser et al. (2011 onwards; Cos Digital Flora). These are further classified as broad or narrow endemics based on the work of Amoroso et al. (2016). Broad endemics have wider distribution in the country, while narrow endemics can be found only in certain regions, and in this study, Mindanao.

These threatened and endemic species in MMPL face the concern on prioritization for conservation interventions. More often, pteridophytes do not get adequate attention in conservation as much as trees and other vascular flora. On the other hand, there are also problem ferns, which need to be controlled, which calls for more studies of its ecophysiological characteristics to effectively regulate their spread in the forest. Moreover, presence of narrow (Mindanao) endemics further strengthens the call for Southern Mindanao stakeholders to prioritize the conservation of these species and integrate this floral group in management plans.

Table 9.5 Threatened Species Categories and their description (DAO 2017–11)

Threatened categories	Description
• Critically Endangered (CR)	Refers to a species, subspecies, varieties, or other infraspecific categories facing extremely high risk of extinction in the wild in the immediate future
• Endangered (EN)	Refers to a species, subspecies, varieties, or forma that is not critically endangered, but whose survival in the wild is unlikely if the causal factors continue operating
• Vulnerable (VU)	Refers to a species or subspecies, varieties, formae or other infraspecific categories of plants that are not critically endangered nor endangered, but are under threat from adverse factors throughout its range and are likely to move to the endangered category in the future. This shall include varieties, formae, or other infraspecific categories
• Other Threatened Species (OTS)	Refers to a species, subspecies, varieties, or other infraspecific categories that are not critically endangered, endangered nor vulnerable but are under threat from adverse factors, such as over collection throughout its range and are likely to move to the vulnerable category in the near future
Non-threatened category • Other Wildlife Species (OWS)	Refers to non-threatened species, subspecies, varieties, or other infraspecific categories that have the tendency to become threatened due to destruction of habitat or other similar causes as may be listed by the Secretary upon the recommendation of the National Wildlife Management Committee

9.9 Conservation Plans for the Pteridophytes in MMPL

As forest degradation is now occurring at higher rates, especially in the tropical regions where forest land areas are converted for agricultural plantations, understanding patterns of species richness and diversity is gaining importance (De Fries et al. 2005). In this way, protection of specious areas within these landscapes may further be pursued, thereby mitigating species' loss.

Prior to every conservation undertaking, it is important to realize its pattern, the environmental factors affecting their distribution as well as those that are necessary to effectively manage their population. From here, it would be possible to see how the patterns might change over temporal scale and under the controlling factors and predict how patterns might change over time under the influence of these controlling factors. Detection of these patterns and understanding their underlying causal mechanisms are important for the future management and conservation of biodiversity (Bhattarai 2017).

Conservation interventions to address the dwindling population of pteridophytes could be a complex puzzle and one successful strategy in one area may not address the problem in the other. The unending search for solution calls for a more emphatic social community which can be achieved through constant information dissemination and education campaigns. Too often, these are neglected and advocacy activities

Table 9.6 Threatened list of Pteridophyte species based on (DAO 2017–11) and relevant literature and their altitudinal range in MMPL

Endangered (EN)	*Athyrium nakanoi Makino*	2000–2100
	Adiantum aff. mindanaoense Copel	1400–1600
	Alsophila hermannii R.M. Tyron	1300–2100
	Cibotium barometz (L.) J.Sm	1800
	Dicksonia mollis Holttum	1600–2200
	Dicksonia amorosoana Lehnert and Coritico	1900–2200
	Ophioglossum pendulum L.	1500–1800
	Phlegmariurus squarrosus (G. Forst.) A. Love & D. Love	1300–1800
	Polystichum moluccense T. Moore	1300–2100
	Sphaeropteris glauca (Blume) R.M. Tyron	1200–2100
	Sphaeropteris lepifera (J.Sm. ex Hook.) R.M. Tryon	900–1100
	Tmesipteris zamorae Gruezo and Amoroso	1500–1900
Vulnerable (VU)	*Asplenium vittaeforme Cav*	1200–1400
	Drynaria aglaomorpha Christenh	1500–1600
	Drynaria cornucopia (Copel) Alderw	1500–1600
	Alsophila fuliginosa Christ.	1800–2100
	Alsophila heterochlamydea Copel.	1300–2200
	Botrychium daucifolium Wall ex Hook & Grev/	1300–1600
	Polystichum elmeri Copel	1800–2000
	Sphaeropteris elmeri (Copel.) Copel	1900–2000
Other Threatened Species (OTS)	*Angiopteris evecta Sw*	1100–1800
	Davallia denticulata (Burm.f.) Mett. ex Kuhn	1400–1800
	Davallia solida (L.f.) Kuhn	1600–1800
	Diplazium calliphyllum M.G. Price	1400–1600
	Diplazium sibuyanense Alderw	1400–1600
	Plenasium banksiifolium C. Presl	1200–1700

appear to be a mere performance of a routine. It is recommended that advocacy interventions transcend boundaries and be understandable instruments in the grassroots level of the society (Aribal and Buot 2009).

In supplementing conservation programs at communities, policy strategies must also be developed. This intervention is very important in conserving protected area as it covers nearly all aspects of conservation at local, national, regional, and international scales (FOS 2006). In MMPL, policies are set towards restrictions in collection of wildlife species, conduct of scientific studies, land use boundaries, and tourism activities (DAO 2017–11). However, the protected area needs a rehash on its policies towards biodiversity and monitoring to include species-specific targets and measurable indicators on desired results (i.e., increase in diversity/richness or reduction in observable threats such as hunting, encroachment of settlements in the protected area) and establishment of permanent biodiversity monitoring stations.

Table 9.7 Endemic species of Pteridophytes and their altitudinal range in MMPL

	Endemic species	Altitude (masl)
Narrow (Midanao)	*Adiantum hosei Baker*	1300–2000
	Alsophila hermannii R.M. Tyron	1300–2100
	Diplazium pseudocyatheifolium	1200
	Dicksonia amorosoana Lehnert & Coritico	1300–1500.
	Hymenophyllum ramosii Copel	1800–2100
	Lindsaea apoensis Copel	1300–1500
	Phlegmariurus delbrueckii (Herter) A.R. Field and Bostock	1700–1800
	Polystichum elmeri Copel	1800–2000
	Pteris whitfordii Copel	1300–1500
Broad (Philippines)	*Alsophila fuliginosa Christ*	1800–2100
	Alsophila hermannii R.M. Tyron	1300–2100
	Alsophila heterochlamydea Copel.	1300–2200
	Alsophila apoensis	1900–2000
	Asplenium apoense Copel	1300–1400
	Asplenium callipteris (Fee) Baker	1300–1400
	Drynaria cornucopia (Copel.) Alderw	1500–1600
	Drynaria descensa Copel	1500–1600
	Drynaria sagitta (Christ) Christenh	1500–1600
	Haplopteris aff. mindanaoensis	1800–1900
	Phlegmariurus bayanicus	1700–1800
	Pneumatopteris laevis (Mett.) Holttum	1000
	Pneumatopteris nitidula (C. Presl) Holttum	1200–1300
	Polystichum elmeri Copel.	1800–2000
	Pteris opposipinnata Fee	1200–1600
	Pteris whitfordii Copel	1200–1300
	Selliguea elmeri (Copel.) Ching	2100–2200
	Sphaeropteris elmeri (Copel.) Copel	1900–2000
	Sphaerostephanos urdanetensis (Copel.) Holttum	1600–1700
	Tmesipteris zamorae Gruezo & Amoroso	1500–1900
	Tomophyllum macrum (Copel.) Parris	2100–2200

Moreover, an integrated resource planning and management of relevant government agencies, local government units (LGUs), PAMB, academe, and civil society organizations (CSOs) to harmonize interventions in MMPL prior to developing policy strategies must be enforced (DAI 2019).

Another plausible intervention is the mass propagation of threatened species through ex situ and in situ mechanisms. In Central Mindanao, there is already ex-situ conservation through botanical gardens, spore banks, and greenhouses. This provides avenues for genetic lines of pteridophytes to be preserved and for the community to develop biophilic interaction which will be crucial for intensifying citizens' involvement in conservation. Conservation efforts along this line could also lead towards local community participation from local livelihood to community involvement in management of targeted species such as pteridophytes.

9.10 Conclusion

The forest landscape in MMPL showed significant pteridophyte flora. Richness, diversity, and evenness showed trends from base to summit, influenced by environmental factors. Middle elevations presented the bulk of richness and diversity which was found to be related to slope. Meanwhile, evenness values were variable and showed association with soil pH. Patches of pteridophytes at the base or lower elevation areas show indication of clearing as marked by the growth of aggressive and ramet-forming individuals as well as other nonindigenous plant groups, thereby decreasing areas for the native ferns and fern allies. The presence of a significant number of threatened species highlights the urgency of conservation interventions, especially that MMPL is a protected area surrounded by growing population of household community and pressed with current economic development interventions.

References

Amoroso VB, Acma FM, Pava HP (1996) Diversity, status and ecology of pteridophytes in selected forests in Mindanao. Central Mindanao University, Musuan, Bukidnon

Amoroso VB, Ruba RP, Demetillo MT, Stortelder A, Lagunzad DA, Aranico EC, Arances JB, Rufila LV, Kessler PJA, Tan BC, Roscom BA, Gorne ND, Opiso GS, Van Den Berg A, Gruezo WS, Co LL, Alava CG (2006) Plant diversity in the northern landscape of Mt. Malindang range and environs, Misamis Occidental, Philippines. SEAMCO SEARCA and PDM Press, College Laguna and Quezon City

Amoroso VB, Aspiras RA, Polizon JJ (2009) Participatory inventory and distribution of endangered, endemic and economically important plants in Hamiguitan range wildlife sanctuary, Davao oriental. Blumea 54:71–76

Amoroso V, Laraga S, Calzada B (2011) Diversity and assessment of plants in Mt. Kitanglad range natural park, Bukidnon, Southern Philippines. Gard Bull Singapore 63(1&2):219–236

Amoroso VB, Amoroso CB, Coritico FP (2012) Diversity and status of plants in three mountain ecosystems in Southern Mindanao, Philippines. Asian J Biodiver 3(1):50–73

Amoroso VB, Coritico F, Fritsch P (2016) Species richness and conservation status of ferns and lycophytes in Mt. Hamiguitan range Wildlife Sanctuary, Davao oriental, Philippines. Philipp J Sci 142(2):127–137

Aribal L, Buot I (2009) The threatened plants species in various regions in Mindanao Island, Philippines. J Nat Stud 8(2):23–33

Banaticla MCN, Buot IE (2005) Altitudinal zonation of pteridophytes on Mt. Banahaw de Lucban, Luzon Island. Philipp Plant Ecol 180:135–151. https://doi.org/10.1007/s11258-004-2494-7

Banaticla MCN, Buot IE Jr (2004) Fern patch structure and species diversity along the altitudinal gradient of Mt. Banahaw de Lucban, Luzon Island, Philippines. Philipp Agric Sci 87(1):49–60

Barcelona JF (2003a) Preliminary report on the ferns and fern allies (pteridophytes) of Mt. Bali-it, Balbalasang, Balbalan National Park, Kalinga, Northern Luzon, Philippines. Sylvatrop 13(1 and 2):81–92

Barcelona JF (2003b) The taxonomy and ecology of the pteridophytes of Mt. Iraya and vicinity, Batan Island, Batanes Province, Northern Philippines. In: Chandra, Srivastava (eds) Pteridology in the New Millenium. Kluwer Academy, Amsterdam, pp 299–325

Barcelona JF (2004) Collection and conservation status of pteridophytes in Panay Island, Philippines. Philipp Sci 41:57–73

Barcelona JF, Dolotina N, Madroñero G, Granert W, Sopot D (2006) The Ferns and Fern Allies of the Karst Forests of Bohol Island, Philippines. Am Fern J 96(1):1–20

Belonias B, Banoc L (1994) Species diversity and distribution of pteridophytes in Mount Pangasugan. Ann Trop Res 15(2):30–38

Bhattarai KR (2003) Variation in species richness at the fine-scale and at the broad-scale in the Himalaya, Nepal. PhD thesis, University of Bergen, Norway

Bhattarai KR (2017) Variation of plant species richness at different spatial scales. Botanica Orientalis. J Plant Sci 11:49–62

Bhattarai KR, Vetaas OL (2003) Variation in plant species richness of different life forms along a subtropical elevation gradient in the Himalayas, East Nepal. Glob Ecol Biogeogr 12:327–340

Bhattarai KR, Vetaas OR, Grytnes JA (2004) Fern species richness along a central Himalayan elevational gradient, Nepal. J Biogeogr 31:389–400

Bickford S, Laffan S (2006) Multi-extent analysis of the relationship between pteridophyte species richness and climate. Glob Ecol Biogeogr 15(6):588–601

BirdLife International (2015) Important bird areas factsheet: Mount Matutum protected landscape

Cadena CD, Hozak KH, Gomez JP, Graham C (2012) Latitude, elevation climatic zonation and speciation in New World vertebrates. Proc R Soc B Biol Sci 279(1726):194–201

Carvajal-Hernandez C, Kromer T, Lopez-Acosta JC, Gomez-Diaz J, Kessler M (2017) Conservation value of disturbed and secondary forests for ferns and lycophytes along an elevation gradient in Mexico. Appl Veg Sci 20(4):662–672

Catapang MV, Reyes PJ, Medecilo MP (2012) Factors influencing species diversity of ferns in Mt. Makulot Cuenca, Southern Luzon, Philippines. In: 2nd International Conference on Environmental and Industrial Innovation (IPCBEI), vol 35. IACSIT Press, Singapore, pp 98–102

Climate-Data.Org (n.d.) Weather in Polomolok and Tupi, South Cotabato. Climate-Data.Org. https://en.climate-data.org/location/20633/

Colwell RK, Lees DC (2000) The mid-domain effects: geometric constraints on the geography of species richness. Trends Ecol Evol 15:70–76

Coritico F, Amoroso V, Acma F, Cariño Y, Fritsch P (2020) Ferns and lycophytes of Mt. Tago range, Bukidnon, Southern Philippines: species richness, distribution and conservation status. Philipp J Sci 149(3–1):773–790

Costa DS, Gerschlauer F, Kiese R, Fischer M, Kleyer M, Hemp A (2018) Plant niche breadths along gradients and their relationship to plant functional traits. Divers Distrib 24(12):1869–1882

DAO 2017–11 Updated national list of threatened Philippine plants and their categories. http://www.denr.gov.ph. Accessed 2 May 2017

DeFries R, Hansen A, Newton AC, Hansen MC (2005) Increasing isolation of protected areas in tropical forests over the past twenty years. Ecol Appl 15:19–26

Delos Angeles MD, Buot IE (2012) Orders and families of Philippine pteridophytes. J Nat Stud 11(1 and 2):19–33

Delos Angeles M, Buot I (2015) Diversity and distribution of pteridophytes along the altitudinal gradient of the Northeastern Slope of a Secondary Forest in Mt. Makiling, Philippines. IAMURE 16:25–46

Department of Environment and Natural Resources (DENR)–Protected Areas and Wildlife Bureau (PAWB) (2006). http://bmb.gov.ph

Duncan R, Webster R, Jensen C (2001) Declining plant species richness in the tussock grasslands of Canterbury and Otago, South Island, New Zealand. N Z J Ecol 25(2):35–37

Fageria NK, Baligar VC, Jones CA (2011) Growth and mineral nutrition of field crops. CRC Press, Boca Raton, FL

Fageria NK, Baligar VC, Li YC (2014) Nutrient uptake and use efficiency by tropical legume cover crops at varying pH of an oxisol. J Plant Nutr 37:294–311

Fernández MH, Vrba ES (2005) A complete estimate of the phylogenetic relationships in Ruminantia: a dated species-level supertree of the extant ruminant. Biol Rev Camb Philos Soc 80(2):269–302

Fernando ES, Co LL, Lagunzad DA, Gruezo WS, Barcelona JF, Madulid DA, Lapis AB, Texon GI, Manila AC, Zamora PM (2008) Threatened plants of the Philippines. Asia Life Sci. Supplement 3:1–52

Fjeldsa J (1999) The impact of human forest disturbance on the endemic avifauna of the Udzungwa Mountains, Tanzania. Bird Conserv Int 9:47–62

Flora Malesiana Series II (n.d.) Naming, describing and inventorying the flora of malesia. www.floramalesiana.org

Foundation for the Philippine Environment (2003) Rapid assessment report on the biodiversity of Mount Matutum protected landscape (unpublished)

Franco CS, Vega IL, Ayala OA, Medina RC (2011) Distributional patterns and biogeographic analysis of ferns in the Sierra Madre Oriental, Mexico. Am Fern J 101(2):81–104

Gao J, Ma K, Feng Z, Qi J, Feng Y (2009) Coupling effects of altitude and human disturbance on landscape and plant diversity in the vicinity of mountain villages of Beijing, China. Acta Ecol Sin 29:56–61

George L, Bazzaz FA (1999) The fern understory as an ecological filter: growth and survival of canopy-tree seedlings. Ecology 80(3):846–856

Gillespie T, Foody G, Rocchini D, Giorgi AP, Saatchi S (2008) Measuring and modelling biodiversity from space. Prog Phys Geogr 32(2):203–221

Gomez PLD (1985) Conservation of Pteridophytes. Proc R Soc Edinb 86B:431–433

Goeltenboth F, Milan PP, Asio VB (eds) (1999) Proceedings of the international conference on applied tropical ecology. ViSCA-GTZ Applied Tropical Ecology Program, Baybay, Leyte

Gonzalez A (2000) Pteridophytes of Mount Matutum. A dissertation presented to the Graduate School Faculty of Central Mindanao University, Musuan Bukidnon. Unpublished

Goswarmi HK, Sen K, Mukhopadhyay R (2016) Pteridophytes: evolutionary boon as medicinal plants. Plant Genet Resour 14(4):328–355

Guignard MS, Leitch AR, Acquisti C, Eizaguirre C, Elser JJ, Hessen DO, Jeyasingh PD, Neiman M, Richardson AE, Soltis PS, Soltis DE, Stevens CJ, Trimmer M, Weider LJ, Woodward G, Leitch IJ (2017) Impacts of nitrogen and phosphorus: from genomes to natural ecosystems and agriculture. Front Ecol Evol 5:70

Hawkins BA, Field R, Cornell HV, Currie DJ, Guegan JF, Kaufman DM, Kerr JT, Mittelbach GG, Oberdorff T, O'Brien EM, Porter EE, Turner JRG (2003) Energy, water, and broad-scale geographic patterns of species richness. Ecology 84:3105–3117

Hill MO (1973) Diversity and evenness: a unifying notation and its consequences. Ecology 54(2): 427–432

Holttum RE (1948) The classification of ferns. Botanic Gardens, Singapore, pp 267–295

Ibars A, Estrelles E (2012) Recent developments in ex-situ and in-situ conservation of ferns. Fern Gazette 19(3):67–86

Ifo SA, Moutsambote JM, Koubouana F, Yoka J, Ndzai SF, Bouetou-Kadilamio LNO, Mampouya H, Jourdain C, Bocko Y, Mantota AB, Mbemba M, Mouanga-Sokath D, Odende R, Mondzali LR, Wenina YEM, Ouissika BC, Joel LJ (2016) Tree species diversity, richness, and similarity in intact and degraded forest in the tropical rainforest of the Congo Basin: case of the Forest of Likouala in the Republic of Congo. Int J For Res 2016:12

Janzen DH (1976) Changes in the Arthropod community along an elevational transect in the Venezuelan Andes. Biotropica 8(3):193–203

Jones A, Berkelmans R, Houston W (2011) Species richness and community structure on a high latitude reed: implications for conservation and management. Diversity 3:329–355

Jost L (2010) The relation between evenness and diversity. Diversity 2(2):207–232

Keller H, Prance G (2015) The ethnobotany of ferns and lycophytes. Fern Gazette 20(1):1–14

Kessler M, Kluge J, Hemp A, Ohlemuller R (2011) A global comparative analysis of elevational species richness patterns of ferns. Glob Ecol Biogeogr 20:868–880

Kim-Bui AT, Kim DD, Kuschk P, Tua TV, Hue NT, Minh NN (2013) Effect of soil on pH on As hyperaccumulation capacity in fern species, Pityrogramma calomelanos. J Environ Biol 34(2): 237–242

Kluge J, Kessler M, Dunn R (2006) What drives elevational patterns of diversity? A test of geometric constraints, climate and species pool effects for pteridophytes on an elevational gradient in Costa Rica. Glob Ecol Biogeogr 15:358–371

Kluge J, Worm S, Lange S, Long D, Bohner J, Yangzom R, Miehe G (2017) Elevational seed plants richness patterns in Bhutan, Eastern Himalaya. J Biogeogr 44:1711–1722

Kozak K, Wiens J (2007) Niche conservatism drives elevational diversity patterns in Appalachian salamanders. Am Nat 176:40–54

Kozak K, Wiens J (2010) Accelerated rates of climatic-niche evolution underlie rapid species diversification. Ecol Lett 13:1378–1389

Lehmann A, Leathwick JR, Overton JMCC (2002) Assessing New Zealand fern diversity from spatial predictions of species assemblages. Biodivers Conserv 11:2217–2238

Leibold MA, Holyoak M, Mouquet N, Amarasekare P, Chase JM, Hoopes MF, Holt RD, Shurin JB, Law R, Tilman D, Loreau M, Gonzalez A (2004) The metacommunity concept: a framework for multi-scale community ecology. Ecol Lett 7(7):601–613. https://doi.org/10.1111/j.1461-0248.2004.00608.x

Lillo EP, Malaki AB, Alcazar SMT, Rosales R, Redoblado BR, Pantinople E, Nuevo RU, Cutillar RC, Almirante A, Buot IE Jr (2020) Diversity and distribution of ferns in forest over limestone in Cebu Island Key Biodiversity Areas (KBAs), Philippines. Biodiversitas 21:413–421

Liu Y, Wujisguleng W, Long C (2012) Food uses of ferns in China: a review. Acta Soc Bot Pol 81(4):263–270

Lou J (2010) The relation between evenness and diversity. Diversity 2:207–232

Magtoto L, Austria C (2017) The pteridophytes of Adams, Northern Luzon, Philippines and their ecosystem services. Philippine J Syst Biol 11(2):43–51

Mannan M, Maridass M, Victor B (2008) A review on the potential uses of ferns. Ethnobot Leafl 2008(1):281–285

McCain CM (2005) Elevational gradients in diversity of small mammals. Ecology 8:366–372

Morin X, Lechowicz MJ (2011) Geographical and ecological patterns of range size in North American trees. Ecography 34:738–750

Nettesheim F, Damasceno E, Sylvestre L (2014) Different slopes of a mountain can determine the structure of ferns and lycophytes communities in a tropical forest of Brazil. An Acad Bras Cienc 86:199. https://doi.org/10.1590/0001-3765201495912

Obemio, CD. And Buot, I. (2021). Enumeration of pteridophytes in Mount Matutum protected landscape, Southern Mindanao, Philippines. Sylvatrop. 31(2): 87–124

Oloyede FA (2012) Survey of ornamental ferns, their morphology and uses for enviromental protection, improvement and management. Lfe J Sci 14(2):245–252

Othman R, Latiff NHM, Tukiman I, Hashim KS (2015) Effects of altitude and microclimate on the distribution of ferns in urban areas. Jurnal Teknologi 77(30):125–131

Page CN (2002) Ecological strategies in fern evolution: a neopteridological overview. Rev Palaeobot Palynol 119:1–33

Pelser PB, Barcelona JF, Nickrent DL (2011) Co's digital flora of the Philippines. http://www.philippineplants.org

Perez H (2004) Diversity of trees along altitudinal gradient Layawn River to North Peak in Mt. Malindang National Park, Barangay Lake Duminagat, Don Victoriano, Misamis Occidental, Samu't Sari 3(2)

PPG I (2016) A community-derived classification for extant lycophytes and ferns. J Syst Evol 54: 563–603. https://doi.org/10.1111/jse.12229

Price MG (1975) The Pteridophytes of Mt. Makiling and vicinity. Institute of Biological Sciences, University of the Philippines Los Baños, College, Laguna

Rahbek C (2005) The role of spatial scale and the perception of large-scale species-richness patterns. Ecol Lett 8:224–239

Robinson RC, Sheffield E, Sharpe J (2010) Problem Ferns: their impact and management. In: Mehltreter K, Walker LR (eds) Fern ecology. Cambridge University Press, Cambridge, pp 255–322

Sen A, Ghosh PD (2011) A note on the ethnobotanical studies of some pteridophytes in Assam. Indian J Tradit Knowl 10(2):292–295

Shannon CE, Weaver W (1949) The mathematical theory of communication. University of Illinois Press

Sharpe J, Mehltreter K, Walker L (2010) Ecological importance of ferns. In: Mehltreter K, Walker L, Sharpe J (eds) Fern ecology. Cambridge University Press, Cambridge

Silva VL, Mehltreterm K, Scmitt JL (2018) Ferns as potential ecological indicators of edge effects in two types of Mexican forests. Ecol Indic 93:669–676

Smith AR, Pryer KM, Schuettpelz E, Korall P, Schneider H, Wolf PG (2008) Fern classification. In: Ranker TA, Haufler CH (eds) Biology and evolution of ferns and lycophytes. Cambridge University Press, New York, pp 417–467

Syfert M, Brummitt N, Coomes D, Bystriakova N, Smith M (2018) Inferring diversity patterns along an elevation gradient from stacked SDMs: a case study on Mesoamerican ferns. Glob Ecol Conserv 16:e00433. https://doi.org/10.1016/j.gecco.2018.e00433

Tolentino DB (1988) The Fern Flora of Mt. Sto.Tomas, Benguet (Luzon). [M.S. Thesis]. College, Laguna, Philippines: University of the Philippines Los Baños. 170 p. (Available at the UPLB Library)

Tu WG, Gao XF, Wu N, Liu SH (2010) Altitudinal patterns of vascular plant richness in high mountains: applying of generative additive model. Pol J Ecol 58(2):231–240

Upreti K, Jalai JS, Tewari LM, Joshi GC, Pangley YPS, Tewari G (2009) Ethnomedical uses of Pteridophytes of Kumuan Himalaya, Uttarakhand, India. J Am Sci 5(4):167–170

Wang C, Zhou X, Guo D, Zhao, Yan L, Feng G, Gao Q, Yu H, Zhao L (2019) Soil pH is the primary factor driving the distribution and function of microorganisms in farmland soils in Northeastern China. Ann Microbiol 69:1461–1473

Yuan TX, Zhang HP, Qu ZY, Tan YB (2014) Effects of topography on the diversity and distribution pattern of ground plants in karst montane forests in Southwest Guangxi, China. J Appl Ecol 25(10):2803–2810

Yumul GP Jr, Dimalanta CB, Maglambayan VB, Marquez EJ (2008) Tectonic setting of a composite terrane: a review of the Philippine island arc system. Geosci J 12(1):7–17

Zhang Y-Y, Wu W, Liu H (2019) Factors affecting variations of soil pH in different horizons in hilly regions. PLoS One 14(6):e0218563

Zhao CM, Chen WL, Tian ZQ, Xie ZQ (2005) Altitudinal pattern of plant species diversity in Shennongjia Mountains, Central China. J Integr Plant Biol 47(12):1431–1449

Zhou Y, Ochola A, Njogu A, Boru B, Mwachala G, Hu G, Xin H, Wang Q (2019) The species richness pattern of vascular plants along a tropical elevational gradient and the test of elevational Rapoport's rule depend on different life-forms and phytogeographic affinities. Ecol Evol 9: 4495–4503

Morphological Diversity of Kopyor Coconut in Indonesia 10

Sukendah and Rossyida Priyadarshini

Abstract

In Indonesia, there is a rare type of coconut known as kopyor coconut which is the result of a mutation expressed in the endosperm. Kopyor coconut trees are found concentrated in several areas on the islands of Java and Lampung. However, there is little information reported about the kopyor coconut in the field, especially its variety and diversity. This chapter describes: (1) the center of kopyor coconut in Java and Sumatera Island, and (2) morphological diversity of kopyor coconut based on type, color, size, and endosperm texture. Kopyor coconut tree is classified into dwarf and tall types, as well as hybrids resulting from dwarf and tall hybridization. Hybrid kopyor coconuts are commonly found in the Pati area, Central Java. In fact, some observations showed that the diversity of kopyor coconuts in terms of color, size, and texture of the endosperm can be found among kopyor coconut population. In the tall kopyor coconut, there are green, brown, and red coconut fruits. However, the color of dwarf kopyor coconut fruit has more variations compared to tall. There are green, red, brown, yellow, brownish green, reddish brown green, greenish brown, and orange (ivory). The endosperm diversity of kopyor coconut was identified based on its texture, the percentage of endosperm volume filling the fruit cavity, and coconut water volume. Generally, kopyor coconut endosperm is classified into 3 types: Type A—clotted endosperm loose from the shell, crumbs, and no coconut water; Type B—some endosperm still attached to the shell, very crumbly, and little coconut water; and Type C—endosperm still attached to the shell and very soft has a lot of

Sukendah (✉) · R. Priyadarshini
Center Innovation of Appropriate Technology for Lowland and Coastal Food, Surabaya, East Java, Indonesia

University of Pembangunan Nasional Veteran Jawa Timur, Surabaya, East Java, Indonesia
e-mail: sukendah@upnjatim.ac.id

S. Ramamoorthy et al. (eds.), *Plant Diversity in Biocultural Landscapes*,
https://doi.org/10.1007/978-981-19-8649-9_10

coconut water. Those information on the diversity of kopyor coconut is very important for researchers, laboratories, and institutions to develop in situ and ex situ conservation, as well as to produce superior kopyor coconut.

Keywords

Mutant coconut · Endosperm · Fruit color · Phenotype diversity

10.1 Introduction

Kopyor coconut is one of the coconut variants which is thought to be the result of natural mutation, having a nut (commonly for coconut fruit) with an abnormal endosperm. The structure of the endosperm is very soft and the flesh separates from the shell into small pieces that fill the entire cavity. Kopyor coconut endosperm abnormalities cause embryos to fail to germinate. Endosperm cannot be utilized by the embryo as an energy source in the germination process, due to the endosperm in kopyor coconut seeds rotting immediately after planting. Thus, pure-bearing kopyor coconut palms that produce 100% kopyor nuts have not been obtained in nature.

Phenotype characters of kopyor coconut are associated with xenia effect, the effect of pollen on seeds and fruit of the fertilized plant, controlled by a single recessive mutant k gene in a K locus (Maskromo et al. 2016; Sukendah et al. 2009). Sudarsono et al. (2013) classified kopyor coconut into three classes based on the phenotype, the genotype of endosperm, and zygotic embryo: (1) Normal phenotype-homozygous KK nuts, with the genotype of endosperm being a homozygous KKK, (2) Normal phenotype-heterozygous Kk nuts, with the genotype of endosperm either KKk or Kkk, and (3) Kopyor phenotype-mutant homozygous kk nuts with the endosperm genotype a homozygous kkk.

Generally, kopyor coconuts are obtained from normal coconut trees which produce normal coconuts and kopyor coconuts in one bunch of inflorescence. This type is called heterozygous kopyor coconut tree. Heterozygous kopyor coconut tree is grown from coconut seeds with normal phenotype, but carries the gene for the kopyor trait on one of the alleles in the locus that regulates the kopyor trait (Kk tree). By this method and depending on where they are planted, also the type of kopyor coconut and its pollination, the percentage of kopyor nuts produced very low, usually only 1–2% fruits per bunch. For tall kopyor coconut type, a cross-pollinated produces 1–2 kopyor fruit per bunch. This is because the possibility of recessive genes in female flower pollinated by recessive gene in male flower is relatively small. Meanwhile, the dwarf type of kopyor coconut produces more kopyor fruit per bunch, sometimes it can reach 50%. Research results Maskromo et al. (2013) showed that the percentage of kopyor fruit production of three varieties of dwarf kopyor coconut (green, yellow and brown) from Pati (Central Java) ranged from 24.8–38.9%. This is because the dwarf type of kopyor coconut is self-pollinating so that the chance of pollination recessive genes between female and male flowers is greater. To develop pure (homozygote) kopyor seedling, the only way is through in vitro culture (embryo rescue technique) (Mashud et al. 2004; Novariantol et al.

2014). This seedling can produce 100% kopyor nuts and called homozygous kopyor coconut tree.

Heterozygous kopyor coconut trees can only be found in certain areas in Java, Madura, and Sumatera Island. In Java, kopyor coconut exists in small-scale community gardens in Jember (East Java), Pati Regency (Central Java), and Ciomas-Bogor (West Java). Kopyor coconut is also grown in farmer plantation in Sumenep (Madura Island), and Kalianda, Lampung (south part of Sumatera Island). Usually, those kopyor coconut trees are tall type. However, in some areas such as Pati, Jember, and Sumenep, there are dwarf and hybrid types. In these areas kopyor hybrid coconut palm is grown among the normal population. Until now, very few researchers have explored the presence and diversity of the kopyor coconut trees on the island of Java or Madura. This chapter aims at (a) describing the diversity of kopyor coconut trees in the farmer's plantation, and (b) describing diversity of kopyor coconut endosperm.

10.2 Centers of Kopyor Coconut in Indonesia

In nature, the population of heterozygous kopyor coconuts is very limited and concentrated in small-scale community gardens on the islands of Java and South Lampung. Until now, there is no clear information about the origin of the kopyor coconut tree. However, the population of kopyor coconut can be found in Jember, Banyuwangi, Sumenep, Tulungagung (East Java), Pati (Central Java), Ciomas, Bogor, and Tangerang (West Java) (Asmah 1999), while in Sumatra it is in Kalianda, South Lampung (Mahmud 2000). Kopyor coconut plants are also often reported to grow sporadically in various areas on the Java island, but the highest population of kopyor coconut is found in five central areas, namely Sumenep, Jember, Pati, Tangerang, and South Lampung.

Heterozygous kopyor coconut centers in Sumenep Regency are located in Dungkek, Batang-Batang, and Gapura Districts. The population of kopyor coconut is scattered sporadically in the three subdistricts. The largest population is in Lapataman Village, Dungkek (Fig. 10.1a, b). Each farmer has an average of 10–20 trees per person. In 2007, the Sumenep Plantation Office has registered the local variety of kopyor coconut, Lapataman Dungkek, to the Department of Agriculture. Aside heterozygous kopyor coconut, the local government has developed a 1 ha homozygous dwarf kopyor coconut plantation consisting of 200 trees (Fig. 10.1c). The seedling came from embryo culture.

In Sumenep, coconut kopyor is known as Kelapa Puan. The number of kopyor coconuts produced from one tree per bunch is an average of 2%. The production of kopyor fruit in the three subdistricts is estimated at an average of 1000 nuts per week (Akuba 2002). Farmers usually sell kopyor coconut in the form of fresh fruit at a price of IDR 50,000 per fruit. This price is ten times compared to normal coconut.

The largest heterozygous kopyor coconuts in Jember are in three subdistricts, namely Wuluhan, Ambulu, and Gumuk Mas. In Jember, there are two types of kopyor coconut, tall and dwarf types. Although generally heterozygous kopyor

Fig. 10.1 Heterozygous kopyor coconut population in Lapataman, Dungkek, Sumenep, East Java. (**a**) Tall heterozygous kopyor coconut, (**b**) Dwarf heterozygous kopyor coconut, and (**c**) Homozygous dwarf kopyor coconut from embryo cultured seedling

coconuts grow scattered among normal coconuts, but in Kesilir Village, Wuluhan District, there is one garden planted with heterozygous kopyor coconuts of approximately 50 trees with a spacing of about 7 × 7 m (Fig. 10.2a). Heterozygous kopyor coconut produced an average of 3–4 fruits per bunch, even up to 7 fruits per bunch per tree. Planting kopyor coconut trees in one area causes an increase in pollination between male flowers and female flowers that carry kopyor traits.

The population of tall kopyor coconut tree in one area was also found in Mayangan Village, Gumuk Mas District. Kopyor coconut trees are planted on bunds surrounding the fish pond (Fig. 10.2b). According to the farmer who planted it, the kopyor coconut seeds planted in Gumuk Mas came from the village of Kesilir Wuluhan. The expansion of the kopyor coconut planting area continues around the Gumuk Mas area, which is now about 4–5 years old (Fig. 10.2c). Meanwhile, the population of dwarf coconut kopyor in Jember is very small and is spread in Ambulu District with orange, yellow, and green fruit colors.

The highest population of kopyor coconut from Pati is found in Tayu, Margoyoso, and Dukuh Seti subdistricts (Fig. 10.3a). As in other areas, the kopyor coconut tree is planted in the gardens around people's homes (Fig. 10.3b). In contrast to Jember, the dwarf type of kopyor coconut in Pati is very numerous and

Fig. 10.2 Heterozygous kopyor coconut population in Jember, East Java. (**a**) Tall heterozygous kopyor coconut plantation in Kesilir Village, and (**b**) Tall heterozygous kopyor coconut in Gumuk Mas Village

widespread in this district. It is suspected that the dwarf kopyor coconut is a native plant of the Pati area. Dwarf kopyor coconut from Pati has grown in other districts in Central Java such as in Jepara, Rembang, Sukoharjo, Magelang, and Jogyakarta. In addition, it has spread to Sumenep, East Java, Berau, East Kalimantan, and Riau Islands (Maskromo et al. 2015). The Association of Indonesian Coconut Farmers (APKI) in Pati Regency, Tayu District, played an important role in the distribution of the heterozygous kopyor coconut seedlings. Association members usually develop dwarf kopyor coconut seedlings from normal coconuts carrying the kopyor gene (Fig. 10.3c).

Aside from dwarf type, hybrid kopyor coconuts are also commonly found in Pati. The hybrid kopyor coconut is the result of a natural pollination between the tall and dwarf kopyor coconuts. Observation of the origin of hybrid kopyor coconut in Pati shows that some hybrid kopyor coconuts in Central Java come from parents in Central Java. For example, the hybrid kopyor coconut from Dukuh Seti has the parents of the tall kopyor coconut from Tayu and the dwarf kopyor coconut from Margoyoso. The Tayu hybrid has parents from Tayu, Dukuh Seti, and Margoyoso (Table 10.1). However, the hybrid kopyor coconut from Margoyoso comes from the parent of tall kopyor coconut Margoyoso and dwarf Tayu.

Fig. 10.3 Heterozygous kopyor coconut in Pati, Central Java. (**a**) Tall heterozygous kopyor coconut, (**b**) Dwarf heterozygous kopyor coconut in Tayu Village, Pati, (**c**) Heterozygous kopyor coconut seedling

Table 10.1 Possible origin of hybrid kopyor coconut in Central Java

No.	Hybrid kopyor coconut (*Offspring*)	Tall kopyor coconut (*Parent* 1)	Dwarf kopyor coconut (*Parent* 2)
1.	Hybrid Dukuh Seti	Tall Tayu	Dwarf Margoyoso
2.	Hybrid Tayu1	Tall Tayu	Dwarf Margoyoso
3.	Hybrid Tayu2	Tall Dukuh Seti	Dwarf Tayu
4.	Hybrid Tayu3	Tall Margoyoso	Dwarf Margoyoso
5.	Hybrid Margoyoso	Tall Margoyoso	Dwarf Tayu

10.3 Diversity of Kopyor Coconut Tree in Farmer Plantation

Kopyor coconuts, like normal coconuts, can be classified into two types—tall and dwarf types (Mashud and Manaroinsong 2007). Meanwhile, Freemond et al. (1966) classified normal coconuts into three varieties, namely tall, dwarf, and hybrid/intermediate (var. Aurantiaca). In the case of coconut kopyor, hybrid species have also been found in certain locations, such as Pati, Central Java and Jember, East Java. In this area, hybrid kopyor coconuts grow among the normal population. Meanwhile, in the Sumenep area, hybrid species have not been found (Sukendah et al. 2011).

Tall kopyor coconut trees have the appearance of trunk size, plant height, and fruit size that are larger than the dwarf type. A distinctive feature is the presence of a bole (the enlarged part at the base of the stem). Tall kopyor coconut begins to flower at the age of 5–7 years with a cross-pollination pattern, so that the variability between individuals and populations is quite high. Naturally, the tall kopyor coconut only produces 1–2 kopyor fruit per bunch.

In addition to the tall and dwarf types, kopyor coconuts are also classified based on the color of the fruit. Heterozygous tall kopyor coconut in Sumenep can be distinguished based on fruit color into 3 types, i.e., green, brown, and red (Fig. 10.4b, c). Meanwhile, the tall kopyor coconut in Kalianda has various fruit colors, ranging from green, brownish green, brown to reddish brown. In general, the fruit color diversity of tall kopyor coconut is relatively the same as the tall normal coconut color variation.

Kopyor coconut fruit diversity is not only in fruit color, but also in fruit size, fruit weight, and endosperm thickness (Sukendah et al. 2011). Besides that, there is also diversity in flower morphology. Kopyor coconut flower color matches the color of the fruit. Inflorescence morphology of brown kopyor coconut is longer with more female and male flowers, but has shorter and thicker flower stalks than green ones.

Diversity of fruit color in dwarf kopyor coconuts is more than the tall one. According to Maskromo et al. (2014), dwarf kopyor coconut in Pati, Central Java, has quite a lot of color variations, namely green, yellow, brown, and orange (Fig. 10.5a–d). Besides that, there are also color gradations, namely brownish green and greenish brown. Until now six kinds of dwarf kopyor coconut in Pati can be identified based on the fruit color: green, brownish green, brown, greenish brown, yellow, and orange (ivory) (Maskromo et al. 2015).

10.4 Diversity of Kopyor Coconut Endosperm

The phenomenon of kopyor coconut endosperm with soft texture or crumbs and lumps (Fig. 10.6a, b) does not only occur in Indonesia. In other countries such as the Philippines, Thailand, Sri Lanka, and India, coconuts with endosperm similar to kopyor coconut are also found. In Indonesia, besides the existing kopyor coconut fruit, endosperm with soft and lumpy texture is also found, known as lilin coconut (Fig. 10.6c). The endosperm texture of lilin coconut from Indonesia is similar to that

Fig. 10.4 Diversity of tall heterozygous kopyor coconut based on fruit color. (**a**) Tall kopyor coconut, (**b**) Diversity of color and size of tall kopyor coconut fruits in Sumenep, (**c**) Diversity of color and size of tall kopyor coconut in Jember

of macapuno from the Philippines, but different from that of kopyor coconut (Figs. 10.6d). The endosperm texture of lilin coconut and makapuno coconut is soft like jelly, not crumbly or not easily crushed, coconut water is viscous like oil. Therefore, the endosperm of makapuno can be made into snacks and mixed in buko-pie cakes. However, the endosperm texture of kopyor coconut is rough crumbs, and kopyor coconut water is not viscous, but like normal coconut water.

The diversity of kopyor coconut is not only seen in the type (tall, dwarf, hybrid), but also in the texture of the endosperm. Based on the texture of the endosperm, the kopyor coconuts in Sumenep are divided into 3 types, namely: Type A which has the following characteristics: the endosperm is lumpy into small flakes and is very crumbly and peels off from the shell, little or no coconut water; Type B: some of the endosperm peeled off into small lumps, some are still attached to the shell, the endosperm is very soft or loose, still has coconut water; and Type C: endosperm still attached to the shell, not peeled off but very loose in texture and has a lot of coconut water (Fig. 10.7).

Akuba et al. (2000) also classify the endosperm of the Sumenep kopyor coconut into 3 groups, namely, the first endosperm that has been broken or totally broken meat is called "Puan" or "Kopyor". In the second type, the endosperm is still intact

Fig. 10.5 Diversity of dwarf kopyor coconut based on fruit color. (**a**) red dwarf kopyor coconut, (**b**) brown dwarf kopyor coconut, (**c**) green dwarf kopyor coconut, and (**d**) yellow dwarf kopyor coconut

but very soft called "Theeri" and the third type is the endosperm slightly harder than "Theeri" which is called "Kapungan". In contrast to the kopyor coconut in Sumenep, the endosperm texture of the kopyor coconut in Kalianda is classified based on the score of the thickness of the endosperm. There are 9 scores of endosperm of Kalianda kopyor fruit, which are ordered based on the percentage of endosperm volume filling the fruit cavity, starting from the thinnest with a score of 1 to the fullest filling the cavity of the coconut fruit with a score of 9. Meanwhile, the dwarf kopyor coconut in Pati obtained a score of 1 to a score of 6 (Maskromo et al. 2014).

10.5 Summary and Conclusion

The diversity of kopyor coconut phenotypes was identified in several kopyor coconut centers in Sumenep (Madura), Jember (East Java), Pati (Central Java), and in Kalianda (Lampung). According to the type of kopyor coconut, it is distinguished by tall, dwarf, and hybrid. Hybrid kopyor coconut is the result of hybridization between tall and dwarf kopyor coconuts. Morphologically, tall and dwarf kopyor coconut vary in fruit color, fruit size, and endosperm texture. Based on the fruit color, both tall and dwarf kopyor coconuts are divided into green, brown, red, and yellow kopyor coconuts. Even in the dwarf kopyor coconut, more variations in fruit

Fig. 10.6 (**a**, **b**) Endosperm kopyor coconut, (**c**) Endosperm lilin coconut (Maskromo et al. 2015), (**d**) Endosperm macapuno Philippines. (http://picssr.com/tags/makapuno)

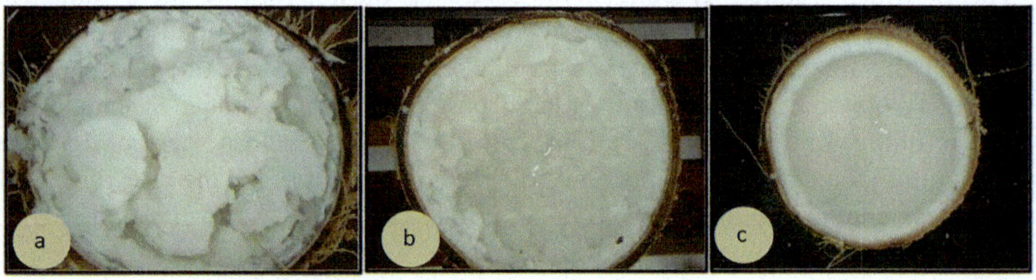

Fig. 10.7 Diversity of kopyor coconut endosperm texture (**a**) Type A: clotted endosperm loose from the shell, crumbs, and little/no coconut water, (**b**) Type B: some endosperm still attached to the shell, very crumbly, and little coconut water. (**c**) Type C: endosperm still attached to the shell and very soft has a lot of coconut water

color were found. The diversity does not only occur in the color of the fruit, but also in the endosperm. Kopyor coconuts in Sumenep are grouped into 3 types of endosperm, while in Pati and Kalianda kopyor coconuts are classified in to a score of 1–6 based on percentage of endosperm volume filling the fruit cavity.

Acknowledgments I am thankful to the Ministry of Research, Technology, and Higher Education who had provided research funding and kopyor coconut team who had collected kopyor coconut data in the field.

References

Akuba RH (2002) Breeding and population genetics studies on coconut (Cocos nucifera L.) composite variety using morphological and microsatellite markers. Disertasi Doctor of Philosophy. University of the Philipines Los Banos

Akuba HR, N Mashud, dan Miftahorracman (2000) Identifikasi plasmanutfah kelapa potensial di Jawa Timur. Laporan Hasil Penelitian Balitka Manado. Belum dipublikasikan

Asmah N (1999) Analisis protein spesifik sebagai penanda sifat kopyor pada kelapa. Skripsi. Jurusan Kimia FPMIPA. IPB. Bogor

Freemond Y, Ziller R, Lamothe ND (1966) The coconut palm. International Potash Institute, Bern

Mahmud Z (2000) Petunjuk teknis budidaya kelapa kopyor. Departemen Kehutanan dan Perkebunan. Dirjen Perkebunan, Jakarta

Mashud N, Manaroinsong E (2007) Embryo culture technology for kopyor coconut development. Buletin Palma 33:37–44

Mashud N, Lumentut N, dan Masing V (2004) Perbanyakan kelapa kenari dan kopyor kopyor memlalui kultur embrio. In: Monograf Agronomi Kelapa, pp 16–23

Maskromo I, Novarianto H, Sukendah, Sukma D, Sudarsono (2013) Productivity of three kopyor dwarf coconut varieties from Pati, Central Java, Indonesia. CORD 29(2):19–28

Maskromo I, Novarianto H, Sukma D, Sukendah, Sudarsono (2014) Keragaman komponen buah dan kuantitas endosperma kelapa Dalam kopyor Kalianda dan kelapa Genjah kopyor Pati. Bul Palma 15(2):75–83

Maskromo I, Tenda ET, Tulalo MA, Novarianto H, Sukma D, Sukendah, Sudarsono (2015) Keragaman fenotipe dan genetik tiga varietas kelapa Genjah kopyor asal Pati, Jawa Tengah. Jurnal Penelitian Tanaman Industri 21(1):75–83

Maskromo I, Larekeng SH, Novarianto H, Sudarsono (2016) Xenia negatively affecting Kopyor nut yield in Kalianda tall Kopyor and Pati dwarf Kopyor coconuts. Emir J Food Agric 28(9): 644–652

Novariantol H, Maskromo I, Dinarti D, Sudarsono (2014) Production technology for Kopyor Coconut seednuts and seedlings in Indonesia. Cord 30(2):31–40

Sudarsono H, Novarianto MLA, Hossang D, Dinarti Y, Sukma D, Maskromo I (2013) Produksi bibit kelapa kopyor *true to type* dengan persilangan terkontrol dan peningkatan produksi buah kelapa kopyor dengan polinator lebah madu. Laporan Tahunan (Annual Report)—Program Hi-Link Kelapa Kopyor. Institut Pertanian Bogor, Bogor, Indonesia

Sukendah, Volkaert H, Sudarsono. (2009) Isolation and analysis of DNA fragment of genes related to kopyor trait in coconut plant. Indian J Biotechnol 14(2):1169–1178

Sukendah, Abidin Z, Wiyatiningsih S, Wijayanti BW (2011) Morphological characters of Kopyor coconut grown in Sumenep, Madura, Indonesia. Isnar C2fs proceeding. pp. 245–254. ISBN: 9786028915939

Ecology of Understory Plants of Forests Over Limestone in Samar Island, Philippines

11

Marjorie D. delos Angeles, Inocencio E Buot, Jr, Ren Divien R. Obeña, and Elaine Loreen C. Villanueva

Abstract

Understory plant species diversity was assessed within forests over limestone across municipalities of Paranas, Taft, and Guiuan in Samar Island, Philippines. Edaphic factors were analyzed and subjected to cluster and ordination analysis. A total of 26 understory plants were recorded, 6 fern species belonging to 6 genera in 6 families and 20 flowering plants belonging to 13 genera in 9 families. Based on the forest floor census, the highest species richness occurs in Paranas, with 18 understory plant species followed by Taft with 12 and Guiuan with 5. Community awareness together with the help of Environmental agencies such as the Department of Environment and Natural Resources and the Local Government Unit is instrumental towards plant species protection and conservation. Lower species diversity in Guiuan may be attributed to generally poor soil characteristics. Soil pH across sites was slightly acidic. Generally, the percent organic matter (OM) was high in all plots but was highest in Guiuan (16.44%) and lowest in Taft (9.61%). Variation in average soil OM across sites can be explained by differences in vegetation cover, amount of leaf litter, and continuous rain-promoting decomposition. Similarly, total N and P are high and above threshold levels in all plots. Total average N was recorded to be highest in Guiuan (0.63) and lowest in Taft (0.47). Interestingly, most of the analyzed edaphic factors are not typical of forests over limestones. Understory plant species recorded from Paranas and Taft are highly affected by Fe and EC. Plant species

M. D. delos Angeles (✉) · I. E. Buot, Jr · R. D. R. Obeña · E. L. C. Villanueva
Institute of Biological Sciences, College of Arts and Sciences, University of the Philippines Los Baños, Laguna, Philippines
e-mail: mddelosangeles1@up.edu.ph

S. Ramamoorthy et al. (eds.), *Plant Diversity in Biocultural Landscapes*,
https://doi.org/10.1007/978-981-19-8649-9_11

from Guiuan, on the other hand, is positively correlated with pH, OM%, N%, P, Cl, K, Cu, Mn, Mg, Na, and Zn. Understanding key environmental factors is important in biodiversity assessment and indicates overall environmental quality.

Keywords

Edaphic factors · Flora · *Kaigangan* · Samar Island

11.1 Introduction

Samar Island together with Leyte is collectively known as Eastern Visayas and is identified as one of the 15 biogeographic zones in the Philippines due to its unique and rich biodiversity (Madulid 2000). Furthermore, it hosts a number of native and endemic species due to its extremely rugged terrain, geographic location, climatic and edaphic conditions, biogeographic conditions, and a variety of habitats and forest formations (Meneses et al. 2018; Tolentino et al. 2020). These forest formations include mangrove forests, beach forests, old and second-growth tropical lowland evergreen forests, forests over ultramafic rocks, and forests over limestone (Chavez et al. 2020). Forests over limestone phenomena are geomorphologic features that result from the dissolution of a layer or layers of soluble bedrock, usually, carbonate rock. Limestone karsts cover 10% of the total land area of the Philippines and are notable for having high species endemism due to their unique flora and fauna (Fernando et al. 2008). Recent expeditions in these forests have led to new species records as well as the discovery and description of new floral species (Meneses et al. 2018; Meneses and Cootes 2019; Adorador 2019; Adorador and Fernando 2019; Adorador et al. 2020; Chavez et al. 2020).

The assessment of floral communities in the epigeic region of forests over limestone ecosystem can generate useful information such as floral diversity, community structure, niche partitioning, as well as the understanding of plant adaptations and plant assemblage patterns as a result of the complexity of environmental conditions. A well-documented and conserved forest over limestone brings in numerous priceless ecosystem services that will sustain ecosystem health, minimizing global warming and drastic climatic changes, aggravating disaster in this vulnerable region as observed during typhoons, particularly Typhoon Haiyan. To understand plant community change and composition as a result of varying soil characteristics, documenting and continuous monitoring is essential.

Generally, this study was conducted to document and examine plant diversity and species composition in forests over the limestone of Paranas, Taft, and Guiuan of Samar Island. Specifically, this study aims to: (1) Document plant understory species in forests over limestones and (2) Describe the pattern of plant distribution regarding ecological parameters using multivariate tools.

11.2 Materials and Methods

11.2.1 Study Site

Assessment of floral diversity in Paranas (Fig. 11.1), Taft (Fig. 11.2), and Guiuan (Fig. 11.3) of Samar Island was conducted from October 1–16, 2019, as part of Project 1: Assessment of Biodiversity in Forest Over Limestone Ecosystem. This is a component of the government-funded research program, Assessment and Conservation of Forest Over Limestone Ecosystem Biodiversity in Selected Municipalities of Samar, Philippines (CONserve-KAIGANGAN). The selection of sampling plots was based on the biophysical characteristics of the area, including (1) plant diversity heterogeneity, (2) topographic attribute (elevation), (3) the extent of forest over limestone area, and (4) the presence of anthropogenic disturbances.

11.2.2 Plant Species Assessment

In documenting understory plant species recorded in established plots from select municipalities, the line intercept technique was used. A total of 54 line transects were established in Paranas (18), Taft (18), and Guiuan (18). Two line transects, 5 meters (m) in length and subdivided with 1 m intervals, were deployed inside each sampling plot. The local name, height, and cover of all plant species traverse in the line were recorded per 1 m interval markings. For species nomenclature, The Plant List (2013) and Co's Digital Flora (Pelser et al. 2011-onwards) was followed.

The dominating plant species in the area were determined by obtaining the following parameters: relative density (RD) and relative frequency (RF). These parameters were then utilized to compute the Importance Values (IV) using the formula adapted from Mueller-Dombois (1974):

$$\mathrm{IV} = \frac{\text{Relative Density (RD)} + \text{Relative Frequency (RF)}}{2}$$

where:

$$\mathrm{RC} = \frac{\text{Cover of the species}}{\text{Total cover of all species}} \times 100$$

$$\mathrm{RF} = \frac{\text{Number of occurrences of the species}}{\text{Total number of occurrences of all species}} \times 100$$

Shannon index of diversity was also computed with the following formula:
Shannon Index of diversity (H'):

$$H = -\sum (p_i \ln p_i)$$

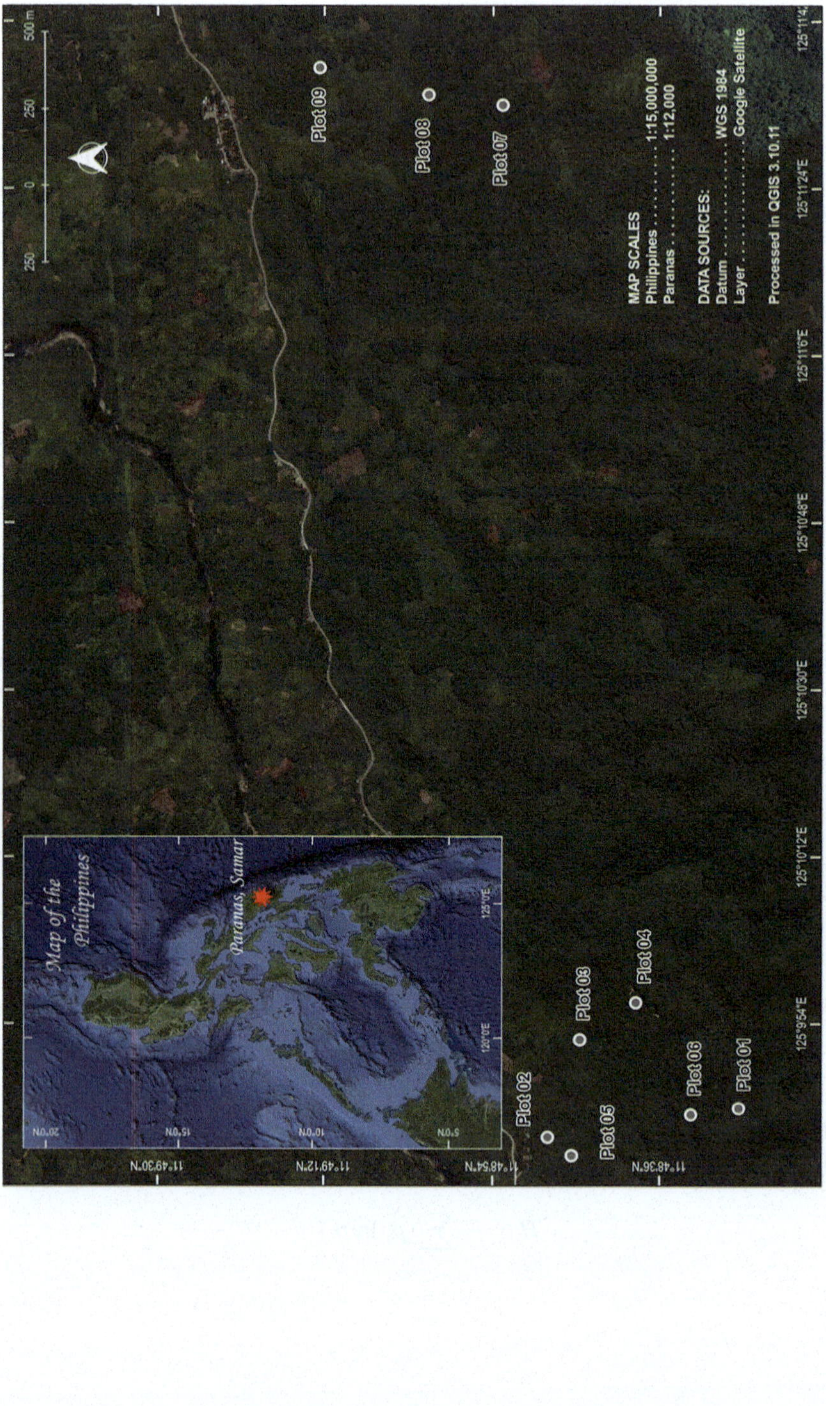

Fig. 11.1 Location of sampling plots in the Municipality of Paranas, Samar Island, Philippines

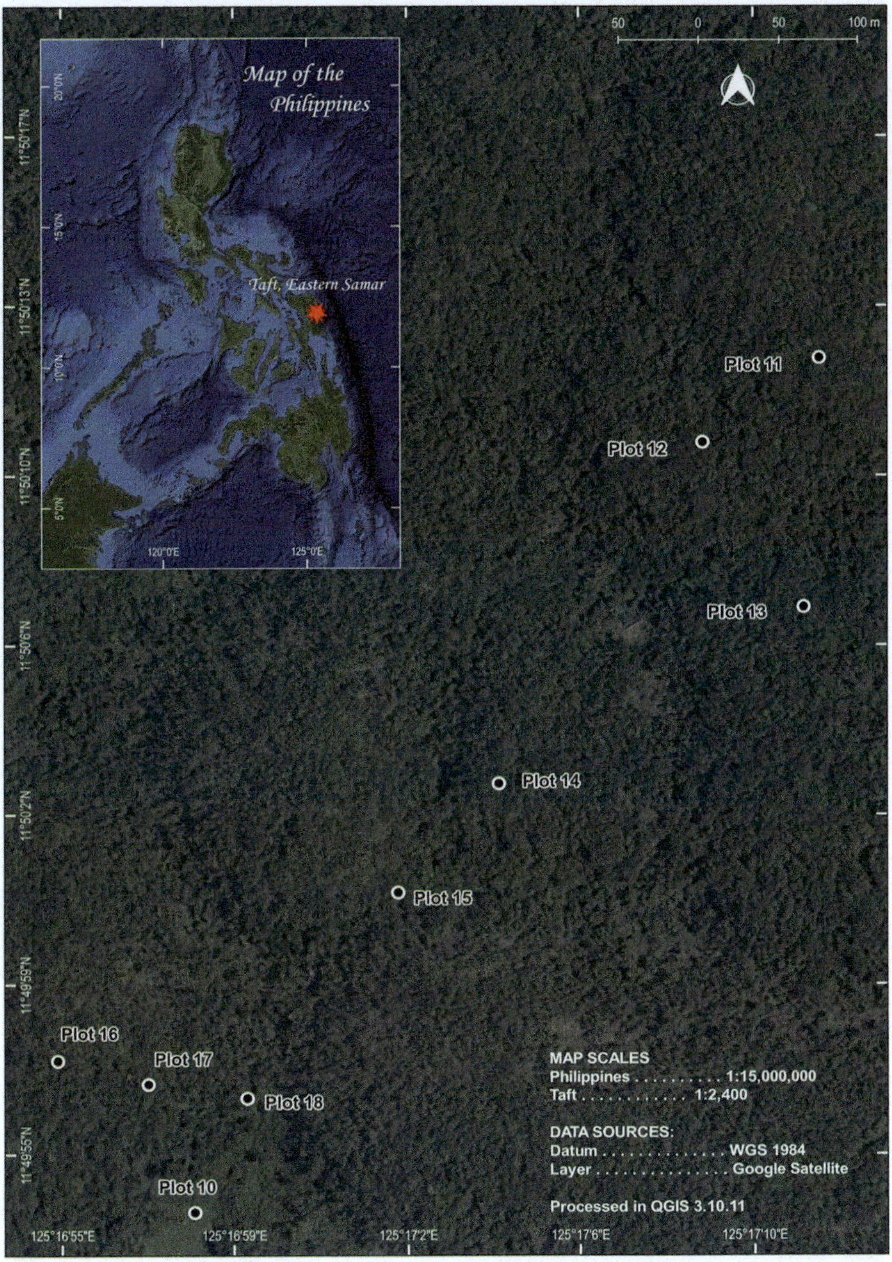

Fig. 11.2 Location of sampling plots in the Municipality of Taft, Eastern Samar, Samar Island, Philippines

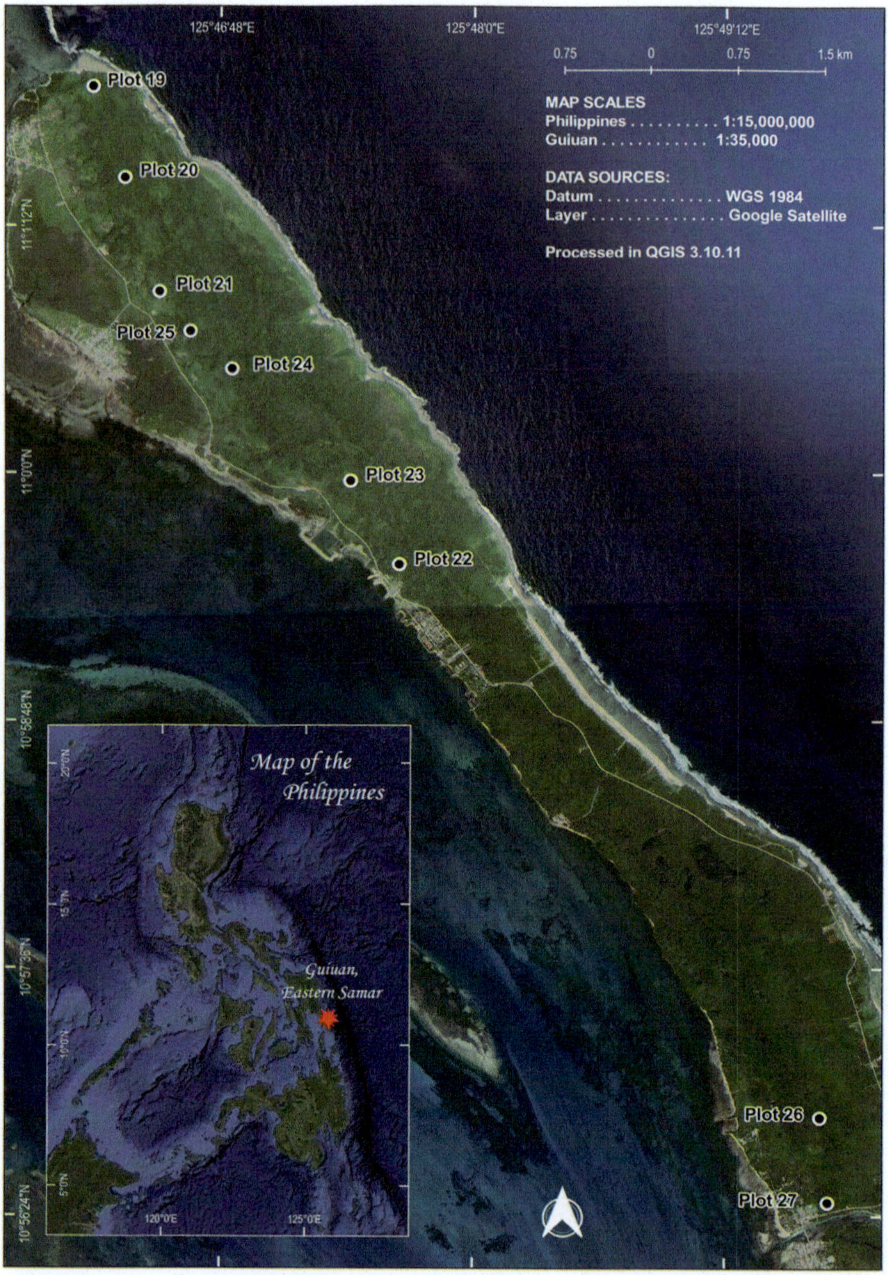

Fig. 11.3 Location of sampling plots in the Municipality of Guiuan, Eastern Samar, Samar Island, Philippines

where $p_i = 10$ relative abundance or the proportion of the total sample belonging to the nth species, and $i = 10$ and ln is the natural logarithm.

11.2.3 Herbarium Specimen

Voucher specimens were collected for proper identification and documentation. These were then processed, and herbarium specimens were deposited in the Plant Biology Division Herbarium (PBDH), Institute of Biological Sciences, University of the Philippines Los Baños. Specimens were identified utilizing original descriptions, taxonomic revisions, and botanical keys. Our specimens were also compared with the collections of PBDH and from available online resources such as "Co's Digital Flora of the Philippines" (Pelser et al. 2011). Because existing databases and updated information regarding the flora of Samar are wanting, experts from the Philippine National Museum were consulted in order to identify the unknown plant specimens.

11.2.4 Data Analysis

Altitude and geographic location were measured using a geographic positioning system (GPS) device. In order to detect significant changes in the floristic composition among plant communities recorded across select municipalities of Samar Island, edaphic factors were analyzed, such as % organic matter (OM), % nitrogen (N), phosphorus (P), electrical conductivity (EC), potassium (K), sodium (Na), calcium (Ca), magnesium (Mg), iron (Fe), zinc (Zn), copper (Cu), manganese (Mn), and chlorine (Cl). A similarity analysis (ANOSIM) was run, which performed 999 randomized permutations on data to calculate the statistic R global with their corresponding probability *p* (Clarke and Gorley 2006). The quantitative data obtained from soil and plant analysis were subjected to Canonical Correspondence Analysis (CCA). The researchers performed principal coordinate analysis (PCoA) of Bray-Curtis similarity of plant species abundance data recorded across select municipalities where square root transformed, and abiotic factors were used without further transformation (% organic matter (OM), % Nitrogen, Phosphorus, Electrical Conductivity, K, Ca, Mg, Fe, Fe, Zn, Cu, Mn, and Cl). To detect significant changes in the plant floristic composition across select municipalities, a similarity index (ANOSIM) was used. All analyses were performed using the Paleontological Statistics Software Package Version 3.25 (2019).

Table 11.1 Recorded plant species recorded from select municipalities of Samar Island, Philippines

Plant group	Order	Family	Scientific name	Common name	Exsiccata
[1]Pteridophyta					
	Selaginalles	Selaginellaceae	*Selaginella* sp.	–	Obeña 7136 (PBDH)
	Marattiales	Marattiaceae	*Angiopteris* sp.	Palko	Obeña 7152 (PBDH)
	Polypodiales	Aspleniaceae	*Aslpenium vittaeforme* Cav.	Lorog/Lurog	Obeña 7063/7091 (PBDH)
		Dryopteridaceae	*Polystichum horizontale* C. Presl	Punit	Obeña 7139/7134 (PBDH)
		Lomariopsidaceae	*Cyclopeltis crenata* (Fée) C. Chr.	Lukdo	Obeña 7111 (PBDH)
		Tectariaceae	*Tectaria dissecta* (G. Forst.) Lellinger	Nito	Obeña 7126 (PBDH)
[2]Spermatophyta/Angiospermae					
	Piperales	Piperaceae	*Piper* sp.	Wenter	Obeña 7137 (PBDH)
	Alismatales	Araceae	*Aglaonema commutatum* Schott	Pilako	Obeña 7089 (PBDH)
			Alocasia sp.	–	Obeña 7100 (PBDH)
			Alocasia zebrina Schott ex Van Houtte	Handuroy	Obeña 7118 (PBDH)
			Rhaphidophora sp.	–	Obeña 7090 (PBDH)
			Homalomena philippinensis Engl.	Payaw	Obeña 7130 (PBDH) /Obeña 7055 (PBDH)
			Schismatoglottis calyptrata (Roxb.) Zoll. & Moritzi	–	Obeña 7140 (PBDH)
	Arecales	Arecaceae	*Saribus rotundifolius* (Lam.) Blume	Anahaw	Obeña 7040 (PBDH)
	Pandanales	Pandanaceae	*Freycinetia* sp.	–	
			Pandanus sp.	Bariw	Obeña 7163 (PBDH)
	Commelinales	Commelinaceae	*Tradescantia zebrina* var. zebrina	–	Obeña 7092 (PBDH)
	Zingiberales	Zingiberaceae	*Alpinia* sp.	–	Obeña 7062 (PBDH)
	Cucurbitales	Begoniaceae	*Begonia* sp.	Kulasiman	Obeña 7052 (PBDH)/Obeña 7161 (PBDH)
	Lamiales	Acanthaceae	*Gymnostachyum affine* Nees	–	–

		Gesneriaceae	*Cyrtandra* sp.	–	–
Unknown			Balukawi/Kawayan	–	Obeña 7142 (PBDH)
			Lukdo Bato	–	Obeña 7163 (PBDH)
			Luta	–	Obeña 7138 (PBDH)
			Marukpurok	–	Obeña 7124 (PBDH)
			No common name (SH 13)	–	Obeña 7059 (PBDH)

Nomenclature based on [1]PPG 1 (2016) and [2]APG IV (2016)

11.3 Results and Discussion

11.3.1 Plant Species Composition

A total of 26 understory plant species were recorded from the different established plots across the municipalities of Paranas, Taft, and Guiuan (Table 11.1). Among these were 6 fern species belonging to 6 genera from 6 families were recorded, whereas a total of 20 flowering plant species belonging to 14 genera and from 9 families were also recorded. Generally speaking, families represented by only one species were numerous compared to those represented by 2 or more species. The family with the greatest species richness was recorded to be Araceae (6), followed by Pandanaceae (2), whereas the remaining seven families were each represented by a single species. Five floral species were unidentified due to the absence of diagnostic features found in the reproductive parts.

Based on the forest floor census, the highest species richness occurred in Paranas, with a total of 18 understory plant species, followed by Taft with 12 species and Guiuan with 5 species (Table 11.2). Paranas was computed to have the highest Shannon values and was deemed to be the most diverse among the sites ($H' = 2.31$), and Guiuan with the lowest Shannon values and was the least diverse among sites ($H' = 1.29$). A lower species diversity in the forests over limestones of Guiuan may be due to the generally poor soil characteristics of the forest as compared to that of Paranas. According to Göltenboth et al. (2006), "Forests over limestone are usually a mosaic of rich and poor growth due to free-draining, steep slopes, water stress, high concentrations of calcium and magnesium, richer soils between jagged peaks and pinnacles." In addition, that may also be attributed to the possibility that the forests over limestone in the area are still young, especially due to the vulnerability of the area to natural calamities such as typhoons. Typhoon Haiyan was a fast-moving and extremely intense tropical cyclone that devastated the island in the year 2013 (Lin et al. 2014; Takagi and Esteban 2016).

The Importance Values (IV) of species were computed for each site and summarized in Tables 11.3, 11.4, and 11.5. Out of the 18 understory species in Paranas, the dominant plant species based on IV was *Pandanus* sp. (26.47%), belonging to the Pandanaceae family (Fig. 11.4a). It was computed to have a relative cover of 29.02% and a relative density of 23.91%. The same species also registered the most number, with 22 individuals. Similarly, in the established transects at Taft, among the 12 species, *Pandanus* sp. (50.41%) was the dominant species based on IV. It was computed to have a relative cover of 53.39% and a relative density of

Table 11.2 Species richness and calculated Shannon index of the diversity of select sites from Samar Island, Philippines

Sites	Richness	Shannon index of diversity
Paranas	18	2.31
Taft	12	1.75
Guiuan	5	1.29

Table 11.3 Importance values and rank of dominant understory species collected from the municipality of Paranas, Samar Island, Philippines

Scientific name	No. of individuals	Relative values (%)		Importance value (IV)	Rank
		Relative cover (RC)	Relative frequency (RF)		
Pandanus sp.	22	29.0180213	23.9130435	26.4655324	1
Cyclopeltis crenata (Fée) C. Chr.	20	23.2806179	21.7391304	22.5098742	2
Polystichum horizontale C. Presl	12	9.63589555	13.0434783	11.3396869	3
Tectaria dissecta (G. Forst.) Lellinger	6	4.70761309	6.52173913	5.61467611	4
Schismatoglottis calyptrata (Roxb.) Zoll. & Moritzi	4	4.52372196	4.34782609	4.43577402	5
Marukpurok	5	1.14012505	7.60869565	4.37441035	6
No common name (SH 13)	7	7.282089	1.08695652	4.18452276	7
Luta	2	2.0228025	5.43478261	3.72879255	8
Lukdo Bato	3	5.22250828	2.17391304	3.69821066	9
Gymnostachyum affine Nees	2	2.94225818	2.17391304	2.55808561	10
Homalomena philippinensis Engl.	1	0.51489518	3.26086957	1.88788237	11
Alocasia sp.	1	2.39058477	1.08695652	1.73877065	12
Balukawi/Kawayan	2	1.25045973	2.17391304	1.71218639	13
Piper sp.	1	2.24347186	1.08695652	1.66521419	14
Angiopteris sp.	1	1.87568959	1.08695652	1.48132306	15
Alocasia zebrina Schott ex Van Houtte	1	1.02979036	1.08695652	1.05837344	16
Selaginella sp.	1	0.58845164	1.08695652	0.83770408	17
Begonia sp.	1	0.33100405	1.08695652	0.70898028	18

47.42% and had the most number with 46 individuals. The dominant plant species in Guiuan based on IV is *Tradescantia zebrina* var. *zebrina* (41.74%), belonging to the family Commelinaceae (Fig. 11.4b). It was computed to have a relative cover of 40.41% and a relative density of 43.06%.

Stakeholders utilize a variety of plant species from the area. Locals have attested to consuming several *Begonia* species. They use it as a souring agent in soups and viands. Whereas other plant species, such as that of *Alocasia* sp. and *Aglaonema* sp. are used for horticultural and decorative purposes. The increasing popularity of plants similarly has led to increased poaching activities. Community awareness together with the help of Environmental agencies such as the Department of Environment and Natural Resources and the Local Government Unit is instrumental towards plant species protection and conservation.

Table 11.4 Importance values and rank of dominant understory species collected from the municipality of Taft, Samar Island, Philippines

Scientific name/ common name	No. of individuals	Relative values (%) Relative cover (RC)	Relative frequency (RF)	IV	Rank
Pandanus sp.	46	53.3887862	47.4226804	50.4057333	1
Cyclopeltis crenata (Fée) C. Chr.	18	17.7141097	18.556701	18.1354054	2
Freycinetia sp.	8	8.10227973	8.24742268	8.1748512	3
Lukdo Bato	3	5.29882933	6.18556701	5.74219817	4
Schismatoglottis calyptrata (Roxb.) Zoll. & Moritzi	4	3.23475046	4.12371134	3.6792309	5
Saribus rotundifolius (Lam.) Blume	6	2.83425755	3.09278351	2.96352053	6
Homalomena philippinensis Engl.	2	3.01910043	2.06185567	2.54047805	7
Angiopteris sp.	2	1.97165742	3.09278351	2.53222046	8
Alpinia sp.	3	0.77017868	3.09278351	1.93148109	9
Aslpenium vittaeforme Cav.	3	1.6327788	2.06185567	1.84731724	10
Cyrtandra sp.	1	1.50955022	1.03092784	1.27023903	11
Kulasiman	1	0.5237215	1.03092784	0.77732467	12

Table 11.5 Importance values and rank of dominant understory species collected from the municipality of Guiuan, Samar Island, Philippines

Scientific name	No. of individuals	Relative values (%) Relative cover (RC)	Relative frequency (RF)	Importance value (IV)	Rank
Tradescantia zebrina var. zebrina	31	40.4145078	43.05555556	41.7350317	1
Pilako	24	26.276832	33.33333333	29.8050827	2
Cyclopeltis crenata (Fée) C. Chr.	10	23.3160622	13.88888889	18.6024755	3
Aslpenium vittaeforme Cav.	5	7.84603997	6.944444444	7.39524221	4
Rhaphidophora sp.	2	2.14655811	2.777777778	2.46216794	5

Further analysis revealed that plant species were similar across different sites (NMDS ordination, Fig. 11.5). Plant species composition was similar among plots established in Paranas and Taft (ANOSIM, R global = −0.02366, $P > 0.05$) and between Taft and Guiuan (ANOSIM, R global = 0.009259, $P > 0.3518$). These similarities can be accounted for by the proximity of sites. A summary of the frequency of the occurrence of understory species is summarized in Table 11.6.

Fig. 11.4 Dominant plant species, based on the calculated Importance Values (IV), collected from Paranas and Guiuan, respectively. (**a**) Vegetative structures of Pandanus sp. and; (**b**) Vegetative structures of Freycinetia sp.

There are five common species between Paranas and Taft which included *Cyclopeltis crenata* (Fée) C. Chr., *Homalomena philippinensis* Engl. Lukdo Bato, *Pandanus* sp., and *Schismatoglottis calyptrata* (Roxb.) Zoll. & Moritzi. Common species between Taft and Guiuan are *Aslpenium vittaeforme* Cav. and *Cyclopeltis crenata* (Fée) C. Chr. On the other hand, plant species composition recorded in plots established in the sites Paranas and Guiuan was found to be different (ANOSIM, R global $= 0.03121$, $P < 0.05$).

11.3.2 Edaphic Factors Affecting Understory Plant Communities Across Select Municipalities of Samar

Environmental factors are equally important in biodiversity assessment and indicate overall environmental quality. The structural and functional characteristics of communities are shaped in accordance with habitat conditions (Protasov et al. 2019). Results from the soil analysis revealed variations in soil chemical characteristics across sites. Soil pH values across sites were identified to be slightly acidic (Table 11.7), and the acidic nature of soils can be explained by the intensive leaching of basic cations brought about by high rainfall (Calubaquib et al. 2016). Samar falls under Type II and Type IV climate according to the Modified Coronas Classification, where the former is characterized by having no dry season with a pronounced maximum rain period throughout the year (Abino et al. 2014). Generally, percent OM was high in all plots. Percent OM was highest in Guiuan (16.44%) and lowest in Taft (9.61%). Variation in average soil OM across sites can be

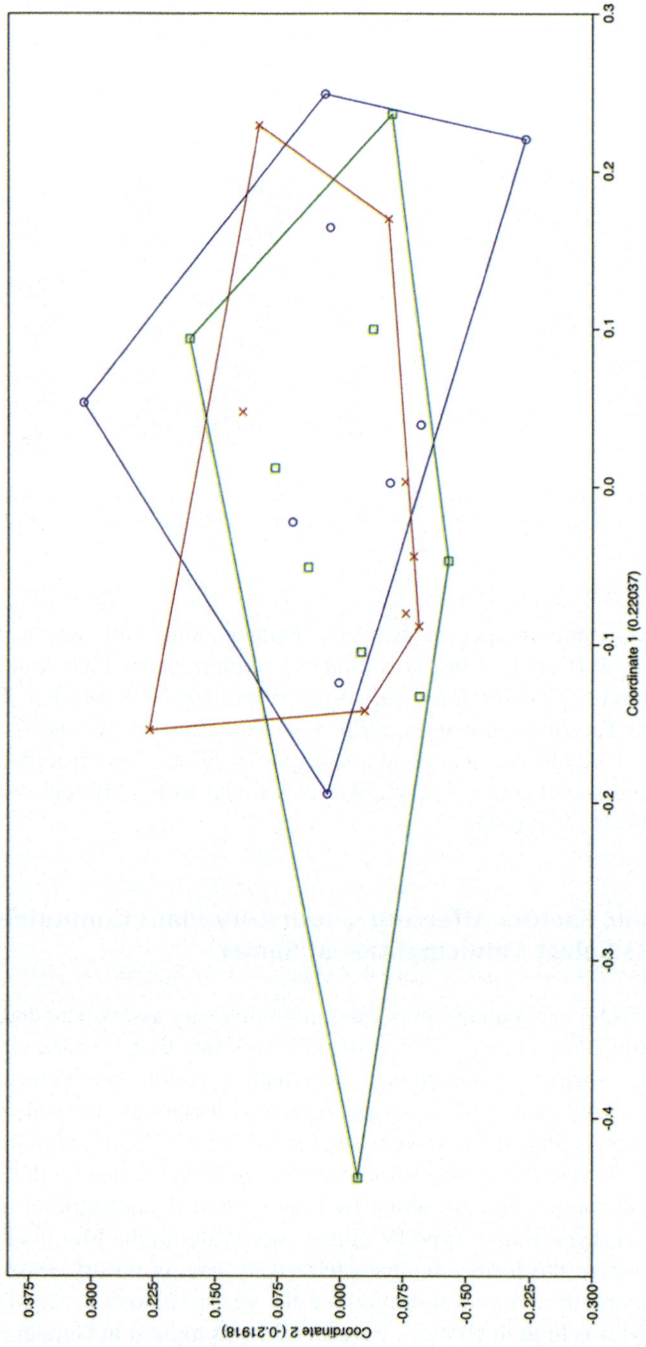

Fig. 11.5 Ordination (NMDS) of understory floral species abundance from 27 plots of 10×10 m^2 across three sites. Plots are symbolized as follows: Paranas (O, blue), Taft (X, red), and Guiuan (□, green). The low-stress value = 0.1475 indicates a good fit of the 2-dimensional representation of distances among plots

Table 11.6 Frequency of occurrence of recorded understory floral species recorded from established plots in select municipalities of Samar Island, Philippines

Scientific name	Paranas	Taft	Guiuan
Alocasia sp.	x	–	–
Alocasia zebrina Schott ex Van Houtte	x	–	–
Alpinia sp.	–	x	–
Angiopteris sp.	x	x	–
Aslpenium vittaeforme Cav.	–	x	x
Balukawi/Kawayan	x	–	–
Begonia sp.	x	–	–
Cyclopeltis crenata (Fée) C. Chr.	x	x	x
Cyrtandra sp.	–	x	–
Freycinetia sp.	–	x	–
Gymnostachyum affine Nees	x	–	–
Homalomena philippinensis Engl.	x	x	–
Kulasiman	–	x	–
Saribus rotundifolius (Lam.) Blume	–	x	–
Lukdo Bato	x	x	–
Luta	x	–	–
Marukpurok	x	–	–
No common name (SH 13)	x	–	–
Pandanus sp.	x	x	–
Payaw	x	–	–
Pilako	–	–	x
Piper sp.	x	–	–
Polystichum horizontale C. Presl	x	–	–
Rhaphidophora sp.	–	–	x
Schismatoglottis calyptrata (Roxb.) Zoll. & Moritzi	x	x	–
Selaginella sp.	x	–	–
Tectaria dissecta (G. Forst.) Lellinger	x	–	–
Tradescantia zebrina var. zebrina	–	–	x

explained by differences in vegetation cover, amount of leaf litter, and continuous rain-promoting decomposition.

Similarly, total N and P are high and above threshold levels in all plots. Percent N was recorded to be highest in Guiuan (0.63%) and lowest in Taft (0.47%). In a study conducted by Calubaquib et al. (2016), they found a positive relationship between OM and total N content in degraded soils collected in Luzon, Philippines. More than 95% of N is bound to soil OM (Scheffer and Schachtschabel 1992). Available P in soil samples was recorded to be highest in Guiuan (95.79 mg/kg) and lowest in Paranas (9.11 mg/kg). Electrical conductivity was highest in Guiuan (368.33 μS/cm) and lowest in Paranas (248.47 μS/cm). Interestingly, most of the analyzed edaphic factors are not typical of forests over limestones. Karst surface deposits are typically

Table 11.7 Average values of edaphic factors (pH, OM%, N%, P, & EC) analyzed from soil samples collected from plots established in Paranas, Taft, and Guiuan

Site	Plot	pH	OM %	N %	P mg/kg (olsen)	EC μS/cm
Paranas	1	6.4	3.12	0.47	5.5	154
	2	6.3	12.15	0.58	7.3	0.207
	3	7	37.97	0.74	12.9	507
	4	7.4	6.86	0.27	5.3	118
	5	6.7	5.87	0.3	7.9	329
	6	6.7	6.28	0.32	7.6	346
	7	6.3	11.81	0.44	11	273
	8	7.6	10.73	0.63	12.3	460
	9	7.1	10.66	0.56	12.2	49
	AVE	6.83	11.72	0.48	9.11	248.47
Taft	10	7.1	16.35	0.63	11.7	523
	11	7	16.35	0.63	11.7	523
	12	6.3	4.66	0.38	12.4	370
	13	5.9	9.05	0.36	*5.9	469
	14	4.9	12.81	0.58	*7.5	204
	15	6.1	8.87	0.37	9.4	535
	16	6.8	7.62	0.49	19.7	0.262
	17	6	6.09	0.24	*6.2	127
	18	6.3	4.7	0.52	12.3	110
	AVE	6.23	9.61	0.47	12.87	317.92
Guiuan	19	7.6	66.55	1.07	626	872
	20	6.7	10.91	0.57	17	547
	21	6.4	9.66	0.5	81.3	451
	22	6.6	10.31	0.62	19.5	209
	23	6.3	7.65	0.43	17	171
	24	6.5	11.48	0.69	17.7	46
	25	6.7	9.78	0.52	27	578
	26	7.1	10.48	0.59	44	314
	27	7.6	11.18	0.7	12.6	127
	AVE	6.83	16.44	0.63	95.79	368.33

*P Bray Method

characterized by a high pH, low organic matter content, and a heterogenous land surface (Abe et al. 2018) (Table 11.8).

The analyzed edaphic factors were found to correlate based on the resulting CCA with the first two axes explaining 36.68% of the variation and patterns among understory plant species (Fig. 11.6). Understory plant species recorded from Paranas and Taft are highly affected by Fe and EC. Plant species from Guiuan, on the other hand, is positively correlated with pH, OM%, N%, P, Cl, K, Cu, Mn, Mg, Na, and Zn.

Table 11.8 Average values of micronutrients analyzed from soil samples collected from plots established in Paranas, Taft, & Guiuan

Site	Plot	K (cmolc/kg)	Na (cmolc/kg)	Ca (cmolc/kg)	Mg (cmolc/kg)	Fe (ppm)	Zn (ppm)	Cu (ppm)	Mn (ppm)	Cl (mg/kg)
Paranas	1	0.17	0.67	11.58	1.05	9.99	1.36	82.53	88.42	161
	2	0.26	0.98	16.38	2.06	21.41	0.75	29.32	30.59	806
	3	0.3	1.46	28.38	3.33	5.96	2.1	6.76	5.94	698
	4	0.03	0.67	10.33	1.2	4.27	0.28	15.26	15.73	376
	5	0.1	0.36	7.32	1.68	7.87	0.48	58.69	63.56	322
	6	0.1	0.36	7.8	2.6	21.24	0.07	24.26	25.9	376
	7	0.13	0.91	13.3	0.13	34.78	0.18	34.52	37.38	537
	8	0.12	1.18	18.27	0.79	2.86	0.24	8.5	7.99	537
	9	0.16	0.8	16.72	1.98	10.31	1.32	36.34	39.18	322
	AVE	0.15	0.82	14.45	1.65	13.19	0.75	32.91	34.97	459.44
Taft	10	0.52	1.04	23.42	4.24	31.84	1.26	8.22	7.67	483
	11	0.1	0.6	7.71	1.08	4.94	ND	27.44	29.41	376
	12	0.12	0.64	7.7	1.34	19.81	1.57	96.22	101.78	376
	13	0.26	0.58	7.8	1.47	46.98	0.42	78.69	84.9	376
	14	0.2	0.42	2.95	1.12	436.94	2.33	15.73	17.34	483
	15	0.15	0.67	9.13	0.56	22.89	0.47	96.42	101.85	430
	16	0.2	1.27	14.06	1.25	33.88	0.56	36.61	39.77	376
	17	0.1	0.44	5.71	1.3	43.36	ND	56.75	61.68	591
	18	0.2	0.71	11.31	0.59	18.09	2.37	81.39	86.37	322
	AVE	0.21	0.71	9.98	1.44	73.19	1.28	55.27	58.97	423.67
Guiuan	19	0.5	2.66	31.02	6.16	6.92	20.33	1.81	8.83	752
	20	0.17	1.07	13.42	2.04	4.98	1.33	81.51	86.78	537
	21	0.11	0.8	11.31	2.43	7.95	6.31	123.29	128.41	376
	22	0.17	0.87	13.05	2.37	15.58	0.93	101.61	107.05	591
	23	0.08	0.51	8.12	1.77	11.06	0.17	96.24	102.08	376

(continued)

Table 11.8 (continued)

Site	Plot	K (cmolc/kg)	Na (cmolc/kg)	Ca (cmolc/kg)	Mg (cmolc/kg)	Fe (ppm)	Zn (ppm)	Cu (ppm)	Mn (ppm)	Cl (mg/kg)
	24	0.25	1.04	13.71	3.34	9.63	0.96	167.94	166.59	537
	25	0.13	1.08	10.07	3.62	14.16	2.36	2.09	160.95	430
	26	0.24	1.94	11.72	3.25	14	2.24	1.45	82.3	537
	27	0.21	1.29	15.23	1.64	2.01	0.09	34.76	37.7	269
	AVE	0.21	1.25	14.18	2.96	9.59	3.86	67.86	97.85	489.4

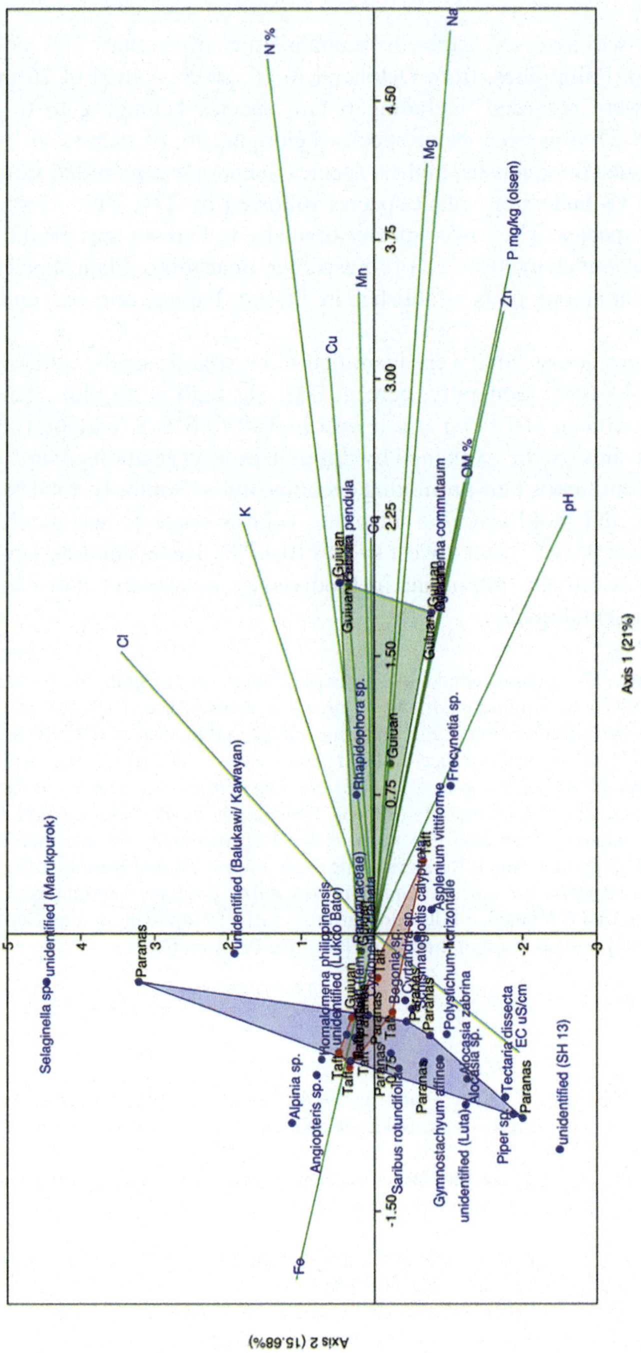

Fig. 11.6 Canonical Correspondence Analysis (CCA) of understory plant species with edaphic factors analyzed from soil samples collected from Paranas, Taft, and Guiuan. Plots are symbolized as follows: Paranas (O, blue), Taft (X, red), and Guiuan (□, green)

11.4 Summary and Conclusion

Floral diversity was assessed across the municipalities of Paranas, Taft, and Guiuan in Samar Island, Philippines, from October 1 to 16, 2019. A total of 26 understory plant species were recorded, including 6 fern species belonging to 6 genera in 6 families, and 21 flowering plant species belonging to 13 genera in 9 families. Based on the forest floor census, highest species richness was recorded from Paranas with a total of 18 understory plant species followed by Taft with 12 species and Guiuan with 5 species. The lower species diversity in Guiuan may be attributed to the poor general soil characteristics of forests over limestones. Plant species composition was similar among plots established in the sites Paranas and Taft and between Taft and Guiuan.

Soil pH values across sites were identified to be slightly acidic with an average pH range of 6.23–6.83. Generally, percent OM was high in all plots. Percent OM was highest in Guiuan (16.44%) and lowest in Taft (9.61%). Variation in average soil OM across sites can be explained by differences in vegetation cover, amount of leaf litter, and continuous rain-promoting decomposition. Similarly, total N and P are high and above threshold levels in all plots. Total average N was recorded to be highest in Guiuan (0.63%) and lowest in Taft (0.47%). Understanding key environmental factors is equally important in biodiversity assessment and can indicate overall environmental quality.

Acknowledgments The authors would like to express their sincere gratitude to Samar Island Natural Park (SINP), City Environment and Natural Resources Office (CENRO) of Catbalogan City, GMRPLS PAMB, Department of Environment and Natural Resources (DENR) Region 8, for the permit to conduct the study and collect soil and plant materials (DENR GP No. 2019-16). Also, our sincere gratitude to the kind people of the *Basaranan nga Organisasyon han San Isidro Samar* (BOSIS), GMRPLS, and the LGUs of Taft and Guiuan for their assistance and hospitality. Furthermore, the authors would also like to thank the Department of Science and Technology (DOST) and the Philippine Council for Agriculture, Aquatic and Natural Resources Research and Development (PCAARRD) through the Grants-in-Aid (GIA) program for funding the program (no. N9A6323), and DOST Science Education Institute (SEI) for allowing the last author to take part in this research program through the Career Incentive Program (CIP).

References

Abe T, Tanaka N, Shimuzu Y (2018) Plant species diversity, community structure and invasion status in insular primary forests on the Sekimon uplifted limestone (Ogasawara Islands). J Plant Res 131(6):1001–1014

Abino AC, Castillo JAA, Lee YJ (2014) Assessment of species diversity, biomass and carbon sequestration potential of a natural mangrove stand in Samar, the Philippines. For Sci Technol 10(1):2–8

Adorador JT (2019) Heterospanthe ferndandoi, a new palm (Arecaceae, Arecoideae, Arecaceae) from the Philippines. Phytotaxa 401(2):117–126

Adorador JT, Fernando ES (2019) Orania zheae (Oranieae, Arecoideae, Arecaceae), a new species from the Philippines, with notes on other slender-stemmed species. Phytotaxa 391(1):69–80

Adorador JT, Meneses ZD, Fernando ES (2020) Pinanga gruezoi (Arecaceae), a new slender clustering palm from the Philippines with notes an amended description of P. samarana. Phtyotaxa 429(2):120–134

APG IV (2016) An update of the Angiosperm Phylogeny Group classification for the orders and families of flowering plants: APG IV. Bot J Linn Soc 181(1):1–20

Calubaquib MAM, Navarrete IA, Sanchez PB (2016) Properties and nutrient status of degraded soils in Luzon, Philippines. Philipp J Sci 145(3):249–258

Chavez JG, Alejandro GJD, Tandang DN, Meve U (2020) Two new species of Timonius (Guettardeae: Rubiaceae) from Samar Island, the Philippines. Bot Lett 167:363–372

Clarke KR, Gorley RN (2006) PRIMER v6: user manual/tutorial. PRIMER-E, Plymouth

Fernando ES, Suh MN, Lee J, Lee DK (2008) Forest formations of the Philippines. ASEAN - Korea Environmental Cooperation Unit (AKECU). GeoBook Publishing Co. www.geobook.co.kr (2001) Securing the future for Asia's stunning karst

Göltenboth D, Langenberger G, Widmann P (2006) Special forest ecosystems. In: Göltenboth F, Timotius KH, Milan PP, Margraf J (eds) Ecology of insular Southeast Asia. Elsevier, pp 385–399. https://doi.org/10.1016/B978-044452739-4/50018-8

Lin II, Pun IF, Lien CC (2014) "Category-6" supertyphoon Haiyan in global warming hiatus: contribution from subsurface ocean warming. Geophys Res Lett 41(23):8547–8553

Madulid DA (2000) A review and assessment of the floristic knowledge of Samar Island. Samar Island Biodiversity Study (SAMBIO) Resources, Environment and Economics Center for Studies, Inc.(REECS). US Agency for International Development, 16 p

Meneses ZD, Cootes JE (2019) Pseuderia samarana (Orchidaceae), a new species and genus records from the Philippines. Taiwania 64(4):353–356

Meneses ZD, Malabrigo PL Jr, Buot IE Jr, Tinio CE (2018) New distribution records of orchids (Orchidaceae) in Samar Island, Philippines. J Nat Stud 17(2):1–9

Mueller-Dombois D (1974) Aims and methods of vegetation ecology. Wiley

Pelser PB, Barcelona JF, Nickrent DL (eds) (2011) Co's digital flora of the Philippines. Retrieved 6 April 2020 from http://www.philippineplants.org

PPG 1 (2016) A community-derived classification for extant lycophytes and ferns. J Syst Evol 54(6):563–603

Protasov A, Barinova S, Novoselova T, Sylaieba A (2019) The aquatic organisms diversity, community structure, and environmental conditions. Diversity 11:190

Scheffer F, Schachtschabel P (1992) Lehrbuch der Bodenkunde. Durchgesehene Auflage. FerdinandEnke Verlag, Stuttgart, 491 p

Takagi H, Esteban M (2016) Statistics of tropical cyclone landfalls in the Philippines: unusual characteristics of 2013 typhoon Haiyan. Nat Hazards 80(1):211–222

The Plant List (2013) Version 1.1. Published on the Internet; http://www.theplantlist.org/. Accessed 1 Jan

Tolentino PJS, Navidad JRL, delos Angeles MD, Fernandez DAF, Villanueva ELC, Obeña RDR, Buot IE Jr (2020) Review: biodiversity of forests over limestone in Southeast Asia with emphasis on the Philippines. Biodiversitas 21(4):1597–1613

Taxo-Ethnobotany of Genus *Ficus* L. in Jammu and Kashmir State (India)

12

H. S. Kirn, Anuradha Pandoh, and Roshan Kumar Yadhav

Abstract

Genus Ficus is one of the most widespread members of the family Moraceae. It comprises ca.800 species in the world and c.70 in India. It is chiefly distributed in tropical and subtropical countries of the world and is of great economic importance. As many as 19 species of this genus are found in Jammu and Kashmir. In the present communication, taxonomy, history, myths, and ethnobotany of the genus in Jammu and Kashmir (India) are presented. A key to identifying and describing different species in detail with the latest nomenclature is provided. Coloured plates of various species are appended herewith to facilitate easy identification.

Keywords

Ficus · Moraceae · Ethanobotany · Jammu and Kashmir · Taxonomy

H. S. Kirn
Shopping Centre Chowk, Bakshinagar, Jammu and Kashmir, India

A. Pandoh
Govt. College, Bishnah, Jammu and Kashmir, India

R. K. Yadhav (✉)
Nepal National Commission for UNESCO, Ministry of Education, Science and Technology, Government of Nepal, Kathmandu, Nepal

S. Ramamoorthy et al. (eds.), *Plant Diversity in Biocultural Landscapes*,
https://doi.org/10.1007/978-981-19-8649-9_12

12.1 History and the Myths

Genus ***Ficus*** is the Latin name for the common edible fig or fig tree, ***Ficus Carica*** **L**, which is native to Asia Minor and the Mediterranean region. The antiquity of the fig tree can be inferred from the use of its sewn leaves by **Adam** and **Eve** as first covering of their bodies and from other numerous references made to it in the holy Bible (Moldenke 1954). In Chinese, fig means a fruit without a flower (Condit 1984). There are different schools of thought regarding the origin and distribution of Ficus species in the world. Fig is grown since the dawn of civilization and the roving tribes probably transplanted it from place to place and thus ***Ficus Carica*** **L** is one of the earliest trees grown by the primitive man more than 4000 years ago. Fig species of genus ***Ficus*** have been named after religion ***(Ficus religiosa)***, distribution (***F. benghalensis***), product (***F. elastica***), the deity (***F. Krishnae***), or to some structures, e.g., ***F. lacor***. The English name Banyan for ***Ficus benghalensis*** is probably derived from the banyas who were favourite to it for running their business under its shade. In Ramayana, Britta is the name given for ***F. benghalensis*** or banyan tree. The name Pippal was given by the Aryans to ***F. religiosa,*** and its sanctity continues today in India (Sastri 1957). Plants of ***F. religiosa*** and ***F. racemosa*** are depicted on several steatite seals of Mohenjodaro (ca. 3000–2000 B.C) and in railings of Bharhut Stupa of first and second century B.C. (Saha 1981; Jain 1981).

There are several mythological stories in Puranas and other sacred literature concerning the ***Ficus religiosa*** and ***F. benghalensis.*** It is believed that Negrtos introduced the worship of ***Pippal* tree in India** (Sengupta 1965)**. It is believed that Lord Krishna used to sit on *pippal*** and ***Cadamb*** tree [***Anthocephalus chinensis*** (Lam.) Rich ex Walp.] after stealing clothes of bathing maidens to tease them. According to another myth, Brahma was transformed into a banyan tree and the latter is thus considered symbol of Brahma. It is also believed that when the demon Bhasmasura wanted to destroy Shiva and marry Parvati, Shiva stood firm and took form of banyan tree and thus it can never be destroyed (Agarwal 1981). The utility of asvatha or pippal is known to the Hindus since the Vedic era. In Kerala, no one dares to tell a lie under this tree. The eminence of the temple in south India is often judged by the height and age of pippal tree associated with it (Nair 1965).

Women desirous of having children often walk around ***Ficus religiosa*** and ***F. benghalensis*** chanting mantras. Some ladies place strips of clothes and having stones or bricks on the branches of these Ficus trees for the birth of a child (Agarwal 1981) The destruction of these trees is considered a sin by Hindus. ***Ficus racemosa*** is also considered sacred. The Sikhs have attachment with ***F. religiosa*** and ***F. benghalensis*** and they believe them as creators (Sengupta 1965). Fig seeds seldom germinate under their parent trees, and probably, the gastrointestinal tracts of birds and animals act as incubators before germination of fig trees in cracks of old walls, tree tops, etc. A famous banyan tree of Andhra Valley was said to be about 520 m in circumference with over 3000 trunks or aerial roots and able to shelter nearly 2000 people under it (Murray and Watt 1890). Trees of ***F. religiosa*** and ***F. benghalensis*** are venerated by Indians, particularly the Hindus, Buddhists, Jains, and other communities and are frequently planted near their houses, temples, villages and roadsides. Stones of various sizes placed under these trees on raised platforms represent Lord Shiva and some other deities (Devi & Devatas) and sometimes small shrines and temples are erected under these trees. Wherever a Hindu family lives, these two trees are often found. By growing upon walls, rooftops, constructions and even as epiphytes on other trees, they cause serious damage to the latter, but still devoted Hindus hesitate to eradicate or check their growth considering it a sin.

12.2 Geography and Physiographic Factors

The state of Jammu and Kashmir lies between **320 and 37 20** north latitude and **73 25 and 80 30** east longitude. It forms a majestic crown in the extreme north of the Indian Union covering nearly 640 km from north to south and 480 km from east to west. It occupies a total area of about 222,236 sq. km in north-western Himalaya. The state is wedged between Afghanistan and north-west frontier province of Pakistan on the west and north-west, Trukmenistan and China on the north and Tibet in the east. It is bordered on the south by Punjab and Himachal Pradesh. The state comprising 14 districts is chiefly rugged and mountainous except for the small tracks adjoining the Punjab plains and the valley of Kashmir. Total population of the state per census of 2001 stands at 10,069,917 with an average population density of about 99 persons per sq. km. Leh with an area of ca 82,665 sq. km is largest of all the districts of state. The entire state can be broadly divided into three main structural groups resulting from the stratified rock formations These are the Panjal, the Zanskar and the Tertiary groups. The Panjal comprises outer hills, outer plains, and middle mountains of lesser Himalaya, The Zanskar comprises the entire tract from Lahul and Spiti in the east to Karakoram in the north, and the Tertiary group is represented by the Valley of Kashmir and other river valleys.

Whole state of Jammu and Kashmir is a three-storeyed entity. The province of Jammu with shivaliks, the outer hills, and ravines on the south of Pir Panjal represent the Ist storey. The Kashmir Valley between the Pir Panjal range on south and south-east and the Great Himalayan range in the north and the north-east represents the second storey. The trans-Himalayan region comprising the territories of Gilgit, Baltistan, and Ladakh in the north of Great Himalayan ranges forms the third storey. The Pir Panjal and the Great Himalayan ranges divide the state almost into three distinct physiographic divisions which are isolated from one another by some lofty mountain barriers and passage from one to the other region is facilitated through many high altitude passes. The area to the south of Pir Panjal receives about 170 cm of average annual rainfall, the Kashmir valley about 105 cm, and the Ladakh region receives the minimum of about 10 cm of rainfall annually. July to September receive nearly 66% of the total rainfall in the foot hills and plains of Jammu province. The Kashmir valley receives the maximum rainfall during the spring season, which lasts from March to May. July to August and December to March are the two rainiest seasons in the ***Ladakh*** region. The total area under forests in Jammu and Kashmir is about 20,441 sq. km constituting 9.2% of the total geographic area of the state

(Anonymous 1956). These forests are in Jammu and in Kashmir Valley. The cold desert areas of Kargil and Leh are devoid of forests. There are 4 national parks and 15 wild life sanctuaries in the state covering about 14,870.22 sq. km thus representing 6.67% of the total area of the state. The forests in Jammu and Kashmir are of the following categories: the tropical dry deciduous forests, Himalayan subtropical pine forests, subtropical dry evergreen forests, dry temperate forests, moist temperate forests, subalpine forests, alpine forests and the dry alpine forests supporting their own type of vegetation.

12.3 Systematic Taxonomic Study and Ethnobotany

The genus ***Ficus*** was founded by Linnaeus in Species Plantarum 1:1059.1753; Lactotype: ***F. carica*** L. N.L Britton Fl. Bermuda from Indian region. Roxburgh's Flora Indica published in 1832 described 55 Indian species including 41 with his name as their author but of these only 15 were new species. King in J.D Hooker's Flora of British India (King 1888) recognized 112 species under this genus. According to Willis 1973 The genus ***Ficus*** comprises 800 species. The genus is represented by ca. 70 species in India (Santapau 1961). According to King (1888), 6 species of Ficus are endemic to eastern Himalaya, and the other 6 are endemic to South India.

Ficus is a large genus of monoecious and dioecious trees, shrubs, root climbers, and epiphytes with milky juice and often with adventitious roots. Leaves stipulate, alternate, spiral or distichous or opposite, entire, lobed or serrate to dentate along the margin. The receptacles usually contain male, female, gall flowers (sterile female flowers). The gall flowers provide shelter to minute wasp-like insects. Special gall insects are associated with the pollination of each species of Ficus. The modified receptacle, possibly a peduncle called syconium is the common fig or fruiting body. Amongst the trees, the most significant in the area of our study are ***F. benghalensis*** L., ***F. religiosa*** L, ***F. elastica*** Roxb. ex Hornem., ***F. Virens*** Ait., ***F. rumphii*** Bl, etc. They are followed by medium or small-sized **F. carica** L., ***F. hispida*** L.f., ***F. racemosa*** L., ***F.palmata***, ***F. cunia*** Buch. Ham., ***F. auriculata*** **Lour**, ***F. nerrifolia*** J.E Sm., ***F. beghalensis*** **var.** ***Krishnae*** (C.DC) Corner etc**.** ***Ficus pumila*** L. and ***F. sarmentosa*** Buch.-Ham ex Sm. are the climbers. The milky latex of some species of Ficus is used for the production of rubber commerce.

Several species of ***Ficus*** produce fruits of variable palatability. **Ficus carica** is chiefly grown for its edible and medicinal fruits. Figs dried and ground into flour are used for making bread and cakes in eastern and northern India. Figs have a laxative property. The chief nutritive value of fig is its rich sugar contents. Fig fruits are used as famine food in east India. The dried fruits of ***Ficus carica*** are sold in the market and are official in Pharmacopeia (Kirtikar and Basu 1935). The arborescent species like ***F. benghalensis***, ***F. religiosa***, ***F. benjamina*** **and** ***F. elastica*** are grown in gardens and along roadside for shadow and avenue purpose. Twigs and leaves of many species provide fodder to domestic animals. Ripe sweet fruits are relished by the visiting birds. The gum of Ficus species is used for trapping birds. The buttress roots are used as planks, but the timber is of poor quality and rarely used for constructions. Latices of several species of Ficus are anthelmintic. Milky juice, bark, leaves buds, seeds and root tips are used as medicines and have been described in many Materia Medica and other literature on the medicinal plant (Biswas and Ghosh 1973) and Bhartacharya (1956). The milky latex dropped over sugar discs locally called ‘Patasas’ eaten to enhance vigour by men.

12.3.1 Taxonomy Key to the Species of Genus Ficus in Jammu and Kashmir State

1.	Leaves opposite	*F. hispida*
	Leaves alternate or spiral	2
2.	Scandent or climbing shrubs	3
	Erect shrubs or trees	4
3.	Plant plants found clinging upon rocks or climbing upon trees in Forests; leaves isomorphous, equal at the base; figs. 1–1.3 cm across, globose	*F. sarmentosa*
	Plants grown in gardens and for covering boundary walls; figs. 4–5 cm long, pyriform	*F. pumila*
4.	Leaves spirally arranged (rarely alternate)	*F. arnottiana*
	Leaves alternate	5
5.	Leaves with prominent caudate –accuminate tip	6
	Leaves obtuse, acute or shortly acuminate at the tip	7
6.	Petioles terete; pinna with 4–8 cm long, linear, acumen at the tip, rounded or truncate at the base, entire; figs purple or black, depressed at the mouth	*F. religiosa*
	Petioles flattened or channeled, the pinna with less than 2.5 cm long at the tip, truncate or subcordate at the base, figs white with dark red or purple spots, not depressed at the mouth	*F. rumphii*
7.	Leaves with a prominent pouch at the base on adaxial surface	*F. benghalensis* var. *krishnae*
	Leaves without any pouch at the base	8
	Leaves equal sided or nearly so at the base, bearing a rounded auricle on one side at the base	*F. semicordata*
9.	Trees with prominent pillar-like aerial prop roots; leaves entire along the margin; figs red when ripe	*F. benghalensis*
	Trees or shrubs without pillar-like aerial prop roots; leaves subentire, toothed or lobed along the margin; figs yellow, purple or black (rarely red)	10
10	Figs yellow when ripe	11
	Figs other than yellow in colour when ripe	12
11.	Leaves distinctly toothed in the upper half, abruptly caudate or acuminate at the tip	*F. subincisa*
	Leaves entire rounded or subacute at the tip subacute at the tip	*F. elastica*
12.	Leaves inequilateral or slightly unequal at the base	13
	Leaves equal sided at the base	14

(continued)

13.	Leaves elliptic-ovate with lateral veins in 5–9 pairs; figs sessile; pink or mauve or purple when ripe	*F. microcarpa*
	Leaves elliptic-lanceolate with lateral veins in 7–11 pairs; Figs stalked, red when ripe	*F. nerriifolia*
14.	Plants without aerial adventitious roots; leaves ovate-orbicular or ovate-oblong	15
	Plants often with aerial adventitious roots; leaves elliptic-oblong or lanceolate-ovate	18
15.	Petioles channeled; leaves glabrous; figs white with red dots when ripe	*F. virens*
	Petioles terete; leaves scabrid or pubescent; figs other than white in colour when ripe	16
16.	Leaves 12–18 cm across, entire, glabrous or nearly so above; figs on special leafless, condensed shoots arising from the main trunk or the main branches	*F. auriculata*
	Leaves less than 12 cm across, dentate or serrate, scabrid above, sometimes lobed; figs solitary, rarely paired, axillary, without ribs	17
17.	Leaves membranous, truncate or subcordate at the base; figs purplish-black when ripe	*F. palmata*
	Leaves thick, cordate at the base; figs greenish purple when ripe	*F. carica*
18.	Leaves 3–7 cm long, elliptic-ovate, abruptly shortly acuminate at the tip, rounded at the base, 1-nerved from the base; figs sessile, solitary	*F. benjamina*
	Leaves 8–12 cm long, ovate-oblong or elliptic-lanceolate, acute at the apex, cuneate or subcordate at the base often infected with insect galls, 3-nerved from the base; figs peduncled, in large clusters on main branches or the trunk, red or purple when ripe	*F. racemosa*

Ficus arnottiana (Miq) Miq in Ann. Mus. Bot. Lugd. Bat 3: 287.1967; king in Ann. Roy. Bot. Gard. Calcutta 1:56, t. 68 A. and 84 V. 1887 and in Hook. f. Fl. Brit. Ind. 5:513.1888; Talb. For. Fl. Bombay Pres. and Sind 2: 516 f. 524.1911; Corner in Dassan and Fosb. Rev. Handb. Fl. Ceylon 3:244.f.10.1981; *Urostigma arnottiana* Miq in London Bot. 6:564.1847.

Vernacular names: **Hindi**-Paras pipal; **Malyalam**-Kallarayal; **Oriya-**Plakhyo; **Sanskrit**-Plaksha; **Tamil**-Kagoli; **Telgu**-Kalaravi.

A glabrous 5–8 m tall tree or a large shrub without aerial roots and pale smooth bark; leaves spirally arranged with pinkish veins and petioles, 5–12 × 4–8 cm, ovate-cordate, up to 2 cm long, the lateral veins in 5–7 paris; figs in axillary pairs and raised below fallen leaves, sessile, yellowish-brown or reddish-purple with green dots when ripe, 6–7 mm across globose, depressed; bracts blunt, caducous; peduncles 2–5 mm long; figs mature in March–July.

Occurrence and distribution: Occasionally found as introduced in drier areas.

Ethnobotany: Leaves and bark used in treatment of skin diseases and leaves are lopped for fodder (Ambasta et al. 1994). Leaves and bark used in skin diseases.

Ficus auriculata Lour., Fl. Cochinch. 666. 1790; Merr. in Trans. Phil. Soc. Philad. n.s. 24:BS. 1955; Corner, The Garden's Bull. Singapore 18: 33. 1960;

Sharma and Kachroo, Fl. Jammu 1: 295. 1981; *F. roxburghii* Wall. ex Miq. Ann. Mus. Lugd. Bat. 3:296.1867; King in Hook. f. Fl. Brit. Ind. 5:534.1888; Parker, Forest Fl. Punjab, Hazara and Delhi, 484. 1924.

Vernacular names: **English**-Even Apron; **Dogri**-Dhusi, Trembal, Trembu; **Hindi**-Timla, tirmal;

Pahari-Tussa, tussi.

An evergreen, small to medium-sized tree upto 8–10 m in height with thick bole, spreading crown, yellowish – grey, with almost smooth bark, and hollow branches; leaves alternate, stipulate, petiolate, 10–20 × 10–25 cm, broadly ovate or orbicular, usually entire, acute or obtuse, sometime repand-toothed, glabrous or sub-glabrous above, softly pubescent beneath, cordate or truncate at the base with 5–7 basal nerved and 3–4 pairs of lateral nerves; petioles 2–15 cm long; figs obovoid in clusters and supported by 3 large ovate-triangular bracts, depressed at the top, umbilicus, large, red or purple-orange when ripe in March–Aug.

Occurrence and distribution: Common in forests of ***Pinus roxburghii*** especially upon rocks. In Jammu, Kathua, Ramban, Udhampur, Reasi, Rajouri and Poonch. Ascending to ca. 1200 m.

Ethnobotany: Leaves are used in making plates ('Pattals') and saucers ('Doonas') commonly used in rituals and for worshiping. Leaves are lopped for fodder, and bark is used for rope making. Ripe fruit is edible and laxative. Also made into jams and jellies (Plate 12.1. 1–9).

Ficus benghalensis L., Sp. Pl. 1059.1753; King in Ann. Roy. Bot. Gard. Calcutta 1: 18, t. 13. 1887 and in Hook. f. Fl. Brit. Ind. 5: 499. 1888; Sharma and Kachroo, Fl. Jammu 1: 294. 1981; Parker, Forest Fl. Punj. Hazara and Delhi 479. 1983 (Repr.)

Vernacular Names: **Bengali**-Bar, bot; **English**-Banyan tree; **Dogri**-Bar, Borh; **Gujrati**-Vad, vadlo, var.; **Hindi**-Bar, bargad, bar**; Kanad**-Ala, alada, mara, vata; **Malyalum**-Ala, vatam; **Marathi**-Vada, vad, var.; **Sanskrit**-Bahupada, vata; **Tamil**-Al, alam; **Telgu**-Marri, peddamarri.

Large, evergreen, tree with grey smooth bark, much spreading branched crown and prop roots giving support to the spreading heavy branches, young part softly pubescent; leaves stipulate petiolate, 10–20 × 4–12 cm, ovate or elliptic, entire obtuse, coriaceous, rounded-subcordate or slightly narrowed at the base, 3–7 nerved, lateral nerves about 5 pairs, prominent; petiole 1–5 cm long, not jointed to the blade; figs. 1–2 cm in diameter, supported by 3 rounded bracts, sessile, in axillary pairs, globose, puberulous, red when ripe in October–November and in April–July.

Occurrence and Distribution: Fairly common in warmer places ascending to ca. 800 m alt. in forests and also planted around habitations. Generally, it is an epiphyte when young and is more common on walls of old buildings and wells. Common in Jammu, Kathua, Udhampur, Reasi, and Rajouri districts.

Ethnobotany: Fruits are eaten in times of scarcity. Leaves lopped for fodder. Latex is locally applied in rheumatism and lumbago. Infusion of bark is considered tonic and astringent, useful in diarrhoea, diabetes and dysentery. Root fibres used in gonorrhea and leaves applied as poultice to abscess. Leaves and seeds are tonic and cooling (Chopra et al. 1956). Wood is suitable for making paper pulp, also used in cheap furniture. Fibre from bark and prop roots is used for ropes. Drops of its latex

Plate 12.1 1—*Ficus carica* 2&3—*F. palmata* 4—*F. benghalensis* var. *krishnae* 5—*F. rumphi* 6—*F. pumila* 7—*F. sarmentosa* 8—*F. arnotiana* 9—*F. hispida*

are put upon sugary discs locally called 'Pattasas' eaten by men to increase their vigor and vitality.

Ficus benghalensis. L. **var.**. Krishnae (C.DC.) Corner in Gard. Bull. Singapore 21:14.1965 & in Dasson and Fosb., Rev. Handb. Fl. Ceylon 3:253.1981; *F. Krishnae* C. DC in Bot. Mag. t. 8092. 1906; Naik, Fl. Marathwada 2:811.1998.

Vernacular Names: **Dogri-**Krishan Borh; **English-**Krishna's butter cup; **Hindi-**Krishan Makhan Katori

A small or medium-sized much-branched tree with dull white bark. Trunk with distant, horizontal wrinkles; leaves alternate, petiolate, 15–25 × 5–10 cm, ovate-lanceolate, acute or bluntly acuminate at the tip, cuneate and forming a pouch at the base, leathery and entire along the margin; figs reported from October to February.

Occurrence and Distribution: Grown in gardens for its peculiar pouched leaves.

Ethnobotany: Grown in gardens.

Ficus benjamina L. Mant. Pl. 129.1767; King in Ann. Roy. Bot. Gard. Calcutta 1: 43, t, 52. 1887 and in Hook. f. Fl. Brit. Ind. 5: 508. 1888; Corner in Dassan. and Fosb. Rev. Handb. Fl. Ceylon 3:257.1981.

Vernacular names: **Assam**-Chillubor; Bengali-Pukar; **Bombay**-Pimpri; **Dogri**-Chilkan, Rondu and Rumbal; **English**-Weeping fig, willow fig; **Malayalam**-Purra-juvi.

Small to medium-sized evergreen trees, with umbrella-like canopy, grey bark and drooping branches, often with few aerial, withering roots; leaves 3–7 × 2–5 cm ovate-elliptic, shining, leathery, rounded or tapering at the base, pointed or abruptly shortly acuminate at the tip, figs axillary, mostly solitary, sometimes in pairs, subglobose or pyriform, 1.5–2 cm across, the figs appear in April–July and are greenish-yellow to orange or dark red when mature and subtended by 3 bracts at the base (Plate 12.2. 1–9).

Occurrence and distribution: Often found grown as an ornamental hedge in gardens and along boundary walls.

Ethnobotany: Decoction obtained from leaves is mixed with oil, which is applied to ulcers. The milky juice is used against the whitening of the cornea.

Ficus carica L.sp.Pl. 1059.1753; Cooke, Fl. Pres. Bombay 3:155.1958 (Repr.); Naik, Fl. Marathwada 2:810.1998; Singh et al. Fl. Maharashtra state 2:947.2001

Vernacular names: **Bombay**-Anjira; **English**-Fig; **Dogri**, **Hindi**-Anjeer; **Punjabi**-Fagari; **Sanskrit**-Anjira; **Tamil-**Simayatti; **Telgu-**Anjuru.

5–10 m tall deciduous tree with pubescent branches and no aerial roots; leaves stipulate, petiolate, 8–15 × 6–15 cm, broadly ovate, thick, cordate at the base, acute or subobtuse at the tip, usually 3–5 lobed, pubescent on both the surfaces; figs. 3–5 cm across, pyriform or obovoid, axillary, solitary green-purple or yellow when mature, finely pubescent upon the surface. The mature figs appear in May–July.

Occurrence and Distribution: Native of Syria and Palestine and grown on a small scale at Batote, Banihal, Thathri, and Mughal maidan and elsewhere in J&K state.

Ethnobotany: Fruit is demulcent, emollient, aperient, and acrid and tonic, used to treat warts.

Ficus elastica *Roxb.* Ex Horn. Hort. Bot. Hatn. Suppl. 7. 1819; King in Ann. Roy. Bot. Gard. Calcutta 1:45,t.54. 1887 and in Hook. f. Fl. Brit. Ind. 5:5081888; Naik, Fl. Marathwada 2: 810.1998.

Vernacular names: **Assam & Bengali**-Bor, Attah bar; **Dogri**-Rabri borh; **English**-Indian rubber, Assam rubber.

A large or medium-sized, evergreen tree with the roots spreading superficially and many exposed above the ground; leaves petiolate, 12–25 × 8–12 cm, oblong or elliptic, rounded or acute at the base, subacute at the tip, coriaceous, glossy, green

Plate 12.2 1—*Ficus benjamina* 2&3—*F. racemosa* 4—*F.auriculata* 5—*F. benghalensis* 6—*F. hispida* 7—*F.microcarpa* 8—*F. semicordata* 9—*F. nemoralis*

above or variously variegated, entire and with numerous, fine parallel nerves, petioles 5–7 cm long; figs axillary and on twigs below leaves, yellow when ripe.

Occurrence and Distribution: Native of eastern Himalaya and Malaya, grown in gardens and parks as ornamental.

Ethnobotany: Planted in gardens as ornamental tree or for avenue purposes along roadsides. Also planted for extraction of rubber.

Ficus hispida L.f. Suppl. 442.1781; King in Roy. Bot. Gard. Calcutta 2:116, t. B4. 1888 and in Hook. f. Fl. Brit. Ind. 5: 522.1888; Cooke, Fl. Pres. Bombay 3: 154.1958 (Repr.); Sharma and Kachroo, Fl. Jammu. 1:192.1981; Parker, Forest Fl. Punj. Hzara and Delhi 482. 1983.(Repr.).

Vernacular names: **Bengali**-Dumoor, Kakodumoor**;** **Dogri**-Kharkhamba; **Gujrati**-Dhedaumara, Jangligangeer; **Hindi**-Daduri, Dagurin, Gobla, Koksha; **Kanad**-Adviatti, Kadatti; **MadhyaPradesh**-Bhuigular, Kathgular; **Malyalam**-Erumnakku, Peyatti; **Marathi**-hokada, Bokaria, Kalumber, Khroti; **Sanskrit-**Kakadhumbura; **Tamil**-Peyatti, Sonatti; **Telgu**-Bodhamamidi, Brammadi.

An evergreen shrub or a small sized tree, generally with hollow, hispidly pubescent branches and greenish-grey, almost smooth bark; leaves opposite, stipulate, petiolate, 10 × 30 × 5–18 cm, ovate-oblong, acute or shortly acuminate, toothed, hispidly scabrid above and hispid beneath, cuneate, rounded or subcordate at the base; stipules 10–15 mm long; ovate-lanceolate; petioles 1–6 cm long; figs usually clustered on special elongated or short leafless branches and from the base of the main trunk, sometimes axillary on young plants, 1–2.5 cm in diam. Obovoid, hispid, supported by 3 small bracts and bearing small brown scales on the surface also, yellow when ripe in July–September.

Ethnobotany: Bark, fruits, and seeds are emetic and purgative. Crushed fruits in warm water given to cattle as lactagogue (Satya and Solanki 2009). Leaf decoction is given in jaundice and fruits are used in diarrhea and dysentery (Khanna and Kumar 2009).

Occurrence and Distribution: Common along canal banks, in waste lands and forests in Jammu, Kathua, Udhampur and Reasi districts ascending to ca. 700 m alt.

Ethnobotany: Fruits edible, also made into jam. Immature fruits are curried but may cause giddiness. Considered tonic, lactagogue, and emetic. Bark contains tannin and yields bark (Ambasta et al. 1994).

Ficus microcarpa L.f, Suppl. 442.1781; Corner in Gardens Bull. Singapore, Singapore 17: 397. 1959 F. retusa sensu King in Ann. Roy. Bot. Gard. Calcutta 1: 50,t. 61,62. 1887 non L. 1767, King in Hook. f. Fl. Brit Ind. 5: 511.1888; Cooke, Fl. Pres. Bombay 3: 146. 1958 (Repr.); Talb., For. Fl. Bombay Pres. and Sind 2: 510, f. 520. 1911, Mathew, 1llus. Fl. Tamil Nadu Carnatic 2:t.676. 1982.

Vernacular Names: **Dogri**-Chilkin; **English**-Chinese banyan;

Evergreen tree with few conspicuous, slender aerial roots and grey bark; leaves petiolate, subdistichous, elliptic –ovate, 6–8 cm long, sub-inequilateral, cuneate at the base, subacute to bluntly subacuminate at the apex, leathery, with 5–8 pairs of lateral veins; petioles 1.5–2.2 cm long, puberulous; figs sessile, in axillary pairs, pink to mauve or purple and black, sometimes white when mature, subtended by 3 spathulate-ovate bracts at the base, ripe in April–July.

Occurrence and Distribution: Occasionally found grown in gardens or an escape from cultivation.

F. nemoralis Wall ex Miq. Hooker's London J. Bot. 7: 453. 1848; King in Hooker. f. Fl. Brit. Ind. 5: 534. 1888 p.p; Parker, Forest Fl. Punj. Hazara and Delhi; 484. 1983 (Repr.) *Ficus nerifolia* J.E. Smith var. nemoralis (Wall. ex Miq.) Corner in Garden Bull. Singapore 17: 426. 1959.

Vernacular Names: **Basohli**–Padola;

A small good looking tree with greyish-brown, glabrous, and almost smooth bark; leaves stipulate, petiolate 6–15 × 2.5 cm, lanceolate, ovate, or elliptic, narrowed to a long acurnen, membranous, entire, usually cuneate and 3-nerved at the base, lateral nerves in 7–14 pairs, wide spreading, the veins prominent, darkly coloured beneath; stipules lanceolate, convolute, 10–15 mm long, reddish, not jointed with the blade; figs. 6–7 mm long, supported by 3 broad connate bracts at the base, axillary, solitary or in pairs, subglobose, umbilicus, glabrous, red when ripe; peduncles 5–6 mm long. The figs ripe in May to October.

Occurrence and Distribution: Sporadic around habitations, along roadsides on rocky slopes and in open forests ascending to ca1300m alt. Basohli, Poonch, Rajouri, Jammu, Kathua, Reasi, Ramban, and Udhampur districts.

Ethnobotany: Figs are edible. Leaves are lopped for fodder but may cause haematurea in cattle.

Ficus palmata Forssk. Fl. Aegypt.–Arab. 179. 1775; King in Hook. f., Fl. Brit. Ind. 5:530.1888; Collet, Fl. Simlensis f. 148. 1921; Parker, Forest Fl. Punjab, Hazara and Delhi 483. 1924; Sharma and Kachroo, Fl. Jammu 1: 293. 1981.

Vernacular names: **Dogri**-Fawara, Phagwara**; Gujrati**-Pepri**; English**-Common figs; **Hindi**-Anjiri, Bedu, Khemri; **Punjabi and Pahari**-Phagwara**; Telgu**-Manjimedi.

A deciduous shrub or a small tree up to 10 m in height with grey bark and grey-tomentose to almost glabrous shoots; leaves alternate, petiolate, variable in shape, 4–15 × 3–12 cm or more, 3–5 lobed or broadly ovate-orbicular, all sometimes born upon the same plant, scabrid above, pubescent or tomentose beneath, rounded or trucate at the base, obtuse or acute at the tip, crenate-dentate along the margin, 3 nerved at the base and with 3–6 pairs of lateral nerves; stipules 5–7 mm long, ovate, acute; petioles 2–5 cm long, figs. 1.2–2.7 cm across supported by 3 or more deciduous basal bracts, axillary, solitary or in pairs, pyriform or globose, purple when ripe in April–August.

Occurrence and Distribution: Fairly common in hedges, along water courses, roadsides, near habitations, and forests in all districts of Jammu and Kashmir but for Ladakh ascending to ca. 2000, alt.

Ethnobotany: Fruit is edible when ripe. It is demulcent and laxative, used in diseases of lungs and bladder. Leaves are lopped of fodder. Wood is used as firewood and in construction. Lates of leaf stalk is put inside the wound caused due pricking by thorns for healing purposes. It also facilitates the extraction of broken spines or thorns from the flesh by the Pahari people in Poonch (H.S.Kirn). In Baderwah of Jammu and Kashmir state the fruits are eaten as laxative against constipation and pulmonary disorders (Kapur and Nanda 1996).

Ficus pumila L., Sp. Pl. 1060.1753; King in Ann. Roy Bot. Gard. Calcutta 2:124. t, 158. 1888; Naik, Fl. Marathwada 2: 812. 1998; Singh et al. Fl. Maharashtra state 2: 948. 2001.

Vernacular names: **English**-Creeping fig.

Perennial evergreen, woody climbing shrub with variable, dimorphic leaves; leaves and branches rooting from the nodes; leaves stipulate, petiolate,

1.3–4 × 1–3 cm, ovate or ovate-oblong, cordate, and unequal sided at the base, acute, entire, the leaves on sterile rooting branches 2–2.5 cm or less long; petioles 6–12 mm or more longer; figs axillary, mostly solitary, 3–5 cm long, pyriform, on spreading branches, purple-black when mature.

Occurrence and Distribution: Native of China and Japan introduced elsewhere.

Ethnobotany: Commonly grown to cover the boundary walls and as epiphyte upon large Palm trees in townships and elsewhere. Fruit is edible and also used for jelly making.

Ficus racemosa L., Sp.Pl. 1060. 1753; Corner in Dansk. Bot. Arkiv. 23: 28. 1963; Nair, Fl. Punjab Plains 243. 1978; Sharma and Kachroo, Fl. Jammu 1: 294. 1981; *F. glomerata* Roxb., Pl. Corom.

2: 18, t.123. 1798; King in Ann. Roy Bot. Gard. Calcutta 2: 173, t. 218. 1888 and in Hook. f., Fl.

Brit. Ind. 5: 535. 1888; Parker, Forest Fl. Punj. Hazara and Delhi. 484. 1924.

Vernacular names: **Bengali**-Dumur, Jagyadhumbar; **Bombay**-Umbar; **English**-Cluster fig; **Dogri**-Rumbal; **Gujrati**-Umar, Gular; **Hindi**-Gular, Umar; **Jharkhand**-Loa; **Kanad and Mayalam**-Athi; **Marathi**-Umbar; **Oriya**-Dimri; **Sanskrit**-Udumbara; **Telgu**-Atti, Bodda.

Large deciduous, 5–12 m tall tree, much branched with a spreading canopy, and greyish-white, smooth bark; leaves alternate, stipulate, petiolate, 8–15 × 5–10 cm, ovate to ovate-oblong or elliptic, subcoriaceous, entire, glabrous, cuneate or blunt at the base and subacute to abruptly blunt at the apex with 3 nerves from the base and 4–8 pairs of lateral nerves, often heavily affected by insect galls; stipules 1–2 cm long, ovate-lanceolate; petioles 2–5 cm long; figs lustered on short leafless warted shoots, rarely axillary, 2.5–4 cm in diam., pyriform or obovoid, white-pilose, with thick wall and supported by 3 small triangular-ovate bracts, ripe in May–August.

Occurrence and Distribution: Common along the water courses in ravines, along roadsides and rocky hill-slopes in Jammu, Kathua, Poonch, Rajouri, Ramban, Reasi, and Udhampur districts ascending to ca. 1800 m alt.

Ethnobotany: Leaves are lopped for fodder. Roasted fruits are made into powder and used as breakfast food. Leaves are also considered useful in bilious infections. Bark is fed to cattle in rinder pest disease. Root is used against diarrhoea and diabetes. Fruits considered stomachic and carminative, used in haemoptysis. Latex is useful in piles and diarrhea, also used for bird-lime. Timber is used for cheap furniture, match boxes. Bark yields 14% tannin. Decoction of bark is used as vulnerary (Ambasta et al. 1994). Bark is astringent and given to cattle suffering from rinder pest, and powder is given for dysentery. The sap of the root is taken to cure diabetes. Powdered leaves mixed with honey are given to bilious infections. Fruit is astringent, carminative, and stomachic given in menorrhagia and haemoptysis. Latex is useful in the treatment of piles and diarrhoea (Chopra et al. 1956). Fruits are pickled in goat milk and taken daily in the morning and evening for about a year to get complete relief from tuberculosis (Goel and Aswal 1990). Fruit juice is given with sugar candy three times per day to cure jaundice (Saren et al. 2009).

Ficus religiosa L., Sp. Pl. 1059. 1753; King in Ann. Roy. Bot. Gard. Calcutta 1: 55, t. 67. 1887; Hook f. Fl. Brit Ind. 513. 1888; Cooke, Fl. Pres. Bombay 3: 149. 1958 (Repr.); Talb. For. Fl. Bombay Pres. and Sind 2: 514. f. 1911; Sharma and Kachroo, Fl. Jammu 1:293.1981; Parker, For. Fl.Punjab, Hazara and Delhi. 480. 1983 (Repr.)

Vernacular names: **Bengali**-Ashathwa; **Bombay**-Asvatha; **English**-Bo tree; **Dogri**, **Hindi**, **Punjabi**-Pipal, Peepal; **Gujrati**-Jari, Pipro, pipul; **kanad**-Arali, ashwattha; **Malyalam**-Arachu, Aryal, Ashwathsm; **Sanskrit**-Ashwatha pipala; **Tamil**-Arasu, Aswatham; **Telgu**-Ashvathamu.

A large deciduous tree up to 15–25 m in height with spreading crown, bark grey, smooth when young, exfoliating in irregular scales when old; leaves alternate, petiolate, 10–18 × 6–12 cm, broadly ovate-abruptly narrowed in a linear-lanceolate, 5–8 cm long acumen at the tip, truncate at the base, entire, red when young, sub-coriaceous, glabrous, shining 5–7 nerved from the base and with 6–8 pairs of lateral nerves; stipules minute, ovate, acute; petioles 7–10 cm long, slender, terete, jointed to the blade; figs sessile, in axillary pairs 5–8 mm in diam, supported by 3 rounded bracts, obovoid-globose, depressed at the mouth, smooth, dark purple or black when ripe in March–November.

Occurrence and Distribution: Common along roadsides, near habitations and temples, along water courses, as epiphyte and on old walls and in forests ascending to ca. 2500 m alt.

Ethnobotany: Fruits and tender buds are eaten during scarcity and they are laxative. Leaves are lopped for fodder. Hardened latex is used to fill cavities in ornaments and is also the source of bird lime. Wood is used for packing cases, and occasionally for making spoons, bowls, wheels and match boxes. Infusion of bark is used against ulcers and skin disorders. Bark contains tannin and also used for the extraction of fibre (Ambasta et al. 1994). Bark is astringent used in the treatment of gonorrhea. Infusion of bark is taken to cure scabies. Leaves and young shoots are purgative. Seeds are alterative and cooling. Fruit is laxative used against constipation (Chopra et al. 1956). Leaves and tender shoots are used externally in the treatment of skin diseases Seeds and fruit have a cooling effect when eaten and are laxative (Kumar and Naqshi 1990).

Ficus rumphii Bl., Bijdr. 437.1825; King in Ann. Roy. Bot. Gard. Calcutta 1:54, t. 67b and 84. 1887 and in Hook.f. Fl. Brit. Ind. 5: 512. 1888; *F. cordifolia* Roxb. Fl. Ind. 3: 561. 1832. non Bl. 1825; Talb. For Fl. Bombay Pres. And Sind 2: 514. 1911; Parker, Forest Fl. Punj. Hazara and Delhi 480.1983 (Repr.); Cooke, Fl. Pres. Bombay 3: 148. 1958 (Repr.);

Vernacular Names: **Assam**-Jakhri, Prap; **Bengali**-Gajaswat; **Hindi**-Gagjaira, Gajna, Kabar, Pokhar, Pilkhan; **Kalakot and Pahari**-Pilkhoie, Piplar, Pipli; **Kanad-**Betyarali; **Marathi**-Asht, Pair, Payar; **Punjabi**-Badha.

Medium sized or Large deciduous glabrous tree, 8–15 m in height, epiphytic in early stages with grey or yellowish – grey, smooth bark; leaves alternate, stipulate, petiolate, 10–15 × 5–10 cm, broadly ovate, truncate or subcordate or slightly narrowed at the base, caudate-acuminate at the apex, the acumen less than 2.5 cm, entire or subundulate along the margin, 3–5 or up to 7 nerved from the base and with

3–6 pairs of lateral irregular nerves, prominent when young; stipules 1–2.5 cm long, ovate, ovate-lanceolate; petioles 5–8 cm long, flat or channeled above, jointed to the blade; figs. 10–12 mm in diam., supported by 3 rounded bracts, sessile, in axillary pairs, globose, smooth, white with dark spots when young, black when ripe supported by 3 rounded bracts at the base in June–August.

Critical note*:* This taxon is often mistaken for ***F. religiosa*** L., which however, differs in having larger leaves with terete petioles, abruptly narrowed into longer (3–8 cm long) acumen and the figs depressed at the tip.

Ethnobotany: Leaves lopped for fodder. Ripe fruits are occasionally eaten. Bark yields a cordage fibre. Fruit juice is used to kill worms and for the treatment of asthma (Ambasta et al. 1994). Bark is used to treat snake bite. Juice of fruit given internally with turmeric, pepper, and ghee in asthma (Chopra et al. 1956).

Ficus sarmentosa Buch.-Ham. ex J.E. Smith in Rees, Encyl. 14: 45, 1810; Corner, Garden Bull. Singapore 18: 6. 1960*; F.foveolata* Wall. Ex Miq. In Ann. Mus. Bot. Lugd. Bat. 3; 294. 1867; King in Hooker,f. Fl. Brit. Ind. 5: 528. 1888 pp. (excluding syn. *F. nipponica* Fr. et. Sav., *F. thunbergii* Maxim., *F.luduca* Roxb. and F. impressa Benth.); Sharma and Kachroo in Fl. Jammu 1: 295.1981; Parker, Forest Fl. Punj. Hazara and Delhi 482. 1983 (Repr.)

Vernacular Names: **Dogri**-Ban-phagaru.

An evergreen climbing often rooting from the nodes and bearing pubescent, brown twigs with rough and raised dots young shoots, petiole and venation beneath rusty-pubescent; leaves on fruiting shoots 4–12 × 1–4 cm, ovate, elliptic or lanceolate, acute or acuminate entire glabrous and shining above, reticulately veined, ultimate nerves prominent beneath with pore-like depressions between them, base 3-nerved, the lateral nerves in 5–8 pairs; leaves on sterile shoots usually 4–5 cm long with ca. 12 mm long petioles; stipules ovate-lanceolate, 10–12 mm long figs pedunculate, axillary, solitary, ca. 1 cm across, globular, bracts, purple when ripe in March –June.

Occurrence and Distribution: Very common creeper over trees, rocks along banks of nallahs in damp forests ascending to ca 2600 m alt.

Occurrence and Distribution: Rare in Kalakot (Rajouri) and Ram Nagar (Jammu) forests.

Ethnobotany: Leaves lopped for fodder.

Ficus semicordata Buch—Ham. ex Sm. In Rees. Cyclop. 14: n.711810; Corner in Garden's Bull. Singapore, Singapore 17: 449. 1959; Bennet, Name changes in Flowering plants of India and adjacent regions 236.1987; Ficus cunia Buch-Ham. ex Roxb., Fl. Ind. 3: 561.1832; King in Hook.

f. Fl. Brit. Ind. 5: 523. 1888 excl. syn. *F. conglomerata* Roxb.; Sharma and Kachroo Fl. Jammu.

294. 1981; Parker, For. Fl. Punjab. Hazara and Delhi 482.1983 (Repr.).

Vernacular names: **Bengali**-Dumbur, Jagyadumur; **Dogri**-Khin KHini, pakh; **Hindi**-Jharphali, khain, khenan, Kheuwnau; **Kanad**-Garagasa; **Malyalam**-Perina, poroh, teregam; **Marathi**-Porodumer; **Tamil**-Taragadi; **Telgu**-Bommamarri; **Jharkhand**-Para.

Small, evergreen, tree of irregular habit without aerial roots, with dark-grey bark and pubescent young shoots; leaves stipulate, petiolate, elliptic-lanceolate or oblong, variable in size, usually 10–20 × 5–7 cm, sometime larger, entire or serrate very unequal and semi-sagitate, with a peculiar 3–4 nerved rounded auricle on one side only at the base, acuminate, pubescent or subglabrous on both the surfaces, occasionally smooth above; stipules 1–2 cm long, linear lanceolate; petioles 5–15 mm long, stout, scabrid, not jointed to the blade; figs in pairs or small clusters on long, leafless pubescent or scally shoots from the old branches or from the main stem near the base, shortly pedunculate, 8–12 mm across, densely hairy, globular or pyriform, reddish-brown when ripe almost throughout the year.

Occurrence and Distribution: Sporadic on steep banks of ravines, roadsides on rocky slopes near Chowki Chowra and Nandani in Jammu, Kathua, Udhampur, Reasi, and along kalkota road in Rajouri ascending to ca. 1000 m alt.

Ethnobotany: Fruits are edible, also made into jam. It is used in apthous complaints. Juice of roots is used in urinary complaints. Leaves lopped for fodder. The bark is the source of fibre, considered suitable for ropes. Bark and fruits are made into a bath used to cure leprosy (Chopra et al. 1956). A garland of green fruits is tied around the neck of the patient suffering from boils and sores. Unripe green fruits are cooked as vegetables. The ripe fruit is sweet and eaten raw (Saren et al. 2009).

Ficus subincisa Buch-Ham ex J.E. Smith in Rees, Ficus Cyclop. 14: 91. 1810; Corner in the Gardens Bull. Singapore 17: 420. 1959; *F. clavata* Wall. ex Mig. In Hooker's London J. Bot. 7: 431.1848; Forest Fl. Punj. Hazara and Dehli 481. 1983. (Repr.)

Vernacular names: **Dogri**-Karedal; **Hindi**-Sansoi, Khankhoi.

A shrub or small tree with scabrid branches; leaves stipulate, petiolate, 6–15 × 2–15 cm, sometime even larger, oblong-lanceolate or oblanceolate, abruptly caudate or acuminate, prominently dentate in the upper half, entire, glabrous, or rough, narrowed and 3-nerved at the base, lateral nerves in 4–6 pairs, distinct beneath; stipules 5–7 mm long, lanceolate; petioles 2–5 mm long, not jointed to the blade; figs axillary, solitary rarely in pairs on 2–5 mm long peduncles, ovoid-globose, supported by minute bracts at the base, strongly umbonate and verucose, 10–15 mm and yellow when ripe.

Occurrence and Distribution: Occasionally found in shady places and ravines.

Ethnobotany: Leaves are lopped for fodder but sometimes do not suit young calves and cause skin problems.

Ficus virens Dyrand in Ait. Hort. Kew. 3: 451. 1789; Corner in The Garden Bull. Singapore 17:

376. 1959; Sharma and Kachroo, Fl. Jammu 1: 293. 1981; *F. infectoria* Roxb., Hort. Beng. 66.1814 and in Fl. Ind. 3: 551. 1832 non Wild. 1806; King in Ann. Roy Bot. Gard. Calcutta 1: 60. 1887 and in Hook f. Fl. Brit. Ind. 5: 515. 1888; Parker, Forest Fl. Punj. Hazara and Delhi 481. 1983 (Repr.)

Vernacular Names: **Bombay**-Pipli; **Dogri**-Palakh, Pakh; **Hindi and Punjabi**-Pakar;

Malyalam-Pepar; **Sanskrit**-Plaksha; **Tamil**-Kuruu; **Telgu**-Badifuvvi.

A small or medium-sized, deciduous glabrous tree without or sometimes with some aerial roots and grey, smooth or scaly bark; leaves stipulate, petiolate, 7-12 × 5–7 cm ovate or oblong-ovate, rounded, truncate or subcordate at the base abruptly shortly acuminate at the tip, entire or subundulate along the margin, membraneous sub-coriaceuous 3-nerved at the base and with 5–7 pairs of lateral nerves, not very prominent; stipules ca. 1 cm long, ovate, acute; petioles, 2.5–5 cm long Channelled, sometimes indistinctly jointed to the blade; figs sessile, in axillary pairs, borne on 1–6 cm long peduncles, globose, 1–1.2 cm long, subtended by 3 minute ovate, glabrous or puberulous bracts at the base, white flushed with red dots when mature in ober-August to October.

Occurrence and Distribution: Often found clinging upon sandy rocks and in cooler ravines of Jammu, Kathua, Udhampur, Rajouri, Ramban, and Reasi districts.

Ethnobotany: Leaves lopped for fodder. The decoction of bark is used for washing of ulcers, as gargle in salivation, and as an injection in leucorrhoea (Chopra et al. 1956).

References

Agarwal SR (1981) Trees, flowers and fruits in Indian folk/songs, folk proverbs and folk tales. In: Jain SK (ed) Glimpses of Indian Ethnobotany 3.12 Oxford and 1BH Publishing Co, New Delhi

Ambasta S, Amachandran SK, Kashyapa K, Chand R (1994) (Rep ed) The useful plants of India CSIR new Delhi

Anonymous (1956) Ficus (moraceae) in wealth of India (raw materials): (F-G) 4:12–40 CSIR New Delhi

Bhartacharya (1956) Cultivation of tree deities of Bengal and growing at the botanical garden Calcutta. J Econ Tax Bot 13(2):191–309

Biswas KP, Ghosh A (1973) *Genus Ficus* L. Baratiya Banasushadi 4: (Rev ed) 1973 Calcutta University, Calcutta (in Bengali)

Chopra RN, Nayar SL, Chopra IC (1956) Glossary of Indian medicinal plants CSIR New Delhi

Condit (1984) The new book of knowledge- F 'Fig' 117.Groller and Co, New York. Condit, 1 and Julius, E 195 A bibliography of the fig Hilgardia 25:1–663 (n.v)

Goel AK, Aswal BS (1990) Less known uses of some plants in indigenous Folk-Lore from Nothern India. J Econ Tax Bot 14(1):185–188

Jain SK (ed) (1981) Glimpses of Indian ethonobotany. Oxford and 1BH Publishing Co, New Delhi

Kapur SK, Nanda S (1996) Traditionally important medicinal plants of baderwah Hills (Jammu Province)- Part II. J Econ Tax Bot Addl Series, 12. Scientific Publishers, Jodhpur, India

Khanna KK, Kumar A (2009) Noteworthy ethnomedicinal uses of plants from the tribals of Betul Distrct, Madhya Pradesh

King G (1888) The genus Ficus L. In: Hooker JD (ed) Fl Brit India 5:494–536. L Reeve and Co, London

Kirtikar KR, Basu RD (1935) *Ficus* L. In: Indian medicinal plants 3:2309–2333 pls 892–904 (Ist d 1918, 2nd revised by E Blatter et al 1935) L M Basu Allahabad

Kumar GM, Naqshi AR (1990) Ethnobotany of Jammu-Banihal. J Econ Tax Bot 14(1):67–74

Moldenke HN (1954) The economic plants of the Bible: food plants. Fig Econ Bot 8:152–153

Murray J, Watt G (1890) Ficus (figs). In: Watt G (ed) A dictionary of economic products of India, vol 3. Superintendent Govt Printing, Calcutta, pp 342–362

Nair PT (1965) Tree symbol worship amonG the Nairs in Kerala. Folklore 6:114–124

Roxburgh (1832) Flora Indica 15:601–602

Saha NC (1981) The role of Ethnobotany in relation to medicinal plants of India. In: Santapau in S. K Jain (ed) Glimpses of ethnobotany. Oxford and IBH Publ Co, New Delhi, pp 69–80
Santapau H (1961) (Publ 1962) Critical notes on the identity and nomenclature of some Indian plants: the genus Ficus in India. Bull Bot Surv India 3:16–17
Saren AM, Halder AC, Singh H (2009) Ethnobotanical plants of Mahilng forest range in Ranchi district, Jharkhand. J Econ Tax Bot 33(4):803–810
Sastri KN (1957) New lights of Indus civilization, religion and chronology Chapter 4. In: Tree worship and tree legends in the Indus valley. Aima Ram and Co, New Delhi, pp 25–35
Satya V, Solanki CM (2009) Indigenous Knowledge of Veterinary medicines among tribals of west Nimar, Madhya Pradesh. J Econ Tax Bot 33940:896–902
Sengupta S (ed) (1965) Tree symbol worship in India. Indian Publication Calcutta
Willis JC (1973) A dictionary of the flowering plants and ferns, 8th edn. Revised by HK Airy Shah. The Cambridge University Press, London

Part III

Ecosystem Processes in Bicultural Landscapes

Food to Medicine: The Impact of Soil and Climatic Factors on the Phytochemical Property of Anahaw (*Saribus rotundifolius* (Lam.) Blume Shoot

13

Dionisio S. Bucao and Xenia Elika N. Bucao

Abstract

Saribus rotundifolius (Lam) Blume has been an important support for humans as food, folkloric uses, material for house construction, and fish pens. Phytochemicals are very useful in health promotion and disease prevention. Their production is greatly affected by various environmental factors. Thus, this study was conducted to determine the effect of soil and climate conditions on the phytochemical properties of the *S. rotundifolius* shoot. Batac and Pagudpud samples were evaluated using various soil laboratory analyses and phytochemical screening. The results of phytochemical screening revealed the presence of flavonoids, saponins, fixed oil, proteins, and condensed tannins for both Batac and Pagudpud *S. rotundifolius* extracts, though there are more flavonoids present and alkaloids in the Batac extract. The variability in phytochemical properties of the two *S. rotundifolius* shoot extracts was found to be affected primarily by soil properties such as soil organic matter content, soil pH, and rainfall. Batac Pedon is characterized by a more developed soil than the Pagudpud pedon, as shown by their subsurface diagnostic horizon with argillic and cambic, respectively. The ustic moisture regime of the Batac pedon suggests a drier condition giving a fine, montmorillonitic, isohyperthermic typic Haplustalf taxonomic classification. Pagudpud pedon, on the other hand, has a udic moisture regime giving a fine, monmorillonitic, acidic typic Haplustalf taxonomic classification. The drier condition of Batac caused the *S. rotundifolius* shoot extract to be higher in flavonoid and alkaloid content than the Pagudpud sample.

D. S. Bucao (✉) · X. E. N. Bucao
Mariano Marcos State University, City of Batac, Ilocos Norte, Philippines
e-mail: dsbucao@mmsu.edu.ph

S. Ramamoorthy et al. (eds.), *Plant Diversity in Biocultural Landscapes*,
https://doi.org/10.1007/978-981-19-8649-9_13

Keywords

Pedon · Soil moisture regime · Temperature regime · Taxonomic classification · Phytochemical screening alkaloids · Flavonoids

13.1 Introduction

IFPs have been used since olden times for health and medical purposes to improve the well-being of people. Edible palm species are among the IFPs. Aside from being nutritious and delicious vegetables, traditional remedies are derived from palms throughout the tropics and subtropics to cure many disorders such as epilepsy, acute hepatitis, and influenza (Gruca et al. 2014).

S. rotundifolius (Lam) Blume is an erect palm reaching a height of 15–20 m and 25 cm in diameter. The trunk is smooth, straight, and marked with close, rather shallow obscure rings, which are the leaf scars. The leaves are crowded at the top of the trunk and ascending. The green, smooth, flattened petiole may have hard, black spines. The circular, fan-shaped, pleated leaf blades are 1 m in diameter and divided into segments 2.5–4 cm wide. The green flowers are 2 mm long. The fruit is 1.5 cm in diameter, fleshy, and yellow with a hard, round, brown seed inside (Department of Environmental and Natural Resources n.d.).

The species is native to Southeast Asia (Malaysia, Indonesia, Java, Moluccas, Philippines, Sulawesi, and Lesser Sunda Islands) but is now cultivated all over the world in tropical and subtropical countries as an ornamental plant (LLIFLE-Encyclopedia of Living Forms 2005). The trunk of *S. rotundifolius* (Lam) Blume is used widely for fish pens. The wood of the trunk is used as pillars and floors for houses in rural areas and is an excellent material for making bows, spear shafts, and canes. Its orbicular, fan-shaped leaves or fronds and pleated leaf blades are used for thatching houses and making hats and fans. Its buds or shoots are eaten as a vegetable (DENR n.d.). It is an ornamentally important plant cultivated extensively with a high demand for both cut foliage and pot plant production in local and foreign markets (Sanjeewani et al. 2013).

Phytochemicals are a wide variety of compounds present in plants, such as fruits and vegetables. They are non-nutritive bioactive secondary compounds that have health promotion and disease prevention potentials (Huang et al. 2015; Neilson et al. 2017). Alkaloids, flavonoids, terpenoids and steroids, saponins, tannins, phenolic compounds, coumarins, and carbohydrates are some examples of phytochemicals (Yin et al. 2013).

Although the production of phytochemicals in plants is guided by genetic processes, it can also be strongly affected by various environmental factors since plants have to produce a specified quantity and quality of phytochemicals to counter environmental stress (Yuan et al. 2020; Pant et al. 2021). Microclimate conditions, location, topography, growing season, soil type and properties (i.e., mineral composition, pH, water content, and temperature), light, and others are among the frequently reported factors (Rao and Rao 2007; Li et al. 2012; Safari et al. 2019;

Elumalero et al. 2021). Thus, this study was conducted to determine the effect of soil and climate conditions on the phytochemical properties of the *S. rotundifolius* shoot.

13.2 Methodology

13.2.1 Study Area

The province of Ilocos Norte is rich in edible plant species, such as indigenous food plants or IFPs, with a total of 46 IFPs identified representing 27 plant families (Antonio et al. 2011). The province has two climate types (Fig. 13.1) based on the Modified Corona Classification (Lantica 2001). The northernmost part (Pagudpud and Adams) is under Type III, with no very pronounced maximum rain period, a short dry season lasting from 1 to 3 months, and an average rainfall of more than 2100 mm. The rest of the province has a Type I climate, with two pronounced seasons, wet (May to October) and dry (November to April), with a maximum rain period from June to September and an average annual rainfall of greater than 2000 mm. Rainfall is being brought about by the Southwest monsoon. Tropical

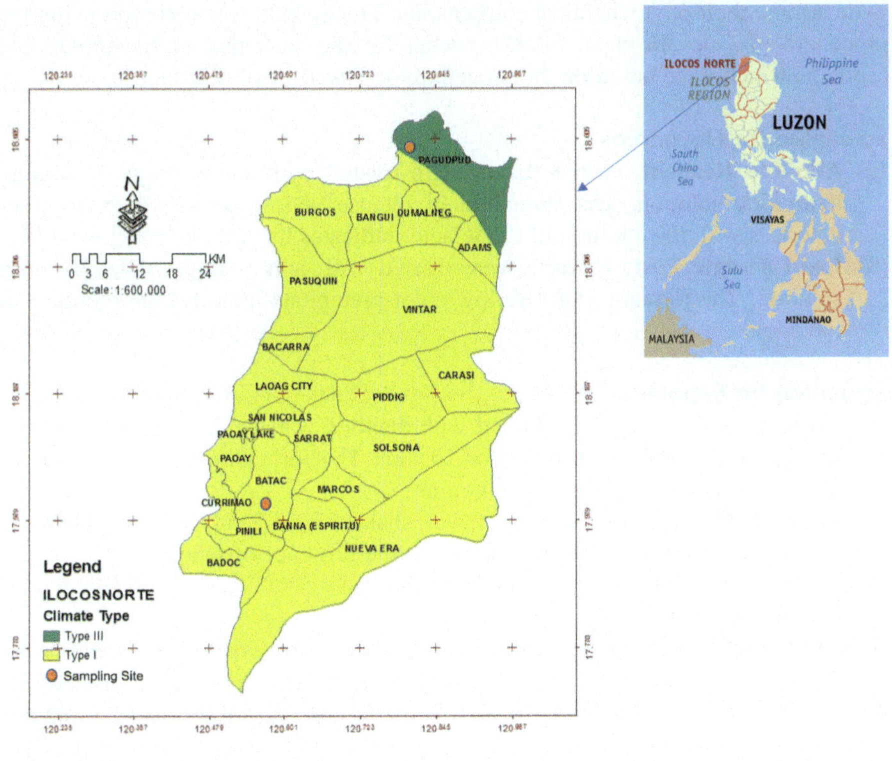

Fig. 13.1 Location and climatic map of the study area

cyclones and thunderstorms also contributed to the total rainfall. The annual mean temperature is 24.73 °C, with December to February as the coolest months and April to June as the hottest months. Annual potential evapotranspiration is about 1314 mm.

13.2.1.1 Plant Sample Collection and Preparation

The *S. rotundifolius* shoots were collected at the City of Batac and Pagudpud, Ilocos Norte. They were placed in a plastic bag that was kept in a Styrofoam box containing ice to retain the sample's freshness. Shortly after this, the *S. rotundifolius* shoots were cut into pieces and oven-dried at 40 °C for 48 h to remove the moisture content. The samples were milled after drying.

13.2.1.2 Preparation of Plant Extract

Powdered samples were extracted for 72 h at room temperature with 80% ethanol and were filtered to separate the marc and the supernatant following the methods used by Sanchooli (2011). The supernatant was evaporated using a water bath set at 50 °C.

13.2.1.3 Phytochemical Screening

Phytochemical screening was performed on the Batac and Pagudpud shoot extracts to determine their phytochemical components. The method was done following the procedure of Guevara et al. (2005), except for the screening of flavonoids and saponins which were based on the procedure of Tiwari et al. (2011).

Screening for Flavonoids

(a) **Alkaline Reagent Test**: Extracts were treated with a few drops of sodium hydroxide solution. The formation of intense yellow color, which becomes colorless with the addition of dilute acid, indicates the presence of flavonoids.
(b) **Lead acetate Test**: Extracts were treated with a few drops of lead acetate solution. The formation of a yellow color precipitate indicates the presence of flavonoids.

Screening for Saponins

(a) **Froth Test**: Extracts were diluted with distilled water to 20 mL, and this was shaken in a graduated cylinder for 15 min. The formation of a 1-cm layer of foam indicates the presence of saponins.
(b) **Foam Test**: 0.5 g of the extract was shaken with 2 mL of water. If foam produced persists for 10 min, it indicates the presence of saponins.

Screening for Alkaloids

An equivalent of 2 g of the leaf extract was evaporated to syrupy consistency in an evaporating dish over a steam bath. Five milliliters of hydrochloric acid was added to the concentrated extract and was heated with constant stirring for about 5 min and was cooled at room temperature. A 0.5 g of powdered sodium chloride was added and filtered. Enough 2 M hydrochloric acid was added to the mixture and was filtered to bring the filtrate to a final volume of 5 mL. Three test tubes were prepared to contain 1 mL each of the filtrate. A few drops of Wagner's reagent were added to 1 mL of filtrate. The formation of brick-red precipitate was an indicative of a positive result. A few drops of Mayer's and Dragendorff's reagents were added to the other two test tubes containing the filtrate. The formation of a white and orange precipitate, respectively, indicates a positive result.

Screening for Steroids: Cardenolides and Bufadienolides

(a) **Liebermann–Burchard Test** was used to determine the presence of unsaturated steroids. An equivalent of 10 g of leaf extract was evaporated to incipient dryness and was allowed to cool to room temperature. Ten milliliters of hexane was added and stirred for a few minutes to the cooled residue. The mixture was allowed to settle to decant off the supernatant liquid until hexane was discarded. Ten milliliters of chloroform was added and stirred for a few minutes to the residue. It was then decanted into a test tube containing 100 mg of anhydrous sodium sulfate. It was shaken and passed through a filter paper. The filtrate was divided into two test tubes, making one portion as a reference. Three drops of acetic anhydride were added and gently mixed. Any change in color was observed. A positive result gives colors ranging from blue to green, red, pink, purple, or violet, because of the steroid/triterpenoid skeleton.

(b) **Keller-Kiliani Test** was used to determine the presence of 2-deoxysugars. An equivalent of 10 g of leaf extract was evaporated to incipient dryness. The residue was defatted by trituration with hexane to remove as much of the colored pigment as possible. The decanting and defatting processes were repeated as required. Hexane was completely decanted and discarded. Over a water bath, the defatted residue was heated to remove the residue of hexane solvent. Three milliliters of $FeCl_3$ reagent were added and stirred to reside, then transferred into a test tube. One milliliter of concentrated sulfuric acid (H_2SO_4) was carefully added to the test tube in an inclined position, letting the acid run along the insides of the test tube. The mixture was allowed to stand upright and was observed for any coloration at the interface of the acid and the aqueous layers. A reddish-brown color which may turn blue or purple indicates the presence of 2-deoxysugars.

Screening for Anthraquinones

The equivalent of 1 g of the shoot extract was evaporated to incipient dryness using a water bath employing the Borntrager's test. The residue was taken up in 10 mL of distilled water and filtered. The filtrate was extracted twice with 5 mL of benzene. The combined benzene extract was divided into two portions. One was reserved as the blank. Five milliliters of ammonia solution was added to the second portion, and the mixture was shaken. The alkaline layer was observed for changes in color. The formation of red color is a positive result for anthraquinone compounds.

Screening of Carbohydrates

Using Fehling's test, 2 mL of the shoot extract was evaporated and 10 mL of distilled water was added to the residue. Freshly prepared Fehling's solution (equal amount of Fehling's A and Fehling's B) was added and heated. The formation of a red precipitate indicates the presence of reducing sugars.

Screening for Fixed Oils and Volatile Oils

(a) **Stain Test**. 2 mL of the shoot extract was boiled in 10 mL hexane. Two drops of the hexane extract were placed on a piece of white paper. The presence of permanent stain indicates the presence of fixed oils.

Screening of Proteins

(a) **Millon's Test.** An equivalent of 2 g of the shoot extract was evaporated, and the residue was taken with 10 mL of water. Ten drops of Millon's reagent were added to 1 mL of the aqueous extract and were placed in a boiling water bath. Any change in color indicates a positive result.

Screening for Tannins

An ethanol extract equivalent to 10 g of plant material was evaporated to incipient dryness over a water bath and then cooled. The residue was extracted with 20 mL of hot distilled water and then cooled. To salt out undesirable constituents, five drops of 10% Sodium Chloride solution were added to the treated extract and filtered.

(a) **Ferric chloride test**. Three drops of $FeCl_3$ solution were added to the test tube containing the extract. The change in color may indicate the presence of hydrolyzable tannins, while a brownish-green color may indicate condensed tannins.

13.2.1.4 Soil Physical Characterization

Soil characterization was done where the *S. rotundifolius* shoot samples were collected through in situ examinations of the soil profile following the standard procedures given by the Soil Survey Division Staff (1993) and Buol et al. (2003, p. 527). Road cuts were cleared large enough for observation of a pedon, and exposed soils were removed using a shovel to show the natural condition of the soil. The exposed profile was probed by hand with the aid of a spade to remove any alterations made while digging and clearing the area. A steel tape was placed on the exposed profile, and soil horizon boundaries were marked using a popsicle stick and photographed. The pedons were examined per horizon in terms of soil color, texture, structures, consistency, root quantity and size, horizon boundary characteristics and continuity, and other features such as concretions, nodules, visible pores, and pH. After a thorough examination of the soil profile, about 2 kg of soil samples were collected from each of the horizons, starting at the bottom up to the surface epipedon. Soil samples were placed in a plastic bag, properly sealed, and stored in an ice chest to avoid direct exposure to sunlight.

13.2.1.5 Soil Sample Preparation

Soil samples taken from the City of Batac and Pagudpud, Ilocos Norte were air-dried in the laboratory and ground using a Willey mill machine at the Crops Research Laboratory, Mariano Marcos State University, City of Batac. Coarse fragments greater than 2 mm diameter were removed from each air-dried sample by sieving, and only the finer fractions 2 mm diameter and below were collected for analyses. Sieved soil samples were placed in a clean, well-sealed, and properly labeled plastic bag.

13.2.1.6 Soil Laboratory Analysis

Soil samples taken from the examined pedons were subjected to various physical and chemical analyses. The kinds of analyses made as well as the methods and references used are presented in Table 13.1.

13.3 Results and Discussion

13.3.1 Phytochemical Screening

The screening was done to determine the phytochemicals present in the Batac and Pagudpud shoot extracts of *S. rotundifolius*. The test results revealed the presence of flavonoids, saponins, fixed oil, proteins, and condensed tannins for both extracts, though there are more flavonoids present in the Batac extract, according to the result of the Alkaline Reagent Test. The result of Wagner's test also revealed that alkaloids are present in the Batac extract (Table 13.2).

Table 13.1 Soil laboratory methods used in the study

Analysis	Reference
Physical analysis	
1. Particle size analysis	Day (1965) and Franzmeier et al. (1977)
2. Bulk density by core method	Philippine Council for Agriculture and Resources Research (1980)
Chemical analysis	
1. Soil reaction (pH)	Philippine Council for Agriculture and Resources Research (1980)
2. Organic matter	Walkley and Black (1934)
3. Exchangeable cations	Chapman (1965) and Thomas (1982)
4. Exchangeable acidity and aluminum	Thomas (1982)
5. Cation exchange capacity (NH_4OAc, pH 7.0)	Chapman (1965)
6. Base saturation percentage	Philippine Council for Agriculture and Resources Research (1980)

Table 13.2 Phytochemical screening of *S. rotundifolius* shoot extracts

Tests	Expected result	Actual result	
		City of Batac sample	Pagudpud sample
Screening for flavonoids			
Alkaline reagent	Formation of intense yellow color which becomes colorless on addition of dilute acid	++	+
Lead acetate	Formation of yellow color precipitate	+++	+++
Screening for saponins			
Froth	Formation of 1 cm layer of foam	+++	+++
Foam	Foam produced persists for 10 min	+++	+++
Screening for alkaloids			
Mayer's	Formation of white precipitate	–	–
Wagner's	Formation of brick-red precipitate	+++	–
Dragendorff's	Formation of orange precipitate	–	–
Screening for steroids: Cardenolides and Bufadienolides			
Liebermann–Burchard	Colors ranging from blue to green, red, pink, purple, or violet	-	–
Keller-Kiliani	Reddish-brown color	–	–
Screening for Anthraquinones	Formation of red color	–	–
Screening for carbohydrates	Formation of red precipitate	–	–
Screening for fixed oils			
Stain	Presence of permanent stain	+++	+++
Screening of proteins			
Millon's	Change in color	+	+
Screening for tannins			
Ferric chloride	Blue-black color (hydrolysable tannins)		
	Brownish-green (condensed tannins)	+++ (condensed)	+++ (condensed)

Note: (–) not detected; (+) low presence; (++) moderate presence; (+++) strong presence

The difference in the phytochemicals presented between the two extracts may be attributed to the variability in the soil properties and the climate conditions where the plant samples were taken. Drought stress increases flavonoid content (Zhang et al. 2017). It should be noted that the Batac site has a ustic moisture regime while Pagudpud has a udic moisture regime. In addition, higher organic matter in soils decreases flavonoid content (Del Valle et al. 2020). Although it was reported in the study of Jansen et al. (2012) that lower alkaloid content was detected in soils with a higher pH, the results of this study are in agreement with the studies of Zhang et al. (2011), Ababaf et al. (2021), Honório et al. (2021), and Pant et al. (2021) that drought stress increases and flooding decrease alkaloid content.

Flavonoids contain phenolic compounds which are usually seen in fruits, vegetables, grains, barks, roots, stems, shoots, and flowers (Sandhar et al. 2011;

Shohaib et al. 2011 as cited by Umesh et al. 2018). These phenolic compounds have anticancer (Batra and Sharma 2013), antibacterial (Ng et al. 1996), and antiviral (Kaul et al. 1985) properties that can cure human diseases, as cited by Umesh et al. (2018). Likewise, alkaloids are organic nitrogen-containing compounds that give dynamic biological activities in the human or animal body, particularly in various diseases like malaria, diabetes, cancer, and cardiac dysfunction, among others (Ain et al. 2016). Research findings also support the benefits of alkaloids and can be applied to discover and design new analogs that could be therapeutically useful for various treatments (Debnath et al. 2018).

13.3.2 Soil Characteristics of the Sampling Sites of *S. rotundifolius (Lam) Blume* Shoots

The characteristics of the pedons (Fig. 13.2) where the *S. rotundifolius* were sampled (Plate 13.1) showed variability in terms of elevation, slope, pH, effective soil depth, subsurface horizon, average annual rainfall, and soil moisture regime (Table 13.3). Despite the variations, however, their taxonomic classification is almost the same

Fig. 13.2 Soil profiles of the sampling sites of *S. rotundifolius* (Lam) Blume shoots

Batac Sampling Site

Pagudpud Sampling Site

Plate 13.1 Photos of the location where *S. rotundifolius* (Lam) Blume shoots were taken

Table 13.3 Pedon characteristics of the sampling sites of *S. rotundifolius* (Lam) Blume shoot

Location	City of Batac, Ilocos Norte	Pagudpud, Ilocos Norte
Elevation	142.5 m	129.3 m
Slope	Rolling (8–18%)	Hilly (18–30%)
Present vegetation	Secondary forest	Secondary forest
Landsurface/ landscape	Colluvial footslope	Colluvial footslope
pH	Neutral	Strongly acid
OM (%)	Moderately low	Moderately high
CEC (cmol+/kg)	Very high	Very high
BS (%)	Very high	Very high
Effective soil depth	Deep	Moderately deep
Texture	Clayey	Clayey
Drainage	Well-drained	Well-drained
Epipedon	Ochric	Ochric
Subsurface diagnostic horizon	Argillic	Cambic
Average annual rainfall	2000 mm	>2100 mm
Mean temperature	26.64 °C (high 28.7; low 23.7)	26.80 °C (high 28.8; low 24.4)
Soil moisture regime	Ustic	Udic
Temperature regime	Isohyperthermic	Isohyperthermic
Taxonomic classification	Fine, montmorillonitic, isohyperthermic, Typic Haplustalfs	Fine, montmorillonitic, acidic isohyperthermic, Typic Hapludalfs
Erosion risk assessment	Moderate	Moderate

only that the Pagudpud pedon is acidic. This can be attributed to the higher average rainfall incurred in the area, which causes the leaching of cations like calcium and magnesium, leaving the anions such as hydrogen, manganese, and aluminum (Zhang 2017; Nathan n.d.).

13.3.3 Taxonomic Classification of the Sampling Sites

Tables 13.4 and 13.5 present the physical properties and chemical properties of the pedons from the Batac and Pagudpud sampling sites.

13.3.3.1 Batac Pedon

This pedon was sampled in the City of Batac in an area with a rolling slope (8–18%) positioned in a colluvial footslope landscape unit with an elevation of 142.5 masl. The area has a molave-type forest and fruit trees as vegetation cover and with slight erosion. The soil is deep about 120 cm, well-drained, dry, and deep water table with an ustic soil moisture regime and isohyperthermic temperature regime. The average annual rainfall is about 2000 mm.

It has a thin surface horizon measuring 16 cm; hue 10 YR; the value of 4, moist and 6, dry; and has chroma of 4 moist, equivalent to dark yellowish brown, moist and light yellowish brown, dry. The soil has a weak fine granular structure; friable, moist, very sticky and very plastic, wet consistency; clear and smooth horizon boundary, with many fine roots. Soil textural class is clay comprising 6.94% sand, 39.51% silt, and 53.55% clay with a bulk density of 1.29 g cm^{-3}. SOM is moderately low (1.31%) with 0.76% OC and a soil reaction of near neutral (7.02), high CEC of 68.07 cmol(+) kg^{-1} soil, and a BS of 101.69%. Due to its thin soil surface and high Munsell color, the pedon falls under ochric epipedon. The subsurface horizons displayed a dark yellowish brown (10 YR 3/4) to very dark brown (10 YR 2/2), moist and yellowish brown to very dark grayish brown (10 YR 3/2), and dry. Soil has a moderate medium subangular blocky to moderate medium prismatic structure; firm, moist, very sticky, and very plastic, wet consistency. The horizon boundary is gradual to clear and smooth, with many fine roots and many fine tubular pores. The clay fraction of the profile comprised 57.81%, 67.25%, 72.27, and 73.84% at the second, third, fourth, and fifth horizons, respectively (Table 29). An increase in clay from the surface to the lowest horizon was observed. SOM and OC of the subsurface horizons were medium to very low. Soil reaction showed near-neutral throughout the subhorizons. CEC (NH_4OAc-pH 7) was high ranging from 54.08 to 66.69 cmol kg^{-1} soil, while ECEC ranged from 55.70 to 69.85 cmol kg^{-1} soil; and average BS of above 100%. It has an argillic subsurface diagnostic horizon.

The pedon has 60.67% clay content (by weight) in the fine-earth fraction between the mineral soil surfaces with a depth of 120 cm after mixing. BS is >35%, while CEC and ECEC are >24 cmol kg^{-1} soil and > 12 cmol kg^{-1} soil, respectively. It has a soil order *Alfisols* and *Ustalfs* soil suborder, which are other *Alfisols* with an ustic moisture regime. It has a great group of *Haplustalfs* and a subgroup *Typic Haplustalfs*.

Table 13.4 Physical properties of Pedons from Batac and Pagudpud sampling sites

Horizon	ISSS System					USDA system				
	Particle size distribution					Particle size distribution				
	C Sand (2–0.2)	F Sand (0.2–0.02)	Silt (0.02–0.002	Clay (<0.002)		Sand (2–0.05)	Silt (0.05–0.002)	Clay (<0.002)		
	%				Texture	%			Texture	Bulk density
City of Batac										
Ap	2.71	6.04	37.71	53.55	HC	6.94	39.51	53.55	C	1.19
Bw	2.69	2.99	36.51	57.81	HC	3.66	38.53	57.81	C	1.20
Bt1	1.32	3.85	27.57	67.25	HC	2.47	30.27	67.25	C	1.24
Bt2	0.39	2.35	24.99	72.27	HC	0.74	26.99	72.27	C	1.27
Bt3	0.30	2.39	23.45	73.86	HC	0.58	25.56	73.86	C	1.27
Pagudpud										
A	0.74	15.05	25.45	58.75	HC	13.12	28.13	58.75	C	1.17
B1	0.98	7.58	32.47	58.97	HC	5.95	35.09	58.97	C	1.20
B2	1.20	5.83	32.91	60.06	HC	4.42	35.52	60.06	C	1.20
BC	1.26	9.03	35.80	53.91	HC	8.12	37.96	53.91	C	1.24

Table 13.5 Chemical properties of Pedons from Batac and Pagudpud sampling sites

	Texture	OC		Exchangeable					CEC (NH_4OAc)	BS
				Ca	Mg	K	Na	Sum		
USDA system	%		pH H_2O (1:1)	Cmol(+) kg^{-1} soil						%
City of Batac										
Ap	1.31	0.76	7.02	35.73	31.24	0.97	1.07	69.01	68.07	101.37
Bw	0.98	0.57	7.29	34.33	33.77	0.79	0.96	69.85	66.69	104.75
Bt1	1.05	0.61	7.32	30.87	33.09	0.71	0.99	65.66	64.82	101.29
Bt2	1.33	0.77	6.83	25.53	28.25	0.82	1.10	55.70	54.08	103.00
Bt3	1.74	1.01	6.67	28.13	28.02	0.71	1.02	57.88	57.24	101.13
Pagudpud										
A	3.17	1.84	5.35	15.47	18.17	0.12	0.45	34.21	48.87	70.00
B1	1.87	1.09	5.42	15.73	19.55	0.12	0.46	35.86	48.40	74.10
B2	1.31	0.76	5.29	18.93	27.60	0.74	0.80	48.08	49.91	96.33
BC	0.94	0.55	5.46	18.93	27.83	0.56	0.71	48.04	49.68	96.70

The pedon has a clay content of 55.68% at the control section with a depth of 43 cm, which is classified as fine. The average CEC to clay ratio is 0.98, which is equivalent to the clay activity class *montmorillonitic*. It has an isohyperthermic moisture regime and average pH of 7.03. Thus, the soil was classified under the family of fine, montmorillonitic, isohyperthermic, Typic Haplustalfs.

13.3.3.2 Pagudpud Pedon

The pedon was taken in Pagudpud, Ilocos Norte in an area with steep slope (30–50%) positioned in a colluvial footslope landscape unit with an elevation of 49.1 masl. The area is hilly and under grasslands. The soil is deep about 100 cm, well-drained, dry, and deep water table with udic soil moisture regime and isohyperthermic temperature regime. The average annual rainfall is >2000 mm.

Pagudpud pedon has a surface horizon with a thickness of 24 cm; hue 10 YR; a value of 3, moist and 6, dry; and chroma of 4 moist, equivalent to dark yellowish brown and light yellowish brown, moist and dry, respectively. Soil has a moderate medium angular blocky structure; firm, moist with very sticky and very plastic, wet consistency. The horizon boundary is clear and smooth; few fine roots; and few fine tubular pores. Soil textural class is clayey comprising 13.12% sand, 28.13% silt, and 58.75% clay with a bulk density of 1.17 g cm^{-3}. SOM is moderately high (3.17%) with 1.84% OC and a soil reaction of strongly acid (5.35), high CEC of 48.87 cmol (+) kg^{-1} soil, and a BS of 70%. Due to its thin soil surface and high Munsell color value and chroma, the pedon falls under ochric epipedon.

Its subsurface horizons displayed a strong brown color (7.5 YR 5/6) to yellowish brown (10 YR 5/6), moist and reddish yellow (7.5 YR 6/6) to brownish yellow (10 YR 6/6), dry. Soil has moderate medium subangular blocky to moderate medium prismatic structure; firm, moist with very sticky and very plastic to plastic, wet consistency. The horizon boundary is clear and smooth; few fine roots; and common fine tubular pores. Clay fraction comprised 58.97%, 60.06%, and 53.91% at the second, third, and fourth horizons, respectively (Table 13.4). The increase in clay was not substantial from the surface horizon to the second and from the second to the third horizon. SOM and OC of the soil decreased with depth (Table 24). SOM at the second horizon was medium (1.87%) and moderately low to low (1.31 and 0.94%) for the succeeding horizons. CEC (NH_4OAc-pH 7) was high ranging from 48.40 to 49.68 cmol kg^{-1} soil with a corresponding ECEC of 36.56 to 50.13 cmol kg^{-1} soil and a BS of 89.04%. Since the increase in clay was not substantial and color development was evident, the subsurface diagnostic horizon of this pedon is cambic.

The pedon has 59.26% clay content (by weight) in the fine-earth fraction between the mineral soil surfaces with a depth of 100 cm after mixing. BS is >50%, while CEC and ECEC are >24 cmol kg^{-1} soil and > 12 cmol kg^{-1} soil, respectively. This pedon qualifies under soil order *Alfisols* and *Udalfs* soil suborder, which are other *Alfisols* with udic moisture regimes. The pedon is under the great group *Hapludalfs,* which are other *Udalfs* and with subgroup *Typic Hapludalfs.*

The pedon has a clay content of 58.86% at the control section with a depth of 24 cm which is classified as fine. The average CEC to clay ratio is 0.85 which is equivalent to clay activity class *montmorillonitic*. The pedon has an isohyperthermic moisture regime and average pH of 5.58. Thus, the soil was classified under the family of fine, montmorillonitic, acidic, isohyperthermic, Typic Hapludalfs.

Pagudpud pedon has moderately high organic matter at the surface (3.17) and subsurface (1.87) compared to Batac pedon with 1.31 and 0.98, respectively. However, Batac pedon has a higher cation exchange capacity (CEC) as a result of the high exchangeable base readings at different horizons. The CEC measures the capacity of the soil to supply and hold nutrients for plant use and prevent leaching (Zhang 2017; Nathan n.d).

Although Pagudpud pedon is more fertile than Batac pedon, the phytochemical properties of the *S. rotundifolius* (Lam) Blume shoots such as flavonoid and alkaloids were higher in the latter due to the differences in climatic factors. The ustic moisture regime in the Batac Pedon characterizes a drought condition that favors the production of flavonoids and alkaloids in *S. rotundifolius* (Lam) Blume shoots. Moreover, the high CEC of Batac pedon supplies the needed nutrients of *S. rotundifolius* (Lam) Blume plant for its growth and development.

13.4 Summary and Conclusion

Shoots of anahaw were taken in the City of Batac and Pagudpud, Ilocos Norte representing the two climatic conditions of the province. Soil Pedons were observed both in situ and ex situ to determine their physical and chemical properties and taxonomic classification. Batac Pedon is characterized by a more developed soil than the Pagudpud pedon as shown by their subsurface diagnostic horizon with argillic and cambic, respectively. The ustic moisture regime of Batac pedon suggests a drier condition giving a fine, monmorillonitic, isohyperthermic typic Haplustalf taxonomic classification. Pagudpud pedon, on the other hand, has a udic moisture regime giving a fine, monmorillonitic, acidic typic Haplustalf taxonomic classification.

Meanwhile, the variability in phytochemical properties of the two anahaw samples was found to be affected primarily by soil properties such as soil organic matter, soil moisture regime, and soil pH and climatic factor. However, the conduct of a study in a controlled environment to confirm the effect of these factors on phytochemicals, and quantitative determination of the phytochemical content of the plant are recommended.

References

Ababaf M, Omidi H, Bakhshandeh A (2021) Changes in antioxidant enzymes activities and alkaloid amount of catharanthus roseus in response to plant growth regulators under drought condition. Ind Crop Prod 167:113505. https://doi.org/10.1016/j.indcrop.2021.113505

Ain Q-U, Khan H, Mubarak MS, Pervaiz A (2016) Plant alkaloids as antiplatelet agent: drugs of the future in the light of recent developments. Front Pharmacol 7:292. https://doi.org/10.3389/fphar.2016.00292

Antonio MA, Utrera RT, Agustin EO, Jamias DL, Badar AJ, Pascua ME (2011) Survey and characterization of indigenous food plants in Ilocos Norte, Philippines. In: SEARCA Discussion Paper Series, vol 2011 No. 2

Batra P, Sharma AK (2013) Anti-cancer potential of flavonoids: recent trends and future perspectives. 3 Biotech 3(6):439–459. https://doi.org/10.1007/s13205-013-0117-5

Buol SW, Southard RJ, Graham RC, Mcdaniel PA (2003) Soil genesis and classification, 5th edn. Iowa State University Press, Ames

Chapman HD (1965) Cation exchange capacity. In: Black CA (ed) Methods of soil analysis. American Society of Agronomy, Madison, pp 891–901

Day FR (1965) Particle fraction and particle-size analysis. Agron Monogr. https://doi.org/10.2134/agronmonogr9.1.c43

Debnath B, Singh WS, Das M, Goswami S, Singh MK, Maiti D, Manna K (2018) Role of plant alkaloids on human health: a review of biological activities. Mater Today Chem 9:56–72. https://doi.org/10.1016/j.mtchem.2018.05.001

Del Valle I, Webster TM, Cheng HY, Thies JE, Kessler A, Miller MK, Ball ZT, MacKenzie KR, Masiello CA, Silberg JJ, Lehmann J (2020) Soil organic matter attenuates the efficacy of flavonoid-based plant-microbe communication. Sci Adv 6(5):eaax8254. https://doi.org/10.1126/sciadv.aax8254

Department of Environmental and Natural Resources (n.d.) "Trees and their management: growing anahaw." New Zealand digital library. https://www.nzdl.org/cgi-bin/library.cgi?e=d-00000-00—off-0hdl–00-0—-0-10-0—0—0direct-10—4——-0-11–11-en-50—20-about—00-0-1-00-0-0-11-1-0utfZz-8-00&cl=CL1.17&d=HASH01df274fb229ae826f6d2355.13&gc=1. Accessed 9 Aug 2022

Elumalero GO, Apenah MO, Ogundele OO, Adesanmi VS, Olalekan OJ (2021) Effect of soil nutrient concentrations on the quantity of tannin and flavonoid deposited in *Gmelina arborea* fruit located in different locations. J Res For Wildl Environ 13(2):2141–1778

Franzmeier DP, Steinhardt GC, Crum JR, Norton LD (1977) Soil characterization in Indiana: I. Field and laboratory procedures. Res Bull. No. 943. Purdue University, Agricultural Experiment Station

Gruca M, Andel T, Balslev H (2014) Ritual uses of palms in traditional medicine in sub-Saharan Africa: a review. J Ethnobiol Ethnomed 10:60. https://doi.org/10.1186/1746-4269-10-60

Guevara R et al (2005) A guidebook to plant screening: phytochemical and Biological Research. Center of the National Sciences, University of Santo Tomas, Manila

Honório ABM, De-la-Cruz-Chacón I, Martínez-Vázquez M, da Silva MR, Campos FG, Martin BC, da Silva GC, Fernandes Boaro CS, Ferreira G (2021) Impact of drought and flooding on alkaloid production in Annona crassiflora Mart. Horticulturae 7:414. https://doi.org/10.3390/horticulturae7100414

Huang Y, Xiao D, Burton-Freeman BM, Edirisinghe I (2015) Chemical changes of bioactive phytochemicals during thermal processing. Elsevier, AmsterdamReference Module in Food Science. https://doi.org/10.1016/B978-0-08-100596-5.03055-9

Jansen G, Jürgens HU, Schliephake E, Ordon F (2012) Effect of the soil pH on the alkaloid content of Lupinus angustifolius. Int J Agron 2012:1–5. https://doi.org/10.1155/2012/269878

Kaul TN, Middleton E Jr, Ogra PL (1985) Antiviral effect of flavonoids on human viruses. J med virol 15(1):71–79. https://doi.org/10.1002/jmv.1890150110

Lantica RM (2001) The science and practice of crop production. Southeast Asian Regional Center for Graduate Study and Research in Agriculture (SEARCA) and the University of the Philippines Los Baños (UPLB), Laguna, p 330

Li H, Tsao R, Deng Z (2012) Factors affecting the antioxidant potential and health benefits of plant foods. Can J Plant Sci 92(6):1101–1111. https://doi.org/10.4141/cjps2011-239

LLIFLE-Encyclopedia of Living Forms (2005) *Livistona rotundifolia*. http://www.llifle.com/Encyclopedia/PALMS_AND_CYCADS/Family/Arecaceae/24545/Livistona_rotundifolia. Accessed 9 Aug 2022

Nathan MV (n.d.) Soils, plant nutrition and nutrient management. University of Missouri Extension. https://extension.missouri.edu/publications/mg4?fbclid=IwAR0CeoG5sbSgFycXmjswUgBkPB4CQl4g9o_fxU_gg1dbwZuSWksycYSeWMs. Accessed 12 Aug 2022

Neilson AP, Goodrich KM, Ferruzzi MG (2017) Bioavailability and metabolism of bioactive compounds from foods. In: Coulston AM, Boushey CJ, Ferruzzi MG, Delahanty LM (eds) Nutrition in the prevention and treatment of disease, 4th edn, pp 301–319. https://doi.org/10.1016/B978-0-12-802928-2.00015-1

Ng TB, Ling JM, Wang ZT, Cai JN, Xu GJ (1996) Examination of coumarins, flavonoids and polysaccharopeptide for antibacterial activity. Gen pharmacol 27(7):1237–1240. https://doi.org/10.1016/0306-3623(95)02143-4

Pant P, Pandey S, Dall'Acqua S (2021) The influence of environmental conditions on secondary metabolites in medicinal plants: a literature review. Chem Biodivers 18(11):e2100345. https://doi.org/10.1002/cbdv.202100345

Philippine Council for Agriculture and Resources Research (1980) Standard methods of analysis for soil, plant tissue, water and fertilizer. los Baños, Laguna

Rao AV, Rao LG (2007) Carotenoids and human health. Pharmacol Res 55(3):207–216. https://doi.org/10.1016/j.phrs.2007.01.012

Safari K, Yadegari M, Hamedi B (2019) Effects of climate and soil properties on phytochemical characteristics of Ferulago angulate (Schltdl.) Boiss. Iranian J Plant Physiol 9(2):2719–2726

Sanchooli N (2011) Antidiabetic potential of *Physalis alkekengi* on biochemical profile in alloxan-induced diabetic rats. Res J Pharm, Biol Chem Sci 2(3):168–173. https://www.researchgate.net/publication/289236133

Sandhar HK, Kumar B, Prasher S, Tiwari P, Salhan M, Sharma P (2011) A review of phytochemistry and pharmacology of flavonoids. Internationale pharmaceutica sciencia 1(1):25–41

Sanjeewani B, Jayasuriya K, Fernando M, Damunupola J (2013) Storage and germination treatments for seeds of an ornamentally important palm, Livistona rotundifolia (lam.) Mart. J Nat Sci Found 41(4):273–277. https://doi.org/10.4038/jnsfsr.v41i4.6257

Shohaib T, Shafique M, Dhanya N, Divakar MC (2011) Importance of flavonoids in therapeutics. Hygeia JD Med 3(1):1–8

Soil Survey Division Staff (1993) Soil survey manual. United States Department of Agriculture, Washington

Thomas GW (1982) Exchangeable cations. Methods of soil analysis, part 2, chemical and microbiological properties. In: Page AL (ed) Agronomy, No. 9, Part 2, 2nd edn. American Society of Agronomy, Soil Science Society of America, Madison, Wl, pp 159–165

Tiwari P, Kumar B, Kaur M, Kaur G, Kaur H (2011) Phytochemical screening and extraction: a review. Int Pharm Sci 1(1):98–106. http://www.ipharmsciencia.com

Umesh CV, Jamsheer AM, Prasad MA (2018) The role of flavonoids in drug discovery—review on potential applications. Res J Life Sci Bioinform Pharm Chem Sci 4(1):70–77. http://www.rjlbpcs.com. https://doi.org/10.26479/2018.0401.06

Walkley A, Black IA (1934) An examination of the different method for determining soil organic matter and a proposed modification of the chromic acid titration method. Soil Sci 37:29–38. https://doi.org/10.1097/00010694-193401000-00003

Yin NS, Abdullah ST, Phin CK (2013) Phytochemical constituents from leaves of *Elaeis guineensis* and their antioxidant and antimicrobial activities. Int J Pharm Pharm Sci 5(4):137–140. ISSN: 0975-1491

Yuan Y, Tang X, Jia Z, Li C, Ma J, Zhang J (2020) The effects of ecological factors on the main medicinal components of Dendrobium officinale under different cultivation modes. Forests 11(1):94. https://doi.org/10.3390/f11010094

Zhang H (2017) Cause and effects of soil acidity. Oklahoma State University. https://extension.okstate.edu/fact-sheets/cause-and-effects-of-soil-acidity.html?fbclid=IwAR34_b9FRp3WfQ_8XEnb7MDDa2en1Ksx0g2pPKd34OQ_xoRcuk4EpBC_QKg. Accessed 12 Aug 2022

Zhang X, Li C, Nan Z (2011) Effects of salt and drought stress on alkaloid production in endophyte-infected drunken horse grass (Achnatherum inebrians). Biochem Syst Ecol 39:471–476. https://doi.org/10.1016/J.BSE.2011.06.016

Zhang W, Cao Z, Xie Z, Lang D, Zhou L, Chu Y, Zhao Q, Zhang X, Zhao Y (2017) Effect of water stress on roots biomass and secondary metabolites in the medicinal plant Stellaria dichotoma L. var. lanceolata Bge. Sci Hortic 224:280–285. https://doi.org/10.1016/j.scienta.2017.06.030

Floral and Fruiting Phenology in the Lowland Forests of Palanan, Isabela, Philippines

14

Analinda C. Manila-Fajardo, Rhollthan A. Tubale, and Jacquilyn L. Estrada

Abstract

The Palanan lowland forests are being utilized in many ways. It has been a research site for determining ecosystem resiliency. To the communities especially the Agtas, it is an area where they get resources from. Among its ecosystem services is pollination and seed dispersal which can be understood by studying the reproductive phenological development of plant species considering there is little knowledge on biotic interactions involved. The collection of reproductive phenological data was performed in the four sites within the lowland forests of the municipality of Palanan: Brgy. Dipogen, PFDP, and Brgy. Alomanay and Mangrove area in Maharlika. A modified reproductive phenology scoring was utilized. Individual tree count in PFDP is high due to the high recruitment rate brought upon by the frequent freeing up of space in the forest canopy every typhoon, which temporarily removes light limitation for seedlings to grow. The majority of the reproducing species during the study period were categorized as supra-annuals. Of these families, Meliaceae, Euphorbiaceae, Lauraceae, Annonaceae, Lamiaceae, Rubiaceae, Arecaceae, and Dipterocarpaceae, all of which are among the most species-rich families inside the plot with flowering and fruiting occurring during the hot dry conditions. Almost forty percent of the species reproduce sub-annually (≈20%) and annually (≈18%). The less rich group of species reproduces continuously. Current data shows flowering was

A. C. Manila-Fajardo (✉)
Institute of Biological Sciences, University of the Philippines Los Baños, Laguna, Philippines
e-mail: amfajardo@up.edu.ph

R. A. Tubale · J. L. Estrada
Institute of Environmental Science and Meteorology, University of the Philippines Diliman, Quezon City, Philippines

S. Ramamoorthy et al. (eds.), *Plant Diversity in Biocultural Landscapes*,
https://doi.org/10.1007/978-981-19-8649-9_14

mildly affected by temperature and moderately affected by rainfall. Fruiting was not really affected by temperature and rainfall.

Keywords

Reproductive phenology · Palanan, Isabela · Lowland forests

14.1 Introduction

With the increasing population of the Philippines, more space is needed for agricultural lands to meet the demand and increase food security. Local practices such as Swidden agriculture (kaingin) clear forest patches for the planting of crops on a smaller scale. Timber, on the other hand, is harvested for the building of houses of locals. Housing and infrastructure for the public such as clinics, hospitals, and schools require space. It is common for developing countries to expand land use to these basic services. Inevitably, a lot of green spaces are removed to accommodate these needs. Along with the removal of forest cover is also the loss of habitat for a part of the wildlife population. Additionally, illegal logging is still a big problem, especially in less monitored forests. With the increasing rate of deforestation in the country, it is of primary concern to replenish the lost green spaces wherever needed and possible. Aside from the lowlands and mountains, the mangrove ecosystems also suffer from indiscriminate logging and conversions. In response to this, reforestation programs have been put up by the government, non-government organizations, and the private sector; through laws that required private companies to help in restoring lost habitats, especially for mining companies, as part of corporate social responsibility. All the aforementioned affects the ecosystem services provided by forest ecosystems. One of the most important of which is pollination and seed dispersal. However, in order to evaluate this, one must understand the reproductive phenological development of plant species. It is, therefore, imperative that floral phenology be considered in literature considering there is little knowledge on biotic interactions involved.

Plant reproductive phenology is the study of the periodicity of repeating biological events in plants, in relation to short-term climatic change (Sakai et al. 2005). Long-term disturbances brought about by anthropogenic factors can directly affect floral phenology and ecology as these may cause a decrease in the population and diversity of plant species (Hussein et al. 2021) and plant community structure (Wolf et al. 2017) which in turn cause habitat loss for the fauna. Climatic conditions also directly affect the floral ecology and in return, if reproduction is successful, it affects the microclimatic environment through vegetation enhancement. Conservation and management can be performed when all aspects are considered.

In the Philippines, a long-term research area has been established in Palanan, Isabela. Palanan, Isabela is a first-class municipality whose population in 2021 is at 17,260 and estimated revenue of PhP2,6M (DTI 2021). It is a remote and isolated coastal town, separated by the Sierra Madre Mountains along its western borders from most of the towns of Isabela. As early as 2020, the Ilagan-Divilacan Road was opened to enable travel by land to the remote towns of Divilacan, Maconacon, Palanan, and Dinapigue.

Palanan's forest is the only forest patch in the Philippines whose monitoring or census is mainly supported by the Smithsonian Tropical Research Institute (STRI) through its Forest Global Earth Observatory Network (ForestGEO). A 16-ha area, the Palanan Forest Dynamics Plot (PFDP), was established in 1994 for this endeavor. Since the Palanan Forests is quite vast, several communities of indigenous Agta live in it. However, their culture has been threatened by socio-economic pressures added to the general ecosystem being affected by illegal logging, resource exploitation, and land-use change (ForestGEO 2022).

This chapter shall discuss the reproductive phenology of plant species within selected sites of the Palanan Forest. Methodologies utilized will also be presented. This data assists in describing the area's long-term dynamics as well as in the understanding of how forests respond to continuous disturbances, such as typhoons and human interventions.

14.2 Gathering Reproductive Phenological Data

This is an offshoot study from the larger objective of assessing the bat pollen diet of phytophagous bats in the lowland rainforests of Palanan, Isabela which is a part of a 2.5 years research program (2016–2018) in understanding fruit bats and their interaction with plants in the said area. The underlying criteria for choosing the sites are based on different populations of phytophagous bats determined through the capture and recapture method which this phenological study is supporting.

The collection of reproductive phenological data was performed in the municipality of Palanan. A total of four sites were utilized for the study. The three study sites in the lowland forests were Brgy. Dipogen, Palanan Forest Dynamics Plot (PFDP), and Brgy. Alomanay (Fig. 14.1). PFDP is a reserved area; Brgy. Alomanay site is at a higher elevation and Brgy. Dipogen is at a mid-elevation of the two. The Agta communities are nearest to Brgy. Dipogen site. The fourth site in the lowland forest was later replaced by a mangrove site in Barangay Maligaya due to the similarity of plant assemblages with PFDP. The selection of the mangrove site was to give an array of information regarding the reproductive phenology of mangrove and mangrove-associated species. These sites were visited one after another. The monitoring of the transects was conducted from April 2016 to November 2017.

Diurnal phenological observation and data collection were performed, in between nocturnal nettings. In each site, three 20 m x 80 m transects were set up in the middle of the 20 nets utilized for the bat research. This rendered a sum of 4800 m^2 of surveyed area for each site and 19,200 m^2 for all four sites. The presence of flowers and fruits was recorded using a reproductive phenology scoring (Table 14.1) devised by Yap (2015), which was used in her previous studies in the PFDP. It includes a recording of flowering and fruiting intensity by eye estimation of its percent canopy cover. This phenological scoring is qualitative, simpler, and more practical in field

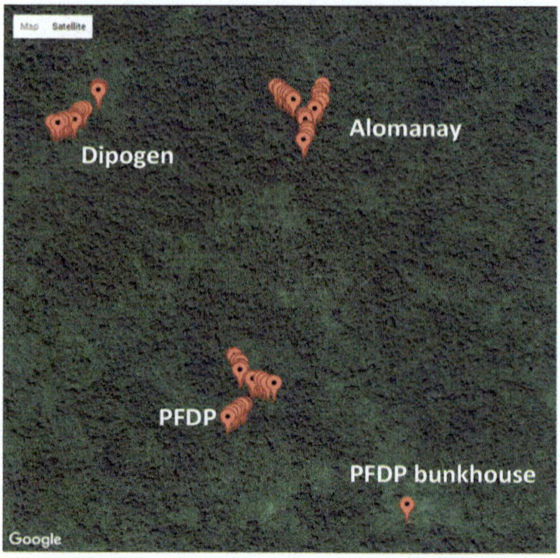

Fig. 14.1 Map of the three study sites. The mangrove site is not represented in the figure. (https://www.darrinward.com/lat-long)

Table 14.1 Modified reproductive phenology scoring (devised by Yap 2015)

Flowering	Intensity	Fruiting
1 if >50% of the buds are closed 2 if >50% of the flowers are opened 3 if >50% of the flowers are in anthesis or dying	0 if ≤25% of the canopy are w/fruits 1 if >25% of the canopy are w/fruits 2 if >50% of the canopy are w/fruits 3 if >75% of the canopy are w/fruits	1 if >50% are unripe and 2 if >50% are ripe

studies without spending long hours gathering absolute counts of flowers and fruits. This was modified from crown estimates of flower and fruit cover within the canopy, methods which were first used in other studies such as Koelmeyer (1959), Sinhaseni (2008), and Elliott et al. (2013). However, the work of Yost et al. (2018) on reproductive phenology scoring using digitized protocol may be utilized later as herbarium specimens have been collected.

Field plant identification was mostly done by long-term field guides and locals. These guides were trained from the previous censuses of the PFDP to identify plants using plant codes representative of the scientific names of the plants in the Palanan plant code guide. Voucher and flower collection were simultaneously done with a gathering of field phenology data (Fig. 14.2). Voucher collections were gathered for identification of individuals that cannot be readily identified in the field, thus were pressed, and brought to the UP Diliman herbarium. Different flora was also utilized for identification and name verification (Pelser et al. 2011; Madulid 2021; Pancho 1983; Pancho and Gruezo 2006; Pancho and Gruezo 2012; Slik 2009). Flower collections were done in conjunction with pollen-type identification.

Data on monthly rainfall and temperature were also gathered through the Davis Weather Station and Vantage Pro™ that was set in the PFDP bunkhouse. This data

Fig. 14.2 Phenology monitoring in the field (**a** and **g**) field guides assist in spotting reproductively active individuals, (**b**) field guides inspecting tag number of trees, (**c** and **d**) climbing trees to get flowers and fruits, (**e**) inspection of buds/fruits, and (**f**) gathering individual flowers/florets for mounting of pollen

was utilized in determining the general conditions in the sites which then is correlated to the reproductive phenology.

14.3 Temperature and Rainfall Readings

Mean temperature together with the total rainfall every month was plotted (Fig. 14.3). Some missing data (e.g., months of February to April 2017) were filled via imputation and interpolation method while data for April 2016 to July 2016 cannot be filled artificially. Examining the timing of reproduction of the species in the lowland forests of Palanan, at least two thirds of the monitored species reproduced during the hot months of April to September. In October 2017, the mean temperature reached 26.06 °C and rainfall hit 500.61 mm from a low of 56.83 mm the preceding month. Almost the same scenario can be seen from the interpolated data for October 2016 (μ temp. = 26.39 °C, total rainfall = 527.03 mm).

14.4 Richness and Diversity of Reproducing Tree Species

Individual tree count in PFDP is high due to the high recruitment rate. The high recruitment in the plot is brought upon by the frequent freeing up of space in the forest canopy every after typhoon, which temporarily removes light limitation for seedlings to grow. This demonstrates good forest resiliency (Yap et al. 2015). The proportion of the reproductive species > and <5.0 cm in DBH from the

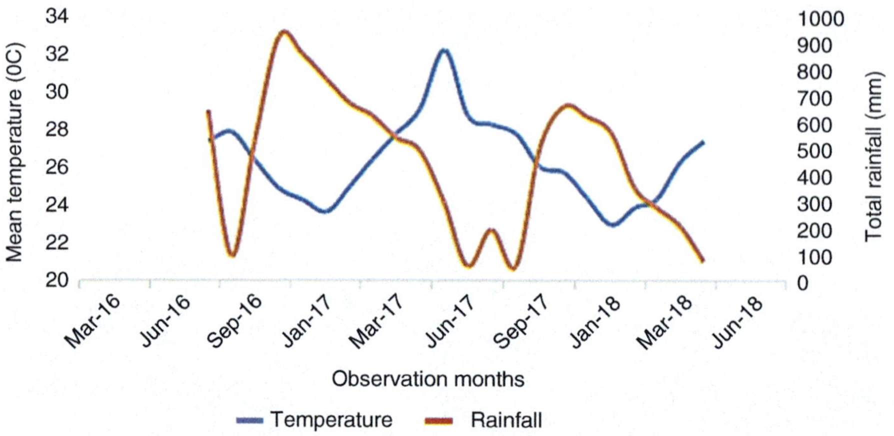

Fig. 14.3 Mean monthly temperature and rainfall in Palanan lowland forest

429 individuals observed to have flowered was examined. This was computed from the data from the first year. It is remarkable that the proportion is 1:1, however even if the individual effort is 1:1, the intensity per individual is greater in bigger conspecifics.

A total of 151 species of fruiting species from 98 genera in 45 families were inside the transect used. Of these, only 107 species from 79 genera in 40 families were observed to have entered at least one episode of reproduction either in the form of flowering, fruiting, or both. Individuals that have been observed to be fruiting but not observed flowering the previous month were considered to have flowered during the 2-week fieldwork break. Of these families, Meliaceae has the most species representation. This was followed by the Euphorbiaceae, Lauraceae, Annonaceae, Lamiaceae, Rubiaceae, Arecaceae, and Dipterocarpaceae, all of which are among the most species-rich families inside the plot. Table 14.2 shows the five most represented families that flowered in PFDP along with their representative species. Their reproductive representation in the phenology data reflect how these families were the most successful and abundant in the lowland forests of Palanan. *Litsea* and *Polyathia* species were observed to be prolific flower producers as they were observed to form flower buds in several months. It is also worth noting that members of Euphorbiaceae such as from the genus *Macaranga* are fast-growing species. Some of them reproduce more than once in a year while members of the family Meliaceae reproduce once or less in a year.

14.5 General Flowering and Fruiting Phenology

The plant species monitored were placed in different phenological classes following the classification system devised by Newstrom et al. (1994). This system standardizes the comparisons on the phenological patterns as tropical plant species

Table 14.2 Five most species represented families that flowered in PFDP

Family	Representative species	
Meliaceae	*Aglaia edulis* (Roxb.) Wall *Aglaia elliptica* (C.DC.) Blume *Aglaia lawii* (Wight) Saldhana & Ramamoorthy *Aglaia oligophylla* Miq. *Chisocheton ceramicus* (Miq.) A. DC. *Chisocheton cumingianus* (C. DC)	*Chisocheton pentandrus* (Blanco) Merr. *Dysoxylum excelsum* Blume *Dysoxylum oppositifolium* F. Muell. *Dysoxylum parasiticum* (Osbeck) Kosterm. *Dysoxylum* sp. 01
Euphorbiaceae	*Endospermum peltatum* Merr. *Macaranga* sp. *Macaranga bicolor* Muell.-Arg. *Macaranga grandifolia* Merr. *Macaranga ovatifolia* Merr.	*Macaranga tanarius* (L.) Muell-Arg. *Omphalea malayana* Merr. *Suregada glomerulata* (Blume) Baill. *Suregada multiflora* (A.Juss.) Baill. *Trigonostemon* sp.
Lauraceae	*Cryptocarya cagayanensis* Merr. *Dehaasia incrassata* (Jack) Kosterm. *Endiandra coriacea* Merr. *Litsea albayana* Vidal *Litsea fulva* (Blume) Fernandez-Villar	*Litsea garciae* Vidal *Litsea tomentosa* Blume *Litsea varians* (Blume) Boerl. *Phoebe sterculioides* (Elmer) Merr.
Annonaceae	*Neouvaria acuminatissima* (Miq.) Airy-Shaw *Haplostichanthus* sp. incert. B *Haplostichanthus lanceolata* (Vidal) Heusden *Goniothalamus elmeri* Merr.	*Enicosanthum grandifolium* (Becc.) airy Shaw *Polyalthia* sp. *Pseuduvaria luzonensis* (Merr.) YCF Su & RMK Saunders
Lamiaceae	*Callicarpa* sp. *Callicarpa* cf. *platyphylla* *Callicarpa cumingiana* Schauer *Callicarpa platyphylla* Merr. *Clerodendrum bracteatum* Wall. ex Walp.	*Clerodendrum macrostegium* Schauer *Clerodendrum mindorense* Merr.

may vary geographically. Of the observed species to have reproduced during the observation period, the majority of the species were categorized as *supra-annuals* meaning that these plants reproduce in a scale greater than a year. Typically, though, the reproductive bouts are less than one per year. Those reproducing yearly and less than a year were labeled as *annual* and *sub-annual*, respectively. Those observed to flowering entirely were labeled *continual*.

However, since the duration of the observation is less than two years, the count of species reproducing supra-annually may indeed still consist of species flowering irregularly or those restricted to flower in general flowering (GF) periods which happens irregularly. Sakai (2000) presented that general flowering is a unique phenomenon occurring in lowland dipterocarp forests in Southeast Asia. In this phenomenon, most dipterocarp species may have profuse flowering in a span of several years in long periods and in synchronicity. Such can contribute to forest diversity because it can disrupt florivory, nectivory, and frugivory due to possible fluctuations in flower and fruit availability. The separation of those reproducing irregularly and strictly in GF events will be made with continuous observation of the species in the lowland forests of Palanan.

Among the monitored species, almost forty percent reproduce sub-annually (≈20%) and annually (≈18%). The less rich group of species reproduce continuously. This includes *Pinanga insignis, Musa textilis, Ficus congesta, F. fistulosa,* and *Caryota cumingii. Musa textilis* and its congeners are known to be visited by phytophagous bats for flowers, such as members of the genera *Rousettus, Macroglossus,* and *Eonycteris* (Marshall 1985). A phenology calendar was also constructed out of the data. Table 14.3 shows the full phenophase calendar for the two-year study which shows the timing of bud appearance, floral anthesis, and fruiting.

From April 2016 to March 2017 (Y1), a general trend of decreasing abundance (Fig. 14.4) in flowering and fruiting individuals was observed for the three sites in Palanan, this includes the sites in Brgy. Dipogen, PFDP in Brgy. Villa Robles and another site in Brgy. Alomanay. As in Table 14.3, most of the species monitored produced flowers and fruits during the observed hot dry conditions. Very few were observed during cold wet conditions.

General continuous (except July 2016 and January 2017) phenological data, and climatic condition observation from April 2016 to March 2017 (Y1) in the three sites, reveal quite interesting patterns (Fig. 14.5). Diversity of flowering species was observed to have at least positive weak ($r = 0.190777816$) correlation with average monthly temperature and negative moderate ($r = -0.512106557$) correlation with total monthly rainfall. The effect of temperature in the diversity of flowering species is very minimal but quite detrimental to rainfall. It is understandable the flowers being sink zones will not be produced at very high average monthly temperatures or when average monthly rainfall is high which will destroy the flowers directly. In both cases, plants are under stress. Similar observations were gathered by Ushio et al. (2020) in their study on GF in Bornean lowland rainforest. Using models and time series data from 1993 to 2011, they found out that GF is synergistic and not independent of cool air temperature and drought affecting it. On the other hand, the opposite was observed with the fruiting diversity ($r_{temp} = -0.30969$, $r_{rain} = 0.671168$). Since most flowering occurred during hot dry conditions, it is more likely that fruiting will come during cold wet conditions. This is not evident, though, as some species were still with fruiting or with fruits during hot dry conditions. Rainfall may have also not affected the presence of fruits. In the research of Polansky and Boesch (2013), rainfall does not explain fruiting phenology and diversity even with long-term changes. Both flowering and fruiting may also be linked to seasonal patterns, pollination syndromes, and dispersal mechanisms (Mohandass et al. 2018) so long-term monitoring is important.

With these insights even without the integration of the current data with previous data gathered by Yap in 2015, some inferences can be deduced. Considering climate change is a major threat to conservation, organizing long-term and multi-taxa data on

Table 14.3 Reproductive phenology of plant species with their phenological classes in select sites of the Palanan Lowland Forest

Scientific name	Family	Monitoring year 1			Monitoring year 2			Phenological class	General observed conditions
		FB	FO	Fruit	FB	FO	Fruit		
Aglaia edulis (Roxb.) Wall.	Meliaceae	NA	Aug	May, Dec	NA	NA	May	Annual	Hot/cold dry
Aglaia elliptica (C. DC.) Blume	Meliaceae	NA	NA	Aug				Supra-annual	Hot dry
Aglaia lawii (Wight) C.J.Saldanha	Meliaceae	NA	Apr	Apr				Supra-annual	Hot dry
Aglaia oligophylla Miq.	Meliaceae	NA	NA	Dec	May	NA	May, June, Aug, Sep, Oct, Nov	Annual	Hot dry
Albizia butarek sp. nov nom. ined.	Fabaceae-Mimosoideae	NA	NA	Oct, Nov, Dec	NA	NA	Sept, Oct, Nov	Annual	Cold wet
Antherostele banahaensis (Elmer) Bremek.	Rubiaceae	NA	Mar	NA				Supra-annual	Hot dry
Antidesma cumingii Müll.Arg.	Phyllanthaceae	NA	Apr, May	June, Aug, Sep, Oct				Supra-annual	Hot dry
Apocynaceae indet Juss.	Apocynaceae	NA	NA	Apr				Supra-annual	Hot dry
Ardisia sp. Sw.	Myrsinaceae	NA	Mar	NA				Supra-annual	Hot wet
Areca sp. L.	Arecaceae	NA	Apr, Aug, Oct, Feb, Mar	Apr, Dec				Supra-annual	Hot dry
Arecaceae Bercht. & J.Presl, nom. Cons.	Arecaceae	NA	Apr	Apr				Supra-annual	Hot dry
Astronia willamsii Merr. ex C.B. Rob. (unresolved name)	Melastomataceae	NA	NA	June, Mar				Supra-annual	Hot dry

(continued)

Table 14.3 (continued)

Scientific name	Family	Monitoring year 1			Monitoring year 2			Phenological class	General observed conditions
		FB	FO	Fruit	FB	FO	Fruit		
Barringtonia pterita Merr.	Lecythidaceae	NA	NA	June				Supra-annual	Hot dry
Callicarpa L.	Lamiaceae	NA	NA	Oct				Supra-annual	Hot dry
Callicarpa cf. *platyphylla* Merr.	Lamiaceae	Oct	June, Oct	June				Supra-annual	Hot dry
Callicarpa cf. *platyphylla* Merr.	Lamiaceae	Oct	June, Oct	June				Supra-annual	Hot dry
Callicarpa pentandra Roxb.	Lamiaceae	May, Oct	June, Oct	Oct				Supra-annual	Hot dry
Callicarpa platyphylla Merr.	Lamiaceae	June, Aug, Oct	June, Oct					Supra-annual	Hot dry
Camellia lanceolata (Blume) seem.	Theaceae	Feb	NA	NA	NA	NA	Sept	Subannual	Hot wet
Caryota cumingii Lodd. Ex Mart.	Arecaceae	Apr, May, Sept	May, Oct, Mar	Apr, May, June, Aug, Oct, Feb, Mar	June, Sept, Oct	May, June, July, Oct, Nov	May, June, July, Sep, Oct	Continual	All conditions
Caryota rumphiana Mart.	Arecaceae	NA	NA	June, Aug, Sept, Oct, Dec, Feb, Mar				Subannual	All conditions
Chisocheton ceramicus Miq.	Meliaceae	June, Aug	June, Aug, Oct	NA	May	NA	Nov	Annual	Hot dry
Chisocheton cumingianus (C.DC.) harms	Meliaceae	June	NA	NA				Supra-annual	Hot dry

Chisocheton pentandrus (Blanco) Merr.	Meliaceae	Aug	NA	Dec	May, June, July, Oct	July, Aug	Oct, Nov	Annual	Hot dry
Clerodendrum bracteatum wall. Ex Walp.	Lamiaceae	NA	May	June, Aug, Oct				Supra-annual	Hot dry
Clerodendrum macrostegium Schauer	Lamiaceae	Aug, Oct	Aug, Oct	NA				Supra-annual	Hot dry
Clerodendrum mindorense Merr.	Lamiaceae	Feb, Mar	Apr, Dec, Feb, Mar	Apr, May, June, Oct, Dec				Subannual	Hot dry
Cryptocarya cagayanensis Merr.	Lauraceae	NA	NA	May				Supra-annual	Hot dry
Cyrtandra oblongata Merr.	Gesneriaceae	Sept, Dec	Sept, Oct, Dec	Oct	NA	May	May	Subannual	Hot and cold dry
Dalrympelea sphaerocarpa (Hassk.) Nor-Ezzaw.	Tapisciaceae	Feb	Feb	Oct				Supra-annual	Cold wet
Dehaasia incrassata (Jack) Kosterm.	Lauraceae	May	May	June, Aug, Sept, Oct, Nov, Dec				Supra-annual	Hot wet and dry
Dendrocnide subclausa (C.B.rob.) chew	Urticaceae	NA	NA	Aug				Supra-annual	Hot dry
Diospyros discolor Willd.	Ebenaceae	Aug, Oct	Aug, Oct	NA				Supra-annual	Hot dry
Diospyros buxifolia Thouars	Ebenaceae	NA	NA	May				Supra-annual	Hot dry
Diplodiscus paniculatus Turcz.	Malvaceae	June	NA	NA	NA	May	May	Annual	Hot dry

(continued)

Table 14.3 (continued)

Scientific name	Family	Monitoring year 1			Monitoring year 2			Phenological class	General observed conditions
		FB	FO	Fruit	FB	FO	Fruit		
Diplospora fasciculiflora Elmer	Rubiaceae	NA	Apr	NA				Supra-annual	Hot dry
Diplospora tinagaoensis (Elmer) S.J.Ali & Robbr.	Rubiaceae	Apr	Apr, Oct	Apr, May				Annual	Hot dry
Dipterocarpus validus Blume	Dipterocarpaceae	NA	NA	Aug				Supra-annual	Hot dry
Discocalyx insignis Merr.	Myrsinaceae	Sept	NA	NA				Supra-annual	Hot dry
Discocalyx micrantha Merr.	Myrsinaceae	June	June	Aug, Sept				Supra-annual	Hot dry
Drypetes grandifolia (C.B.rob.) pax & K. Hoffm.	Putranjivaceae	NA	NA	Apr, May, June, Aug, Nov, Dec, Feb	NA	Apr, July	Apr, May, July, Oct	Annual	Hot wet and dry
Dysoxylum excelsum Blume	Meliaceae	Feb	NA	NA				Supra-annual	Cold wet
Dysoxylum oppositifolium F. Muell.	Meliaceae	NA	NA	Apr, May, Aug, Oct	NA	NA	May	Annual	Hot dry
Dysoxylum parasiticum (Osbeck) Kosterm.	Meliaceae				Mar	NA	May, June, July	Supra-annual	Hot dry
Dysoxylum sp. 01 Blume	Meliaceae	NA	NA	Apr, Aug, Oct, Nov				Supra-annual	Hot dry
Endiandra coriacea Merr.	Lauraceae				NA	Apr, Oct	Apr, Oct	Supra-annual	Hot dry

Endospermum peltatum Merr.	Euphorbiaceae	NA	NA	Dec	May, Sept	May	NA	Annual	Hot dry
Monoon grandiflorum (Becc.) B.Xue & R.M.K. Saunders	Annonaceae	Aug	Apr	Apr				Supra-annual	Hot dry
Ficus congesta (H. Lév. & Vaniot) H. Lév.	Moraceae	Apr, Aug, Nov, Dec	Apr, Dec, Feb, Mar	Apr, May, June, Aug, Sept, Oct, Dec, Feb, Mar	NA	May	Apr, May, June, July, Sept, Oct, Nov	Continual	All conditions
Ficus fistulosa Reinw. Ex Blume	Moraceae	May, Aug, Dec, Feb	Apr, May, Aug, Dec, Feb	Apr, May, June, Aug, Sep, Oct, Nov, Dec, Feb, Mar	May	May, July	Apr, May, June, July, Aug, Sep, Oct, Nov	Continual	All conditions
Garcinia macgregorii Merr.	Clusiaceae	May	NA	NA				Supra-annual	Hot dry
Phyllanthus albus (Blanco) Müll.Arg.	Phyllanthaceae	NA	NA	Mar				Supra-annual	Hot dry
Phyllanthus lutescens (Blume) Müll.Arg.	Phyllanthaceae	NA	NA	Mar				Supra-annual	Hot dry
Glycosmis greenei Elmer	Rutaceae	Sept, Oct, Feb	Oct	Dec	June, Oct	NA	NA	Subannual	Warm and cold wet
Gomphandra cumingiana (Miers) Fern.-Vill.	Stemonuraceae	Sept, Oct, Dec, Feb	Feb	May, Sept, Oct, Dec	Oct, Nov	Oct, Nov	Oct	Annual	Cold wet
Gomphandra luzoniensis (Merr.) Merr.	Stemonuraceae	Apr	Apr	Apr, May, June, Aug, Sept, Oct, Feb, Mar				Subannual	Hot dry

(continued)

Table 14.3 (continued)

Scientific name	Family	Monitoring year 1			Monitoring year 2			Phenological class	General observed conditions
		FB	FO	Fruit	FB	FO	Fruit		
Goniothalamus elmeri Merr.	Annonaceae	Dec	Dec	Dec, Mar	Nov	NA	NA	Annual	Cold wet
Gonocaryum calleryanum (Baill.) Becc.	Cardiopteridaceae	Na	Feb	Na				Supra-annual	Cold wet
Gynotroches axillaris Blume	Rhizophoraceae	Sept, Dec, Feb	May, Dec, Feb	May, Oct, Nov, Mar				Subannual	Warm and cold wet
Polyalthia lanceolata S.Vidal	Annonaceae	Apr	Oct	Oct	Na	Apr, Sep	Apr, Sept	Subannual	Hot dry
Polyalthia sp. incert. B. Blume	Annonaceae	Oct, Feb	Sept	May, Aug, Sept, Oct, Dec, Feb, Mar				Subannual	Warm and cold dry
Knema glomerata Merr.	Myristicaceae	May	NA	NA	NA	NA	June, July	Annual	Hot dry
Knema sp. lour.	Myristicaceae	Aug	NA	May, June, Aug				Supra-annual	Warm/ cold dry
Leea congesta Elmer	Vitaceae	Apr, Mar	Apr, May	May, Oct				Subannual	Hot dry
Leea guineensis G. Don	Vitaceae	Aug, Sep, Oct, Nov	Sept, Oct	Nov, Dec, Feb, Mar				Subannual	All conditions
Leea "hairy" D. Royen ex L.	Vitaceae	Oct	Oct	NA				Supra-annual	Hot dry
Leea indica (Burm. f.) Merr.	Vitaceae	Oct	Oct	NA				Supra-annual	Hot dry
Leea philippinensis Merr.	Vitaceae	NA	June	June				Supra-annual	Hot dry

Lepisanthes tetraphylla Radlk.	Sapindaceae	Apr	May	May, Oct				Annual	Hot dry
Leptonychia banahaensis (Elmer) Merr.	Malvaceae	NA	Apr, May, Feb, Mar	Apr, Mar	NA	Apr	Apr	Subannual	Hot dry
Litsea albayana S. Vidal	Lauraceae	May	NA	June, Feb, Mar	NA	Apr	Sept, Oct, Nov	Subannual	Hot dry
Litsea fulva Fern.-Vill.	Lauraceae	Apr, May, June, Sept	Apr, May, June, Sept	Oct, Nov, Dec, Feb, Mar				Subannual	Warm wet and dry
Litsea garciae S. Vidal	Lauraceae	Apr, Aug, Dec, Mar	NA	Apr, Aug				Subannual	Hot dry
Litsea tomentosa Blume	Lauraceae	Apr	Apr	Apr, Mar				Supra-annual	Hot dry
Litsea varians Boerl.	Lauraceae				Oct	NA	NA	Supra-annual	Hot dry
Macaranga Thouars	Euphorbiaceae	Aug	Apr, May	June, Sept, Oct, Mar				Subannual	Hot dry
Macaranga bicolor Müll.Arg.	Euphorbiaceae				May	May	June	Supra-annual	Hot dry
Macaranga grandifolia Merr.	Euphorbiaceae	Feb	Feb	NA	May, June, July	May, June, July	June	Annual	Hot dry
Macaranga ovatifolia Merr.	Euphorbiaceae	May, June, Sept	May, June	May, June, Aug				Subannual	Hot dry
Macaranga tanarius Müll.Arg.	Euphorbiaceae	Apr, May, Sept, Feb, Mar	Apr, Sept, Oct, Feb, Mar	Apr, Oct, Mar	NA	NA	May, Oct	Subannual	Hot dry
Magnolia acuminata (L.) L.	Magnoliaceae				NA	NA	July	Supra-annual	Hot dry

(continued)

Table 14.3 (continued)

Scientific name	Family	Monitoring year 1			Monitoring year 2			Phenological class	General observed conditions
		FB	FO	Fruit	FB	FO	Fruit		
Matthaea heterophylla Quisumb. & Merr.	Monimiaceae	NA	Apr	Apr				Supra-annual	Hot dry
Melastoma malabathricum L.	Melastomataceae	May	May	June, Feb				Supra-annual	Hot dry
Melicope confusa (Merr.) T.S.Liu	Rutaceae	Dec	Dec	Na				Supra-annual	Cold wet
Memecylon ramosii Merr.	Melastomataceae	May	NA	Oct				Supra-annual	Hot dry
Microcos triflora (Blanco) R.C.K. Chung	Malvaceae	NA	NA	Apr, may				Annual	Hot dry
Morinda sp. L.	Rutaceae	NA	NA	May, June				Supra-annual	Hot dry
Musa textilis Née	Musaceae	Apr, May, Aug, Sept, Oct	Apr, May, June, Aug, Oct	Apr, May, June, Aug, Sept, Oct, Nov, Dec				Continual	All conditions
Mussaenda anisophylla S. Vidal	Rubiaceae	May, Aug, Sept, Dec, Feb, Mar	May, Aug, Oct, Dec, Feb, Mar	May, June, Aug, Oct, Nov				Subannual	All conditions
Myristica philippensis lam.	Myristicaceae	Aug	NA	May, June, Aug				Subannual	Hot dry
Neo-uvaria acuminatissima (Miq.) airy Shaw	Annonaceae				May, Sept, Oct, Nov	NA	NA	Supra-annual	Hot dry

Nephelium ramboutan-ake (Labill.) Leenh.	Sapindaceae	NA	Apr	Oct	May	May	May	Annual	Hot dry
Octomeles sumatrana Miq.	Tetramelaceae	NA	Oct	NA				Supra-annual	Hot dry
Omphalea malayana Merr.	Euphorbiaceae	Apr, June, Oct, Feb	May, June, Oct Feb	NA				Supra-annual	Hot dry
Palaquium bataanense Merr.	Sapindaceae	NA	NA	Sept				Supra-annual	Hot dry
Palaquium tenuipetiolatum Merr.	Sapotaceae	NA	NA	Sept				Supra-annual	Hot dry
Phoebe sterculioides Merr.	Lauraceae	NA	Apr	May, Oct				Subannual	Hot dry
Phyllanthus ramosii Quisumb. & Merr.	Phyllanthaceae	NA	NA	Dec				Supra-annual	Cold wet
Pinanga insignis Becc.	Arecaceae	Continuous			Continuous			Continual	All conditions
Pinanga maculata Porte ex Lem.	Arecaceae	Continuous			NA	Jun	Apr, June, July, Sept, Oct	Subannual	Warm and cold dry
Pipturus asper Wedd.	Urticaceae	NA	NA	Sept				Supra-annual	Hot dry
Planchonia spectabilis Merr.	Lecythidaceae	Oct, Dec	Aug, Sept	Aug				Supra-annual	Hot dry
Polyalthia lanceolata S. Vidal	Annonaceae	Apr	NA	Apr				Supra-annual	Hot dry
Polyalthia sp. Blume	Annonaceae	Apr, May, Aug, Dec, Feb, Mar	June, Sept, Mar	Apr, May, June, Aug, Sept, Oct, Mar				Supra-annual	Hot dry

(continued)

Table 14.3 (continued)

Scientific name	Family	Monitoring year 1			Monitoring year 2			Phenological class	General observed conditions
		FB	FO	Fruit	FB	FO	Fruit		
Praravinia sablanensis (Elmer) Bremek.	Rubiaceae	Apr, Aug, Mar	Feb	Sept, Oct, Nov				Subannual	Hot dry
Pseuduvaria luzoniensis (Merr.) Y. C.F.Su & R.M.K. Saunders	Annonaceae				NA	NA	May	Supra-annual	Hot dry
Psychotria pallidifolia Merr.	Rubiaceae	Dec	NA	Apr, Oct, Nov, Dec	Apr	Apr, July	Apr, July, Aug, Oct	Subannual	Hot dry
Psychotria sp. L.	Rubiaceae	NA	NA	June				Supra-annual	Hot dry
Pterocarpus indicus Willd.	Fabaceae - Papilionoideae	NA	NA	Feb				Supra-annual	Warm and cold wet
Saurauia klemmei Merr.	Actinidiaceae	Mar–June	Mar–June	Mar–June, Aug–Feb	Apr, May, June	Apr, May, June	Apr, June, July, Sept, Oct	Annual	Hot dry
Semecarpus sp. big leaf L.f.	Urticaceae	Oct	Oct	NA				Supra-annual	Hot dry
Atalantia racemosa Wight & Arn.	Rutaceae	NA	Aug	Apr, Dec				Supra-annual	Hot dry
Shorea assamica dyer	Dipterocarpaceae	NA	May	May				Supra-annual	Hot dry
Shorea contorta S. Vidal	Dipterocarpaceae				NA	Aug	NA	Supra-annual	Hot dry
Shorea palosapis Merr.	Dipterocarpaceae	Mar	Mar	NA				Supra-annual	Hot dry
Shorea polysperma Merr.	Dipterocarpaceae	Mar	NA	NA				Supra-annual	Hot dry

Strombosia philippinensis S. Vidal	Erythropalaceae	Aug	Apr, Aug	Apr, June, Aug	NA	NA	Oct	Annual	Hot dry
Suregada glomerulata Baill.	Euphorbiaceae	Feb	Feb	NA				Supra-annual	Hot dry
Suregada multiflora Baill.	Euphorbiaceae	Nov	NA	Apr				Subannual	Cold wet
Syzygium cf. *longisimum* P. Browne ex Gaertn.	Myrtaceae	NA	NA	May, Sept, Oct				Annual	Hot dry
Syzygium everettii (C.B.rob.) Merr.	Myrtaceae				NA	May	May	Supra-annual	Hot dry
Syzygium nitidum Benth.	Myrtaceae	Feb	NA	NA	NA	NA	June, Oct	Annual	Hot dry
Syzygium sp. P. Browne ex Gaertn.	Myrtaceae				NA	July	Oct	Annual	Hot dry
Tabernaemontana pandacaqui lam.	Apocynaceae	Dec	May	May, Aug, Sept, Feb, Mar				Supra-annual	All conditions
Trigonostemon sp. Blume	Euphorbiaceae	Feb	NA	Na				Supra-annual	Cold wet
Voacanga globosa Merr.	Apocynaceae	Sept, Nov	Apr, Aug, Oct, Feb	Apr, Aug, Oct, Feb, Mar				Subannual	All conditions
Xanthophyllum palawanense Elmer	Polygalaceae	Apr, May	May	May				Annual	Hot dry

FB floral bud; *FO* flower open

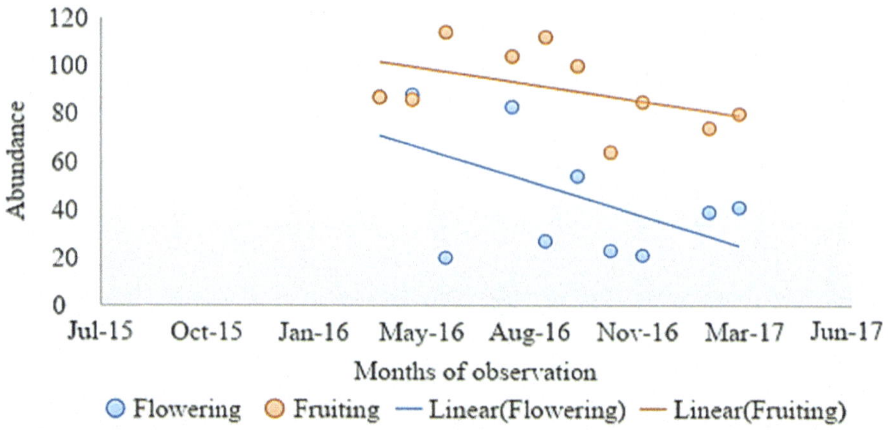

Fig. 14.4 Abundance of flowering/fruiting individuals per month from April 2016 to March 2017 in Palanan, Isabela

phenology is important to developing proper conservation strategies (Rosemartin et al. 2014), especially on reproductive phenology data as it will give information for seedling recruitment. Borghi et al. (2019) further presented that disruptions in plant–pollinator interactions reduce crop yields when drastic environmental conditions occur during flowering periods. Additionally, close monitoring may provide useful information on the behavior of Palanan lowland forests in phenomena such as general flowering.

14.6 Summary and Conclusion

Individual tree count in PFDP is high due to the high recruitment rate. A total of 151 species of fruiting species from 98 genera in 45 families were inside the transect used. Of these, only 107 species from 79 genera in 40 families were observed to have entered at least one episode of reproduction either in the form of flowering, fruiting, or both. The most reproductively active families were Meliaceae, Euphorbiaceae, and Lauraceae. The integration of previous phenological data with what has been gathered in this study is needed. The current information generated can be amended to the phenological schedules generated from February 2014 to March 2015. In general, the majority of the species reproduce during the relatively hot period of the year from April to September. The phenophase calendar constructed has shown that most of the plant species are supra-annuals with flowering and fruiting occurring during the hot dry conditions. This calendar can be continuously revised with previous and new incoming data.

The tropical lowland forests of the Philippines are indeed one of the most diverse in the region. The population of species, dynamics, and resiliency of these forests are well documented for some designated areas, but most are still under the pressure of

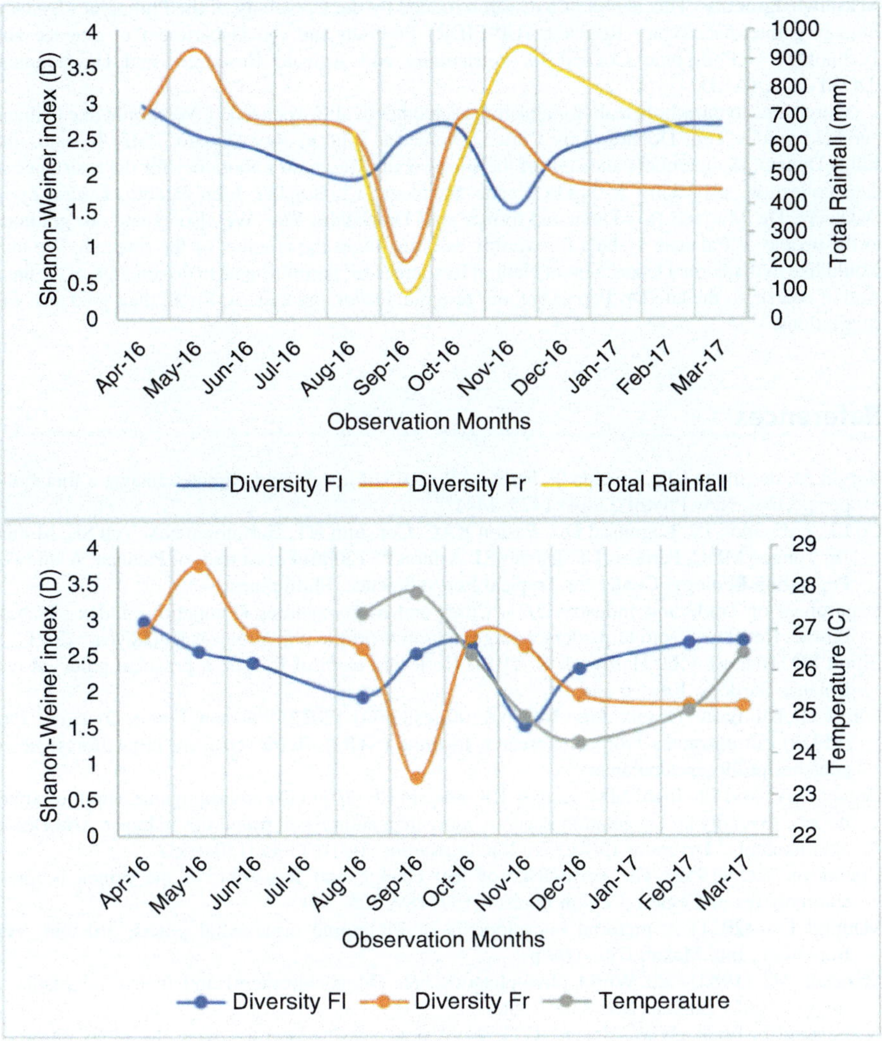

Fig. 14.5 Patterns of flowering and fruiting in the lowland forests of Palanan, Isabela with respect to mean monthly temperature and total monthly rainfall

conversion for human use. Furthermore, little is known of its reproductive phenology (Co et al. 2006), from species to community levels. With the ever-changing climate, gradual change in global temperatures, and increased fluctuations from established climatic patterns, it is imperative to study the responses and adaptability of these species. Palanan, a unique lowland forest, a frontier in the Philippines, diverse and resilient, may enable us to understand how to not replace but to "restore forests" (Ong 2017).

Acknowledgments This is part of a research funded by the University of the Philippines System-Emerging Interdisciplinary Research (UP-EIDR) Program and the Department of Science and Technology – Philippine Council for Agriculture and Aquatic Resources and Development (DOST-PCAAARD).

Our thanks to the whole team for assisting each other in all ways needed. We would like to thank our field guides Toni Donato, Emy Cortez, Diomedes Salazar, Rey Donato, Raffy Donato, and Julius Donato; the members and staff of the Biodiversity Research Laboratory for the assistance in data collection, especially to Nikki Yvette Mendoza, Christopher John Pueblo, Lystra Zyrill Dayapera, Dr. Mariano Roy Duya and former staff Dr. Sandra Yap. We also extend our gratitude to the people of Palanan, Isabela for warmly assisting us in the conduct of the research. We also would like to thank the Department of Natural Resources for permitting us to do research in the area. And of course, to the late Dr. Perry Ong, our program leader and mentor, for all their guidance and suggestions.

References

Borghi M, de Souza LP, Yoshida T, Fernie AR (2019) Flowers and climate change: a metabolic perspective. New Phytol 224(4):1425–1441

Co LL, LaFrankie JV, Lagunsad DA, Pasion KAC, Consunji HT, Bartolome NA, Yap SL, Molina JE, Tongco MDC, Ferreras UF, Davies SJ, Ashton PS (2006) Forest trees of Palanan. A study of Population Ecology. Center for Tropical Forest Science, Philippines

Department of Trade and Industry 2021. Cities and Municipalities Competitive Index—Isabela Profile. Retrieved from https://cmci.dti.gov.ph/prov-profile.php?prov=Isabela&year=2021

Elliott SD, Blakesley D, Hardwick K (2013) Restoring tropical forests: a practical guide. Royal Botanic Gardens, Kew, p 344

Forest Global Earth Observatory Network (ForestGEO) (2022) Palanan Forest Dynamic Plot (PFDP). Smithsonian Tropical Research Institute (STRI). Retrieved from https://forestgeo.si.edu/sites/philippines/palanan

Hussein EA, Abd El-Ghani MM, Hamdy RS, Shalabi LF (2021) Do anthropogenic activities affect floristic diversity and vegetation structure more than natural soil properties in hyper-Arid Desert environments? Diversity 13(4):157, 23p. https://doi.org/10.3390/d13040157

Koelmeyer KO (1959) The periodicity of leaf change and flowering in the principal forest communities of Ceylon. Ceylon Forest 4:157–189, 308–364

Madulid DA (2021) A pictorial encyclopedia of Philippine ornamental plants, 3rd edn. The Bookmark, Inc, Makati City, 699 p

Marshall AG (1985) Old World phytophagous bats (Megachiroptera) and their food plants: a survey. Zool J Linnean Soc 83:351–369

Newstrom LE, Frankie GW, Baker HG (1994) A new classification for plant phenology based on flowering patterns in lowland tropical rain Forest trees at La Selva. Costa Rica Biotropica 26(2): 141–159

Mohandass D, Campbell MJ, Chen XS, Li QJ (2018) Flowering and fruiting phenology of woody trees in the tropical-seasonal rainforest, Southwestern China. Curr Sci 114(11):2313–2322

Ong P (2017) Personal communication

Pancho JV (1983) Vascular Flora of mount Makiling and vicinity (Luzon: Philippines), part 1. In: Kalikasan, the Philippine journal of biology, suppl 1. Kalikasan Press, Manila

Pancho JV, Gruezo WS (2006) Vascular Flora of mount Makiling and vicinity (Luzon: Philippines), Part 2. Vascular Flora of Mount Makiling and Vicinity (Luzon:Philippines), Part 2. National Academy of Science and Technology (NAST) Philippines, Department of Science and Technology, Bicutan, Taguig City and Institute of Biological Sciences, University of the Philippines Los Baños, College, Laguna, Philippines (Publishers), 626 p

Pancho JV, Gruezo WS (2012) Vascular Flora of mount Makiling and vicinity (Luzon: Philippines), part 4. National Academy of Science and Technology (NAST) Philippines, Department of Science and Technology, Bicutan, Taguig City and Institute of Biological Sciences, College of Arts and Sciences, University of the Philippines Los Banos, College, Laguna, Philippines (Publishers), 405 p

Pelser PB, Barcelona JF, Nickrent DL. 2011 onwards. Co's Digital Flora of the Philippines. www.philippineplants.org

Polansky L, Boesch C (2013) Long-term changes in fruit phenology in a west African lowland tropical rain Forest are not explained by rainfall. Biotropica 45(4):434–440

Rosemartin AH, Crimmins TM, Enquist CAF, Gerst KL, Kellermann JL, Posthumus EE, Denny EG, Guertin P, Marsh L, Weltzin JF (2014) Organizing phenological data resources to inform natural resource conservation. Biol Conserv 173:90–97

Sakai S (2000) Plant reproductive phenology in tropical forests: implications of general flowering in a lowland dipterocarp forest. Japanese J Ecol 50(1):23–39

Sakai S, Momose K, Yumoto T, Nagamitsu T, Nagamasu H, Hamid Karim AA, Nakashizuka T, Inoue T (2005) Chapter 4. Plant reproductive phenology and general flowering in a mixed dipterocarp Forest. In: Roubik DW, Sakai S, Hamid Karim AA (eds) Pollination ecology and the rain Forest. Ecological studies, vol 174. Springer, New York, NY. https://doi.org/10.1007/0-387-27161-9_4

Sinhaseni K. 2008. Natural establishment of tree seedlings in Forest restoration trials at ban Mae Sa Mai, Chiang Mai Province. MSc thesis, Chiang Mai University, Thailand

Slik JWF 2009 Onwards. Plants of Southeast Asia. Retrieved from: http://www.asianplant.net/Lythraceae/Duabanga_moluccana.htm

Ushio M, Osada Y, Kumagai T, Kume T, Pungga RAS, Nakashizuka T, Itioka T, Sakai S (2020) Dynamic and synergistic influences of air temperature and rainfall on general flowering in a Bornean lowland tropical forest. Ecol Res 35:17–29. https://doi.org/10.1111/1440-1703.12057

Wolf AA, Zavaleta ES, Selmants PC (2017) Flowering phenology shifts in response to biodiversity loss. PNAS 114(13):3463–3468. https://doi.org/10.1073/pnas.1608357114

Yap SL (2015) Phenology and seed-seedling dynamics of the Palanan Forest Dynamics Plot. In: Research and Development Terminal Report 2015. Institute of Biology, University of the Philippines, Diliman, Quezon City. Unpublished

Yap SL, Davies SJ, Condit R (2015) Dynamic response of a Philippine dipterocarp forest to typhoon disturbance. J Veg Sci 27:133–143

Yost J, Sweeney P, Gilbert E, Nelson G, Guralnick R, Gallinat A, Ellwood ER, Rossington N, Willis CG, Blum SD, Walls RL, Haston EM, Denslow MW, Zohner CM, Morris AB, Stucky BJ, Carter JR, Baxter DG, Bolmgren K, Denny EG, Dean E, Pearson KD, Davis CC, Mishler BD, Soltis PS, Mazer SJ (2018) Digitization protocol for scoring reproductive phenology from herbarium specimens of seed plants. Appl Plant Sci 6(2):e1022. https://doi.org/10.1002/aps3.1022. Retrieved from https://escholarship.org/uc/item/3s12x2p1

A Survey of Understory Vegetation in the Biocultural Landscape of Mount Makiling, Luzon Island, Philippines: Implications for Sustainable Management

15

Damasa B. Magcale-Macandog, Vachel Gay V. Paller, Cleofas R. Cervancia, Inocencio E Buot, Jr, Nerissa K. Torreta, Ivy Amor F. Lambio, Analinda C. Manila-Fajardo, Alejandro C. Fajardo, Jr., Marjorie D. delos Angeles, Hazel Marie O. Chua, Christopher Alexis A. Belison, Fatima A. Natuel, and Jeoffrey M. Laruya

Abstract

Mount Makiling is known to be one of the megadiverse tropical rainforests and dormant volcanoes in the Philippines. As one of the ASEAN heritage parks, Mount Makiling showcases its biocultural landscapes that provide numerous ecosystem goods and services, especially to its nearby communities. Located at the foothills of Mount Makiling are several scientific institutions including UPLB, IRRI, PCAARRD, DOST Region IV, and Boy Scout of the Philippines. This isolated and mystical mountain is an important watershed that shelters numerous species of flora and fauna which are very popular for hikers, birdwatchers, and campers alike. Aside from its mega biodiversity, Mount Makiling abounds in natural resources like flatrocks, mud springs, hotsprings,

D. B. Magcale-Macandog (✉) · V. G. V. Paller · C. R. Cervancia · I. E. Buot, Jr · N. K. Torreta · I. A. F. Lambio · A. C. Manila-Fajardo · A. C. Fajardo, Jr. · M. D. delos Angeles · H. M. O. Chua · C. A. A. Belison
Institute of Biological Sciences, College of Arts and Sciences, University of the Philippines Los Baños, Laguna, Philippines
e-mail: dmmacandog@up.edu.ph

F. A. Natuel
College of Arts and Sciences, Laguna State Polytechnic University, San Pablo City, Laguna, Philippines

J. M. Laruya
Department of Social Forestry and Forest Governance, College of Forestry and Natural Resources, University of the Philippines Los Baños, College, San Pablo City, Laguna, Philippines

S. Ramamoorthy et al. (eds.), *Plant Diversity in Biocultural Landscapes*, https://doi.org/10.1007/978-981-19-8649-9_15

and natural spring water. Mount Makiling has provided an environment for the development of different vegetation types and floral species that have several economic uses such as food, medicinal, source of raw materials, construction, and ornamental. More so, it has been serving as the living laboratory of natural science courses of the University of the Philippines system since the 1960s.

A vegetation survey was conducted to present and analyze the richness and diversity of understory species along the Northeastern slopes of Mount Makiling. The mountain is also a habitat for a number of host plants and a diversity in bee forage for the Philippine giant honey bees. As the integrity of the great biodiversity of Mount Makiling is threatened by the activities of its biocultural landscape across history, institutions and activities within Mount Makiling were discussed to determine their potential impacts on the sustainable management of such forest reserve. The UPLB administration, in collaboration with the surrounding local government units of Los Baños, Calamba City, Sto. Tomas City, and Bay, should uphold the implementation of the existing Mount Makiling Forest Reserve (MMFR) management plan to safeguard the future of the species in Mount Makiling and ensure sustainable ecology and economy for both present and future generations.

Keywords

Mount Makiling Forest Reserve · Tropical mountain · Ecosystem services · Biocultural landscape · Sustainable ecology

15.1 Introduction

Mount Makiling is known to be one of the megadiverse tropical rainforests and dormant volcanoes in the Philippines (PHIVOLCS 2008). Located within Laguna and Batangas provinces, about 65 km away from Metro Manila, this low mountain spans a total of 4244 hectares and rises to an elevation of 1090 meters above sea level (m a.s.l.) (ASEAN Centre for Biodiversity 2013). Recognized as one of the first national parks in the country by Proclamation No. 552 in 1933, Mount Makiling embodies a variety of species and forest ecosystems. However, it was decommissioned as a national park when it was transferred under the management of the University of the Philippines (UP) for forestry education purposes in 1963 by Republic Act 3523 (ASEAN Centre for Biodiversity n.d.). Since 1990, it was reclassified as Mount Makiling Forest Reserve (MFR) under Republic Act 6967 and was managed by the Makiling Centre for Mountain Ecosystems under the UP Los Baños College of Forestry and Natural Resources (Makiling Center 2014).

As one of the ASEAN heritage parks, Mount Makiling showcases its biocultural landscapes that provide numerous ecosystem goods and services, especially to its nearby communities. Biocultural landscapes refer to unique land designs or observable patterns which are by-products of human adaptations of natural settings for the purpose of securing shelter, food, and/or pleasure (Ciftcioglu et al. 2016). It is for this reason that more often than not, biodiversity with significant relevance to particular village culture can be protected and conserved. Hence, many scholars would like to look into biocultural landscapes if this is still true, despite the burgeoning human population living in a disturbed world frequented by disasters as aggravated by climate change in the Anthropocene Epoch.

A number of studies had been reported on the biocultural landscape in many parts of the world: in Korea (Hong 1998, 2001, 2007, 2010, 2011, 2014; Hong et al. 1995, 2014; Choi and Kim 2014); in China (Min and He 2014; He et al. 2014; Jiao et al. 2014); in Japan (Hong et al. 1995; Fukamachi et al. 2001, 2003; Naito et al. 2014; Kim et al. 2014); in Europe (Bogaert et al. 2014); in Slovakia (Spulerova et al. 2014); in England (Rackham 1990, 1994; Fukamachi et al. 2003). These studies have confirmed that there is rich biodiversity in biocultural landscapes, and indeed, this is the case because people want to protect and conserve what they need most of the time, be it for food, medicine, construction material, or firewood purposes.

In Japan, satoyama landscapes, considered biocultural landscapes, are found to have biodiversity that is rapidly declining. They hypothesize that this is due to a lack of human activity which the landscape had been accustomed to for several decades (Fukamachi et al. 2001, 2003). Japanese studies reported that many inhabitants of the rural satoyama landscapes have migrated to urban metropolises to enjoy the conveniences of city life (Fukamachi et al. 2001, 2003; Naito et al. 2014; Buot Jr and Osumi 2004). Hence, the utilization of biodiversity has remarkably decreased. The proliferation of unutilized biodiversity may result in ecological interaction which might not be benign to both the ecosystem and the biodiversity.

The case of the Philippine biocultural landscapes is different. Floral and faunal biodiversity have been endangered due to too much human activity, alongside several natural and human-induced calamities. Studies conducted in many national parks and protected areas reported patches of human activities in various areas of the national monumental landscapes (Buot Jr and Okitsu 1998; Buot Jr and Okitsu 1999; Buot Jr. 2007, 2008a, b, c; Banaticla and Buot Jr. 2004; Banaticla and Buot Jr. 2005; Buot Jr and Osumi 2004; Buot Jr and Osumi 2011; Malaki and Buot Jr. 2011; de los Angeles and Buot Jr 2015; de los Angeles and Buot Jr 2019; de los Angeles et al. 2020; Chanthavong and Buot Jr. 2017; Chanthavong and Buot Jr. 2019a, b; Vallena and Buot Jr. 2018; Martinez and Buot Jr. 2018a, 2018b; Sulistiyowati and Buot Jr. 2013; Sulistiyowati and Buot Jr. 2016; Cadiz and Buot Jr. 2009; Sopsop and Buot Jr. 2009; Sopsop and Buot Jr. 2011a, b; Doydee and Buot Jr. 2011; Santiago and Buot Jr. 2018a, b; Caringal et al. 2019, 2020, 2021; Villanueva and Buot Jr. 2020; Obemio and Buot Jr. 2021; Lillo et al. 2020, 2021; Tobias et al. 2021; Villanueva et al. 2021a, b). Humans have gone to the uplands to build houses and clear forests to establish farms and earn a living. Widespread shifting cultivation occurs even in protected areas.

This paper presents the diversity of understory vegetation in the biocultural landscape of Mount Makiling. Specifically, this paper aims to describe and locate biocultural institutions and activities in the entire Mount Makiling forest landscape and vicinities; record the richness and diversity of understory species along the Northeastern slopes of Mount Makiling; present the host plants and pollen diversity of forage for the Philippine giant honey bees; and discuss the implications of these biocultural institutions and activities on the sustainable management of Mount Makiling.

15.2 Biocultural Landscape of Mount Makiling

Mount Makiling is a small mountain mass located south of Manila. Mount Makiling was the country's first national park before it became a state-owned forest reserve. It has three peaks of which Peak II is the most accessible and most popular among the public. Toward Mount Makiling's Peak II is an 8-kilometer "Mariang Makiling Trail" trek, where many can observe different vegetation types. Just a two-hour drive from Manila, the country's capital, this legendary mountain also offers a multitude of cultural ecosystem services that can be explored both by nearby locals and tourists. The biocultural landscape of Mount Makiling is shown in Fig. 15.1. Described below are the institutions, natural resources, and most notable places which offer several outdoor recreation activities within or around Mount Makiling (Makiling Center 2014).

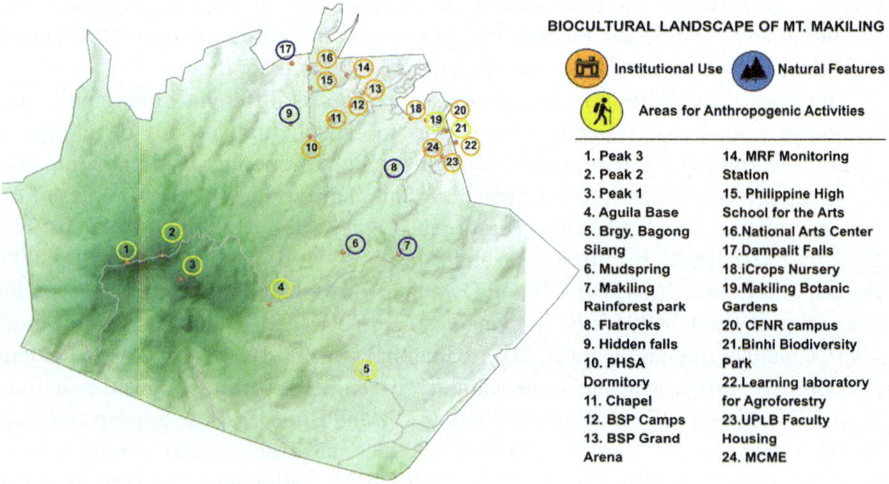

Fig. 15.1 Biocultural landscape of Mt. Makiling featuring the location of (1) Peak 3, (2) Peak 2, (3) Peak 1, (4) Aguila base, (5) Brgy. Bagong Silang, (6) Mudsprings, (7) Makiling Rainforest Park, (8) Flatrocks, (9) Hidden Falls, (10) PHSA Dormitory, (11) Chapel, (12) BSP Camps, (13) BSP Grand Arena, (14) MRF Monitoring Station, (15) Philippine High School for the Arts, (16) National Arts Center, (17) Dampalit Falls, (18) iCrops Nursery, (19) Makiling Botanic Gardens, (20) CFNR campus, (21) Binhi Biodiversity Park, (22) Learning laboratory for Agroforestry, (23) UPLB Faculty Housing, and (24) Making Center for Mountain Ecosystems

15.3 Institutions in Mount Makiling

15.3.1 University of the Philippines Los Baños

In 1909, the University of the Philippines Los Baños was originally established as the University of the Philippines College of Agriculture (UPCA) by the UP Board of Regents. Its first Dean, Dean Edwin Copeland, together with four other American teachers built the college and started conducting classes in their houses and in tents. The first structure was built on 73-hectare abandoned farmland below Mount Makiling. After World War II, UPCA started to become a training center for technical agriculture in Southeast Asia. Finally, on November 20, 1972, UPLB was granted complete autonomy, thereby making it a full-fledged university, not just an agricultural college by virtue of Presidential Decree No. 58. In 1977, the UPLB College of Agriculture was awarded "Ramon Magsaysay Award for International Understanding" in recognition of its world-class quality of research and teaching, which foster collaboration and modernize agriculture in Southeast Asia (UPLB Internet Archive n.d.).

Establishments found at the foothills of Mount Makiling around and within the UPLB campus are the buildings of the nine colleges, UPLB Faculty and Staff housing, IRRI Staff housing, and student dormitories. During the school year, the campus is very much alive with the presence of around 14,160 resident students, 964 faculty members, and more than 500 research, extension, and administrative personnel.

15.3.2 The UPLB Museum of Natural History, Makiling Botanic Gardens, and ASEAN Center for Biodiversity

The UPLB Museum of Natural History (MNH), established on September 30, 1976, is located on the upper campus of the College of Forestry and Natural Resources. The museum exhibits voucher specimens of Philippine biodiversity. Curators and research staff of the MNH are actively promoting biodiversity and natural resource conservation through education and information dissemination. The Museum is a popular science field trip destination for elementary and high school students in the region. The regional office of the ASEAN Center for Biodiversity (ACB) is located close to the UPLB MNH.

Situated at the lower elevations of the Molawin-Dampalit watershed, the Makiling Botanic Gardens (MBG) hosts a number of native, endemic, and exotic plants that visitors can enjoy. The MBG has a pavilion where social events like weddings and parties are catered. Guests attending these events can also enjoy the beautiful landscape and refreshing sceneries. The Molawin and Maralas creeks join within the vicinity of the park, pass through the UPLB campus, and are drained to Laguna Lake.

15.3.3 Los Baños: The Science City

The municipality of Los Baños, located at the foothills of Mount Makiling, is also known as the Science City. This is attributed to the presence of various research institutions and government science agencies located in the area. Among these are the Philippine Council for Agriculture, Aquatic, and Natural Resources Research and Development (PCAARRD), the Department of Science and Technology (DOST)—Region 4, the Local Government Training Center, the Philippine Carabao Research Center, the International Rice Research Institute (IRRI), and other CGIAR Offices.

15.3.4 Center for the Arts

Since the 1970s, Mount Makiling has been hosting the National Arts Center and the Philippine High School for the Arts. The Philippine High School for the Arts offers special courses on music, dancing, performance, and visual arts. Students had to undergo a rigid selection process before they were admitted to the school. The students and teachers had to stay in the dormitories and have their lessons up in the mountains. This environment provided them opportunities to focus on their studies and be creative in their respective arts.

15.4 Natural Resources in Mount Makiling

15.4.1 Flatrocks

Aside from its lush greenery, Mount Makiling also has a small river, Molawin Creek, where people can swim or dip their feet in. One of the most popular places to visit when in Mount Makiling, Flatrocks is located nearby the Mount Makiling Trail Station 2, and close to Station 4. Backpackers say it is rewarding to have a side trip at the Flatrocks while trekking at Mount Makiling.

15.4.2 Mudspring, Hotsprings, and Natural Spring Water

Situated within the Molawin-Dampalit watershed, mud springs are known to be formed due to Mount Makiling volcanic history. It is one of the mountain's mud pots or hot springs which were created due to the breaking down of rocks into clay by volcanic heat and sulfuric acid. It is also one of the most visited places in Mount Makiling, the same as flatrocks. Mount Makiling is the source of energy for the Makban Geothermal Power Plant located in Bitin, Bay Laguna.

Hot spring resorts abound in the downstream municipalities of Calamba, Los Baños, and Bay surrounding the foothills of Mount Makiling. These hot spring swimming pools are popular weekend and summer destinations and getaways for

Metro Manila residents. Mount Makiling is also a bountiful source of spring water that is being tapped by bottling companies like Viva Spring Water and Summit Spring water.

15.5 Anthropogenic Activities in Mount Makiling

15.5.1 Tourist Destination for Hiking, Camping, and Relaxation

This isolated and mystical mountain is an important watershed that shelters numerous species of flora and fauna which are very popular for hikers, birdwatchers, and campers alike. It is always a challenge for active hikers to climb up to Peak 1 (Fig. 15.1). Most hikers reach up to Peak 2, which is more accessible. Along the way, hikers may camp overnight at Aguila Base and enjoy the biodiversity as well as the nearby mudspring.

Mount Makiling is also a popular destination for devotees during the Lenten season. They would spend the Lenten season up in the mountains to rejuvenate themselves. Notable pilgrimage sites in the mountain include Saint Marc's Chapel and the grottos found along the Jamboree road. They can also go hiking along Magnetic Hill and observe the flow of rainwater defying the direction of gravity. Another exciting activity is the trek going to Dampalit Falls passing through a previous quarry site.

One of this mountain's rare gems is the rare, endo-holoparasitic species, the *Rafflesia lagascae* (commonly known as malaboô), which blooms from late February to June and attracts pollinators with its giant rotten-smelling flower (UPLB Museum of Natural History 2019; Pelser et al. 2017; Tolod et al. 2020). A lot of tourists, campers, and hikers have been enjoying the lush green and clean spring water of the forest especially because of its proximity to urban regions in Southern Tagalog and Metro Manila (Lapitan et al. 2013; Makiling Center 2014).

The Jamboree Site and Boy Scout of the Philippines Hotel are also located within the Makiling Forest Reserve. After World War II, the Jamboree Site was the official national camping site for the Boy Scouts of the Philippines. This is a huge event as boy scouts from various regions in the country congregate for their annual camping convention in Mount Makiling. In 2010, the Boy Scouts of the Philippines Hotel was established at the foothills of Mount Makiling to provide accommodation for those who want to relax and enjoy the tropical forest environment of Mount Makiling. The hotel is also a great venue for conferences.

15.5.2 Biodiversity Learning Laboratory

Owing to its nearness to the country's capital, scientists all over the world have been visiting and studying the biodiversity of tropical Mount Makiling. In the early 1900s, Mount Makiling was used primarily for forestry education by the University of the Philippines Los Baños. Nowadays, Mount Makiling serves as the living laboratory of natural science courses of the University of the Philippines system. Permanent biodiversity monitoring plots were established inside the Mount Makiling Forest Reserve for research purposes. Even social science courses make use of Mount Makiling as a social laboratory since there are communities living within the forest reserve.

15.5.3 Human Settlements in Mount Makiling

During the Japanese occupation in the 1940s, migrants from neighboring areas arrived in Mount Makiling as they sought refuge. They cleared the forest and grassland in building their houses and growing their crops, which unfortunately led to soil erosion and a decrease in soil productivity (Magcale-Macandog et al. 2011). Aside from upland rice cultivation for home consumption, migrant settlers cultivated garlic during summer and hunted wild honey bees for income generation. However, upland rice cultivation was halted due to the high cost of fertilizer. Another reason why the cultivation ceased was the rat and bird pest problems, especially in Brgy. Lalakay. Hence, farmers shifted to agroforestry by planting coffee, lanzones, citrus, and coconut in their cropping areas. Farmers also planted perennial crops which later provided soil cover that improved soil fertility and controlled soil erosion. In the 1970s, the logging ban policy was implemented to control the cutting of trees and clearing of lands for cropping purposes in the MFR. Another issue is flooding because of the establishment of the Makiling Heights Subdivision in the lower areas of Brgy. Lalakay in 1970. Generally, meeting the demand for food subsistence and livelihood, as the population increases, was the prime driver of converting grasslands and cutting trees in Mount Makiling throughout the years.

15.5.4 Honey Hunting

In Mount Makiling, honey hunting has been a practice of some locals as a source of their livelihood or supplementary farming income. Most of these locals were inhabitants of nearby communities of Bay, Calauan, Sto. Tomas and Los Banos or informal settlers within the forest reserve. These local honey hunters get honey from the Philippine Giant Honey Bees, *Apis breviligula* Fabr. colonies. These colonies are found in diverse host plants which have significantly changed since the study of Morse and Laigo in 1969. Considering that plant-pollinator relationships play a key role in maintaining and enhancing forest diversity, knowledge of host plants for bee pollinators as affected by anthropogenic activities occurring in Mount Makiling becomes imperative.

15.6 Floral Diversity of Mount Makiling

Established as one of the most significant biodiversity conservation areas in the country, Mount Makiling is rich in endemic, indigenous, and introduced species (Mallari et al. 2001). Mossy forests are found in the forest's highest elevation above 750 m a.s.l. while numerous hardwood species and forest plantations are found below 750 m a.s.l. (Combalicer et al. 2011). Based on altitude, there were four types of vegetation recognized in Mount Makiling: upper montane rainforest (>1000 m a.s.l.), lower montane rainforest (750 to 1000 m a.s.l.), lowland evergreen rainforest (100 to 500 m a.s.l.), and Parang vegetation consisting of grassland combined with second-growth forest (<100 m a.s.l.) (Fernando et al. 2004). Based on the information in Makiling Biodiversity Information System (MakiBIS) 2.0, an online database of Mount Makiling biodiversity, there are around 434 plant species in the upper montane rainforest, 110 species in lower montane rainforest, 474 species in lowland evergreen rainforest, and 28 species in grassland vegetation (Magcale-Macandog et al. 2021; Pancho and Gruezo 2006, 2009, 2012).

Among its flora, approximately 940 genera, 2038 species, 19 subspecies, 167 varieties, 225 families of ferns and flowering plants, and 15 endemic species of angiosperms have been recorded from the forest reserve (Pancho 1983; LLDA 2005; Fernando et al. 2004). Meanwhile, 148 tree species with 87 native species and 44 endemic species were reported during the floral biodiversity surveys conducted in Mount Makiling (Malabrigo et al. 2016; Castillo et al. 2018; Abraham et al. 2010). In a survey conducted by Alcala and colleagues in Alcala et al. 2019, a total of 77 moss species were discovered along the altitudinal gradients in the Sipit and Makiling trails of the forest reserve. Mount Makiling has indeed provided an environment for the development of different vegetation types and floral species that have many economic uses such as food, medicinal, source of raw materials, construction, and ornamental (Fernando et al. 2008; Magcale-Macandog et al. 2021).

15.7 Survey of Bryophytes, Pteridophytes, Vascular Epiphytes, and Understory Woody Species along the NE Slopes of Mount Makiling

As a National Center of Excellence in biological science, the Institute of Biological Sciences (IBS) received a research grant from the Commission on Higher Education (CHED) to conduct a biodiversity study in Mount Makiling. Thus, IBS faculty members and thesis students embarked on a biodiversity survey of flora and fauna including fungi, algae, and microorganisms. In this paper, vegetation survey of bryophytes, pteridophytes, vascular epiphytes, and understory woody species will be presented.

15.8 Methodology

Vegetation survey was conducted along a trail on the Northeastern slopes of Mount Makiling at various elevational gradients of every 100 m. Nine sampling sites, located 20–50 m from the trail, were established at alternating sides of the trail except at an altitude of 950 m a.s.l. due to the steepness of the region and thus, a lack of suitable sites for establishing the sampling quadrats.

15.8.1 Mosses

Within the predetermined sampling site used by the understory and pteridophytes group, moss patches were scanned on soil banks, tree trunks, and rock surfaces. Moss samples were collected from these patches and were placed carefully inside labeled collection paper bags. The specimens were brought back to the laboratory for drying, further observation, and identification. Identification was done in two ways: physical examination and microscopic examination for very small specimens that it is difficult to identify the parts even under the dissecting microscope. For the microscopic observations, Hoyer's Mounting Solution was used in mounting the specimens. Hoyer's Mounting Solution was made by mixing 25 cm^3 of distilled water, 15 g of Gum Arabic (U.S.P. Flake), 100 g of Chloral Hydrate, and 10 cm^3 of Glycerine. Hoyer's Solution was used since it is a rapid permanent mounting medium that is highly recommended when mounting mosses and/or liverworts. Photos of the specimens/samples were also taken for documentation purposes and also to aid in the identification process.

15.8.2 Understory Woody Species and Pteridophytes

Understory woody species and pteridophytes were surveyed following the quadrat method. Three random 2 m × 5 m sampling subplots were laid out within a 20 m × 20 m plot (Cadiz and Buot Jr. 2009) except at 950 m a.s.l. due to limits of available area for sampling. At each sampling plot, the species name and total cover of each species were determined. Voucher specimens of each plant species were collected and identified by taxonomists from the National Museum of the Philippines and UPLB College of Forestry and Natural Resources. All the understory woody plants were measured for Diameter at Breast Height (DBH) and height. Woody plants with diameters ranging from 3 to 20 cm and heights of less than 1 m were considered in the data set for further diversity analysis.

15.8.3 Vascular Epiphytes

Quadrats at each altitudinal site were also surveyed for epiphytic species and recorded. Identification was done on-site for species with floral appendages present.

Voucher specimens of non-reproductive specimens were collected and later identified.

15.8.4 Host Plant Diversity and Pollinator Ecology

15.8.4.1 Methodology

A survey of the bee colony sites previously visited and listed by Morse and Laigo (1969) was conducted to reverify the *Apis breviligula* population in Mount Makiling in 2011. The site locations were obtained from published maps. During the survey, pertinent information like colony height, host plant identification, angle of branch, size of comb, and direction of colony opening were determined. Honey and pollen samples were collected to determine colony forage. Finally, the nest location was noted using GPS. The survey was also conducted to determine species utilization and determine potential threats to biodiversity conservation. To determine foraging preference, honey and bee bread were collected from harvested colonies and analyzed for pollen load. Similarly, phenological and floral visitor surveys were conducted in select plant species in Mount Makiling to evaluate pollinator populations. Evaluation trials for managed pollinators were also conducted to assess the feasibility of their utilization in forest conservation.

15.8.5 Results

15.8.5.1 Mosses

A total of 24 species of mosses belonging to 19 families were collected in various altitudinal gradients of Mount Makiling. Appendix 1 shows a list of moss species recorded along different elevations of the Northeastern slope of Mount Makiling. Four species were found at the lowest elevation (100–200 m a.s.l.) including *Himantocladium, Mnium* sp., *Rhizogonium distichum,* and *Sphagnum cristatum.* At elevations between 200–300 m a.s.l., five species were identified including *Cratoneuropsis relaxa, Hylocomium* sp., *Mnium* sp., *Rosulabryum wightii,* and *Thuidium furfurosum.* The number of species recorded at every 100 m elevational gradient ranged from three to five species (Appendix 1). The species found at high elevations (>1000 m a.s.l.) include *Camptochaete arbuscula, Ctenidium pubescens, Isopterygiopsis pulchellum, Leucobryum candidum,* and *Pseudohypnella verrucosa.* Species found at mid-elevational gradients (from 400 to 800 m a.s.l.) include *Ctenidium pubescens, Hennediella heimii, Himantocladium* sp., *Hypnum triquetrum, Leucobryum glaucum, Orthorrhynchium elegans, Papillidiopsis* sp., and *Polytrichum formosum.* Data showed that varying moss species were found at different elevational gradients.

A comparison of the identified species was done for the different sampling sites. In most cases, there were one to two species that were found to be common in some of the sampling sites. Indices of similarity showed that elevations between 700–800 and 800–900 m a.s.l. as well as >1000 and 900–1000 m a.s.l. exhibited greater similarity indices (0.5–0.67) compared to the other zones. These zones were found in higher altitudes where moisture is most abundant to support the growth and development of mosses.

15.8.5.2 Pteridophytes

Appendix 2 shows a list of pteridophyte species recorded during the dry season and Appendix 3 shows those recorded during the wet season both along different elevations of the Northeastern slope of Mount Makiling. There are more fern species recorded during the wet season (38 species) than in the dry season (29 species). This is due to the higher moisture content during the wet season that is favorable for the growth of various fern species. Fern species recorded at low elevation (100–200 m a. s.l.) include *Bolbitis heteroclita* (C. Presl) Ching, *Blechnum egregium* Copel. In, and *Selaginella involvens* (Sw.) Spring. The lone fern species that has a wide range of distribution in the lower elevations from 100 to 400 m a.s.l. was *Selaginella involvens* (Sw.) Spring. Most of the fern species were commonly found at mid-elevation (500–800 m a.s.l.) including *Microsorum heterocarpum (Blume)* Ching, *Tectaria beccariana* (Cesati) C Chr, *Nephrolepis biserrata* (Sw.) Schott, *Bolbitis heteroclita (*C. Presl) Ching, *Bolbitis sinnuata* (C Presl) Hennipman, *Davallia hymenophylloides* (Blume) Kuhn, *Sphaerostephanos lobatus* (Copel.) Hotumm, *Asplenium tenerum* G Forst., *Pteris blumeana* J Agardh, *Lindsaea obtusa* J Sm. Ex Hook, *Cyathea* sp. 1, *Marattia sylvatica* Blume, and *Selaginella cupressina* (Willd.) Spring. At elevations greater than 1000 m a.s.l., fern species found include *Oleandra maquilensis* Copel., *Histiopteris incisa* (Agardh) J Smith, *Cyathea* sp. 1, and *Selaginella* sp.

During both the dry and wet seasons, the elevation of 650 m a.s.l. had the highest value for the Shannon index of the diversity of 1.81 and 1.79, respectively. This is most likely because of a favorable environmental condition for fern growth in this zone. The established sub-quadrats were located within the wilderness zone which is characterized by a rocky substrate and is enclosed by canopies of trees. In addition, a stream was present at the site which provided sufficient moisture for ferns to grow. Among the sampling sites, the presence of rocks in the area provides a suitable substrate for ferns to attach to. Since the area is enclosed by a canopy, low evaporation keeps the soil moist. High humidity, wet rocks, and moist soils provide favorable environmental conditions for the growth and proliferation of ferns.

In a study conducted by Zuquim et al. (2009), most fern species were concentrated in areas with low-light plots. Furthermore, the opening of gaps in a forest causes a local input of light, driving microclimate changes and shifts in the composition of species. During the dry season, site 9 at 950 m a.s.l. had the lowest value for the Shannon index of diversity (0.257319). Even if ferns have strategies that allow them to adapt to disturbed areas and invest in rapid gametophyte growth and recruitment (Watkins Jr et al. 2007), their growth and reproduction are negatively affected by disturbance and gaps. Site 9 is considered to be a disturbed site since it is close to the trail and the site was located close to a cliff.

The Shannon index of diversity was higher during the dry than during the wet season at sampling sites 3 (350 m a.s.l.) and 5 (550 m a.s.l.). At 350 m a.s.l. elevation, the sampling site was covered by a thick canopy which tended to lessen the evaporation of moisture from the soil into the atmosphere. Ferns are moisture-loving organisms; consequently, fern diversity was still high during the dry season at site 3. Likewise, at 550 m a.s.l., the area was also covered by a thick canopy and a stream flowed within the area of the sampling site. The presence of moisture greatly contributed to the diversity of ferns.

Significant floral species identified included fern species *Oleandra maquilingensis*, an endemic fern species that can only be found at 1050 m a.s.l. in the mossy forest of Mount Makiling. Among the threatened fern species is *Microsorum membranifloium* (R.Br.) Ching which can be found at an elevation of 500 to 600 m a.s.l. during both the wet and dry seasons. The Tree fern or *Cyathea* sp. is another group of fern species that are threatened. It can be found at an altitude of 600–1000 m a.s.l. The fern species *Histiopteris incisa* is found nowhere else in the mountain but at an altitude of 1000 m a.s.l.

15.8.5.3 Vascular Epiphytes

Epiphytes are plants that live on the surface of another plant primarily for physical support and not necessarily for nutrition. The vascular epiphytes recorded in this study belong to six (6) fern families (Aspleniaceae, Davalliaceae, Grammitidaceae, Lomariopsidaceae, Polypodiaceae, and Vittariaceae), three monocot families (Araceae, Orchidaceae, and Pandanaceae), and six dicot families (Apocynaceae, Araliaceae, Gesneriaceae, Melastomataceae, Nepenthaceae, and Piperaceae).

At an altitude of 100 m a.s.l. up to the peak of Mount Makiling, different fern species were encountered. The ferns can be said to make the most successful epiphytes since they possess characteristics that enable them to survive stressful conditions like low light intensity and the ability to withstand drying. *Drynaria quercifolia*, a common fern epiphyte under the family Polypodiaceae, makes use of its cabbage-like sterile leaf to collect humus. This humus serves to absorb water and allow the species to survive water stress conditions. *Asplenium nidus* (Family Aspleniaceae) has whorled bowl-like fronds that similarly collect and retain the humus. Like in *Drynaria*, this humus acts as a compost pile that later on releases the nutrients needed by the plants. Epiphytic angiosperms like members of the family Orchidaceae (*Dendrochilum* sp., *Bulbophyllum* sp., and *Dendrobium* sp.) at higher altitudes (from 900 m a.s.l. up to the peak) have thick fleshy stems that can store water.

Based on the index of similarity, each zone is taxonomically different from the others. The index of similarity ranges from 0.12 (between 400 and 800 m.a.s.l. sampling sites) to 0.53 (between 200 and 300 m a.s.l. sampling sites). The higher index of similarity among sampling sites with closer proximity (e.g., between 200 and 300 m a.s.l.) might be due to similarity in environmental conditions and the extent of dispersal of propagules within the vicinity.

15.8.5.4 Understory Woody Species

Appendix 4 shows a list of understory woody species recorded along different elevations of the Northeastern slope of Mount Makiling. There were 101 species belonging to 45 families identified during the survey. The 2 most represented families with 10 species each are Meliaceae and Araceae. Family Moraceae was represented by 8 species, followed by Euphorbiaceae with 5 species while Annonaceae and Commelinaceae families were each represented by 4 species. All the other 39 families were represented by one, two, or three species only. Of the 101 species identified, more than half (54 species) were native while 23 species were endemic in Mount Makiling. The 4 species introduced in the Northeastern slopes of Mount Makiling include *Laurentia longiflora*, *Zebrina pendula*, *Swietenia macrophylla* and *Coffea arabica*.

Species richness (S) of forest understory woody species and epiphytes along the altitudinal gradient shows a decreasing value from lower to higher elevations (Table 15.1 and Fig. 15.2). The highest species richness was observed at 250 m a.s.l. during both dry (29 species) and wet (25 species) seasons while the least species richness was observed at 950 m a.s.l. again for both dry (4 species) and wet (6 species) seasons. The decrease in temperature with increasing altitude might be a primary factor that inhibits the growth and proliferation of many fern species at higher altitudes. Further, mountain peaks may act as "island formations" of the theory of island biogeography limiting the movement of propagules (MacArthur and Wilson 1967).

Sorensen's Community Coefficient (Index of Similarity) of understory woody species was also compared between altitudinal gradients and between seasons within gradients. Generally, it was observed that species similarity between adjacent stands was high since propagules were easier dispersed over short distances, but in the case of the observed understory species, very low similarity indices were noted between altitudes. This could be attributed to the degree of shade given by the overstory forest canopy in the area. Forest overstory vegetation dictates the type of understory vegetation. If the overstory canopy has some gaps allowing the penetration of light

Table 15.1 Number of species (S) and species diversity index (H′) of understory woody species and epiphytes along different elevational gradients of the Northeast slope of Mount Makiling during the wet and dry seasons. Total number of species found in all sites: 108

	Dry		Wet	
Elevation (m a.s.l.)	S	H′	S	H′
100–200	15	2.41	16	2.39
200–300	29	2.74	25	2.3
300–400	17	1.86	19	2.56
400–500	13	1.67	15	2.34
500–600	24	2.51	22	2.69
600–700	8	1.3	12	2.09
700–800	12	1.3	12	1.35
800–900	10	1.31	6	1.2
900–1000	4	0.24	6	1.19

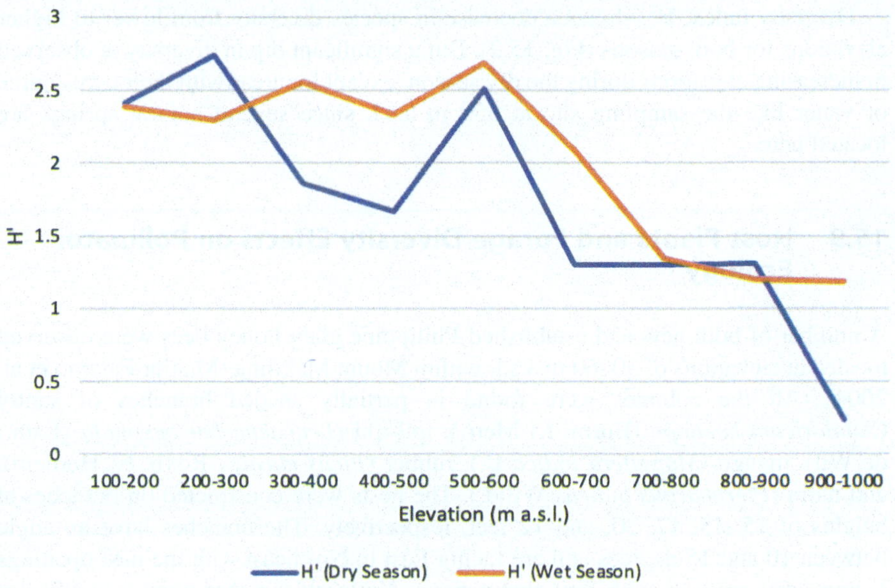

Fig. 15.2 Diversity index (H′) of understory plants and epiphytes along different elevational gradients of the Northeast slope of Mount Makiling during the dry and wet seasons

flecks, then understory species needing more light may be able to grow such as those plants with elongated leaves. However, if the overstory canopy species does not allow gaps for light penetration, and become a close canopy, then the understory vegetation will be composed of broad-leaved species (e.g., aroids) that are more adapted to low light conditions. There is a very high competition between understory species for the limited light resources on the forest floor, thus competitive exclusion might be working for the success of some species in the forest understory. In Mt. Makiling, the most abundant understory plants belong to the aroid family, with their dark green-colored and wide-bladed leaves adapted to trap the limited light energy reaching the forest floor. The dark green color of leaves indicate the abundance of chlorophyll pigments that are active in capturing the limited sunflecks on the forest floor while the wide-bladed leaves increase the surface area of the leaves thus increasing their efficiency to capture ephemeral sunflecks. Similarities in species composition between seasons per site were also low, indicating the importance of water as a resource and condition for the dominance of a few species per season.

Dendrogram analysis of the community coefficients between gradients shows that there are two distinct zones of understory species in Mt. Makiling. The lower montane forest covering 150 to 650 m a.s.l. sampling areas and the upper montane portion from 750 to 950 m a.s.l. and possibly covering up to the peak.

Diversity index, H′, shows a decrease in species diversity from lower to higher elevations for both seasons (Fig. 15.2). But a significant dip in diversity is observed in increasing elevations during the dry season, except for areas with their own source of water like the sampling site at 550 m a.s.l. since several natural springs are located here.

15.9 Host Plants and Forage Diversity Effects on Pollinator Ecology

A number of both new and established Philippine giant honey bees were observed located at elevations of 30–60 m a.s.l. within Mount Makiling (Manila-Fajardo et al. 2004). All the colonies were found in partially shaded branches of santol (*Sandoricum koetjape* (Burm. f.) Merr.), ipil-ipil (*Leucaena leucocephala* (Lam.) de Wit), mango (*Mangifera indica* L.), rubber (*Ficus elastica* Roxb. ex Hornem), and narra (*Pterocarpus indicus* Willd.). The nests were constructed on branches at heights of 15, 15, 17, 30, and 12 feet, respectively. The branches have an angle between 10 and 15 degrees and are facing East to Northeast with the nest openings facing open areas or with few obstructions. Noticeable is that quite a number of colonies were found on *Pterocarpus indicus* Willd. and *Ficus elastica* Roxb. ex Hornem within the lower and upper campus of the university reserve. However, 2–3 km outside the university, the colonies were found mostly in *M. indica* trees. This was observed at the onset of the pandemic during the summers of 2021 and 2022 suggesting tree nesting preference for the species. The results of the survey performed in 2011 were significantly different from the observations of Morse and Laigo in 1969 (Table 15.2). In 1969, Morse and Laigo observed 30 *A. breviligula* colonies which are typically nested close to the ground, preferred trees that are smooth-barked, do not nest in the open, and are not aggregated. While in the 2011 survey, the 18 Philippine wild honey bee colonies (observed during the same time frame of observation by Morse and Laigo) nested in high places on trees that are either smooth or rough barked, nest in the open, and typically show apparent aggregation of colonies (Paller et al. 2012). It can be deduced that the change in nesting behavior observed as well as the reduction in numbers is brought about by the land-use changes, deforestation, and other human disturbances occurring in the lower elevations of the Mount Makiling forest reserve.

Results of pollen load analysis have shown that there is no predominant pollen source. However, a number of pollen types were identified as belonging to Family Leguminosae which along with *Sesbania roxburgii* Merr. and an unidentified species were considered as secondary sources comprising 42.09% of the total pollen load of bee bread and honey. Amongst the Leguminosae family, *Alnus japonica* (Thunb.) Steud, a species that thrives in higher elevations, was identified as an important minor source (3.87%). This was followed by *Bombax malabaricum* DC (2.53%) and *Cassia fistula* L. (2.19%). Other species comprised the minor sources

Table 15.2 Comparison of host plants and nesting behavior of *Apis breviligula* Fabr. in Mount Makiling vicinity

Characteristics	1969 (Morse and Laigo)	2011 survey
Nest trees/host plants	*Caryota cuminggii* Lodd* *Gliricidia sepium* (Jacq.) Steud *Leucaena leucocephala* (Lam.) de Wit Others undescribed	*Acacia senegal* (L.) Willd. *Caryota cuminggii* Lodd *Celtis philippinensis* Blanco *Cocos nucifera* L. *Ficus elastica* Roxb. ex Hornem *Leucaena leucocephala* (Lam.) de Wit *Mangifera indica* L. *Nephelium lappaceum L.* *Parashorea malaanonan* Blanco *Pterocarpus indicus* Willd. *Sandoricum koetjape* (Burm. f.) Merr. *Schefflera actinophylla* (Endl.) harms** *Shorea contorta* Vidal
Height of nest	12 cm to 36 ft	25–62 ft
Angle of branch	25–45 degree	25–45 degree
Elevation (a.s.l.)	33–300	25–300
Comb size	65.90 × 51.23 cm	55.75 × 74.16 cm
Size of nest branch	12–30 cm	10–16 cm
Distribution	Putho, Tranca, Maahas Lower UPLB	Putho, Tranca***, Maahas, Silangan, lower UPLB, Umali subdivision

*Felled by typhoon or with low lying branch; **Ornamental; ***Reports, no visuals

group including *Mangifera indica* L. and a Cucurbitaceae species. In all, 20 pollen types from 14 families were identified.

In a related study, Manila-Fajardo and Cervancia (2003) similarly observed that members of Leguminosae family were heavily foraged on by *Apis mellifera* L. (European honey bees) in Mount Makiling despite the colonies lasting only several weeks on the site due to predation pressure, presence of aggressors and the seasonal flowering of the species in the reserve, that significantly degraded colony strength. The bees foraged on 23 plant species among which are *Mimosa pudica* L., *Mimosa diplotricha* C. Wright ex Sauvalle var. *diplotricha*, *Terminalia catappa* L., and *Callistemon citrinus* L., several species from Leguminosae (like *Acacia* sp., *Albizzia* sp.), Gramineae, Palmae (i.e. *Veitchia merillii* L.), Rosaceae, Solanaceae, Compositae, and Acanthaceae. This number was lower than those observed in colonies that were placed in agricultural and industrial sites indicating that the exotic species are really not adapted to the Philippine tropical forest ecosystem as they cannot forage on more species.

The Makiling reserve was also home to numerous solitary bee and wasp species that were also essential pollinators. These species were likewise dependent on the range of plant hosts for reproduction and feeding. Quakenbush (2018) studied the morphology and distribution of *Medinilla multiflora* Merr. on the mountain, the documentation of reproductive phenological patterns, a study of the mating system, and observations of biotic interactions. In the investigation of its reproductive biology, he has observed its pollen to be collected by four bee species, *Halictus* sp1 and *Halictus* sp2, *Amegilla cingulata* Fabr., and *Xylocopa* sp. in the morning generally under good weather conditions. These solitary species were however susceptible to habitat loss especially that of nesting native trees which could have profound consequences for the *Medinilla* populations.

Plant-pollinator interactions hinge not only on the availability of rewards for floral visitors but also on the wide variety of functions provided by both. Plant hosts may be especially for nidification and breeding but not for feeding. Others play the role of alternate forage and nesting species in the absence of their preferred host. The pollinators may also play as predator or parasites for other species as in the case of wasps and solitary bees. Or even play vector for plant diseases. Either way, their activities play a crucial check and balance role in maintaining the biodiversity of forests' ecosystems. The preservation and conservation of pollinator species, therefore, underscores the importance of biodiversity management.

15.10 Implications for Sustainable Management of Mount Makiling

Recognized as a significant biodiversity conservation area and as one of the 170 conservation priority areas by the Philippine government, Mount Makiling has grasslands, agricultural farms, agroforestry, leased areas, and many plantations that continually serve its surrounding communities (Mallari et al. 2001). In 2012, numerous anthropogenic activities were determined to cause changes in the species richness and diversity of the forest reserve. Examples include the conversion of lands and the expansion of illegal settlements. In fact, households in Mount Makiling were observed to increase from 280 in 2003 to more than a thousand in 2012. More so, some informal settlers were found to do illegal activities in the forest like hunting and bird trading (Paller et al. 2012). Not only the locals are adding pressure to the increasing number of threats that the forest is experiencing but the visitors are as well. According to the local communities in the forest, some tourists usually leave their marks and trash in the forest, which has resulted in pollution and vandalism.

In terms of flora species, about 70 species in Mount Makiling were found to be threatened according to the category of DENR Administrative Order 2007-01 and 72 species as per the IUCN threatened category. Based on the country's Department of Environment and Natural Resources (DENR 1997 and the International Union for Conservation of Nature (IUCN 2020), about 40 plant families have species that are already threatened, of which the highest number was recorded for Dipterocarpaceae (16 species). These are hardwood tree species that were cut down for construction purposes (Magcale-Macandog et al. 2021). Thus, the illegal collection of specimens and samples from the forest is strictly prohibited by UPLB. Interestingly, although there have been many expeditions and comprehensive surveys conducted in Mount Makiling regarding plant species, much of the biology and ecology of its richness still remain unknown.

As the decision-making on biodiversity conservation and sustainable development must be based on comprehensive and accurate data, it is crucial to manage all the biodiversity information of a particular ecosystem (Xiao-Ting 2012). In the case of Mount Makiling, an online repository known as MakiBIS or Makiling Biodiversity Information System was created to store baseline data on the forest's species composition, their morphological and molecular identification, as well as economic use and conservation status (Lapitan et al. 2013). This initiative could support various conservation plans and manage the sustainable utilization of species that have potential medicinal, industrial, and economic uses (Magcale-Macandog et al. 2021).

In addressing issues that affect the resources in the MFR, a Presidential Commission to formulate a master plan for the development of the reserve was created and signed under Executive Order 121 in 1994. This is also the same Commission that was crafted in managing and developing Laguna Lake, the largest lake in the Philippines. In 2014–2015, a management plan for the Mount Makiling Forest Reserve in the context of an ASEAN Heritage Park was also developed (Bantayan 2015) through a project funded by the German Development Corporation. Since UPLB has exclusive administration and jurisdiction over the MFR, especially for academic, research, and extension purposes; its administration has been developing plans and implementing ways forward to control any human-related pressures on the forest's sustainable preservation and conservation. With the help of the Makiling Center for Mountain Ecosystems (MCME), the UPLB administration has been closely monitoring and limiting the number of tourists visiting the forest reserve; strictly forbidding the collection of biodiversity samples from the reserve; and regularly guarding the boundaries of the reserve against possible encroachment of illegal settlers. Several biodiversity information awareness campaigns and participatory consultation meetings with upland communities were also conducted to protect and rehabilitate the biodiversity of Mount Makiling. The UPLB administration, in collaboration with the surrounding local government units of Los Baños, Calamba City, Sto. Tomas City, and Bay, should uphold the implementation of the existing Mount Makiling Forest Reserve (MMFR) management plan to safeguard the future of the species in Mount Makiling and ensure sustainable ecology and economy for both present and future generations.

Appendix 1

List of Moss Species Found along Elevational Gradients of the Northeastern Slope of Mount Makiling Forest Reserve and their Distribution Ranges

Species recorded			Elevation (m a.s.l.)									
Family	Scientific name	Distribution range	100–200	200–300	300–400	400–500	500–600	600–700	700–800	800–900	900–1000	1000–UP
Hookeraceae	*Calyptochaeta* sp.	Tropical				X						
Lembophyllaceae	*Camptochaete arbuscula*	Australasian, Malesia and Pacific									X	X
Amblystegiaceae	*Cratoneuropsis relaxa*	Tropical		X								
Hypnaceae	*Ctenidium pubescens*	Australasian, Malesia and Pacific					X			X		X
Dicranaceae	*Dicranella heteromalla*	North America, Europe, and Asia									X	
Pottiaceae	*Hennediella heimii*	Worldwide except Africa				X						
Neckeraceae	*Himantocladium*	Africa, subtropical to tropical Asia, and Oceania	X					X		X		
Hylocomiaceae	*Hylocomium* sp.	Worldwide		X								
Hypnaceae	*Hypnum chrysogaster*	Worldwide except Africa			X							
Hypnaceae	*Hypnum triquetrum*	Worldwide except Africa				X		X	X			
Hypnaceae	*Isopterygiopsis pulchellum*	Worldwide except Antarctica and South America										X

Dicranaceae	*Leucobryum candidum*	Worldwide except Antarctica								X		X
Dicranaceae	*Leucobryum glaucum*	Worldwide except Antarctica						X			X	
Mniaceae	*Mnium* sp.	Northern hemisphere	X	X				X				
Orthorrhynchiaceae	*Orthorrhynchium elegans*	Malesia							X			
Sematophyllaceae	*Papillidiopsis* sp.	Tropical			X	X	X					
Polytrichaceae	*Polytrichum formosum*	Asia, Oceania, and North America						X	X			
Hypnaceae	*Pseudohypnella verrucosa*	Asia, Malesia, and Australia										X
Rhacocarpaceae	*Rhacocarpus purpurascens*	Asia, Oceania, and North America			X							
Rhizogoniaceae	*Rhizogonium distichum*	Asia and Oceania	X		X							
Bryaceae	*Rosulabryum wightii*	Subtropics to tropics		X								
Sphagnaceae	*Sphagnum cristatum*	Worldwide	X									
Sematophyllaceae	*Taxithelium novaeguineae*	Worldwide				X						
Thuidiaceae	*Thuidium furfurosum*	Worldwide		X								

Appendix 2

List of pteridophyte species found along elevational gradients of the Northeastern slope of Mount Makiling Forest Reserve during the dry season and their conservation status

Species recorded					Elevation (m a.s.l.)									
Family	Scientific name	Common name	Conservation status	Distribution range	100–200	200–300	300–400	400–500	500–600	600–700	700–800	800–900	900–1000	1000–up
Polypodiaceae	*Microsorum heterocarpum* (Blume) Ching			Widespread					X	X	X	X		
	Microsorum membranifolium (R. Br.) Ching		Threatened							X				
Tectariaceae	*Pleocnemia* sp.						X			X				
	Tectaria beccariana (Cesati) C chr									X	X	X		
	Tectaria siifolia (wild.) Copel.								X					
	Tectaria sp.						X							

Lomariopsidaceae	*Nephrolepis biserrata* (Sw.) Schott	Forked sword fern		Widespread					X		X			
	Oleandra maquilensis Copel.	Kaliskis Ahas	Endemic to mount Makiling											X
Dryopteridaceae	*Bolbitis heteroclita* (C. Presl) Ching	El Nino fern Asian water fern			X				X	X	X			
	Bolbitis sinnuata (C Presl) Hennipman									X				
	Davallia hymenophylloides (Blume) Kuhn										X			
	Polystichum obtusum J Sm. Ex C Presl			Throughout The Philippines								X		
Blechnaceae	*Blechnum egregium* Copel. In Perkins				X							X		

(continued)

List of pteridophyte species found along elevational gradients of the Northeastern slope of Mount Makiling Forest Reserve during the dry season and their conservation status (continued)

Species recorded					Elevation (m a.s.l.)									
Family	Scientific name	Common name	Conservation status	Distribution range	100–200	200–300	300–400	400–500	500–600	600–700	700–800	800–900	900–1000	1000–up
Thelypteridaceae	*Christella parasitica* (L.) lev.						X	X	X					
	Sphaerostephanos lobatus (Copel.) Hotumm									X	X	X	X	
Aspleniaceae	*Asplenium tenerum* G Forst.										X	X		
Pteridaceae	*Pteris blumeana* J Agardh						X			X				
Dennstaedtiaceae	*Histiopteris incisa* (Agardh) J Smith													X
Lindsaeaceae	*Lindsaea obtusa* J Sm. Ex Hook									X				
Cyatheaceae	*Cyathea* sp. *1*	Tree fern, "Natong"	Threatened							X		X		X
	Cyathea sp. 2	Tree fern, "Natong"	Threatened									X		

Lygodiaceae	*Lygodium circinnatum* (Burm.) Sw	Nitong puti				X	X							
Marattiaceae	*Marattia sylvatica* Blume										X			
Lycopodiaceae	*Selaginella cupressina* (Willd.) spring									X	X			
	Selaginella involvens (Sw.) Spring				X	X	X	X				X		
	Selaginella sp. *1*												X	
	Selaginella sp. 2													X
Unknown	Fern species. 1							X						
	Fern species. 2								X					

Appendix 3

List of pteridophyte species found along elevational gradients of the Northeastern slope of Mount Makiling Forest Reserve during the wet season and their conservation status

Species recorded					Elevation (m a.s.l.)									
Family	Scientific name	Common name	Conservation status	Distribution range	100–200	200–300	300–400	400–500	500–600	600–700	700–800	800–900	900–1000	1000–up
Polypodiaceae	*Microsorum heterocarpum* (Blume) Ching			Widespread					X			X		
	Microsorum membranifolium (R. Br.) Ching		Threatened							X				
Tectariaceae	*Pleocnemia* sp.						X							
	Tectaria beccariana (Cesati) C chr								X	X				
	Tectaria siifolia (wild.) Copel.								X					
	Tectaria sp.						X							

Lomariopsidaceae	*Oleandra maquilensis* Copel.	Kaliskis ahas	Endemic to mount Makiling											X
Dryopteridaceae	*Bolbitis heteroclita* (C. Presl) Ching	El Nino fern, Asian water Fern			X					X				
	Bolbitis sinnuata (C Presl) Hennipman									X				
	Bolbitis sp.				X									
	Davallia hymenophylloides (Blume) Kuhn										X	X		
	Polystichum stenophyllum Christ.										X			
	Polystichum obtusum J Sm. Ex C Presl			Throughout the Philippines								X		

(continued)

List of pteridophyte species found along elevational gradients of the Northeastern slope of Mount Makiling Forest Reserve during the wet season and their conservation status (continued)

Species recorded					Elevation (m a.s.l.)									
Family	Scientific name	Common name	Conservation status	Distribution range	100–200	200–300	300–400	400–500	500–600	600–700	700–800	800–900	900–1000	1000–up
Blechnaceae	*Blechnum egregium* Copel. In Perkins										X	X		
Thelypteridaceae	*Christella parasitica* (L.) Lev.						X	X	X					
	Sphaerostephanos heterocarpus (Blume) Holttum								X					
	Sphaerostephanos lobatus (Copel.) Hotumm									X	X	X	X	
	Diplazium esculentum (Retz.) Sw	Vegetable fern, "Pako"							X	X				
Aspleniaceae	*Asplenium affinii* Sw.											X	X	
	Adiantum sp.								X					

Pteridaceae	*Pteris blumeana* J Agardh						X	X						
Dennstaedtiaceae	*Dennstaedtia* sp.												X	
Dennstaedtiaceae	*Histiopteris incisa* (Agardh) J smith			Throughout the Philippines										X
Lindsaeaceae	*Lindsaea obtusa* J Sm. Ex hook									X				
Cyatheaceae	*Cyathea* sp. *1*	Tree fern, "Natong"	Threatened		X								X	X
	Cyathea sp. *2*	Tree fern, "Natong"	Threatened							X				
Lygodiaceae	*Lygodium circinnatum* (Burm.) Sw	Nitong puti		Throughout the Philippines			X					X		
Marattiaceae	*Marattia* sp.											X		
	Marattia sylvatica Blume							X			X			
Lycopodiaceae	*Selaginella cupressina* (Willd.) spring									X	X	X		

(continued)

List of pteridophyte species found along elevational gradients of the Northeastern slope of Mount Makiling Forest Reserve during the wet season and their conservation status (continued)

Species recorded					Elevation (m a.s.l.)									
Family	Scientific name	Common name	Conservation status	Distribution range	100–200	200–300	300–400	400–500	500–600	600–700	700–800	800–900	900–1000	1000–up
	Selaginella involvens (Sw.) spring									X	X	X		
	Selaginella nummularia								X					
	Selaginella sp.*1*											X	X	
	Selaginella sp.2									X				X
Unknown	Fern species. 1					X								
	Fern species. 2						X							
	Fern species. 3				X									
	Fern species. 4				X									

Appendix 4

List of understory woody species (tree and shrub species) found along elevational gradients of the Northeastern slope of Mount Makiling Forest Reserve, their residency status in the Philippines, and conservation status

Species recorded					Elevation (m a.s.l.)								
Family	Scientific name	Common name	Residency status	Conservation status	100–200	200–300	300–400	400–500	500–600	600–700	700–800	800–900	900–1000
Acanthaceae	*Strobilanthes pluriformis*	"Tuhod-manok"	Endemic						X	X	X	X	X
Anacardiaceae	*Buchanania nitida*	"Balitantan"			X								
	Dracontomelon dao	"Dao"	Native				X						
Annonaceae	*Anaxagorea luzonensis*	"Bagang-Aso"	Native		X		X						
	Haplostichanthus lanceolata		Luzon Endemic						X				
	Platymitra arborea	"Bolon"	Endemic		X								
	Polyalthia flava		Luzon endemic		X				X				
Araceae	*Aglaonema* sp.		Native			X	X	X	X		X		
	Alocasia zebrina	Striped Alocasia	Luzon Endemic	Vulnerable				X					
	Amorphophallus konjac	"Pongapong"	Native			X	X				X	X	X

(continued)

List of understory woody species (tree and shrub species) found along elevational gradients of the Northeastern slope of Mount Makiling Forest Reserve, their residency status in the Philippines, and conservation status (continued)

Species recorded					Elevation (m a.s.l.)								
Family	Scientific name	Common name	Residency status	Conservation status	100–200	200–300	300–400	400–500	500–600	600–700	700–800	800–900	900–1000
	Amorphophallus phaeniifolius				X	X	X	X	X				
	Dieffenbachia sp.	"Bakya"						X					
	Epipremnum aureum	"Lukmoy", "Tibatib"				X							
	Homalomena philippinensis		Endemic					X	X	X	X	X	
	Schismatoglottis sp.	Painted tongue, "Alapayi"							X	X			
	Arenga sp.	"Kaong"						X					
	Caryota sp.	Fishtail palm			X	X	X		X				
Burseraceae	*Canarium* sp.	"Pili"	Native				X		X		X		
Campanulaceae	*Laurentia longiflora*	Star of	Introduced			X	X			X			
		Bethlehem											
Capparaceae	*Capparis* sp.	Tinikan	Native						X	X			
Celastraceae	*Salacia chinensis*	Lolly vine, Lolly berry			X	X							
Chloranthaceae	*Chloranthus elatior*		Native		X								
	Sarcandra glabra		Native			X			X		X	X	X
Clusiaceae	*Garcinia dulcis*		Native							X			

Combretaceae	*Terminalia microcarpa*	"Kalumpit"	Native			X							
	Terminalia nitens		Endemic		X								
Commelinaceae	*Amischotolype hispida*				X								
	Commelina diffusa								X	X			
	Pollia secundiflora		Native					X					
	Zebrina pendula	Wandering jew	Introduced						X				
Datiscaceae	*Octomeles sumatrana*		Endemic						X				
Dipterocarpaceae	*Parashorea malaanonan*	White lauan		Critically Endangered		X							
Ebenaceae	*Diospyros pilosanthera*	"Bolongeta"	Native			X							
Euphorbiaceae	*Antidesma ghaesembilla*		Native			X			X				
	Bridelia insulana		Native			X							
	Drypetes maquilingensis		Endemic									X	
	Mallotus sp.		Native			X							
	Neotrewia cumingii		Endemic		X	X							
Fagaceae	*Lithocarpus ovalis*		Luzon Endemic		X								

(continued)

List of understory woody species (tree and shrub species) found along elevational gradients of the Northeastern slope of Mount Makiling Forest Reserve, their residency status in the Philippines, and conservation status (continued)

Species recorded					Elevation (m a.s.l.)								
Family	Scientific name	Common name	Residency status	Conservation status	100–200	200–300	300–400	400–500	500–600	600–700	700–800	800–900	900–1000
Flacourtiaceae	*Hydrocarpus alcalae*		Luzon Endemic		X	X		X	X		X		
Gesneriaceae	*Cyrtandra* sp.		Native									X	
Hypoxidaceae	*Curculigo capitulata*		Native			X							
Icacinaceae	*Stemonurus luzoniensis*		Endemic							X	X		
Lauraceae	*Beilschmiedia* sp.		Native										X
	Neolitsea villosa		Native								X		X
Lecythidaceae	*Planchonia spectabilis*		Endemic		X		X						
Leeaceae	*Leea philippinensis*		Native									X	
Lobeliaceae	*Hippobroma* sp.											X	
Loganiaceae	*Strychnos* sp.		Native			X			X				
Marantaceae	*Calathea* sp.								X	X			
	Donax cannaeformis		Native		X								
	Phyrnium sp.								X				
Meliaceae	*Aglaia argentea*		Native		X	X							
	Aglaia edulis		Native	Vulnerable	X				X				

	Aglaia elliptica		Native				X	X					
	Aglaia rimosa		Native	Vulnerable				X	X				
	Aphanamixis polystachya		Native			X	X	X	X				
	Chisocheton pentandrus		Native					X					
	Dysoxylum arborescens		Native			X	X						
	Dysoxylum gaudichaudianum		Native				X						
	Dysoxylum mollissimum		Native				X	X	X				
	Swietenia macrophylla		Introduced			X	X						
Mimosaceae	*Parkia roxburghii*		Native					X		X			
	Wallaceodendron celebicum		Endemic			X			X				
Moraceae	*Artocarpus rubrovenius*		Native	Vulnerable		X	X						
	Ficus congesta		Native		X								
	Ficus pseudopalma		Endemic			X	X						
	Ficus suberula				X	X	X	X	X				
	Ficus ulmifolia		Endemic						X			X	
	Ficus nervosa pubine rvis		Native				X		X	X			

(continued)

List of understory woody species (tree and shrub species) found along elevational gradients of the Northeastern slope of Mount Makiling Forest Reserve, their residency status in the Philippines, and conservation status (continued)

Species recorded					Elevation (m a.s.l.)								
Family	Scientific name	Common name	Residency status	Conservation status	100–200	200–300	300–400	400–500	500–600	600–700	700–800	800–900	900–1000
	Parartocarpus venenosa		Native		X								
	Streblus asper		Native						X				
Myristicaceae	*Myristica glomerata*		Native		X		X						
Myrsinaceae	*Ardisia* sp.		Native			X							
Myrtaceae	*Syzygium calubcob*		Endemic						X				
	Syzygium nitidum		Native		X	X							
Olacaceae	*Strombosia philippinensis*		Endemic			X							
Orchidaceae	*Habenaria* sp.		Native			X			X				
	Rhombusa sp.				X		X	X	X				
Papilionaceae	*Erythrina subumbrans*		Native			X							
	Pterocarpus indicus *Forma echinatus*		Native			X							
Rubiaceae	*Coffea arabica*		Introduced						X				
	Ixora sp.		Native				X						
	Pavetta sp.		Native		X	X	X						

	Psychotria sp.		Native			X	X						
	Melicope triphylla		Endemic								X		
Sapindaceae	*Harpullia arborea*		Native					X					
	Litchi chinensis philippinensis		Native			X	X	X					
	Mischoscarpus pentapetalus		Native					X					
	Nephelium ramboutan-ake		Luzon endemic			X							
Sapotaceae	*Palaquium foxworthyi*		Luzon endemic		X								
Sterculiaceae	*Heritiera sylvatica*		Native					X					
Tiliaceae	*Diplodiscus*		Endemic			X							
	Paniculatus												
Ulmaceae	*Celtis luzonica*		Luzon endemic	Vulnerable					X				
Urticaceae	*Dendrocnide* sp.		Native			X							
	Elatostema angustatum		Native				X						
	Elatostema *Longifolium*		Native			X			X	X			
	Villebrunea sp.		Native			X	X		X				X
Xanthorrhoeaceae	*Dianella* sp.				X								
Zingiberaceae	*Alpinia elegans*				X		X		X				

References

Abraham ERG, Gonzalez JCT, Castillo ML, Lit IL Jr, Fernando ES (2010) Forest cover and biodiversity profile of the crater area of mt. Makiling, Luzon, Philippines. Asia Life Sci Suppl 4:48–82

Alcala AA, Angeles MDD, Buot IE Jr (2019) Fern species diversity across various land use types of Mt. Makiling, Luzon Island, Philippines. Biodivers J 209:2437–2445

ASEAN Center for Biodiversity (n.d.). https://www.aseanbiodiversity.org/

ASEAN Center for Biodiversity. 2013. Mt. Makiling Forest Reserve. Retrieved from https://asean.chm-cbd.net/mt-makiling-forest-reserve

Banaticla MCN, Buot IE Jr (2005) Altitudinal zonation of pteridophytes on mount Banahaw de Lucban, Luzon Island, Philippines. Plant Ecol 180:135–151

Banaticla MCN, Buot IE Jr (2004) Fern patch structure and species diversity along the altitudinal gradient on mount Banahaw de Lucban, Luzon Island, Philippines. Philipp Agric Sci 87:49–60

Bantayan N. 2015. An ASEAN Heritage Park: Mount Makiling Forest Reserve. Retrieved from https://makiling.center/category/news/page/41/?ak_action=accept_mobile

Bogaert JB, Vranken I, Andre M (2014) Anthropogenic effects in landscapes: historical context and spatial pattern. In: Hong SK, Bogaert K, Min Q (eds) Biocultural landscapes: diversity, functions and values. Springer, Berlin

Buot IE Jr (2008a) Vertical distribution and zonation pattern of woody vegetation on the north-western slope of Mt. Mayon, Philippines. Asia Life Sci 17:189–205

Buot IE Jr (2008b) A new way of looking at environmental health: focused on man and his environment. Asia Life Sci Suppl 2:1–5

Buot IE Jr (2008c) Sustaining environmental health in Philippine Satoyama landscapes. Asia Life Sci Suppl 2:129–138

Buot IE Jr (2007) Vegetation zonation of Mt. Akiki, Cordillera Mountain range, Benguet Province, Philippines. Asia Life Sci 16(1):53–70

Buot IE Jr, Osumi K (2011) Land use type pattern and woody species composition near human disturbed landscapes on mount Makiling, Luzon Island. Am J Environ Sci 7(4):306–315

Buot IE Jr, Osumi K (2004) Satoyama landscapes in the Philippines. J Nat Stud 3:1–11

Buot IE Jr, Okitsu S (1999) Leaf size zonation pattern of woody species along the altitudinal gradients of Mt. Pulag, Philippines. Plant Ecol 145:197–208

Buot IE Jr, Okitsu S (1998) Vertical distribution and structure of the tree vegetation in the montane forest of Mt. Pulag, Cordillera Mountain range, the highest mountain in Luzon Is., Philippines. Veg Sci 15:19–32

Cadiz GO, Buot IE Jr (2009) An enumeration of the woody plants of Cantipla forest fragments, Cebu Island, Philippines. Philipp J Syst Biol 3:1–7

Caringal AM, Buot IE Jr, Villanueva ELC. (2021). Endemic Philippine teak (*Tectona philippinensis* Benth. & hook. (f.) and associated flora in the coastal landscapes of Verde Island passage, Luzon Island, Philippines. Curr Sci, 120 (6): 1057–1065

Caringal AM, Buot IE Jr, Villanueva ELC (2020) Analysis of human and Philippine teak forest interaction in the lasang-Baybay landscape along Verde Island passage marine corridor, Batangas Province, Philippines. J Mar Isl Cult 9(1):1–19

Caringal AM, Buot IE Jr, Villanueva ELC (2019) Woody plant communities in the Philippine teak forest landscape along Verde Island passage, Batangas, Luzon, Philippines. Biodiversitas 20(11):3189–3198

Castillo ML, Castillo LA, Canceran MS, Gonzalvo KJP, Barua LD, Alegre AC, ... , Bantayan NC. 2018. Distribution, diversity and spatial analysis of tree species in a long-term ecological research plot in Molawin-Dampalit watershed, mount Makiling Forest reserve Asian J Biodiver, 91 12–36

Chanthavong S, Buot IE Jr (2017) Plant diversity in dong Na tard provincial protected area, Lao People's Democratic Republic (Lao PDR): species structure and Forest zonation. Environment Asia 10:52–62

Chanthavong S, Buot IE Jr (2019a) Priority Areas for Conservation Planning in Dong Na Tard Provincial Protected Area, Lao People's Democratic Republic (Lao PDR). Environment Asia 12(2):116–125

Chanthavong S, Buot IE Jr (2019b) Conservation status of plant diversity at dong Na tard provincial protected area, Lao, PDR. Int J Conserv Sci 10:393–402

Choi JU, Kim DY (2014) Characteristics and Management of old and Sacred Dangsan Forests in Korea. In: Hong SK, Bogaert K, Min Q (eds) Biocultural landscapes: diversity, functions and values. Springer, Berlin

Ciftcioglu GC, Uzun O, Nemutlu FE (2016) Evaluation of biocultural landscapes and associated ecosystem services in the region of Suğla Lake in Turkey. Landsc Res 41(5):538–554. https://doi.org/10.1080/01426397.2016.1173659

Combalicer MS, Kim D, Lee DK, Combalicer EA, Cruz RVO, Im S (2011) Changes in the forest landscape of mt. Makiling forest reserve, Philippines. Forest Sci Technol 72:60–67

de los Angeles MD, Buot IE Jr, Cuevas VC, Sanchez PB (2020) Phytoremediating Capacity of Copper Tolerant Plants in Mine Tailing Soil Materials with Compost Amendment in Mankayan Benguet, Philippines. Environment Asia 13(1):86–98

de los Angeles MD, Buot IE Jr (2019) A checklist of Fern species in disturbed areas of mount Makiling, Philippines. Thailand Nat Hist Mus J 13(2):163–172

de los Angeles MD, Buot IE Jr (2015) Diversity and distribution of pteridophytes altitudinal gradient of a secondary forest in mt Makiling, Philippines. Int J Ecol Conser 16:25–46

Department of Environment and Natural Resources 1997. National list of Threatened Flora. Retrieved from https://bmb.gov.ph/index.php/facts-and-figures-wild/national-list-of-threatened-flora

Doydee P, Buot IE Jr (2011) Mangrove vegetation zones in Ranong coastal wetland ecosystem, Thailand. KU Fish Research Bull 35:14–28

Fernando ES, Co LC, Lagunzad DA, WSm G, Barcelona JF, Madulid DA, Zamora PM (2008) Threatened plants of the Philippines: a preliminary. Asia Life Sci 3:1–52

Fernando ES, Sun BY, Suh MH, Kong HY, Koh KS (2004) Flowering plants and ferns of mt. ASEAN-Korea Environmental Cooperation Unit, Makiling, p 368

Fukamachi K, Oku H, Rackham O (2003) A comparative study on trees and hedgerows in Japan and England. In: Palang H, Fry G (eds) Landscape interfaces. Kluwer Academic Publishers, London, pp 53–69

Fukamachi K, Oku H, Nakashizuka T (2001) The change of a Satoyama landscape and its causality in Kamiseya, Kyoto prefecture, Japan between 1907 and 1995. Landsc Ecol 16:703–717

He Q, Wu Z, Xu F, Guo X (2014) Human-elephant relations becoming crisis in Xishuangbanna, Southwest China. In: Hong SK, Bogaert K, Min Q (eds) Biocultural landscapes: diversity, functions and values. Springer, Berlin

Hong SK (1998) Changes landscape patterns and vegetation process in the far-eastern cultural landscapes: human activity on pine-dominated secondary vegetations in Korea and Japan. Phytocoenologia Band 28 Heft 1:45–66

Hong SK (2001) Factors affecting landscape changes in Central Korea: cultural disturbance on the forested landscape systems. In: Zonneveld IS, van der Zee D (eds) Landscape ecology applied in land evaluation, development and conservation. ITC Publisher, Enschede

Hong SK (2007) Linking man and nature landscape systems: landscaping blue-green network. In: Hong SK et al (eds) Landscape ecological applications in man-influenced areas: linking man and nature systems. Springer, Dordrecht, pp 505–523

Hong SK (2010) Island ecology on biological-cultural diversities and human adaptation in seascapes. J Ecol Field Biol 32(2):115–120

Hong SK (2011) Biocultural diversity and traditional ecological knowledge in island regions of southwestern Korea. J Ecol Field Biol 34(2):137–147

Hong SK (2014) Philosophy and background of biocultural landscapes. In: Hong SK, Bogaert K, Min Q (eds) Biocultural landscapes: diversity, functions and values. Springer, Berlin

Hong SK, Maffi L, Oviedo G, Mastuda H, Kim JE (2014) Development and vision of island biocultural diversity initiative. In: Hong SK, Bogaert K, Min Q (eds) Biocultural landscapes: diversity, functions and values. Springer, Berlin

Hong SK, Nakagoshi N, Kamada M (1995) Human impacts on pine-dominated vegetation in rural landscapes in Korea and western Japan. Vegetatio 116:161–172

Jiao Y, Liang L, Okuru T, Takeuchi K (2014) Ecosystem services and biodiversity of traditional agricultural landscapes: a case study of the Hani terraces in Southwest China. In: Hong SK, Bogaert K, Min Q (eds) Biocultural landscapes: diversity, functions and values. Springer, Berlin

Kim JE, Hong SK, Nakagoshi N (2014) Biocultural landscape dynamics in Japanese rural regions. In: Hong SK, Bogaert K, Min Q (eds) Biocultural landscapes: diversity, functions and values. Springer, Berlin

Laguna Lake Development Authority (LLDA) 2005. Los Baños-Mt. Makiling micro-watershed ecoprofile. Laguna De Bay Environmental Action Planning

Lapitan FRG, Magcale-Macandog DB, Raymundo MDS, dela Cruz AG (2013) Makiling biodiversity information system (MakiBIS): development of an online species information system for mount Makiling. J Nat Stud 12(1):1–11

Lillo EP, Malaki AB, Alcazar SMT, Rosales R, Redoblado BR, Diaz JLB, Pantinople EM, Buot IE Jr (2021) Inventory of native and mother trees in key biodiversity areas of Cebu Island, Philippines for species selection in local reforestation programs. Biodiversitas 22:4740–4749

Lillo EP, Malaki AB, Alcazar SMT, Rosales R, Redoblado BR, Diaz JLB, Pinote JP, Rosales R, Buot IE Jr (2020) Native trees in Nug-as forest key biodiversity area, Cebu Philippines. Biodiversitas 21(9):4162–4167

MacArthur RH, Wilson EO (1967) The theory of island biogeography. Princeton University Press

Magcale-Macandog DB, Balon JL, Engay KG, Nicopior OBS, Luna DA, dela Cruz CP. (2011) Assessment of the environmental degradation and proposed solutions in the Los Baños subwatershed through participatory approaches. J Nat Stud 10(2):1–15

Magcale-Macandog D, Lapitan FRG, Laruya JM, Valerio JIF, Aguila JCD, Mesina CAL, Santos TMF, Cuevas ANT, Baylon KD, Silapan IM, Dimalibot R, Edrial JD, Larida NJA, Natuel FA, Perez MGLD, Quinones SGL (2021) Updating of Makiling biodiversity information system (MakiBIS) and analysis of biodiversity data. Biodiversitas 22(1):211–226

Makiling Center for Mountain Ecosystems [MCME] (2014, November 23) Laws governing MMFR. An ASEAN Heritage Park. Retrieved January 10, 2023, from https://makiling.center/about/

Malaki ABB, Buot IE Jr (2011) Conservation status of indigenous trees in Argao River watershed reserve, Cebu Island, Philippines. Asia Life Sci Suppl 6:45–59

Malabrigo PL Jr, Umali AGA, Tiburan CL Jr, Pampolina NM, Balatibat JB, Abasolo WP, Luna AC, Boncodin JC (2016) Tree diversity and stand structure or permanent biodiversity monitoring area in Mt. Makiling. Asian J Biodivers 7. https://doi.org/10.7828/ajob.v7i1.885

Mallari NA, Tabaranza B, Crossby MJ (2001) Key conservation sites in the Philippines. Bookmark, Inc., Makati City, Philippines, p 485

Manila-Fajardo AC, Fajardo ACJR, Cervancia CR (2004) Pollen sources of Apis dorsata F. in mt. Makiling, Luzon island, Philippines. Bees for new Asia. In: Proc. of the 7th Asian apicultural association conference and 10th BEENET symposium and techno-fora. University of the Philippines Los Baños, College, Laguna, pp 63–66

Manila-Fajardo AC, Cervancia CR (2003) Performance of honey bees Apis mellifera Fabricius in three ecosystems in Laguna. Phil Agric Sci 86(2):146–157

Martinez MR, Buot IE Jr (2018a) Mangrove assessment in Manamoc Island for coastal retreat mitigation. J Mar Island Cult 7(1):65–83. https://doi.org/10.21463/jmic.2018.07.1.05

Martinez MR, Buot IE Jr (2018b) Altitudinal distribution of some Hoya species in the Philippines. Thailand Nat Hist Mus J 12(1):9–18

Min Q, He L (2014) Agro-cultural landscapes in China: types and significances. In: Hong SK, Bogaert K, Min Q (eds) Biocultural landscapes: diversity, functions and values. Springer, Berlin

Morse RA, Laigo FM 1969. Apis Dorsata in the Philippines: (including an annotated bibliography). Philippine Association of Entomologists—Apis (Insects). Monograph. 96 pp.

Naito K, Kikuchi N, Ohsako Y (2014) Role of the oriental white stork in maintaining the cultural landscape in the Toyooka Basin, Japan. In: Hong SK, Bogaert K, Min Q (eds) Biocultural landscapes: diversity, functions and values. Springer, Berlin

Obemio CD, Buot IE Jr (2021) Checklist of pteridophytes in mount Matutum protected landscape, southern Mindanao, Philippines. Sylvatrop. Techn J Philippine Ecosyst Nat Res 31(2):87–124

Paller VGV, Cervancia CR, Macandog DM, Raymundo AK, De Guia APO, Pagaduan DC, Dimalibot JC, Lacaste AV, de Chavez ERC, MCB L, Hadsall AS, Torreta NK, Nacorda OS, Buot IE Jr, IAF L, Mendoza BC, Lit II Jr, dale Cruz CPP, DAP F (2012) Biodiversity along elevational gradients of the Northeastern slope of Mt. Makiling Forest Reserve. UPLB, College, Laguna, Laguna, Philippines, p 4031

Pancho JV (1983) Vascular Flora of mount Makiling and vicinity (Luzon: Philippines), part 1. In: Kalikasan, the Philippine journal of biology, suppl 1. Kalikasan Press, Manila

Pancho JV, Gruezo WSM 2006. Vascular Flora of mount Makiling and vicinity (Luzon: Philippines), part 2. National Academy of science and technology (NAST) Philippines, Department of Science and Technology, Bicutan, Taguig City and Institute of Biological Sciences, University of the Philippines Los Baños, College, Laguna, Philippines (Publishers), 626 p

Pancho JV, Gruezo WSM 2009. Vascular Flora of mount Makiling and vicinity (Luzon: Philippines), part 3. National Academy of science and technology (NAST) Philippines, Department of Science and Technology, Bicutan, Taguig City and Institute of Biological Sciences, University of the Philippines Los Baños, College, Laguna, Philippines (Publishers), 496 p

Pancho JV, Gruezo WSM 2012. Vascular Flora of mount Makiling and vicinity (Luzon: Philippines), part 4. National Academy of science and technology (NAST) Philippines, Department of Science and Technology, Bicutan, Taguig City and Institute of Biological Sciences, University of the Philippines Los Baños, College, Laguna, Philippines (Publishers), 405 p

Pelser PB, Nickrent DL, Gemmill CEC, Barcelona JF (2017) Genetic diversity and structure in the Philippine *rafflesia lagascae* complex (Rafflesiaceae) inform its taxonomic delimitation and conservation. Syst Bot 42(3):543–553. https://doi.org/10.1600/036364417X696186

Philippine Institute of Volcanology and Seismology [PHIVOLCS] (2008, July 30) Inactive Volcanoes. Retrieved January 11, 2023, from https://www.phivolcs.dost.gov.ph/index.php?option=com_content&view=article&id=78&Itemid=500024

Quakenbush JP (2018) Pollination, mating system, phenology and characterisation of *Medinilla multiflora* Merr. (Melastomataceae) on mt. Makiling. Philippines Sibbaldia 16:121–139

Rackham O (1990) Trees and woodland in the British landscape, 2nd edn. Dent, London, p 234

Rackham O (1994) The illustrated history of the countryside. Weidenfeld and Nicolson, London, p 240

Santiago JO, Buot IE Jr (2018a) Assessing the status of pinuchu as indicator of socio-ecological resilience of Chaya socio-ecological production landscape, Ifugao. Philippines Biodiversitas 19(6):2010–2019

Santiago JO, Buot IE Jr (2018b) Conceptualizing the socio-ecological resilience of the Chaya Rice terraces, a socio-ecological production landscape in Mayoyao, Ifugao, Luzon Island, Philippines. J Mar Isl Cult 7(1):107–126. https://doi.org/10.21463/jmic.2018.07.1.07

Spulerova J, Dobrovodsak M, Stefunkova D, Baca A, Lieskovsky J (2014) Biodiversity of traditional agricultural landscapes in Slovakia and their threats. In: Hong SK, Bogaert K, Min Q (eds) Biocultural landscapes: diversity, functions and values. Springer, Berlin

Sopsop LB, Buot IE Jr (2011a) The importance of non-wood Forest products in the household economy of the direct users of Aborlan Guba system, Palawan Island, Philippines. J Environ Sci Manag 14(2):50–59

Sopsop LB, Buot IE Jr (2011b) Human forests interaction in Aborlan-Guba system, Palawan Island, Philippines: implications for conservation and management. Asia Life Sci 20(1):153–171

Sopsop LB, Buot IE Jr (2009) The endangered plants of Palawan Island, Philippines. Asia Life Sci 18(2):251–279

Sulistiyowati H, Buot IE Jr (2016) Ecological valuation tools to appraise biomass and soil organic matter in a natural Forest ecosystem. J Wetl Biodiversity 6:97–108

Sulistiyowati H, Buot IE Jr (2013) Integrated biodiversity valuation framework: ecological approach. J Wetl Biodiversity 3:7–16

The IUCN 2020 Red List of Threatened Species. Version 2015-4. www.iucnredlist.org. Downloaded on 17 October 2020

Tobias A, Baltazar AM, Taguinod JJ, Buot IE Jr (2021) The role of home gardens in conserving threatened plants of the Philippines. Asian J Biodiver 12:87–103

Tolod JR, Galindon JMM, Atienza RR, Duya MV, Fernando ES, Ong PS (2020) Flower and fruit development and life history of *rafflesia consueloae* (Rafflesiaceae). Philippine J Sci 150(S1): 321–334. Retrieved from https://www.researchgate.net/publication/349731300_Flower_and_Fruit_Development_and_Life_History_of_Rafflesia_consueloae_Rafflesiaceae

UPLB Internet Archive: Wayback Machine (n.d.). UPLB History. Retrieved January 10, 2023, from https://web.archive.org/web/20090720100734/http://www.old.uplb.edu.ph/about/uplb-history

UPLB Museum of Natural History (2019). Makiling's rare gems and the flies that come with them. Retrieved from https://mnh.uplb.edu.ph/14-content/news/382-makiling-phis-rare-gems-and-the-flies-that-come-with-them

Xiao-Ting D (2012) Research Progress on biodiversity information system. Int Conf Comput Sci Serv Syst Nanjing 2012:296–299. https://doi.org/10.1109/CSSS.2012.81

Vallena MG, Buot IE Jr (2018) Checklist and conservation status of Hoya species in Bicol peninsula, Luzon Island, Philippines. Thailand Nat Hist Mus J 12(2):91–102

Villanueva ELC, Fernandez DAP, de los Angeles M, PJS T, RDR O, Buot IE Jr (2021a) Checklist of the Flora and Fauna of the karst forests in Basey, Samar, Philippines. Thailand Nat Hist Muse J 15(2):147–160

Villanueva ELC, Fernandez DAP, de los Angeles M, PJS T, RDR O, Buot IE Jr (2021b) Biodiversity in forests over limestone in Paranas, Samar Island Natural Park (SINP), a UNESCO world natural heritage site nominee. Trop Nat Hist 21(1):119–145

Watkins JE Jr, Mack MC, Sinclair RR, Mulkey SS (2007) Ecological and evolutionary consequences of desiccation tolerance in tropical fern gametophytes. New Phytol 176(3): 708–717. https://doi.org/10.1111/j.1469-8137.2007.02194

Zuquim G, Costa FRC, Prado J, Braga-Neto R (2009) Distribution of pteridophyte communities along environmental gradients in Central Amazonia, Brazil. Biodivers Conserv 18:151–166

Floral Diversity and Carbon Stock Assessments of Montane Forests along the Tri-Boundaries of Benguet, Ifugao, and Mountain Province, Philippines

16

Diomedes A. Racelis, Pastor L. Malabrigo, Jr., and Elenita L. Racelis

Abstract

Floral diversity and carbon stock assessments were conducted within the tri-boundaries of Benguet, Ifugao, and Mountain Province, the Philippines. The floral diversity assessment revealed that 32 morpho-species belonging to 26 genera and 24 families can be found inside the 43 sample plots used in forest carbon assessment. The additional 94 species that were recorded from the opportunistic survey conducted in the designated study sites brought the total count to 126 morpho-species. The dominant families identified include Pinaceae, Fagaceae, Myrtaceae, Pentaphylacaceae, Podocarpaceae, and Ericaceae. Thirty-six (36) species recorded were found to be endemic to the Philippines which includes rare species of orchids. Six (6) species recorded were found to be threatened under DAO 2017-11 and/or the IUCN Red list of Threatened Species (2018-2) which includes Yew (*Taxus wallichiana* Zucc.). Further, the five (5) species with the highest Importance Value or the most important species are Benguet Pine (*Pinus kesiya* ssp. *langbianensis*), Palayen (*Lithocarpus* sp.), Boltik (*Syzygium* sp.), Igem (*Dacrycarpus imbricatus*), and Bini (*Eurya buxifolia*). In

D. A. Racelis (✉)
Institute of Renewable Natural Resources, College of Forestry and Natural Resources, University of the Philippines, Los Baños, College, Los Baños, Laguna, Philippines
e-mail: daracelis@up.edu.ph

P. L. Malabrigo, Jr.
Department of Forest Biological Sciences, College of Forestry and Natural Resources, University of the Philippines, Los Baños, College, Los Baños, Laguna, Philippines

E. L. Racelis
Training Center for Tropical Resources and Ecosystems Sustainability, College of Forestry and Natural Resources, University of the Philippines, Los Baños, College, Los Baños, Laguna, Philippines

S. Ramamoorthy et al. (eds.), *Plant Diversity in Biocultural Landscapes*,
https://doi.org/10.1007/978-981-19-8649-9_16

terms of biodiversity indices, Sample Plot 1 had the highest Shannon and Simpson index wherein 18 species were recorded. Most of the plots have low to very low diversity based on the H′ diversity classification range.

Further, results of forest carbon baseline measurement show that the 1195.0-ha montane forest that straddles the tri-boundaries of Mt. Province, Benguet, and Ifugao contains a total of 300,973.7 Mg or tons of C which is equivalent to 1,103,570.25 Mg of CO_2 contained in 247,636.46 Mg of forest biomass. The land cover type which contains the highest C on a per hectare basis is Closed Forest Broadleaved (CFB) with 688.72 Mg C which is equivalent to 2525.29 Mg of CO_2. This is followed by Open Forest Broadleaved (OFB) (363.52 Mg ha^{-1} C or 1332.90 Mg ha^{-1} CO_2), Open Forest Conifer (OFC) (238.92 Mg ha^{-1} C or 876.03 Mg ha^{-1} CO_2), Other Wood Land Shrubs (OWLS) (164.43 Mg ha^{-1} C or 602.90 Mg ha^{-1} CO_2), Other Land, Cultivated/Farm Land/Annual Crops (CFL) (161.67 Mg ha^{-1} or 592.80 Mg ha^{-1}, and Other Wooded Land, Wooded Grassland (147.48 Mg ha^{-1} or 540.77 Mg ha^{-1}), CFL contains more C than OWLG owing to the comparatively high C content in soil component of the former. Moreover, among the vegetation types only Closed Forest Broadleaved exhibited the tree C sink with the highest C and CO_2 stored. This is expected particularly for broadleaved species based on previous studies conducted but mostly in the lowland areas. The rest of the land cover types; OFB, OFC, OWLS, and OWLG, significantly contributed to the carbon accumulation of this vegetation. Trees and roots followed next with a mean share of 33% and 5%, respectively out of the total C density. Necromass or litter also gave a good contribution to carbon stored in this study (4.86 Mg ha^{-1}) as compared to other studies conducted where only minimal or as low as half a ton per hectare was recorded.

Keywords

Montane forest · Floral diversity · Forest carbon stock · Cordillera Administrative Region

16.1 Introduction

The tri-boundaries of Benguet, Ifugao, and Mountain Province are host to one of the few remaining heavily forested areas in the country. It also serves as the major headwater of the Abra and Chico River Basins that constitute almost two-thirds of the entire Cordillera Administrative Region (CAR). However, the area is currently under threat from destructive human activities such as illegal logging, kaingin, and wildlife poaching which are in turn the results of worsening poverty in the surrounding upland communities (Cordillera Peoples Alliance 2016).

One of the strategies to ensure the effective management of the forests and natural resources in the area is to harness their potential to provide sustainable sources of livelihood to the local people to wean them out from destructive activities. To realize this, one must first assess the extent of resources present in the area, particularly their level of biodiversity and carbon storage potential, and in view of the current initiatives of the government to promote climate change mitigation and biodiversity conservation. The information to be generated could be the basis for operationalizing the Payment for Environmental Services (PES) in the area. In turn, PES can be used to incentivize the protection and conservation of forest resources by the local communities.

Thus, the Department of Environment and Natural Resources, Cordillera Administrative Region (DENR-CAR) initiated a collaborative research and extension project with the University of the Philippines Los Baños, College of Forestry and Natural Resources to conduct forest biodiversity and carbon stock baseline assessments in the area. This study aimed to determine the level of floral diversity and measure the amount of various carbon pools including aboveground, ground, and belowground biomass including soil organic carbon, within the 78-ha forest area along the tri-boundaries of Benguet, Ifugao, and Mt. Province. The project was funded by the Global Environment Facility (GEF) as a partner agency of the Asian Development Bank in the implementation of the Integrated Natural Resources Management Project (INREMP) in the Upper Chico River Basin.

Furthermore, the Cordillera Administrative Region is the most forested region in the Philippines with more than 81% of its total land area still covered with verdant forests. This is equivalent to almost 10% of the remaining forest cover in the country as reported by the Philippine Forestry Statistics (DENR-FMB 2020). The high-elevation forests of CAR provide a conducive environment for a wide array of endemic species to survive (Villanueva 2005). Out of the 101 Key Biodiversity Areas in the Philippines, only 27 are protected. Three of these are located in CAR namely, Banao Protected Landscape in the province of Kalinga; Mt. Pulag in Benguet, Ifugao, and Nueva Vizcaya; and the Apayao Lowland Forest bordered by Apayao and Cagayan (BMB-DENR 2016). Aside from the exceptionally rich biodiversity, CAR is also home to many indigenous peoples who play an important role in biodiversity conservation and management (Magcale-Macandog 2010). The upland biodiversity conservation of the people of Cordillera is often linked to their indigenous knowledge (IK), particularly in agroecosystem management. The best example of this IK is the *Muyong*, a known tradition among Ifugaos that promotes sustainable forest management and biodiversity conservation (Camacho et al. 2016). Their value of conservation is also rooted in their strong appreciation that biodiversity is a precious asset to present and future generations for it provides supporting, regulating, provisioning, and cultural services unique to the Cordillera (Jianchu and Ruscoe 1993). This close link between the indigenous peoples and the ecosystem services resulted in reduced integrity of the forests and the endangerment of the biodiversity living within (Conservation International Philippines 2007), making it more critical to do biodiversity-related research.

Biodiversity studies in Cordillera are best summarized in the paper of Baoanan et al. (2020). According to their comprehensive review, there were a total of 226 biodiversity reports for the Cordillera mountains. More than half (130) of the studies were on flora, 66 (30%) on fauna, and 30 (13%) on the ecosystem. However, 145 (64%) of the literatures are unpublished project reports including local research of the Department of Environment and Natural Resources, Ecosystems Research and Development Service (DENR-ERDS), terminal reports of funded projects, graduate, and undergraduate theses. Moreover, of the 82 published reports, only 54 have international circulation. Most (131/226) of the biodiversity studies in Cordillera were concentrated in the province of Benguet (58%). The provinces of Abra, Apayao, and Ifugao can be considered understudied with only 3 (1%), 11 (5%), and 14 (6%) biodiversity studies, respectively (Baoanan et al. 2020).

Among the published flora studies in Cordillera, the plant diversity assessment in Balbalasang Balbalan National Park (now Banao Protected Landscape) is one of the most comprehensive (Malabrigo 2013) where the author documented a total of 319 species of plants, including a new species, *Rafflesia banaoana Malabrigo*, a very rare parasitic plant. The same paper also reported 106 Philippine endemics (38 of which are Luzon endemics) and 18 threatened species. One of the recent flora studies that were not included in the study of Baoanan et al. (2020) is the plant survey conducted by Malabrigo et al. (2021) in the tri-boundaries of Ifugao, Benguet, and Mountain Province. This study recorded a total of 125 species including 36 endemic and six (6) threatened species. Some of the most noteworthy species recorded are the high-elevation orchid species such as *Ceratostylis ramosa*, *Dendrochilum cinnabarinum*, *D. unicorne*, *Liparis philippinensis*, *Pinalia philippinensis*, and *Pinalia tridens*.

More recently, Mendoza et al. (2022) and Habito et al. (2022) conducted studies to determine the species of wood used by Ifugaos in their traditional houses. Thirty-two (32) species of mostly indigenous and endemic wood tree species were identified to be used in the construction of traditional houses. Some of the tree species that were consistently utilized by the Ifugaos for traditional house construction were Molave or Amugawon (*Vitex parviflora*), Narra or Udyo (*Pterocarpus indicus*), and Itangan (*Weinmannia luzoniensis*). The same tree species were also identified in the earlier works of Rondolo (2000), Alberto et al. (2011), and Taguiling (2013) as the wood species commonly used in the construction of the traditional Ifugao houses. However, all the above studies revealed a noticeable shift to the use of commonly available but lesser quality wood species as the premium hardwood trees traditionally utilized by the indigenous house makers were no longer available due to over-utilization. This is supported by the timber inventory study conducted by DENR-CAR (2011) in the municipality of Hungduan, Ifugao which revealed that the diameter of the premium wood species used for traditional houses ranged from 5 to 10.7 cm. Despite the exceptionally rich biodiversity brought about by its unique topography and physiognomy and considering the very few published biodiversity reports in CAR, there is a dearth of biodiversity data available to the public (Baoanan et al. 2020).

A study by Dedicatoria et al. (2010) examined the environmental changes in the nearby Mt. Data National Park over the last 50 years and analyzed how such changes affected the sustainable management of the area. Results of the key informant interviews done in the study revealed that the conditions in the park had drastically changed over the last 50 years. It was determined in the study that the area was once thickly forested but the forest cover had since been converted to vegetable farms with only few patches remaining. Specifically, the study found that the place was once thickly forested and served as a hunting ground for indigenous peoples but had been turned into mixed patches of pine forests and grasslands due to logging and commercial farming. This resulted in forest fragmentation and habitat loss that eventually led to decrease in species number and diversity and an increase in the number of tolerant species such as exotic rats and other fruit bats. Local practices of indigenous peoples that include *anop* (practice of hunting mammals) and *ikik* (lighting a tree during the night to attract and trap avian species) also contributed to decline in wildlife population in the area. Diverse species of birds and mammals used to abound in the watershed but locals reported fewer wildlife sightings with some species of birds believed to have already become locally extinct.

Moreover, a number of forest carbon stock assessment-related studies have been conducted in the area. Doyog et al. (2018) used a combined Landsat image and forest cover map to delineate the spatial distribution of the aboveground biomass (AGB, Mg C ha^{-1}) storage of the Pinus kesiya Royle ex Gordon (Benguet pine) forest of Sagada, Mt. Province, Philippines. The AGB was estimated using the Normalized Difference Vegetation Index (NDVI) values (with filter and with no filter). The data was obtained from 66 plots that were established in the different Benguet pine stands in Sagada. The estimated aboveground biomass (AGB) density of the Benguet pine forest was 249.66 Mg ha^{-1} corresponding to 112.35 Mg C ha^{-1}.

A study by Racelis et al. (2017) evaluated the 12,185-ha Benguet Pine (*Pinus kesiya* Royle ex Gordon) mining forests in Philex, Padcal, Benguet Province for its climate change mitigation potential. It adopted a single stratum stratified sampling to determine the amount of biomass, carbon, and equivalent CO_2 stored in trees, understorey and herbaceous (UH), necromass and litter (NL), coarse woody debris (CWD), roots, and soil organic carbon (SOC) under three stand density types. The stand types were delineated using Google Earth image and GIS-generated map based on crown stocking density (CSD): high (HDS) = 2/3 CSD, medium (MDS) $\geq$ 1/3 to <2/3 CSD and low (LDS) $\geq$ 1/3 CSD. A non-destructive sampling was applied to all trees with a diameter of 5 cm and above. Destructive sampling of UH and NL samples were done in subplots inside the main sample plots. Composite soil samples were collected per soil depth (0–10 cm; 10–20 cm, and 20–30 cm) and replicated per elevation type (high, mid, and lower elevation). Results showed a mean biomass density of 176.83 Mg ha^{-1} with a C content of 169.74 Mg ha^{-1}. The C content per stand type is as follows: HDS = 199.06 Mg ha^{-1}; MDS = 188.91 Mg ha^{-1}, and LDS = 123.03 Mg ha^{-1}. This amounted to a total of 8,277,474.88 Mg of CO_2.

Lumbres and Lee (2014) estimated the aboveground biomass (AGB) of the La Trinidad forests in Benguet, Philippines, using Landsat Thematic Mapper, forest map, field data, and k-nearest neighbor method. Results showed that the mean AGB of the La Trinidad forests was 240.46 Mg ha^{-1} with an overall accuracy of 56% determined using a confusion matrix. Avtar et al. (2020) used field-based information in the *muyong* forest management system to estimate the total carbon pools in aboveground biomass (AGB), belowground biomass (BGB), forest floor biomass, and soil carbon in Ifugao Province, Philippines. The study revealed a difference in the total carbon stock of the private woodlots (*muyong*) and that of the communal forest (*bilid*). The average carbon stock per unit area is higher in *muyong* (150.8 Mg ha^{-1}) than in *bilid* (126.1 Mg ha^{-1}). The study was conducted to determine whether Ifugao's *muyong* forest system should be included under the REDD+ framework.

16.2 Study Site

The study was conducted along the tri-boundaries of Benguet, Ifugao, and Mountain Province (Fig. 16.1). Results of the analysis of the GIS map of the study area and on-site observation revealed six (6) land cover types. They include the following:

- **Closed Forest Broadleaved (CFB)**: Forest with a predominance (more than 75% of tree crown of trees) of broadleaved species. Formation where tress in various storey and undergrowth cover a high proportion (>40%) of the ground and do not have a continuous dense grass layer.
- **Open Forest Broadleaved (OFB)**: Formations composed of broadleaved species with discontinuous tree layers with coverage of at least 10% and less than 40%.
- **Open Forest Conifer (OFC)**: Forest with predominance (more than 75% of tree crown cover) of trees of coniferous species with discontinuous tree layer with coverage of at least 10% and less than 40%.
- **Other Wood Land Shrubs (OWLS)**: Land where the dominant woody vegetation are shrubs, generally of more than 0.5 meters and less than 5 meters in height in maturity and without a definite crown. The growth habit can be erect, spreading, or prostate. The height limits for trees and shrubs should be interpreted with flexibility, particularly the minimum tree and maximum shrub height, which may vary between 5 and 7 meters approximately.
- **Other Wooded Land, Wooded Grassland (OWLG)**: Areas predominantly vegetated with grasses such as *Imperata*, *Themada*, *Saccharum* spp., among others.
- **Other Land, Cultivated/Farm Land/Annual Crops (CFL)**: Land not classified as forest or other wooded land used by man for agriculture or pastures.

The remaining forest cover is approximately 795 ha which is almost 66% of the total area. These forests are highly fragmented and occur in relatively small patches, an indication of intense anthropogenic disturbance. The remaining forests fall under either tropical lower montane rainforest (with the presence of broad-leaf forest and pine forest types) or tropical upper montane forest (mossy forest).

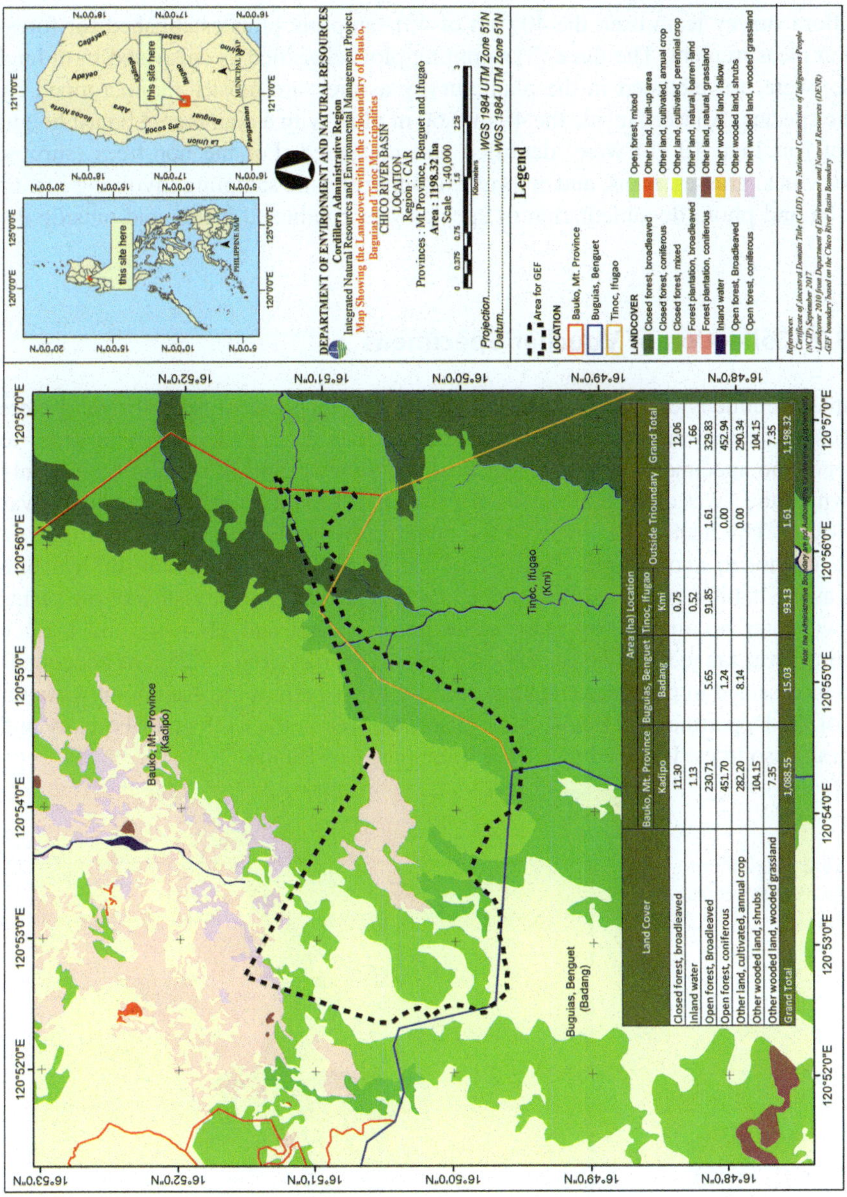

Land Cover	Area (ha) Location				
	Bauko, Mt. Province Kadipo	Buguias, Benguet Badang	Tinoc, Ifugao Kmi	Outside Trioundary	Grand Total
Closed forest, broadleaved	11.30		0.75		12.06
Inland water	1.13		0.52		1.66
Open forest, Broadleaved	230.71	5.65	91.85	1.61	329.83
Open forest, coniferous	451.70	1.24		0.00	452.94
Other land, cultivated, annual crop	282.20	8.14		0.00	290.34
Other wooded land, shrubs	104.15				104.15
Other wooded land, wooded grassland	7.35				7.35
Grand Total	1,088.55	15.03	93.13	1.61	1,198.32

Fig. 16.1 The study site

16.3 Floral Diversity Assessment

16.3.1 Survey Method

The flora survey team used the 40 (out of 43) sampling plots established for forest carbon measurement. The three (3) sampling plots classified as cultivated farm land (CFL) were not included in the plant survey as they are mostly monocropped by exotic species. All trees inside the 40 m × 60 m plots with a diameter at breast height of not less than 10 cm were identified and measured. For the non-trees (shrubs, herbs, vines, grasses, ferns, and epiphytes), opportunistic sampling involving identification and photo documentation of species was conducted inside and outside the plots.

16.3.2 Collection of Voucher Specimens

Sample specimens of each species that were difficult to identify in the field were collected. These were processed at the end of each day of the survey in order to preserve the specimens prior to identification. Schweinfurth's method (commonly known as the "wet collection" method), a standard plant collection technique was employed. This technique involves the soaking of properly labeled specimens in ethyl alcohol to avoid rapid wilting and crumpled drying. Important information such as the habitat, physiognomy, slope and aspect, and characteristics of the plants that will later become unobservable after drying; DBH and TH were noted. Each specimen was tagged using proper coding prior to storage (Fig. 16.2). The specimens were then sealed in polyethylene bags for further analysis at the Taxonomy Laboratory of the Department of Forest Biological Sciences in the College of Forestry and Natural Resources, University of the Philippines Los Baños.

Fig. 16.2 Herbarium specimens with proper tagging

16.3.3 Data Analysis

16.3.3.1 Diversity Indices

Diversity indices (Shannon, Simpson's, and Evenness) of monitoring plots were computed using the Paleontological Statistical software package for education and data analysis (PAST version 3.20). Data on the occurrence of each species and their richness in surveyed plots were primarily used in the computation of indices.

16.3.3.2 Importance Value

The relative density, relative dominance, and relative frequency values for each tree species were determined to obtain their Importance Value (IV)—a standard measurement in forest ecology to determine the rank relationships of species. The relative values indicate different aspects of species importance in a community. Importance values were computed using the following formula:

$$\text{Density} = \frac{\text{Number of individuals}}{\text{Area sampled}}$$

$$\text{Relative density} = \frac{\text{Density for a species}}{\text{Total density for all species}} \times 100$$

$$\text{Frequency} = \frac{\text{Number of plots in which species occur}}{\text{Total number of plots sampled}}$$

$$\text{Relative frequency} = \frac{\text{Frequency value for a species}}{\text{Total frequency for all species}} \times 100$$

$$\text{Dominance} = \frac{\text{Basal area or volume for a species}}{\text{Area sampled}}$$

$$\text{Relative dominance} = \frac{\text{Dominance for a species}}{\text{Total dominance for all species}} \times 100$$

$$\text{Importance value} = \text{Relative density} + \text{relative frequency} + \text{relative dominance}$$

16.4 Forest Carbon Stock Baseline Assessment

16.4.1 Forest Vegetation Sampling

16.4.1.1 General Vegetation

The study area is a mosaic of different land uses consisting of closed forests, open forests, grassland, and agricultural lands. The remaining forest cover is approximately 795 ha which is almost 66% of the total area. The remaining forests are highly fragmented and occur in relatively small patches, an indication of intense anthropogenic disturbance. The remaining forests fall under either tropical lower montane rainforest (with the presence of broad-leaf forest and pine forest types) or tropical upper montane forest (mossy forest).

It conducted sampling of biomass and carbon content in the five (5) major carbon pools covering the forest vegetation and different land use types in the 78-ha study sites for the carbon project. These are located within the tri-boundaries of Benguet, Ifugao, and Mt. Province. The five (5) carbon pools inventoried include tree biomass, understory herbaceous (UH), necromass litter (NL), roots, and soil. They can be further grouped into aboveground biomass (tree, UH, and NL), ground biomass (roots), and soils. The sampling was done through the establishment of main plot, bigger plot, and subplots within the main plot. These sampling plots were established within the six major land cover types in the study area as described earlier.

The required number of sample plots was computed by determining first the standard deviation derived from the results of pre-sampling done representing each land cover type within the study sites. An initial of ten (10) sample plots were established for the pre-sampling. Using the formula recommended by Pearson et al. (2005), the number of sampling plots per vegetation type was determined by factoring in the computed standard deviation at 10–20% allowable error or desired precision, set at 95% confidence level and the total area covered. The results are shown in Table 16.1. The number of sample plots was also estimated at 10% sampling intensity out of the area covered per land use type.

The study followed more than one stratum/land use test group that employed a stratified sampling based on forest vegetation/land-cover type. Across the study areas, sample plots and sub-plots were established for each vegetation type for a total of 43 sample plots including cultivated/farm lands. Table 16.2 shows the elevation and coordinates of each sample plot.

Table 16.1 Computed number of sample plots established within the carbon sampling area of the project

Land cover classification	DENR carbon plot (ha)	Actual area (ha)	Computed number of sample plots to be established	Actual no. of sample plots/ established
Closed forest, broadleaved (CFB)	**8**	7.46	**4**	**4**
Open forest, broadleaved (OFB)	**60**	16.73	**8**	**8**
Open forest, coniferous (OFC)	**60**	33.07	**17**	**17**
Other wooded land, shrubs (OWLS)	**10**	17.49	**5**	**8**
Other wooded land, wooded grassland (OWLG)	**10**	3.28	**2**	**3**
Other land, cultivated/ farm land, annual crops (CFL)				**3**
TOTAL		**78.03**	**39**	**43**

Table 16.2 Elevation and coordinates per sample plot

Plot no.	Land cover type	Elevation (masl)	Longitude	Latitude
1	Open forest, broadleaved (OFB)	2384.57	120.8918	16.8326
2	Open forest, broadleaved (OFB)	2395.74	120.8923	16.8325
3	Open forest, broadleaved (OFB)	2346.63	120.8939	16.8313
4	Open forest, broadleaved (OFB)	2424.09	120.8911	16.8312
5	Other wooded land/shrubs (OWLS)	2469.15	120.8887	16.8289
6	Other wooded land/wooded grassland (OWLG)	2329.00	120.8911	16.8334
7	Open forest, broadleaved (OFB)	2494.37	120.8878	16.8288
8	Open forest, broadleaved (OFB)	2385.63	120.8885	16.8298
9	Open forest, broadleaved (OFB)	2405.04	120.8888	16.8311
10	Open forest, coniferous (OFC)	2256.34	120.8928	16.8374
11	Open forest, coniferous (OFC)	2188.08	120.8936	16.8372
12	Open forest, coniferous (OFC)	2045.30	120.8944	16.8383
13	Open forest, coniferous (OFC)	2070.35	120.8950	16.8405
14	Open forest, coniferous (OFC)	2077.12	120.8957	16.8408
15	Open forest, coniferous (OFC)	2029.37	120.8965	16.8413
16	Open forest, coniferous (OFC)	1852.22	120.8980	16.8471
17	Open forest, coniferous (OFC)	1937.04	120.8992	16.8466
18	Open forest, coniferous (OFC)	2051.55	120.9011	16.8458
19	Open forest, coniferous (OFC)	2058.72	120.9013	16.8452
20	Open forest, coniferous (OFC)	1935.80	120.8936	16.8434
21	Open forest, coniferous (OFC)	1912.12	120.8944	16.8435
22	Other wooded land/wooded grassland (OWLG)	1900.30	120.8802	16.8544
23	Other wooded land/wooded grassland (OWLG)	1863.24	120.8807	16.8548
24	Open forest, coniferous (OFC)	2479.01	120.8866	16.8291
25	Open forest, coniferous (OFC)	1821.83	120.8914	16.8443
26	Open forest, coniferous (OFC)	1785.58	120.8909	16.8432
27	Open forest, coniferous (OFC)	1965.67	120.8903	16.8409
28	Open forest, coniferous (OFC)	1989.36	120.8911	16.8410
29	Open forest, coniferous (OFC)	1940.67	120.8897	16.8416
30	Other wooded land/shrubs (OWLS)	1770.21	120.8953	16.8462
31	Other wooded land/shrubs (OWLS)	1792.25	120.8972	16.8464
32	Other wooded land/shrubs (OWLS)	1769.44	120.8953	16.8456
33	Other wooded land/shrubs (OWLS)	1814.11	120.8990	16.8440
34	Other wooded land/shrubs (OWLS)	1795.58	120.8990	16.8422
35	Other wooded land/shrubs (OWLS)	1901.18	120.8984	16.8409
36	Other wooded land/shrubs (OWLS)	1876.23	120.8985	16.8399
37	Closed forest, broadleaved (CFB)	2637.92	120.9408	16.8428
38	Closed forest, broadleaved (CFB)	2635.33	120.9410	16.8422
39	Closed forest, broadleaved (CFB)	2611.27	120.9400	16.8426

(continued)

Table 16.2 (continued)

Plot no.	Land cover type	Elevation (masl)	Longitude	Latitude
40	Closed forest, broadleaved (CFB)	2591.15	120.9420	16.8401
41	Cultivated/farm land (CFL)	2363.60	120.8905	16.8328
42	Cultivated/farm land (CFL)	2341.60	120.8901	16.8317
43	Cultivated/farm land (CFL)	2376.34	120.8893	16.8302

16.4.1.2 Sampling and Measurement of Biomass, C, and CO_2 Stored in the C Pools

Revegetation is considered as the most cost-effective carbon offset project or mitigating measure. Trees or plants sequester carbon from the atmosphere through the process of photosynthesis. The carbon that plants absorb and store in their biomass is estimated to be about 50% of their total weight. However, IPCC (2006) recommended 47% as the average default value and 44%-49% as the minimum and maximum default values, respectively. This study used 45% wood C content as recommended by Lasco and Pulhin (2001). This is an average C content for tropical trees based on their studies conducted within the country.

Carbon stock was measured both in the aboveground and belowground biomass. Aboveground biomass includes both live and dead biomass. Standing live biomass comprises all woody stems, branches, and leaves of living trees, creepers, climbers, and epiphytes as well as herbaceous undergrowth. Crop and weed biomass are considered in agricultural lands. Live tree biomass is also referred to by other studies as aboveground biomass (AGB) (Lumbres et al. (2012) and Lasco and Pulhin (2001)). Dead organic material or necromass covers dead standing and fallen trees and logs, other coarse woody debris, and litters. Underground biomass and C content, includes roots and soil organic matter (SOC). For roots alone, other studies referred to it as belowground biomass (BGB), Lumbres et al. (2012).

Estimation of biomass within a certain vegetation was either done through destructive and non-destructive sampling. Non-destructive sampling was applied to standing trees with DBH $\geq$ 5 cm. Destructive sampling was done on trees <5 cm in diameter and other vegetation like creepers, climbers, and herbaceous undergrowth. This study used the nested plot sampling method developed by Hairiah et al. (2011) also known as the Alternative to Slash and Burn or ASB technique. Nested plots consist of smaller sub-units of various shapes and sizes which is considered as the most cost-efficient and practical method of sampling (Fig. 16.3). It is applicable to stands with a wide range of tree diameters or stands with changing diameters and stem densities (Pearson et al. 2005). A total of 400 samples each for understorey/herbaceous vegetation and litter were collected from forty (40) sample plots each consisting of ten (10) nested plots within the main plot.

Trees

A complete enumeration of trees with a diameter at breast height (DBH) of 5–30 cm that falls within the 5 m × 40 m main plot was done by measuring its individual DBH

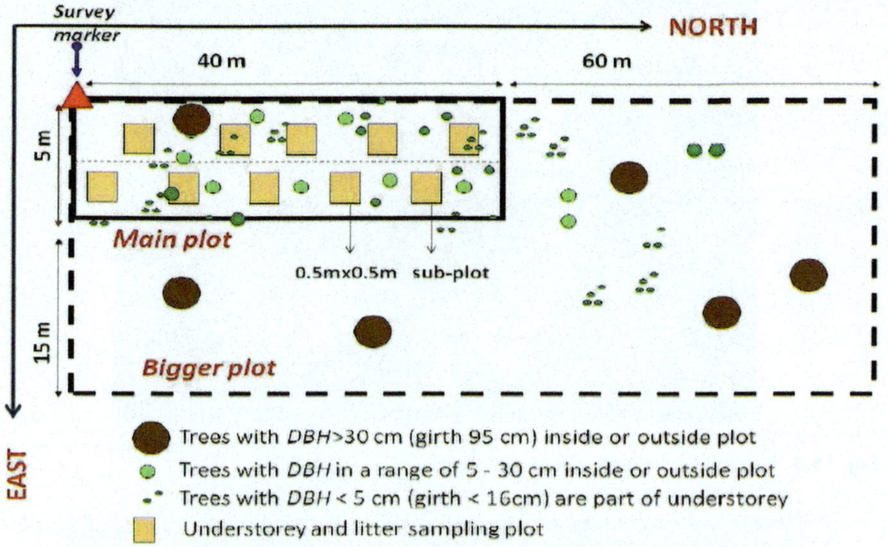

Fig. 16.3 Diagram of nested plot design for vegetation sampling. (Source: (Hairiah et al. 2011))

Fig. 16.4 Measuring the tree DBH

(measured at 1.3 m above the soil surface). Each tree was identified by its local name and scientific name. The main plot was extended to form a bigger plot measuring 20 m × 60 m to cover all larger trees with DBH > 30 cm found within both plots (Fig. 16.4).

Understorey and Herbaceous (UH) Vegetation

This is also referred to as non-tree vegetation according to Pearson et al. (2005). This includes all herbs and woody plants with diameters less than 5 cm found inside the

Fig. 16.5 UH and necromass or litter sampling

Fig. 16.6 Sampling of necromass or litter on the ground

0.5 m × 0.5 m plots. All these vegetative samples found within the quadrat were clipped to the surface of the ground using pruning shears (Fig. 16.5). Collected samples per sub-plot were put inside labeled plastic bags. The samples were transported to the Philex Padcal Assay Laboratory for oven-drying at 80 ° C until a constant weight was achieved. A total of 400 samples each for understorey/ herbaceous vegetation and litter were overdried representing forty (40) sample plots each consisting of ten (10) nested plots within the main plot.

Necromass and Litter (NL)

Litters or dead woody debris and branches were sampled within the 0.5 m × 0.5 m subplot (Fig. 16.6). All samples were collected from the soil surface and their fresh weight was measured before sub-sampling was done per subplot. Fresh weight of the

Fig. 16.7 Processing and oven drying of vegetation samples at the Philex, Padcal Assay Laboratory

Fig. 16.8 Actual sampling of SOC and bulk density

sub samples was also determined before air drying. Samples collected per plot were placed in a labelled paper bag ready for oven drying. Subsequent oven drying of samples was done at the Philex Padcal Assay Laboratory (Fig. 16.7). Ten (10) samples were collected each from the 40 main plots.

Roots

Roots or belowground biomass measurement is a relatively costly and time-consuming method. Instead of conducting actual field sampling, the study used the Pearson et al. (2005) regression model for tropical forests using individual tree biomass. This is considered as the most practical and effective method of determining root biomass.

Soil Organic Carbon and Bulk Density

Soil samples were collected from each of the three (3) soil layers within the plot: 0–10 cm; 10–20 cm, and 20–30 cm depth per elevation type: high, middle, and lower elevation to determine soil organic matter content (Fig. 16.8). Another sample was taken from each of the three (3) elevation types per plot for bulk density analysis. This was done by choosing an undisturbed spot near the plot and positioning the soil core sampler in the middle of the 0–30 cm soil layer. Surface litter was removed before carefully pushing the soil core sampler with a metal cylinder into the designated spot. Each core sample was carefully put in labeled plastic bags. Thus, twelve (12) samples were collected from each plot totaling 516 samples from the 43 plots.

The samples were then transported to the Soils Laboratory of the Institute of Renewable Natural Resources, UP Los Baños College of Forestry and Natural Resources for analysis. To reduce the number of samples, some soil samples collected from the same soil layer from nearby plots with more or less the same vegetation type were mixed into composite samples prior to analysis to determine the soil organic matter and % C. Thus, only 387 samples were subjected to laboratory analysis.

16.4.2 Biomass and Carbon Computation

16.4.2.1 Trees

Tree biomass is commonly termed as aboveground biomass or AGB (Lasco and Pulhin 2001). It is computed by getting the diameter at breast height (DBH) of each tree and applying an appropriate allometric equation, a non-destructive method to determine tree biomass. It is ideal to come up with a good estimate using equations that were developed for each location, species, or group of species, and for trees of similar sizes and ages. However, such equations are not yet available in our country although some species type has but still limited. Several studies used the Brown (1997) equation. The equation is based on diameter at breast height-DBH (1.3 m from the ground) or height of tree (H); and the density of the wood(s). For this study, three (3) equations were used to calculate tree biomass density. Specific for coniferous species at DBH limit from 2 to 52 cm developed by Brown (1997), the equation is as follow:

$$\mathrm{TB} = \exp.\{-1.170 + 2.119 \times \ln\ (\mathrm{DBH})\}$$

For broad-leaf species, the study used allometric equations formulated by Brown (1997) specific for a certain DBH limit. The equations are applied for trees in tropical humid regions with an annual rainfall that ranges from 1400 to 4000 mm suited for the climatic condition in the CAR region:

For trees with DBH $<$ 60 cm.

$$\mathrm{TB} = \mathrm{EXP}\ (-2.134 + 2.530 \times \mathrm{LN}\ (\mathrm{DBH}))$$

While for trees with 60–148 cm DBH.

$$\mathrm{TB} = 42.69 - 12.800 \times \mathrm{DBH} + 1.242 \times (\mathrm{DBH})^2$$

where TB $=$ biomass in kg; DBH $=$ in cm.

Biomass in kg per tree was converted to Mg measurement unit which is also equivalent to ton.

Total tree biomass in t ha^{-1} was derived using the formula below:

$$T_{\mathrm{TB}} = \frac{\sum \text{Biomass of all trees in a sample plots (ton)}}{\text{Area of the sample plots } (\mathrm{m}^2)} \times \frac{10{,}000\ \mathrm{m}^2}{1\ \mathrm{ha}}$$

Carbon Content:

$$\mathrm{TC}_{\mathrm{TB}} = T_{\mathrm{TB}}\ \left(\mathrm{t\ ha}^{-1}\right) \times \%\mathrm{C\ content}$$

where: % C = 45% (as suggested by Lasco et al. 2007).

16.4.2.2 Understorey and Herbaceous (UH) Vegetation

Biomass density for UH was computed using the formula below:

$$\mathrm{UH\ density} = \frac{\mathrm{Oven\ dry\ weight/area}(\mathrm{g/m}^2) \times 10,000\ \ \mathrm{m}^2/\mathrm{ha}}{10^6\ \ (\mathrm{g/ton}^2)}$$

where: UH = biomass density of UH vegetation.

$$\text{Carbon content : Total C UH} = \text{biomass density of the samples} \times \%\text{C content}$$

16.4.2.3 Necromass and Litter (NL)

Biomass for necromass or litter is computed using the formula:

$$\mathrm{Ground\ litter} = \frac{\mathrm{Oven\ dry\ weight/area}(\mathrm{g/m}^2) \times 10,000\ \ \mathrm{m}^2/\mathrm{ha}}{10^6\ \ (\mathrm{g/ton}^2)}$$

16.4.2.4 Roots

Computation for the root biomass was based on the tree biomass of individual trees measured using the formula of Pearson et al. (2005) as presented below:

$$\text{Root Biomass} = \text{EXP}\ (-\ 1.0587 + 0.8836 \times \text{LN (TB)})$$

where TB = Tree Biomass.

16.4.2.5 Soil Organic Carbon (SOC) and Bulk Density

To determine the soil organic carbon stored (SOC) in the stand in Mg ha^{-1}, the study computed first the bulk density:

Bulk density (BD) was derived using the formula:

$$BD\ (g\ cm^3) = \text{Dry wt.of soil (g)/Vol.of cylinder}\ (cm^3)$$

The dry weight of soil and the equivalent SOC was computed using the following formula:

$$\text{Volume of 1 ha soil} = 100\ m \times 100\ m \times 0.3\ m\ \text{(or 0.1 m per soil level/depth)}$$

$$\text{Weight of soil (t)} = \text{Bulk Density} \times \text{Soil Volume}$$

$$SOC = \text{Weight of soil (t)} \times \%C$$

16.4.2.6 Computation of Carbon Dioxide (CO_2) Equivalent

The CO_2 density in Mg ha^{-1} per carbon sink was derived using the formula below:

$$CO_2 = C\ \text{stored in Mg}\ ha^{-1} \times 44/12$$

where.

CO_2 = 1 molecule of Carbon and 2 molecules of Oxygen.
Atomic wt. C = 12 O = 16.
Wt. of CO_2 is C + 2*O = 43.999915 or 44.
The ratio of CO_2 to C is 44/12.
Total CO_2 = Total carbon per sink × 44/12.

16.4.2.7 Biomass Content, C, and CO_2 Stored

The study covered the 78.0 hectares of forested lands and other land cover types designated as carbon sites within the tri-boundaries of Benguet, Ifugao, and Mt. Province. Table 16.3 presents the total area for each land cover type of the entire study sites.

The study followed the recommendation by Hairiah et al. (2011) that the process of disaggregating the total area into classes of land cover and zones can make a substantial difference to the final estimates as well as affect the certainty level of the estimate. It will further avoid overestimation or underestimation of values in the final analysis.

Table 16.3 Total area covered per land cover type

Land cover classification	Designated sampling area (ha)
Closed forest, broadleaved (CFB)	**7.46**
Open forest, broadleaved (OFB)	**16.73**
Open forest, coniferous (OFC)	**33.07**
Other wooded land, shrubs (OWLS)	**17.49**
Other wooded land, wooded grassland (OWLG)	**3.28**
Other land, cultivated/farm land, annual crops (CFL)	
Total	**78.03**

Table 16.4 lists the individual sample plots according to vegetation type. It should be noted that Sample plots/Plot Numbers 37–40 were established outside the identified sampling site for closed forest broadleaved (CFB). This was the result of the consultation with the DENR personnel together with one GIS technician to identify alternative sites with more or less similar land cover classifications. This was in response to the account of the field inventory team that the pre-identified site is highly inaccessible. The inventory team reported that they spent the whole day trying to reach the site to no avail. Thus, the UPLB team recommended applying the proxy sampling approach in consultation with local residents who are familiar with the area. This involves an inventory of an area whose land cover characteristics are similar to the target site.

16.4.2.8 Profile of the Sampled Forest Vegetation

The study measured 1330 trees from the 40 sample plots established within the CFB, OFB, OFC, OWLS, and OWLG land cover types excluding the CFL. Trees found in the main and bigger plots laid-out within each sample plot were measured in terms of their DBH and recorded in a tally sheet per sample plot. Out of these, 26% were from the CFB, 36% OFB, 31% OFC, 5% OWLS, and only 1% from OWLG. There was not much difference in tree population between OFB and OFC except OWLS and OWLG as expected which have the least number of trees listed.

The majority of the smaller diameter trees (with DBH range: 5–30 cm) come from the OFB (388 trees) followed by CFB (151 trees) and OFC (120 trees). However, larger diameter trees (with DBH range above 30 cm) were mostly found in OFC (296 trees) followed by CFB (194 trees) and OFB (97 trees). Comparing further the two forest types in terms of stocking density for bigger trees, CFB is more densely stocked than OFC at an average of 38 trees/plot for CFB while OFC is only 17 trees/plot. CFB has the most number of trees per plot with larger DBH (19-60 trees/plot) and with the highest average DBH of 52.06 cm followed by OFB of 42.99 cm. OFB mean DBH is slightly higher than OFC (40.66 cm).

CFB scored high far above the rest of the forest types with the highest number of trees with DBH $\geq$ 60 cm. CFB has about 80 trees as compared to only 11 trees from OFC and eight (8) trees from OFB. This means that CFB is highly dominated by large/matured and even overmatured trees which resembled an old-growth forest. These trees significantly contributed to the high biomass content and C/CO_2 equivalent of the CFB stand. Table 16.5 shows the profile of the different forest types inventoried.

The DBH composition of the trees inventoried is further summarized in Table 16.6. Most of the trees measured (44%) from the main plots belong to the DBH range of 11–20 cm followed by DBH of 5–10 cm of about 31% and about 25% are within 21–30 cm DBH. It has an overall mean DBH of 15.4 cm. The majority of the trees in the bigger plots fall within the DBH bracket of >30-50 cm (77%) followed by the DBH range of 51–70 cm of about 17%. Trees in bigger plots have an average DBH of 53.6 cm.

Among the trees measured, Benguet pine (*Pinus kesiya* ssp. *langbianensis*), a coniferous species, tops the list of species found in the project sites. This species is

Table 16.4 List of sample plots per land cover type

Sample plot no.	Land cover type
1	Open forest, broadleaved (OFB)
2	Open forest, broadleaved (OFB)
3	Open forest, broadleaved (OFB)
4	Open forest, broadleaved (OFB)
5	Other wooded land/shrubs (OWLS)
6	Other wooded land/wooded grassland (OWLG)
7	Open forest, broadleaved (OFB)
8	Open forest, broadleaved (OFB)
9	Open forest, broadleaved (OFB)
10	Open forest, coniferous (OFC)
11	Open forest, coniferous (OFC)
12	Open forest, coniferous (OFC)
13	Open forest, coniferous (OFC)
14	Open forest, coniferous (OFC)
15	Open forest, coniferous (OFC)
16	Open forest, coniferous (OFC)
17	Open forest, coniferous (OFC)
18	Open forest, coniferous (OFC)
19	Open forest, coniferous (OFC)
20	Open forest, coniferous (OFC)
21	Open forest, coniferous (OFC)
22	Other wooded land/wooded grassland (OWLG)
23	Other wooded land/wooded grassland (OWLG)
24	Open forest, coniferous (OFC)
25	Open forest, coniferous (OFC)
26	Open forest, coniferous (OFC)
27	Open forest, coniferous (OFC)
28	Open forest, coniferous (OFC)
29	Open forest, coniferous (OFC)
30	Other wooded land/shrubs (OWLS)
31	Other wooded land/shrubs (OWLS)
32	Other wooded land/shrubs (OWLS)
33	Other wooded land/shrubs (OWLS)
34	Other wooded land/shrubs (OWLS)
35	Other wooded land/shrubs (OWLS)
36	Other wooded land/shrubs (OWLS)
37	Closed forest, broadleaved (CFB)
38	Closed forest, broadleaved (CFB)
39	Closed forest, broadleaved (CFB)
40	Closed forest, broadleaved (CFB)
41	Cultivated/farm land (CFL)
42	Cultivated/farm land (CFL)
43	Cultivated/farm land (CFL)

Table 16.5 Summary profile of trees sampled inside the carbon project sites

Profile	Land cover type CFB	OFB	OFC	OWLS	OWLG	CFL	Total
Total no. of trees	345	485	416	70	14	0	1330
• Main plot	151	388	120	29	14	0	702
Ave. no. of trees/main plot	38	49	7	4	5	0	
• Bigger plot	194	97	296	41	0	0	628
Ave. no. of trees/bigger plot	38	12	17	5	0	0	
Range: No. of trees/plot							
• Main plot	32–43	24–94	1–20	1–16	14	0	
• Bigger plot	19–60	4–17	8–25	6–9	0	0	
Ave. DBH							
• Main plot	16.46 cm	12.98 cm	22.16 cm	12.01 cm	4.56 cm	0.00	
• Bigger plot	52.06 cm	42.99 cm	40.66 cm	23.27 cm	0.00	0.00	
Highest DBH	120 cm	74 cm	79.3 cm	74 cm	27.5 cm	0.00	
Lowest DBH	5 cm	5 cm	5.5 cm	6.3 cm	5.5 cm	0.00	
No. of trees with DBH ≥ 60 cm	80	8	11	1	0	0.00	100

Table 16.6 Frequency of trees was measured within the DBH range

DBH range	No. of trees sampled
Small plot	
5–10	217
11–20	308
21–30	117
Total	702
Ave. DBH	16.40
Bigger plot	
>30–50	483
51–70	105
71–90	34
91–110	3
>110	3
Total	628
Ave. DBH	37.96

very common in highland areas like Benguet and Mt. Province. It comprises about 36% of the tree species as shown in the tree inventory list. Among the top 5 species next to Benguet Pine which is all broadleaved species are: Paleyan/Paleyen/Palayen (*Lithocarpus* sp.) 18%; Boltik/Boltek (*Syzygium* sp.) 13%; Igem (*Dacrycarpus imbricatus* (Blume) de Laub.) 5%, and Bini (*Eurya buxifolia* Merr.) 3%. These species are found in the closed broadleaved forest (CFB) which commonly belong to large DBH classes (Table 16.7).

16.5 Vegetation Structure and Plant Diversity

16.5.1 Species Diversity and Stand Structure

A total of 32 morpho-species belonging to 26 genera and 24 families were found inside the 40 sampling plots. Dominant families were Pinaceae, Fagaceae, Myrtaceae, Pentaphylacaceae, Podocarpaceae, and Ericaceae. The 10 most abundant species were listed in Table 16.8. This includes Benguet pine (*Pinus kesiya* ssp. *langbianensis*), palayen (a species of *Lithorcarpus*), boltik (a species of *Syzygium*), papat-ek (*Eurya coriacea*), and igem (*Dacryocarpus imbricatus*).

The average number of trees per quadrat (20 m × 100 m) is approximately 33 trees or 165 trees ha^{-1}. The average diameter for all trees inside the plot is 30.05 cm which can be classified as medium-sized trees. Trees with the largest average diameter are gmelina (*Gmelina arborea*), gatiley (*Cryptocarya tomentosa*), igem (*Dacrycarpus imbricatus*), Benguet pine (*Pinus kesiya* ssp. *Langbianensis*), and *Schefflera* sp. (*Schefflera glabra* Merr.). The summary of average DBH of recorded trees is listed in Table 16.9.

16.5.2 Opportunistic Sampling

Apart from the species recorded inside the plots, additional 94 species were recorded from the opportunistic survey. Hence a total of 126 morpho-species were encountered during the survey. Forty-six (46) specimens were not identified to the species level and have therefore been tentatively assigned to the most probable family and/or genus. Most of these are in their juvenile stage (seedlings/saplings) and/or are sterile specimens (without flowers or spores), as identification of flora species is very much dependent on the variation exhibited by the reproductive structures.

16.5.3 Noteworthy Species

16.5.3.1 Endemic Species

Thirty-six (36) species recorded were found to be endemic to the Philippines (Pelser et al. 2011) (Table 16.10). These includes rare species of orchids including *Ceratostylis ramosa* Rolfe ex Ames, *Dendrochilum* cf. *cinnabarinum* Pfitzer in

Table 16.7 Summary of trees inventoried per species according to its local and scientific name

Common name	Scientific name	Main plot	Big plot	Grand total	%
Apih-it	*Symplocos* sp.	1	0	1	0.07
Apiit/Pedped	*Clethra canescens var. luzonica* (Merr.*) Sleum.*	4	0	4	0.30
Attipen	*Melicope sessilifoliola* (Merr.*) T.G. Hartley*	6	4	10	0.75
Bangtelen/ Bongtelin/ Danupong	*Meliosma pinnata* subsp. *pendula* (Merr.*) Beus.*	8	1	9	0.67
Bini/Damdame-eg	*Eurya buxifolia* Merr.	27	17	44	3.29
Boltik	*Syzygium* sp.	101	60	161	12.02
Bosleg	*Platea excelsa var. boornensis (Heine) Sleumer*	2	2	4	0.30
Dalung	*Dendrochnide* sp.	0	1	1	0.07
Damaan	*Schefflera glabra* Merr.	1	1	2	0.15
Dumiplas	*Eurya* sp.	20	1	21	1.57
Ficus	*Ficus* sp.	1	0	1	0.07
Ganaba	*Polyosma philippinensis* Merr.	2	0	2	0.15
Gatiley	*Cryptocarya tomentosa Blume.*	0	2	2	0.15
Gmelina	*Gmelina arbroea Roxb.*	0	1	1	0.07
Gotmo	*Vaccinium cumingianum var. cumingianum S.Vidal*	14	13	27	2.02
Haling-hingen	*Endiandra coriacea* Merr.	1	0	1	0.07
Igem	*Dacrycarpus imbricatus (Blume) de Laub.*	30	39	69	5.15
Kandaruma	*Astronia* sp.	1	0	1	0.07
Kawasi	*Theaceae* sp.	3	2	5	0.37
Ladew	*Syzygium* sp. *4*	2	0	2	0.15
Lobso	*Ficus ampelas Burm.f.*	2	0	2	0.15
Lomte	*Syzygium* sp. *5*	1	0	1	0.07
Lusong	*Medinilla* sp.	4	0	4	0.30
Madmadanum	*Polyosma verticillata* Merr.	13	7	20	1.49
Masmasiyu	*Skimmia japonica Thunb.*	7	0	7	0.52
Odyew	*Taxus wallichiana Zucc.*	1	1	2	0.15
Otkulan	*Bhesa* sp.	7	4	11	0.82
Palansinsingen	*Eurya* sp. *2*	12	1	13	0.97
Palayen	*Lithocarpus* sp.	130	108	238	17.77
Panabugen	*Neolitsea microphylla* Merr.	19	1	20	1.49
Papat-ek	*Eurya coriacea* Merr.	69	8	77	5.75
Saleng	*Pinus kesiya* ssp. *langbianensis (A. Chev.) Gaussen ex Bui*	166	401	567	42.35
Supek	*Tasmannia piperita (Hook.f.) Miers*	9	0	9	0.67
Total		664	675	1339	100.00

Table 16.8 Top 10 most abundant species surveyed

Local name	Scientific name	Abundance
Benguet pine	*Pinus kesiya* ssp. *langbianensis (A. Chev.) Gaussen ex Bui*	567
Palayen	*Lithocarpus* sp.	238
Boltik	*Syzygium* sp.	163
Papat-ek	*Eurya coriacea* Merr.	77
Igem	*Dacrycarpus imbricatus (Blume) de Laub.*	69
Bini	*Eurya buxifolia* Merr.	44
Gutung	*Vaccinium cumingianum var. cumingianum S. Vidal*	27
Dumiplas	*Eurya* sp.	21
Panabugen	*Neolitsea microphylla* Merr.	20
Madmadanum	*Polyosma verticillata* Merr.	20

Engl, *Dendrochilum unicorne* (Ames) L.O.Williams, *Liparis philippinensis* (Ames) Schltr, *Pinalia philippinensis* (Ames) W.Suarez & Cootes, and *Pinalia tridens* (Ames) W.Suarez & Cootes. These rare species are documented during the opportunistic survey.

It should be emphasized that categorizing species as endemic is very much dependent on the availability of published biodiversity data, recent taxonomic revisions, nomenclatural changes, and new evidences from various disciplines used in systematics among others. Thus, estimates of endemism should be interpreted within the context of the methodologies and limitations imposed by contributing factors. In this study, a number of specimens were not identified at the species level. This will influence the percent endemism estimates.

16.5.3.2 Threatened Species

Six (6) species recorded were found to be threatened under (DAO 2017-11) and/or the IUCN Red list of Threatened species (2018-2). Noteworthy among the list is the endangered species which can only be found in higher elevation, Yew (*Taxus wallichiana* Zucc.), recorded during an opportunistic survey (Table 16.11).

16.5.4 Importance Value

The relative density (RD), relative dominance (RDom), and relative frequency (RF) values of each tree species in the whole area were determined to calculate their respective importance values (IV), a standard measure in ecology that determines the rank relationships of species. A high IV among species indicates a composite score for high relative species dominance, density, and frequency. Based on the computed IV (Table 16.12) the five species with the highest IV or the most important species are Benguet pine (*Pinus kesiya* ssp. *langbianensis*), palayen (*Lithocarpus* sp.), boltik (*Syzygium* sp.), igem (*Dacryocarpus imbricatus*), and bini (*Eurya buxifolia*). These five species accounts to almost 40% of the total importance value of all species surveyed. More importantly, the computed IV coincide with the species which are commonly found in higher elevation.

Table 16.9 Summary table of the average diameter of trees recorded

Local name	Scientific name	Average DBH (cm)
Gmelina	*Gmelina arbroea* Roxb.	54.0
Gatiley	*Cryptocarya tomentosa* Blume.	41.0
Igem	*Dacrycarpus imbricatus* (Blume) de Laub.	36.0
Saleng	*Pinus kesiya* ssp. *langbianensis* (A.Chev.) Gaussen ex Bui	35.3
Schefflera sp.	*Schefflera glabra* Merr.	33.5
Gutung	*Vaccinium cumingianum var. cumingianum* S. Vidal	33.4
Dendrochnide sp.	*Dendrochnide* sp.	31.5
Palayen	*Lithocarpus* sp.	31.1
Bosleg	*Platea excelsa var. boornensis* (Heine) Sleumer	31.0
Odyew	*Taxus wallichiana* Zucc.	31.0
Bini/Damda-eg	*Eurya buxifolia* Merr.	29.1
Boltik	*Syzygium* sp.	27.1
Theaceae sp.	*Theaceae* sp.	24.8
Madmadanum	*Polyosma verticillata* Merr.	22.8
Bhesa sp.	*Bhesa* sp.	21.7
Attipen	*Melicope sessilifoliola* (Merr.) T.G.Hartley	19.9
Astronia sp.	*Astronia* sp.	19.0
Lobso	*Ficus ampelas* Burm.f.	18.0
Ganaba	*Polyosma philippinensis* Merr.	16.2
Halinghingen	*Endiandra coriacea* Merr.	16.0
Ficus sp.	*Ficus* sp.	16.0
Papat-ek	*Eurya coriacea* Merr.	13.9
Eurya sp.	*Eurya* sp.	13.9
Bangtilen/ Danupong	*Meliosma pinnata* subsp. *pendula* (Merr.) Beus.	12.5
Apih-it	*Symplocos* sp.	12.5
Eurya sp. *2*	*Eurya* sp. *2*	12.4
Lomte	*Syzygium* sp. *5*	12.0
Bohian-liitan	*Neolitsea microphylla* Merr.	11.8
Ameweg/Pidpid	*Clethra canescens var. luzonica* (Merr.) Sleum.	11.4
Medinilla sp.	*Medinilla* sp.	8.6
Skimmia sp.	*Skimmia japonica* Thunb.	8.4
Supek	*Tasmannia piperita* (Hook.f.) Miers	7.9

16.5.5 Diversity Indices

Based on the number and abundance of the recognized morpho-species across the plots surveyed, the diversity indices such as Shannon (H′), Simpson (D), and Evenness index (J′) were computed. Shannon index gives an estimate of species richness and distribution while Simpson's index gives the probability of getting

Table 16.10 List of endemic species surveyed

Common name	Family	Scientific name
Batinai	Pentaphylacaceae	*Adinandra macgregorii* Merr.
Benguet Dungao	Melastomataceae	*Astronia benguetensis* Maxw.
Berberis	Berberidaceae	*Berberis barandana* S. Vidal
Ming Ramos	Orchidaceae	*Ceratostylis ramosa* Rolfe ex Ames
Sandkuhi cinnamon	Lauraceae	*Cinnamomum sandkhulii* Merr.
Cleyera	Pentaphylaceae	*Cleyera japonica var. montana* (Merr.) Kobuski
Cyrtandra	Gesneriaceae	*Cyrtandra oblongata* Merr.
Igem	Podocarpaceae	*Dacrycarpus imbricatus* (Blume) de Laub.
Dendrobium	Orchidaceae	*Dendrobium unicorne* (Ames) A.D.Hawkes
Dendrochilum	Orchidaceae	*Dendrochilum* cf. *cinnabarinum* Pfitzer in Engl
Diplycosia sp.	Ericaceae	*Diplycosia luzonica var. pubens* Sleumer
Eurya	Pentaphylacaceae	*Eurya amplexicaulis* Moore
Eurya	Pentaphylacaceae	*Eurya coriacea* Merr.
Mistletoe	Loranthaceae	*Lepidaria quadriflora* Tiegh.
Orchid	Orchidaceae	*Liparis philippinensis* (Ames) Schltr
Curran kulilisiao	Lauraceae	*Machilus curanii* Merr.
Medinilla	Melastomataceae	*Medinilla cordata* Merr.
Medinilla	Melastomataceae	*Medinilla pendula* Merr.
Aroking-bitin	Sabiaceae	*Meliosma pinnata* subsp. *pendula* (Merr.) Beus.
Bohian-Liitan	Lauraceae	*Neolitsea microphylla* Merr.
Pilea	Urticaceae	*Pilea luzonensis* Merr.
Gremlins orchid	Orchidaceae	*Pinalia philippinensis* (Ames) W.Suarez & Cootes
Gremlins orchid	Orchidaceae	*Pinalia tridens* (Ames) W.Suarez & Cootes
Pipturus	Urticaceae	*Pipturus dentatus* (C.B.rob.) C.B.rob.
Magbut	Escalloniaceae	*Polyosma philippinensis* Merr.
Buduan	Escalloniaceae	*Polyosma verticillata* Merr.
Rhododendron	Ericaceae	*Rhododendron quadrasianum var. quadrasianum* S. Vidal
Saurauia	Actinidiaceae	*Saurauia elegans* (Choisy) Fern.-Vill.
Saurauia	Actinidiaceae	*Saurauia panduriformis* Elmer.
Schefflera	Araliaceae	*Schefflera blancoi* Merr.
Schefflera	Araliaceae	*Schefflera glabra* Merr.
Schefflera	Araliaceae	*Schefflera simplicifolia* Merr.
Lauisanan	Symplocaceae	*Symplocos glabriramifera* Noot.
Baltik	Myrtaceae	*Syzygium acrophilum* (C.B.rob.) Merr.
Gutung	Ericaceae	*Vaccinium cumingianum var. cumingianum* S.Vidal
Itangan	Cunoniaceae	*Weinmannia luzoniensis* S.Vidal

different species when two individuals were drawn (with replacement) inside a plot. On the other hand, the Evenness index tells us how evenly species and/or individuals are distributed inside a quadrat.

Table 16.11 List of threatened species surveyed

Species	Scientific name	Conservation status	
		IUCN 2018-2	DAO 2017-11
Berberis	*Berberis barandana* S. Vidal		VU
Igem	*Dacrycarpus imbricatus* (Blume) de Laub.		OTS
Mistletoe	*Lepidaria quadriflora* Tiegh.		EN
Belvisia	*Lepisorus platyrhynchos* (Kunze) li Wang		EN
Petroleum nut	*Pittosporum resiniferum* Hemsl.		OTS
Yew	*Taxus wallichiana* Zucc.	EN	VU

VU Vulnerable, *OTS* Other threatened species, *EN* Endangered

Table 16.12 Top 10 species with the highest importance value

Species	RD	Rdom	RF	IV
Pinus kesiya ssp. *langbianensis* (A.Chev.) Gaussen ex Bui	0.07	82.41	0.63	83.12
Lithocarpus sp.	0.82	11.37	2.52	14.71
Syzygium sp.	0.30	4.06	1.89	6.25
Dacrycarpus imbricatus (Blume) de Laub.	0.15	1.28	1.26	2.69
Eurya buxifolia Merr.	5.15	0.34	7.55	13.04
Eurya coriacea Merr.	0.07	0.24	0.63	0.94
Vaccinium cumingianum var. cumingianum S.Vidal	0.07	0.17	0.63	0.87
Polyosma verticillata Merr.	3.29	0.04	7.55	10.88
Eurya sp.	5.75	0.02	8.18	13.94
Bhesa sp.	1.57	0.01	3.14	4.72

RD relative density, *Rdom* relative dominance, *RF* relative frequency, *IV* importance value

Table 16.13 Top 10 plots with the highest computed diversity index

Plots	No. of species	Simpsons index (D′)	Shannon index (H′)	Evenness
1	18	0.73	1.97	0.40
24	13	0.72	1.80	0.47
4	13	0.71	1.76	0.45
3	11	0.72	1.75	0.52
8	10	0.71	1.66	0.52
7	14	0.70	1.66	0.37
2	11	0.69	1.58	0.44
9	10	0.69	1.57	0.48
5	8	0.70	1.55	0.59
38	12	0.68	1.52	0.38

Value interpretation for H′: Very High $\geq$ 3.5 above, High = 3.0–3.49, Moderate = 2.5–2.99, Low = 2.0–2.49, Very Low $\leq$ 1.9 and below

The computed Shannon index of the different plots varies from as low as 0.69 and as high as 1.97 (Table 16.13). Quadrat 1 had the highest Shannon and Simpson index wherein 18 species were recorded. Most of the plots have low to very low diversity based on the H′ diversity classification range developed by Fernando et al. in 1996. The lower diversity is expected as it is the general elevational trend in the tropics, diversity decreases with increasing elevation. In addition, most of the plots surveyed were dominated by Benguet pine.

16.6 General Assessment of Biomass Density of the Six Major Land Cover Types

Among the four vegetation types, CFB is characterized as closed and densely populated forest vegetation which has the highest total biomass density of 1299.44 Mg ha^{-1} followed by OFB (436.06 Mg ha^{-1}). The top two forest stands are both dominated by broadleaved species they only differ in terms of stocking density. CFB is 66% higher in biomass density than OFB. They are followed by OFC (181.41 Mg ha^{-1}, OWLS (61.67 Mg ha^{-1}), OWLG (36.35 Mg ha^{-1}), and CFL which is no longer reflected because of its insignificant biomass content or considered as zero biomass density. The land cover types with the least biomass density are those which are poorly vegetated with trees or no trees at all. They are mostly dominated by shrubs, grasses, and cultivated crops. In particular, the CFL whose vegetation/crops are not retained in the stand because they are regularly harvested.

Table 16.14 shows the mean biomass content of each vegetation type from their individual plot and per carbon sink. The result of this study shows the great potential of an undisturbed forest like CFB to store huge amount of carbon from its biomass which is more than that of a matured stand or an old-growth forest as compared to the findings of Lasco et al. (2011). It contains a diverse and very rich amount of biomass. The stand has considerably accumulated high amounts of biomass through time. This supports further the general rule that the more biomass produced the greater the amount of C sequestered. Trees continue to accumulate carbon as they get older although at varying rates depending on the species and age. However, the rate of accumulation declines as the age of the tree. This can be attested by the presence of larger trees in CFB and other forest types with DBH ≥ 60 cm and whose biomass is greater than 2 tons. Though these trees are usually few, according to Brown and Lugo (1982) and Brown et al. (1992) as cited by Racelis et al. (2008), they can account for more than 40% of the carbon, particularly for old-growth stands. This finding is evident in the sampled big plots in this study which covered trees whose DBH is >30 cm that already accounts for half (51%) of the total mean biomass

Table 16.14 Summary of mean biomass density per land cover type and C pool

	Land cover type mean biomass density (Mg ha^{-1})				
Carbon pools	CFB	OFB	OFC	OWLS	OWLG
Trees	1121.46	360.38	147.72	46.05	23.80
UH	2.42	3.86	3.22	3.55	3.63
NL	10.47	8.17	5.26	4.10	4.32
Roots	165.09	63.66	25.20	7.98	4.60

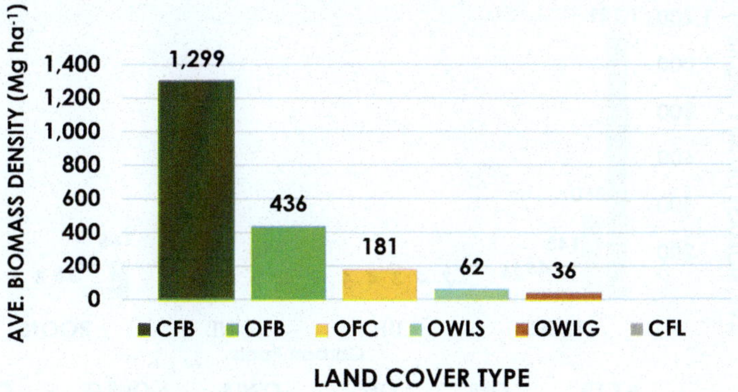

Fig. 16.9 Comparing the mean biomass stored per vegetation type

accumulated among vegetation types. Its species richness and diversity can also be attributed to its inaccessibility and far from human disturbances.

There is an observed empirical relationship that possibly exists between and among different land cover types in terms of biomass stocking potential. CFB is about 66% higher than OFB and 86% higher than OFC. Comparing OFB with OFC in terms of biomass stocking rate, the difference is still high. OFB is 58% higher than OFC but much higher than OWLS (86%) and OWLG (92%). For OFC as compared to the least vegetated areas like OWLS and OWLG the rates are still significant, 66% and 80%, respectively. In between OWLS and OWLG, the rate of biomass density change has lowered, OWLS 41% greater than OWLG. As a general observation, the contribution of tree biomass especially those or large diameter trees affect much on the increase of biomass content on each vegetation type. The above results are demonstrated in a graphical form. Fig. 16.9 presents the comparative biomass accumulation among and between land cover types.

On the performance of each carbon pool in accumulating biomass, aboveground biomass like trees is also enhanced by its root system. Root biomass corresponds to the amount of tree biomass. Its biomass is proportional to the tree biomass density value. CFB is densely populated by matured trees as manifested by its large DBH. The presence of these trees somehow prohibits the growth of younger regenerations because sunlight can no longer penetrate down to the forest floor. This could be the reason why UH has the lowest biomass density among the vegetation types studied particularly on CFB.

The considerable supply of necromass or litter (ranges from 4.10 to 10.47 Mg ha^{-1}), which is common to all stands, contributed to the higher biomass content. The results on necromass production in this study support the findings of several authors. Lugo and Brown (1992) estimated the biomass density of fine litter that ranges from about 2 to 16 t ha^{-1} (average of 6 t ha^{-1} or less than 5% of aboveground biomass), with higher values generally in moist environments. Osman (2013) reported that in tropical regions, forest floor mass has a wide range variation: 3.3–10.0 t ha^{-1} for

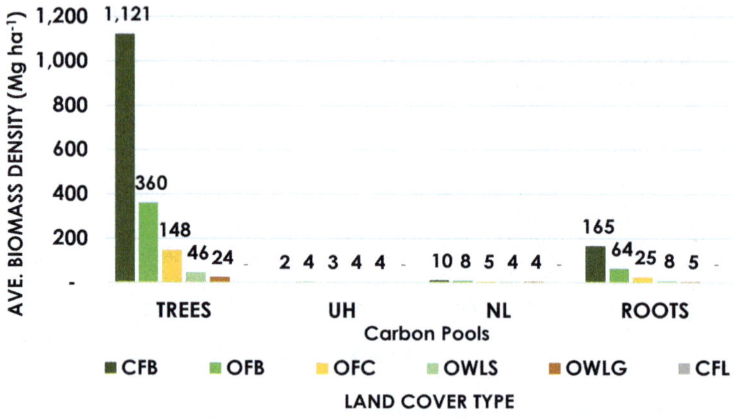

Fig. 16.10 Graph showing the mean biomass content per C pool among vegetation types

lowland rainforest and 5.1–16.5 for the montane rainforest. He also cited the forest floor mass of Malaysia at the range of 3.2–7.1 t ha^{-1}. On the other hand, in the temperate region, Zhao et al. (2014) quantified the necromass density of a Chinese pine species that ranges from 14.9 to 23.0 t ha^{-1}.

Figure 16.10 summarizes the C storage of C pools between and among vegetation types. It shows an overall descending order in biomass accumulation except for UH vegetation. UH is lower in CFB compared to the rest of the vegetation types that obtained an almost equal UH biomass density of 3–4 Mg ha^{-1}. The CFB exhibited the highest tree biomass stored among the vegetation types. The downward trend was also significant in roots and necromass or litter. But the understory and herbaceous C sink exhibited an opposite trend where OFB, OWLS, and OWLG exceeded by about 2 Mg higher than the CFB and 1 Mg higher for OFC.

Figure 16.11 summarizes the mean biomass density of different carbon pools for each vegetation type. This is supported by the percentage share of each C sink in the overall stocking potential of each land cover type, Fig. 16.12. Generally, the biomass storage capacity of the five (5) land cover types is in the order of tree, roots, necromass, and litter and UH. The tree component of the forest stand represented by the standing live trees shared the highest biomass build-up at an average of 78%. This is followed by roots (14%), necromass or litter (5%), and UH (4%) (Fig. 16.11). Similar findings were also reported by several studies, particularly on the major contribution of trees in biomass accumulation both in the lowland and high-elevated areas. Trees constituted 80% of *P. kesiya* plantation in the Mindanao area (Patricio and Tulod 2010); 82% of mossy forest in MFR while 90% of mossy forest in Pagbilao (Lasco et al. 2000); 79–99% of secondary forestry in Subic Zambales (Lasco et al. 2001), MFR (Lasco et al. 2004), Agusan del Sur (Lasco et al. 2000); 82% pine forest in Baguio (Lasco et al. 2000); 85% from tree plantations in Nueva Ecija (Lasco et al. 2000); and 78% from old growth forest in Atimonan, Quezon (Lasco et al. 2001).

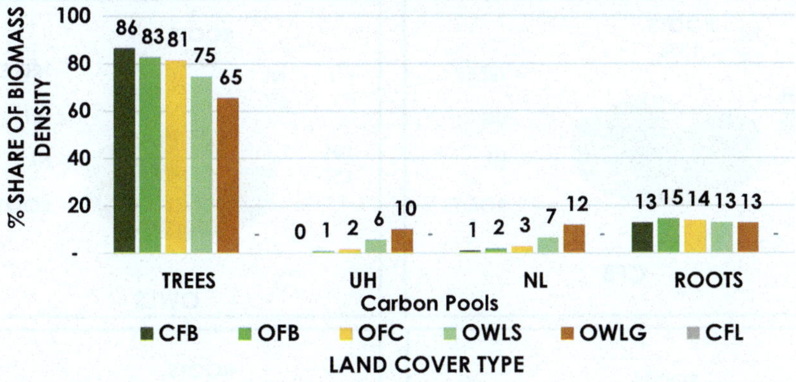

Fig. 16.11 A total average % share of different C pools in the biomass density of the six land cover types

The significant contribution of necromass/litter can be attributed to the findings of Barbosa and Wagner (1989) that necromass or litter component in a coniferous stand continues to build up on the forest floor as the species shows a slow decomposition rate due to its nature and the surrounding environment where it thrives. The decomposition rate is also affected by some climatic conditions wherein low-temperature limits microbial activity and low moisture and available nitrogen concentration as well. This is true for the study sites with the cooler environment as they are situated in high elevated areas. Litters and other dead materials take more time to decompose unlike in a warmer environment.

A number of studies in the country focused on the aboveground (AGB) or tree biomass estimation of different forest types. This is primarily due to the cost and time involved for a detailed study and some other considerations like rate of change, magnitude and direction of the change and availability and accuracy of methods to quantify change. For LULUCF projects which include reforestation, afforestation and forest management, biomass estimation of the five major carbon pools is generally required as the rate of change is high and essential for management purposes (IPCC 2003 as cited Lasco et al. 2004).

Local studies on biomass estimation on different vegetation types in high-elevation areas are limited particularly for closed broadleaved species. Rarely one can find a closed broadleaved forest in a lower elevation. Most of them can only be found in high elevated areas like the case of Mt. Province and Tinok. So far, carbon studies specific to Benguet Pine stand both pure and mixed stands gradually gained in number. Benguet Pine biomass and C studies were conducted in Luzon (Sakurai et al. 1994; Lasco et al. 2004, 2005; Lumbres 2009; Lumbres et al. 2012; Racelis et al. 2017) and only one was conducted in the Mindanao area (Patricio and Tulod 2010).

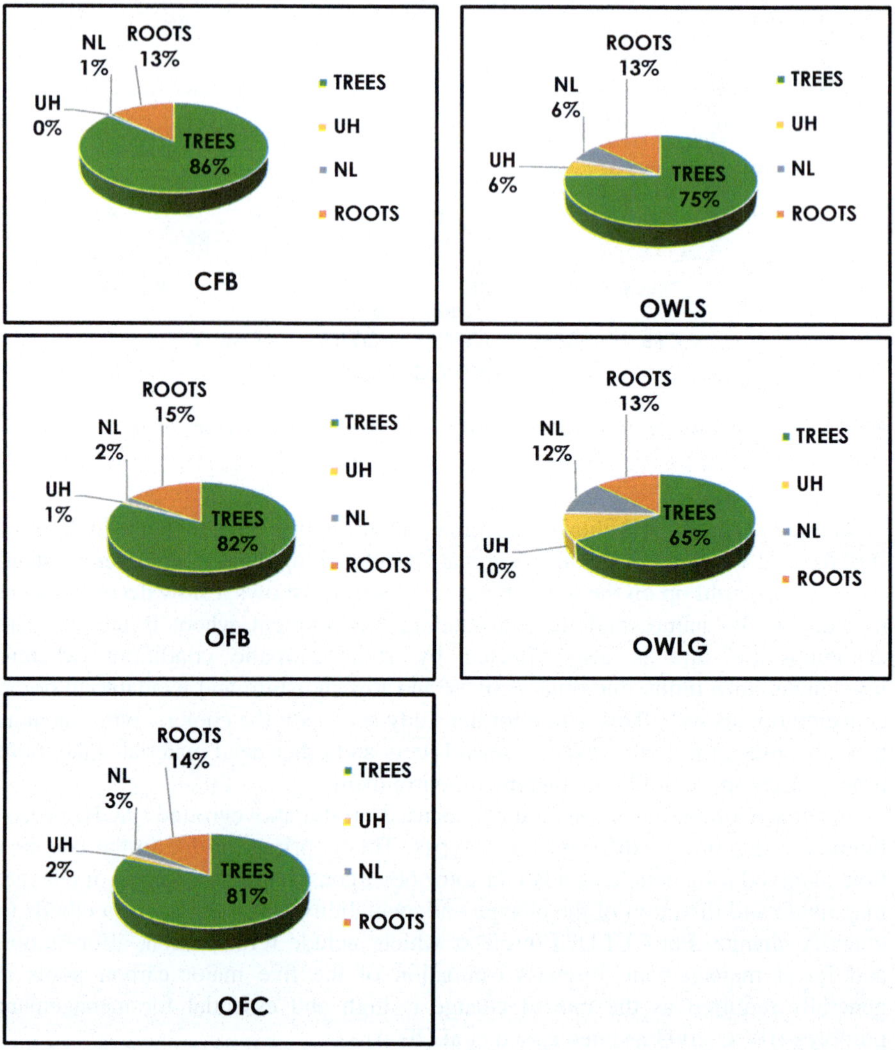

Fig. 16.12 A pie chart of the percent share of biomass stored among C pools per land cover type

16.6.1 General Assessment of C and CO_2 Stock among Land Cover Types

In terms of carbon and CO_2 storage capacity of the six vegetation types, Table 16.15 reflects the mean C and CO_2 in every vegetation type per carbon source/pool. Computation for C and CO_2 densities per vegetation type already includes SOC which is excluded in the computation for biomass. The carbon sources include the trees, UH, NL, roots, and soils (Fig. 16.12).

Table 16.15 Summary of C and CO_2 stored in the different carbon pools per vegetation type

	Land cover type mean C/CO_2 stored (Mg ha^{-1})					
Carbon pools	CFB	OFB	OFC	OWLS	OWLG	CFL
Trees	504.66	162.17	66.48	20.72	10.71	0
UH	1.09	1.74	1.45	1.60	1.63	0
NL	4.71	3.68	2.37	1.84	1.94	0
Roots	74.29	28.65	11.34	3.59	2.07	0
Soils	103.97	167.29	157.28	136.67	131.12	161.67
Total C	**688.72**	**363.52**	**238.92**	**164.43**	**147.48**	**161.67**
Total CO_2	**2525.31**	**1332.90**	**876.03**	**602.90**	**540.77**	**592.80**

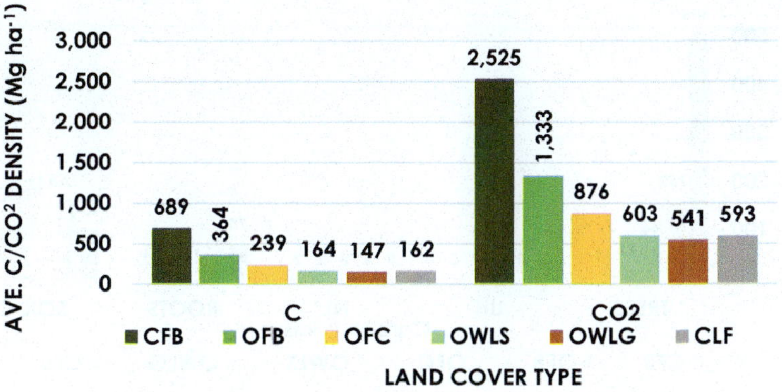

Fig. 16.13 Graph showing average C and CO_2 stock among land cover types

Comparing further each vegetation type, CFB vegetation has the highest carbon stock in its biomass, totaling to 688.72 Mg ha^{-1} or 2525.29 Mg ha^{-1} of CO_2 equivalent. This is followed by OFB (363.5.74 Mg ha^{-1} C or 1332.90 Mg ha^{-1} CO_2), OFC (238.92 Mg ha^{-1} C or 876.03 Mg ha^{-1} CO_2), and OWLS (164.43 Mg ha^{-1} C or 602.90 Mg ha^{-1} CO_2). CFL outdo OWLG in its potential to sequester carbon at a difference of 14.19 Mg ha^{-1} C. CFL carbon content is mainly contributed by its soil component. The amount of carbon or CO_2 stored by CFB as compared to other vegetation types supersedes to about 46–76% high. The change between and among vegetation types excluding CFB is below 38%, particularly in the case of OWLS, OWLG, and CFL which is less differentiated. Fig. 16.13 presents a general decreasing trend of C and CO_2 accumulation among the land cover types.

Table 16.16 and Figs. 16.13 and 16.14 show that among the vegetation types only CFB exhibited the tree C sink with the highest C and CO_2 stored. This is expected particularly for broadleaved species based on previous studies conducted but mostly in the lowland areas. The rest of the land cover types; OFB, OFC, OWLS, and OWLG, soils significantly contributed to the carbon accumulation of this vegetation.

Table 16.16 Ranking of C and CO_2 stored per carbon pools per vegetation type

Carbon pools	Land cover type (Mg ha^{-1})					
	CFB	OFB	OFC	OWLS	OWLG	CFL
Trees	504.66 (1)	162.17 (2)	66.48 (2)	20.72 (2)	10.71 (2)	0
Uh	1.09 (5)	1.74 (5)	1.45 (5)	1.60 (5)	1.63 (5)	0
NL	4.71 (4)	3.68 (4)	2.37 (4)	1.84 (4)	1.94 (4)	0
Roots	74.39 (3)	28.65 (3)	11.34 (3)	3.59 (3)	2.07 (3)	0
Soils	103.97 (2)	167.29 (1)	157.28 (1)	136.67 (1)	131.12 (1)	161.67 (1)
Total C	**688.72**	**363.52**	**238.92**	**164.43**	**147.48**	**161.67**
Total CO_2	**2525.31**	**1332.90**	**876.03**	**602.90**	**540.77**	**592.80**

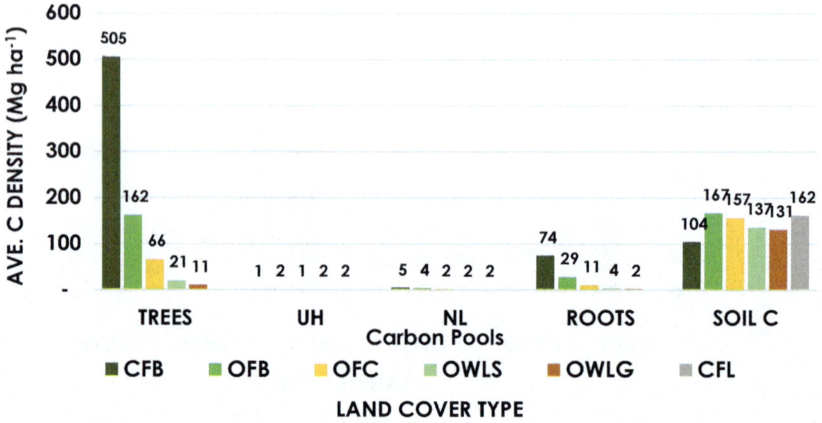

Fig. 16.14 Comparing the C stocking potential of each C pool between and among vegetation types

Trees and roots followed next with a mean share of 33% and 5%, respectively, out of the total C density. Necromass or litter also gave a good contribution to carbon stored in this study (4.86 Mg ha^{-1}) as compared to other studies conducted where only minimal or as low as half a ton per hectare was recorded.

Figure 16.15 further exhibits the graphical relationship of C pool accumulation between and among land cover types. Generally, the accumulation potential of each carbon sink per vegetation type is in decreasing order from CFB to OWLG as highlighted by the tree and soil components. There is a slight difference between UH and NL but with a marked difference in the tree pools, soil, and roots. On the soil aspect, it follows an irregular trend with the CFB got the lowest soil C.

Comparing further the sequestration potential of each carbon pool per land cover type in terms of percentage share, Figs. 16.15 and 16.16 show the individual share of C stock for each C pool per vegetation type. Ranking its percentage share of C/CO_2 accumulation, only CFB follows this order of ranking; trees >soils> roots > UH >

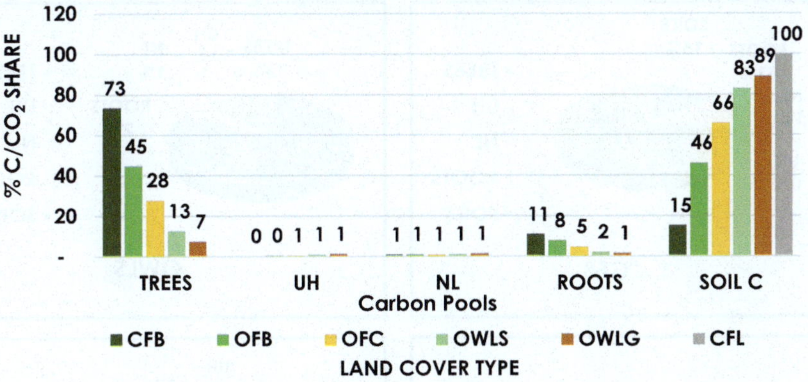

Fig. 16.15 Comparison of C and CO_2 percentage share of different C pools among vegetation types

and necromass and litter. The OFB, OFC, OWLS, and OWLG vegetation interchange soils with trees, with soil C as the highest followed by trees C. Its ranking is in the order of soils > trees > roots > NL > and UH. Roots, NL, and UH are consistent in their rank as third, fourth, and fifth, respectively in all vegetation types. Table 16.16 reflects the order of C and CO_2 accumulation by each carbon pool ranked from 1 to 5 with 1 as the highest.

The major contribution of soil in the carbon storage of other vegetation types can be accredited to the nature of the decomposition process that occurs in a conifer stand or elevated areas, particularly in this study. The decomposition process in a conifer forest is slow due to unfavorable microclimatic conditions in the site like low temperature that limits microbial activity and low moisture and available nitrogen concentration as well (Barbosa and Wagner 1989). Van den Driessche (1991) espoused that cold climates generally favor the accumulation of organic matter, although decomposition is slowed, the yield of humus substances is also increased during the break down of plant litter in soils of cold climate. Because of slow decomposition, organic matter, existing in various stages of decomposition remains at the surface of the mineral soil forming the so-called forest floor.

The above findings corroborate the results of this study. It showed that the amount of carbon from the biomass was doubled by the presence of soil C. Lugo and Brown (1992) reported that forest soils contain twice the amount of carbon present in the biomass. Moreover, this was more than the proposition of Moura-Costa (1996) as cited by Racelis et al. (2008) and Lasco and Pulhin (2001) that 30% and 40% C is tied up in the soil. Brown et al. (1992) as cited by Racelis et al. (2008) reported that variations in the rate of SOCs accumulation can be attributed to the differences in species, environmental factors, and stand age (Zhao et al. 2014). Some species produce more litters and roots than others do, thus producing more organic inputs which eventually influence SOC. This is true with the conifer species, like Benguet Pine. It produces considerable amounts of litters, especially pine needles, which are very abundant on the forest floor. These types of litter decompose slowly although

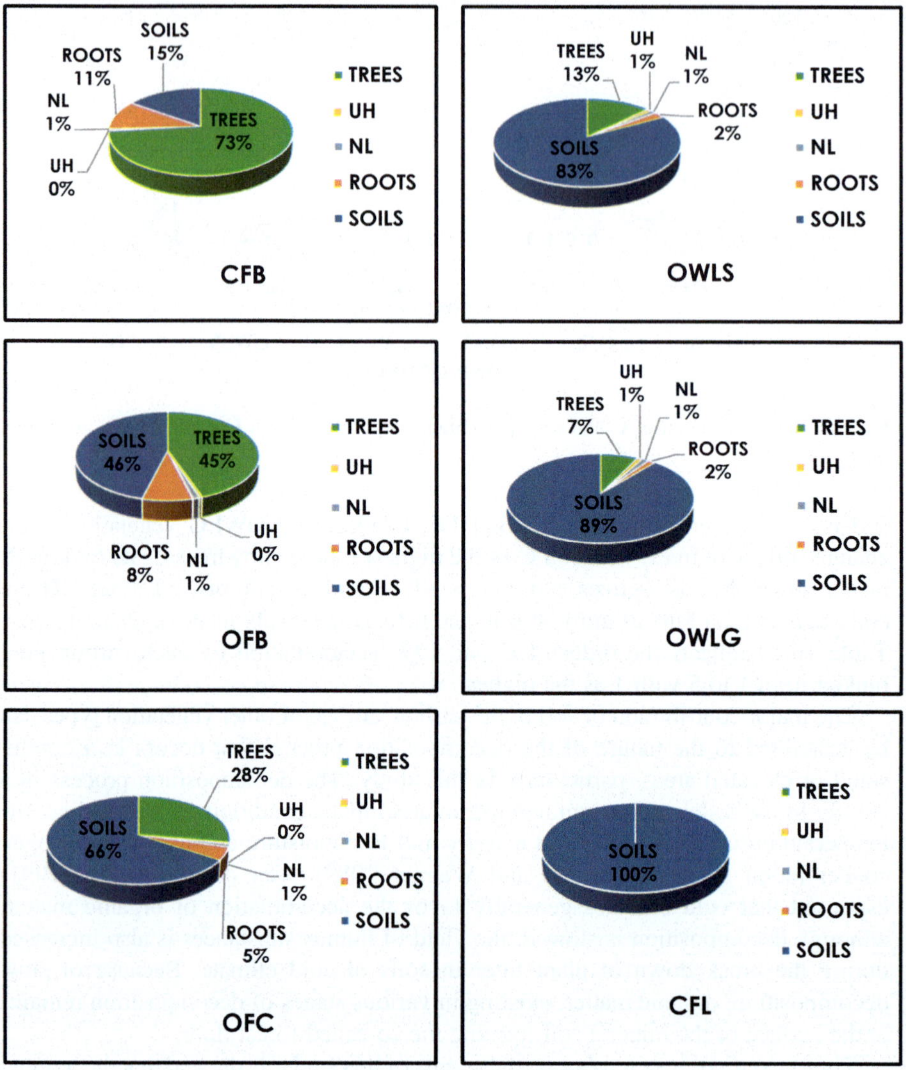

Fig. 16.16 Percent stocking rate per C pool per vegetation type

they constitute a fire hazard. The slow decomposition process helps in the gradual but continuous build-up of soil C on the forest floor. Soil organic matter does not easily leach since the fresh litter on top of the old one serves as a trapping mechanism for the decomposed litter. Hence, it is regarded as a long-term storage compartment for atmospheric carbon. It can continue to accumulate soil C over hundreds of years. It was also found that the amount of SOC increases over time. So as the vegetation matures, naturally, its soil carbon content also increases. Soil C continuous to pile up and eventually percolates to the lower depth or layer of the soil.

Necromass also plays an important role in the C sequestration potential of the study which had some bearing on the increased soil C. Necromass is regarded as a long-term carbon sink in the forest. Some studies refer to it as forest floor mass (Zhao et al. 2014; Osman 2013). It constitutes the litter and wood debris found on the forest floor. The study's necromass or litter C value validates the findings of Lugo and Brown (1992) that necromass value could reach from two (2) t ha^{-1} to as high as 16 t ha^{-1} or about 0.9–7.2 t ha^{-1} C. The results of the study found a total C density of necromass that ranges from 1.84 to 4.71 Mg ha^{-1}. Several authors have already documented the importance of necromass to carbon storage, particularly for pine species in temperate forests. In China, in particular, the study by Zhao et al. (2014) of a Chinese pine forest registered a litter biomass C ranging from 5.1 to 7.9 t ha^{-1} for a young to a mature stand. Other pine species in China have litter C content as low as 0.8 to 16.6 t ha^{-1}. This condition was observed to be unique to a conifer forest located in highly elevated areas as compared to most broad-leaf dominated species in lowland areas where necromass or litter C value is usually insignificant and even null. The reason is more of the warmer environment in the lowland areas that favors a faster decomposition process.

Sampson (1992) claimed that biomass contained in the litter on the forest floor and soil C is perhaps the least understood and least appreciated. According to Grierson et al. (1991), as cited by Racelis et al. (2008), the estimation of the total carbon stored in litter and detritus will vary according to stand age, season, climate, forest type, and fire history.

Generally, it was observed that the carbon storage in different forest ecosystem components (tree, necromass/litter, and CWD) increases with age. As the stand gets older, the more C it has accumulated except for understorey vegetation. The responsiveness of UH to biomass accumulation was claimed to be dependent more on forest management, stand-specific canopy, and soil conditions, which affect light, water, and nutrient availability to favor the growth of UH instead of stand age. Soil C stock, on the other hand, follows an opposite trend as the stand aged. It was observed in pine species in China that its share in the total ecosystem C decreased gradually from young stand to mature stand of about 61.3% to 48.4%, respectively. Mindanao *P. kesiya* species also followed the same pattern, the soil C % share decreased out of the total carbon density as the stand matured from 72% (for a 4-year-old plantation) to 28% (for a 75-year-old stand). It was concluded that aboveground C storing capacity would ultimately exceed the soil C accumulation through time or as the stand ages. This finding could be applied in this study, particularly in the CFB vegetation which is considered as matured stand composed of large-diameter trees.

16.6.2 Overall Computation of the Biomass Density, C, and CO_2 Stored in the Entire Project Site

Table 16.17 presents the total amount of biomass, C, and CO_2 equivalent of each vegetation/land cover type in Mg per hectare. It serves as a basis for the computation of the total biomass, C, and CO_2 captured by the entire project ecosystem. The total

Table 16.17 Total mean biomass, C, and CO_2 stored per land cover type (Mg ha^{-1})

Land cover type	Total ave. biomass density	Total ave. C stock	Total ave. CO_2 stored
CFB	1299.44	688.72	**2525.29**
OFB	436.06	363.52	**1332.90**
OFC	181.41	238.92	**876.03**
OWLS	61.67	164.43	**602.90**
OWLG	36.35	147.48	**540.77**
CFL		161.67[a]	**592.80[a]**

[a]Present in soil

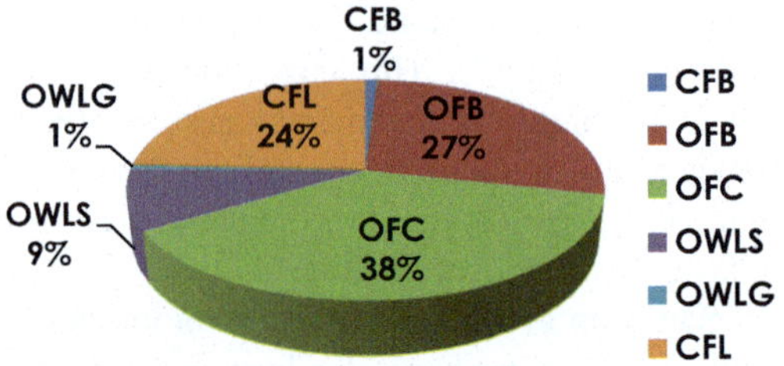

Fig. 16.17 Percent area coverage of each land use

density per vegetation is multiplied by the total area covered for each land cover type. Fig. 16.17 shows the map of the project sites indicating the total area covered by each land cover type.

The study computed a total amount of 243,801.91 Mg of biomass, 298,422.97 Mg of C, and 1,094,217.56 Mg of equivalent CO_2 stored in the 1195.04-ha project area inside the tri-boundaries of Mt. Province, Benguet, and Ifugao (Table 16.18). The area is further subdivided into different land cover types. About 38% of the total area (452.94 ha) is colonized by conifers, particularly Benguet pine species followed by OFB (328.21 ha) which is 27%, CFL 24%, and OWLS (9%). Only 1% of the project area is covered either by CFB or OWLG vegetation (Fig. 16.18).

In terms of C/CO_2 stocking potential, though OFB has less area covered than OFC it got the highest C stored (119,310.02 Mg) and CO_2 (437,470.08 Mg) build-up of about 10% higher than OFC (Table 16.18). OFC has 108,215.71 Mg C and 396,790.94 Mg of CO_2 which is 36% of the total C/CO_2. CFL also covers a large portion of the area that contributed to C and CO_2 stored of about 46,939.27 Mg C or 172,113.15 CO_2 coming solely from soil C. This is followed by OWLS (17,125.00 Mg C/62,791.67 CO_2), CFB (8299.02 Mg C/30,429.73 CO_2), and the least is OWLG (1084.01 Mg C/3974.69 Mg CO_2). Figure 16.18 shows the percent overall inputs of each vegetation type in the C/CO_2 stocks.

Table 16.18 Overall computation of the total amount of biomass, C, and CO_2 stored in each land cover type inside the tri-boundaries of Mt. Province, Benguet, and Ifugao

Land cover type	Area covered (ha)				Total amount of biomass (Mg)	Total C stored (Mg)	Total CO_2 stored (Mg)
	Mt. Province	Benguet	Ifugao	Total area (ha)			
CFB	11.30		0.75	**12.05**	15,658.26	8299.02	30,429.73
OFB	230.71	5.65	91.85	**328.21**	143,120.72	119,310.02	437,470.08
OFC	451.70	1.24		**452.94**	82,166.90	108,215.71	396,790.94
OWLS	104.15			**104.15**	6423.38	17,125.00	62,791.67
OWLG	7.35			**7.35**	267.19	1084.01	3974.69
CFL	282.20	8.14		**290.34**	0	46,939.95	172,113.15
Grand total				**1195.04**	**247,636.46**	**300,973.71**	**1,103,570.25**

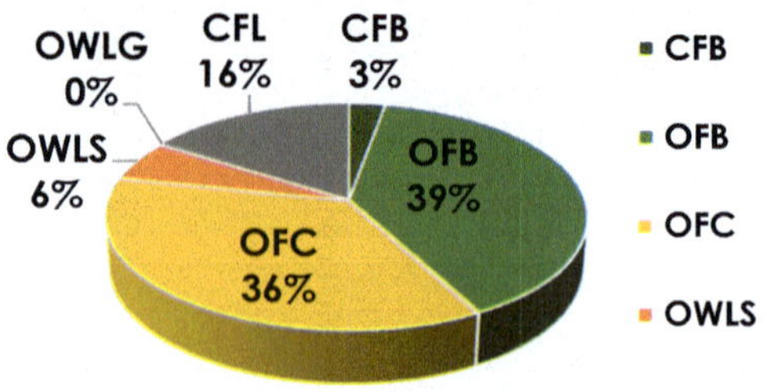

Fig. 16.18 Overall percent C/CO_2 contribution of each land cover type

16.7 Threats to Floral Biodiversity

Based on field observations obtained from literature reviews, the flora in the study site is threatened by the following:

16.7.1 Land Conversion

The conversion of forests into farmlands and residential lots poses the biggest threat to the local flora and fauna. Patches of cultivated lands are common along the fringes of forested areas which show incipient signs of continuing expansion. This is brought about by the high demand for vegetable crops in the city. Likewise, the increasing population due to high birth rates and upland migration put pressure on the forest as indicated by the growing number of settlements in the area.

16.7.2 Poaching

Illegal hunting and gathering of forest products contribute to the degradation of forest biological diversity in the study site. Medium to low biodiversity indices indicate that there is a declining trend in the population of flora and fauna species in the area.

16.7.3 Forest Fire

Periodic episodes of forest fire contribute to the decline in forest cover. Patches of burnt-out areas are common in the study sites. Most of the fire incidents were reported to be anthropogenic in origin, particularly from kaingin areas. There were also cases of arson committed by unscrupulous persons. The situation is made worse by occasional El Nino episodes.

16.7.4 Changing Climate

Changing weather patterns and extreme climatic events also contribute to forest degradation. Gradually increasing average temperature threatens the ecology of forest ecosystems attuned to low temperatures. Likewise, prolonged drought and powerful storms wreak havoc on the local vegetation and consequently on wildlife habitats.

16.7.5 Pests and Diseases

Pests and diseases could potentially decimate certain flora and fauna species. For example, some mistletoes (*Amyema* spp.) were observed nestled on branches of a number of trees encountered during the survey. They also noticed a number of pine trees with signs of bark beetle (*Ips calligraphus*) infestation. Changes in climatic regimes could weaken certain individuals and make them susceptible to pests and diseases.

16.8 Summary and Conclusion

Forest carbon stock baseline measurement and biodiversity assessments within the tri-boundaries of Benguet, Ifugao, and Mountain Province were conducted. Specifically, the study sought to: (a) measurement of various carbon pools including aboveground, ground, and belowground biomass including soil organic carbon; (b) assess biodiversity survey of forest flora and macro-fauna; (c) provide a list of flora and fauna present in the study site and determining their conservation status using IUCN red list; and (d) identify existing threats to floral diversity in the study site.

The initial sampling plots established were utilized for the pre-sampling survey to determine the minimum number of plots that needed to be established for the forest carbon baseline study. Specifically, a total of 43 sample plots were established across the study area for each vegetation type including cultivated/farm lands. For the non-trees (shrubs, herbs, vines, grasses, and ferns), opportunistic sampling involving identification and photo documentation of species were conducted inside and outside the plots. This revised methodology extended the coverage of biodiversity assessment since all the plots for carbon will be included in the survey. In addition, this synchronized the carbon measurement with the tree inventory.

Further, results of forest carbon baseline measurement show that the 1195.04-ha GEF area that straddles the tri-boundaries of Mt. Province, Benguet, and Ifugao contains a total of 300,973.71 Mg or tons of C which is equivalent to 1,103,570.25 Mg of CO_2 contained in 247,636.46 Mg of forest biomass. The land cover type which contains the highest C on a per hectare basis is Closed Forest Broadleaved (CFB) with 688.72 Mg C which is equivalent to 2525.29 Mg of CO_2. This is followed by Open Forest Broadleaved (OFB) (363.52 Mg ha^{-1} C or

1332.90 Mg ha^{-1} CO_2), Open Forest Conifer (OFC) (238.92 Mg ha^{-1} C or 876.03 Mg ha^{-1} CO_2), Other Wood Land Shrubs (OWLS) (164.43 Mg ha^{-1} C or 602.90 Mg ha^{-1} CO_2), Other Land, Cultivated/Farm Land/Annual Crops (CFL) (161.67 Mg ha^{-1} or 592.80 Mg ha^{-1}and Other Wooded Land, Wooded Grassland (147.48 Mg ha^{-1} or 540.77 Mg ha^{-1}), CFL contains more C than OWLG owing to the comparatively high C content in soil component of the former. Moreover, among the vegetation types only Closed Forest Broadleaved exhibited the tree C sink with the highest C and CO_2 stored. This is expected particularly for broadleaved species based on previous studies conducted but mostly in the lowland areas. The rest of the land cover types; OFB, OFC, OWLS, and OWLG, significantly contributed to the carbon accumulation of this vegetation. Trees and roots followed next with a mean share of 33% and 5%, respectively, out of the total C density. Necromass or litter also gave a good contribution to carbon stored in this study (4.86 Mg ha^{-1}) as compared to other studies conducted where only minimal or as low as half a ton per hectare was recorded.

The floral biodiversity assessment revealed that 32 morpho-species belonging to 26 genera and 24 families can be found inside the 40 sample plots used in forest carbon assessment. The additional 94 species that were recorded from the opportunistic survey conducted in the designated study sites brought the total count to 126 morpho-species. The dominant families identified include Pinaceae, Fagaceae, Myrtaceae, Pentaphylacaceae, Podocarpaceae, and Ericaceae. Thirty-six (36) species recorded were found to be endemic to the Philippines which includes rare species of orchids. Six (6) species recorded were found to be threatened under (DAO 2017-11) and/or the IUCN Red list of Threatened Species (2018-2) which includes Yew (*Taxus wallichiana* Zucc.). Further, the five species with the highest Importance Value or the most important species are Benguet Pine (*Pinus kesiya* ssp. *langbianensis*), Palayen (*Lithocarpus* sp.), Boltik (*Syzygium* sp.), Igem (*Dacryocarpus imbricatus*), and Bini (*Eurya buxifolia*). In terms of biodiversity indices, Sample Plot 1 had the highest Shannon and Simpson index wherein 18 species were recorded. Most of the plots have low to very low diversity based on the H′ diversity classification range.

Acknowledgments The project team wishes to acknowledge the Department of Environment and Natural Resources –Cordillera Administrative Region and Integrated Natural Resources and Environmental Project for the opportunity to assess the forest carbon and biodiversity resources in the tri-boundary of Benguet, Ifugao, and Mountain Province. In particular, our sincere gratitude to Engr. Ralph C. Pablo, DENR-CAR Regional Director, For. Augusto D. Lagon, Assistant Regional Director, For. Moises H. Butic, INREMP Deputy Project Director, For. Rose Y. Apilis, INREMP Regional Focal Person, For. Florentino Yan, INREMP Assistant Regional Focal Person, and the rest of DENR-CAR staff who in one way or another contributed to the successful conduct of the study. We also would like to extend our appreciation for the valuable comments and suggestions made by the members of the DENR-CAR Technical Working group.

Recognition is also due to Philex Mines for allowing the use of their laboratory for oven-drying of vegetation samples. The same goes for the Institute of Renewable Natural Resources for the use of its Soils Laboratory. We are also grateful to LGU Bauko and Tinoc for allowing the safe conduct of the study in their municipalities. We also would like to thank PNP Bauko for giving the team the safe conduct pass to ensure its security during the survey. Our sincere gratitude to the Barangay

Council of Monamon Sur, Bauko, Mt. Province for the resolution endorsing the conduct of the study in their barangay.

The representatives of local communities in Bauko, Mt. Province, and Tinoc, Ifugao also merit recognition for their active participation and support to the project. They contributed a lot in locating study sites, identification of local names of species, and collection of samples for the project.

Lastly, the project would like to extend its sincere thanks to the University of the Philippines Los Baños for allowing the conduct of the study.

References

Alberto EG, Alcayna EA, Babang MB, Labhat MB, Benohlan JO (2011) In: Ngohayon SL (ed) Ifugao indigenous knowledge (IK) workbook. Ifugao State University, Ifugao, Philippines

Avtar R, Tsusaka K, Herath S (2020) Assessment of forest carbon stocks for REDD+ implementation in the muyong forest system of Ifugao, Philippines. Environ Monit Assess 192:571. (2020). https://doi.org/10.1007/s10661-020-08531-8

Baoanan ZG, Abansi CL, Abalos CFS (2020) A review of biodiversity-related reports in the cordillera highlands, northern Luzon, Philippines. J Nat Stud 19(2):84–103

Barbosa P, Wagner MR (1989) Introduction to Forest and shade tree insects. Academic Press, Inc, San Diego, California

Biodiversity Management Bureau (BMB) Department of Environment and Natural Resources (DENR) (2016) In: Cabrido C (ed) Philippine biodiversity and strategy and action plan (2015–2028): bringing resilience to Filipino communities. BMB-DENR, United Nations Development Program, Quezon City, Philippines

Brown S (1997) Estimating biomass and biomass change of tropical forest: A primer. Forestry paper 134. FAO, Rome

Brown SA, Lugo AE (1982) The storage and production of organic matter in tropical forests and their role in the global carbon cycle. Biotropica 14:161–167

Brown SA, Lugo AE, Iverson L (1992) Processes and lands for sequestering carbon in the tropical forest landscape. Water Air Soil Pollut 64:139–155. Kluwer Academic Publishers. Netherlands

Camacho LD, Gevaña DT, Carandang AP, Camacho SC (2016) Indigenous knowledge and practices for the sustainable management of Ifugao forests in cordillera, Philippines. Int J Biodiver Sci Ecosyst Ser Manag 12(1–2):5–13. https://doi.org/10.1080/21513732.2015.1124453

Conservation International-Philippines (2007) Human wellbeing and conservation framework and strategy. Quezon City, Philippines, p 91

Cordillera Peoples Alliance (CAP). 2016. Cordillera forests in peril. Retrieved: August 23, 2022, from https://cpaphils.wordpress.com/2016/03/21/cordillera-forests-in-peril/

Dedicatoria RMM, Rebancos CM, Afuang LE, Espaldon MVO (2010). Identifying environmental changes in Mt. Data Watershed, Bauko, Mt. Province, Northern Philippines: Implications to sustainable management. Proceedings 4th Asian Rural Sociology International Conference. ISBN 978-971-8778-87-6. 545 pp.

DENR-FMB. (2020). Philippine Forestry Statistics. Retrieved from: https://drive.google.com/file/d/1AYk7J3jo7hFlshFcXgcQwhe_ewQ2kRLZ/view

DENR-CAR (2011) Assessment of agricultural biodiversity of pilot sites of globally important agricultural heritage systems, Ifugao Rice terraces (GIAHS—IRT) at Hungduan, Ifugao, Philippines. Department of Environment and Natural Resources, Ifugao, Philippines

Doyog ND, Lumbres RIC, Young JL (2018) Mapping of the spatial distribution of carbon storage of the Pinus kesiya Royle ex Gordon (Benguet pine) forest in Sagada, Mt. Province, Philippines. J Sustain For 37(7):661–677. https://doi.org/10.1080/10549811.2018.1450155

Grierson PF, Polglase PJ, Attiwill PM, Adams MA (1991) Carbon storage in Victoria's forests. Climate Change Impacts, Australia

Habito CD, Mendoza RC, Dong-e R, Malabrigo PL Jr (2022) Identification and conservation of culturally important trees used for traditional houses in Ifugao, Philippines. In: Ramamoorthy S, Buot IJ, Chandrasekaran R (eds) Plant genetic resources, inventory, collection and conservation. Springer, Singapore. https://doi.org/10.1007/978-981-16-7699-4_7

Hairiah K, Dewi S, Agus F, Velarde S, Ekadinata A, Rahayu S, Van Noordwijk M (2011) Measuring carbon stocks across land use systems. A manual. World Agroforestry Centre (ICRAF), SEA Regional Office, Bogor, Indonesia, p 154

IPCC. (2006). Guidelines for National Greenhouse gas inventories. Volume 4: agriculture, forestry and other land uses. Chapter 4: forestland

Jianchu X, Ruscoe MT (1993). The use of indigenous knowledge in agroecosystem Management for Conservation: A case study in barangay Haliap, Kiangan, Ifugao central cordillera, Philippines. University of the Philippines Los Baños, Institute of Environmental Science and Management and Ford Foundation

Lasco RD, Pulhin FB, Bugayong LA, Mendoza MA (2011) An assessment of potential benefits to smallholders of REDD+ components in the Philippines. Ann Trop Res 33(1):31–48. https://doi.org/10.32945/atr3313a.2011

Lasco RD, Pulhin PB, Cruz RVO (2007) Carbon stocks assessment of Forest land uses in the Kaliwa watershed, Philippines. J Environ Sci Manag 10(2):40–48

Lasco RD, Pulhin FB, Cruz RVO, Pulhin JM, Roy SSN (2005). Carbon budgets of terrestrial ecosystems in the Pantabangan-Carranglan watershed. Assessments of Impacts and Adaptations to Climate Change (AIACC) Working Paper No. 10:1-23

Lasco RD, Pulhin FB, Roshetko JM, Banaticla MRN (2004) Land use, land use change and forestry climate change mitigation projects: A primer. World Agroforestry Centre. Southeast Asia Regional Research Programme, p 43

Lasco RD, Pulhin FB (2001) Climate change mitigation activities in the Philippine forestry sector: application of the COMAP model. Mitig Adapt Strateg Glob Chang 6:313–334. https://doi.org/10.1023/A:1013335320992

Lasco RD, Pulhin FB, Visco RG, Racelis DA, Guillermo IQ, Sales RF (2000). Carbon stocks assessment of Philippine forest ecosystems. Paper presented at the science-policy workshop on terrestrial carbon assessment for possible carbon trading, Bogor, Indonesia. 28–20 February 2000

Lasco RD, Visco RG, Pulhin JM (2001) Formation and transformation of secondary forests in the Philippines. J Trop Forest Sci 13:652–670

Lugo AE, Brown S (1992) Tropical forests as sinks of atmospheric carbon. For Ecol Manag 54: 239–255

Lumbres RC (2009). Carbon stock assessment of trees in Burnham Park, Baguio City, Philippines: a tool for urban environmental management. Master thesis (unpublished). Benguet State University. 80 pp

Lumbres RIC, Lee YL, Seo YO (2012) Development of height-dbh growth model and biomass estimation of Pinus kesiya Royle ex Gordon in La Trinidad, Benguet. Philippines Asia Life Sci 21(1):177–188

Lumbres RIC, Lee YL (2014) Aboveground biomass mapping of La Trinidad forests in Benguet, Philippines, using Landsat thematic mapper data and k-nearest neighbor method. For Sci Technol 10(2):104–111. https://doi.org/10.1080/21580103.2013.866171

Magcale-Macandog D (2010) Conserving indigenous knowledge: information system towards food information system towards food security and sustainable NRM in the security and sustainable NRM in the cordilleras. In: SEARCA agriculture and development seminar series. University of the Philippines Los Baños, Laguna, Philippines

Malabrigo PL (2013) Vascular flora of the tropical montane forests in Balbalasang-Balbalan National Park, Kalinga Province, northern Luzon, Philippines. Asian J Biodiver 4(1):1–22. https://ejournals.ph/article.php?id=1465

Malabrigo PL, Eduarte GT, Racelis DA (2021). Floral diversity assessment within the tri boundaries of Benguet, Ifugao and Mountain Province. Paper presented during the 3rd international conference on cordillera studies, 05 July 2021

Mendoza RC, Romano AD, Dulnuan ED, Bulong MP, Malabrigo PL, Habito CDL (2022) Wood identification: A tool for preservation of indigenous architecture of traditional houses in Ifugao, Philippines. Int J Wood Culture:1–36. https://brill.com/view/journals/ijwc/aop/article-10.1163-27723194-bja10004/article-10.1163-27723194-bja10004.xml

Moura-Costa P (1996) Tropical forestry practices for carbon sequestration. In: Schulte A, Schone D (eds) Dipterocarp forest ecosystems: towards sustainable management. World Scientific, Singapore, pp 308–334

Osman KT (2013) Nutrients dynamics in Forest soil. In: Forest soils: properties and management. Springer International Publishing, Switzerland. ISBN 978-3-319-02541 (eBook), pp. 114

Patricio JH, Tulod A (2010) Carbon sequestration potential of Benguet pine (Pinus kesiya) plantations in Bukidnon, Philippines. J Nat Stud 9(1):99–104. Retrieved May 25, 2015, from http://www.researchgate.net/publication/266896815_Carbon_Sequestration_Potential

Pearson T, Walker S, Brown S (2005) In: Schlamadinger B, Emmer I, Kagi W, Noble I, Bosquest B, Ringius L (eds) Sourcebook for land use, land-use change and forestry projects. Bio CF WI Winrock International, Arlington, VA

Pelser PB, Barcelona JF, Nickrent DL. 2011 Onwards. Co's Digital Flora of the Philippines. www.philippineplants.org. As cited by Co's Digital Flora of the Philippines. www.philippineplants.org

Racelis EL, Carandang WM, Lasco RD, Racelis DA, Castillo ASA, Pulhin JM (2008) Assessing the Carbon Budgets of Large-leaf Mahogany (*Swietenia macrophylla* King) and Dipterocarp Plantations in the Mt. Makiling Forest Reserve, Philippines. J Environ Sci Manag 11(1): 40–55. (June 2008). ISSN 0119-114. UPLB, College, Laguna

Racelis EL, Racelis DA, Villanueva TR, Florece LM, Carandang MG, Lapitan RL (2017) Carbon stock potential of Benguet pine (*Pinus kesiya* Royle ex Gordon) stands within a mining site in Padcal, Benguet Province, Philippines. Ecosyst Develop J 7(2):28–36, Oct. 2017. ISSN 2012-3612.College, Laguna

Rondolo MT. (2000). The changing Ifugao woodlots; implications for indigenous Ifugao plant knowledge and biodiversity. PhD dissertation. Australian National University

Sakurai S, Ragil RB, de la Cruz LU (1994) Tree growth and productivity in degraded forest land. In: Rehabilitation of degraded lands in the tropics. JIRCAS international symposium series No. 1, Tsukuba, Japan, pp 64–71

Sampson RN (1992) Forestry opportunities in the United States to mitigate the effects of global warming. Water, air and soil pollution, vol 64. Kluwer Academic Publishers, Netherlands, pp 157–180

Taguiling NK (2013) Macrofloral biodiversity conservation in Ifugao. Eur Sci J 9(10):469–482. https://doi.org/10.19044/esj.2013.v9n10p%25p

Van Den Driessche R (1991) In mineral nutrition of conifer seedling. CRC Press Inc, Boca Raton

Villanueva TR (2005) Upland ecosystem management, 2nd edn. UP Open University, Laguna

Zhao J, Kang F, Wang L, Yu X, Zhao W, Song X, Zhang Y, Chen F, Sun Y, He T, Han H (2014) Patterns of biomass and carbon distribution across a Chronosequence of Chinese pine (*Pinus tabulaeformis*) forests. PLoS One 9(4):e94966

The Dynamics of Soil Microbiome Upon Anthropogenic Changes in Plant Diversity and Land Management Practices

17

Rossyda Priyadarshini, Saefur Rohman, and Amir Hamzah

Abstract

Plants and microorganisms are part of ecosystems. Soil microbes have a crucial role in providing nutrition and thus influence plant growth. Plants through their root exudates affect the microbiome structures. The soil biogeochemical process in the rhizosphere is controlled by soil microbial activity that directly or indirectly supports the plant performance. The mechanisms of this interaction and the alteration between plant-rhizosphere-microbiome are bidirectional and poorly understood. Plants release the exudates through the rhizodeposition that become a nutrient source for soil microbes to grow. Land use change will affect the soil biota structure and composition through the soil biomass and nutrient input. Land use change will change the plant aboveground biomass then it will the shifted of the bacterial and fungal communities. Land use change due to anthropogenic activities positively correlated with the soil chemical characteristics and it will have a negative effect on Proteobacteria, and positive effect on Acidobacteria and Actinobacteria. We focus on the plants and root-associated bacteria interaction that enhance soil nutrition. This interaction results in plant traits changes, including the members of the plant-associated community at the trophic levels.

R. Priyadarshini (✉) · S. Rohman
Center Innovation of Appropriate Technology for Lowland and Coastal Food, Surabaya, East Java, Indonesia

University of Pembangunan Nasional Veteran Jawa Timur, Surabaya, East Java, Indonesia
e-mail: rossyda_p@upnjatim.ac.id

A. Hamzah
University of Tribhuwana Tunggadewi, Malang, East Java, Indonesia

S. Ramamoorthy et al. (eds.), *Plant Diversity in Biocultural Landscapes*,
https://doi.org/10.1007/978-981-19-8649-9_17

Keywords

Plant diversity · Anthropogenic changes · Land-use change · Soil microbiome

17.1 Introduction

Plants are utterly important to support the survival of human and other living beings on Earth due to their ability to conduct photosynthesis. Plants are photoautotrophs that acquire their energy via the conversion of carbon dioxide (CO_2) and water (H_2O) into six-carbon molecule of glucose with the help of sunlight energy. Glucose as the final form of energy storage would further be used to produce plentiful plant parts which some of them are used by humans either as food (grains, fruits, bulbs, etc.) or for other needs (cottons, woods, dyes, etc.) (Checchetto et al. 2013).

A current estimate counts 400,000 known plant species on Earth. This number includes vascular plants that represent 383,671 species, while the rest belongs to bryophytes, liverworts, and hornworts (Brummitt et al. 2021). Among vascular plants, seed plants and flowering plants account for 370,492 and 369,434 species, respectively (Lughadha et al. 2016). Yet, plants are not uniformly spatially distributed. Their distribution are mainly driven by biotic (ecology or vegetation) and abiotic (topographic and climate) variations. Several areas harbor higher species richness than others particularly on tropical areas such as Amazonian rainforest, Congo basin in Africa, Indian subcontinent, and Malay Archipelago, which are due to the proximity to the equator while vertically plant richness decreases as the increase in altitude (Brummitt et al. 2021).

The major threat to plant diversity loss, by far, is caused by human and human-driven changes. Since their emergence on Earth, humans have manipulated the environment to satisfy their needs which cause several disturbances to the natural ecosystem. For instance, the domestication of plants has caused major plant genetic diversity loss due to deliberately selecting a few plants with superior characteristics than others (Bourguiba et al. 2012). This is followed by the cultivation of plants, i.e., agricultural production that results in huge conversion of natural ecosystem into cropping land which causes habitat loss, fragmentation, and land degradation in the process. Also, pollution, introduction of invasive alien species, overexploitation, and climate changes are other anthropogenic disturbances that contribute to the loss of biodiversity, particularly on plants (Corlett 2016).

Plants are subjected to their environment, particularly soil. Soil serves as the anchorage of roots and the main source of nutrients (Roy and Bassham 2014). Further, plant–soil interaction is heavily mediated by soil biological property, i.e., soil microbes. These microbial communities occupy a niche in the proximity of plant roots and are fed on nutrient-rich exudates secreted by the roots (Sasse et al. 2018). In turn, the microbes perform various key soil biogeochemical processes such as nutrient cycling and plant-growth promotion that contribute to the overall plant health and growth status (Fierer 2017).

It can be said that plants shape their own soil microbiome (i.e., soil microbial community and the process they perform) while the structure and functioning are heavily influenced by the aboveground plants diversity (Sasse et al. 2018). If any disturbances occur on the aboveground due to anthropogenic-driven activities, for instance land use changes, they would also surely alter the belowground microbiome status (Cornell et al. 2022). Unfortunately, our knowledge on how the changes on the aboveground would affect the belowground microbiome is still lacking.

Here in this chapter we try to explain the current progress on: (1) the interrelationship of plant, soil, and soil microbiome within the spatial and temporal framework of plant rhizosphere, (2) plant diversity as a function of anthropogenic intervention upon natural ecosystem, (3) how anthropogenic-driven disturbances upon plant diversity cause shift in soil microbiome structure and function.

17.2 The Bidirectional Relationship of Plants and Soil Microbiome

17.2.1 Plant Rhizosphere

The role of soil on plants is very important. Soil serves as the anchorage for plant roots and the main source of water and nutrients for their growth (Roy and Bassham 2014). Soil is further divided into three constituents: physical, chemical, and biological properties. Any changes on any of these properties, e.g., change in pH, soil organic content, liming practice, etc., surely affect plant growth and health status (Adedayo et al. 2022).

The soil rhizosphere, which has different characteristics from other parts of the soil, plays an important role in plant growth. The rhizosphere is a narrow body of soil that surrounds plant roots and is under the influence of plant roots activity (Adedayo et al. 2022; Philippot et al. 2013). Plants continuously secrete certain organic compounds (carbohydrates, organic acids, and amino acids) from their roots to the surrounding soils which is called as rhizodeposition (Lu et al. 2017). This activity provides a nutrient-rich niche where soil microbes can grow and thrive. The microbes in turn influence the rhizosphere through various biogeochemical processes such as nutrient cycling, soil fertility maintenance, and carbon sequestration (Jacoby et al. 2017).

The influence of plant rhizodeposition activity upon soil microbes and the biogeochemical process they undertake further shapes the rhizosphere which is distinct with the rest of the bulk soil (Philippot et al. 2013). Soil characteristics may differ greatly with the bulk soil even if they are just few millimeters apart. Hence, it is thoughtful to consider soil not as a single, but a complex and dynamic, environment depending on the soil in question (Fierer 2017).

17.2.2 Soil Microbiome

Soil is a dynamic and rich community that contains numerous microbial taxa from all three domains of life (prokaryotic bacteria, archaea, and eukaryotic fungi) (Chen et al. 2003; Fierer 2017; Kalayu 2019). They perform many important soil ecosystem and biogeochemical processes. Soil microbes can produce or consume atmospheric gasses (oxygen, nitrogen, carbon dioxide), assist in rock weathering and soil formation process, influence soil acidity, stabilize the soil by aggregate formation, and mediate carbon and other elements cycling. In addition, they can also facilitate or suppress pathogen infection, degrade xenobiotic compounds, and chelate heavy metals in bioremediation. The structure of soil microbes community composition together with the dynamic of their activity is known as soil microbiome (Fierer 2017).

Soil cannot be simplified as a single and uniform entity, but a diverse and wide range of environment that contains distinct microbial communities. Even within just a few millimeters apart, soil can differ considerably in terms of their abiotic and microbial characteristics (structure, abundance, and activity). It is now known that soil microbial biomass rivals the aboveground plants and animals biomass with a recent estimate calculating more than 1000 kg of microbial biomass carbon per hectare. The diversity of soil microbial taxa is so huge that even 1 g soil can contain thousands of microbial taxa and up to 10 billion cells of bacteria (de Vrieze 2015; Fierer 2017).

To date, it is almost impossible to deliberately culture all soil microbial taxa. Only a few has been successfully cultured in vitro. Hence, the culture-based method to assess soil microbes would obviously underestimate their actual diversity and function (Fierer 2017). To answer this problem, culture-independent metagenomic (also metatranscriptomic, and metaproteomic) approaches are being developed.

17.2.3 Soil Microbiome Function

Soil biogeochemical process that shapes the rhizosphere is mainly mediated by microbial activity in soil. Soil microbes perform various processes that directly or indirectly contribute to the aboveground plants health and growth status. For example, soil microbes promote plant growth directly via nitrogen fixation, mobilization of certain nutrients (phosphorus and potassium) to increase their availability, synthesis of plant phytohormone (indole-3-acetic acid, IAA), etc. (Aasfar et al. 2021; Egamberdieva et al. 2017; Etesami et al. 2017; Rawat et al. 2021) whereas some examples of indirect roles of soil microbiome are mediating nutrient cycling, soil aggregate formation, and litter decomposition (Cania et al. 2019; Ravn et al. 2020). Table 17.1 lists some notable examples of soil biogeochemical processes mediated by soil microbes.

Table 17.1 Some notable examples of soil biogeochemical processes mediated by soil microbes

No.	Name of process	Microbes responsible	Mechanism/modes of action
Processes that directly affect plant growth			
1	Nitrogen fixation	Nitrogen-fixing bacteria, e.g., *Rhizobium, Azotobacter, Bacillus,* etc.	The conversion of atmospheric nitrogen (N_2) into a usable form of nitrogen, ammonium (NH_4^+), facilitated by *nif* genes encoded nitrogenase enzymes complex (Aasfar et al. 2021)
2	Phosphate solubilization	Phosphate solubilizing bacteria (*Pseudomonas*, *Bacillus,* and *Micrococcus*) and fungi (*Penicillium, Aspergillus,* and *Rhizopus*)	Secretion of organic acids, enzymes, or siderophores that can chelate metal into ions complexes and hence release phosphate and increase its availability in soil (Rawat et al. 2021)
3	Potassium solubilization	*Acidithiobacillus*, *Paenibacillus*, and *Bacillus*	Increase available K by dissolving silicate minerals, lowering the pH, and enhancing the chelation of the cations bound to K (Etesami et al. 2017)
4	Plant growth promotion and stress alleviation	Numerous bacteria and fungi such as *Acinetobacter, Bacillus, Rhodopseudomonas, Aspergillus*, and *Streptomyces*	Synthesis of plant growth hormone and stress alleviation related compounds, i.e., auxin (indole acetic acid), cytokinin, and proline (Egamberdieva et al. 2017)
Process that indirectly affect plant growth			
5	Decomposition of organic matter	Numerous saprotrophic bacteria and fungi	Enzymes assisted physical breakdown and biochemical transformation of large and complex organic molecules into smaller and simpler organic and inorganic molecules (Ravn et al. 2020)
6	Biological rock weathering in soil formation	Numerous bacteria and fungi	Mineral dissolution facilitated by ligands complex (siderophore and organic acids), altering the pH, or performing redox reaction (Ribeiro et al. 2020)
7	Soil aggregate formation	Bacteria (e.g., *Chitinophagaceae, Nitrospiraceae,* and *Bradyrhizobiaceae*) and fungi particularly mycorrhizal fungi	By producing exopolysaccharides and lipopolysaccharides that stabilize soil aggregates by "gluing" soil particles in bacteria (Cania et al. 2019) or by interaction between mycelium and soil particles in fungi (Lehmann et al. 2020)

(continued)

Table 17.1 (continued)

No.	Name of process	Microbes responsible	Mechanism/modes of action
8	Bioremediation of xenobiotic compounds and heavy metals	Numerous soil bacteria and fungi	By binding, oxidizing, volatilization, immobilization, or chemical transformations of pollutants into less harmful forms of compounds (Reddy and Osborne 2022)

17.2.4 How Plants Shape Their Microbiome

Plants live in association with a large variety of microbes that can live inside (endosphere) or outside (episphere) of plant tissues. Among these microorganisms, bacteria and fungi are predominant. Bacterial and fungal taxa play important roles such as nutrient availability, plants uptake, and increased plant stress tolerance. Therefore, plant growth and survival is not only the result of physical and physiological functions of plants but it is also associated with the microbiome or it was known as plant holobiont.

The microbial community associated with roots was proposed to be assembled in two steps: first, the rhizosphere is colonized by a subset of the bulk soil community and, second, the rhizoplane and the endosphere are colonized by a subset of the rhizosphere community. Intriguingly, a set of recurring plant-associated microbes has emerged (core microbiome). The microbial community associated with roots was proposed to be assembled in two steps: first, the rhizosphere is colonized by a subset of the bulk soil community and, second, the rhizoplane and the endosphere are colonized by a subset of the rhizosphere community. Interestingly, a set of recurring plant-associated microbes has emerged (core microbiome) (Bulgarelli et al. 2013).

Rhizobiomes are influenced by their spatial orientation toward roots in two ways. First, the radial proximity of microbial communities to roots defines community complexity and composition. Second, the lateral position of microbes along a root shapes the community. Root tips are the first tissues that make contact with bulk soil: root tips are associated with the highest numbers of active bacteria compared with other root tissues, and likely select microbes in an active manner. The root elongation zone is specifically colonized by *Bacillus subtilis*, which suggests a particular role of this zone in plant–microbe interactions. Mature root zones feature a microbial community distinct from root tips (Massalha et al. 2017). Their community includes decomposers, which could be involved in the degradation of dead cells shedding from old root parts. Similarly, lateral roots are associated with distinct microbial communities, differing between tips and bases, as well as between different types of lateral root (Kawasaki et al. 2016).

17.3 Soil Microbiome of Several Land Use Types

Land use types are a crucial factor influencing soil microbial structure and composition. Land use types are closely related with the β-diversity of soil biota. Diversity and richness of soil biota vary depending on the land use types. Generally, terrestrial ecosystems are more diverse than aquatic ecosystems while, on a more acidic environment, fungi are more abundant than bacterial or protist communities (Wang et al. 2021).

Land use conversion affects all trophic levels and significantly changes the microbial community structure and composition. Typically, land use conversion causes diminishing effect on bacterial and archaea richness and abundance, otherwise increases fungal richness and abundance. The shifts on soil microbial community structure and composition are the results of changes in soil ecosystem and properties (Berkelmann et al. 2020). As demonstrated by Berkelmann et al. (2020) the conversion of forest into agroforestry and monoculture croplands had negative effects on *Proteobacteria* (especially *Rhizobiales* and *Burkholderiales*), but it had positive effects on *Acidobacteria* and *Actinobacteria* abundance. These changes were positively correlated with the soil chemical characteristics, such as pH, C:N ratio, C, and N content of the sites studied.

Land use change is a stronger driver for bacterial and fungal communities shift. Soil from specific sites tends to harbor unique bacterial communities. For instance, microbial taxa of *Burkholderiaceae, Oxalobacteraceae, Sphingomonadaceae, Sphingobacteriaceae, Xanthomonadaceae, Flavobacteriaceae, Micrococcaceae, and Streptomycetaceae* tend to be more prevalent on cultivated soil while taxa of *Solirobrobacteraceae, Rubrobacteraceae, Propionibacteriaceae, Opitutaceae, and Cytophagaceae* are more prominent in soil with conservation management. Moreover, there were also significant interactions between land use and root proximity for *Oxalobacteraceae*, *Pseudomonadaceae*, and *Phyllobacteriaceae* (Schlatter et al. 2020).

17.3.1 Forest

Forests cover a third of the earth's surface area and are characterized by a large number of trees. There are three types of forest: temperate, tropical, and boreal with different characteristics. Temperate forests have varying temperatures throughout the year, high rainfall, high fertility, and a wide variety of fauna while tropical forests have temperatures between 20 and 31 °C. This type of forest is known as a symbol of biodiversity because of its high biodiversity. Tropical mangrove forests are characterized by trees and shrubs that are able to grow in salt water. The last forest type, boreal forest (taiga), is one of the largest biomes and has a significant role in carbon sequestration. The temperature in the boreal forest is below freezing. Conifers, spurce, spruce, and pine trees are the predominant plant species at this site.

Biodiversity and biome's types are closely related to the soil microbial diversity. Biome's types affect the plant–microbial relations rather than latitude or elevation gradients, and soil types. Tropical forests have the strongest correlations between plant diversity and microbial diversity compared with the subtropical forest and shrublands (Liu et al. 2020). Distribution of microbial biogeography was related to the types of ecosystems. Less environmental pressure in the tropics causes a higher population of plant and soil microbes. The variety of litter inputs and the quality of the litter increases the diversity of microbial communities in tropical forests (Bradford et al. 2017).

Microbial richness was also positively correlated with plant diversity. Variability between taxonomic groups due to plant diversity is more prominent in latitude, elevation cover, and tropical forests, which could be explained by high variations in microbial diversity and plant diversity. The plant–microbial correlations declined with increased environmental differences in most cases. Total Organic Carbon (TOC) and Total Nitrogen (TN) content has a positive effect on microbial communities in the tropical forests.

Commonly, soils contain dominant bacterial phyla, mainly Acidobacteria, Actinobacteria, Bacteroidetes, and Proteobacteria that have been found consistently in both cultivated and natural environments. The Rhizobiales order within Alphaproteobacteria were more abundant in lime and hornbeam mono stands at total community and active community level than in beech and oak mono stands while Betaproteobacteria that were represented by Burkholderiales and Nitrosomonadales were more abundant in lime and hornbeam mono stands compared to beech and oak mono stands. Both orders form part of the nitrogen-fixing bacterial community in forests soils and participate in symbiotic relationships with plants.

17.3.2 Agroforestry

Agroforestry is an agricultural cropping system that integrates trees into the cropping system. The perennial, i.e., woody trees are combined with annual crops and, often with, livestock. Agroforestry systems are developed and advocated to answer the major issues arising from conventional monoculture farming system including food security, climate change, and biodiversity loss (Beule et al. 2022; Isaac and Borden 2019). This system are based on the hypothesis that yield benefits are to occur if resources are used in a complementary manner. In this sense, the combination of trees and crops within a farming system creates a balanced and noncompetitive resource sharing particularly nutrients uptake via the roots. Tree roots are likely to uptake the leachable nutrients unused by the crops and incorporate them into the trees biomass which can be returned to the soil through tree litter. This ensures effective resource use and low-input sustainable agricultural practice (Isaac and Borden 2019).

The benefits of agroforestry are the results of increasing aboveground plant diversity. The incorporation of trees or shrubs into the cropping system along with their interaction results in increase of ecosystem services and environmental benefits (Beule et al. 2022; Kuyah et al. 2019). Agroforestry brings efficient use of nutrient and resources by partitioning the resources between trees and crops, reducing nitrate leaching through nitrate uptake by trees root, and increasing soil fertility through consistent tree-litter input (Beule and Karlovsky 2021). Further, agroforestry raises carbon stocks, assists in the regulation of soil moisture content, controls erosion, enhances pollination, and supplies more resources (fruits, nuts, woods, fodder, etc.) (Kuyah et al. 2019). The various benefits derived from agroforestry systems are due to the high diversity of plants, animals, and particular microbes either above and below the ground (Beule and Karlovsky 2021).

Increasing the aboveground plant diversity by incorporating trees or shrubs to the farming land in agroforestry system also affects the belowground soil microbiome. Several researches have shown shifts of microbial community composition and abundance in agroforestry (Zhang et al. 2019). Agroforestry increased soil microbial biomass, particularly microbial biomass C and N (Zaia et al. 2012). In addition to the increase of biomass, specific impacts are also experienced by specific groups particularly bacteria and fungi. Tree rows harbor the highest soil bacterial activity than crop row or monoculture crop land (Beule and Karlovsky 2021). However, it seems that this applies only to specific bacterial taxa. For instance, a study from Beule et al. (2022) found that only six of nine taxa (Acidobacteria, Actinobacteria, Alpha- and Gammaproteobacteria, Firmicutes, and Verrucomicrobia) are affected while the other three (Bacteroidetes, Betaproteobacteria, and Gemmatimonadetes) remained unaffected. The same pattern is also find in fungi which the highest fungal abundance and activity occurring in the proximity of the tree row than of crop row or monoculture cropland. The effect also varies among a fungal group, for example, Basidiomycota is seem to be particularly promoted among others.

It seems that the impact of trees incorporated on the agroforestry system on soil microbiome structure and activity is not uniform but spatially distributed in a gradual manner following the distance from trees row. For instance, the highest accumulation of fungal biomass is found in the vicinity of trees row and diminishes toward the crop rows. This applies also to other parameters such as abundance and activity in both bacteria and fungi. This phenomenon seems to be related to exudate activity of trees roots which are stronger than the crop roots (Beule et al. 2022; Beule and Karlovsky 2021).

17.3.3 Dryland

Dryland constitutes 41% of the terrestrial surface and is vulnerable to land degradation. The restoration of degraded dryland is currently being under concern. Dryland is characterized by high temperatures, low and scarce precipitation, low nutrient availability, and low vegetation cover which all contribute to its vulnerability (Havrilla et al. 2020). Vegetation has a crucial role in maintaining soil microbial community activities. Vegetation has a crucial role in maintaining soil microbial community activities. The microbe was an ecological engineer and provider of ecosystem services that could function normally under vegetated systems. The lack of vegetation cover on dryland greatly affects soil microbial community and functioning (van der Heijden et al. 2008); thus the restoration of aboveground vegetation would simultaneously improve soil microbiome and is the key factor on the restoration of dryland ecosystem (Hart et al. 2020).

Plants are a key factor of land restoration particularly in dryland. Vegetation types control both the soil microbial community and the process they mediate related to the ecosystem services provided. Plant diversity contributes to soil fertility which is associated with the amount and diversification of plant biomass and biochemical characteristics of nutrients (Banning et al. 2011). Nutrient pools, especially soil organic matter supporting the diversity of soil microbes, mean that plant species affect the growth of soil microbiome (Prober et al. 2015) due to the function of each species and the diversity and composition of root exudates (Steinauer et al. 2016).

Williams et al. (2022) observed that diversifying the cropping systems on dryland ecosystems affect soil microbe communities that is associated with abiotic changes in the environment. They found that on fallow or winter period, where drought preventing new crops, changes in organic matter led to small changes in copiotrophic microbes that utilize the labile organic matter which would be followed by more persistent changes in fungal communities that are able to decompose a more recalcitrant organic matter. While in the summer where more crops were present, the copiotrophic microbes would continue their presence and support the differences of bacterial and fungal operational taxonomic (OTU) richness and community composition. Persistent fallow means no vegetation cover, hence hiving the smallest bacterial and fungal OTUs. It correlated with the fresh input of organic matter and nutrient input. Soil microbiome diversity and composition are closely related with the environment. In the low nutrient environment only few species could survive, and the species would be dominated by the species that could decompose recalcitrant organic matter (Tardy et al. 2015).

Under dry ecosystems which is characterized by water deficits, the members of Actinobacteria and Proteobacteria seem to increase while under agricultural land use, the unknown Acidobacteria phylum ("Candidatus Solibacter" and "Candidatus Koribacter") is found in the dry season. In many drylands the presence of Actinobacteria and Cyanobacteria is increased while the Proteobacteria, Acidobacteria, and Bacteroidetes are decreased (Lacerda-Júnior et al. 2019). Meanwhile, the increase of Actinobacteria under drought periods in dryland is quite interesting. They have the ability to thrive under high temperatures, high concentration of salt, as well as radiation. Hence they are a potential target for bioprospecting (Mohammadipanah and Wink 2015). Similarly, Cyanobacteria can grow well in arid environments by developing specific strategy for ecological adaptation, including extracellular polysaccharides synthesis for water retention (Tamaru et al. 2005).

17.4 Soil Microbiome of Several Agricultural Management Practices

Agronomic soil management practices are a critical factor in determining short- and long-term soil health. Intensive tillage enhances soil compaction and leads to soil loss and diminishes the crop residue on the soil surface; therefore fertilization is needed due to the decrease of carbon sequestration, and soil organic matter. Tillage will also diminish the exchangeable nitrogen, phosphorus, potassium, and calcium. Thus, reducing tillage or even no tillage application is needed.

No tillage can broadly increase soil microbial diversity (microbiome composition and diversity) and abundance, as well as reduce the amounts of fungal pathogens, compared to conventional practices. Arbuscular mycorrhizal fungi (AMF) are also negatively impacted by tillage intensity. Reduced tillage will increase AMF root colonization in some plant hosts. Less soil disturbance would promote the niche differentiation that it reflected with the higher microbial diversity. Tillage will structure more fungal communities than bacteria while the cover type has significant effect only for bacteria.

Tillage and cover type are expected to influence a number of soil characteristics that can impact microbial fitness, including soil pore size, nutrient bioavailability, and moisture. We therefore hypothesized that treatment would significantly structure the beta diversity of hosted microbial communities. In addition, we predicted that tillage type would impact fungi to a greater extent than bacteria, due, in part, to their formation of extensive hyphae networks.

17.5 Summary and Conclusion

Apart from what have been mentioned above, it is noteworthy to consider that what happens belowground on soil microbiome structure and functioning upon trees row introduction in cropping land in agroforestry system may vary considerably depending on the site in question. Soil is a complex systems with a multitude of intertwined biotic and abiotic factors affected. Though the general pattern can be concluded, i.e., higher microbial activity in agroforestry compared to the monoculture cropping system, the underlying detail may differ from site to site. The scope of parameter studied also affects the extent of our knowledge since it is hard to conduct studies thoroughly on soil microbial status under agroforestry.

Acknowledgment I am thankful to the Ministry of Research, Technology and Higher Education who provided research funding related to plant diversity and soil metagenomics.

References

Aasfar A, Bargaz A, Yaakoubi K, Hilali A, Bennis I, Zeroual Y, Meftah Kadmiri I (2021) Nitrogen fixing azotobacter species as potential soil biological enhancers for crop nutrition and yield stability. Front Microbiol 12:628379. https://doi.org/10.3389/fmicb.2021.628379

Adedayo AA, Babalola OO, Prigent-Combaret C, Cruz C, Stefan M, Kutu F, Glick BR (2022) The application of plant growth-promoting rhizobacteria in *Solanum lycopersicum* production in the agricultural system: a review. PeerJ 10:e13405. https://doi.org/10.7717/peerj.13405

Banning NC, Gleeson DB, Grigg AH, Grant CD, Andersen GL, Brodie EL, Murphy DV (2011) Soil microbial community successional patterns during forest ecosystem restoration. Appl Environ Microbiol 77(17):6158–6164. https://doi.org/10.1128/AEM.00764-11

Berkelmann D, Schneider D, Meryandini A, Daniel R (2020) Unravelling the effects of tropical land use conversion on the soil microbiome. Environ Microbiome 15(1):5. https://doi.org/10.1186/s40793-020-0353-3

Beule L, Karlovsky P (2021) Tree rows in temperate agroforestry croplands alter the composition of soil bacterial communities. PLoS One 16(2):e0246919. https://doi.org/10.1371/journal.pone.0246919

Beule L, Vaupel A, Moran-Rodas VE (2022) Abundance, diversity, and function of soil microorganisms in temperate alley-cropping agroforestry systems: a review. Microorganisms 10(3):616. https://doi.org/10.3390/microorganisms10030616

Bourguiba H, Audergon J-M, Krichen L, Trifi-Farah N, Mamouni A, Trabelsi S, D'Onofrio C, Asma BM, Santoni S, Khadari B (2012) Loss of genetic diversity as a signature of apricot domestication and diffusion into the Mediterranean Basin. BMC Plant Biol 12(1):49. https://doi.org/10.1186/1471-2229-12-49

Bradford M, Veen C, Bonis A, Bradford E, Classen A, Cornelissen J, Crowther T, De Long J, Freschet G, Kardol P, Manrubia M, Maynard D, Newman G, Logtestijn R, Viketoft M, Wardle D, Wieder W, Wood S, Putten W (2017) A test of the hierarchical model of litter decomposition. Nat Ecol Evol 1(12):1836–1845. https://doi.org/10.1038/s41559-017-0367-4

Brummitt N, Araújo AC, Harris T (2021) Areas of plant diversity—what do we know? Plants People Planet 3(1):33–44. https://doi.org/10.1002/ppp3.10110

Bulgarelli D, Schlaeppi K, Spaepen S, Themaat E, Schulze-Lefert P (2013) Structure and functions of the bacterial microbiota of plants. Annu Rev Plant Biol 32:64. https://doi.org/10.1146/annurev-arplant-050312-120106

Cania B, Vestergaard G, Krauss M, Fliessbach A, Schloter M, Schulz S (2019) A long-term field experiment demonstrates the influence of tillage on the bacterial potential to produce soil structure-stabilizing agents such as exopolysaccharides and lipopolysaccharides. Environ Microbiome 14(1):1. https://doi.org/10.1186/s40793-019-0341-7

Checchetto V, Teardo E, Carraretto L, Formentin E, Bergantino E, Giacometti GM, Szabo I (2013) Regulation of photosynthesis by ion channels in cyanobacteria and higher plants. Biophys Chem 182:51–57. https://doi.org/10.1016/j.bpc.2013.06.006

Chen G, Zhu H, Zhang Y (2003) Soil microbial activities and carbon and nitrogen fixation. Res Microbiol 154(6):393–398. https://doi.org/10.1016/S0923-2508(03)00082-2

Corlett RT (2016) Plant diversity in a changing world: status, trends, and conservation needs. Plant Divers 38(1):10–16. https://doi.org/10.1016/j.pld.2016.01.001

Cornell CR, Zhang Y, Ning D, Wu L, Wagle P, Steiner JL, Xiao X, Zhou J (2022) Temporal dynamics of bacterial communities along a gradient of disturbance in a U.S. Southern Plains Agroecosystem. mBio 13(3):e03829-21. https://doi.org/10.1128/mbio.03829-21

de Vrieze J (2015) The littlest farmhands. Science 349(6249):680–683. https://doi.org/10.1126/science.349.6249.680

Egamberdieva D, Wirth SJ, Alqarawi AA, Abd Allah EF, Hashem A (2017) Phytohormones and beneficial microbes: essential components for plants to balance stress and fitness. Front Microbiol 8:2104. https://doi.org/10.3389/fmicb.2017.02104

Etesami H, Emami S, Alikhani HA (2017) Potassium solubilizing bacteria (KSB): mechanisms, promotion of plant growth, and future prospects - a review. J Soil Sci Plant Nutr 17(4):897–911. https://doi.org/10.4067/S0718-95162017000400005

Fierer N (2017) Embracing the unknown: disentangling the complexities of the soil microbiome. Nat Rev Microbiol 15(10):579–590. https://doi.org/10.1038/nrmicro.2017.87

Hart MM, Cross AT, D'Agui HM, Dixon KW, Van der Heyde M, Mickan B, Horst C, Grez BM, Valliere JM, Rossel RV, Whiteley A, Wong WS, Zhong H, Nevill P (2020) Examining assumptions of soil microbial ecology in the monitoring of ecological restoration. Ecol Solutions Evid 1(2):e12031. https://doi.org/10.1002/2688-8319.12031

Havrilla CA, Munson SM, McCormick ML, Laushman KM, Balazs KR, Butterfield BJ (2020) RestoreNet: an emerging restoration network reveals controls on seeding success across dryland ecosystems. J Appl Ecol 57(11):2191–2202. https://doi.org/10.1111/1365-2664.13715

Isaac ME, Borden KA (2019) Nutrient acquisition strategies in agroforestry systems. Plant Soil 444(1–2):1–19. https://doi.org/10.1007/s11104-019-04232-5

Jacoby R, Peukert M, Succurro A, Koprivova A, Kopriva S (2017) The role of soil microorganisms in plant mineral nutrition—current knowledge and future directions. Front Plant Sci 8:1617. https://doi.org/10.3389/fpls.2017.01617

Kalayu G (2019) Phosphate solubilizing microorganisms: promising approach as biofertilizers. Int J Agron 2019:1–7. https://doi.org/10.1155/2019/4917256

Kawasaki A, Donn S, Ryan P, Mathesius U, Devilla R, Jones A, Watt M (2016) Microbiome and exudates of the root and rhizosphere of Brachypodium distachyon, a model for wheat. PLoS One 11:e0164533. https://doi.org/10.1371/journal.pone.0164533

Kuyah S, Whitney CW, Jonsson M, Sileshi GW, Öborn I, Muthuri CW, Luedeling E (2019) Agroforestry delivers a win-win solution for ecosystem services in sub-Saharan Africa. A meta-analysis. Agron Sustain Dev 39(5):47. https://doi.org/10.1007/s13593-019-0589-8

Lacerda-Júnior GV, Noronha MF, Cabral L, Delforno TP, de Sousa STP, Fernandes-Júnior PI, Melo IS, Oliveira VM (2019) Land use and seasonal effects on the soil microbiome of a Brazilian dry forest. Front Microbiol 10:648. https://www.frontiersin.org/articles/10.3389/fmicb.2019.00648

Lehmann A, Zheng W, Ryo M, Soutschek K, Roy J, Rongstock R, Maaß S, Rillig MC (2020) Fungal traits important for soil aggregation. Front Microbiol 10:2904. https://doi.org/10.3389/fmicb.2019.02904

Liu L, Zhu K, Wurzburger N, Zhang J (2020) Relationships between plant diversity and soil microbial diversity vary across taxonomic groups and spatial scales. Ecosphere 11:e02999. https://doi.org/10.1002/ecs2.2999

Lu H, Sun J, Zhu L (2017) The role of artificial root exudate components in facilitating the degradation of pyrene in soil. Sci Rep 7(1):7130. https://doi.org/10.1038/s41598-017-07413-3

Lughadha EN, Govaerts R, Belyaeva I, Black N, Lindon H, Allkin R, Magill RE, Nicolson N (2016) Counting counts: revised estimates of numbers of accepted species of flowering plants, seed plants, vascular plants and land plants with a review of other recent estimates. Phytotaxa 272(1):82. https://doi.org/10.11646/phytotaxa.272.1.5

Massalha H, Korenblum E, Malitsky S, Shapiro O, Aharoni A (2017) Live imaging of root-bacteria interactions in a microfluidics setup. Proc Natl Acad Sci U S A 114:4549–4554. https://doi.org/10.1073/pnas.1618584114

Mohammadipanah F, Wink J (2015) Actinobacteria from arid and desert habitats: diversity and biological activity. Front Microbiol 6:1541. https://doi.org/10.3389/fmicb.2015.01541

Philippot L, Raaijmakers JM, Lemanceau P, van der Putten WH (2013) Going back to the roots: the microbial ecology of the rhizosphere. Nat Rev Microbiol 11(11):789–799. https://doi.org/10.1038/nrmicro3109

Prober SM, Leff JW, Bates ST, Borer ET, Firn J, Harpole WS, Lind EM, Seabloom EW, Adler PB, Bakker JD, Cleland EE, DeCrappeo NM, DeLorenze E, Hagenah N, Hautier Y, Hofmockel KS, Kirkman KP, Knops JMH, La Pierre KJ et al (2015) Plant diversity predicts beta but not alpha diversity of soil microbes across grasslands worldwide. Ecol Lett 18(1):85–95. https://doi.org/10.1111/ele.12381

Ravn NR, Michelsen A, Reboleira ASPS (2020) Decomposition of organic matter in caves. Front Ecol Evol 8:554651. https://doi.org/10.3389/fevo.2020.554651

Rawat P, Das S, Shankhdhar D, Shankhdhar SC (2021) Phosphate-solubilizing microorganisms: mechanism and their role in phosphate solubilization and uptake. J Soil Sci Plant Nutr 21(1): 49–68. https://doi.org/10.1007/s42729-020-00342-7

Reddy S, Osborne J (2022) An insight on the advancements of biological technologies in the bioremediation of textile effluents. Urban Water J 19(5):468–480. https://doi.org/10.1080/1573062X.2022.2030369

Ribeiro IDA, Volpiano CG, Vargas LK, Granada CE, Lisboa BB, Passaglia LMP (2020) Use of mineral weathering bacteria to enhance nutrient availability in crops: a review. Front Plant Sci 11:590774. https://doi.org/10.3389/fpls.2020.590774

Roy R, Bassham DC (2014) Root growth movements: waving and skewing. Plant Sci 221–222:42–47. https://doi.org/10.1016/j.plantsci.2014.01.007

Sasse J, Martinoia E, Northen T (2018) Feed your friends: do plant exudates shape the root microbiome? Trends Plant Sci 23(1):25–41. https://doi.org/10.1016/j.tplants.2017.09.003

Schlatter DC, Yin C, Hulbert S, Paulitz TC (2020) Core rhizosphere microbiomes of dryland wheat are influenced by location and land use history. Appl Environ Microbiol 86(5):e02135-19. https://doi.org/10.1128/AEM.02135-19

Steinauer K, Chatzinotas A, Eisenhauer N (2016) Root exudate cocktails: the link between plant diversity and soil microorganisms? Ecol Evol 6(20):7387–7396. https://doi.org/10.1002/ece3.2454

Tamaru Y, Takani Y, Yoshida T, Sakamoto T (2005) Crucial role of extracellular polysaccharides in desiccation and freezing tolerance in the terrestrial cyanobacterium nostoc commune. Appl Environ Microbiol 71(11):7327–7333. https://doi.org/10.1128/AEM.71.11.7327-7333.2005

Tardy V, Chabbi A, Charrier X, de Berranger C, Reignier T, Dequiedt S, Faivre-Primot C, Terrat S, Ranjard L, Maron P-A (2015) Land use history shifts in situ fungal and bacterial successions following wheat straw input into the soil. PLoS One 10(6):e0130672. https://doi.org/10.1371/journal.pone.0130672

van der Heijden MGA, Bardgett RD, van Straalen NM (2008) The unseen majority: soil microbes as drivers of plant diversity and productivity in terrestrial ecosystems. Ecol Lett 11(3):296–310. https://doi.org/10.1111/j.1461-0248.2007.01139.x

Wang C, Masoudi A, Wang M, Yang J, Yu Z, Liu J (2021) Land-use types shape soil microbial compositions under rapid urbanization in the Xiong'an New Area, China. Sci Total Environ 777:145976. https://doi.org/10.1016/j.scitotenv.2021.145976

Williams A, Birt HWG, Raghavendra A, Dennis PG (2022) Cropping system diversification influences soil microbial diversity in subtropical dryland farming systems. Microb Ecol. https://doi.org/10.1007/s00248-022-02074-w

Zaia F, Gama-Rodrigues A, Gama-Rodrigues E, Moço M, Fontes A, Machado RC, Baligar V (2012) Carbon, nitrogen, organic phosphorus, microbial biomass and N mineralization in soils under cacao agroforestry systems in Bahia, Brazil. Agroforestry Syst 86:197–212. https://doi.org/10.1007/s10457-012-9550-4

Zhang X, Gao G, Wu Z, Wen X, Zhong H, Zhong Z, Bian F, Gai X (2019) Agroforestry alters the rhizosphere soil bacterial and fungal communities of moso bamboo plantations in subtropical China. Appl Soil Ecol 143:192–200. https://doi.org/10.1016/j.apsoil.2019.07.019

Plants' Anatomical and Genetic Responses to Anthropogenic Climate Change and Human-Induced Activities

18

Lerma S. J. Maldia, Marilyn S. Combalicer, and Crusty E. Tinio

Abstract

Anthropogenic climate changes have many impacts on plants and these changes leave imprints in their key anatomical and genetic characteristics. The first section of this chapter presents the roles of morpho-anatomical traits in shaping plant diversity patterns and forest landscapes, focusing on the trade-offs in plant morpho-anatomy, physiological functions, and structural diversity were discussed. The functional implications of morpho-anatomical traits will help in understanding the mechanisms of plants' resistance strategies that underpin plant resilience amid frequent anthropogenic climate change and human-induced activities. These implications are demonstrated in the different type of plants according to their water relations: mesophytes, xerophytes, halophytes, hydrophytes, and heliophytes. The morpho-anatomical trait–functioning relationships were also discussed in terms of the aboveground and belowground traits and plant and ecosystem functions. Variations of morpho-anatomical traits at multi-scales (forest landscape structure) along different environmental gradients, which will help in predicting the likely responses of plant species to climate change, were also presented. The second section accounted the coevolution of plants and human and highlighted how the different anthropogenic activities, centering on domestication and introductions resulting to biological invasion, caused genetic and molecular changes in plants. In particular, the genetic effects of human-driven population bottlenecks and selection pressures brought about by domestication processes and streamlined agricultural and silvicultural practices and interventions alongside climate change were discussed.

L. S. J. Maldia (✉) · M. S. Combalicer · C. E. Tinio
Department of Forest Biological Sciences, College of Forestry and Natural Resources, University of the Philippines Los Baños, Los Baños, Laguna, Philippines
e-mail: lsmaldia@up.edu.ph

S. Ramamoorthy et al. (eds.), *Plant Diversity in Biocultural Landscapes*,
https://doi.org/10.1007/978-981-19-8649-9_18

Genetic responses were centered on two significant climate change-induced responses in plant populations: changes in flowering and fruiting patterns and changes in the patterns of the population genetic diversity and structure. Lastly, various molecular and potential biotechnological interventions for plants to adapt to climate change, revive and make ecosystems more climate-resilient, sustain food, and climate change mitigation for the attainment of the greater good of human were synthesized.

Keywords

Anthropogenic climate change · Morpho-anatomical traits · Morpho-anatomical trait–functioning relationships · Plant genetic responses · Population bottlenecks

In this chapter, we explain the changes in plants in response to various anthropogenic activities through anatomical and molecular perspectives.

The first section of this chapter discusses plant morpho-anatomy in a changing environment. It presents the impacts of anthropogenic activities on the plant key anatomical characteristics amid changing climates. We discussed the roles of morpho-anatomical traits in shaping plant diversity patterns and forest landscapes, focusing on trade-offs in plant morpho-anatomy, physiological functions, and structural diversity. The morpho-anatomical traits of plants classified according to their water relations were discussed and their functional implications were presented.

The second section presents significant accounts on the co-evolution of plants and humans. It also discusses the different anthropogenic activities (domestication, introductions resulting to biological invasion) and anthropogenic climate change causing genetic changes in plants. We presented the genetic effects of human-driven population bottlenecks and selection pressures brought about by domestication processes and streamlined agricultural and silvicultural practices and interventions. Along with domestication, we discussed the plant's genetic responses to human-mediated species introductions, leading to bioinvasion, particularly in the light of climate change. We focused on the aspect of plant adaptation, and the discussion is centered on two significant changes in plant populations amid climate change, namely changes in plant phenology patterns (specifically flowering and fruiting behaviors) and changes in population genetic diversity patterns (structure and levels). We lastly cite various molecular and biotechnological interventions for plants to adapt to climate change that, in turn, will help sustain the human population's need for food and other raw materials.

18.1 Plant Morpho-Anatomy in a Changing Environment

18.1.1 Anthropogenic Climate Change Impacts on Plant Key Anatomical Characteristics

Anthropogenic climate change has many impacts on plants, be it stress brought by heat (Stangler et al. 2017; Yuan et al. 2017), flooding (Jia et al. 2021; De Oliveira et al. 2015), drought (Hernandez et al. 2021; Wong 2018), elevated carbon dioxide (CO_2) concentrations (Huang et al. 2020; Innocente et al. 2020; Anderegg and Meinzer 2015), and temperature (Calleja-Cabrera et al. 2020; Kadipaya et al. 2017), which directly affect plant growth, reproduction, and resilience (Parmesan and Hanley 2015) and consequently plant key anatomical characteristics (Pandey 2021; Wong 2018; Yuan et al. 2017; De Oliveira et al. 2015).

Many changes in plant species, particularly in the anatomical characteristics, have already been well documented due to anthropogenic climate change (i.e., elevated atmospheric CO_2 concentrations, warming climate, and changing precipitation regimes) (Pandey 2021). Some of these plants' anatomical characteristics evolved as their means of adaptation. For instance, with the effect of heat, Yuan et al. (2017) observed thickness in leaf lamina, upper epidermis, and palisade mesophyll of *Brassica campestris*. In the study of Wong (2018) *Syzygium cumini* demonstrated exceptional drought tolerance with no mortality and only slight wilting; however, the lack of water and the energy storage starch in the pith cells indicates that the plants were stressed. Similarly, Huang et al. (2020) observed the most striking anatomical variations in *Ottelia cordata* responding to high CO_2, such as the enlarged upper epidermal cells and the decreased area of epidermal chloroplasts. During flooding, it was reported by De Oliveira et al. (2015) that *Tabebuia rosea* (Bignoniaceae) in Brazil developed aerenchyma in the root cortex.

Some negative impacts of anthropogenic climate change were also observed in plants' morpho-anatomical characteristics. Stangler et al. (2017) revealed that sugar maple (*Acer saccharum* Marsh.), a species that currently dominates northern hardwood forests, became vulnerable to heat waves during leaf expansion. With the effects of flooding, *Kielmeyera coriacea* (Calophyllaceae) and *Aspidosperma macrocarpon* (Apocynaceae) showed a decrease in survival and necrosis of the leaves and roots (De Oliveira et al. 2015). Innocente et al. (2020) mentioned in their study the anatomical changes in the stems of young *Eucalyptus urophylla* brought by elevated CO_2 and observed a decrease in the lignification content in the newly formed tissues of the plant. In the case of higher temperatures, plant anatomy of *Triticum aestivum* L was also adversely affected (Kadipaya et al. 2017). With these studies and observations, it is certain that anthropogenic climate change and human-induced activities will have continuous effects on plant anatomical characteristics over the coming years and decades.

18.1.2 Role of Morpho-Anatomical Traits in Shaping Plant Diversity Patterns and Forest Landscapes

Trade-Offs in Plant Morpho-anatomy, Physiological Functions, and Structural Diversity

Plant species' habitat affinities along environmental gradients are determined by a combination of morpho-anatomical and physiological traits (Belluau 2018). According to Silvertown et al. (2015), in their developed idea with trade-offs in water economy at different organizational levels, at a morpho-anatomical level, the trade-off is related to the safety or efficiency of water transport, while at a physiological level, the trade-off is related to leaf gas change. Leaf photosynthesis is largely determined by anatomical characteristics (Xie et al. 2019). Muir (2020) explained that stomatal pores control leaf gas exchange and are one route for infection of internal plant tissues by many foliar pathogens, setting up the potential for trade-offs between photosynthesis and pathogen colonization. Xiao et al. (2022) also mentioned in their study that leaf thickness and stomatal density are the anatomical characteristics most closely related to photosynthesis.

The leaf economics spectrum (LES) is an eco-physiological concept describing the trade-offs of leaf structural and physiological traits that have been widely investigated across multiple scales (Xiong and Flexas 2018), including construction costs per unit leaf area, nutrient concentrations, and rates of carbon fixation and tissue turnover (Onoda 2017). Xiong and Flexas (2018) observed that the photosynthetic traits of domesticated rice (*Oryza* sp.) responded differently to leaf structural and biochemical changes while mesophyll conductance was the most sensitive to leaf N status. Consequently, in the study of Yuan et al. (2017), the thickness of leaf lamina, upper epidermis, and palisade mesophyll of *Brassica campestris* increased with the presence of heat that might contribute to the higher assimilation of photosynthates. In a different study conducted by Wang et al. (2022), it was observed that needle anatomical structures change across the five studied Picea species as affected by needle size (needle width). This is substantially related to the differences in net photosynthetic capacity (Pnmax) and photosynthetic nitrogen use efficiency (PNUE). Understanding morpho-anatomical and physiological traits of plants will provide a better understanding of the existing relationship between different biophysical structures and will be beneficial for identifying and selecting species for a specific site condition.

Vegetative aerial organs, which are more exposed to environmental conditions, also reflect specific adaptations of plants to their local environment. In the study by Muniz et al. (2018), it was found that environmental factors, including climate and soil from different localities, have a significant influence on assessing the morpho-anatomical characteristics of the *Aldama grandiflora*. Plant roots also exhibit inter-specific variation in root functional traits, such as changes in root morphology, root exudation, and mycorrhizal symbioses, for the acquisition of nutrients (Wen et al. 2019).

18.1.3 Morpho-Anatomical Traits and Their Functional Implication

Understanding the mechanisms of plants' resistance strategies can help us explain plant resilience amid frequent anthropogenic climate change and human-induced activities, such as hydroponics, use of fertilizers, urban landscaping, and mining, among others. Functional traits can be grouped into ensembles possessing hierarchical relationships: morpho-anatomical traits and physiological traits (Silvertown et al. 2015; Chapin III et al. 1993). Morpho-anatomical traits form the group at the top of the hierarchy, which are less plastic and have slow response times to fluctuating environments compared to physiological traits (Belluau 2018).

Plants are usually classified according to their water relations: mesophytes, xerophytes, halophytes, hydrophytes, and heliophytes. Mesophytes require abundant soil water and a relatively humid atmosphere (Kirkham 2005), while xerophytes are adapted to dry habitats (Lefor 1999). Hydrophytes (hygrophytes) depend on a large supply of moisture or grow partly or completely submerged in water (Kirkham 2005). Halophytes possess efficient salt-tolerance mechanisms and can complete their life cycles in naturally saline soils with NaCl contents exceeding 200 mM (Yuan et al. 2019). Lastly, heliophytes require full sunlight (Hale et al. 2005).

Plants are functionally diversified due to the many ways they respond to varying environmental gradients (i.e., climate change, heat, drought, temperature, etc.) (Huang et al. 2020; Calleja-Cabrera et al. 2020; Rundel et al. 2018; Stangler et al. 2017) and as exposed to human stress factors (i.e., land use change, invasiveness, fire, etc.) (Galapon et al. 2020; Rundel et al. 2018; Gratani 2014).

18.1.4 Mesophytes

Leaves play a key role in plant function and long-term environmental adaptation (Galviz and Valerio 2021). The characteristics that distinguish plants of the various habitats are most striking in the leaves (Kirkham 2005). Moreover, leaves show differences between plant species due to phylogenetic relationships and their adaptation to specific environments (Tian et al. 2016). Plants adjust their morphological, anatomical, and/or physiological traits in response to the deficit of resources. These adjustments are "plastic responses" and can facilitate plant acclimation to new environmental conditions (Schlichting 1986). Anthropogenic climate change exacerbated by human-induced activities is one of the causes of morphotype differentiation observed in plants responsible for inducing leaf-trait variation (Dong et al. 2020). Due to frequent flooding (as a result of deforestation), soil waterlogging reduces gas exchange between the soil and the atmosphere, leading to oxygen deprivation in the rhizosphere. In the study of Cardoso et al. (2013), *Brachiaria humidicola* was observed having aerenchyma in shoots and roots under drained conditions but were further increased under waterlogging. Superior growth of one accession of *B. humidicola* under waterlogged soil conditions is probably a result of morpho-anatomical traits acting together to enhance root aeration and shoot ventilation. A better understanding of plant drought responses was explained by Hernandez et al. (2021) in their study, which is essential to improve plant water use efficiency, productivity, and resilience to ever-changing climatic conditions. Growth, morpho-anatomical, physiological, and biochemical responses of *Quercus acutissima* Carruth., *Q. serrata* Murray, and *Betula schmidtii* Regel to progressive water-stressed (WS) were investigated. Larger vessels (i.e., >100 to ≤130) and tyloses were more frequent at WS after 8 weeks. Analysis showed that specific leaf area and stomatal pore are associated with WS seedlings of *Q. serrata* and *B. schmidtii*, while the tylosis frequency, total soluble sugar, and vessel diameter are associated with WS seedlings of *Q. acutissima*.

One of the human-induced activities that affect plants' morpho-anatomical characteristics is the practice of hydroponic systems. The use of hydroponic systems for cultivation in controlled climatic conditions for the specific environment helps improve crop growth and yield (Paradiso et al. 2017). De Sousa et al. (2020) cultivated *Hymenea courbaril* seedlings in a static hydroponic system and evaluated morpho-anatomical, physiological, and growth parameters over the course of seedling development. They observed an interesting relationship between the increase in the density and stomatal conductance, which reflected higher rates of photosynthesis. Similarly, leaves of *Glycine max* (L.) Merr. inoculated with plant growth-promoting organisms had changes in leaf anatomical traits (higher density of smaller stomata, thicker palisade parenchyma, and larger intercellular spaces in the mesophyll) that enhanced plant's photosynthetic activity, compared to non-inoculated one. Another human-induced activity that affects the mopho-anatomy of the plants is the foliar applications of fertilizer as an alternative for traditional fertilization. For example, foliar application of 50 ppm concentration of silver nanoparticles (AgNPs) had stimulatory effect on *Lactuva sativa* L. as shown on its improved stem epidermis thickness (44%), hypodermis (130%), and cortex cell area (372%). The morpho-anatomy of lettuce was also induced by the use of vegetal-based biostimulant and enhanced its photosynthetic activity in the early phase of growth using 5.0 mL L^{-1} (Carillo et al. 2022).

18.1.5 Xerophytes

Xerophytes exhibit certain adaptive mechanisms to complete their life cycle in dry environments (Atia et al. 2014). An endemic species in the Philippines, *Tectona philippinensis* Benth. & Hook.f., possessed anatomical characteristics typical of a xerophyte, such as the presence of four types of trichomes, extended and well-developed vascular system, multiple layers of palisade and sclerenchyma cells, and extension of vascular bundles to both non-glandular hairs on the adaxial surface and glandular hairs on the abaxial surface of the leaf (Hernandez et al. 2016). In the investigation of the leaf anatomy of *Jacquinia armillaris* plants in two different Venezuelan xeric shrublands, it was found to contain many traits of xerophytes, among them are thick cuticles, thick epidermis in both leaf surfaces, stomata only in the abaxial side, and epidermal depressions lodging a glandular trichome. Furthermore, the hypodermis is composed of one or two cell layers below the upper epidermis (Galviz and Valerio 2021). Hence, this kind of species must be considered for reforestation.

Many anatomical characteristics have been recognized as protective mechanisms that allow the plants to survive against various levels of environmental pressure (Hernandez et al. 2016). Variations in the root, xylem, and leaf anatomy, among others, have been observed in response to environmental stresses (Hernandez et al. 2016; Richardson et al. 2002; Stephanou and Manetas 1997). Morphological changes in response to various stresses may appear in the form of reduced growth of internodes, leaf size, leaf surface area, branching pattern, and shoot and root growth (Ojo and Shoele 2022; Park et al. 2022; Wang et al. 2020; Klem et al. 2019; Bano et al. 2019). Alteration at the anatomical level is mainly due to decreased cell elongation, restricted stimulation of cell division, and changes in differentiation conditions of cells, which result in varied anatomical characteristics of plant organs like roots, vascular tissues, and leaves (Bano et al. 2019). Stresses cause the production of reactive oxygen species, which cause oxidative stress, thereby inducing growth retardation (Zhou et al. 2020), disturbing photosynthetic apparatus (Pospišil 2016), and alteration in the ultrastructure of plant tissues (Wu et al. 2009), the permeability of plasma membrane (Demidchik et al. 2003), and regulation of stomatal movements (Sierla et al. 2016).

18.1.6 Halophytes

Halophytes are plant species with particular morpho-anatomical characteristics that allow them to complete their biological cycles in an environment with very stringent conditions (Pérez Cuadra et al. 2021). Morpho-anatomical adaptations or functional traits prevent water loss (Sánchez et al. 2021) with the dilution of salts absorbed in excess and with the excretion of salts (Lombardi et al. 2022; Rančić et al. 2019). These characters are mostly found in leaves and stems since they are the plant organs most rigorously exposed to environmental conditions. Specifically, among the most important characters are reduced leaves or their absence, a large number of glandular and/or glandular trichomes, presence of salt glands (Rančić et al. 2019), greater development of support tissues, presence of aqueous tissue, and development of atypical secondary growth, for example in the case of Cholistan flora (Wahid 2003). With few structural differences, they are similar functionally and associated with ions' secretion. Beyond these generalizations, it should be taken into account that no morpho-anatomical characters define all the halophytes; each species has adapted to the environment from the characteristics that developed within the evolutionary lineage to which it belongs (Lombardi et al. 2022).

As a result of anthropogenic climate change, the effects of elevated CO_2 are being studied due to its effect in offsetting physiological damages due to salinity stress in the plants, indicating that halophytic plants may gain from rising atmospheric CO_2 concentration in the scenario of climate change (Jothiramshekar et al. 2018; Yadav et al. 2018). Acclimation to variable CO_2 was studied by Huang et al. (2020) in floating leaves of the freshwater monocot *Ottelia cordata*. The most striking

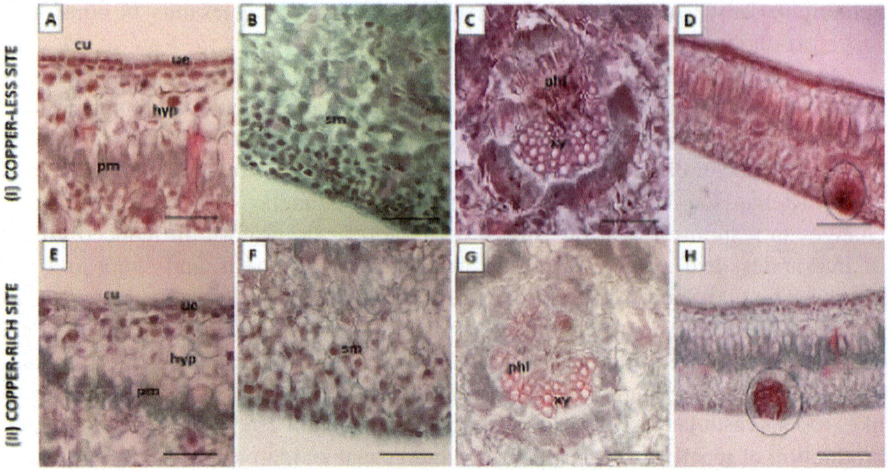

Fig. 18.1 Leaf cross-section of *Rhizophora mucronata* in (I) copper-less site and (II) copper-rich site showing the cuticle (cu), the upper epidermis (ue), hypodermis (hyp), and palisade mesophyll (pm) (I-A, II-E); the spongy mesophyll (sm) (I-B, II-F); vascular bundle (phl, phloem; xy, xylem) (I-C, II-G); and salt excretion cell (I-D, II-H), respectively. The bar represents 100 μm (source: Perez et al. 2021)

anatomical variations responding to high CO_2 included the enlarged upper epidermal cells and the decreased area of epidermal chloroplasts implying that *O. cordata* do not just rely on the atmospheric CO_2 for its inorganic carbon but also depend on CO_2 and bicarbonate in the water.

Mining is considered a human-induced activity that threatens the mangrove plant species because of heavy metal pollution. However, plants develop certain adaptations and mechanisms. For example, *Rhizophora mucronata* was found to contain thicker cuticular, upper epidermis, and hypodermal layers; thin mesophyll layers and vascular bundle; smaller and fewer xylary elements; and larger salt secretion cells indicating functional and adaptive traits in mangrove areas with copper (Fig. 18.1) (Perez et al. 2021). These structures may help plants tolerate the high concentration of heavy metals in the plant body and soil.

18.1.7 Hydrophytes

Aquatic plants are constantly subjected to a wide variety of environmental factors (Lusa et al. 2011), be it on a continental scale (e.g., latitude, longitude, and altitude range) (Chapuis et al. 2012; Santamaría et al. 2003) or at a local scale (e.g., light, salinity, flow velocity, area, rainfall, water resource, evapotranspiration, and temperature) (Chapuis et al. 2012; Santamaría 2002). Many widely distributed aquatic species can respond plastically to diverse conditions (Santamaría 2002). Studies showed that even clonal populations of aquatic plants with low genetic variability can exhibit compensatory plastic responses to environmental gradients (Pazzagalia et al. 2021; Bricker et al. 2018; Grace 1993). The ability of an organism to change its morphology and physiology in response to environmental conditions is known as phenotypic plasticity (Via et al. 1995).

In the study of Lusa et al. (2011), aquatic macrophytes (*Potamogeton polygonus* Cham. and Schltdl) showed great phenotypic plasticity and can occupy environments with different physicochemical conditions, such as a higher number of fiber bundles in the leaves to increase resistance to hydrodynamic forces, aerenchyma in the thicker central region to prevent water from damaging other parts of the plant, presence of Casparian strips to control the entry and exit of water, and thickening of the outer periclinal wall of the epidermal cells to support the plasmalemma area. In the Philippines, *Calophyllum inophyllum* L. (bitaog) with the natural habitat in the coastal area was found to contain thickest mesophyll. But when it is grown outside of its natural habitat, e.g., urban area for landscaping, its morpho-anatomical characteristics are altered by having smaller leaf length, leaf width, palisade mesophyll, and spongy mesophyll as affected by the climatic variables (Calibo et al. 2021).

18.1.8 Heliophytes

Since heliophytes prefer full sunlight, they have thicker leaves filled with palisade mesophyll (Shipunov 2021; Gotoh et al. 2018). Sun leaves are also smaller or more slender and have a greater mass per area and higher nitrogen content per unit area (Rozendaal et al. 2006), have higher volume of photosynthetic machinery per unit leaf area (Elser et al. 1996), and have higher growth rates (Freschet et al. 2011) than shade leaves. Gaps increase the diversity of forests across the landscape and present local environments that encourage the establishment and growth of new species, including bioinvasive species. A study in *Thunbergia grandiflora* Roxb. (Acanthaceae) or "skyflower," now engulfing natural vegetation and some forest plantations in Mount Makiling, Philippines, showed that it could survive intense light conditions but is suppressed under partial to fully shaded treatments (Galapon et al. 2020). The leaf and stem anatomical structures of *T. grandiflora* have characteristics typical of invasive species that can adapt to certain harsh conditions. This includes the presence of dense non-glandular trichomes, cystolith, multi-layered collenchyma cells in the hypodermis, sclerenchymatous tissues, wider xylem pores, and closely packed parenchyma cells in the stem.

Due to extensive mining activities, strip vegetation and soil contamination with heavy metals have become common which affect the morpho-anatomy of some species thriving in the area. In the study of Interior-Hallegado et al. (2022), six species of grasses—namely, *Alloteropsis semialata* (R.Br.) Hitchc., *Cymbopogon citratus* (DC.) Stapf, *Dinochloa acutiflora* (Munro) Soenarko, *Imperata cylindrica* (L.) Raeusch., *Machaerina disticha* (C.B. Clarke) T. Koyama, and *Panicum repens* L. Leaf anatomical features in the transverse section—such as thickened epidermal cell walls, presence of sclerenchyma caps, extended vascular bundles, and bulliform cells—were observed in the leaves. Root anatomical features observed were thickened epidermal cell walls, thick endodermis, and a large cortex filled with intercellular spaces (*P. repens, D. acutiflora*, and *M. disticha*), as well as black spots in the cortex and pericycle (*A. semialata*), suggesting *A. semialata*, *D. acutiflora*, and *P. repens* suitable for phytostabilization.

18.1.9 Morpho-Anatomical Traits and Plant Community Assembly Along Environmental Gradients

Environmental gradients, defined as gradual changes in biotic or abiotic environmental factors, are essential determinants of the structure and functioning of ecological systems and their components. Traits whose values vary predictably in response to such gradients are called "response traits"; these differ according to the gradient considered (Garnier et al. 2015). On the other hand, community assembly processes are thought to shape the mean, spread, and spacing of functional trait values within communities (Cornwell and Ackerly 2009). According to Subedi et al. (2019), a combination of plant architecture, wood density, and leaf traits (SLA, leaf phosphorus, and leaf nitrogen) comprised a key set of functional traits that are important for understanding the community assembly process in dry tropical forest. In the coastal California landscape with topographically mediated gradient in water availability, strong shifts in both the plot-level mean trait values and the variation in trait values within communities were observed by Cornwell and Ackerly (2009).

In terms of morpho-anatomical traits, leaf anatomical traits are considered sensitive and adaptable to environmental changes. In the study of Liu et al. (2021), different elevational patterns of leaf anatomical traits occurred between the species and community levels. Similarly, variation of functional (morphological and physiological) traits in different tree species in the Fandoghlo forest region (NW Iran) was observed as affected by altitude and light regime (Jahdi et al. 2020). Plant anatomy is also known to vary in time and space (Anderegg and Meinzer 2015). A large number of studies have quantified anatomical and physiological differences in species in contrasting environments, especially along moisture gradients (Alder et al. 1996; Maherali and DeLucia 2000). Ciccarelli and Bona (2022) emphasized the usefulness of anatomical traits (including root system traits) in understanding the functional strategies adopted by plants.

18.1.10 Morpho-Anatomical Trait–Functioning Relationships

18.1.10.1 Links Between Aboveground Traits and Plant and Ecosystem Functioning

The trait composition and trait diversity of plant communities are globally applicable predictors of ecosystem functioning (Happonen et al. 2022). Aboveground traits of plant species and the functional composition of plant communities have been found to vary with LES (Wright et al. 2004), including as SLA, leaf nitrogen content and leaf dry matter content (LDMC) (Bruelheide et al. 2018; Díaz et al. 2016), and plant size, characterizing the trade-offs between investments to light competition, and photosynthesis and reproduction, which can be measured with, for example, plant vegetative height or stem-specific density (Díaz et al. 2016).

According to Happonen et al. (2022), plant height, SLA, and LDMC have clear effects on tundra ecosystem functions differently. Variability in SLA and LDMC increased CO_2 fluxes and soil organic carbon stocks, while variability in height increased the aboveground carbon stock. In the Northern Tibetan alpine grasslands (Zhu et al. 2016), it was found that both species and functional diversity were positively related to the aboveground biomass production.

18.1.10.2 Links Between Belowground Traits and Plant and Ecosystem Functioning

The effects of plants on the biosphere, atmosphere, and geosphere are key determinants of terrestrial ecosystem functioning (Freschet et al. 2021). However, despite substantial progress made regarding plant belowground components (cite references here), we need to explore the complex relationships between root traits and functions. According to Freschet et al. (2021), belowground traits with the widest importance in plant and ecosystem functioning are not necessarily those that are the most commonly measured.

According to Laliberté (2017), trait-based approaches have led to significant advances in plant ecology, but are currently biased toward aboveground traits. A stronger emphasis on belowground traits is needed to better predict future changes in plant biodiversity and their consequences for ecosystem functioning. He proposed "belowground frontiers" in trait-based plant ecology, with an emphasis on traits governing soil nutrient acquisition: redefining fine roots; quantifying root trait dimensionality; integrating mycorrhizas; broadening the suite of root traits; determining linkages between root traits and abiotic and biotic factors; and understanding ecosystem-level consequences of root traits.

18.1.10.3 From Morpho-Anatomical Traits to Forest Landscape Structure

Understanding variation of morpho-anatomical traits at multi-scales (forest landscape structure) along different environmental gradients can help predict the likely responses of dominant plant species to climate change (Lin et al. 2021). For example, plant leaf traits at the cell level (e.g., stomatal traits, mesophyll cells, etc.), tissue level (anatomical traits), and organ level (morphological traits) may be especially useful in relation to climate as they reflect aspects of carbon acquisition, water use, and gaseous exchange (Liu et al. 2021; Baillie and Fleming 2020; Muir 2020). At the cell level, stomata are the essential structures for the leaf to exchange gasses with the atmosphere (Lin et al. 2021; Baillie and Fleming 2020; Muir 2020). The stomatal movement and conductance directly regulates the balance of carbon gain (photosynthesis) and water loss (transpiration) (Ainsworth and Rogers 2007). At the tissue level, palisade cell shape affects the light-induced chloroplast movements and leaf photosynthesis (Gotoh et al. 2018). Also, thicker spongy mesophyll enhances the drought and cold resistance of plants in temperate forests, thus adapting to low temperature and moisture (Liu et al. 2021). At the organ level, specific leaf area is well known to positively associate with the plant's relative growth rate (Lambers and Poorter 1992).

Several studies have been conducted to study the relationship between morpho-anatomical traits and reproduction (Glubsheva et al. 2021), phylogenetics diversity, and ecosystem dynamics (Fagundes 2019), and environmental gradients (Lin et al. 2021; Fagundes 2019). The revealed morpho-anatomical features of the structure of the *Tulipa biebersteiniana* flower in the south of the Central Russian Upland suggest that it retains the characteristics that support seed reproduction, despite the low real seed productivity during several years of observations. Fagundes (2019) concluded in her study that Leguminosae structure was strongly related to altitude as well as phylogenetic distance. In the study of Lin et al. (2021), variations in morpho-anatomical traits were found correlated with temperature, precipitation, and soil pH. Further, dominant oak species in tropical forests respond to environmental change by modulating traits at multiple levels, from that of the individual cell, through tissue and up to the whole leaf scale.

18.2 Anthropogenic Activities and Genetic Changes in Plants

18.2.1 Plant-Human Co-evolution Through Domestication

The co-evolution of plants and animals can be traced as early as the Holocene, around 12,000 years ago (kya) after the last glacial period, through agriculture and the consequent existence of domesticated plants (Fuller et al. 2014; see also Altman and Mesoudi 2019; Przelomska et al. 2020). Accordingly, the process of crop domestication has been a long-standing topic of active research not only among biologists/agriculturist and the like but even among anthropologists. Larson et al. (2014) stated that it is difficult to overstate the cultural and biological impacts of plants and animal domestication on species.

The origin of crop domestication, including whether there is evidence of multiple, geographically distinct domestication events, and the tempo of the domestication event(s) can be traced using both archeological and evolutionary genetic techniques. These combined techniques have revolutionized our understanding of the pattern and process of domestication as well as the origins of agriculture that paved the way to our modern way of life (Larson et al. 2014). Specifically, archeological science has successfully complemented the traditional knowledge and explored in-depth evolutionary and cultural processes underlying domestication, and has been valuable for the chronology and composition of past agroecosystems, domestication syndrome, and elucidating past diets (Purugganan and Fuller 2009; Smýkal et al. 2018; reviewed in Przelomska et al. 2020).

In this section, however, our main aim is to relate anthropogenic activities and plants' responses to these activities, particularly to domestication, in a molecular and genetic perspective. With the ancient DNA revolution in the 1980s, archeological plant remains and herbarium specimens have been analyzed with molecular techniques to probe the evolutionary interface of plants and humans (Przelomska et al. 2020). Since then, it has offered tremendous insights into the co-evolution of plants and people.

Although agriculture is believed to have begun as an attempt to modify the landscape, encouraging the growth of edible wild plants at the expense of less useful ones (Smith 1998 as cited in Doebley et al. 2006), human's intention to domesticate crops has led to the so-called domestication syndrome (Smýkal et al. 2018)—a condition by which traits can be developed or altered as a result of human preferences to meet their needs (e.g., increased palatability and recently for nutrition for food crops, increased yield or growth advantages, ease of harvest, among many others) such that these domesticated crops can thrive best only under human care (Doebley et al. 2006). "The pathways to agriculture were prolonged episodes of co-evolution, genetic adaptations on the part of the plants, and cultural shifts and innovations on the part of people" (Fuller et al. 2014). By far, the number of domesticated plant species have been estimated at 160 taxonomic families, of which approximately 2500 species have undergone some degree of domestication and 250 species are considered fully domesticated (Smýkal et al. 2018; see Fuller et al. 2014).

Altman and Mesoudi (2019) argued that the emergence and impact of agriculture can be understood within new theoretical frameworks arising within the evolutionary human sciences. "The study of agriculture can benefit from being situated within a set of new evolutionary approaches to human behavior–cultural evolution (CE), gene-culture co-evolution (GCC), and cultural niche construction (CNC)." These approaches, accordingly, incorporate cultural change and individual agency within a rigorous scientific and multidisciplinary evolutionary framework. The authors also pointed out that changes in agricultural knowledge and practices are a prime example of cumulative CE (CCE) and GCC, and agriculture, in part, is associated with an extensive CNC and an excellent case of CCE since it involves the long sequential improvement of agricultural knowledge and practices through repeated cycles of innovation and cultural transmission. They furthered that due to the fact that agricultural practices transform the environment, agriculture can modify the biotic and abiotic and the social environments (e.g., social norms, regulation, markets)—and these changes alter the selection pressures on agricultural CCE, making agriculture one of the prime causes of the Anthropocene. They also emphasized that recognizing domestication, breeding, and other agricultural advancements as outcomes of plant gene-human culture co-evolution is essential to public acceptance of newly developed genetically modified plants. Furthermore, molecular breeding is also a case of GCC.

18.2.2 Plants' Genetic Responses to Domestication, Breeding, and Improvement

"The study of domesticated species has led to increased interest in several important issues in genetics and evolutionary biology, including the underlying genetic architecture of adaptations and parallel evolution" (Larson et al. 2014). Let us first look at how domestication affects the genetic processes in plants by reviewing the basic concepts on how genetic variations arise in a population by the actions of evolutionary factors. This will help us understand how selection for the desired trait during domestication and breeding or any human intervention can cause significant effects on the genetic structure of the population of a species. Genetic data have been powerful in drawing out inferences on the geographic origins of domesticated crops, the numbers of independent domestication events for a certain species, the specific molecular changes underlying domestication traits, and the nature of artificial selection during domestication and subsequent improvement (Gross and Olsen 2010). Assuming there is an absence of several forces that disturb the equilibrium acting on a population, a population in Hardy–Weinberg equilibrium has balanced gene frequencies such that the progenies will have the same allele and genotypic ratios as the parent population. However, when an evolutionary force such as genetic drift, a change in the population's allele frequency resulting from a random variation in the distribution of alleles from one generation to the next, acts upon a population particularly in small ones, rare alleles can disappear completely, reducing the overall genetic variation in the population. In the same way, phenotypic selection (natural and artificial through domestication and breeding) can also lead to the rapid loss or fixation of alleles within a population (see also Flint-Garcia 2013). Although it can be counterbalanced by gene flow due to the migration of individuals between populations by bringing together alleles from each population, it will negate the primary objectives of domestication and breeding for the desired traits.

Selection driven by human preferences, cultivation practices, and agricultural environments, as well as other population genetic processes that reduce effective population size have shaped the outcomes of plant domestication. In general, it is a known theory that any selection, natural or human-induced, imposes a reduction in the diversity, favoring genotypes of traits that are heavily selected, for instance, toward economic goals. Genome sequencing has provided evidence that human-induced selection is evident during and after domestication. Molecular level analyses of population genetic structure have also elucidated the demographic histories of the process of domestication, illuminating the origins of crops. For some species, this was not detected using morphological or archeological evidence.

The early agricultural practices of clearing and burning of lands and re-planting of favored plant species such as food crops (or even higher plants such as trees) have left their genetic signatures on the patterns of genetic diversity in the genomes of the species (Doebley et al. 2006). Due to the limited number of individuals used as starting materials in the domestication, reduced genetic diversity has been generally traced in domesticated crop populations, while a large amount is left behind in the progenitor species. This is because the allelic diversity in the domesticated crop is expected to be a subset of that found in the wild population(s) from which it was derived. Genetic studies have examined that if populations of the wild progenitor existed in the area where domestication occurred, the geographic origin(s) of the domesticated crop can be tracked to a particular population or region.

Modern agricultural practices have greatly reduced the effective population sizes of crops, allowing genetic drift to alter genotype frequencies. Further reduction in diversity throughout the genome of domesticated crops results from bottleneck due to the subsequent use of seeds from the best plants in the succeeding generation throughout the domestication process (Doebley 1989). However, the rate of reduction in genetic diversity is highly dependent on how much population size or how long a period is involved in the domestication (Eyre-Walker et al. 1998; see also Doebley et al. 2006). Although population bottlenecks are a common important demographic event during domestication as it reduces diversity in neutral genes (those that do not influence favored genotypes), selection forces tend to lose more genes so that one or a few genes influencing the desired phenotype is retained in the plant after domestication (e.g., weaker defense response to wounding and insect feeding in domesticated tomato, Liu et al. 2018). In plants several domestication genes have been already identified compared to animals due to the relative ease of discovering the genetic basis of domestication in plants (Larson et al. 2014).

Two approaches to understanding the genetic changes underlying crop domestication and improvement have been examined in the review of Doebley et al. (2006) (see also Ross-Ibarra et al. 2007). One is a classical genetic approach of starting with phenotype and working back to the gene and the other is a population genetic approach of starting with genes and determining whether these genes were targets of selection. Based on these approaches the review arrived at four major conclusions: (1) the major genes controlling morphological and structural changes during domestication are transcriptional regulators, suggesting their central role in domestication, (2) many of the different sets of genes that contribute to varietal differences harbor early stop codons or other lesions that disrupt the coding sequence, suggesting loss of function mutations, (3) transcriptional regulators are over-represented for morphological traits and flowering time, while basic enzymatic genes in the biosynthetic pathways predominate for the nutrient composition of seeds and fruits; and (4) some of the genes under selection for morphology or physiology may have never been considered as such during domestication and improvement.

Relative to the earlier mentioned review, the identification of the specific genes that control some of the most important morphological changes associated with domestication traits as well as varietal differences started since the 1990s with the mapping of the quantitative trait loci (QTL) in progenitor crop hybrid populations. The application of the QTL mapping and the subsequent characterization of functional mutations at some domestication genes had shown that the suite of changes associated with plant domestication can have a relatively simple genetic basis. QTLs are discussed extensively in numerous reviews and original papers (e.g., Doebley 2004; see also review of Doebley et al. 2006; Meyer and Purugganan 2013). Among the QTLs found are for economic traits of major agricultural crops, while most traits of economic interest in forest trees do not fall into discrete phenotypic classes but instead result from the collective action of multiple genes which exhibit quantitative variation.

Aside from QTL mapping, recent developments on genome-wide association studies (GWAS) and whole-genome resequencing studies (WGS) have identified genes associated with domestication traits and subsequent diversification of crops (Meyer and Purugganan 2013). Accordingly, "these studies, altogether, reveal the functions of genes that are involved in the evolution of crops that are under domestication, the types of mutations that occur during this process, and the parallelism of mutations that occur in the same pathways and proteins, as well as the selective forces acting on these mutations and are associated with geographical adaptation of crop species."

18.2.3 Plants' Genetic Responses to Introduction, Invasion, and Climate Change

Scheffers et al. (2016) had comprehensive accounts of the broad footprint of climate change from genes to biomes to people. They demonstrated that most ecological processes recently show responses to anthropogenic climate change. Whether in terrestrial or marine ecosystems, species are changing genetically, physiologically, morphologically, and phenologically, and these changes can lead to shift in species distribution and consequently affect food webs, which may later form new interactions. With climate change, species are able to undergo evolutionary adaptation and climate change has substantial impacts on species physiology. This includes changes in tolerances to high temperatures, changes in sex ratios in species with temperature-dependent sex determination, and increased metabolic costs of living in warmer ecosystems. On a broader scale, plants' responses to climate change include phenological changes, abundance, and species distribution. For example, temperate plants are found to initiate early bud formation and flowering during spring and later in autumn. There is also documented widespread evidence of range expansion in warm-adapted species and range contraction in cold-adapted ones. Scheffers et al. (2016) argued that while these changes are taking place, many cryptic dynamics such as genetic changes operate simultaneously. Hence, an understanding of these changes in ecological processes can help humans strategize adaptation from climatic changes.

Natural populations are under threat from a variety of anthropogenic stresses, including habitat fragmentation and degradation, pollution, and invasive species. These could impede dispersal (Kremer et al. 2012; Anderson et al. 2012), change selection patterns on complex traits (Franks and Hoffman 2012), and reduce genetic variation (Jump and Peñuelas 2006; Lowe et al. 2005), thereby limiting evolutionary responses to climate change. Climate change will continue to alter distribution patterns; it will directly cause the introduction of exotics into new geographic areas by driving shifts in species ranges; invasive species will become more prevalent, and phenological changes will alter species interactions (Parmesan and Hanley 2015; Bennett et al. 2015; Revilla et al. 2015).

Natural plant populations face extinction if they do not respond to climate change by migrating to more favorable climates, adapting to novel conditions, or shifting their phenotypes (plasticity) (Aitken and Whitlock 2013; Nicotra et al. 2010). Historically, strategies that allow species to persist in changing environments were classified as ecological (distribution shifts and phenotypic plasticity) or evolutionary (adaptation and gene flow) (Anderson et al. 2012). To mitigate the extinction risks posed by rapid contemporary climate change, most species will almost certainly need to employ a combination of strategies. The ability of species to track favorable conditions by shifting distribution ranges (Chen et al. 2011), acclimate via phenotypic plasticity (Boyd et al. 2022; Parmesan and Hanley 2015; Nicotra et al. 2010), and evolve adaptations to novel stresses (Hoffmann and Sgrò 2011) is critical to their persistence in the face of contemporary climate change.

Climate change has an impact on many organisms and, in some cases, can cause evolutionary change, altering the gene and allele frequencies in populations (Franks and Hoffman 2012). Allele frequencies in populations of species can fluctuate rapidly at relatively short generation times (Hoffmann and Willi 2008). Neutral genetic markers can detect the changes especially when stressful conditions cause a population decline. However, fluctuations in allele frequencies triggered by specific conditions (stress) are dependent on isolating sets of genes that are associated with adaptive responses. Some of these are related to various environmental stresses such as pollution, global warming, and others. Recent research has identified a number of candidate genes involved in climate change adaptation, as well as evidence that genetic regulatory networks and epigenetic effects (Miryeganeh 2022) may be particularly important for climate-change-driven evolution. Examining evolutionary responses to climate change can help determine the genetic basis of adaptive evolution. Genome scans (Prunier et al. 2011), transcriptome comparisons (Wu et al. 2021), candidate gene comparisons (Haas et al. 2021), and genetic mapping (Brachi et al. 2010) are some of the genetic strategies used to investigate the genetic basis of climate change adaptation. Furthermore, the methods could make use of a wide range of possible source material, such as sampling along latitudinal or altitudinal gradients and before and after a natural or artificial environmental change (Franks and Hoffman 2012).

18.2.4 Changes in Phenological Patterns in Plants

Climate change allows researchers to investigate the genetic basis of adaptation. For instance, the timing of flowering is an important research area for advancing our understanding of the genetic basis of climate change adaptation. Flowering time is a complex trait in plants that is tightly controlled by a genetic network that incorporates environmental cues (Fernandez-Calleja et al. 2021). It is well understood at the molecular level and is highly relevant for genetic adaptation to climate change. It is considered to be ideal for studying the genetic basis of evolution under climate change because of the large amount of information already known about the genes, pathways, and processes involved in determining flowering time (Michaels

2009; Wellmer and Riechmann 2010). According to Michaels (2009), flowering time in *Arabidopsis* is controlled by a genetic regulatory network that includes four major pathways: photoperiod, temperature, gibberellin, and autonomous. Internal signals (such as gibberellins) and external signals (such as photoperiod and temperature) influence these pathways, preventing plants from flowering too early or too late. The regulatory network is quite complex, with hundreds of genes, transcription factors, and receptor proteins, as well as gene expression variation and epigenetic regulation. Despite the complex regulatory network, only a small number of genes appear to vary in the wild and influence the flowering time phenotype in natural populations; these are considered prime candidate genes for investigating the genetic basis of flowering time variation and evolution across taxa in natural conditions (Lempe et al. 2005). Many species have shifted their phenology (the seasonal timing of reproduction and other life history events) (Parmesan and Yohe 2003). Franks et al. (2007) discovered that flowering time in the annual plant, *Brassica rapa,* was heritable. When ancestors and descendants were grown together under common conditions, early flowering plants had greater fitness under drought conditions than late flowering plants, and average flowering time was earlier in plants following a natural drought than in plants before the drought. Nonmodel plants, such as the pearl millet, *Pennisetum glaucum*, have also been found to have flowering time genes that are linked to climate conditions (Mariac et al. 2011). A genome scan was initially used to identify a signature of selection along a rainfall gradient in a MADS-box gene, PgMADS11, and then association mapping was used to link polymorphism in this gene to flowering time variation in a separate population. It was discovered that in many other species, flowering time is heritable, is influenced by climatic factors, varies across climatic clines, and can evolve in response to changes in climatic conditions (Franks and Hoffman 2012).

Fruit tree phenology in temperate regions has changed in various parts of the world as a result of global warming. Chilling hours, a major determinant in tree phenology in temperate regions, have decreased, resulting in significant yield reductions in several species (Ramirez and Kallarackal 2015). Furthermore, phenological observations revealed that bud break, flowering, and fruiting have all advanced by several days in most fruit trees. Also, in the subtropics, most studies have shown that flowering time in fruit trees is advancing. On the other hand, in the tropics, precipitation is a major factor regulating the phenology and yield in fruit trees. Phenology in tropical ecosystems may be less sensitive to temperature and photoperiod and more sensitive to seasonal shifts in precipitation (Kallarackal and Roby 2012; Kallarackal and Renuka 2014). Many fruit trees' reproductive behavior has been disrupted as a result of the change in rainfall pattern. Climate change, according to Sthapit et al. (2012), will have both positive and negative effects on tropical fruits. Further temperature increases will have a negative impact on fruit yield and quality in regions where temperatures are already high. Temperature increases may be beneficial in regions where cold temperatures are one of the primary factors limiting crop production (Sthapit et al. 2012).

Higher temperatures caused by global warming are responsible for a decrease or increase in phenological cycles in trees. Global warming has impacted horticultural fruit tree phenology in the past, as evidenced by apple (Hoffmann and Rath 2013), pear (Guédon and Legave 2008), peach (Luedeling et al. 2009), cherries (Primack et al. 2009); mangosteen (Boonklong et al. 2006), coffee (Paes de Camargo 2010), and mango (Ramírez and Davenport 2012). According to Ramírez et al. (2014), mango trees do not have a distinct phenology, but instead exhibit a series of easily identifiable stages that occur independently on individual stems and are closely related to temperature, the age of the previous vegetative flush of that stem, precipitation, and dry seasons, all of which are important events in the tropics. Climate change, on the other hand, is known to have an impact on mangosteen (*Garcinia mangostana* L.) production in Thailand. According to Boonklong et al. (2006), mangosteen production in Thailand's eastern region increased as the drought period before flowering lengthened. As a result, mangosteen production should be higher in a year with a longer drought period (Boonklong et al. 2006) and this is most likely due to the minimal temperature increase that has occurred in the eastern region over the years.

Plant phenology changes have significant implications for ecosystem carbon cycles and ecosystem feedback to climate. Climate change has shifted the timing of major phenological events, such as the widely reported spring advancement and autumn postponement (Piao et al. 2019; Gill et al. 2015), and has caused phenological mismatches across trophic levels (Renner and Zohner 2018). Some tree species' seed production is reduced as a result of their different phenological responses to warming (Kudo and Ida 2013).

Genetic diversity could influence phenological responses to climate change. This was exemplified in the study of Doi et al. (2010) on 11 species across Japan where they estimated the regional variation across Japan in flowering and leaf budburst dates of plants based on a dataset of phenological timings from 1953 to 2005. The observed plants' genetic diversities varied according to human cultivation. The within-species variations of phenological response to temperature as well as regional variations in the responses of flowering and leaf budburst of plant species with lower genetic diversity are remarkably smaller than those with higher genetic diversity. Thus, genetic diversity influences the variation in phenological responses of plant populations. Under increased temperatures, low variation in phenological responses may allow drastic changes in the phenology of plant populations with synchronized phenological timings. This is a first step toward considering the relationships between genetic diversity and the effects of climate change on populations. It is imperative to maintain genetic diversity within species to sustain phenological resilience (Doi et al. 2010).

18.2.5 Changes in Patterns of Genetic Diversity in Plants

While many efforts are made towards addressing the impacts of climate change on biodiversity, most studies considered species as a unit, ignoring the intraspecific genetic variation, when in fact the short-term and long-term survival of the species is heavily dependent on how much genetic diversity is maintained within a species.

Among the documented responses of species to climate change are local adaptations, range shift, range reduction, or a combination of these (reviewed in Alsos et al. 2012). The loss of range for many species is one great impact of climate change. Range shift may alter the genetic diversity within species while range reduction is most likely to result in loss of genetic diversity, thus severely limiting the ability of the species to adapt to a changing climate. This tendency has been observed in 1200 populations of 27 northern plant species (Alsos et al. 2012). Loss of genetic diversity varied considerably among species, and this variation could be explained by dispersal adaptation (up to 57%) and by genetic differentiation among populations (up to 61%).

Plant populations are expected to face strong directional selection pressures as a result of climate change (Jump and Penuelas 2005; Anderson et al. 2012). By permitting evolutionary responses, genetic diversity may play an important role in allowing individual species to resist climate change. Plant populations may adapt in situ through selection on standing genetic variation (Jump et al. 2008a), and this adaptive response may be an important component of species' resistance to climate change because it provides an in situ "option" for persistence despite environmental change (Jump et al. 2008a).

Climate-related genetic responses have been identified using genomic and quantitative genetic approaches, for example, on the Mediterranean shrub *Fumana thymifolia* (Jump et al. 2008b) as well as the annual plant *Brassica rapa* (Franks 2011). Both studies have found that, despite potentially high levels of gene flow, climatic factors can act as potent forces of selection, driving adaptive differentiation between and within populations at fine spatial scales (Jump et al. 2008b; Franks 2011). Ravencroft et al. (2015), on the other hand, investigated species' genetic responses of two co-existing, and the most abundant perennial plant study species (*Festuca ovina* and *Plantago lanceolata*) within the community to 15 years of simulated climate change treatments (involving summer drought, increased temperature, and enhanced rainfall). The aims were to determine the extent to which grassland plants can respond genetically to climate change and whether these responses are consistent with a process of evolution in response to climatic selection. The analysis of DNA markers in plants revealed that climate change treatments had changed the genetic composition of plant populations. The findings also revealed an evolutionary change process in *P. lanceolata*, implying that genetic diversity may be able to protect plants from the harmful effects of climate change, allowing for an "evolutionary rescue" (Ravencroft et al. 2015).

Recently, the study of Franks et al. (2016) demonstrates how a plant species can rapidly change at the genetic level because of harsh environmental conditions caused by climate change. It was the first documented case of a natural plant population rapidly evolving in response to a natural climate change, and the changes occurred in just 7 years (Goshier 2016). The study discovered that field mustard, *Brassica rapa*, evolved to flower earlier in response to a drought that lasted from 1997 to 2004 in Southern California. To determine how the post-drought generation differs genetically, the researchers sequenced the genomes of pre-drought and post-drought plants from the first study. They discovered that in response to the drought, hundreds of genes had rapidly evolved, many of which were involved in traits such as flowering time and stress response. Plants that flowered earlier were either more likely to survive the drought or produced more offspring. The study demonstrated the value of comparing generations of species over a shorter time period to study genetically based changes. The study also demonstrates how climate change can either hasten genetic changes in species or endanger other species that cannot evolve as quickly. The fragmentation of plant habitats resulting from greater urbanization may impede evolution, since smaller plant populations have less of the genetic variation that allows hardier members of the species to emerge (Franks et al. 2016).

Furthermore, the anthropization of the Cerrado biome's landscape in recent decades has fragmented its natural environments, affecting the connectivity of plant populations and altering gene flow (Vitorino et al. 2020). Plant species may also reduce population size in response to climatic and environmental conditions that are less than ideal. The scientists investigated whether landscape features and climate characteristics affect the genetic diversity and inbreeding coefficients of Cerrado plant species, and whether the distribution of genetic diversity of these plants is consistent with the center-periphery model. They examined 121 populations of 31 plant species using generalized linear mixed models (GLMM) and multiple matrix regressions (MMRRs) to assess the effects of landscape features and climatic variables on the observed (HO) and expected heterozygosity (HE), allelic richness (AR), and inbreeding (Fis) patterns of the local populations. The landscape was assessed in terms of the percentage land cover of agriculture (AG), forestry (FO), remnant vegetation (RV), urban areas (UA), pasture (PA), and water (WA) within 1 km, 3 km, and 5 km buffers around the study populations. The study found that the genetic diversity of many Cerrado plant species is distributed in a typical center-periphery pattern, with greater heterozygosity and lower levels of inbreeding in populations closer to the biome's center. These plants are directly influenced by the landscape matrix in which they grow, specifically the percentage of forestry cover, water, and urban areas. Because anthropogenic changes to the landscape can affect gene flow, they will ultimately determine the levels of genetic heterozygosity and inbreeding found in a population. The mean annual temperature also influenced genetic diversity in Cerrado plant species, indicating the influence of the optimal temperature range for the occurrence and genetic diversity of the species. Overall, the study indicates the need for systematic management of the effects of both climate change and shifts in landscape characteristics to minimize the impacts on the genetic diversity of Cerrado plant species (Vitorino et al. 2020).

Lasky et al. (2014) investigated how plants adapt to harsh climate conditions in order to better predict and manipulate how plants will respond to climate change. *Arabidopsis thaliana* (Brassicaceae) provides an opportunity to investigate ecological drivers of natural variation from molecular to continental scales because it thrives in a wide range of environments, from northwestern Africa to Scandinavia and Siberia. *Arabidopsis* has a relatively small genome (about 25,000 genes) and was the first plant to be fully sequenced in 2000 (The Arabidopsis Genome Initiative 2000). Lasky et al. (2014) investigated gene expression or how snippets of material in the plant's DNA create proteins and other chemical instructions that help regulate cold tolerance, and how gene expression can change in response to certain conditions (Funkhouser 2015). They discovered evidence that evolutionary changes in gene expression allow populations to perform well in harsh environments, and that genes sensitive to cold and drought help plants adapt to environmental changes. Researchers may be able to manipulate or improve plants in an agricultural setting if they have a better understanding of how plants have evolved tolerances to natural environmental stresses (Lasky et al. 2014).

In many species, including soybean, wheat, rice, peanut, and bean, the increase in shoot biomass is associated with significant increases in seed yield (reviewed in Hatfield et al. 2011). Soybean (*Glycine max*) is a well-studied model of whole plant growth and reproductive output responses to elevated CO_2. Gray and Brady (2016) created a comprehensive summary of the effects of climate change factors such as elevated CO_2, elevated temperature, and drought on soybean development stages such as leaf development, root development, flowering, seed yield, and genotypic variation. Improving our mechanistic understanding of plant developmental responses to multiple, interacting climate change factors will be critical for forecasting effects on agricultural and natural systems.

Despite the recent focus on the potential effects of climate change on genetic diversity, little research has been conducted on the effects of climate change on genetic connectivity and the relationship between climate stability and genetic divergence. Guan et al. (2021) use a combination of population connectivity and genetic data to investigate the effects of future climate change on genetic connectivity. They used POPS software, which predicts population genetic structure using latent regression models, along with climatic variables and genetic data to create spatially explicit simulations and predict the dynamics in genetic clusters in response to climate change. A generalized additive model was used to investigate the relationship between climatic stability and genetic diversification. Guan et al. (2021) showed how to forecast the genetic vulnerability of species and the persistence of local populations in the face of future environmental change by combining landscape connectivity predictions with genetic simulations. Their findings suggest that the current accelerated climatic changes may cause the range of the small deciduous tree *Cornus kousa subsp.chinensis* to contract, affecting genetic connectivity across the landscape and potentially leading to a significant loss of genetic variation. Because of the loss of fundamental evolutionary potential caused by climate change, the displacement of climatic genetic clusters will make species adaptation to new environmental conditions more difficult (Guan et al. 2021).

Genetic studies are being used to analyze the effects of climate change and to aid in the prediction and development of climate change mitigation measures. It is particularly interesting to learn whether the effects of climate change have penetrated deep into organisms, to the level of their genes. Plants and animals are not defenseless against climate change because they can respond in three ways: migration, acclimatization, and adaptation (Rejon 2021). First, there is migration from areas with increasing temperature levels to colder areas in terms of latitude or altitude. Raf et al. (2018) used nuclear microsatellite markers to clarify genetic evidence of human-mediated historical seed transfer of larch (*Larix decidua*) forest from Alps to Carpathian. Iwaizumi et al. (2018), on the other hand, used nuclear microsatellite markers to assess the genetic diversity of cultivated Japanese black pine populations spread across Japan, and suggested potential extension from southwest refugia to northeast following glaciation. Red pine populations' potential range expansion from glacial refugia to current range in North America's north east has also been reported (Boys et al. 2005).

Acclimatization and adaptation are two other ways that organisms respond to global warming (Rejon 2021). The role of omics techniques derived from genetics (i.e., genomics, transcriptomics, proteomics, etc.) in dividing phenotypic changes (i.e., morphology, physiology, behavior, etc.) caused by climate change into acclimatization (differences in the number of certain RNAs and/or proteins) and adaptation (there are also DNA differences, and new types of RNAs and proteins may appear). Most of the research indicates that possible changes in organisms caused by climate change are acclimatization; there is little evidence of true adaptation (Rejon 2021). The appearance of these possible genetic adaptations as a result of climate change is being studied in organisms with short life cycles, such as insects like flies, mosquitoes, and butterflies, whose populations have shown possible changes in genes related to heat stress, heat shock, or cell death apoptosis, though the changes appear to be polygenic (Rejon 2021).

Climate change's impact is frequently assessed using expected changes in species' geographical distributions. Genomic diversity, local adaptation, and migration are rarely considered when projecting future species. Aguirre-Liguori et al. (2019) investigated the effects of climate change on the populations of two wild relatives of maize, the *teosintes Zea mays* ssp. *mexicana* and *Z. mays* ssp. *parviglumis*. Throughout their evolutionary history, cultivated landraces of maize (Zea mays ssp. mays) have had a close relationship with their wild relatives, the teosintes. The researchers used ecological niche modeling and climate averages from 1960 to 1990 to create geographical distribution models for the two-teosinte subspecies based on eight climate change projections (Aguirre-Liguori et al. 2019).

Despite high levels of genetic diversity within populations and widespread future habitat suitability, they predict that climate change will alter patterns of local adaptation and reduce migration probabilities in more than two-thirds of current teosinte populations. These changes are geographically diverse, implying that the potential impacts of climate change will vary greatly among populations. Climate change has population-specific effects in maize landraces, implying that climate change may result in maize landraces becoming maladapted to the climates in which they are currently cultivated. The fact that future climate change predicts significant changes in habitat distribution, migration potential, and patterns of local adaptation raises a red flag for the future of wild and cultivated maize populations. The heterogeneous nature of predicted population responses demonstrates that the selective impact of climate change varies across populations and is influenced by a variety of processes, including past adaptation (Aguirre-Liguori et al. 2019).

Plants have played a critical role in shaping the Earth's atmosphere by producing oxygen and sequestering carbon. In turn, historical atmospheric and climatic changes have shaped plant diversity (Xu and Weng 2020). Plants respond by evolving new developmental and metabolic traits. It produces an incredible suite of specialized metabolites, also called secondary metabolites (Kortbeek et al. 2019; Singh and Sharma 2015), which they use for their own immunity and growth-defense balance (i.e., internal and external signaling, attracting pollinators and seed dispersers, and defending against herbivores and pathogens). Humans use these metabolites for food, medicine, cosmetics, and other purposes (Xu and Weng 2020). However, the anthropogenic acceleration of climate change poses an increasing threat to the earth's plants, such as forest fires, because it not only destroys the plants that produce oxygen, but also releases a significant amount of sequestered carbon back into the atmosphere. Other plant stressors include rising temperatures, which release large amounts of sequestered methane (Moskvitch 2014), a more potent greenhouse gas than carbon dioxide (CO_2), back into the atmosphere; rising atmospheric carbon dioxide concentrations (Kahn 2017); and damage to the stratospheric ozone layer (Oltmans et al. 2006).

Many important commercial crops, including coffee (*Coffea arabica*), sugar cane (*Saccharum* spp.), rubber trees (*Hevea brasiliensis*), and oil palms (*Elais* spp.), are sensitive to growing conditions and, as a result, are threatened by rapid climate change. Humans have also used artificial selection to favor certain growth traits at the expense of defense and stress responses in these species. Mangroves (*Rhizophora* spp.) have also adapted to niche environments and thus face habitat loss (reviewed in Xu and Weng 2020). Climate change has the potential to influence plant growth, thereby influencing the quantity and quality of specialized metabolites produced. For instance, the richness of coffee's flavor profile correlates strongly with altitude, but ideal altitude ranges have been shown to change and become more inconvenient for farmers in recent years (Jonsson et al. 2015; Avelino et al. 2005).

Given that a few crop species have emerged as the most successful terrestrial plants as a result of human selection and facilitation, it is foreseeable that new generations of engineered plants with various desirable traits—whether bred, edited, or even created from scratch—will emerge to accompany future human life on Earth (Xu and Weng 2020).

18.2.6 Molecular and Biotechnological Intervention for Plants to Adapt to Climate Change

Climate change, combined with rapid population growth, significantly raises global food demand, and achieving the goal of food security for current and future generations is critical. It has affected millions of people worldwide, particularly the most vulnerable, by increasing the frequency and virulence of extreme weather events that cause crop yield reductions and population displacement (Begna 2021). Agriculture and climate are inextricably linked and interdependent. Climate modeling studies in many tropical and mid-latitude countries suggested that climate change will have a negative impact on food crop yield potential (Knox et al. 2012). As a result, increased climate change will reduce both the quantity and quality of food crops and forage (Myers et al. 2014). It is expected to have an impact on all of the factors that influence food security, including availability, access, stability, and utilization. Because land is shrinking while world population is rapidly increasing, modern agricultural practices are struggling to meet the level of primary productivity required to feed approximately 10 billion people by 2050 (Begna 2021).

In this challenging scenario of extreme weather events, increasing incidence of weeds, pests, and pathogens that are heavily influencing major cropping systems, proactive strategies are required to accelerate the rate of genetic gains with the goal of developing novel varieties. Large plant breeding populations, efficient high-throughput technologies, big data management tools, and downstream biotechnology and molecular techniques are the foundations for next generation breeding (Taranto et al. 2018). Plant breeding has always played an important role in human history, from revolutionizing agriculture to feeding the world's ever-growing population. It is primarily based on the presence of significant genetic variations in order to address the crops' maximum genetic yield potential and to exploit these variations through effective selection for improvement. Intensive efforts are currently being made to develop desirable crop cultivars in order to meet food demand while supporting sustainable agricultural productivity for climate change adaptation (Munaweera et al. 2022). Current efforts are being made to develop climate resilient fruit crop varieties using smart breeding tools to ensure food security in adverse climatic conditions, for instance, in almond, apple, bananas, citrus, coconut, grapevine, peach, and Rosaceous berries varieties (Kole 2020). Climate change, including extreme temperature and drought, is expected to have a negative impact not only on the agronomic conditions of the aforementioned fruit crops, but also on soil nutrients, pathogens, and pests. Therefore, there is a need to develop climate-resilient varieties with broad spectrum and durable resistance to both biotic and abiotic stresses.

Creating new plant genotypes is one of the most important options for agricultural climate change adaptation. Very different genotypes may be required to perform in the modified environments of protected agriculture. Improving the nutritional value and taste of foods, as well as ensuring healthy, safe food that is also sustainably and ethically produced and producing climate-resilient crops are significant challenges amid climate change. The plant genetic innovation will be enabled by genomic sequences of relevant germplasm and an understanding of the functional role of alleles controlling key traits (Henry 2020). According to the author, there are three strategies that can be used for agricultural adaptation to climate change. This necessitates the implementation of one or more of a variety of complementary strategies: developing technology (genotypes and production systems); relocating agriculture to new locations in response to environmental change; or adopting protected agriculture by partially or completely controlling the environment. Direct selection of all desirable alleles and/or gene editing will be required to deliver genotypes with the targeted alleles to provide the required yield and food with the necessary nutritional and functional traits for the new environments. However, for species with long-lived genotypes such as trees a more proactive approach to climate adaptation is required. Genomics provides a critical infrastructure for understanding the genetic potential of plants to respond to a wide range of environments and breeding better adapted crop varieties that can anticipate future climate changes (Abberton et al. 2015). As Alston (2000) emphasized, translating the strategies and genomic tools into climate resilient crops will necessitate additional large investments because agricultural initiatives can take 15–30 years to yield maximum returns.

To meet the ever-increasing demands of the human population, climate-resilient crops must be grown. Plant breeding, which entails crossing and selecting superior gene pools, has made significant contributions to this goal over the last few decades. Plant breeding technology is as old as agriculture itself, dating back more than 10,000 years to the origins of crop domestication (Allaby et al. 2017). To accelerate domestication two main avenues have been generally discussed in many literatures: one is through genetic modification (genetic engineering), by introducing new genes into already existing elite genotypes, and the other is through the exploitation of native genetic diversity in breeding programs. These strategies both largely benefit from an in-depth understanding of gene–function relationships.

18.2.7 Genetic Engineering, Gene Editing, and "Omics" Technologies for Accelerated Crop Production and Trait Improvement

Genetic tools to uncover genes that will help crops like rice to withstand harsh conditions are therefore necessary. Biotechnology allows for significant changes in crops to withstand the stress of anthropogenic climate change. Biotechnological approaches such as genetic engineering (GE) or transgenic technology, genome editing, RNA-mediated gene silencing armored with next-generation sequencing, genome mapping, and many more have paved the way for precise and faster genetic modifications of plants (Munaweera et al. 2022; Gaba et al. 2021).

GE could help move genes into different varieties to produce more resilient crops. GE allows for the transfer of the desired gene(s) into crops, making this method precise and unbound by interspecies incompatibility, free of linkage drag, and, most significantly, less time-consuming than conventional breeding methods (Gaba et al. 2021). The use of GE methods can overcome the majority of the disadvantages associated with breeding approaches, and this is relatively simple and targets can be met quickly. For perennial species like trees GE methods can overcome the long juvenile periods, circumvent phenotypic variation observed in progeny as a result of recombination and segregation, restore elite genotypes which may exhibit susceptibility to biotic stresses after many years of breeding and selection, and allow for the use of genes at present in the genome of the genetically engineered plant.

Many genes involved in plant adaptations to drought stress and desiccation tolerance also confer an improved salinity tolerance. New genes for salt tolerance can be introduced into crop cultivars; thus identification of salt-tolerant germplasm is very important. Water use efficiency (WUE) is one measure of the ability of a plant to perform well under incipient drought. Accordingly, three key processes can be exploited in breeding for higher WUE: (1) moving more of the available water through the crop rather than it being wasted as evaporation from the soil surface or drained beyond the root zone or left behind in the root zone during harvest; (2) acquisition of more carbon (biomass) in exchange for the water loss through plant transpiration; and (3) partitioning more of the achieved biomass into the harvested product (Condon et al. 2004). Plant epigenetics has recently gained interest for plant breeding (Mirouze and Paszkowski 2011), and knowledge on epigenetics will be useful for identifying candidate genes for the selection of promising genotypes and as a source of new useful characters for developing new varieties.

However, plant scientists are hopeful that the recently developed gene or genome editing technique, CRISPR (clustered regularly interspaced short palindromic repeats), will allow them to precisely integrate desirable traits while removing undesirable ones (reviewed in Gaba et al. 2021). It has emerged as a method for either assisting organisms in adapting to climate change-driven increases in droughts, flooding, temperature, soil salinity and acidity, and the number of crop-destroying insects, or helping to mitigate the effects of climate change on agriculture (Karavolias et al. 2021). It employs a molecular tool such as CRISPR to make targeted improvements to a plant's DNA, typically working only within the plant's own family. For example, scientists used gene editing to identify a gene in tomatoes that could be rendered inactive to make the tomatoes more heat tolerant. CRISPR-Cas-based gene-editing technologies have given hope and charged the old plant breeding machine with the most energetic and powerful fuel to address the challenges involved in feeding the world (Vu et al. 2022). CRISPR-Cas technology also aims to increase yields while using less fertilizer, pesticide, or water.

Plant and soil scientists also hope to largely increase and accelerate carbon storage using gene editing on a few major common crops such as rice and sorghum. For instance, the Innovative Genomics Institute in San Francisco Bay aimed to create gene-edited rice and sorghum seeds that when planted globally could remove more than a billion extra tons of carbon from the atmosphere each year (Stone 2022). The researchers embarked to develop new crop varieties that can efficiently photosynthesize and channel more carbon into the soil using CRISPR genome editing, which was attempted for years using traditional GE technology—introducing chunks of DNA from bacteria, or other plants, with desirable traits, into the genes encoding light-harvesting proteins and other biochemical machinery.

The precision and efficiency of gene editing are two of its most significant advantages as it enables plant breeders to achieve the same goals as traditional breeding methods in relatively shorter time and with higher precision. It is a good alternative to plant breeding and GE because it can precisely modify the genomes at the target site. As with GE it has been seen to provide revolutionary plant breeding solutions to sustain food production and reach zero hunger by 2050 as envisioned under the UN Sustainable Development Goals. Compared to GM crops, however, the biosafety regulations in many countries for commercialization of gene-edited crops are simpler and relaxed (Vu et al. 2022). Since gene editing techniques like CRISPR is like a pair of molecular scissors that make cut-and-paste edits throughout an organism's genome without introducing any foreign DNA, gene-edited products have become more popularly recognized worldwide (Stone 2022). Hence, it guarantees better safeguards for food supplies, which are critical to communities all over the world.

Rice, maize, wheat, and soybean are known to dominate the global food system, accounting for 90% of the world's food energy intake (Vogel et al. 2019). Some efforts have been made on major cereal and noncereal crops to make them "climate-ready" using advanced biotechnological approaches (see Munaweera et al. 2022). *Arabidopsis thaliana*, barley, potato, tomato, cassava, and banana are other climate-ready crops improved through genetic engineering. *Colocasia esculenta* (Taro), on the other hand, is an underutilized crop that has been identified as a potential source of antioxidants as well as a means to combat chronic malnutrition and hunger. Taro's genomic sequencing recently revealed that 17,097 genes may be functional proteins. Because Taro is distributed globally, it requires significant attention to modern biotechnology techniques in order to become one of the "climate-ready crops." Similarly, crops with relatively low commercial potential, such as cassava, cowpea, and yams, are considered to grow in most developing countries in order to meet food requirements (Munaweera et al. 2022).

Accelerating the domestication process in generally all trees has the potential to improve yields, pest resistance, adaptation to managed environments and biomass properties for various ends while also reducing our reliance on various inputs. However, compared with agricultural crops, timely domestication of trees is virtually unachievable through traditional genetic improvement methods alone, because of the long breeding cycles and rotation times. Conventional breeding coupled with modern genetic and genomic techniques will accelerate tree domestication (Harfouche et al. 2012). Genetic engineering and recently gene editing have gained considerable attention for tree breeding. For example, transgenesis has led to the development of cold-tolerant hybrid eucalyptus trees (*Eucalyptus grandis, Eucalyptus urophylla*) and lignin-modified hybrid poplar trees (*Populus. tremula, P. alba*). A single cycle of genetic improvement can span several decades. Marker-assisted selection (MAS) has long been considered to accelerate tree domestication in trees based on QTL analysis. However, an abundance of minor frequency alleles in forest populations have resulted in the identification of only a few polymorphisms associated with certain phenotypes. Prospects of omics technologies (genomics, proteomics, transcriptomics, and metabolomics), accordingly, will play an increased importance in trait improvement in trees because of their great potential to uncover the molecular basis of valuable traits for tree breeding and for early selection in breeding programs.

18.3 Synthesis

Human-induced activities in addition to anthropogenic climate change influence the morpho-anatomical and physiological characteristics as well as the genetic responses of plants in various ways. These also help the plants to develop adaptive mechanisms. Plant morpho-anatomical traits at the cell, tissue, and organ levels are very useful in understanding their variations and can help predict the responses to climate change and human-induced activities. It is important to develop an understanding of how these structural and genetic changes allow plants to cope with environmental constraints and anthropogenic stress.

The various biotechnological breakthroughs for plants as well as animals have enormous potentials to revive ecosystems or make ecosystems more climate-resilient, protect food systems, mitigate further effects of climate change, and ultimately attain the greater good of human society. However, there are also risks and unknowns. The use of these technologies and their associated health and ecological risks has been recently the focus of discussion, but it should be made easily available (interpreted in the layman's terms) to the people for safeguard.

References

Abberton M, Batley J, Bentley A, Bryant J, Cai H, Cockram J, Costa de Oliveira A, Cseke LJ, Dempewolf H, De Pace C et al (2015) Global agricultural intensification during climate change: a role for genomics. Plant Biotechnol J 14:1095–1098

Aguirre-Liguori JA, Ramirez-Barahona S, Tiffin P, Eguiarte LE (2019) Climate change is predicted to disrupt patterns of local adaptation in wild and cultivated maize. Proc Biol Sci 286:20190486. https://doi.org/10.1098/rspb.2019.0486

Ainsworth EA, Rogers A (2007) The response of photosynthesis and stomatal conductance to rising [CO_2]: mechanisms and environmental interactions. Plant Cell Environ 30:258–270

Aitken SN, Whitlock MC (2013) Assisted gene flow to facilitate local adaptation to climate change. Annu Rev Ecol Evol Syst 44:367–388

Alder NN, Sperry JS, Pockman WT (1996) Root and stem xylem embolism, stomatal conductance, and leaf turgor in *Acer grandidentatum* populations along a soil moisture gradient. Oecologia 105(3):293–301

Allaby RG, Stevens C, Lucas L, Maeda O, Fuller DQ (2017) Geographic mosaics and changing rates of cereal domestication. Philos Trans R Soc Lond B Biol Sci 372(1735):20160429

Alsos IG, Ehrich D, Thuiller W, Eidesen PB, Tribsch A, Schoenswetter P, Lagaye C, Taberlet P, Brochmann C (2012) Genetic consequences of climate change for northern plants. Proc Biol Sci 279:2042–2051

Alston JM (2000) A meta-analysis of rates of return to agricultural R&D: ex Pede Herculem? International Food Policy Research Institute, Washington, DC

Altman A, Mesoudi A (2019) Understanding agriculture within the frameworks of cumulative cultural evolution, gene-culture co-evolution, and cultural niche construction. Hum Ecol 47: 483–497

Anderegg WRL, Meinzer FC (2015) Wood anatomy and plant hydraulics in a changing climate. In: Hacke U (ed) Functional and ecological xylem anatomy. Springer International Publishing Switzerland, Cham. https://doi.org/10.1007/978-3-319-15783-2_9

Anderson JT, Panetta AM, Mitchell-Olds T (2012) Evolutionary and ecological responses to anthropogenic climate change. Plant Physiol 160:1728–1740

Atia A, Rahbi M, Debez A, Abdelly C, Gouia H, Haouari C, Smaoui A (2014) Ecophysiological aspects and photosynthetic pathways in 105 plants species in saline and arid environments of Tunisia. J Arid Land 6:762–770

Avelino J, Barboza B, Araya JC, Fonseca C, Davrieux F, Guyot B, Cilas C (2005) Effects of slope exposure, altitude and yield on coffee quality in two altitude terroirs of Costa Rica, Orosi and Santa María de Dota. J Sci Food Agric 85:1869–1876

Baillie AL, Fleming AJ (2020) The developmental relationship between stomata and mesophyll airspace. New Phytol 225:1120–1126

Bano C, Amist N, Singh B (2019) Morphological and anatomical modifications of plants for environmental stresses. In: Roychoudhury A, Tripathi D (eds) Molecular plant abiotic stress: biology and biotechnology. Wiley, Hoboken, NJ, p 454

Begna T (2021) Global role of plant breeding in tackling climate change. Int J Agric Sci Food Technol 7(2):223–229

Belluau M (2018) Linking hard and soft traits: physiology, morphology and anatomy interact to determine habitat affinities to soil water availability in herbaceous dicots. PLoS One 13(3): e0193130. https://doi.org/10.1371/journal.pone.0193130

Bennett NL, Severns PM, Parmesan C, Singer MC (2015) Geographic mosaics of phenology, host preference, adult size and microhabitat choice predict butterfly resilience to climate warming. Oikos 124(1):41–53

Boonklong B, Jaroensutasinee M, Jaroensutasin K (2006) Climate change affecting mangosteen production in Thailand. In: Proceedings of the 5th WSEAS international conference on environment, ecosystems and development, Venice, Italy

Boyd JN, Odell J, Cruse-Sanders J, Rogers W, Anderson JT, Baskauf C, Brzyski J (2022) Phenotypic plasticity and genetic diversity elucidate rarity and vulnerability of an endangered riparian plant. Ecosphere 13(4):e3996

Boys J, Cherry M, Dayanandan S (2005) Microsatellite analysis reveals genetically distinct populations of red pine (*Pinus resinosa*, Pinaceae). Am J Bot 92(5):833–841

Brachi B, Faure N, Horton M, Flahauw E, Vazquez A, Nordbord M, Bergelson J, Cuguen J, Roux F (2010) Linkage and association mapping of *Arabidopsis thaliana* flowering time in nature. PLoS Genet 6(5):e1000940

Bricker E, Calladine A, Virnstein R, Waycott M (2018) Mega clonality in an aquatic plant - A potential survival strategy in a changing environment. Front Plant Sci 9:435. https://doi.org/10.3389/fpls.2018.00435

Bruelheide H, Dengler J, Purschke O, Lenoir J, Jiménez-Alfaro B, Hennekens SM, Botta-Dukát Z, Chytrý M, Field R, Jansen F, Kattge J, Pillar V, Schrodt F, Mahecha MD, Peet RK, Sandel B, van Bodegom P, Altman J, Alvarez-Dávila E, Jandt U (2018) Global trait–environment relationships of plant communities. Nat Ecol Evol 2(12):1906–1917

Calibo MKT, Tinio CE, Combalicer MS, Maldia LSJ (2021) Leaf traits of *Calophyllum inophyllum* L. (Calophyllaceae) in different locations suggest suitability for planting outside of natural habitat. Philippine J Sci 150(4):643–655

Calleja-Cabrera L, Boter M, Oñate-Sánchez L, Pernas M (2020) Root growth adaptation to climate change in crops. Front Plant Sci 11:544. https://doi.org/10.3389/fpls.2020.00544

Cardoso JA, Rincón J, Jiménez JLC, Noguera D, Rao IM (2013) Morpho-anatomical adaptations to waterlogging by germplasm accessions in a tropical forage grass. AoB Plants 5:plt047. https://doi.org/10.1093/aobpla/plt047

Carillo P, De Micco V, Ciriello M, Formisano L, El-Nakhel C, Giordano M, Colla G, Rouphael Y (2022) Morpho-anatomical, physiological, and mineral composition responses induced by a vegetal-based biostimulant at three rates of foliar application in greenhouse lettuce. Plants (Basel) 11(15):2030. https://doi.org/10.3390/plants11152030

Chapin FS III, Autumn K, Pugnaire FI (1993) Evolution of suites of traits in response to environmental stress. Am Nat 142:278–292

Chapuis E, Ballesteros E, Garcia E (2012) Distribution and richness of aquatic plants across Europe and Mediterranean countries: patterns, environmental driving factors and comparison with total plant richness. J Veg Sci 2012. https://doi.org/10.1111/j.1654-1103.2012.01417.x

Chen IC, Hill JK, Ohlemueller R, Roy DB, Thomas CD (2011) Rapid range shifts of species associated with high levels of climate warming. Science 333:1024–1026

Ciccarelli D, Bona C (2022) Exploring the functional strategies adopted by coastal plants along an ecological gradient using morpho-functional traits. Estuarines Coasts 45:114–129

Condon AG, Richards RA, Rebetzke GJ, Farquhar GD (2004) Breeding for high water-use efficiency. J Exp Bot 55(407):2447–2460

Cornwell WK, Ackerly D (2009) Community assembly and shifts in plant traits distributions across an environmental gradient in coastal California. Ecol Monogr 79(1):109–126

De Oliveira AS, Ferreira CS, Graciano-Ribeiro D, Franco AC (2015) Anatomical and morphological modifications in response to flooding by six Cerrado tree species. Acta Bot Brasilica 29(4). https://doi.org/10.1590/0102-33062014abb0035

De Sousa DDCV, Bessa LA, Silva FG, Rosa M, Filho SDCV, Vitorino LC (2020) Morpho-anatomical and physiological responses can predict the ideal period for the transplantation of hydroponic seedlings of *Hymenaea courbaril*, a neotropical fruit tree. Plants (Basel) 9(6):721

Demidchik V, Shabala SN, Coutts KB, Tester MA, Davies JM (2003) Free oxygen radicals regulate plasma membrane Ca^{2+-} and K^{+}-permeable channels in plant root cells. J Cell Sci 116(Pt 1): 81–88

Díaz S, Kattge J, Cornelissen JHC, Wright IJ, Lavorel S, Dray S, Reu B, Kleyer M, Wirth C, Colin Prentice I, Garnier E, Bönisch G, Westoby M, Poorter H, Reich PB, Moles AT, Dickie J, Gillison AN, Zanne AE, Gorné LD (2016) The global spectrum of plant form and function. Nature 529(7585):167–171

Doebley J (1989) Isozymic evidence and the evolution of crop plants. In: Soltis D, Soltis P (eds) Isozymes in plant biology. Dioscorides Press, Portland, OR, pp 165–191

Doebley J (2004) The genetics of maize evolution. J Ann Rev Genet 38:37–59

Doebley JF, Gaut BS, Smith BD (2006) The molecular genetics of crop domestication. Cell 127 (7):1309–1321

Doi H, Takahashi M, Katano I (2010) Genetic diversity increases regional variation in phenological dates in response to climate change. Glob Chang Biol 16:373-379

Dong N, Prentice IC, Wright IJ et al (2020) Components of leaf-trait variation along environmental gradients. New Phytol 228(1):82–94

Elser JJ, Dobberfuhl DR, MacKay NA, Schampel JH (1996) Organism size, life history, and N:P stoichiometry: towards a unified view of cellular and ecosystem processes. Bioscience 46:674–684

Eyre-Walker A, Gaut RL, Hilton H, Feldman DL, Gaut BS (1998) Investigation of the bottleneck leading to the domestication of maize. Proc Natl Acad Sci U S A 95:4441–4446

Fagundes NCA (2019) Morphoanatomical, functional and evolutionary traits conditioning the adaptability of leguminosae in an edaphoclimatic gradient. Thesis, Universidade Federal De Lavras, 154 p

Fernandez-Calleja M, Casa AM, Igartua E (2021) Major flowering time genes of barley: allelic diversity, effects, and comparison with wheat. Theor Appl Genet 134:1867–1897

Flint-Garcia SA (2013) Genetics and consequences of crop domestication. J Agric Food Chem 4(61):8267–8276

Franks SJ (2011) Plasticity and evolution in drought avoidance and escape in the annual plant *Brassica rapa*. New Phytol 190:249–257

Franks SJ, Hoffman AA (2012) Genetics of climate change adaptation. Annu Rev Genet 46:185–208

Franks SJ, Sim S, Weis AE (2007) Rapid evolution of flowering time by an annual plant in response to a climate fluctuation. Proc Natl Acad Sci U S A 104:1278–1282

Franks SJ, Kane NC, O'hara NB, Tittes S, Rest JS (2016) Rapid genome-wide evolution in *Brassica rapa* populations following drought revealed by sequencing of ancestral and descendant gene pools. Mol Ecol 25:3622–3631

Freschet GT, Dias ATC, Ackerly DD, Aerts R, van Bodegom PM, Cornwell WK, Dong M, Kurokawa H, Liu G, Onipchenko VG, Ordoñez JC, Petzer DA, Richardson SJ, Shidakov II, Soudzilovskaia NA, Tao J, Cornelissen JHC (2011) Global to community scale differences in the prevalence of convergent over divergent leaf trait distributions in plant assemblages. Glob Ecol Biogeogr 20:755–765

Freschet GT, Roumet C, Comas LH, Weemstra M, Bengough AG (2021) Root traits as drivers of plant and ecosystem functioning: current understanding, pitfalls and future research needs. New Phytol 232(3):1123–1158

Fuller DQ, Denham T, Arroyo-Kalina A, Lucasa L, Stevens CJ, Qin L, Allaby RG, Purugganan MD (2014) Convergent evolution and parallelism in plant domestication revealed by an expanding archaeological record. Proc Natl Acad Sc U S A 111(17):6147–6152

Funkhouser D (2015) Study finds genetic clues to how plants adapt to climate. Columbia Climate School, State of the Planet. https://news.climate.columbia.edu/2015/01/30/study-finds-genetic-clues-to-how-plants-adapt-to-climate/

Gaba Y, Pareek A, Singla-Pareek SL (2021) Raising climate-resilient crops: journey from the conventional breeding to new breeding approaches. Curr Genomics 22:450–467

Galapon MMF, Quimado MO, Tinio CE, Combalicer MS (2020) Growth response of *Thunbergia grandiflora* (Roxb. ex Rottler) Roxb. (Skyflower) on shading and anatomical characteristics under various soil series. Philippine J Sci 149(3-a):935–946

Galviz YC, Valerio R (2021) Leaf morphoanatomical traits of *Jacquinia armillaris* Jacq. (Theophrastoideae–Primulaceae) in two xeric shrublands from Venezuela. Neotrop Biodivers 7(1):364–375

Garnier E, Navas ML, Griculis K (2015) Gradients, response traits, and ecological strategies. In: Plant functional diversity: organism traits, community structure, and ecosystem properties (Oxford, 2015; online edn, Oxford Academic, 24 Mar 2016). https://doi.org/10.1093/acprof:oso/9780198757368.003.0004

Gill AL, Gallinat AS, Sanders-Demott R, Rigden AJ, Short Gianotti DJ, Mantooth JA, Templer PH (2015) Changes in autumn senescence in northern hemisphere deciduous trees: a meta-analysis of autumn phenology studies. Ann Bot 116:875–888

Glubsheva T, Chernuavskih V, Dumacheva E, Vorobyova O, Grigorenko S (2021) Morpho-anatomical features of the flower *Tulipa biebersteiniana* Schult. et Schult. in the South of the Central Russian Upland. BIO Web Conf 40:01011. https://doi.org/10.1051/bioconf/20214001011

Goshier C (2016) Climate change drives fast genetic changes in plants, study finds. Fordham News. https://news.fordham.edu/science/climate-change-drives-fast-genetic-changes-in-plants/

Gotoh E, Suetsugu N, Higa T, Matsushita T, Tsukaya H, Wada M (2018) Palisade cell shape affects the light-induced chloroplast movements and leaf photosynthesis. Sci Rep 8:1472. https://doi.org/10.1038/s41598-018-19896-9

Grace JB (1993) The adaptive significance of clonal reproduction in angiosperms: a perspective. Aquat Bot 44:159–180

Gratani L (2014) Plant phenotypic plasticity in response to environmental factors. Adv Bot 2014: 208747. https://doi.org/10.1155/2014/208747

Gray SD, Brady SM (2016) Plant developmental responses to climate change. Dev Biol 419:64–77

Gross BL, Olsen KM (2010) Genetic perspectives on crop domestication. Trends Plant Sci 15(9): 529–537

Guan B, Gao J, Chen W, Gong X, Ge G (2021) The effects of climate change on landscape connectivity and genetic clusters in a small subtropical and warm-temperate tree. Front Plant Sci 12:671336

Guédon Y, Legave JM (2008) Analyzing the time-course variation of apple and pear tree dates of flowering stages in the global warming context. Ecol Model 219:189–199

Haas JC, Vergara A, Serrano AR, Mishra S, Hurry V, Street NR (2021) Candidate regulators and target genes of drought stress in needles and roots of Norway spruce. Tree Physiol 41(7): 1230–1246

Hale WG, Saunders VA, Mangham JP (2005) Dictionary of biology, 3rd edn. Oxford University Press, Oxford

Hatfield J, Boote KJ, Kimball BA, Ziska L, Izaurralde R, Ort D, Thomson AM (2011) Climate impacts on agriculture: implications for crop production. J Agron 103:351–370

Harfouche A, Meilan R, Kirst M, Morgante M, Boerjan W, Sabatti M, Scarascia Mugnozza G (2012) Accelerating the domestication of forest trees in a changing world. Trends Plant Sci 17:64–72

Happonen K, Virkkala AM, Kemppinen J, Niittynen P, Luoto M (2022) Relationships between above-ground plant traits and carbon cycling in tundra plant communities. J Ecol 110(3): 700–716

Henry RJ (2020) Innovations in plant genetics adapting agriculture to climate change. Curr Opin Plant Biol 56:168–173

Hernandez JO, Malabrigo PL Jr, Quimado MO, Maldia LSJ, Fernando ES (2016) Xerophytic characteristics of *Tectona philippinensis* Benth. & Hook.f. Philippine J Sci 145(3):259–269

Hernandez JO, An JY, Combalicer MS, Chun JP, Oh SK, Park BB (2021) Morpho-anatomical traits and soluble sugar concentration largely explain the responses of three deciduous tree species to progressive water stress. Front Plant Sci 12:738301. https://doi.org/10.3389/fpls.2021.738301

Hoffmann H, Rath T (2013) Future bloom and blossom frost risk for *Malus domestica* considering climate model and impact model uncertainties. PLoS One 8:e75033. https://doi.org/10.1371/journal.pone.0075033

Hoffmann A, Sgrò C (2011) Climate change and evolutionary adaptation. Nature 470:479–485

Hoffmann AA, Willi Y (2008) Detecting genetic responses to environmental change. Nat Rev Genet 9(6):421–432

Huang W, Han S, Xing Z, Li W (2020) Responses of leaf anatomy and CO_2 concentrating mechanisms of the aquatic plant *Ottelia cordata* to variable CO_2. Front Plant Sci 11:1261. https://doi.org/10.3389/fpls.2020.01261

Innocente PS, Teixeira SP, Balbuena TS (2020) Elevated atmospheric CO_2 increases *Eucalyptus urophylla* S. T. Blake stem diameter by stimulating cell proliferation and reducing lignin deposition. Acta Bot Brasilica 34(3):589–591. https://doi.org/10.1590/0102-33062020abb0107

Interior-Hallegado ID, Tinio CE, Luna AC, Quimado MO, Combalicer MS (2022) Morpho-physiology and anatomy of the six grass species growing on lateritic soil: diagnosis of characters for phytostabilization of soils depleted by mining. Philippine J Sci 151(4):1541–1556

Iwaizumi MG, Miyata S, Hirao T, Tamura M, Watanabe A (2018) Historical seed use and transfer affects geographic specificity in genetic diversity and structure of old planted Pinus thunbergii populations. For Ecol Manag 408:211–219

Jahdi R, Arbai M, Bussotti F (2020) Effect of environmental gradients on leaf morphological traits in the Fandoghlo forest region (NW Iran). iForest 13(6):523–530

Jia W, Ma M, Chen J, Wu S (2021) Plant morphological and anatomical adaptation to flooding stress and the underlying molecular mechanisms. Int J Mol Sci 22(3):1088. https://doi.org/10.3390/ijms22031088

Jonsson M, Raphael IA, Ekbom B, Kyamanywa S, Karungi J (2015) Contrasting effects of shade level and altitude on two important coffee pests. J Pest Sci 88(2):281–287

Jothiramshekar S, Benjamin JS, Krishnasamy R, Pal AK, George S, Swaminathan R, Parida AK (2018) Responses of selected C3 and C4 halophytes to elevated CO_2 concentration under salinity. Curr Sci 115(1):129–135

Jump AS, Penuelas J (2005) Running to stand still: adaptation and the response of plants to rapid climate change. Ecol Lett 8:1010–1020

Jump AS, Peñuelas J (2006) Genetic effects of chronic habitat fragmentation in a wind-pollinated tree. Proc Natl Acad Sci U S A 103:8096–8100

Jump AS, Marchant R, Penuelas J (2008a) Environmental change and the option value ~ of genetic diversity. Trends Plant Sci 14:51–58

Jump AS, Penuelas J, Rico L, Ramallo E, Estiarte M, Martinez-Izquierdo JA, Lloret F (2008b) Simulated climate change provokes rapid genetic change in the Mediterranean shrub *Fumana thymifolia*. Glob Chang Biol 14:637–643

Kadipaya KB, Singh C, Bhalara RL, Kandoliya UK, Dabhi KH (2017) Effect of higher temperature on leaf anatomy of heat tolerance and heat susceptible wheat genotypes (*Triticum aestivum* L.) by scanning electron microscopy. J Pharmacogn Phytochem 6(5):2270–2277

Kahn B (2017) We just breached the 410 ppm threshold for CO2. Scientific American. https://www.scientificamerican.com/article/we-just-breached-the-410-ppm-threshold-for-co2/

Kallarackal J, Renuka R (2014) Phenological implications for the conservation of forest trees. In: Kapoor R, Kaur I, Koul M (eds) Plant reproductive biology and conservation. I.K. International, Delhi, pp 150–168

Kallarackal J, Roby TJ (2012) Response of trees to elevated carbon dioxide and climate change. Biodivers Conserv 21:1327–1342

Karavolias NG, Horner W, Abugu MN, Evanega SN (2021) Application of gene editing for climate change in agriculture. Front Sustain Food Syst 5:685801

Kirkham MB (2005) Leaf anatomy and leaf elasticity. In: Kirkham MB (ed) Principles of soil and plant water relations. Elsevier, Amsterdam, pp 357–378

Klem K, Gargallo-Garriga A, Rattanapichai W, Oravec M, Holub P, Veselá B, Sardans J, Peñuelas J, Urban O (2019) Distinct morphological, physiological, and biochemical responses to light quality in barley leaves and roots. Front Plant Sci 10:1026. https://doi.org/10.3389/fpls.2019.01026

Knox J, Hess T, Daccache A, Wheeler T (2012) Climate change impacts on crop productivity in Africa and South Asia. Environ Res Lett 7:034032

Kole C (ed) (2020) Genomic designing of climate-smart fruit crops. Springer Nature Switzerland, Cham, p 418. https://doi.org/10.1007/978-3-319-97946-5

Kortbeek RWJ, van der Gragt M, Bleeker PM (2019) Endogenous plant metabolites against insects. Eur J Plant Pathol 154(1):67–90

Kremer A, Ronce O, Robledo-Arnuncio JJ, Guillaume F, Bohrer G, Nathan R, Bridle JR, Gomulkiewicz R, Klein EK, Ritland K (2012) Long-distance gene flow and adaptation of forest trees to rapid climate change. Ecol Lett 15:378–392

Kudo G, Ida TY (2013) Early onset of spring increases the phenological mismatch between plants and pollinators. Ecology 94:2311–2320

Laliberté E (2017) Below-ground frontiers in trait-based plant ecology. New Phytol 213(4): 1597–1603

Lambers H, Poorter H (1992) Inherent variation in growth-rate between higher-plants—a search for physiological causes and ecological consequences. Adv Ecol Res 23:187–261

Larson G, Piperno DR, Allaby RG, Purugganan MD, Andersson L, Arroyo-Kalin M, Barton L, Climer Vigueira C, Denham T, Dobney KM, Doust AN, Gepts P, Gilbert MT, Gremillion KJ, Lucas L, Lukens L, Marshall FB, Olsen KM, Pires JC, Richerson PJ, Rubio de Casas R, Sanjur O, Thomas MG, Fuller DQ (2014) Current perspectives and the future of domestication studies. Proc Natl Acad Sci 111:6139–6146

Lasky JR, David L, Marais D, Lowry DB, Povolotskaya I, McKay JK, Richards JH, Keitt TH, Juenger TE (2014) Natural variation in abiotic stress responsive gene expression and local adaptation to climate in *Arabidopsis thaliana*. Mol Biol Evol 31(9):2283–2296

Lefor MW (1999) Xerophyte. In: Environmental geology. Encyclopedia of earth science. Springer, Dordrecht. https://doi.org/10.1007/1-4020-4494-1_371

Lempe J, Balasubramanian S, Sureshkumar S, Singh A, Schmid M, Weigel D (2005) Diversity of flowering responses in wild *Arabidopsis thaliana* strains. PLoS Genet 1(1):e6

Liu MJ, Sugimoto K, Uygun S, Panchy N, Campbell MS, Yandell M, Howe GA, Shiu SH (2018) Regulatory divergence in wound-responsive gene expression between domesticated and wild tomato. Plant Cell 30:1445–1460

Lin Y, Kuan L, Tang S, Mou Z, Phillips OL, Lambers H, Liu Z, Sardans J, Peñuelas J, Lai Y, Lai Y, Lin M, Chen D, Kuan Y (2021) Leaf traits from stomata to morphology are associated with climatic and edaphic variables for dominant tropical forest evergreen oaks. J Plant Ecol 14(6): 1115–1127

Liu X, Chen H, Sun T, Li D, Wang X, Mo W, Wang R, Zhang S (2021) Variation in woody leaf anatomical traits along the altitudinal gradient in Taibai Mountain, China. Glob Ecol Conserv 26:e01523. https://doi.org/10.1016/j.gecco.2021.e01523

Lombardi T, Bertacchi A, Pistelli L, Pardossi A, Pecchia S, Toffanin A, Sanmartin C (2022) Biological and agronomic traits of the main halophytes widespread in the Mediterranean region as potential new vegetable crops. Horticulturae 8(3):195. https://doi.org/10.3390/horticulturae8030195

Lowe AJ, Boshier D, Ward M, Bacles CFE, Navarro C (2005) Genetic resource loss following habitat fragmentation and degradation; reconciling predicted theory with empirical evidence. Heredity 95:255–273

Luedeling E, Zhang M, Girvetz EH (2009) Climatic changes lead to declining winter chill for fruit and nut trees in California during 1950–2099. PLoS One 4:e6166

Lusa MG, Boeger M, Cecilia M, de Chiara Moço MC, Bona C (2011) Morpho-anatomical adaptations of *Potamogeton polygonus* (Potamogetonaceae) to lotic and lentic environments. Rodriguesia 62(4):927–936

Maherali H, DeLucia EH (2000) Xylem conductivity and vulnerability to cavitation of ponderosa pine growing in contrasting climates. Tree Physiol 20:859–867

Mariac C, Jehin L, Saidou AA, Thuillet AC, Couderc M, Sire P, Judge H, Adam H, Bezancon G, Pham JL, Vigouroux Y (2011) Genetic basis of pearl millet adaptation along an environmental gradient investigated by a combination of genome scan and association mapping. Mol Ecol 20: 80–91

Meyer R, Purugganan M (2013) Evolution of crop species: genetics of domestication and diversification. Nat Rev Genet 14:840–852

Michaels SD (2009) Flowering time regulation produces much fruit. Curr Opin Plant Biol 12:75–80

Mirouze M, Paszkowski J (2011) Epigenetic contribution to stress adaptation in plants. Curr Opin Plant Biol 14(3):267–274

Miryeganeh M (2022) Mangrove forests: natural laboratories for studying epigenetic and climate changes. Front Plant Sci 13:85151518

Moskvitch K (2014) Mysterious Siberian crater attributed to methane. Nature News. https://doi.org/10.1038/nature.2014.15649

Muir CD (2020) A stomatal model of anatomical tradeoffs between gas exchange and pathogen colonization. Front Plant Sci 11:518991. https://doi.org/10.3389/fpls.2020.518991

Munaweera TIK, Jayawardana NU, Rajaratnam R, Dissanayake N (2022) Modern plant biotechnology as a strategy in addressing climate change and attaining food security. Agric Food Security 11:26

Muniz LF, Bombo AB, Filartiga AL, Appezzato-da-Glória B (2018) Can climate and soil conditions change the morpho-anatomy among individuals from different localities? A case study in *Aldama grandiflora* (Asteraceae). Braz J Biol 78(4):706–717. https://doi.org/10.1590/1519-6984.174636

Myers SS, Zanobetti A, Kloog I, Bloom AJ, Carlisle EA, Dietterich LH et al (2014) Rising CO2 threatens food quality. Nature 510:139–142. https://bit.ly/3hHQ39y

Nicotra AB, Atkin OK, Bonser SP, Davidson AM, Finnegab EJ, Mathesius U, Poot P, Purugganan MD, Richards CL, Valladres F, van Kleunen M (2010) Plant phenotypic plasticity in a changing climate. Trends Plant Sci 15:684–692

Ojo O, Shoele K (2022) Branching pattern of flexible trees for environmental load mitigation. Bioinspir Biomim 17(5):056003

Oltmans SJ, Lefohn AS, Harris JM, Galbally I, Scheel HE, Bodeker G, Brunke E, Claude H, Tarasick D, Johnson BJ, Simmonds P, Shadwick D, Anlauf K, Hayden K, Schmidlin F, Fujimoto T, Akagi K, Meyer C, Nichol S, Davies J, Redondas A, Cuevas E (2006) Long-term changes in tropospheric ozone. Atmos Environ 40(17):3156–3173

Onoda Y (2017) Physiological and structural tradeoffs underlying the leaf economics spectrum. MS Thesis, Graduate School of Agriculture, Kyoto University. 57 p

Paes de Camargo MB (2010) The impact of climatic variability and climate change on Arabic coffee crop in Brazil. Bragantia 69:239–247

Pandey S (2021) Climatic influence on tree wood anatomy: a review. J Wood Sci 67:24. https://doi.org/10.1186/s10086-021-01956-w

Paradiso R, Arena C, De Micco V, Giordano M, Aronne G, De Pascale S (2017) Changes in leaf anatomical traits enhanced photosynthetic activity of soybean growth in hydroponics with plant growth-promoting microorganisms. Front Plant Sci 8:674. https://doi.org/10.3389/fpls.2017.00674

Park SH, Pauli CS, Gostin EL, Staples SK, Seifried D, Kinney C, Heuvel V (2022) Effects of short-term environmental stresses on the onset of cannabinoid production in young immature flowers of industrial hemp (*Cannabis sativa* L.). J Cannabis Res 4(1):1. https://doi.org/10.1186/s42238-021-00111-y

Parmesan C, Hanley ME (2015) Plants and climate change: complexities and surprises. Ann Bot 116(6):849–864

Parmesan C, Yohe G (2003) A globally coherent fingerprint of climate change impacts across natural systems. Nature 421:37–42

Pazzagalia J, Reusch TBH, Terlizzi A, Guirao LM, Procaccini G (2021) Phenotypic plasticity under rapid global changes: the intrinsic force for future seagrasses survival. Evol Appl 14(5): 1181–1201

Pérez Cuadra VP, Verolo M, Cambi V (2021) Morphological and anatomical traits of halophytes. In: Grigore MN (ed) Handbook of halophytes. Springer, Cham. https://doi.org/10.1007/978-3-030-57635-6_120

Perez KLDJ, Quimado MO, Maldia LSJ, Tinio CE, Hernandez JO, Combalicer MS (2021) Effects of copper on the leaf morpho-anatomy of *Rhizophora mucronata*: implications for mangrove ecosystem restoration. Biodiversitas 22(4):2058–2065

Piao S, Liu Q, Chen A, Janssens IA, Fu Y, Dai J, Liu L, Lian X, Shen M, Zhu X (2019) Plant phenology and global climate change: current progresses and challenges. Glob Chang Biol 25: 1922–1940

Pospišil P (2016) Production of reactive oxygen species by photosystem II as a response to light and temperature stress. Front Plant Sci 7:1950. https://doi.org/10.3389/fpls.2016.01950

Primack RB, Higuchi H, Miller-Rushing AJ (2009) The impact of climate change on cherry trees and other species in Japan. Biol Conserv 142:1943–1949

Prunier J, Laroche J, Beaulieu J, Bosquet J (2011) Scanning the genome for gene SNPs related to climate adaptation and estimating selection at the molecular level in boreal black spruce. Mol Ecol 20(8):1702–1716

Przelomska NAS, Armstrong CG, Kistler L (2020) Ancient plant DNA as a window into the cultural heritage and biodiversity of our food system. Front Ecol Evol 8(74):1–8

Purugganan M, Fuller D (2009) The nature of selection during plant domestication. Nature 457: 843–848

Raf H, Konrad H, Curtu LA, Geburek T (2018) Genetic evidence of human mediated, historical seed transfer from the Tyrolean Alps to the Romanian Carpathians in *Larix decidua* (Mill) forests. Ann Forest Sci 75(4):98–111. https://doi.org/10.1007/s13595-018-0776-9

Ramírez F, Davenport TL (2012) Mangoes in Colombia. In: Valavi SG, Rajmohan K, Govil JN, Peter KV, Thottappilly G (eds) The mango. Studium Press, USA, pp 346–358

Ramirez F, Kallarackal J (2015) Responses of fruit trees to global climate change. Springer Briefs in Plant Science. pp 47

Ramírez F, Davenport TL, Fischer G, Pinzón JCA, Ulrichs C (2014) Mango trees have no distinct phenology: the case of mangoes in the tropics. Sci Hortic 168:258–266

Rančić D, Pećinar I, Aćić S, Stevanović ZD (2019) Morpho-anatomical traits of halophytic species. In: Hasanuzzaman M, Shabala S, Fujita M (eds) Halophytes and climate change: adaptive mechanisms and potential uses. CABI Publisher, Wallingford, pp 152–178

Ravencroft CH, Whitlock R, Fridley JD (2015) Rapid genetic divergence in response to 15 years of simulated climate change. Glob Chang Biol 21:4165–4176

Rejon MR (2021) Is there genetic evidence of climate change? Science environment. BBVA Open Mind. https://www.bbvaopenmind.com/en/science/environment/is-there-genetic-evidence-of-climate-change/

Renner SS, Zohner CM (2018) Climate change and phenological mismatch in trophic interactions among plants, insects, and vertebrates. Annu Rev. Ecol Evol Syst 49:165–182

Revilla TS, Encinas-Viso F, Loreau M (2015) Robustness of mutualistic networks under phenological change and habitat destruction. Oikos 124:22–32
Richardson DA, Duigan SP, Berlyn GP (2002) An evaluation of noninvasive methods to estimate foliar chlorophyll content. New Phytol 153(1):185–194
Ross-Ibarra J, Morrell PL, Gaut BS (2007) Plant domestication, a unique opportunity to identify the genetic basis of adaptation. Proc Natl Acad Sci U S A 104(1):8641–8648
Rozendaal DMA, Hurtado VH, Poorter L (2006) Plasticity in leaf traits of 38 tropical tree species in response to light; relationships with light demand and adult stature. Funct Ecol 20:207–216
Rundel PW, Arroyo MTK, Cowling RM, Keeley JE, Lamont BB, Pausas JG, Vargas P (2018) Fire and plant diversification in Mediterranean-Climate Regions. Front Plant Sci 9:851. https://doi.org/10.3389/fpls.2018.00851
Sánchez AR, Pineda JE, Casas XM, Calderón. (2021) Influence of edaphic salinity on leaf morphoanatomical functional traits on juvenile and adult trees of red mangrove (*Rhizophora mangle*): implications with relation to climate change. Forests 12:1586. https://doi.org/10.3390/f12111586
Santamaría L (2002) Why are most aquatic plants widely distributed? Dispersal, clonal growth and small-scale heterogeneity in a stressful environment. Acta Oecol 23:137–154
Santamaría L, Figuerola J, Pilon JJ, Mjelde M, Green AJ, De Boer T, King RA, Gornall RJ (2003) Plant performance across latitude: role of plasticity and local adaptation in aquatic plant. Ecology 84:2454–2461
Scheffers BR, De Meester L, Bridge TCL, Hoffmann AA, Pandolfi JM, Corlett RT, Butchart SHM et al (2016) The broad footprint of climate change from genes to biomes to people. Science 354: 6313
Schlichting CD (1986) The evolution of phenotypic plasticity in plants. Annu Rev. Ecol Evol Syst 17(1):667–693
Shipunov A (2021) Introduction to botany. June 7, 2021 version. 192 p
Sierla M, Waszczak C, Vahisalu T, Kangasjärvi J (2016) Reactive oxygen species in the regulation of stomatal movements. Plant Physiol 171(3):1569–1580
Silvertown J, Araya Y, Gowing D (2015) Hydrological niches in terrestrial plant communities: a review. J Ecol 103:93–108
Singh B, Sharma RA (2015) Plant terpenes: defense responses, phylogenetic analysis, regulation and clinical applications. Biotechnology 5(2):129–151
Smith BD (1998) The emergence of agriculture. W. H. Freeman, New York
Smýkal P, Nelson MN, Berger JD, Von Wettberg EJ (2018) The impact of genetic changes during crop domestication. Agronomy 8:119
Stangler DF, Hamann A, Kahle HP, Spiecker H, Mäkelä A (2017) A heat wave during leaf expansion severely reduces productivity and modifies seasonal growth patterns in a northern hardwood forest. Tree Physiol 37(1):47–59
Stephanou M, Manetas Y (1997) The effects of seasons, exposure, enhanced UV-B radiation and water stress on leaf epicuticular and internal UV-B absorbing capacity of *Cistus creticus*: a Mediterranean field study. J Exp Bot 48:1977–1985
Sthapit BR, Ramanatha RV, Sthapit SR (2012) Tropical fruit tree species and climate change. Bioversity International, New Delhi
Stone M (2022) Can we hack DNA in plants to help fight climate change? National Geographic. https://www.nationalgeographic.com/environment/article/can-we-hack-plant-dna-to-suck-up-more-carbon-emissions
Subedi SC, Ross MS, Redwine JP, Baraloto C (2019) Trait-based community assembly pattern along a forest succession gradient in a seasonally dry tropical forest. Ecosphere 10(4):e02719. https://doi.org/10.1002/ecs2.2719
Taranto F, Nicolia A, Pavan S, De Vita P, Agostino ND (2018) Biotechnological and digital revolution for climate-smart plant breeding. Agronomy 8:277. https://doi.org/10.3390/agronomy8120277

The Arabidopsis Genome Initiative (2000) Analysis of the genome sequence of the flowering plant Arabidopsis thaliana. Nature 408:796–815

Tian M, Yu G, He N (2016) Leaf morphological and anatomical traits from tropical to temperate coniferous forests: mechanisms and influencing factors. Sci Rep 6(1):19703

Via S, Gomulkiewicz R, De Jong G, Schlichting CD, Van Tienderen PH (1995) Adaptative phenotypic plasticity: consensus and controversy. Trends Ecol Evol 10:212–217

Vitorino LC, Reis MNO, Bessa LA, de Souza UJB, Silva FG (2020) Landscape and climate influence the patterns of genetic diversity and inbreeding in Cerrado plant species. Diversity 12(11):421. https://doi.org/10.3390/d12110421

Vogel E, Donat MG, Alexander LV, Meinshausen M, Ray D, Karoly D, Mainhausen N, Frieler K (2019) The effects of climate extremes on global agricultural yields. Environ Res Lett 14: 054010

Vu TV, Das S, Hensel G, Kim JY (2022) Genome editing and beyond: what does it mean for the future of plant breeding? Planta 255(6):130. https://doi.org/10.1007/s00425-022-03906-2

Wahid A (2003) Review: physiological significance of morpho-anatomical features of halophytes with particular reference to Cholistan flora. https://www.researchgate.net/publication/242186882_Review_Physiological_Significance_of_Morpho-Anatomical_Features_of_Halophytes_with_Particular_Reference_to_Cholistan_Flora

Wang M, Zhang J, Guo Z, Guan Y, Qu G, Guo Y, Yan X (2020) Morphological variation in Cynodon dactylon (L.) Pers., and its relationship with the environment along a longitudinal gradient. Hereditas 157(1):4. https://doi.org/10.1186/s41065-020-00117-1

Wang J, Ouyang F, An S, Wang L, Xu N, Ma J, Wang J, Zhang H, Kong L (2022) Variation, coordination, and trade-offs between needle structures and photosynthetic-related traits across five *Picea* species: consequences on plant growth. BMC Plant Biol 22:242. https://doi.org/10.1186/s12870-022-03593-x

Wellmer F, Riechmann JL (2010) Gene networks controlling the initiation of flower development. Trends Genet 26(12):519–527

Wen Z, Li H, Shen Q, Tang X, Xiong C, Li H, Pang J, Ryan M, Lambers H, Shen J (2019) Trade-offs among root morphology, exudation and mycorrhizal symbiose for phosphorus-acquisition strategies of 16 crop species. New Phytol 223(2):882–895. https://doi.org/10.1111/nph.15833

Wong A (2018) The art of botany: the effect of drought on plant anatomy. Horizons 3. https://scholarspace.manoa.hawaii.edu/items/cea49aed-7d43-401e-b13d-217660a3616f/full. Accessed 09 Aug 2022

Wright IJ, Reich PB, Westoby M, Ackerly DD, Baruch Z, Bongers F, Cavender-Bares J, Chapin T, Cornelissen JHC, Diemer M, Flexas J, Garnier E, Groom PK, Gulias J, Hikosaka K, Lamont BB, Lee T, Lee W, Lusk C, Villar R (2004) The worldwide leaf economics spectrum. Nature 428(6985):821–827

Wu Z, Chen L, Long YL (2009) Analysis of ultrastructure and reactive oxygen species of hyperhydric garlic (*Allium sativum* L.) shoots. In Vitro Cell Dev Biol Plant 45(4):483–490

Wu W, Liao T, Du K, Wei H, Kang X (2021) Transcriptome comparison of different ploidy reveals the mechanism of photosynthetic efficiency superiority of triploid poplar. Genomics 113(4): 2211–2220. https://doi.org/10.1016/j.ygeno.2021.05.009

Xiao Z, Ma G, Bai X, Li J, Zhao M, Su L, Ma H, Zhou H (2022) The influence of leaf anatomical traits on photosynthetic in coffee leaves in Yunnan of China. BioRxiv. https://doi.org/10.1101/2022.02.18.481015

Xie K, Lu Z, Pan Y, Gao L, Hu P, Wang M, Guo S (2019) Leaf photosynthesis is mediated by the coordination of nitrogen and potassium: the importance of anatomical-determined mesophyll conductance to CO_2 and carboxylation capacity. Plant Sci 290:110267. https://doi.org/10.1016/j.plantsci.2019.110267

Xiong D, Flexas J (2018) Leaf economics spectrum in rice: leaf anatomical, biochemical and physiological trade-offs. J Exp Bot 69(22):5599. https://doi.org/10.1093/jxb/ery322

Xu SY, Weng JK (2020) Climate change shapes the future evolution of plant metabolism. Adv Genet 1:e10022

Yadav S, Mishra A, Jha B (2018) Elevated CO_2 leads to carbon sequestration by modulating C4 photosynthesis pathway enzyme (PPDK) in *Suaeda monoica* and *S. fruticosa*. J Photochem Photobiol B Biol 178:310–315

Yuan L, Tang L, Zhu S, Hou J, Chen GH, Liu F, Liu S, Wang C (2017) Influence of heat stress on leaf morphology and nitrogen-carbohydrate metabolisms in two wucai (*Brassica campestris* L.) genotypes. Acta Soc Bot Pol 86(2):3554. https://doi.org/10.5586/asbp.3554

Yuan F, Gui J, Shabala S, Wang B (2019) Reproductive physiology of halophytes: current standing. Front Plant Sci 9:1954. https://doi.org/10.3389/fpls.2018.01954

Zhou X, Xiang Y, Li C, Yu G (2020) Modulatory role of reactive oxygen species in root development in model plant of *Arabidopsis thaliana*. Front Plant Sci 11:485932. https://doi.org/10.3389/fpls.2020.485932

Zhu J, Jiang L, Zhang Y (2016) Relationships between functional diversity and aboveground biomass production in the Northern Tibetan alpine grasslands. Sci Rep 6:34105. https://doi.org/10.1038/srep34105

Impact of Anthropogenic Compounds on Biodiversity: A Comprehensive Analysis 19

Mariadoss Alphonse, Kalaivani Thiagarajan, Nagaraj Nallakaruppan, William Raja Tharsius Raja, Sushmita Paul, Sangamithirai Jaichandran, Aravind Mukundan, Inocencio E Buot, Jr, Michael Pillay, Sunil Nautiyal, Siva Ramamoorthy, and Rajasekaran Chandrasekaran

Abstract

Anthropogenic chemicals are substances that are created mainly in industries for agriculture, medicine, and other commercial purposes. When exposed to the environment at moderate to high levels, they become toxic to plants, humans, and animals. The continuous addition of toxins to the environment causes their accumulation in the tissues of the organisms that consume the lower trophic levels. Gradually the levels increase in concentration as they reach the higher

Kalaivani Thiagarajan and Rajasekaran Chandrasekaran contributed equally with all other contributors.

M. Alphonse · K. Thiagarajan · N. Nallakaruppan · S. Paul · S. Jaichandran · A. Mukundan · S. Ramamoorthy · R. Chandrasekaran (✉)
Department of Biotechnology, School of Bio Sciences and Technology, Vellore Institute of Technology, Vellore, Tamil Nadu, India

W. R. T. Raja
Division of Ethnopharmacology, Entomology Research Institute, Loyola College, Chennai, Tamil Nadu, India

PG and Research Department of Biotechnology, Bishop Heber College (Autonomous), Tiruchirappalli, Tamil Nadu, India

I. E. Buot, Jr
Institute of Biological Sciences, College of Arts and Sciences, University of the Philippines Los Baños, Laguna, Philippines

M. Pillay
Erudite Editing and Science Writing Solutions, Vanderbijlpark, South Africa

S. Nautiyal
Centre for Ecological Economics and Natural Resources, Institute for Social and Economic Change, Bangalore, India

S. Ramamoorthy et al. (eds.), *Plant Diversity in Biocultural Landscapes*,
https://doi.org/10.1007/978-981-19-8649-9_19

trophic levels (food chain energetics). Anthropogenic chemicals become major environmental pollutants, affecting biodiversity and ecosystem functions by habitat destruction. Therefore, it is necessary to monitor and assess. Anthropogenic activities lead to the degradation of land and water bodies. This chapter discusses the major sources and ecological effects of anthropogenic chemicals in plants and animals and the mechanism of biomagnification on the environment. We also discuss various measures required to mitigate the effects of anthropogenic chemicals or find alternatives to the chemicals to conserve our ecosystem with favorable conditions for organisms to grow and evolve.

Keywords

Anthropocene chemicals · Climate change · Biodiversity · Ecosystem · Soil food web

19.1 Introduction

Paul Crutzen coined the phrase "Anthropocene" to describe the anthropogenic influence on the Earth (Crutzen and Brauch 2016). It refers to outcomes brought about by the effect of human activity on nature. The rise of agriculture, deforestation, and extensive species trade are characteristics of the Anthropocene (Zhu and Penuelas 2020). Anthropogenic pollution is a term that describes a particular kind of contamination due to human activity, such as mining, industrial combustion, burning fossil fuels, etc. To comprehend and manage these Earth's fluctuations, humanity and the scientific community, in particular, have thus been active for nearly 30 years. It is critical to consider how chemistry affects the Anthropocene in this setting. Chemistry is fundamentally the science of change. Chemists can change matter, and this power to alter reality can be frightening. Irrational chemophobia is currently widely disseminated by traditional and social media in the contemporary climate of widespread fear. Therefore, it is critical to restore chemistry's status for both the present and future generations (Lutz 2017). Due to these artificial contaminants, the environment has undergone dramatic changes that have changed the ecosystem's functioning, endangering animals and plant life and the interactions among organisms. The recent changes in global climate have made the detrimental effects of man-induced environmental degradation quite apparent. Unexpected rainstorms, glacier melting that causes floods, and other unforeseen weather phenomena are all prevalent. Anthropogenic pollutants include greenhouse gases, sulfur oxides, nitrogen oxides, heavy metals, and aerosols such as volatile organic compounds, which have significantly and abruptly disrupted the ecosystem. The number and concentration of anthropogenic chemical sources have risen, posing a serious threat to the environment due to the growing global population and demand for energy. The main sources include transportation, industries like metallurgy or building, and the combustion of fossil fuels for energy (Vigiak et al. 2021). One of the primary characteristics of the Anthropocene from the beginning of the industrial revolution in Europe in the nineteenth century is the detectable pollution fingerprint that resulted from it (Logemann et al. 2022). Manufactured chemicals, including heavy metals, polyfluoroalkyl compounds, polychlorinated biphenyls (PCBs), and other conventional, emerging, and unidentified inorganic and organic pollutants can all be classified as pollution fingerprints (Gałuszka et al. 2020).

Anthropogenic pollution has been a burden to the environment since the dawn of industries. The industrial revolution resulted in increased pollution in the atmosphere (burning of fossil fuels), water bodies (draining of effluents), and soil (excessive use of pesticides, fungicides, etc.), giving rise to adverse health effects on humans, animals, plants, and marine organisms. Biomagnification plays a role where the toxicity concentration progressively increases on every trophic level until it reaches the top of the food chain. The toxins accumulate in the tissues of the organisms and cause moderate to adverse health defects depending on the level of toxicity. Reduced biodiversity could result from the effects of toxic artificial chemicals on the environment's living things (Rhind 2009).

According to the International Union for Conservation of Nature's (IUCN) Red List of threatened and endangered species, a quarter of mammal species, an eighth of bird species, and more than 40% of amphibian species are threatened; thousands of species of invertebrates and plants are also in danger (McGill et al. 2015; McBeath and Leng 2006). According to the Living Planet Index, there are 52% fewer vertebrate populations today than forty years ago (McGill et al. 2015). A sixth major mass extinction, comparable to the preceding five verified mass extinctions, is being discussed. Many ecologists are examining problems of the implications of this predicted reduction of biodiversity on humans because the significant negative impact is widely acknowledged (Hance 2015).

Using synthetic chemical pesticides has negatively impacted the abiotic and biotic aspects of the ecosystem. Thus, this review aims to assess the major sources of Anthropocene chemicals and their effect on the environment.

19.2 Major Sources of Anthropogenic Chemicals

With the rise in global population and ongoing need for energy, the quantity and concentration of anthropogenic chemical sources have increased, posing a grave danger to the environment. Primary sources include industries like metallurgy construction, transportation, and burning fossil fuels to produce energy (Lewis et al. 2011). Two sources, stationary and mobile, can be used to classify anthropogenic chemical emissions. The sources that release anthropogenic chemicals from a permanent location are known as stationary sources. Examples of stationary sources include traditional biomass burning and the smoke from factories, power plants, and waste incinerators. Emissions from ships, aircraft, and automobiles all count as mobile sources (Popescu and Ionel 2010). According to concentration and environmental consequences, the five biggest anthropogenic chemicals are carbon monoxide (49.1%), sulfur oxides (16.4%), nitrogen oxides (14.8%), volatile organics (13.6%), and particulates (6%).

Carbon Monoxide The transport sector contributes 75% of the CO produced, residential wood burning 10%, and industrial processes 15%. Carbon monoxide is mainly produced as an intermediate by-product of combustion in piston engines and boilers. Each CO molecule undergoes oxidation, producing one ozone molecule and a carbon dioxide molecule. The combustion chamber must be kept at a minimum temperature of 990 K for an extended period for carbon monoxide to be converted to carbon dioxide. If the temperature is not high enough or the chamber's homogeneity is disturbed, CO will not oxidize (Rosen and Van Metre 2010).

Sulfur Oxides Approximately 67% of Sox pollution is due to natural sources, and the remaining is due to fossil fuel burning, transportation, and industrial activities. Due to the burning of fossil fuels, mostly coal, most of the anthropogenic sulfur content is accounted for and completely oxidized into sulfur dioxide during complete combustion. When there is insufficient oxygen, partial combustion occurs, forming sulfur and hydrogen sulfide. Additionally, hydrogen sulfide can be created when low-quality coal is burned, as well as in piston engines with low air-to-fuel ratios (Battista et al. 2021).

Nitrogen Oxides According to a recent estimate, the burning of fossil fuels, road transport, public power, and the heating sector are the main global sources of NOx emissions, accounting for 33 Tg annually. When the nitrogen component of the fuel and the combustion air undergo oxidation, NOx is generated (Copley 2009). Only NO is generated in the initial phase. Then, NO2 is formed after the combustion when oxygen is more available in the atmosphere. There is a risk of ozone depletion, which might result in global warming, as NOx levels in the atmosphere rise.

Particulates In the atmosphere, solid and liquid droplets combine to form particulate matter. They can be acquired either naturally or through human activity. The natural sources include biological sources, ocean spray, volcanic eruptions, and forest fires. Transportation, power plants, woodstoves, fuel combustion, industrial and agricultural operations are all examples of anthropogenic sources of particulates.

Volatile Organic Chemicals (VOCs) Given the large number of VOC species generated by natural and anthropogenic sources, it is quite challenging to distinguish between them. The main source of VOCs is the use of fossil fuels in the chemical and transportation industries. Mobile sources include exhaust emissions and emissions from fuel distribution and transportation, such as from ships, tankers, and gas stations. Refining and chemical-related businesses are examples of stationary sources. Soon after being released into the atmosphere, the VOCs are oxygenated by photochemical reactions, which eventually cause their elimination from the atmosphere. Atmospheric radicals such as hydroxides, ozone, nitric oxides, and chlorine are necessary to initiate this reaction.

19.2.1 Agriculture

A significant portion of naturally occurring land has been transformed into human-influenced biomes during the Anthropocene, which comprises around 75% of the total land. The pressure on the remaining land is only worsening due to the rapidly increasing human population and the increasing ecological impact per person (Ellis and Ramankutty 2008). In addition, most land use activities affect the ecosystem quality. For instance, industrial and mining activities, pest management, and chemical treatment of the land and infrastructure are prevalent practices. Applying fertilizers and plant protection materials is a frequent and growing agricultural practice that mostly contaminates the land (Hakoun et al. 2017).

Pesticides are created to be poisonous to fungi, insects, or weeds that hinder the growth of plants. Although pesticides make food production more efficient, they also have many unintended consequences since they eradicate various wildlife species, including bees, earthworms, and mammals. Over the ages, pesticides (e.g., DDT, dieldrin, agent orange) have remained dangerous and long-lasting pollutants affecting biodiversity. Increased use of pesticides for several decades causes the accumulation of toxic chemicals in soil, which influence the food chain. Therefore, these pesticides have been crucial in destroying natural resources, ecosystems, and biodiversity (Boatman et al. 2007).

Our pesticide-intensive agriculture paradigm has been identified as a significant contributor to biodiversity loss. Pesticides can remain in the environment for decades and constitute a global threat to the entire ecological system, thus affecting food production worldwide. Using inappropriate or excessive pesticides contaminates neighboring soil and water sources, destroys beneficial insect populations that act as pests' natural enemies, and reduces the nutritional value of local food sources. A loss of biodiversity is another effect of pesticides (Pesticides and the loss of biodiversity 2010).

19.2.2 Pharmaceutical

A global issue that presents more difficulties for nations with rapid population growth is the rise in pharmaceutical levels in the environment and their possible negative effects on biological systems. Research suggests that pharmaceuticals endanger genetic variety, species diversity, and community diversity when incorporated into biological organisms and ecosystems (Néstor and Mariana 2017).

Both human health and biodiversity are in danger when pharmaceuticals are released into the environment. Pharmaceutical product residues have polluted the environment since the second half of the twentieth century. These residues can adversely impact the growth and physiology of many unicellular and multicellular species and cellular metabolism. In many ecosystems and geographical locations worldwide, their presence has been detected in microbes, plants, invertebrates, and vertebrates (Daughton and Ternes 1999; Küster and Adler 2014).

Pharmaceuticals are emerging pollutants. Despite being less common, they are widely dispersed globally. Contrary to other contaminants, pharmaceutical compounds come in a variety of forms. Around 4000 different kinds are available globally for human and veterinary use, and 600 have been distributed in terrestrial and aquatic ecosystems (Arnold et al. 2014). Although this kind of pollution is not physically obvious, there is evidence that it can negatively affect biodiversity at many scales, endangering the long-term sustainability of genetic diversity, populations, and ecosystems (Brodin et al. 2014).

Among the significant factors affecting biodiversity, pharmaceutical compounds top the list. Globally, civilization has hastened the extinction of species, populations, and the destruction of environments. The pressures on biodiversity are increasing rather than decreasing, and new deteriorating agents like contaminants of pharmaceutical origin are also starting to appear (Dirzo et al. 2014).

19.2.3 Industrials

Natural ecosystems have a complicated structure because they typically contain diverse groups of organisms. Due to their location and the uniqueness of layering processes, these ecosystems generally are distinguished by distinct spatial features and temporal aspects brought on by the successional patterns of the current populations. All populations of organisms that live in a certain habitat develop a variety of interactions with one another and the abiotic environment. Natural phenomenon dynamics and human-made interventions have a significant impact on these interactions (Cismasiu et al. 2017).

Rapid industrialization and aggressive economic expansion severely influence biodiversity, causing a loss of green cover, the gradual or sudden extinction of many species, and global environmental deterioration. The five industrial chemicals most frequently emitted include sulfuric acid, ethylene, sodium hydroxide, propylene, and nitrogen gas. The number of heavy metals and other pollutants produced has a detrimental effect on the environment, poisoning flora and fauna (Sahu and Lal 2021). Heavy metal contamination by industrial activity reduces microbial biomass. Even if it does decrease the species population, it also lowers biodiversity or disturbs community structure. Industrialization has a huge negative influence on biodiversity, which has caused species extinction over time and sometimes quite quickly (Chen et al. 2000).

The diversity and activity of species populations have been negatively impacted by high concentrations of heavy metals in the soil from industrial contaminants. When processing operations like scouring, bleaching, mercerizing, dyeing, printing, and finishing in the textile industries, a variety of chemicals, including acids, alkalies, oil detergents, dyes, aromatic amines, etc. are used, producing significant amounts of pollutants that harm biodiversity and ecosystem (Sahu and Lal 2021) (Table 19.1).

Table 19.1 Effect of different pollutants on the ecosystem

Category	Potential contaminants	Effect on ecosystem	Reference
Agriculture	Fertilizer (nitrates and phosphates)	Exceeds uptake capacity of plants	Burri et al. (2019)
	Pesticides and herbicides (2,4-dichlorophenoxyacetic acid, pentachlorophenol, DDT, BHC, and parathion)	Life cycle change of organisms, effects on the structure and function of ecosystem	Burri et al. (2019), Katayama et al. (2015)
	Antibiotics and hormones	Significant risk to the receiving environment	Burri et al. (2019)
Pharmaceuticals	Anti-inflammatory drugs (17b-oestradiol and 17a-ethinyloestradiol)	Pharmaceutical chemicals caused a population-level impact on nontarget wildlife	Arnold et al. (2014)
	Antibiotics, B-blockers, X-ray, contrast media, cosmetic, personal care products, surfactants, and phytosanitary products	Alterations in behavior, reproduction system, stress, community composition, and structure	Néstor and Mariana (2017)
Industrials	Organic pollutants (polycyclic aromatic hydrocarbons)	Life cycle changes of organisms and decline in the quantity and density of the population	Yuan et al. (2020)
	Spills and leaks (benzene, toluene, ethylbenzene, and xylene)	Safety threats to the aquatic and land ecosystem	Burri et al. (2019)
	Heavy metals (Cd, Pb, Cu, As, Zn, and Cr)	Transport and enrichment along the food chain, inhibit the functions of producers, consumers, and decomposers and affect the stability of the entire ecosystem	Ashraf et al. (2019)

19.3 Ecological Effects of Anthropogenic Chemicals

Understanding anthropogenic change is essential for knowing both historical and current changes in our world. Anthropogenic changes may be an unnoticed or poorly understood side effect of human activity, as with the declining biodiversity that comes with increased urbanization or with much of the pollution brought on by industrialization and the technological advancements of the twentieth century. The rising production of carbon dioxide and other greenhouse gases and the subsequent alteration in the global climate are two human impacts that have been gradually revealed over the past few decades. Since humans learned how to process metals, they have been adding to the environmental burden. Anthropogenic pollution is not a recent occurrence. However, the composition and distribution of contaminants in the environment have changed as new compounds have been discovered (Vigiak et al.

2021). Due to the urbanization of the population, there has been increased air and water pollution from burning fossil fuels and organic pollutants like sewage. These have harmed human health, ruined river ecosystems, and killed off fish and other species (Rhind 2009).

While heavy metal pollutants, such as lead and copper, are naturally occurring elements that are concentrated through industrial processes, such as smelting, some organic pollutants, such as PAHs, are mostly caused by the incomplete burning of fossil fuels, though they can also be caused by natural forest fires (Younis and Younis 2018). On the other hand, several synthetic organic substances, such as PCBs, PBDEs, and various organochlorine insecticides, can interfere with endocrine function. Recently, nitrates, widely used as fertilizers, have been linked to animal physiological disruption. Potential toxic element (PTE) accumulation caused by mining activities results in unnatural enrichment, ecological contamination, and environmental degradation. Anthropogenic activities like mining are well known for contaminating ecological resources. The effluents produced significantly impact the physical, chemical, and biological aspects of the resources involved (Jabbar Khan et al. 2020).

19.3.1 Effect on Animals

It is recognized that typical anthropogenic agents like pesticides can depress amphibians' immune systems. Atrazine, a herbicide, lowers leukocyte production and raises host susceptibility to Ranavirus, while chlorpyrifos, an insecticide, boosts Ranavirus infection rates in tiger salamanders (North et al. 2015).

These recently created compounds, such as phthalates, polybrominated diphenyl ethers (PBDEs), polychlorinated biphenyls (PCBs), and polycyclic aromatic hydrocarbons (PAHs), which are present in the environment, have a variety of chemical properties. Organochlorine compounds, such as DDT, are also very effective insecticides. Many of these compounds have been discovered to impact animal physiology, highlighting the significance of this change in the kind and quantity of pollutants. They are referred to as endocrine-disrupting chemicals (EDCs) because they can interfere with normal endocrine function (Fox 2004; Rhind 2012).

Diurnal rhythms have a crucial role in both plants and animals. The circadian rhythm provides a 24-h day/night cycle, an internal timekeeping mechanism that enables animals to coordinate behavioral, physiological, and molecular activities chronologically. After being first identified in *Drosophila*, its molecular architecture was later found to be substantially conserved in various other species (Zheng et al. 2021). A disruption of the circadian rhythm system leads to physiological malfunction. The impact of pollutants on circadian relationships in people and animals is poorly understood because past research has mostly focused on figuring out fundamental ideas and concepts. The regulation of the circadian rhythm in aquatic species is shown to be disrupted by several environmental pollutants, both natural and artificial, which are discharged into surface waters by industries, hospitals, and

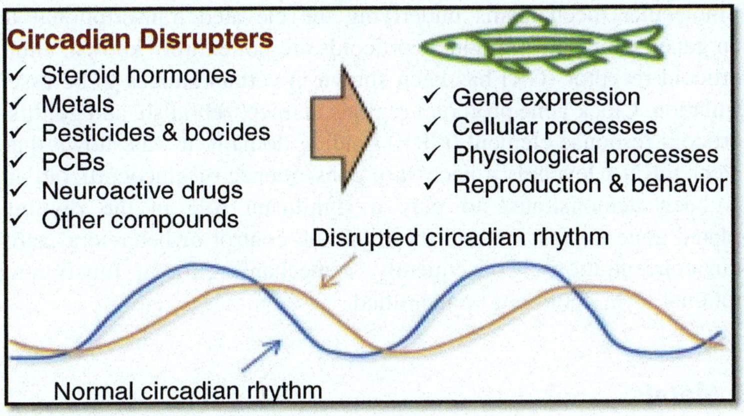

Fig. 19.1 Environmental chemicals affect circadian rhythms (Zheng et al. 2021)

domestic/municipal wastewater. These substances are circadian disrupters (Fang et al. 2017) (Fig. 19.1).

There are more than 40 environmental circadian disruptors in fish, both natural and artificial. They fall into six broad categories: neuroactive pharmaceuticals, cyanobacterial toxins, metals, pesticides and biocides, polychlorinated biphenyls, pesticides and biocides, and steroid hormones. Metals and steroid hormones are the subjects of most research. At the molecular and physiological levels, zebrafish revealed circadian dysregulation, particularly for progestins and glucocorticoids, with major behavioral abnormalities.

Anthropogenic effects, such as aquatic animals being exposed to environmental toxins or cyanobacterial toxins that are promoted because of eutrophication and climatic warming, were recognized as variables influencing circadian rhythms and natural ecological influences. Numerous aquatic creatures, particularly fish, showed circadian disruption after exposure to various environmental chemicals.

19.3.2 Steroid Hormones

In zebrafish exposed to fludrocortisone, the circadian rhythm network in the brain underwent the most significant alterations due to the transcriptional upregulation of per1a and nr1d1. Zebrafish eggs prematurely hatched, F1 embryos developed faster, and the swimming patterns of larvae were altered because of exposure.

Progestins, a significant subgroup of steroid hormones, have also been intensively researched for their circadian-disrupting effects. After being exposed to progesterone for 14 days, core circadian genes such as nr1d2b and per1b were markedly downregulated in the adult zebrafish brain, while nr1d1 was dramatically downregulated in the ovary (Zucchi et al. 2013).

The molecular mechanisms underlying the elevated transcriptional levels of circadian genes in response to glucocorticoids are not entirely known, although the glucocorticoid receptor (GR) has been shown in certain studies to be essential for such regulation. Clock gene promoter regions in mice, zebrafish, and goldfish have a glucocorticoid response element (GRE) binding domain. It was shown that a rapid rise in Perl mRNA levels is a necessary consequence of glucocorticoid signaling. GR has been demonstrated to play a significant role in the rhythmicity of clock-related gene expression and the rhythmic control of behavioral activity and melatonin concentrations. Consequently, a mechanism for gr functioning at the transcriptional level could not be identified.

19.3.3 Metals

Another major group of environmental circadian disrupters is metals, particularly transition metals. Lithium was found in the first study to affect the circadian rhythm in goldfish, which was published in 1981. Goldfish circadian rhythms were slowed and disturbed by long-term (30 days) lithium administration, similar to the effects seen in mammals. Lithium treatment significantly increased the expression of the circadian genes clock1a and bmal1b in zebrafish (Xiao et al. 2017).

The circadian rhythm is also disrupted by cadmium. Zebrafish larvae exposed to 5 ppm of cadmium chloride showed altered clock1a, bmal1b, and per2 transcript levels, which showed considerable upregulation during the light phase. Additional metabolic pathways were impacted, including proteolysis, amino acid metabolism, and autophagy (Xiao et al. 2016).

19.3.4 Pesticides and Biocides

The herbicide atrazine caused fish to exhibit circadian disturbance. It induces lipid peroxidation, alters liver metabolism, and is an endocrine disruptor. The average behavioral intensity of goldfish significantly decreased after 6 days of exposure to 0.003 mg/L, and the circadian rhythm phase switching pattern was modified. When atrazine was administered, the zebrafish's activity level decreased, and their circadian rhythm's periodicity showed a noticeable 1-h delay (Zheng et al. 2021).

19.3.5 Effect of Anthropogenic Chemicals on Plants

Some pollutants discharged into the environment due to human activities include phthalate esters, petroleum hydrocarbons, heavy metals, carbon monoxide, sulfur oxides, and nitrogen oxides. As a result of the build-up of these pollutants, the global climate changes, which in turn influence rainfall patterns, droughts, and other aspects of climatic mean and variation, significantly impact terrestrial ecosystems (Fuchs et al. 2018). When these anthropogenic chemicals build up in plants, it might result in chromosomal anomalies such as bridges, dispersion, and premature movements. The concentration of anthropogenic chemicals rises along with population growth due to emissions from industrial activities, vehicle combustion, and agricultural practices using pesticides, herbicides, additives, and other chemicals for commercial purposes (Jürgens and Bischoff 2017). New chemicals and physical agents introduced to plant populations may alter their development. The environment is frequently impacted by xenobiotic substances released from human-made sources and atmospheric and terrestrial contaminants, which affect plants' meiotic cells. Reactive oxygen species, like ozone, may stress out sensitive plants physically and environmentally, which can lead to impacts such as decreased flowering and seed production in legumes. They can also destroy molecules that are important for plant signaling. The photosynthetic efficiency, chlorophyll content, plant density, green leaf biomass, and plant health have all been proven to be impacted by exposure to anthropogenic pollutants. Anthropogenic pollution has resulted in a major loss in plant variety, which has allowed weed species to dominate. The supply of nitrogen oxides to the terrestrial ecosystem has increased due to recent industrial upgrades. This deposition can affect the soil's pH and deplete calcium and other cations that plants require. Anthropogenic volatile pollutants from industry and transportation significantly influence plant signaling and their interference with the plant's chemical metabolisms and affect communication between pollinators. Plants emit volatile chemical molecules that function as olfactory cues. The communication between plants and pollinators is hampered by anthropogenic pollutants, although these VOCs also aid in defense against viruses and folivores.

Heavy metals naturally occur in the environment, but when released in large concentrations by anthropogenic causes, they may have devastating impacts. Because of different human activities, including mining, industries, excessive use of hazardous chemicals in fertilizers, and pesticides in agricultural fields, the soil may be detected with heavy metals such as cadmium, copper, lead, chromium, etc. These heavy metals have a significant impact on the metabolic processes of plants.

Plants uptake these heavy metals, which are soluble components in the soil solution. If they are below a certain concentration, they are beneficial. When an excessive amount is absorbed, it causes toxic effects such as damage to cell structure due to oxidative stress and inhibition of cytoplasmic enzymes, reduced photosynthetic activities, plant mineral nutrition, and reduced activity of some enzymes. As is known, microbial growth in the soil is also essential for the development of the plant. The presence of these heavy metals negatively influences these microorganisms, leading to a decrease in the decomposition of organic matter and low soil fertility. Even at extremely low concentrations, metals, including Pb, Cd, and Hg, have negatively impacted plant development. It is well known that cadmium adversely affects the shoot and root development in wheat plants.

Overexposure to copper causes oxidative stress and ROS in plants, which lowers plant mortality, biomass production, seed production, and root deformity. Damage to macromolecules and disruption of metabolic pathways result from oxidative stress. The younger leaves develop chlorosis as a result of zinc poisoning. A surplus of zinc can result in a copper and manganese shortage. Another indication of Zn poisoning is the development of a purplish red tint in the leaves, which is related to phosphorus deficiency. Metal poisoning can change the plasma membrane's permeability, leading to a drop in water content; Cd has been proven to affect the water balance in plants (Pichhode and Nikhil 2016).

As a typical sign of Mn poisoning, necrotic brown staining on leaves, petioles, and stems occurs. Too much Mn prevents a Fe-related mechanism that would otherwise impede chlorophyll production. Chlorosis due to Mn build-up in the shoot and root shortens these organs. The excess of Fe+2 creates free radicals, which permanently alter the cellular structure and harm proteins, DNA, and membranes. Plants raised on soil with high Ni+2 levels displayed nutrient imbalance impairment, and abnormal cell membrane function. According to research on *Oryza sativa* shoots, Ni+2 impacted the plasma membrane's lipid content and H-ATPase function. Adding calcium and phosphorus to the soil can decrease the lead concentrations that plants take up via their roots.

19.4 Effect of Anthropogenic Chemicals on Biodiversity

According to Rockström et al. (2009), chemical pollution is one of the planetary boundaries that should not be overlooked to protect humanity. One in six deaths is premature, with over nine million people dying from exposure to contaminated food, water, air, houses, workplaces, or consumer goods each year. The annual death toll from chemical exposure is far higher than that during World War II, and it currently represents the leading preventable cause of mortality. Additionally, it causes catastrophic damage to wildlife, particularly to insects and the animals that depend on them, as well as to ecosystems and the services they provide, such as clean water and pollination, which are necessary for human survival. This emphasizes how chemical contamination contributes to the possibility of global ecological collapse (David and Juyuan 2013). Evidence of cognitive, reproductive, and developmental abnormalities, as well as early deaths brought on by chemical contamination of human habitation, has grown in recent decades (Diamanti-Kandarakis et al. 2009).

Numerous organisms of various sizes and functions make up the soil food web. This includes a variety of higher trophic level invertebrate eaters and predators, root-associated biota-like pathogens or mutualists, and saprotrophs that decompose decaying organic matter.

Soil food webs impact the cycling of nutrients and carbon (C) in the soil, which has repercussions for C storage and, eventually, mitigation of growing atmospheric carbon dioxide concentrations. One way soil food webs influence how much nitrogen (N) is available to plants is by mineralizing organic nitrogen. The soil food web does mineralize nitrogen, but this does not always lead to the free availability of nutrients for plants. Through the seclusion of their living or dead biomass or indirectly by changing the chemistry or structure of the soil, soil food webs can assist the soil system in maintaining nitrogen levels by preventing leaching and denitrification. It has been shown that altering soil food webs' composition, network structure, and connectivity can vary on how quickly crucial ecosystem processes occur (Kardol and Jonathan 2018). The soil food web is vital for disease prevention and plant defense against root pathogens.

Inadequate risk assessments are frequently used when releasing novel compounds (Sala and Goralczyk 2013; Wang et al. 2020), and their mixes result in new chemical environments with very ambiguous toxicity. Almost all major industries are experiencing chemical intensification. For instance, modern agriculture uses about five million tons of pesticides and 200 million tons of concentrated nitrogen, phosphorus, and potassium (NPK) fertilizers annually to produce crops and livestock intensively enough to feed a large portion of the world. The statistics maintained by the Food and Agriculture Organization of the United Nations (FAO) show that 3,835,826 tons of pesticides were used worldwide in 2008, a rise of about 7% over the following 10 years (FAOSTAT 2019). There are more than 400 active chemical ingredients in different pesticides in the USA alone (Bexfield et al. 2020). Agrichemical use is also growing in newly industrializing nations like China, which is currently the world's largest producer and consumer of industrial chemicals and accounts for 36% and 25%, respectively, of global demand for chemical fertilizers and pesticides (Guo et al. 2010).

19.4.1 Channels Through Which Plant Communities Affect Soil Food Webs

Recent research has shown that various plant species affect the soil ecosystems they sustain. Anthropogenic modifications to the plant community's composition may substantially impact the soil web's structure, particularly in urban green spaces. Here, we outline three mechanistic pathways by which alterations in the composition of the plant community can impact the soil food webs (Fig. 19.2). First, the amount and type of leaf and root litter that plants return to the soil fluctuate significantly depending on their species (i.e., chemical composition) (Fig. 19.3). Plant litter's suitability as a resource for bacteria that feed on detritus and degrade it depends on its chemical composition.

Soil food webs can be affected by changes in the amount and chemical composition of litter (dead plant tissue, shoots, and roots), changes in root morphology, tissue chemistry, and exudate composition, or changes in the soil's abiotic conditions, such as the amount of resources (such as nutrients and water) and microclimate (such as temperature). Changes in the structure of the soil food web, which are mediated by variations in the composition of the plant community, have a substantial impact on the activities of the soil ecosystem, such as the cycling of nutrients and carbon and the management of diseases (Fig. 19.4).

The numerous feeding types of soil nematodes include omnivores, predators, bacterial and fungal feeders, and plant feeders (both those that live in and feed inside plant root tissue and those that feed externally from outside the root).

The structure and operation of soil food webs and the services they provide are expected to be significantly impacted by anthropogenic changes in the variety and composition of plant communities.

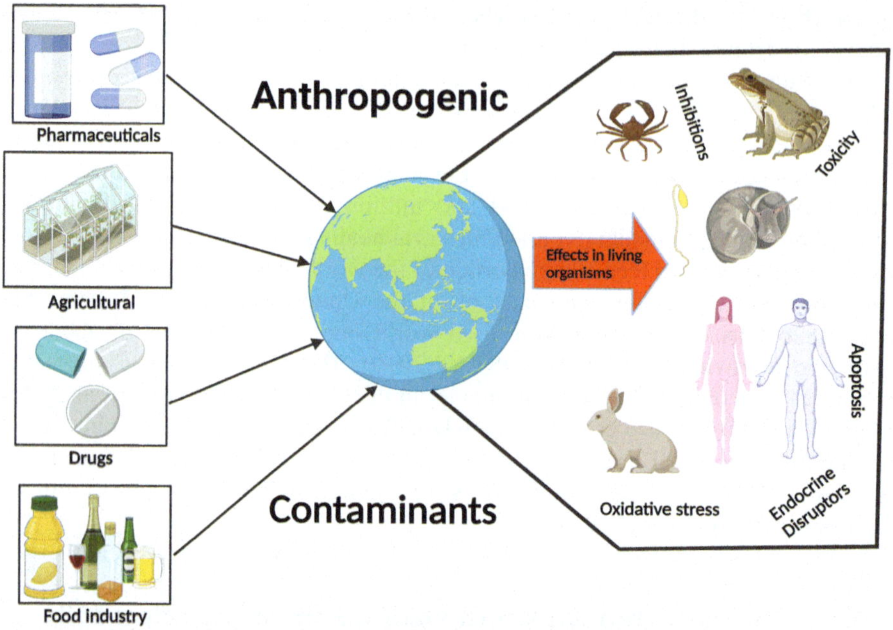

Fig. 19.2 Adverse effects of anthropogenic compounds (López-Pacheco et al. 2019)

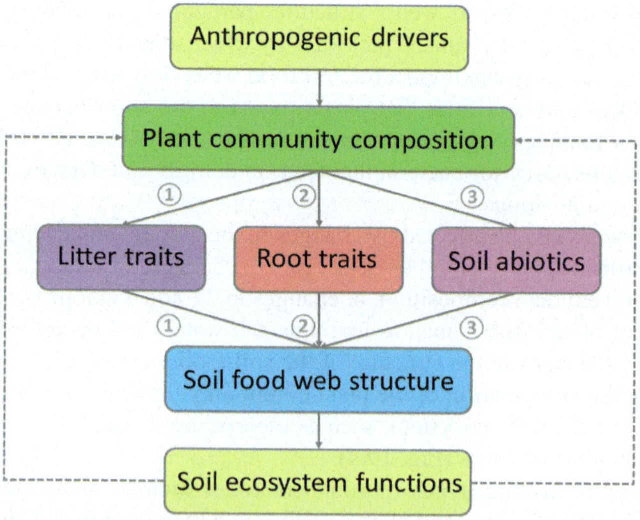

Fig. 19.3 Conceptual map of the paths on the composition of the plant community that impact the soil food webs (Kardol and Jonathan 2018)

Fig. 19.4 Plant root characteristics affect the formation and operation of the soil food web. (**a**) Plant communities frequently include several coexisting and interacting species with varied values for functional features, such as the volume and kind of litter they return to the soil. (**b**) The composition and diversity of root-associated organisms, as well as their consumers and predators, are influenced by root nutrient acquisition traits (such as associations with nitrogen-fixing bacteria or mycorrhizal fungi), architectural habitat traits (such as root depth, diameter, and branching pattern), and chemical traits (such as the carbon-to-nitrogen ratio and defense compounds). (**c**) The structure and operation of the soil food web are frequently assessed using the taxonomically and functionally varied soil nematode groups (Kardol and Jonathan 2018)

Crop species differ in their attractiveness to soil pests and diseases, such as host-specific nematodes, due to their distinctive root chemistry properties. Given these inter-specific differences in crop characteristics, converting from monoculture cropping to mixed cropping would infuse traits into the system, boosting trait packing and favorably affecting the variety and functionality of the soil food web (Fig. 19.5).

Changes in the components of plant communities affect the composition, variety, and connectedness (i.e., the ratio of observed pair-wise interactions represented as a percentage of the total number of interactions imaginable) of the soil food web. These modifications to the soil food web may have a major effect on the activities of the soil ecosystem, including nutrient retention and disease control. A simple plant community, like an early successional monoculture, can change into a complex, diversified plant community, like a late successional grassland or shrubland.

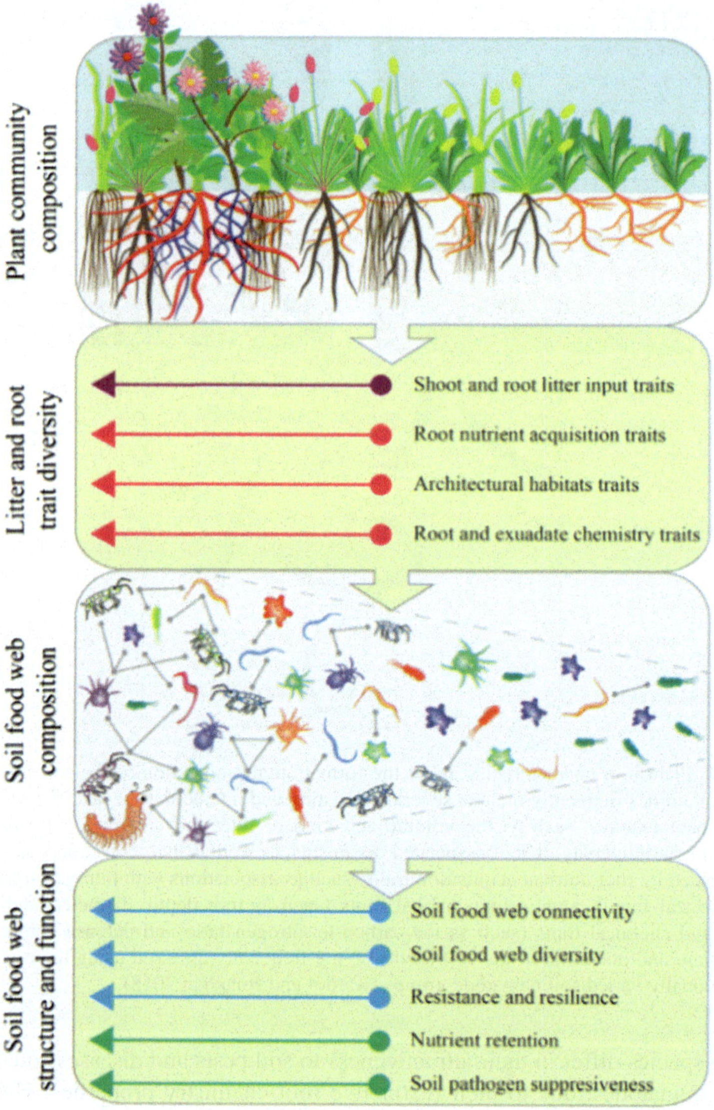

Fig. 19.5 Hypothetical relationships between changes in plant community trait composition and the structure and operation of the soil food web (Kardol and Jonathan 2018)

19.5 Biomagnification

Biomagnification or biological accumulation is when toxic substances, such as pesticides, accumulate in the environment, including water bodies or soil, and gradually move up the food chain in much higher concentrations. The toxic substances ultimately amass into the tissues of the organisms that consume the organisms of the lower trophic levels. The basic concept of biomagnification depends on the substance concentration increasing gradually up the trophic levels (food chain energetics) and causing adverse health effects as the toxin levels rise (Szynkowska et al. 2018).

In the agricultural sector, the chemicals used as fertilizers and for other purposes are one of the main causes of biomagnification. These substances—pesticides, herbicides, inorganic fertilizers, etc.—permeate the soil, build up to dangerous concentrations, and eventually make their way into water bodies during run-offs brought on by potential overwatering or storms (Kumar 2012). The effluents from the agricultural and other manufacturing industries also contribute to biological accumulation since the chemicals used in the farms or released from industries contain heavy metals like lead, mercury, arsenic, etc. (Kelly et al. 2007). After being consumed by other consumers, the harmful compounds and heavy metals are taken up by zooplankton and plants, respectively, and released into the bodily tissues of the primary consumers. Heavy metals are emitted by the combustion of fossil fuels in industries, residences, and transportation. Ecosystems on land and water both face a serious threat from toxic trace metals. After being discharged from both natural and anthropogenic sources, heavy metals contaminate natural water bodies, sediments, and soils (Rhind 2009).

DDT (Dichloro Diphenyl Trichloroethane) was introduced to India in 1948 to control pests and curb levels and gained wide popularity as a pesticide. However, it was banned in 1989 due to scientific evidence proving its harmful effects on health and the environment. DDT is a class II insecticide which means it is moderately toxic. But studies have demonstrated that DDT seriously affects human and animal health. Low to moderate exposure (8–10 mg/kg) may result in nausea, diarrhea, and irritation of the eyes, nose, or throat, while higher exposure (13–16 mg/kg) can lead to tremors and convulsions. In some cases, increased doses can also cause chromosomal damage. Moreover, DDT can remain in the soil or water body for many years due to its non-biodegradable characteristics, leading to biomagnification. Governments work to limit or eliminate the environmental and human health dangers posed by substances that build up to dangerous levels in living things. Bioaccumulative compounds are defined by regulatory agencies as hydrophobic, fat-soluble molecules with high n-octanol/water partition coefficients (KOW—100,000). In all three food webs, calculated and observed concentrations of recalcitrant substances exhibited good agreement in bioaccumulation modeling studies (Kelly et al. 2007). The model also shows that the elimination rate controls the relationship between the BMF (biomagnification factor) and chemical properties. If the KOW (octanol–water partition ratio) of the chemical surpasses 105, elimination becomes slow enough to produce biomagnification in water-respiring organisms. This also occurs in air-breathing species for compounds with a high KOA (octanol–air partition ratio) (106), which causes slow respiratory elimination, and a KOW $>$ 102, which causes sluggish elimination in urine or nitrogenous wastes. KOW and the BCF (bioconcentration factor) in fish are not reliable indicators of biomagnification in air-breathing animals for compounds with a KOA of 106 and a

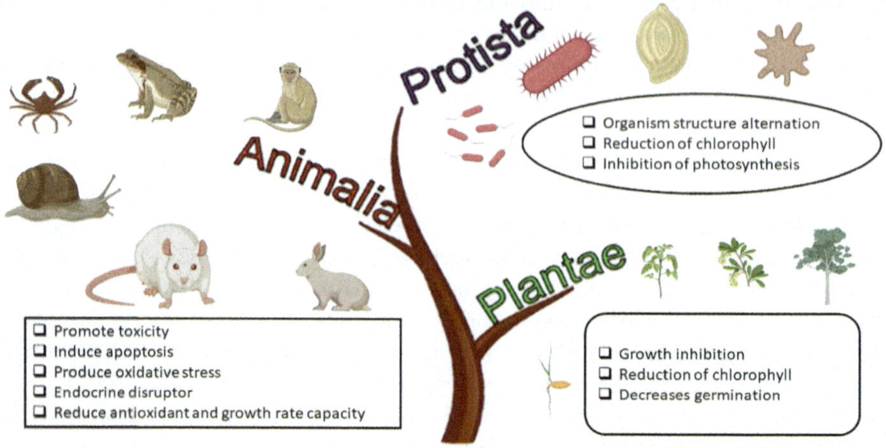

Fig. 19.6 Adverse effects of anthropogenic chemicals on living organisms in a biological kingdom (López-Pacheco et al. 2019)

KOW > 102, according to the difference in biomagnification behavior between water- and air-breathing species (Fig. 19.6).

Due to the concentration and variety of pollutants emerging from anthropogenic processes, wastewater treatment procedures are extremely important since chemical dissemination primarily occurs through water sources (Bilal et al. 2017). The food chain is disturbed by the chemicals and pollutants released into the water bodies. Larger animals ingest toxins after being absorbed by small organisms. As a result, these toxins build up in higher-level species.

19.6 Effect of Anthropocene Chemicals on the Microbiome

Human activity is also causing significant changes in the microbial world, in addition to changes in temperature, land cover, biodiversity, and chemical composition.

19.6.1 Antimicrobial Resistance (AMR)

The human, animal, plant, and environmental sectors are all under severe threat from AMR. Compared to AMR in human or animal health, the environmental component of antimicrobial resistance has received less attention. Antimicrobial resistance is, nonetheless, widely distributed in the natural world. In the transmission of AMR to the environment, human activities are a key factor. The main factors influencing the emergence of bacteria that are resistant to antibiotics are their misuse and excessive use. The lack of hygienic circumstances, clean water, and effective infection prevention and control all contribute to the spread of pathogens, some of which may be resistant to antibiotic therapy (Mendelson and Matsoso 2015).

19.7 Biodiversity Loss and Damage to Crops and Livestock

The principal source of services for the sustenance of life on Earth is provided by biodiversity throughout the Earth's surface layer, from bedrock to the vegetation canopy (Banwart et al. 2019). A significant loss of the Earth's biodiversity is caused by the acute and chronic effects of excessive use of agriculture and other industrial pollutants in the present and in the past. Neonicotinoid insecticides have caused the collapse of honeybee communities worldwide, which has resulted in a global disaster for crop pollination (David and Juyuan 2013). According to accounts, pesticide pollution in some areas has resulted in the loss of more than 40% of the total taxonomic pools of stream invertebrates (Beketov et al. 2013).

In heavily polluted locations (Gevao et al. 2000), residues of more persistent chemicals, including many pesticides, pose a serious hazard to groundwater and marine life (Arias-Estévez et al. 2008; Jamieson et al. 2017). From 290 sites in Germany, losses of up to 78% of insect species have been documented (Seibold et al. 2019). Such ecological effects and their persistence may significantly affect biological processes in natural settings, such as decomposition and soil formation, creating unfavorable or difficult conditions for human food production. At both a regional and global level, diminishing biodiversity is caused by reactive nitrogen pollution of the atmosphere and its deposition. Phytoremediation focuses on using plants and the associated rhizosphere microorganism to transform, eliminate, or suppress the toxic chemicals in the soil, groundwater, surface water, and atmosphere. Phytoremediation focuses on using microalgae, cyanobacteria, and macroalgae to eliminate or transform the toxic chemicals present in the water and absorb atmospheric CO_2 (Bauddh and Korstad 2022).

19.8 Alternatives to Anthropogenic Chemicals

The most important causes of biodiversity loss have typically been identified as climate change and habitat loss. However, as was already mentioned, biodiversity can also be seriously harmed by exposure to manufactured pollutants. The fact that chemicals are prevalent everywhere is significant because they can have unpredictable and disproportionate effects on all species. Additionally, the requirement to adapt to chemical exposure might make it harder for an organism to cope with additional stressors, such as those brought on by climate change. By altering the pH and, in turn, the absorption and bioavailability of chemicals, climate change can impact the destiny and toxicity of chemicals in the environment.

Another key finding is that some creatures may gradually adjust to the changing environment when climate change happens if the time scales necessary for recovery and adaptation of stressor-exposed organisms are taken into account. Contrarily, the considerably more rapid shifts in chemical exposure might not provide enough time for recovery and adaptation, resulting in the species' irrevocable extinction. Most existing methods for producing and managing synthetic chemicals are known to significantly impact habitat destruction and climate change. This occurs directly, as in the case of greenhouse gas emissions, highly energy- and water-intensive processing of fossil fuels, or mining operations, and indirectly, as in the widespread distribution of manufactured synthetic chemicals and materials, which may eventually result in habitat destruction.

When seen from these perspectives, anthropogenic chemicals may represent a highly significant variable in the biodiversity loss equation. If we are to ever fully comprehend the current situation and discover viable remedies, the chemical pollution driver needs to receive equal awareness and attention when investigating biodiversity loss.

19.9 Conclusion

The biosphere and the larger Earth system are now being shaped by human activity as a global force. The symptoms include biodiversity loss and climate change. Human well-being, prosperity, and maybe even the survival of societies and civilizations are currently under threat due to the growing spread of human activities that have weakened the resilience of the biosphere and the Earth system. Humanity is challenged by the Anthropocene's chaotic times and new linked dynamics of people and earth, where rapid and gradual change interact in surprising and unanticipated ways.

Anthropogenic changes in the biodiversity and composition of plant communities are anticipated to significantly impact the structure and functionality of soil food webs and the services they offer. This book chapter provides useful information, but to fully understand the role of soil food webs in how changes in plant community composition affect soil ecosystem functioning, we still need to look at the entire soil food webs, including organisms at higher trophic levels. Due to the quantity and diversity of contaminants emerging from human activities, wastewater treatment procedures are crucial since chemical distribution mostly occurs through water sources. People, businesses, societies, and cultures urgently need to take control of nature's positive contributions to our well-being and develop a robust biosphere for future generations.

Acknowledgments The authors acknowledge the Vellore Institute of Technology Management and the Dean, School of Bio Sciences and Technology, for providing the infrastructure, constant support, and encouragement. We sincerely thank DST–SPLICE–Climate Change Programme, Govt. of India, for providing financial support to Mr. A. Mariadoss & Mr. N. Nagaraj—Grant No. DST/CCP/NCC&CV/133/2017 (G).

References

Arias-Estévez M, López-Periago E, Martínez-Carballo E, Simal-Gándara J, Mejuto JC, García-Río L (2008) The mobility and degradation of pesticides in soils and the pollution of groundwater resources. Agric Ecosyst Environ 123(4):247–260

Arnold KE, Brown AR, Ankley GT, Sumpter JP (2014) Medicating the environment: assessing risks of pharmaceuticals to wildlife and ecosystems. Philos Trans R Soc B Biol Sci 369(1656): 20130569

Ashraf S, Ali Q, Zahir ZA, Ashraf S, Asghar HN (2019) Phytoremediation: environmentally sustainable way for reclamation of heavy metal polluted soils. Ecotoxicol Environ Saf 174: 714–727

Banwart SA, Nikolaidis NP, Zhu YG, Peacock CL, Sparks DL (2019) Soil functions: connecting Earth's critical zone. Annu Rev Earth Planet Sci 47:333–359

Battista G, de Lieto Vollaro E, de Lieto Vollaro R (2021) Analysis of the main anthropogenic sources' contribution to pollutant emissions in the Lazio region, Italy. Appl Sci 11(17):7936. https://doi.org/10.3390/app11177936

Bauddh K, Korstad J (2022) Phycoremediation: use of algae to sequester heavy metals. Hydrobiology 1(3):288–303

Beketov MA, Kefford BJ, Schäfer RB, Liess M (2013) Pesticides reduce regional biodiversity of stream invertebrates. Proc Natl Acad Sci 110(27):11039–11043

Bilal M, Asgher M, Parra-Saldivar R, Hu H, Wang W, Zhang X, Iqbal HMN (2017) Immobilized ligninolytic enzymes: an innovative and environmental responsive technology to tackle dye-based industrial pollutants - a review. Sci Total Environ 576:646–659

Boatman ND, Parry HR, Bishop JD, Cuthbertson AG (2007) Impacts of agricultural change on farmland biodiversity in the UK. Issues Environ Sci Technol (25):1–32

Bexfield LM, Belitz K, Lindsey BD, Toccalino PL, Nowell LH (2020) Pesticides and pesticide degradates in groundwater used for public supply across the United States: occurrence and human-health context. Environ Sci Technol 55(1):362–372

Brodin T, Piovano S, Fick J, Klaminder J, Heynen M, Jonsson M (2014) Ecological effects of pharmaceuticals in aquatic systems—impacts through behavioural alterations. Philos Trans R Soc B Biol Sci 369(1656):20130580

Burri NM, Weatherl R, Moeck C, Schirmer M (2019) A review of threats to groundwater quality in the anthropocene. Sci Total Environ 684:136–154

Chen H, Zheng C, Wang S, Tu C (2000) Combined pollution and pollution index of heavy metals in red soil. Pedosphere 10(2):117–124

Cismasiu CM, Cioboiu O, Pacesila I (2017) An integrative synthesis about the biodiversity research related to the decontamination process of the natural ecosystem affected by the industrial pollution from Romania. Oltenia J Stud Nat Sci 32(2):164–172

Copley SD (2009) Evolution of efficient pathways for degradation of anthropogenic chemicals. Nat Chem Biol 5(8):559–566

Crutzen PJ, Brauch HG (eds) (2016) Paul J. Crutzen: a pioneer on atmospheric chemistry and climate change in the Anthropocene, vol 50. Springer

Daughton CG, Ternes TA (1999) Pharmaceuticals and personal care products in the environment: agents of subtle change? Environ Health Perspect 107(Suppl 6):907–938

David H, Juyuan J (2013) A study of smog issues and $PM_{2.5}$ pollutant control strategies in China. J Environ Prot 4(7):746–752

Diamanti-Kandarakis E, Bourguignon JP, Giudice LC, Hauser R, Prins GS, Soto AM, Zoeller RT, Gore AC (2009) Endocrine-disrupting chemicals: an Endocrine Society scientific statement. Endocr Rev 30(4):293–342

Dirzo R, Young HS, Galetti M, Ceballos G, Isaac NJ, Collen B (2014) Defaunation in the Anthropocene. Science 345(6195):401–406

Ellis EC, Ramankutty N (2008) Putting people in the map: anthropogenic biomes of the world. Front Ecol Environ 6(8):439–447

Fang M, Kang H-G, Park Y, Estrella B, Zarbl H (2017) In vitro bioluminescence assay to characterize circadian rhythm in mammary epithelial cells. J Vis Exp (127):55832. https://doi.org/10.3791/55832

Faostat FAO (2019) Production crops. http://www.fao.org/faostat/en/#data

Fox JE (2004) Chemical communication is threatened by endocrine-disrupting chemicals. Environ Health Perspect 112(6):648–653. https://doi.org/10.1289/ehp.6455

Fuchs LK, Jenkins G, Phillips DW (2018) Anthropogenic impacts on meiosis in plants. Front Plant Sci 9:1429

Gałuszka A, Migaszewski ZM, Rose NL (2020) A consideration of polychlorinated biphenyls as a chemostratigraphic marker of the Anthropocene. Anthrop Rev 7(2):138–158

Gevao B, Semple KT, Jones KC (2000) Bound pesticide residues in soils: a review. Environ Pollut 108(1):3–14

Guo JH, Liu XJ, Zhang Y, Shen JL, Han WX, Zhang WF, Christie P, Goulding KWT, Vitousek PM, Zhang FS (2010) Significant acidification in major Chinese croplands. Science 327 (5968):1008–1010

Hakoun V, Orban P, Dassargues A, Brouyère S (2017) Factors controlling spatial and temporal patterns of multiple pesticide compounds in groundwater (Hesbaye chalk aquifer, Belgium). Environ Pollut 223:185–199

Hance J (2015) How humans are driving the sixth mass extinction. The Guardian, 20

Jabbar Khan A, Akhter G, Gabriel HF, Shahid M (2020) Anthropogenic effects of coal mining on ecological resources of the Central Indus Basin, Pakistan. Int J Environ Res Public Health 17(4): 1255. https://doi.org/10.3390/ijerph17041255

Jamieson AJ, Malkocs T, Piertney SB, Fujii T, Zhang Z (2017) Bioaccumulation of persistent organic pollutants in the deepest ocean fauna. Nat Ecol Evol 1(3):1–4

Jürgens A, Bischoff M (2017) Changing odor landscapes: the effect of anthropogenic volatile pollutants on plant-pollinator olfactory communication. Funct Ecol 31(1):56–64. https://doi.org/10.1111/1365-2435.12774

Kardol P, Jonathan R (2018) How anthropogenic shifts in plant community composition alter soil food webs. F1000Res 7:4

Katayama N, Baba YG, Kusumoto Y, Tanaka K (2015) A review of post-war changes in rice farming and biodiversity in Japan. Agric Syst 132:73–84

Kelly BC, Ikonomou MG, Blair JD, Morin AE, Gobas FAPC (2007) Food web-specific biomagnification of persistent organic pollutants. Science 317(5835):236–239

Kumar S (2012) Biopesticides: a need for food and environmental safety. J Biofertil Biopestic 3(4). https://doi.org/10.4172/2155-6202.1000e107

Küster A, Adler N (2014) Pharmaceuticals in the environment: scientific evidence of risks and its regulation. Philos Trans R Soc B Biol Sci 369(1656):20130587

Lewis M, Pryor R, Wilking L (2011) Fate and effects of anthropogenic chemicals in mangrove ecosystems: a review. Environ Pollut 159(10):2328–2346

Logemann A, Reininghaus M, Schmidt M, Ebeling A, Zimmermann T, Wolschke H, Friedrich J, Brockmeyer B, Pröfrock D, Witt G (2022) Assessing the chemical anthropocene–development of the legacy pollution fingerprint in the North Sea during the last century. Environ Pollut 302: 119040

López-Pacheco IY, Silva-Núñez A, Salinas-Salazar C, Arévalo-Gallegos A, Lizarazo-Holguin LA, Barceló D, Iqbal HMN, Parra-Saldívar R (2019) Anthropogenic contaminants of high concern: existence in water resources and their adverse effects. Sci Total Environ 690:1068–1088

Lutz JF (2017) Catalyst: being a chemist in the Anthropocene. Chem 2(2):155–156

McBeath GA, Leng TK (2006) Governance of biodiversity conservation in China and Taiwan. Edward Elgar Publishing

McGill BJ, Dornelas M, Gotelli NJ, Magurran AE (2015) Fifteen forms of biodiversity trend in the Anthropocene. Trends Ecol Evol 30(2):104–113

Mendelson M, Matsoso MP (2015) The World Health Organization global action plan for antimicrobial resistance. SAMJ. S Afr Med J 105(5):325–325

Néstor MC, Mariana C (2017) Impact of pharmaceutical waste on biodiversity. In: Ecopharmacovigilance, pp 235–253

North AC, Hodgson DJ, Price SJ, Griffiths AGF (2015) Anthropogenic and ecological drivers of amphibian disease (ranavirosis). PLoS One 10(6):e0127037

Pesticides and the loss of biodiversity (2010) https://www.paneurope.info/issues/pesticides-and-loss-biodiversity

Pichhode M, Nikhil K (2016) Effect of heavy metals on plants: an overview. Unpublished. https://doi.org/10.13140/RG.2.2.27583.87204

Popescu F, Ionel I (2010) Anthropogenic air pollution sources. In: Air quality. Sciyo

Rhind SM (2009) Anthropogenic pollutants: a threat to ecosystem sustainability? Philos Trans R Soc Lond Ser B Biol Sci 364(1534):3391–3401

Rhind SM (2012) Anthropogenic pollutants – an insidious threat to animal health and productivity? Acta Vet Scand 54(S1):S2. https://doi.org/10.1186/1751-0147-54-s1-s2

Rockström J, Steffen W, Noone K, Persson A, Chapin FS, Lambin E, Lenton TM, Scheffer M, Folke C, Schellnhuber HJ, Nykvist B, de Wit CA, Hughes T, van der Leeuw S, Rodhe H, Sorlin S, Snyder PK, Costanza R, Svedin U, Falkenmark M, Karlberg L, Corell RW, Fabry VJ, Hansen J, Walker B, Liverman D, Richardson K, Crutzen P, Foley J (2009) Planetary boundaries: exploring the safe operating space for humanity. Ecol Soc 14(2):32. http://www.ecologyandsociety.org/vol14/iss2/art32/

Rosen MR, Van Metre PC (2010) Assessment of multiple sources of anthropogenic and natural chemical inputs to a morphologically complex basin, Lake Mead, USA. Palaeogeogr Palaeoclimatol Palaeoecol 294(1-2):30–43. https://doi.org/10.1016/j.palaeo.2009.03.017

Sahu N, Lal ASN (2021) Industrial pollutants: hamper or enhance soil biodiversity. Actinomycetes 1012(1013):107–108

Sala S, Goralczyk M (2013) Chemical footprint: a methodological framework for bridging life cycle assessment and planetary boundaries for chemical pollution. Integr Environ Assess Manag 9(4): 623–632

Seibold S, Gossner MM, Simons NK, Blüthgen N, Müller J, Ambarlı D, Ammer C, Bauhus J, Fischer M, Habel JC, Linsenmair KE (2019) Arthropod decline in grasslands and forests is associated with landscape-level drivers. Nature 574(7780):671–674

Szynkowska MI, Pawlaczyk A, Maćkiewicz E (2018) Bioaccumulation and biomagnification of trace elements in the environment. In: Chojnacka K, Saeid A (eds) Recent advances in trace elements. Wiley, pp 251–276. https://doi.org/10.1002/9781119133780.ch13

Vigiak O, Udias A, Pistocchi A, Zanni M, Aloe A, Grizzetti B (2021) Probability maps of anthropogenic impacts affecting ecological status in European rivers. Ecol Indic 126:107684

Wang Z, Walker GW, Muir DC, Nagatani-Yoshida K (2020) Toward a global understanding of chemical pollution: a first comprehensive analysis of national and regional chemical inventories. Environ Sci Technol 54(5):2575–2584

Xiao B, Chen T-M, Zhong Y (2016) The possible molecular mechanism underlying cadmium-induced circadian rhythms disruption in zebrafish. Biochem Biophys Res Commun 481(3-4): 201–205

Xiao B, Cui L-Q, Ding C, Wang H (2017) Effects of lithium and 2,4-dichlorophenol on zebrafish: circadian rhythm disorder and molecular effects. Zebrafish 14(3):209–215

Younis MH, Younis YMH (2018) Monitoring the anthropogenic and geochemical environment surrounding the Butana (Sudan) drinking water sources via the determination of heavy metals composition of the soil, streams sediments, and gold mining tailings in the wet season (III). Biomed J Scient Tech Res 9(4). https://doi.org/10.26717/bjstr.2018.09.001821

Yuan J, Lu Y, Wang C, Cao X, Chen C, Cui H, Du D (2020) Ecology of industrial pollution in China. Ecosyst Health Sustain 6(1):1779010

Zheng X, Zhang K, Zhao Y, Fent K (2021) Environmental chemicals affect circadian rhythms: an underexplored effect influencing health and fitness in animals and humans. Environ Int 149: 106159

Zhu YG, Penuelas J (2020) Changes in the environmental microbiome in the Anthropocene. Glob Chang Biol 26(6):3175–3177

Zucchi S, Castiglioni S, Fent K (2013) Progesterone alters global transcription profiles at environmental concentrations in the brain and ovary of female zebrafish (*Danio rerio*). Environ Sci Technol 47(21):12548–12556

Part IV

Best Practices in Plant Diversity Conservation

Documentation of the Phenocalendar of *Allaeanthus luzonicus* (Blanco) Fern.-Vill. (Family Moraceae) to Sustain Its Utilization

20

Menisa A. Antonio and Evangeline S. Galacgac

Abstract

This chapter introduces an innovation leveraging on the phenological calendar of the growth and development of *Allaeanthus luzonicus* (Blanco) Fern.-Vill. to ensure a sustainable inflorescence supply for household consumption. *A. luzonicus* is a favorite indigenous vegetable in Northern Philippines, but it is wild-growing and seasonal in availability; hence edible inflorescences are available only during the regular flowering season in the months of January to March each year. So, we sought to establish and explore the species' phenocalendar to have longer supply of inflorescence. We first documented the species' growth and development stages and compared its phenology under four climate types in three provinces in Northern Philippines where the indigenous vegetable is most utilized. We also discovered that there are off-season varieties of *A. luzonicus* which flower outside the regular flowering period of January to March. Thus, a production system that integrates the documented phenocalendar and off-season varieties is recommended for year-round inflorescence supply.

Keywords

Allaeanthus luzonicus · Indigenous vegetable · Phenocalendar · Growth and development · Sustainable utilization

M. A. Antonio (✉) · E. S. Galacgac
Mariano Marcos State University, City of Batac, Ilocos Norte, Philippines
e-mail: maantonio@mmsu.edu.ph

S. Ramamoorthy et al. (eds.), *Plant Diversity in Biocultural Landscapes*,
https://doi.org/10.1007/978-981-19-8649-9_20

20.1 Introduction

Many local communities in the Philippines harness the abundance of nature for household food sufficiency and nutritional security. Dubbed as indigenous foods (IF), these plant species are usually gathered from the wild; and thus sometimes called foods from the wild; other times, *manna* from heaven. Biblically, *manna* pertains to the miraculously supplied food to the Israelites in the wilderness.

Through an R&D Project which surveyed and characterized the indigenous food plants (IFPs) in Ilocos Norte, Philippines, 46 indigenous plant species are consumed across the province, constituting a significant part of the rural diet (Antonio et al. 2011). Sourcing these species from the wild and serving them in their dining tables have rooted in the Ilocano life, food culture, and plant heritage. They are served either as vegetable viand, salad, sauteed dish, delicacy and snack, among others, adding new flavor and color for diet diversity. While these plant resources are generally underutilized and neglected, they are gaining importance in transforming our food system to become sustainable and climate resilient with raised bar for diet quality and diversity.

One of the identified indigenous vegetables is *Allaeanthus luzonicus* (Blanco) Fern.-Vill., Family Moraceae, locally called himbabao (Tag.), alokon (Ilk.), alibabag (Itawis), or birch flower (Engl.) (Antonio et al. 2011). *A. luzonicus* is native to the Philippines and Sulawesi (POWO 2022). It is found throughout the Philippines. Its distribution is likewise cited in Indonesia, and spread to some parts of Hawaii, USA, but it is considered nonnative there.

It is a popular and much-loved indigenous vegetable in the Province of Ilocos Norte, including the rest of the provinces in Northern Philippines. The staminate inflorescence added with some leaftops are the edible plant parts, and favorite for various dishes such as vegetable stews, viand, sauteed or salad prepared singly or in combination with other vegetables (Antonio et al. 2016). The plant's wild-growing habit and the edible parts' seasonal availability (usually in the months of January to March every year) have hampered the sustainable utilization of the plant.

So, we collected, analyzed, and integrated data from projects done on the plant which were implemented by the Mariano Marcos State University to develop a phenological calendar (phenocalendar in short) to extend the supply of the edible inflorescence. Specifically, we analyzed data to answer the following questions: (a) what are the species' biological stages?; (b) how does the plant's phenology vary under Climate types I, II, III, and IV (Modified Corona's Classification of Climate based on average monthly rainfall) in three provinces in Northern Philippines?; (c) what are the components of the plant that support its traditional use for food?; (d) what plant behaviors can be harnessed to contribute to a sustainable inflorescence supply?; (e) what other milestones can be integrated into a production system which we will propose to establish or enhance a commercial market niche for the crop?

20.2 Body

20.2.1 Plant Parts Used and Flower Characteristics

A. luzonicus is a dioecious tree (Fig. 20.1a) growing wild to semi-wild (Pancho 1979). The staminate plant produces an elongated catkin (Fig. 20.1b), measuring 3–58 cm long, while the pistillate plant produces globose to sub-globe head (Fig. 20.1c), measuring 2–3 cm in diameter (Antonio et al. 2018). Both types of inflorescences, together with tender leaves, are edible. But in Northern Philippines, the staminate inflorescence is more preferred than the pistillate inflorescence for vegetable stew. It produces a peculiar inviting aroma, adds texture, and provides additional roughage to any vegetable dish. When cooked, it turns vibrant green, soft, and a bit slick or slimy.

In terms of flowering behavior, *A. luzonicus* is annual, seasonal, and indeterminate (flowers emerge and open from proximal to distal orientation). Its flowering habit is either: (a) synchronous, characterized by distinct vegetative and reproductive phases, following the developmental cycle described below; or (b) asynchronous, characterized by two or more stages occurring simultaneously, i.e., simultaneous vegetative, leaf falling, and flowering all together.

It flowers once a year, usually from January to March every year, coinciding with the summer months in the Philippines. Since a tree, inflorescences are harvested by

Fig. 20.1 *A. luzonicus* (Blanco) Fern.-Vill. (**a**) Growing habit, (**b**) staminate inflorescence, and (**c**) pistillate inflorescence. (Photo credit: Menisa A. Antonio, Mariano Marcos State University)

pruning the whole tree or major branches. More information on plant flowering will be discussed in the succeeding sections of this chapter.

20.2.2 Nutrient and Nonnutrient Components

In a phytochemical screening done by Antonio and Vivit (2017), *A. luzonicus* inflorescence contains the secondary metabolites flavonoid, phenol, steroid, coumarin, triterpene, tannin, cardiac glycoside, and xanthoprotein. The identified secondary metabolites or phytochemicals offer several medicinal and pharmacological applications providing chemo-protection against various biotic and abiotic factors as these secondary metabolites are often associated with improving nutrient absorption and boosting the immune system (Saxena et al. 2013).

In a related work, the antioxidant potential of *A. luzonicus* inflorescence was evaluated through DPPH radical scavenging activity (RSA) (Antonio and Vivit 2017). *A. luzonicus* gives a good antioxidant food, affording total efficacy of about 89% at 200 μg/ml. When compared with four other indigenous vegetables from Ilocos Norte, *A. luzonicus* inflorescence is most potent, affording 87% RSA at a concentration as low as 150 μg/ml.

In addition, the edible parts also contain minerals such as calcium, phosphorus, iron, sodium, and vitamins such as beta-carotene, thiamin, riboflavin, niacin, and ascorbic acid (Table 20.1) (DOST-FNRI FCT. Released Dec 1, 2019).

20.2.3 Growth and Development Cycle

We selected and observed *A. luzonicus* trees that grow independently in the four ecological zones (central lowland, northern coastal, southern coastal, eastern interior) and mountainous area in Ilocos Norte (Fig. 20.2). We monitored them for 3 years (2016–2018), done weekly during the start of reproductive stage to end of flowering (January to March), and bimonthly during the vegetative phase. Two to three trees from each area were considered. Four branches (one branch on each side of the tree) were tagged which served as samples for monitoring purposes. Phenological events (Table 20.2) and flowering periods were recorded. Phenophase descriptions are adapted from the Project Budburst Phenophase Field Guide (http://www.windows.ucar.edu) and the methods were adapted from the National Phenology Network (Haggerty and Mazer 2008).

Additionally, weather data consisting of minimum and maximum temperatures, precipitation, and adverse weather condition were obtained from the two weather stations of Ilocos Norte (MMSU-PAGASA Agrometeorological Station and Laoag Synoptic Station). Other weather data from automatic weather stations installed in different areas of Ilocos Norte were also considered. The climate in the study areas was extracted from NAMRIA.

Table 20.1 Food composition table of *A. luzonicus*. DOST-FNRI FCT. Released Dec 1, 2019

Components	*Flower (boiled)*	*Leaves (boiled)*
Proximates (per 100 g edible portion)		
Water (g)	86.8	88
Energy, calculated (kcal)	52	49
Protein (g)	2.9	1.6
Total fat (g)	0.9	0.5
Carbohydrate, total (g)	8.1	9.6
Total ash, total (g)	1.3	0.3
Other carbohydrates (per 100 g edible portion)		
Sugars, total (g)	1	0.9
Minerals (per 100 g edible portion)		
Calcium (mg)	278	116
Phosphorus (mg)	75	54
Iron (mg)	4.3	1
Sodium (mg)	13	12
Vitamins (per 100 g edible portion)	5.1	–
Retinol, vitamin A (μg)	0	0
Beta-carotene (μg)	300	925
Retinol activity equivalent (μg)	25	77
Thiamin, vitamin B1 (mg)	0.06	0.03
Riboflavin, vitamin B2 (mg)	0.13	0.07
Niacin (mg)	0.8	0.5
Ascorbic acid, vitamin C (mg)	10	9
Lipids (per 100 g edible portion)		
Fatty acids, saturated, total (g)		
Fatty acids, monounsaturated, total (g)	–	–
Fatty acids, polyunsaturated, total (g)	–	–
Cholesterol (mg)	–	–

20.2.3.1 Observed Growth and Development Phases

Plant growth and development generally consists of vegetative and reproductive phases. In *A. luzonicus*, there is no clear delineation of the end of the vegetative phase and the start of the reproductive phase. In this chapter, we call this phase as the intermediate phase, which corresponds to the post-vegetative and pre-reproductive changes that occur in the plant.

The vegetative phase in *A. luzonicus* lasts for 7–8 months, usually from late March to October of every year. This period consists of leaf bud emergence, full leaf development, and leaf maturity and senescence (Fig. 20.3). The reproductive phase is usually from 4 to 5 months, from November to March. It is characterized by floral bud emergence, inflorescence market maturity, and inflorescence physiological maturity.

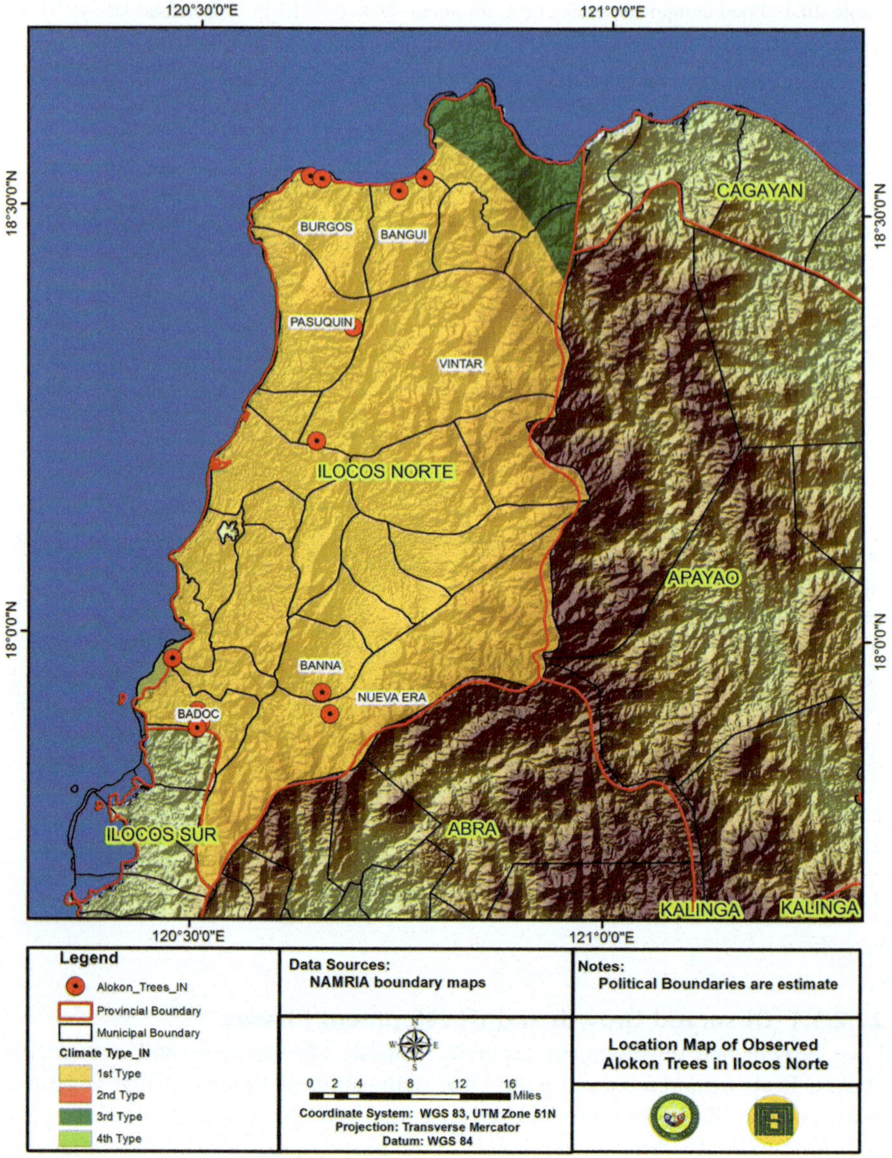

Fig. 20.2 Location of observed *A. luzonicus* trees in the different ecological zones of Ilocos Norte. Map by Engr. Rodel T. Utrera, Mariano Marcos State University

The flowering period marked by floral bud emergence to last inflorescence harvest (Fig. 20.3) lasts for 2–3 months, generally from January to March, peaking in February. The intermediate phase is characterized by the deciduous changes which consists of leaf yellowing and abscission, and leafless period. These periods

Table 20.2 Phenophases monitored in *A. luzonicus*

	Phenophase	Description
1	Emerging leaves or leaf budding	In at least 4 locations on the plant, an emerging leaf is visible. A leaf is considered "emerging" once the green tip is visible
2	95% full leaves or vegetative stage	Report the date when nearly all (at least 95%) of the growing leaf buds have already reached its mature size (majority of leaves have elongated or fully expanded)
3	Leaf maturity	Report the date when nearly all (at least 95%) of the leaves began to dry up or turn yellow
4	Leaf abscission initiation	At least 10% of the leaves fell
5	Leaf abscission completion	All leaves fell
6	Leafless period	Record the duration of leafless period
7	Emerging flower buds or inflorescence bud initiation	In at least 4 locations on the plant, an emerging flower bud is visible. A flower is considered "emerging" once the green tip is visible
8	Harvestable inflorescence	Report the date when inflorescence reached market maturity
9	Physiologically mature inflorescence	Report the date when the inflorescence become light yellow
10	End of flowering	Report the date when no more flower buds emerged and the matured inflorescences fell

usually last for one to 1 ½ months, and 3–4 weeks, respectively. This phase is in the months of November to December.

Several sample trees exhibited extreme cases, which do not conform to the above general growth and development cycle. These cases were as follows: (a) Early flowering in the month of December to January, stop flowering, then 2° floral flushing in May, June, July (the tree is found in the Municipality of Badoc); (b) Late flowering in April to May (tree also found in Badoc); (c) Long-duration flowering from November or December to May or June but peaking in the regular season then gradually declines until June or July (trees found in Brgy 4 and 6 in Vintar; Brgy Labut in Badoc; and Brgy Saoit in Burgos).

20.2.4 Phenology of *A. luzonicus* Under Four Climate Types

In addition to the observed trees in the Province of Ilocos Norte (Fig. 20.2), several trees in different towns of Cagayan and Apayao (Fig. 20.4) were also identified and monitored to compare the effect of different climate types on the phenological cycle of the species. In addition to the sample trees in the identified study towns, observations were also made on trees in the rest of the towns in Cagayan; however this was done only once in 2018. Similar phenological events (Table 20.2) and flowering periods were observed in Cagayan and Apayao.

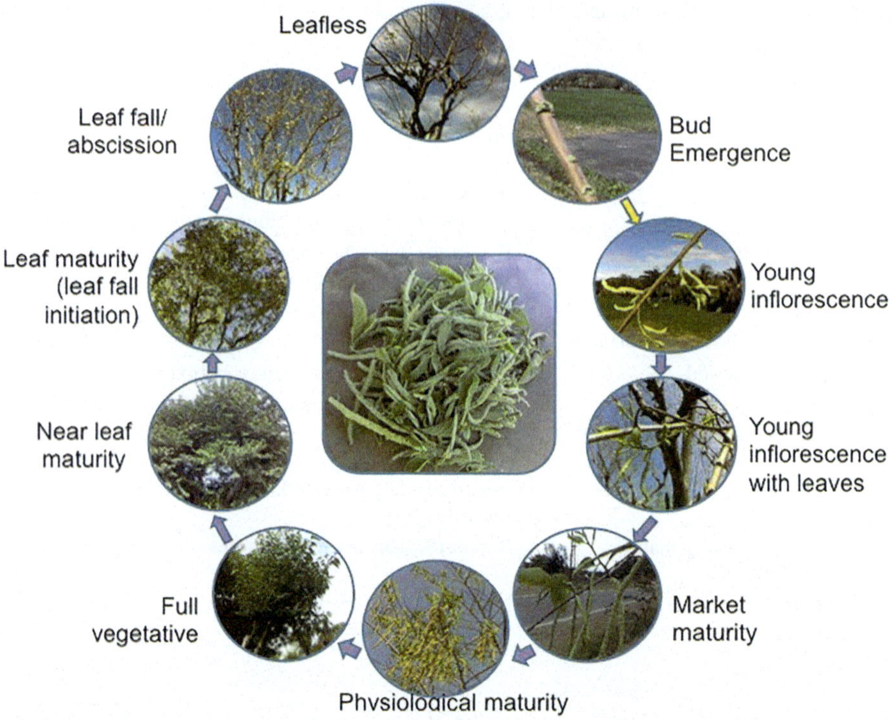

Fig. 20.3 The morphological structures of *A. luzonicus* at various phases in its growth and development cycle

20.2.4.1 Phenophases in Different Ecological Zones of Ilocos Norte

During the start of the three-year documentation in January 2016, some of the sample trees from the different ecological zones have entered varied stages. Some were already on the leaf fall initiation to completion stage, while others were on leafless stage, i.e., all leaves have fallen off. Leaf abscission occurs in a staggered manner. Still, other trees were on leaf or floral bud emergence.

The following are observations from the ecological zones of Ilocos Norte.

1. *Southern Coastal in Ilocos Norte.* Figure 20.5 shows the phenology of *A. luzonicus* in the southern coastal area of Badoc, Ilocos Norte. In 2016, flowering started in late January, peaked in mid-February, and ended in early March. The tree completed the phenophases of leaf maturity, leaf abscission, leafless, and flower bud initiation in chronological order (synchronous). In 2017, flowering started in late March, peaked in late April, and ended in early June. Leaf and flower bud initiation occurred simultaneously. In 2018, the phenophase pattern was the same as in year 2016 but the flowering period was until the later part of March. The flowering period in 2017 was late compared to 2016 and 2018. In October 19–20, 2016, Super Typhoon Lawin devastated Ilocos Norte.

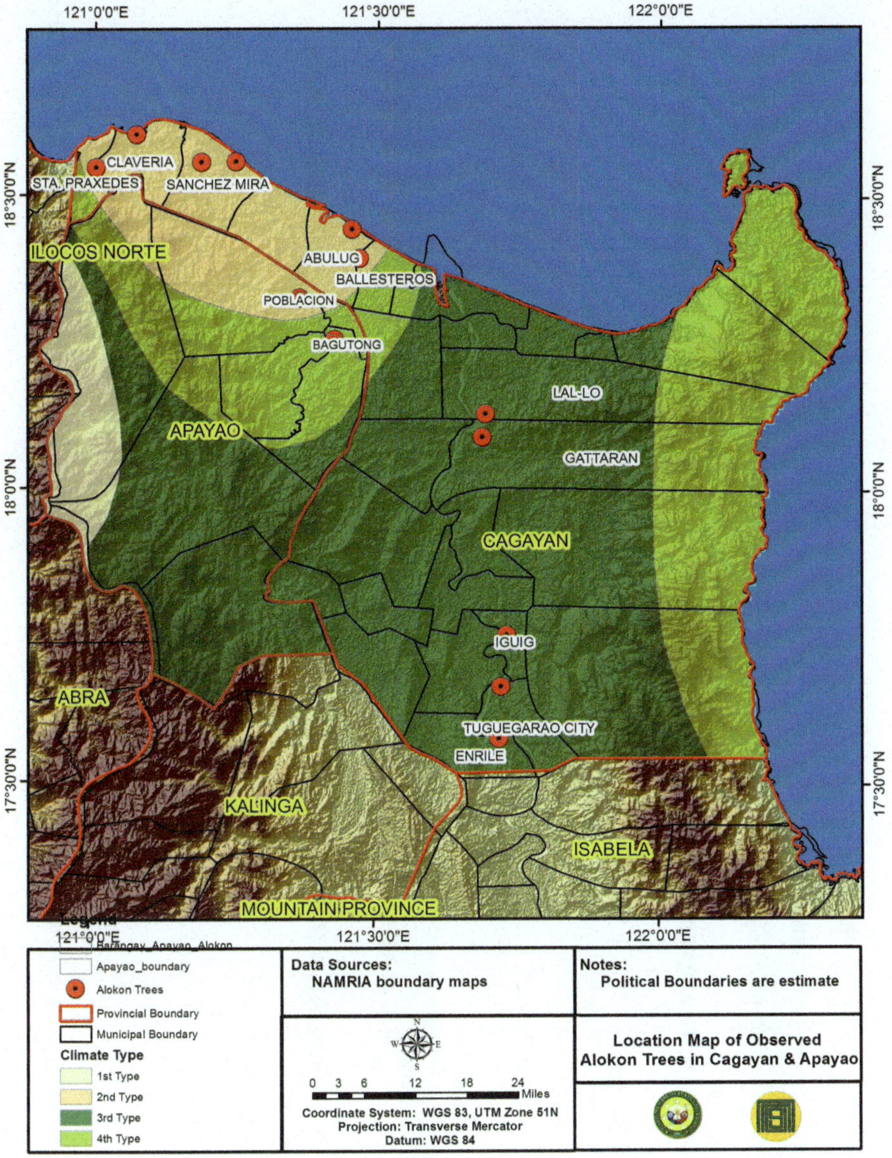

Fig. 20.4 Location of observed *A. luzonicus* trees in the Provinces of Cagayan and Apayao. Map by Engr. Rodel T. Utrera, Research Directorate, Mariano Marcos State University

The trees were affected and defoliated forcedly, and after some time they sprouted again and entered vegetative phase. The weather disturbance might have caused the prolonged vegetative period of the trees (November to December) and the late flowering time (March) in 2017.

5 Badoc (Southern Coastal)

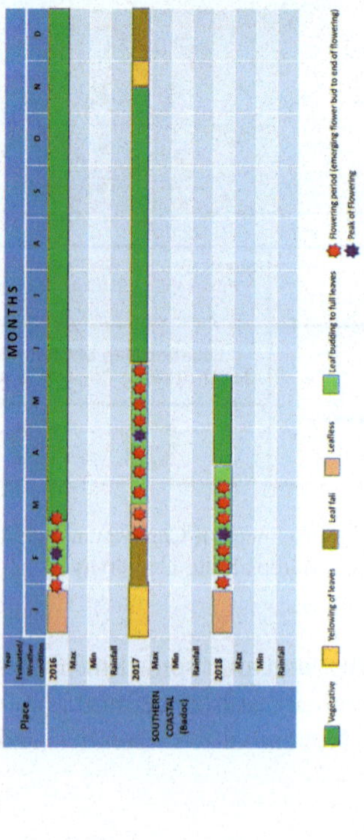

6 Badoc (Central Lowland 1)

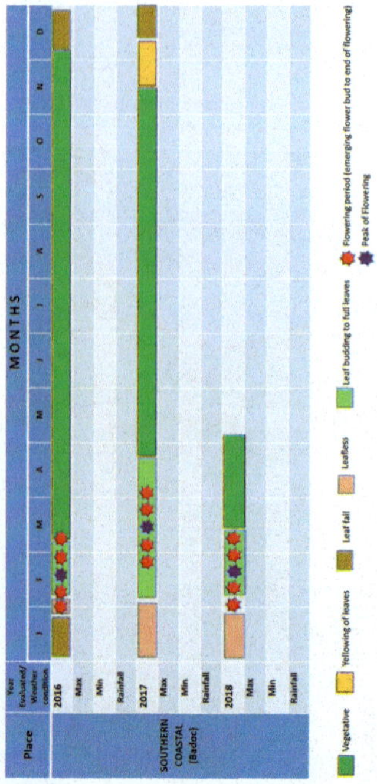

7 Badoc (Central Lowland 2)

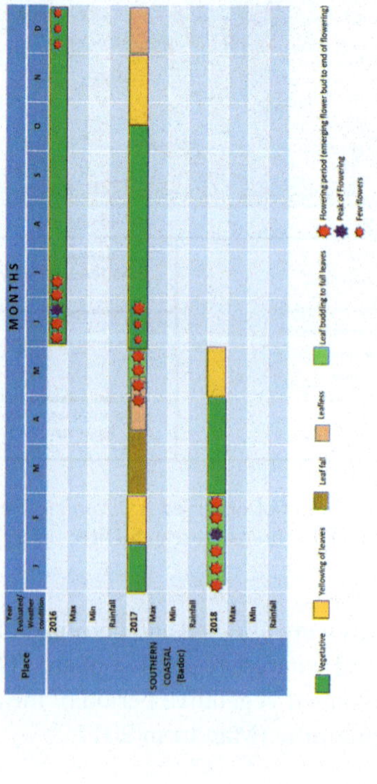

8 Badoc (Central Lowland 3, Rosita Yasay)

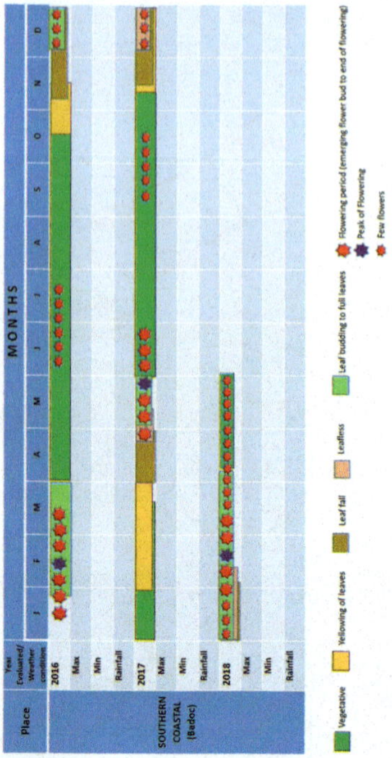

Figs. 20.5–20.8 Phenology of *A. luzonicus* in the southern coastal (**5**) and three sites in the central lowland (**6, 7, 8**) in the municipality of Badoc, Ilocos Norte

2. *Central Lowland in Ilocos Norte*. Figures 20.6, 20.7, and 20.8 show the phenology of *A. luzonicus* in the central lowlands of Badoc, Ilocos Norte. Phenophases and flowering behavior of the trees differ from each other. The flowering time of one of the sample trees (Fig. 20.6) followed the same pattern as in Fig. 20.5 except in 2017. In 2016, flowering started in late January, peaked in mid-February, and ended in March. In 2017, flowering was a month later (February) than in 2016, and the flowering period was from February to early April. In 2018, the trees followed the same pattern and time as in 2016. In 3 years, all the trees completed the phenophases in chronological manner (yellowing, leaf abscission, then leafless) prior to bearing flower buds.
   Figures 20.7 and 20.8 show the trees with different flowering behavior in the central lowland of Badoc. Flowers were abundant during the rainy season (June to July) in 2016, and started to bear few flowers in December (Fig. 20.7). In 2017, the flowering period started in late April and ended in late June. In 2018, the flowering period started in early January and ended in late February. Another sample tree, which is found also in the same community in Brgy. Labut, Badoc, exhibited another flowering behavior (Fig. 20.8). In 2016, the flowering period started in mid-January when the tree was totally leafless and ended in mid-May. It started to bear few flowers again in June to July (2^0 floral flush) when the tree was in its vegetative stage. Matured leaves started to fall in early November, then bore few flowers in leafless branches in December. In 2017, the flowering period started in late April and ended in June. It bore flowers again in September to October, then in December. In 2018, flowers were abundant from late January to early March, although flowering continued until May. Leaf maturity, leaf abscission, leafless, leaf formation, and flowering occurred simultaneously (asynchronous). The tree owner-grower revealed that his tree indeed exhibits long-duration flowering (almost all throughout the year). The "variety's" behavior coupled with the cultural technique of no pruning employed by the grower-owner contributed to the prolonged flowering period in the tree.
   Figure 20.9 shows the phenology of *A. luzonicus* in Vintar. Documentation started in December 2016 when the tree was in its leaf and flower budding stage. In 2017, the tree bore flowers from late January to mid-June. It bore few flowers in November to February 2018. It was observed that the tree continuously bears flowers but not all the time abundant. The peak of flowering was in April in the year 2017, while in January in the year 2018.
3. *Northern Coastal in Ilocos Norte*. Two trees were documented and monitored in the Municipality of Burgos, and one in the Municipality of Bangui. One of the sample trees in Burgos and the sample tree in Bangui had the same flowering pattern (Figs. 20.10 and 20.11). In 2016, flowers emerged when the trees were at their leaf abscission stage in January, but flowering occurred a little earlier in Burgos and both ended at the same time. In 2017, these two trees started to bear flowers in February and ended in April. The tree in Bangui had prolonged flowering period because the branches were not pruned during the peak of flowering. The usual practice of the owner is to top-prune the branches during its peak of flowering, so the flowering period was ended there. In 2018, the

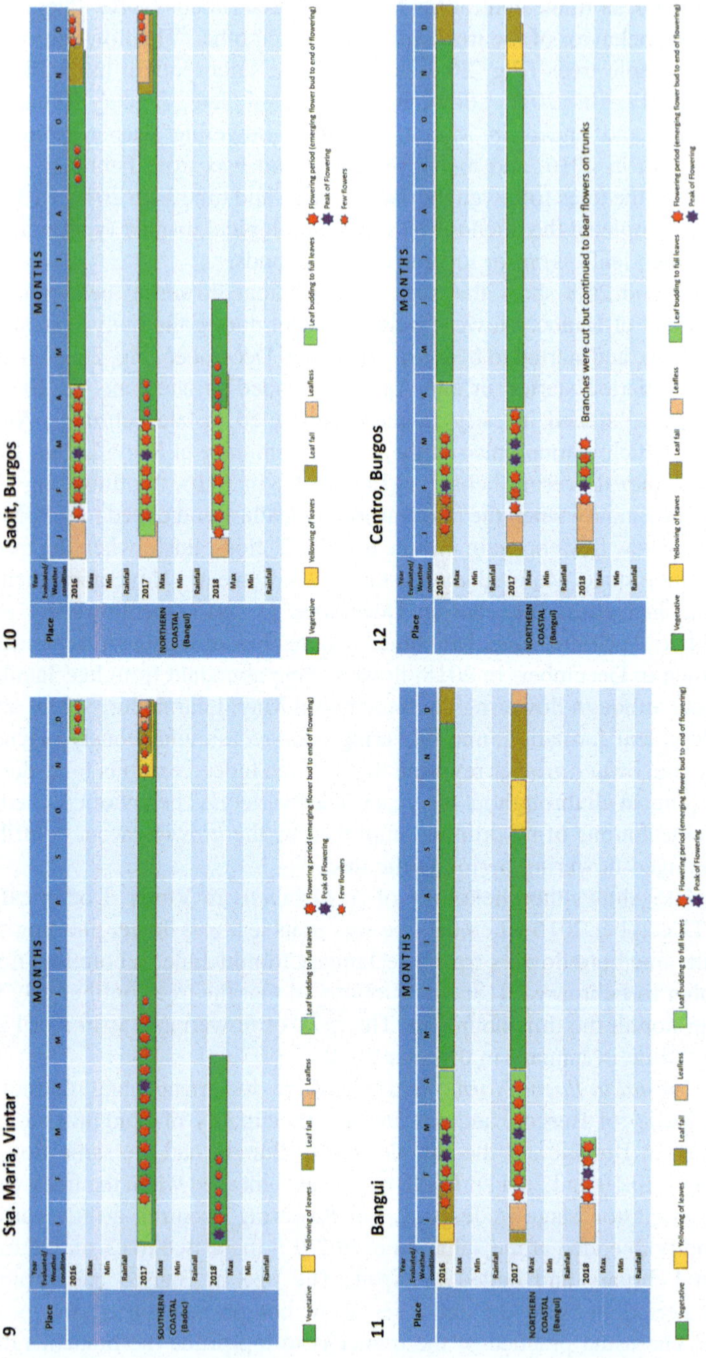

Fig. 20.9–20.12 Phenology of *A. luzonicus* in the central lowland in the Municipality of Vintar (**9**) and southern coastal in the Municipality of Burgos (**10, 12**) and Municipality of Bangui (**11**), Ilocos Norte

flowering period of these two trees started in February and ended in early March because the branches were cut but continued to bear flowers on trunks and sprouts. These trees completed the phenophases (leaf yellowing, abscission, and leafless) before flowers emerged in 2017 and 2018.
   Another tree in Burgos had a different flowering pattern (Fig. 20.12). This tree had the same growth and flowering pattern as the long-duration flowering trees in the Municipalities of Badoc and Vintar. It produced flowers almost all throughout the year but not as abundant as the yield during its regular flowering period (January to April). This tree exhibited the usual phenological pattern, i.e., leafless first before the flower emerged.
4. *Mountainous area in Ilocos Norte*. In a village in Tadao in the Municipality of Pasuquin (Fig. 20.13), the tree was on its leafless stage when the monitoring period was started in January 2016. Flowering period started in late January and ended in mid-March. Leaf maturity started in December and leafless period ended in late February. Flower buds immediately occurred after 3 weeks of leafless period and ended in April. Leaf maturity occurred earlier (late August) in 2017 than in 2016 (December). In 2018, leaf buds emerged first before the flower buds. Flowering started in the later part of January and ended in the early part of March. The flowering season in the year 2016 and 2018 had the same pattern.
5. *Eastern Interior in Ilocos Norte*. In the Municipality of Nueva Era (Fig. 20.14), the tree was on its leaf fall stage when the monitoring period was started in 2016. Flower buds emerged when the leaf falling stage was about to end. Flowering period started in mid-March and ended in late April. In 2017, flowering started 1 month earlier than in 2016. Flower buds also emerged when the leaf falling stage was about to end. In 2018, leaf maturity and leaf abscission stages occurred simultaneously, in different parts or branches of the tree. Flowering period occurred in early March, the branches were cut when there were abundant flowers, but still continued to bear flowers on trunks.
   Another tree (Fig. 20.15) in Nueva era exhibited a different pattern when compared to the tree in Fig. 20.14. The regular flowering period was in February but frequently cut during its peak of flowering but still continued to bear flowers on trunks. Few flowers were also present in January (Fig. 20.15).

20.2.4.2 Phenophases in the Province of Cagayan and Apayao

A. luzonicus in Cagayan and Apayao also follow the same flowering behavior based on phenophases observed in Ilocos Norte. Phenophases such as leaf abscission, leafless period, floral bud emergence and burst, leaf emergence, secondary floral burst, etc. were documented through interview with local residents. *A. luzonicus* flower earlier (as early as November/December) in coastal barangays in Sanchez Mira, i.e., Brgy Marzan, than in the rest of the province.

According to key informants, there are two general flowering periods in the Province of Cagayan, namely: (a) January to March, flowering peak in March, in the Municipalities of Sta. Praxedes, Claveria, Sanchez Mira, Ballesteros, and Abulug; and (b) April to June in the Municipalities of Gattaran, Iguig, and Tuguegarao. A few cases of early flowering varieties, asynchronous habit, and

Fig. 20.13–20.15 Phenology of *A. luzonicus* in a mountainous village in the Municipality of Pasuquin (13), and in the eastern interior Municipality of Nueva Era (14, 15), Ilocos Norte

double flowering (i.e., in the Municipalities of Lal-lo and Enrile) were likewise noted in some areas.

Ocular observation started in June 2017 in almost all the municipalities of Cagayan. Figures 20.16, 20.17, and 20.18 show the phenology of *A. luzonicus* in the province based on ocular observation. Most of the trees in June 2017 were in their vegetative stage (Figs. 20.16 and 20.17) except for some trees in a part of Lal-lo and Gattaran which were on leafy flowering stage. If we follow the flowering season mentioned by the informants, the June flowering could be the regular season in that part of Lal-lo and Gattaran.

A. luzonicus trees monitored in November 8–9, 2017, were on their leaf shedding stage in the Municipalities of Claveria, Sanchez Mira, Ballesteros, and Lallo while others were on leaf maturity stage in Gattaran, Alcala, and Enrile. A tree in Sanchez Mira was seen on full flowering stage on November 8, 2017, which was probably due to the removal of its bark.

In March 2018, most of the trees visited are now on their peak of flowering. Bud emergence could have started in February. The trees in Lal-lo (near Magapit Bridge) which flowered in June are again flowering in February and March (Fig. 20.18), which may now coincide with the mentioned second flowering.

In the Province of Apayao, the sample trees being monitored in Luna and Flora generally follow the flowering period of trees observed in the Municipalities of Claveria, Abulug, and Ballesteros.

20.2.5 Off Season Varieties of *A. luzonicus*

Since there is no identified and registered variety of *A. luzonicus* yet, preliminary identification of ecotypes or varieties was done to identify plants with excellent flowering behavior and gives a desired consumer appeal and acceptability at the same time. Trees were selected based on flowering habit and maturity, identifying preferably those trees that flower outside the general flowering period as observed above. Subsequently, the selected varieties were evaluated for the sensory qualities of the edible part, particularly texture, aroma, taste, slick/sliminess, tenderness, still-green trait after cooking, and general acceptability.

From the accessions monitored, six accessions were selected for their desired flowering behavior (Table 20.3). These include three accessions (Accn Yasay1, Yasay2, Badoc1-CL) from the Municipality of Badoc, two accessions (Accn 4Vintar and Accn 6Vintar) from the Municipality of Vintar, and one accession (Accn Salucag1) from the Municipality of Burgos, all situated in the Province of Ilocos Norte. The specific flowering period together with the consumer acceptability rating for each accession is indicated in Table 20.3. These accessions are geo-tagged and prioritized for mass propagation.

The six accessions were selected for having early flowering (as early as November or December), double flushing, and long-duration flowering. In addition, their desirable flowering behavior is complemented by the excellent sensory qualities earning a general acceptability rating of liked moderately to liked very

Fig. 20.16–20.18 Phenology of *A. luzonicus* in the Municipalities of Claveria (16), Sanchez Mira (17), and Lal-lo (18), Province of Cagayan

Table 20.3 *A. luzonicus* varieties with desired flowering behavior and sensory qualities

Accession/ location	Inflorescence length (cm)	General acceptability	Flowering period	Special character
Yasay1 Badoc, Ilocos Norte	3.2–10-4	Liked very much; slick	January–March June–July, December	Early flowering in December; double flushing in June/July; long-duration flowering Asynchronous
(Yasay2) Badoc, Ilocos Norte	3.2–10-4	Liked very much; slick	January–May	Long-duration flowering Asynchronous
Badoc1 Badoc, Ilocos Norte	6–11	Liked moderately; slightly slick	April–June/ June–July, December	Early December; Late or double flushing in April–July
4 Vintar, Ilocos Norte	6–10.8	Liked moderately; slightly slick	November–December; February–June	Early and long-duration flowering Asynchronous
6 Vintar, Ilocos Norte	6–11	Liked moderately; slick; still green	February–July; September–December*	Long-duration flowering *flowers at this period are significantly lesser than during February–June
Salucag1 Saoit, Burgos, Ilocos Norte	3.6–8.7	Liked very much; slick	April–June; September–October; December	Early & long-duration flowering Asynchronous

much (Table 20.3). All six accessions are moderately slick to slick. Slickness, similar to the slickness of *Corchorus* sp., is a desired sensory quality of *A. luzonicus*. One of the accessions (Accn 6Vintar) has a still-green feature, retaining a vibrant green after at least one hour from cooking. This indicates less oxidation in this accession.

20.2.6 Phenocalendar in Climate Types I, II, III, and IV

The study areas fall under four climate types (Fig. 20.19). The study areas in the Province of Ilocos Norte belong to climate Type 1, to which majority of areas in Ilocos Norte belong. In the Modified Corona's Climate Classification based on rainfall distribution, Type I climate is characterized by two pronounced seasons: dry from November to April and wet during the rest of the year (https://www.pagasa.dost.gov.ph). The maximum rain period is from June to September. On the other hand, the Province of Cagayan belongs to three climate conditions (Type II, III, and IV). Type II climate prevails in the Municipalities of Sta. Praxedes, Claveria, Sanchez Mira, Pamplona, Abulug, and part of Ballesteros while Type III climate prevails in majority of the municipalities of Cagayan. Type II climate is

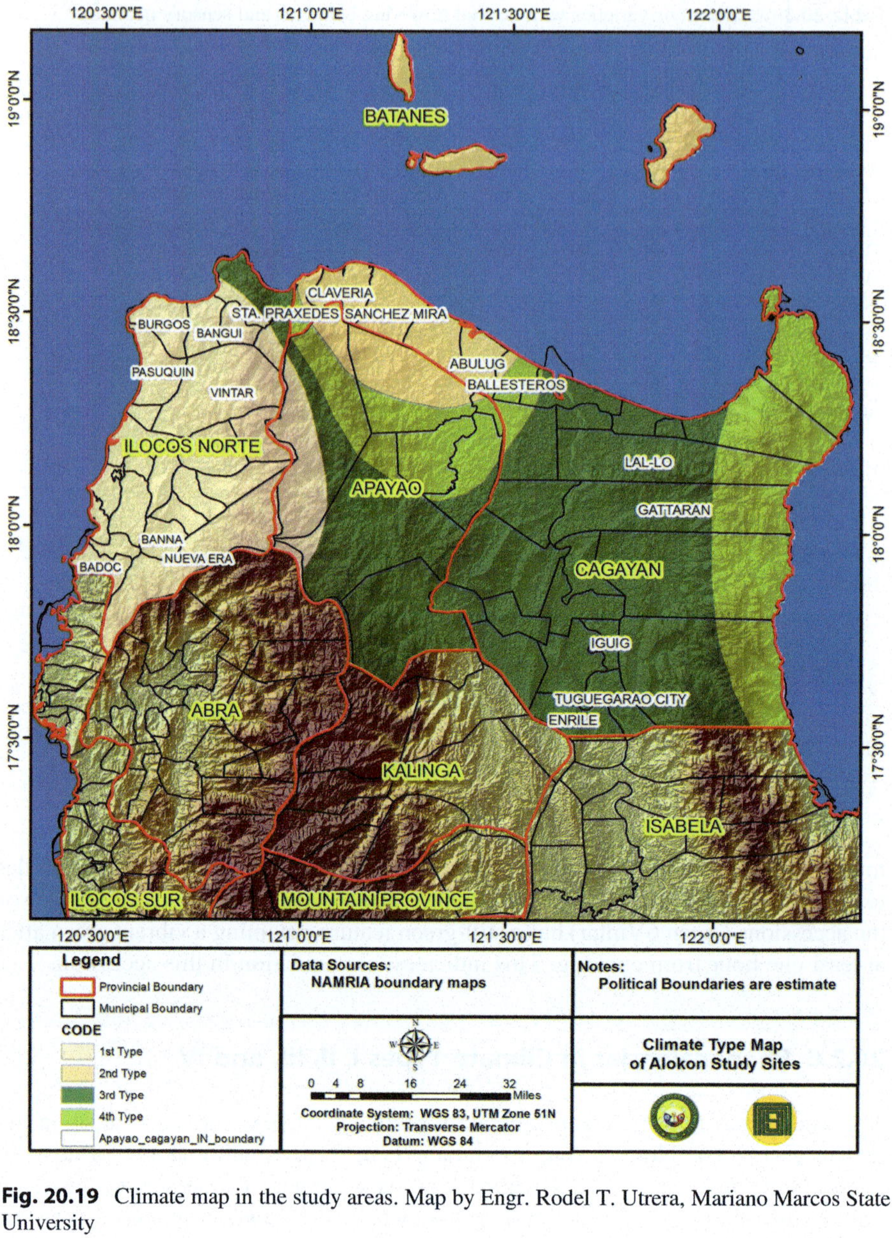

Fig. 20.19 Climate map in the study areas. Map by Engr. Rodel T. Utrera, Mariano Marcos State University

characterized by no dry season with a very pronounced maximum rain period from December to February. Type III climate, on the other hand, is characterized by no very pronounced maximum rain period, with a very short dry season lasting only from 1 to 3 months (https://www.pagasa.dost.gov.ph).

The Municipality of Sta. Ana and portions of Gonzaga, Lal-lo, Gattaran, Baggao, and Penablanca belong to Type IV. They are characterized by more or less even distribution of rainfall throughout the year (https://www.pagasa.dost.gov.ph). Meanwhile, four climate types prevail in the Province of Apayao. The study sites in Luna and Flora belong to Types II and IV, respectively.

Flowering period in Ilocos Norte under Type 1 climate is generally from January to March, peaking in February (Fig. 20.20). With the identified off-season varieties, flowering is almost year round, except in the month of August.

Meanwhile, flowering in northwestern Cagayan with Type II climate, and northern Apayao both with Type II (i.e., Municipality of Luna) and Type IV climates (i.e., Municipality of Flora) is generally from January to March, too. However, flowering is behind Ilocos Norte by about 2–3 weeks; hence, peak of flowering is in the month of March (Fig. 20.21).

Flowering in most towns of Cagayan, which have Type III climate is generally from April to June (Fig. 20.21). Despite this general flowering pattern observed per climate type, there exist few to some cases of early flowering, double flowering, and late flowering in above areas.

20.2.6.1 Factors Affecting Phenophases and Flowering Behavior

The development of *A. luzonicus* appeared to be influenced by climate type in the study areas. This is evident in the different flowering periods observed in the three provinces (Figs. 20.20 and 20.21).

Additionally, phenophases were influenced by temperature and daylength (photoperiodism) and moisture before and during the phase occurrence. In the Province of Ilocos Norte, the usual low temperature (Fig. 20.20) and short daylength in the months of November to December appeared to favor leaf abscission and flowering. Additionally, weather disturbance such as Typhoon Lawin in October 2017 caused delay in the flowering period, especially in the Municipality of Badoc where its effect was most pronounced.

The low temperature (as low as 12 °C) usually started in November to December, extending up to February (as low as 9 °C) of the succeeding year, is likened to vernalization, which makes a plant receptive to flowering stimulus. Although it has no direct role in inducing flowering, vernalization stimulus is perceived by dividing cells, e.g., shoot tips and embryo, in temperate and some tropical species (http://www.biologydiscussion.com). Vernalization treatment which includes exposure to about 0–5 °C can substitute the role of gibberellin in flowering and its effect is quite demonstrated in a number of species such as *Arabidopsis thaliana*.

A. luzonicus is also a photoperiodic species. Like garlic whose bulbing is favored by short days, *A. luzonicus* flowering seemed to be induced by the short-day length in November to December. The floral-stimulating action of photoperiodism is well reported in many species. The reported transfer of photoperiodic induction stimulus to non-induced plants is observed during propagation studies done in *A. luzonicus*. Once induced to flower (cell determined), new cuttings continue to flower even if they haven't yet developed roots and regenerated into new plants.

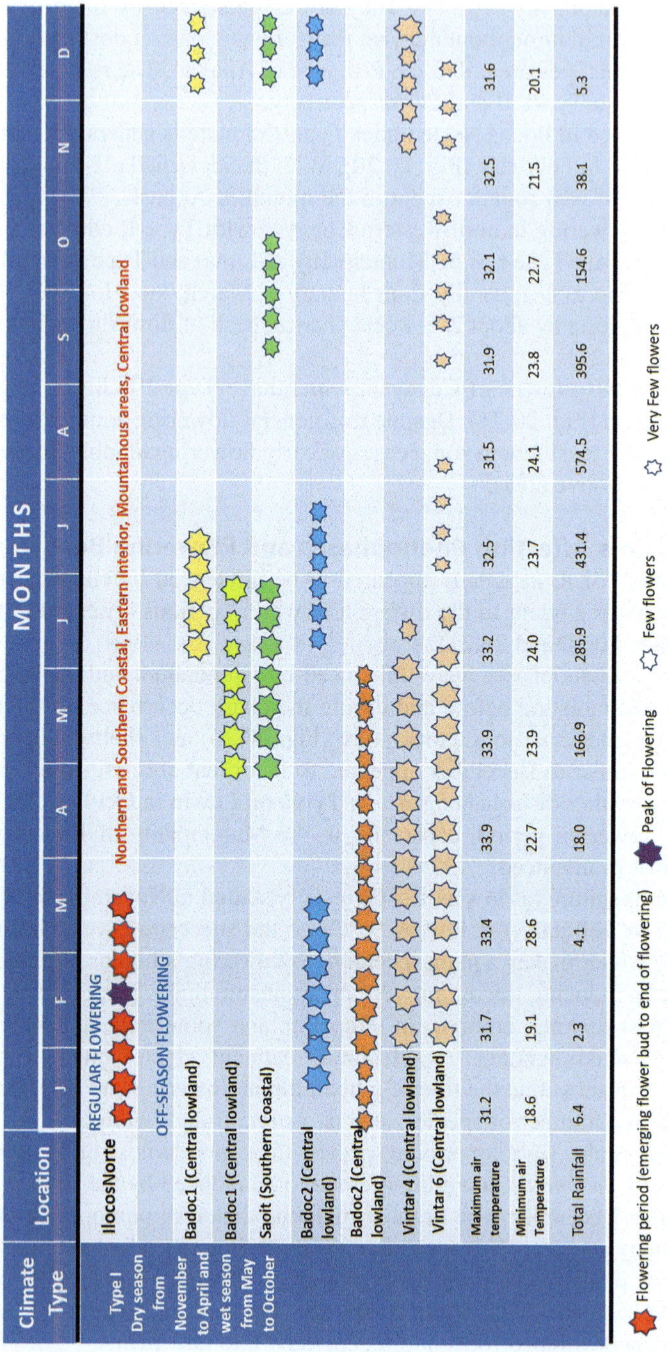

Fig. 20.20 Flowering of *A. luzonicus* under Climate Type I of Ilocos Norte

Climate Type	Location	J	F	M	A	M	J	J	A	S	O	N	D
Type II - No dry season with very pronounced maximum rain period from Dec- Feb **Type IV-** Rainfall is more or less evenly through out the year	**Cagayan: Claveria, Sanchez Mira , Lal-lo**		REGULAR FLOWERING										
	Maximum Air Temperature	27.8	29.0	30.6	32.7	33.4	34.0	33.5	32.9	32.5	31.5	30.2	28.2
	Minimum Air Temperature	20.7	21.5	22.5	24.1	24.8	25.1	25.1	24.8	24.4	23.8	23.3	21.6
	Rainfall	91.7	65.0	34.3	44.8	128.5	165.7	207.1	172.4	246.8	321.2	306.2	256.9
Type III - No very pronounced maximum rain period, with a short dry season lasting only from one to three months	**Cagayan: Enrile**					REGULAR FLOWERING							
	Maximum Air Temperature	28.2	31.1	32.0	35.2	35.5	34.9	34.4	33.1	32.7	31.5	29.9	27.9
	Minimum Air Temperature	18.1	19.7	20.3	22.0	23.8	22.9	23.2	22.6	22.2	21.1	20.6	18.9
	Rainfall	42.7	72.2	47.4	49.6	130.6	179.7	197.9	236.7	238.7	237.1	296.8	137.5

Flowering period (emerging flower bud to end of flowering) Peak of Flowering Few flowers

Fig. 20.21 Flowering of *A. luzonicus* under Climate Types II and IV, and Type III in Cagayan and Apayao

The effect of temperature is also observed on the rate of development in *A. luzonicus* inflorescence. An important assumption here is that plant growth is a linear function of temperature. An increase in temperature results in an increase in development rate (Sivertsen et al. 1999). On flower characters, low temperature is not favorable to inflorescence elongation as manifested in the stunted growth of inflorescence as observed in trees in the Municipality of Bangui. Observing the same trees, the increase in temperature in the month of March favored inflorescence development as indicated in the longer and faster rate of elongation. The usual period it takes from bud emergence to market maturity is 2 weeks. But this was shortened to 1 week during the hotter month of March.

In the regular flowering season during the summer months (starting March), during which temperature, solar radiation, and evapotranspiration rise while rainfall becomes nil, there are distinct features observed on flowering, as follows: (a) inflorescence emerge starting on the first axil and every axil thereafter, (b) many inflorescence arise directly from the stem, somewhat similar to cauliflory, (c) inflorescence is plenty (1 to 9 per spike), with few or no leaf at all, and iv) spikes have shorter internodes or distances between axils.

During the rainy season when sufficient moisture becomes available for plant growth, second flush of flowers occurs (but in few varieties only). Inflorescence produced is only few (1 to 3 per spike), but longer than inflorescence produced in summer or regular season, had big and numerous leaves, and usually start emerging at third or fourth axil. Likewise, spikes were observed to be longer, thus a factor of longer internodes. This is presumably a direct effect of the relative abundance of water which favors vegetative growth. Similar hypothesis is drawn for the slightly extended vegetative phase in January in Cagayan as the months of December and January usually have continuous rain.

Biotic factors were also inferred to affect the flowering behavior of *A. luzonica*. The sample trees did not manifest similar flowering behaviors. One possible reason could be attributed to the age of the trees. Trees pass through specific developmental phases as they age, including juvenile to adult, and vegetative to reproductive phases (Groover 2017). The timing of these transitions is regulated genetically but is also highly influenced by the environment. Transition to reproduction is influenced by conserved mechanisms in angiosperm trees, which determine when to flower by integration of information from physiological changes associated with size and carbohydrate metabolism with measurement of seasonal and environmental factors.

Variety is another factor in the observed differences on flowering behavior such as the occurrence of long-duration (year-round) flowering, second floral flushing, and early and late flowering. At present, there is no identified and registered variety of *A. luzonicus* yet.

20.3 Summary, Conclusions, and Recommendations

The growth and development cycle in *A. luzonicus*, also referred to as phenological cycle in this context, consists mainly of vegetative, intermediate, and reproductive phases. Under Ilocos Norte conditions, the vegetative phase lasts for 7–8 months (March to October) while the reproductive phase lasts for 4–5 months (late November to March). The specific phases are accompanied by morphological changes in the plant such as leaf bud emergence, full leaf development, leaf maturity, leaf yellowing and senescence, leaf abscission, leafless, floral bud emergence, flower development, and flower physiological maturity. Regardless of ecological zone in Ilocos Norte, trees follow a similar phenological cycle and period of occurrence over the three-year observation period. But a few trees exhibit extreme cases, not conforming with the regular cycle (hence called asynchronous) and period of occurrence.

Comparing trees under four climate types in the Provinces of Ilocos Norte, Cagayan, and Apayao, the phenological cycles are similar but flowering period varies. The flowering period under Type I climate of Ilocos Norte is generally from January to March, peaking in February. Flowering in Type II (northwestern Cagayan and northern Apayao, i.e., Municipality of Luna) and Type IV climates (northern Apayao, i.e., Municipality of Flora) is generally from January to March but behind Ilocos Norte flowering by about 2 to 3 weeks; hence, peak of flowering is in the *month of March*. Flowering under Type III climate in Cagayan is generally from April to June. Climate, weather factors such as temperature and daylength (photoperiodism), and biological factors such as age and variety of tree seemed to affect the occurrence and duration of the different phases, and some flower characteristics.

There exist six accessions which exhibit early flowering, double flowering, and year-round flowering (but still peaks during the regular flowering season).

Flower induction is employed in some high value crops, such as mango and dragonfruit, for off-season crop production. While cultural intervention can be tested also in *A. luzonicus*, a production system integrating the species' phenological cycle and off-season varieties is an alternative for year-round inflorescence production. Thus, a phenocalendar for a sustainable crop production is hereby developed (Fig. 20.22). The Provinces of Ilocos Norte, Cagayan, and Apayao are viable sources of raw materials to meet consumption requirements in the months of January to June. For the remaining months of the year (excluding August), the off-season varieties will supply the inflorescence requirement. The raw material requirement for the month of August can be met by using frozen, freeze-dried, or ready-to-cook formulations of the inflorescence. Freezing technique is already in place. Blanching and freeze-drying duration are currently being standardized for freeze-dried *A. luzonicus*. Additionally, research on the development of ready-mix or ready-to-cook viands of *A. luzonicus* is also underway.

The phenocalendar-based crop production system is simple and economical, not requiring cultural interventions or chemical inputs. The following recommendations are hereby put forward in support of the developed phenocalendar: (a) Validate the consistency of the phenocalendar in response to climate change. (b) Mass propagate

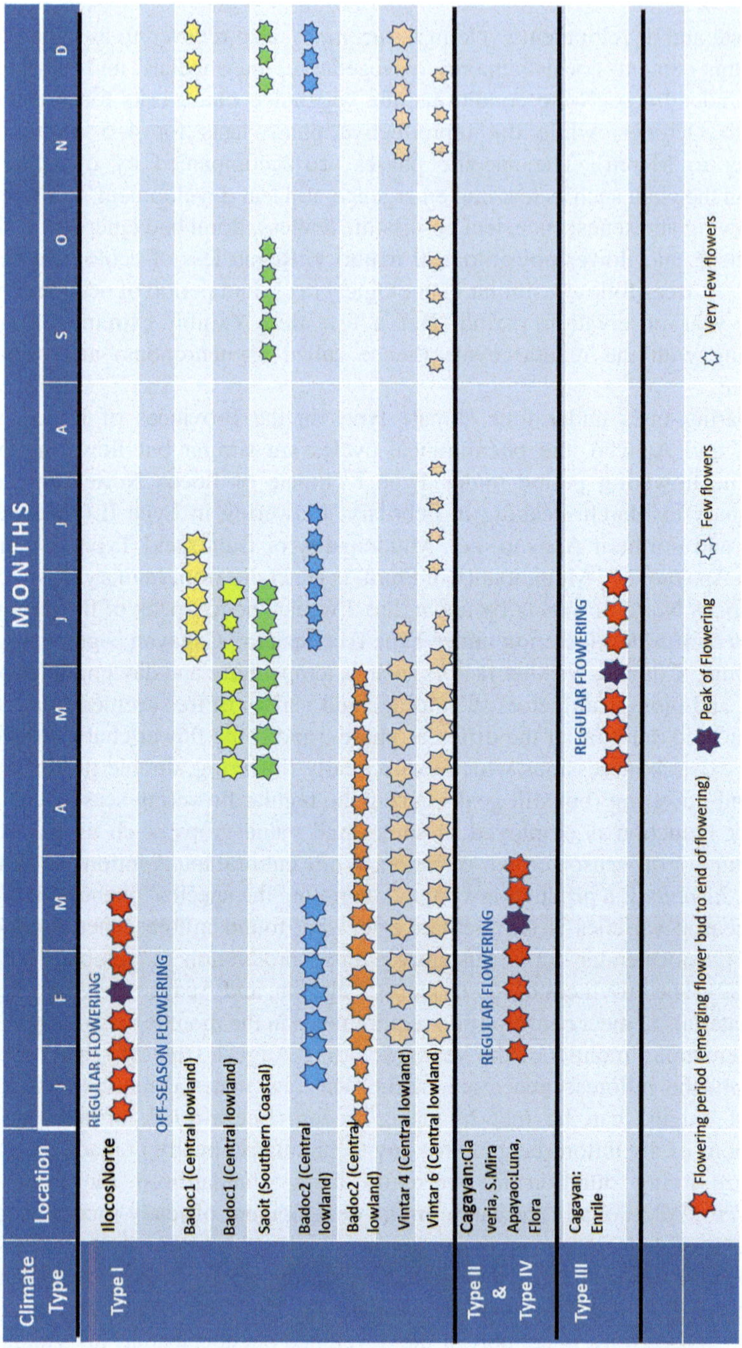

Fig. 20.22 Phenocalendar-based inflorescence production in *A. luzonicus*

the identified varieties and promote additional planting. (c) Accelerate proactive research on product development to address future need and supply gaps (i.e., market glut, supply shortage).

Acknowledgments Special thanks to the Mariano Marcos State University for the funding support in the implementation of the various R&D projects from which information included in this chapter has been taken.

References

Antonio MA, Vivit MB (2017) Phytoconstituents and in vitro anti-oxidant activity of selected indigenous vegetables in the Ilocos. MMSU Sci Tech J 7(2):16–27

Antonio MA, Utrera RT, Agustin EO, Jamias DL, Badar AJ, Pascua ME (2011) Survey and characterization of indigenous food plants in Ilocos Norte, Phil. SEARCA Agr & Devt Discussion Paper Series. SEARCA, Los Baños, Laguna

Antonio MA, Agustin EO, Badar EO (2016) Catalog of indigenous food plants in Ilocos Norte, 2nd edn. Mariano Marcos State University, City of Batac

Antonio MA, Galacgac ES, Utrera RT (2018) The biology and phenology of *Allaeanthus luzonicus* (Moraceae), a promising indigenous vegetable. Paper presented in the ILARRDEC regional symposium on R&D highlights. DMMMSU, Bacnotan, La Union. 27–29 November 2018

Groover A (2017) 201er/psw_2017_groover001.pdf.7. Agre-related changes in tree growth and physiology. In: eLs. Wiley, Chichester. https://doi.org/10.1002/9780470015902.a0023924., www.eld.nethttps://www.fs.fed.us.psw.publications/groov. Accessed 13 Aug 2018

Haggerty BP, Mazer SJ (2008) The phenology handbook: a guide to phenological monitoring for students, teachers, families and nature enthusiasts. Accessed 14 Aug 2018

Pancho JV (1979) Kalikasan, The Philippine Journal of Biology No. 1. Vascular flora of Mt. Makiling and vicinity (Luzon:Phil) Part 1. UP Los Banos. College, Laguna

POWO (2022) Plants of the World Online. Facilitated by the Royal Botanic Gardens, Kew. Published on the Internet; http://www.plantsoftheworldonline.org/. Retrieved 15 August 2022

Saxena M, Saxena J, Nema R, Singh D, Gupta A (2013) Phytochemistry of medicinal plants. J Pharmacogn Phytochem 1(6) http://www.phytojournal.com/archives/2013/vol1issue6/PartA/26.pdf

Sivertsen TH, Nejedlik P, Oger R, Sigvald R (1999) The phenology of crops and the development of pests and diseases. www.bioforsk.no/ikbViewer/Content/16776/Publikasjon1999.pdf. Accessed 14 Aug 2018

Analyzing the Spatio-temporal Changes of the Biocultural Landscape of Banaue, Philippines Using GIS

21

Consuelo DI. Habito, Rosalyn D. Sontillanosa, and Dana Lorraine C. Faurillo

Abstract

The Ifugao Rice Terraces (IRT) of the Philippines are listed among UNESCO's World Heritage Sites and are considered an outstanding example of an evolved, living cultural landscape. Dating back to several centuries, the rice terraces were carved out by the Ifugao, a group of indigenous people who inhabit the Cordillera mountains. To this day, the rice terraces have been tilled using age-old sustainable farming practices and planted with traditional rice varieties. Fittingly, the IRT has been recognized as a Globally Important Agricultural Heritage Site by the FAO. However, the FAO has drawn attention to the "very critical stage of deterioration" of the IRT. The continued survival of this eminent biocultural landscape is threatened by combined economic and tourism growth, environmental degradation, unregulated housing development, and neglect. This can result to plant diversity degradation and even extinction.

Geographic Information System (GIS) and Remote Sensing (RS) can significantly contribute to the decision-making and management of cultural heritage sites such as the IRT. This study used digitized thematic maps and satellite images from different time periods to analyze land use changes in the IRT. This analysis

C. D. Habito (✉)
Faculty of Management and Development Studies, University of the Philippines Open University, Los Banos, Laguna, Philippines
e-mail: cdhabito@up.edu.ph

R. D. Sontillanosa
GDMD, Geospatial Information System Management Branch, National Mapping Resource Information Authority, Taguig City, Philippines

D. L. C. Faurillo
Department of Trade and Industry, Construction Industry Authority of the Philippines, Makati City, Philippines

S. Ramamoorthy et al. (eds.), *Plant Diversity in Biocultural Landscapes*,
https://doi.org/10.1007/978-981-19-8649-9_21

can potentially contribute to informed decision-making by national and local government officials and planners toward IRT plant diversity protection and management. This study recommends a comprehensive plan for the heritage towns that will take into consideration the conservation and preservation of the bio-cultural landscape of Ifugao in combination with sustainable tourism development.

Keywords

Biocultural landscapes · Urban sprawl · Cultural heritage sites · GIS

21.1 Introduction

The Ifugao Rice Terraces (IRT) is a living bio-cultural landscape located in the Northern Island of Luzon, Philippines. This site stands out as part of the unique history and heritage of the Ifugao, a culturally distinct indigenous people that have inhabited the mountainous communities of the Philippine Cordilleras for centuries. The hardy Ifugao are known for having carved out the rice terraces along rugged mountain terrain. To this day, a dwindling number of Ifugao farmers have continued to preserve and till the IRT. In 1995, five distinct rice terrace clusters were collectively recognized as a World Heritage Site by the United Nations Educational, Scientific, and Cultural Organization. In 2005, the Food and Agriculture Organization of the United Nations also designated the IRT as a Globally Important Agricultural Heritage System (GIAHS). The majestic landscape and the Ifugao people are inextricably linked—influenced by strong ethnic, sociocultural, supernatural, and traditional beliefs and practices. For a visitor, the richness of the Ifugao cultural heritage is immediately highlighted by the majestic vista of the rice terraces as a tangible proof of the people's hardiness and resilience. But beyond the captivating scenery, there are the heritage towns, the characteristic architecture of their native houses, the traditional customs and practices of the local people, and the traditional Ifugao knowledge and expertise on farming, agriculture, arts, wood carving, and weaving, to name a few.

Over the past 10 years, there has been increasing recognition of the importance of strengthening linkages between biological and cultural diversity initiatives (Agnoletti and Rotherham 2015). In particular, the Florence Declaration (ICOMOS 2014) has put a spotlight on the concept of biocultural diversity and has identified four key issues related to research activities for its promotion. Although the main ideas put forward were largely centered on biological and cultural diversity in Europe, the key issues identified by Agnoletti and Rotherham (2015) resonate with and are highly relevant to contexts and environments in other parts of the globe. In Southeast Asia, the IRT stands out as an example of such a biocultural landscape. Collectively, the massive land area covered by the IRT and the distinct cultural heritage of the Ifugao people are highly integral to the Philippine Cordillera biocultural landscape. The incalculable value of the Ifugao heritage towns has been described by Macapagal and Bermejo (2015) as a "jewel of Filipino architectural patrimony." From a broader perspective, historical and heritage sites such as the IRT are of immense value for a "nation's history, identity, and regional diversity" (El Menchawy et al. 2011). Within the IRT, there has been a dynamic and evolving mix of environmental, social, economic, biological, and cultural processes over time and space—elements that add to the complexity of managing and sustaining both the biological and cultural richness of the landscape.

In recent decades, mounting challenges have come from an increasing population, demographic shifts, changing attitudes and beliefs, especially among the younger population, environmental degradation, economic factors, and political priorities, all of which have put tremendous pressure on the preservation of the IRT (Zialcita et al. 2015). A dwindling number of aging Ifugao farmers, social change, and modernized attitudes have led to neglect and abandonment of some once-productive rice terraces (Calderon et al. 2009).

From the time that five agricultural clusters of rice terraces were declared as a World Heritage Site by UNESCO, there have been several initiatives from the government and from donor agencies for the heritage conservation of the Ifugao Rice Terraces. However, these were mostly focused on the maintenance of the physical structure of the rice terraces, with minimal or rare participation from the local communities (Acabado et al. 2017). These conservation efforts were mostly driven by the potential income to be generated from the influx of tourists without regard to the communities directly involved in the culture of rice in the terraces. In fact, some people considered the inscription as a World Heritage Site a burden since there were strict policies that had to be implemented as prescribed by heritage experts and conservationists (Nozawa et al. 2008). There should be respect and recognition of the living traditions and way of life of local communities who expect their fair share in the development and conservation of this important bio-cultural landscape. They can share equal responsibility with local, provincial, national, and international institutions in the conservation of the outstanding universal value of the Ifugao Rice Terraces.

Urban and tourism pressure, in addition to the inadequate or lack of a management plan, has created challenges in the precarious balance of people and rice terraces in Ifugao. This is particularly true in the town of Banaue which is located at the center of the road network connecting the four WHS rice terraces of Hungduan, Bangaan, Battad, and Mayoyao. Urban sprawling has been evident. Sprawl is defined as "unrestricted growth with little or no planning." Dulnuan (2014) mentioned the tourism problems and concerns for all the heritage towns of Ifugao. In Banaue's Comprehensive Land Use Plan (CLUP), four barangays, namely Poblacion, Bocos, Tam-an, and Viewpoint were classified as urban barangays. The location of the four barangays is strategic as they are all located along the national highway with the major cross-road leading to the four WHS rice terraces clusters. While being classified as urban barangays, it is important to note that extensive mapping of the other barangays since 1963 in addition to four urban barangays by Conklin (1980) showed extensive rice terraces. In addition to the absence of a building code, the area is also geologically unstable and prone to earthquakes and landslips (Nozawa et al. 2008). Its proximity to Hapap Fault may cause some problems with buildings not following codes for earthquake safety. Unregulated building construction has marred the rice terraces view with buildings unsuitable for heritage towns such as Banaue. Moreover, multi-story structures can cause collapse if they do not follow building codes.

Ifugao tourist arrivals reached their peak from 2004 to 2011 and have slowly declined (Dulnuan 2014). Despite this decline, there were 70,000 tourists[1] and 180,000 visitors in Ifugao in 2017 (Geminiano 2018). Domestic tourism at the moment has caused the construction of tourist inns, informal settlements, and shanties on roadsides, completely marring or blocking the view of rice terrace clusters. The lack of parking spaces for tourist vehicles and local transport has caused chaos in the Banaue Poblacion proper coupled with narrowed roads leading out of Poblacion. While Covid-19 pandemic has closed all tourist travel since 2020. This however has not stopped the construction of tourist establishments. There is a need for sustainable tourism development in the area that will benefit both the conservation of the Ifugao biocultural landscape and the tourism revenues that it can generate.

Objectives
This study has the following general objectives:

1. Analyze the spatio-temporal changes of the biocultural landscapes of IRT using geographic information systems (GIS);
2. Determine the various factors causing spatio-temporal changes;
3. To recommend policy and future research implications that are relevant to the management, preservation, or protection of biocultural landscapes of the IRT.

21.1.1 Description of Banaue, the Study Site

The study was conducted in the Municipality of Banaue, Ifugao which is about 320 km north of Metro Manila, the capital of the Philippines. It can be reached after a 10-h bus drive from Manila. It lies at the foot of Mt. Amuyao, the highest peak in the area. It is bounded on the north by the municipality of Barlig, Mountain Province, on the east by the municipality of Mayoyao, on the west by the municipality of Hungduan, and on the south by the municipality of Hingyon.

Banaue is a fourth class municipality[2] consisting of 18 barangays (Fig. 21.1). Based on the Provincial Planning Development Office report (2006), out of these 18 barangays, 4 are considered as urban barangays (Poblacion, Viewpoint, Tam-an,

[1]Tourists as those who booked in accommodation establishments for at least 24 h, and visitors as the day tourists.

[2]Municipalities that have obtained an average total revenue of 300,000 pesos or more but less than 500,000 pesos per annum.

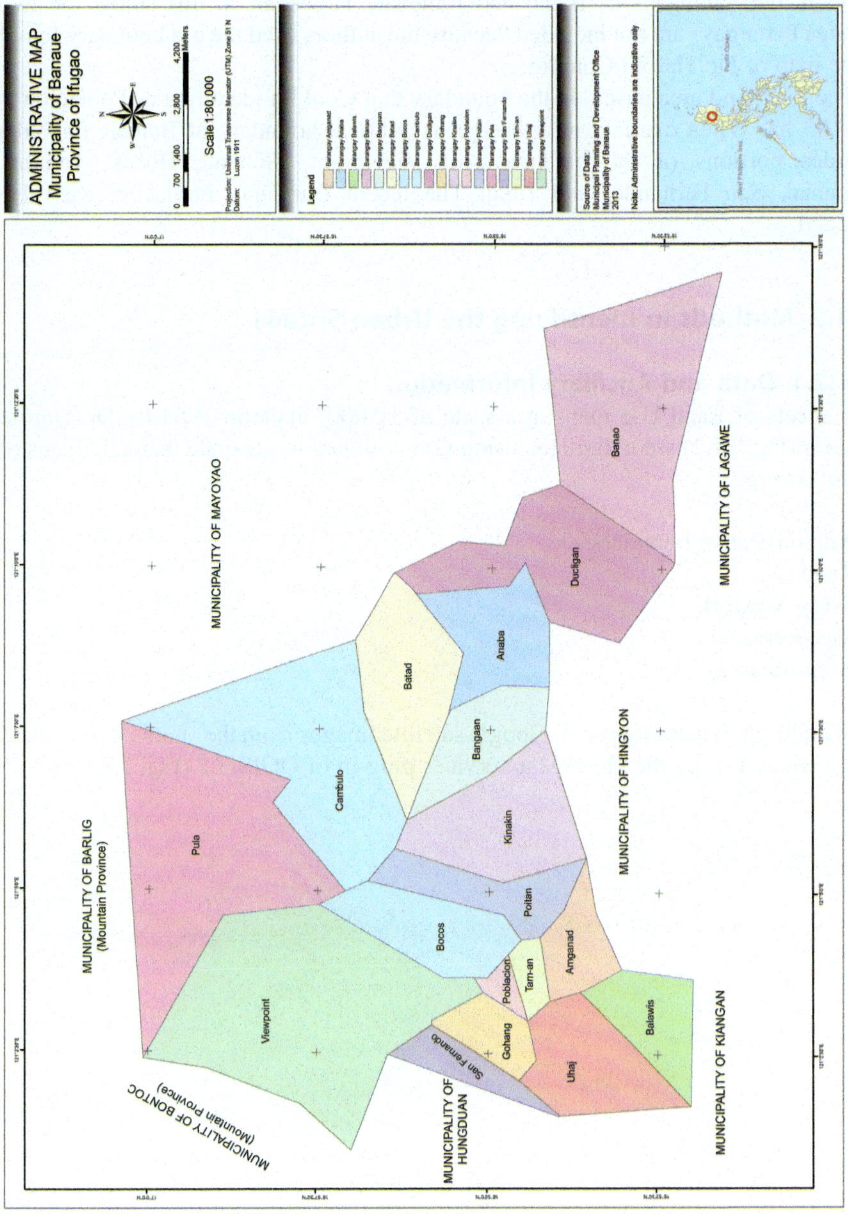

Fig. 21.1 Barangay map of Banaue

and Bocus). The remaining 14 barangays (Anaba, Amganad, Balawis, Bangaan, Batad, Gohang, Poitan, Pugo, Uhaj, Ducligan, Kinakin, San Fernando, Cambulo, and Banao) are considered rural areas. Two UNESCO World Heritage Sites are found in the barangays of Batad and Bangaan. However, in this study, the two heritage barangays are not included because the authors used the old boundary found in the map of Dr. Harold Conklin.

The total land area based on the boundary that Conklin identified as Poblacion is 1353.83 ha. When overlaid into the recent political boundary of Banaue this area includes portions of the barangays of Viewpoint, Gohang, Bocos, Tam-an, Amganad, San Fernando, and Uhaj. The recent Barangay Poblacion was also included.

21.1.2 Methods in Identifying the Urban Sprawl

21.1.2.1 Data and Ancillary Information

Two sheets of Land Use maps at a scale of 1:5000, made in 1963 by Dr. Harold Conklin (Fig. 21.2) were digitized using GIS software to generate thematic maps of the following:

1. Administrative boundaries
2. Road
3. River Network
4. Rice terraces
5. Built-up areas

In addition to these maps, 30 Google satellite images from the years 2010 to 2015 were captured using the QuickMapServices plug-in of QGIS 3.2 (Fig. 21.3).

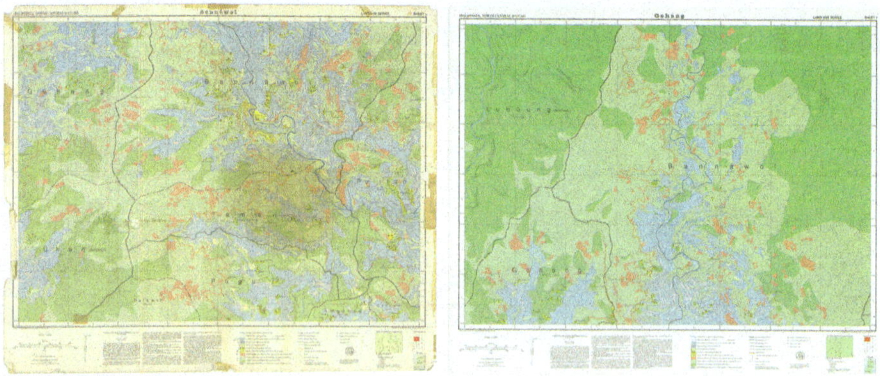

Fig. 21.2 1963 land use maps of Dr. Harold Conklin

Fig. 21.3 Google Earth Images from QuickMapService plug-in of QGIS

21.1.2.2 Methods

In order to identify the urban sprawl, information in the land use change is required. In this study, land use maps produced by Harold Conklin in 1963 were converted first to digital format before any analysis was done. All collected images and thematic maps were pre-processed (image restoration and rectification) before they can be used in Geographic Information System for further analysis. This process is important to ensure that the relationships of different features on the map are in the right location on the ground.

Pre-processing of data involved image enhancement, image clipping, and image georeferencing. In image enhancement, the brightness, tone, and colors were adjusted to see clearly the details of different feature objects. On the other hand, image clipping simply cropped the data based on the required boundary of the area of study. This also reduced the data file size. The next pre-processing step was image georeferencing to ensure that all the spatial data used the same coordinate reference system in mapping. For the purpose of this study, the maps were georeferenced using the Universal Transverse Mercator (UTM) coordinate system based on Luzon 1911 Datum, to be consistent with what Conklin used.

After georeferencing, the maps were digitized. This process in GIS involved tracing the feature on the map and converting it to points, lines, and polygons using QGIS software. The features that were digitized were: administrative boundary, road, river network, rice terraces, and built-up areas.

Since there are no available satellite images with the same spatial resolution as Conklin's map, the 30 Google Earth map images used were taken from the QuickMapService plug-in in QGIS 3.2 software. Built-up areas and roads were then digitized on satellite images. The projection of the Google Map satellite images is in the World Geodetic System of 1984 (WGS 84). The digitized feature layers were converted to UTM Luzon 1911 datum. In GIS, all feature layers were converted to have the same projection and coordinate system before overlay analysis was carried out. The built-up areas digitized from Google Maps were overlaid on the map of Conklin to identify the expansion of land use and land conversion.

21.1.3 Discussion

21.1.3.1 Biocultural Landscapes

Hong (2014) defines biocultural landscapes as the interaction between man and nature. This relationship between man and nature is expressed as culture and biological diversity interaction in the landscape. The Ifugao rice terraces are recognized as a bio-cultural landscape as it is peopled by several ethnolinguistic groups that have uniquely managed the environment, both built and natural. The wet-rice culture in the rice terraces is very much linked to the culture of the indigenous people of Ifugao. In addition, their management of the surrounding forests and wood lots has helped maintain its rich biodiversity. Other similar terminologies have described the same bio-cultural landscape such as socio-ecological production landscapes (SEPLs), agro-cultural landscapes, cultural heritage sites, satoyama, and geo-parks.

The next two sections will describe how GIS and Remote Sensing are used in cultural heritage sites and the urbanization of bio-cultural heritage sites.

21.1.3.2 The Application of Remote Sensing and GIS to Demonstrate Urban Growth in Cultural Heritage Sites

The emergence of geospatial technology like Geographic Information System, Remote Sensing, and Spatial Modelling has made it possible for these to be used as tools in mapping and managing cultural heritage sites. The capabilities of these tools to overlay and combine different information provide a rich opportunity in analyzing the factors that can protect and conserve cultural heritage sites. A few examples where these technologies have been applied in cultural heritage conservation are mentioned below.

Cultural Heritage Monuments in Paphos District in Cyprus

The study by Agapiou et al. (2015) in cultural heritage sites in Cyprus presents an example of a multi-temporal spatial analysis to demonstrate growth processes in urbanization. Landsat data were classified using four land cover types: four main land cover types (built areas; water bodies; vegetation and land). As the study covered an extensive period (1984–2010) where spatial analysis was done at four time points (the years 1984, 1990, 2000, and 2010), this study presents a good example of how Geographic Information System (GIS) and spatial analysis can be used to demonstrate urban sprawl over time. Moreover, the study went further to create a 2020 projection map of the areas. The results of their analysis showed a significant increase in urbanization increase in Paphos district that could put strong pressure on historically important archeological sites.

Historic City of Georgetown in Penang, Malaysia

Mohd Noor et al. (2014) analyzed the application of geospatial technology in the study of urban morphology of the historic city of Georgetown in Penang, Malaysia. It was established in 1786 by the British and is considered to be a heritage area due to the numerous heritage buildings and living monuments that can be found here. To analyze and identify the expansion of land use development in Georgetown, Penang the authors used Remote Sensing and Geographic Information System (GIS) to conduct a temporal analysis using four satellite images taken from different years (2004–2014).

Three different land use categories were identified namely: built-up, non-built-up areas, and water bodies. The result of the study showed that there was an increase in built-up areas. This gave the urban planners an idea on the kind of urban growth that exists in the area. The use of geospatial technology like GIS and Remote Sensing can help assist urban planners and decision-makers identify the problems and possible solutions in urban planning and management issues, especially in a heritage site.

The Heritage City of Lahore in Pakistan

Sherwani (2017) used GIS to map the historic city of Lahore in Pakistan. Urban expansion affected the heritage boundaries which resulted in the shrinking and deterioration of precious structures in Lahore. Using GIS, complex structures and heritage sites were accurately delineated on its exact location. Old photos and maps were analyzed and overlayed on each other in order to delineate the heritage boundary. A 3-dimensional map (3D) was also used to record and visualize the topology of the historical sites to be used in buffer analysis.

The result of the analysis generated different maps that showed the changes that took place in Lahore over time, especially in terms of the boundary. The study also provided different applications that used geospatial technology in the documentation and analysis of the impacts of urban development on heritage conservation and preservation.

The Historical Cities of Perm and Usolie in the Territory of Perm Region (Russia)

The rich cultural heritage cities of Perm and Usolie in the territory of the Perm Region (Russia) are threatened because of urban development. The study conducted by Bushmakina et al. (2017) showed how they used GIS to develop a geospatial model of the two cities which included the historical monuments of history, architecture, and urban development according to the national cultural heritage register of Russia.

Using historical data like old maps, several buildings, monuments, and plans were digitized and developed into a geodatabase. A geodatabase is crucial especially in the study of cultural heritage sites because it allows you to edit and manage a large number of datasets. Historical data were stored in the geodatabase in an orderly manner for easy accounting of cultural heritage objects and spatial monitoring. The result of the study produced two geospatial models which contain information about the urban planning development that happened in the two heritage cities of Usolie and Perm. The study provides opportunities for further research and monitoring of the heritage sites.

Architectural Heritage of Buddhist Monasteries in Qinghai-Tibet Plateau

A recent study by Zhang et al. (2018) also used GIS to assess changes related to urbanization around 152 historically important Buddhist monasteries in Qinghai-Tibet. Similar to the results of the Cyprus study cited above, this analysis demonstrated the steady increase in growth and urbanization around the historical sites. More importantly, the analysis provided maps and tangible data as a basis for policy formulation and more focused interventions that are expected to help in the management and preservation of the monasteries.

These are just a handful of cases that highlight the usefulness of GIS and spatial analysis in the study of urban sprawl in heritage sites in different parts of the world. In essence, these tools and approaches present an opportunity for better and informed decision-making that could possibly strike a balance between the seemingly inevitable onslaught of urbanization and the preservation of culturally invaluable heritage sites.

21.1.3.3 Urbanization in Biocultural Heritage Sites: Some Experiences from Different Settings

The International Council on Monuments and Sites (ICOMOS) in 2005 conducted an analysis of the threats to World Heritage Sites (WHS) from 1994–2004 and their

Table 21.1 Summary of threats to World Heritage Sites in Asia/Pacific region (ICOMOS 2014)

Asia/ Pacific	46%	**F6** Inadequate/lack of management strategies/priorities/plan/monitoring/ mechanisms (conservation included)
	31%	**B1** Urban pressure (destruction of traditional building, construction of large buildings, high rise, modern houses, incinerator, demographic growth)
	25%	**E4** Over-visiting/tourism pressure
	22%	**A6** Natural deterioration
	22%	**F11** Unclear boundaries
	20%	**F26** Inadequate/lack of maintenance/restoration
	20%	**F24** Use of inadequate material for restoration/inadequate techniques

findings showed that in all WHS around the world, there is an increasing threat from development. In the Asia-Pacific Region, the top three threats were inadequate or lack of management strategies/priorities/plans/monitoring/mechanisms and conservation followed by urban pressure and over-visiting/tourism pressure (Table 21.1).

According to the United Nations (2018), with urbanization becoming a global phenomenon, a little more than half of the world's population now lives in urban areas. Urbanized city areas, and even traditional heritage towns, now face both benefits as well as threats of urban sprawl characterized by uncontrolled and unplanned growth. In the case of many heritage towns that hold unique aesthetic, architectural, historical, and cultural significance (Kiruthiga and Thirumaran, 2019), unplanned growth poses numerous threats to the heritage sites and alters the way of life of residents and communities. Despite the potential of urbanization to make cities and communities more "developed" (UN Habitat 2016), unsustainable environmental, social, and economic practices may contribute to the erosion and commercialization of indigenous culture and unhinge the delicate balance between the important bio-cultural landscapes and the communities living within.

In analyzing the extent and impact of sprawl in Banaue, experiences in Japan, India, and Afghanistan are illustrative. Like the heritage towns in Ifugao, mountainous villages enlisted in UNESCO World Heritage sites are also found in Japan. For example, the heritage villages of Shirakawa-go and Gokoyama are known for the unique architecture of their traditional farmhouses called "*Gassho-zukuri*" (Singh 2017). These cultural heritage sites are facing different challenges and are constantly finding ways to address these problems and ensure the preservation of their heritage. The difficulty of access to the outside world of these villages years ago has enabled the birth of the unique cultural heritage of these traditional villages that is perfectly adapted to their natural mountainous environment (Singh 2017).

Shirakawa-Go, located in Shirakawa in Gifu Prefecture, faces problems in balancing tourism with landscape preservation. Shirakawa-go has problems managing severe traffic congestion as many tourists in their own private vehicles flock to the village (The Japan Times 2010). Due to the rise of tourists visiting Shirakawa-go since its enlistment as a World Heritage Site, many restaurants, and souvenir shops have sprung up in the area. Previous farmlands have been turned into parking lots to accommodate the flock of tourists.

Meanwhile, Gokoyama which is located in Nanto in Toyama Prefecture is faced with problems with its declining and aging population, making it difficult for the preservation and maintenance of the heritage site (The Japan Times 2010). Unlike Shirikawa-go, Gokoyama does not face issues with traffic congestion as their tourism did not increase as much as with Shirakawa-go. However, the depopulation of the village was seen as their most urgent concern as their elderly passed away and the younger ones choose to leave the village to find jobs away from their village. They cannot increase the number of residents, which increased the number of vacant and unmaintained houses.

To address these problems, Shirakawa-go has limited bus access to the village and built new parking areas outside the heritage site to ease traffic congestion (The Japan Times 2010). They are also considering limiting access to private vehicles, but are also apprehensive about the idea since it may reduce and impede tourism in the area which serves as the source of livelihood for most of the residents of the village (The Japan Times 2010). Limiting tourism may hinder the revitalization of the village since it may result in the loss of jobs, hence, more residents leaving the village to find jobs outside the village. Meanwhile, in Gokoyama, the problem of the lack of successors is seen to be difficult to address since increasing the number of residents is already a problem. Though Gokoyama is fairly well preserved, the need for successors that will maintain the heritage site in the future is a major challenge for the village.

In Tamil Nadu, India, the temple town of Kumbakonam is known for its historic structures and cultural heritage. Grand palaces and residences across the town were built during the reign of their legendary kings. Despite being colonized during the British colonial period in the country, the heritage town was able to maintain its architecturally unique style and was able to hold a good balance between development and the preservation of its heritage character (Kiruthiga and Thirumaran 2019). The unique architectural style during the time of the ruling of the kings was still exhibited through the town's temples while the residential houses possess the features of the vernacular architecture of the region. Commercial buildings, on the other hand, express the colonial style (Kiruthiga and Thirumaran 2019).

However, because of the strong pressure of rapid urbanization, the temple town and its cultural heritage are in danger. The development promised to be brought by urbanization was said to have no regard for the importance of the town's cultural and historical heritage, and rapidly replacing the historical structure and style with "amorphous, contemporary, and global designs" (Kiruthiga and Thirumaran 2019). In recent decades, the number of modern and commercial buildings are slowly surpassing the number of traditional designs, while the existing traditional establishments are being modified based on global trends and modern design (Kiruthiga and Thirumaran 2019). These interventions are now resulting in the loss of the built heritage setting of this heritage town.

To address these issues, Kumbakonam implemented its Local Development Plan (LDP) for sustainable urban growth and preservation of its heritage character (Kiruthiga and Thirumaran 2019). They are continuously monitoring the significant changes in the heritage characteristics of the town, to further contribute to the implementation of its LDP in the hopes of the preservation of this heritage site.

Finally, Afghanistan has an example that shows cities are also drastically affected by urbanization and unplanned urban development. Herat, the provincial capital of Afghanistan is already at least two millennia old and holds significant heritage and history through hundreds of years—surviving since the time of Alexander of Macedon and the invasion of the Mongols in the country to the present time (Kristy 2018). The city has experienced rapid and unplanned urban development, causing damage to the city's heritage sites and its historical roots. The risk of heritage sites being overtaken by unregulated urban sprawl is very high (Kristy 2018). Like what is described in Banaue in the succeeding section, analysis of the threats of rapid urbanization and investigation of the patterns of urban sprawl are ongoing to determine the impact of urbanization on the cultural heritage of Herat.

Urban Sprawl in Banaue

Banaue is a heritage town experiencing the same threats as other heritage towns cited in the previous section, as well as in other locations around the world. When Conklin produced these hand-made cartographic maps of the rice terraces cluster in Banaue, he probably never thought that it would serve as an important base map for the areas of Banaue now considered as urban barangays. From traditional houses and granaries in 1963 occupying a total of 28,907.37 sqm or 2.89 ha (Fig. 21.4), that number has almost doubled with a total land area of 199,704.19 sqm or 19.9 ha (Fig. 21.5), to include multi-story buildings such as trade center, market, and multi-level parking at Barangay Poblacion.

In a recent flash flood last July 07, 2022, Brgy. Poblacion, Banaue experienced the consequences of poor planning that resulted in urban sprawl. In the 50 years since Conklin drew his cartographic maps, Brgy. Poblacion has expanded its built-up areas by encroaching on areas that used to be wet rice terraces or on very steep slopes. As a consequence of poor planning, it experienced the worst flooding and landslide (Fig. 21.6a, b).

Main Drivers of Urban Sprawl in Banaue

Roads Expansion and Improvement

In terms of the road network, about 14.2 km of roads existed in Banaue in 1963. Moreover, the roads going to Banaue were not yet paved during that time. The government's Cordillera Road Improvement Project (CRIP) that started in 2001 has resulted in a steady increase in the proportion of roads paved with concrete or asphalt across the region. With just 20% of national roads in CAR paved in 1987, over 80% of roads in the region have now been paved (NEDA-CAR 2021).[3] In 2015, most roads in the municipality of Banaue have subsequently been paved and there have

[3]The Cordillera Roads Improvement Project: Making the Cordillera dream a reality | NEDA Cordillera.

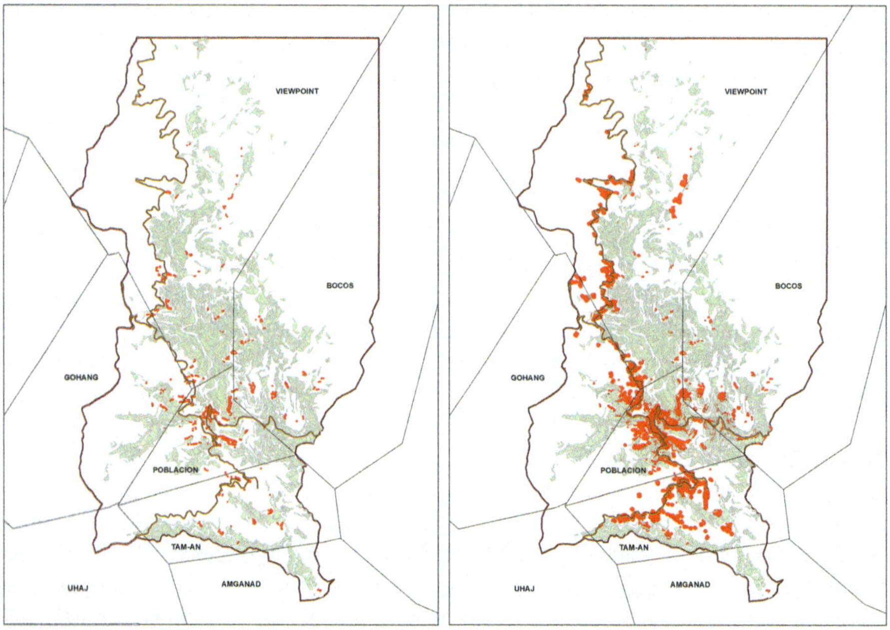

Fig. 21.4 (**a**) Built-up areas 1963. (Source Conklin 1980); (**b**) Built-up areas 2015. (Source Google Satellite)

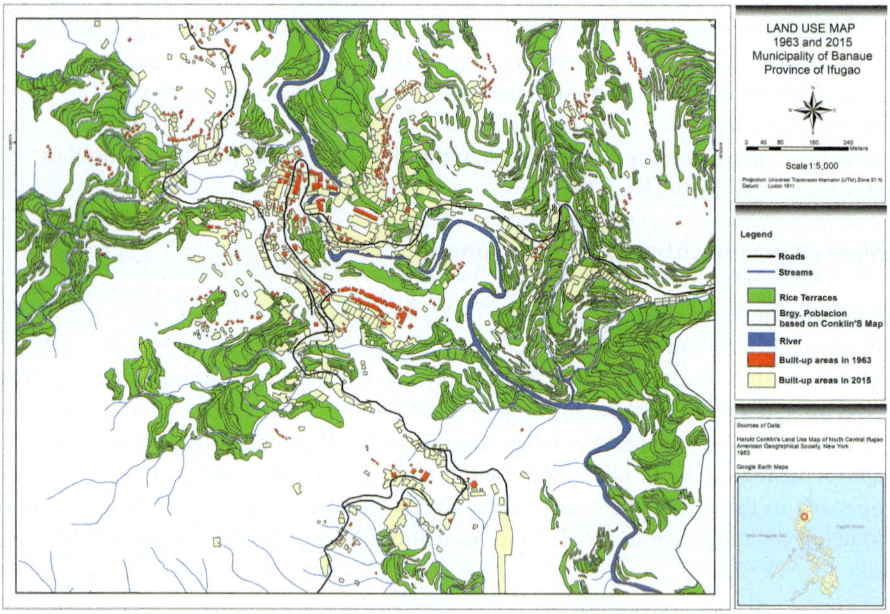

Fig. 21.5 Combined land use map of built-up areas (in red) in 1963 with built-up areas (in yellow) in 2015 and rice terraces in Barangay Poblacion. (Scale 1:5000)

Fig. 21.6 (**a**) Recent flash flood in Brgy. Poblacion, Banaue Ifugao on July 7, 2022. Photo credit to Peewee Bacuño and (**b**) Landslide in Brgy. Poblacion, Banaue, Ifugao on July 7, 2022. Photo credit Daniel Mariano Jr

been several road improvements and widening projects. In addition, the roads leading to the neighboring towns of Bontoc, Hungduan, and Mayoyao are now also fully paved. It is well established that areas with good roads are considered as growth corridors that hasten local growth and development. In the context of Ifugao province, they not only allow better accessibility for remote mountainous communities in the conduct of economic and other human activities, but they have also consequently lead to an increase in land valuation, especially in areas within easy access to the road network.

Conversion of Rice Terraces to Other Land-Uses

Another threat is land conversion. Comparing images from Google Earth and Conklin's map, significant changes were observed. Some terraces especially along the road in Barangay Bocos had been converted into residential areas (Fig. 21.7).

The three succeeding maps show some examples of land use conversion in the urban barangays of Poblacion and Bocos in Banaue (Fig. 21.7a–c). In these examples, Conklin's land use map was compared with a Google Earth map of the same area to check for land use changes. In all cases, the rice terraces were converted into built-up areas.

Conflicting Laws and Legal Instruments

Aside from road network expansion and land conversion in the urban barangays of Banaue as presented in the previous section using GIS and RS contributing to the problem of urban sprawl, several conflicting laws and legal instruments in the study area have also exacerbated the problem. Based on personal interviews with some local officials, it was mentioned that the increasing number of houses used as tourist

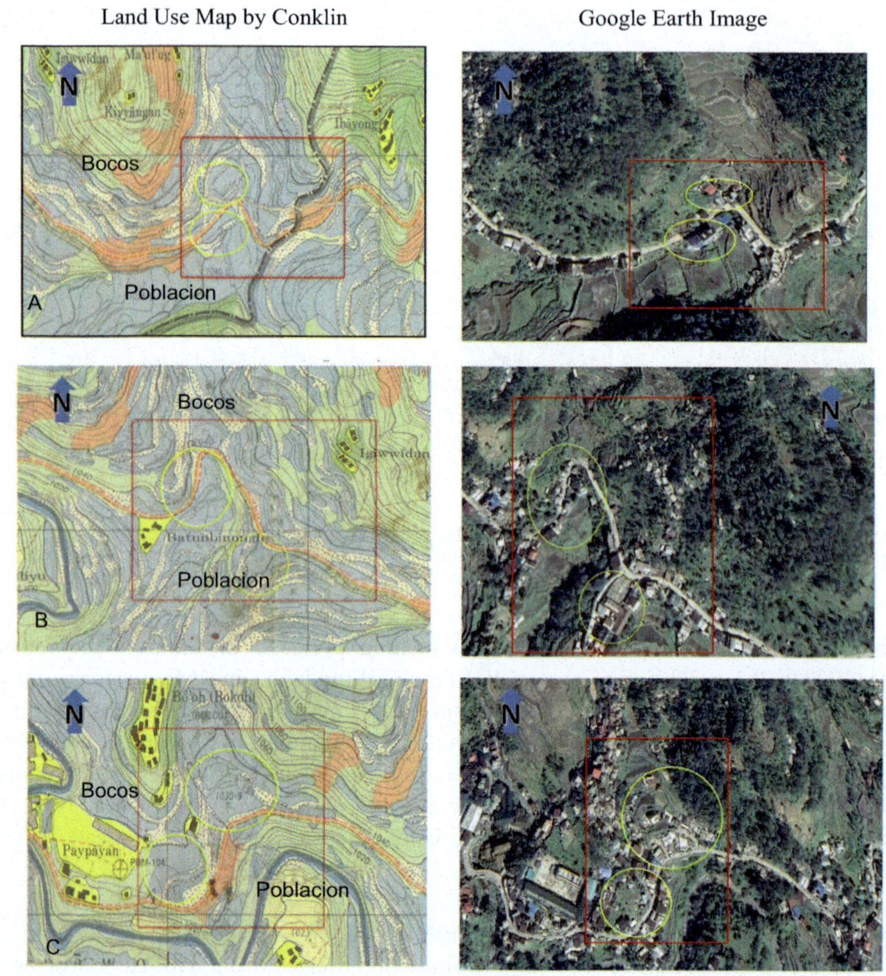

Fig. 21.7 (**a**–**c**) Comparison of land use change from rice terraces to built-up area using Google earth map and Conklin's map of Barangay Poblacion and Bocos. Conklin's map land use code: road = orange; rice terraces = ash blue and villages/houses = yellow with black square as compared to the present landuse in the same area

inns has affected the water allocation and distribution to the rice terraces. Some terraces have been abandoned due to a lack of water. Despite being a famous tourist spot and a Heritage Site, the whole Municipality of Banaue is considered to be highly susceptible to landslide due to its topography. This is based on the Geohazard map created by the Mines and Geosciences Bureau (Fig. 21.8).

While based on the Land Classification Map of the National Mapping & Resource Information Authority, Banaue is classified as Forestland. In Presidential Decree 705 or "Revised Forestry Code," forestlands are lands with topography that is more

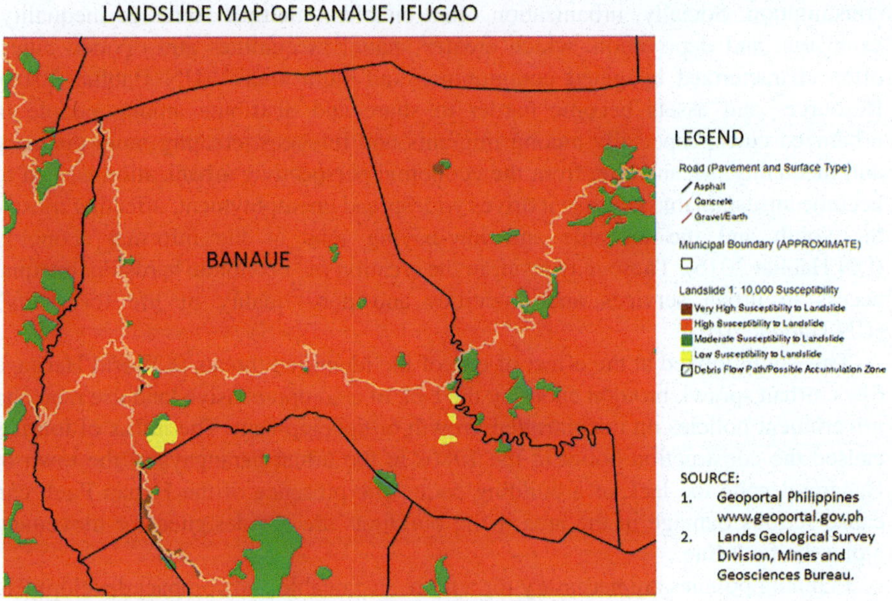

Fig. 21.8 Geohazard map of Banaue, Ifugao, Philippines, Mines and Geosciences Bureau, Philippines

than 18% slope and it includes the public forest, the permanent forest, forest reserves, and forest reservations. All of these are owned by the state which means people living in these areas cannot own their lands. This is in conflict with Republic Act 8371 or otherwise known as the Indigenous People's Right Act (IPRA Law), which recognizes the rights of the indigenous people over their land.

These different laws and instruments can create conflicting interpretations and confusion as to what relevant and appropriate rules and regulations can apply in terms of land use. Paper copies of maps are still used in the study area. While land cover data are now available in digital format, the analysis of these maps will require intensive capacity training in GIS and Remote Sensing. Based on the interviews with some staff of the LGU, they also lack the software and hardware technology to handle such a huge amount of data.

21.2 Summary and Conclusion

Environmentally, urbanization enables low-density suburbanization which may potentially threaten the environment and may contribute negatively to climate change (UN Habitat 2016). As more people sprawl out to the suburbs, energy and fuel consumption, household driving, and land consumption increase. It promotes dependence on car ownership and pushes for more intensive energy and fuel consumption. Socially, urbanization often enables "multiple forms of inequality, exclusion, and deprivation, which creates spatial inequalities and divided cities, often characterized by gated communities and slum areas" (UN Habitat 2016). Resources and assets become harder to share and distribute among people in urbanized cities, especially among migrants and refugees, creating more disparity and inequality. Meanwhile, from the economic perspective, urbanization is seen to become unsustainable because it drives widespread unemployment, affecting mostly the youth and the unstable and low-paying jobs in the informal economy (UN Habitat 2016). These may result in the creation of "economic hardship, unequal access to urban services and amenities and poor quality of life for many" (UN Habitat 2016).

This study looked at the effect of one of the identified threats to World Heritage Sites, urban sprawl, brought about by the lack of a comprehensive land use map and government policies on uncontrolled growth or development. The influx of tourists caused the construction boom in the 1980s in the urban barangays of the town of Banaue despite the lack of a building code. The presence of the Hapap Fault can cause serious damage in the area if the buildings are not designed for movement along the fault line.

Planning policies are necessary for a more sustainable and compact development pattern (Tian et al. 2017). The existence of urban sprawl in Banaue is indicative of poor planning and poses a serious threat to the bio-cultural landscapes of Ifugao. In the absence of an enforced comprehensive land use plan, the local government and the communities residing in the urban barangays will slowly bring about the demise of the rise terraces clusters. In addition, the lack of effective management strategies may not bode well for the rice terraces clusters as they are slowly being converted to built-up areas and other uses. While communities evolve and change with the times, there should be a conscious and concerted effort of all stakeholders for the preservation of the rice terraces clusters both outside and inside the World Heritage Sites.

References

Acabado S, Martin M, Datar F (2017) Ifugao archaeology: collaborative and indigenous archaeology in the northern Philippines. Adv Archaeol Pract 5(1):1–11

Agapiou A, Alezakis DD, Lysandrou V, Sarris A, Cuca B, Themistocleous K, Hadjimitsis DG (2015) Impact of urban sprawl to cultural heritage monuments: the case study of Paphos area in Cyprus. J Cult Herit 16(1):671–680. https://doi.org/10.1016/j.culher.2014.12.006

Agnoletti M, Rotherham I (2015) Landscape and biocultural diversity. Biodivers Conserv 24:3155–3165. https://doi.org/10.1007/s10531-015-1003-8

Bushmakina YV, Balyberdina PA, Dmitrieva MK, Gogoleva MV (2017) The use of GIS for studying cultural heritage and historical urban landscape: the case of Pern and Usolie (Russia). Ge-conservacion 11:264–271

Calderon MM, Dizon JT, Sajise AJ, Andrada RT II, Bantayan NC, Salvador MG (2009) Towards the development of a sustainable financing mechanism for the conservation of the Ifugao rice terraces in the Philippines (research report). IDRC Regional Office for Southeast and East Asia, Nairobi

Conklin HC (1980) Ethnographic atlas of Ifugao. Yale University Press, London

Dulnuan ED (2014) The Ifugao rice terraces tourism: status, problems and concerns. IAMURE Int J Ecol Conserv 10(1) http://ejournals.ph/form/cite.php?id=3046
El Menchawy A, Aly SS, Hakim MA (2011) The impact of urban sprawl on the heritage areas through the urban fabric of cities. WIT Trans Ecol Environ 150(1):299–314. https://doi.org/10.2495/SDP110261
Geminiano P (2018) Ifugao sets up homestays, as PRRD eyes it as next tourist hub. Philippine News Agency, Quezon City. https://www.pna.gov.ph/articles/1039149
Hong SK (2014) Philosophy and background of biocultural landscapes. In: Hong SK, Boagaert J, Min Q (eds) Biocultural landscapes. Springer Science+Business Media, Dordrecht. https://doi.org/10.1007/978-94-017-8941-7_1
International Council on Monuments and Sites (ICOMOS) (2014) The Florence declaration on heritage and landscape as human values. ICOMOS, Manila
Kiruthiga K, Thirumaran K (2019) Effects of urbanization on historical heritage buildings in Kumbakonam, Tamil Nadu, India. Front Archit Res 8(1):94–105. https://doi.org/10.1016/j.foar.2018.09.002
Kristy G (2018) The impact of urban sprawl on cultural heritage in Herat, Afghanistan: a GIS analysis. Digit Appl Archaeol Cult Herit 10(1):e00086. https://doi.org/10.1016/j.daach.2018.e00086
Macapagal RA, Bermejo RA III (2015) The Batad Kadangyan ethnic lodges project: community-based indigenous tourism in a UNESCO world heritage rice terrace cultural landscape. In: Mena AC (ed) Second international conference on best practices in world heritage: peoples and communities. UCM, Madrid, pp 793–810. https://eprints.ucm.es/34899/1/ActasMenorca15.pdf
Mohd Noor N, Mohd Nor M, Abdullah A, Zahari RK (2014) Geospatial technology approaches in urban morphology for resilient urban governance. Plan Malaysia III:1–14
Nozawa C, Malingan M, Plantilla A, Ong J (2008) Evolving culture, evolving landscapes: the Philippine rice terraces. In: Amend T, Brown J, Kothari A, Phillips A, Stolton S (eds) Protected landscapes and agrobiodiversity values. Kasparek Verlag, Heidelberg, pp 71–86
Sherwani RK (2017) Technological advancement in mapping of heritage by using GIS (mapping of heritage: preservation for the future). Int J Syst Model Simul 2(4):11–16. (ISSN Online: 2518-0959)
Singh RPB (2017) The world heritage villages of Shirakawa-Gō and Gokayama. Continuing Culture and Meeting Modernity, Tokyo. https://www.researchgate.net/publication/320002994
The Japan Times (2010) UNESCO villages in battle with modernity. The Japan Times. [online] https://www.japantimes.co.jp/news/2010/12/11/national/unesco-villages-in-battle-with-modernity/#.XL0cJ-gzaM-. Accessed 17 Apr 2019
Tian, L. , Li, Y. , Yan, Y. and , Wang, B. (2017). Measuring urban sprawl and exploring the role planning plays: a Shanghai case study, Land Use Policy 67, 426–435
United Nations Human Settlements Programme (UN-Habitat) (2016) Urbanization and development: emerging future (world cities report 2016). United Nations Human Settlements Programme (UN-Habitat), Nairobi. http://wcr.unhabitat.org/main-report/
Zhang Y, Zhang H, Zheng S (2018) Effects of urban growth on architectural heritage: the case of Buddhist monasteries in the Qinghai-Tibet plateau. Sustainability 10(1):1–18. https://doi.org/10.3390/su10051593
Zialcita FN, Canilao MP, Roldan SR, Santos MW, Stanyukovich MV, Toohey A (2015) Introduction: issues in the Literature. In: Zialcita FN et al (eds) Preserving the Ifugao Terraces: a literature review. UNESCO National Commission in the Philippines, Pasay City. 175p

22 Plant Diversity in Selected Agro- and Forest Ecosystems in Indigenous Cultural Communities (ICCs) in the Cordillera Region, Northern Philippines

Romeo A. Gomez, Jr, Jason M. Leung, Joyce N. Paing, Elizabeth T. Dom-ogen, Aldrin V. Baniaga, and Rey D. Napoleon

Abstract

The description of the nature, observations, and findings relating to biocultural landscapes in the Cordillera Region is the theme elucidated in this chapter as a paper review. It brings to light the documentation, experiences, and analysis in select ecosystems studied, both agro- and forest ecosystems, accentuated quite strongly by the historical and cultural influences of indigenous cultural communities (ICCs) on these ecosystems. It centers around the relevance of the important biological resources especially plant diversity, contained therein. The emerging framework being proposed in this paper is indigenous knowledge systems and practices (IKSP) in ICCs and its interface with traditional rice-based ecosystems (TRBEs), protected areas, and communal forests. This model is seen to capitalize on and is strongly influenced by other models. These include the UNESCO World Heritage Listings, Globally Important Agricultural Heritage Systems (GIAHS), socio-ecological production landscapes (SEPL), and *satoyama-satoumi* concepts—all bring about strong environmental conservation and management value. This paper provides an overall treatment of the agro-ecosystems in the region, particularly those where cultural values, customs, and traditions are markedly in operation for a significant period of time. In essence, it depicts TRBEs as a focal model across various sites, and the importance of other essentially locally available plant resources, such as *Coix* and other indigenous vegetables. With the same frame of thought as mentioned earlier, discussions on the forest ecosystems are presented via select studies in protected areas in two case sites: the Mount Data National Park (MDNP) and Mount Pulag National Park (MPNP) in Mountain Province and Benguet-Nueva Vizcaya-Ifugao

R. A. Gomez, Jr (✉) · J. M. Leung · J. N. Paing · E. T. Dom-ogen · A. V. Baniaga · R. D. Napoleon
Benguet State University, La Trinidad, Benguet, Philippines
e-mail: r.gomezjr@bsu.edu.ph

S. Ramamoorthy et al. (eds.), *Plant Diversity in Biocultural Landscapes*,
https://doi.org/10.1007/978-981-19-8649-9_22

tri-boundary, respectively. This is followed by a presentation on the dynamics in select communal forests in various sites in Benguet. Toward the end, future challenges are posited to give sense to possible trajectories in the medium and long terms.

Keywords

Indigenous cultural communities (ICCs) · Traditional rice-based ecosystems (TRBEs) · UNESCO World Heritage · GIAHS · SEPL · *Satoyama-satoumi* · Protected areas · Communal

22.1 Introduction

A recent surge of interest in the theorization of biocultural landscapes has become apparent. As articulated by Dublin et al. (2018), the term refers to landscapes that are the result of the interplay of ecological processes and human industry; it can be seen roughly as within the purview of human–environment interaction. While humans and their activities have oftentimes been conventionally seen as having caused many anthropogenically driven environmental degradations and resource depletion, there have been a number of cases, where they have resulted to the contrary. That is, a number of systems have in fact exemplified a good level of promise and sustainability. The term 'socio-ecological production landscape' (SEPL) is a closely associated term that connotes ecological sustainability and biological conservation while addressing human security and well-being.

A description of the Japanese model of *satoyama* (associated with the forests) and *satoumi* (associated with coastal ecosystems) as applied in the Philippines (Buot 2008; Gomez 2013) can be found in a number of publications. Basically, these are ecosystems described to be characteristically high in biodiversity which provide ecological and economic flows to community people. These systems are fundamentally 'biocultural' that reflect how traditional culture and values interface to bring about benefits to the community in terms of food and other forms of security, which may collectively address the so-called human security. It is postulated that this is happening within the epoch of the 'Anthropocene,' where human influences have extraordinarily risen to unprecedented levels.

Conceptual framework: The discourse on biocultural landscapes presupposes a strong reciprocal interaction between humans and the environment, both physically and biologically. The interest in this particular paper is to offer an analysis of the influence of indigenous cultural communities (ICCs) on phyto-diversity—be it from the nature of the pedon and associated vegetation, historical, and cultural interfaces with traditional societies, taking into account some insights on climate change trajectories and scenarios. In the Cordilleras where ICCs abound, such influence is demonstrably exemplified in various natural or man-made ecosystems and landscapes. Particularly in this chapter is the focus on traditional agro-ecosystems and forest ecosystems (including areas declared as 'protected areas and landscapes').

Integral in this discourse is the evolutionary and developmental interactions of these physical templates, with the attendant plant diversity as the central focus, as influenced and nurtured by traditional cultures or societies. The discourse contained herein is a review paper in key research output in the offering a synthesis of Cordillera Region.

Much of the discussions offered to describe the dynamics in agroecosystems are drawn principally from experiences in traditional rice-based ecosystems (TRBEs) in general (Gomez et al. 2022; Paing and Gomez 2021; Gomez 2020; Batani et al. 2019; Napoleon 2018; Napoleon and Gomez 2022; Gomez and Baniaga 2018; Tad-awan et al. 2015; Ingosan et al. 2008). Many such ecosystems abound in the Cordilleras where numerous exemplifications of human–environment interactions are centered in the cultivation of traditional rice. As a staple food, traditional rice varieties (TRVs; also called native or heirloom rice) have been defined and also reciprocally define the particular culture that has nourished them through several generations. Such interactions have also led to the evolution and emergence of a myriad of traditional customs and practices, even to the extent of permeating into religion, mysticism, and the like. Invariably, these interactions have defined to some extent, the sustainability of human-ecological systems, by virtue of the operationalization of ecosystems akin to mimick very closely the 'laws of nature,' thus making it consistent with the workings of 'Mother Nature.'

Present and recently emerged frameworks of the Food and Agricultural Organization (FAO) based on Italy, such as the UNESCO Heritage Sites and the Globally Important Agricultural Heritage Systems (GIAHS), the traditional concept of the Japanese *satoyama-satoumi* concept, the German model of the ecosystem-based adaptation (EbA), among others, have extensively lent an enrichment of human influences on ecosystems that capitalize on biodiversity, of course, to include phytodiversity.

The so-called monument of sustainability in the Cordilleras, the world-renowned Ifugao Rice Terraces (IRT) truly demonstrates the universality of these concepts. Various empirical evidences, from several scientific literatures, have pointed to this fact. To a lesser extent, there are equally important similar TRBEs spread in different parts of the region-Palina in Kibungan, Poblacion in Bakun, all in the Province of Benguet, as well as in other provinces, such as Mountain Province, Kalinga, and Abra. These agroecosystems in the latter provinces may not be necessarily within the purview of UNESCO or GIAHS, but definitely, as TRBEs, the element of cultural influence on their management is a very strong one. All of them exemplify a close link between humans and the environment and are fundamentally characterized by a good level of biodiversity, including plant diversity.

On the other hand, forest ecosystems and other protected areas or landscapes, especially those within the vicinities or confines of ICCs, are also influenced largely by how these communities perceive, use, and manage them. Though it may "cut both ways," the traditional values' communities accrue toward the forest ecosystems, as resources bequeathed by an omnipotent, as well as their revered ancestors, have positively given impetus for protecting and conserving such community resources. On the other end, the loss of such values by the community, or at least to some segment of the constituents, may also impinge on forest ecosystem integrity. Sadly, the latter observation is also beginning to be observed even in protected areas; such is attributed to sociocultural changes that unduly impinge on the commodification of traditionally revered resources. This is largely due to the growing interface with the market economy, similar to what happens in the IRT (Gomez and Pacardo 2005).

In both cases, whether the description refers to agro- or forest ecosystem, it attempts to cite reasonable empirical data on plant diversity from the floral inventories conducted. In the agro-ecosystems, for instance, different varieties (TRVs, as a form of morphological variation) and related characteristics are presented. Moreover, other species of indigenous food plants are corollary mentioned as associated species in the TRBEs. All these put the context of the existing status of plant diversity in these ecosystems. On the other hand, basically, the more common tree species (or the more dominant species, technically 'more correct' terminology) are enumerated. Other relevant characteristics are also discussed.

22.2 The Agro-ecological System and the ICC Nexus

TRBEs are generally considered home to a higher level of diversity of plants thriving in the area, owing to the fact that traditional organic production systems maintain the natural flows, in as much as they closely mimic the laws of nature or ecological laws. Moreover, these low-input agroecosystems are not typically receiving exogenous substances in the cultivation of locally available plant resources. This is fundamentally the characteristic nature of the farms planted to TRVs. The Ifugao Rice Terraces is one documented example of such an ecosystem. Thus, plant diversity in this kind of system is generally regarded as high. TRVs as genetic resources are conserved; in a study in Kiangan, Ifugao, for instance, more than 20 varieties of TRVs are known to be in active cultivation (Gomez and Pacardo 2005). Given the typical nature of TRBEs in the Cordilleras, Fig. 22.1, which share more or less closely related or very similar culture, traditions, indigenous knowledge, and practices, it is reasonable to assume that such is widespread in rice-producing areas in the region—be it in Benguet, Mountain Province, Abra, Kalinga, or Apayao, apart from Ifugao.

The unique advantage of TRBEs in Ifugao is the massive preponderance of such in the entire province, thus earning recognition as UNESCO Heritage Sites (Hungduan, Banaue, Mayoyao, and Kiangan) as well as Globally-Important Agricultural Heritage System (GIAHS) (Gomez 2013). These recognitions bestowed on the IRT have afforded formal financial, legal, political, and institutional mechanisms to support the conservation of these landscapes and ecosystems. Furthermore, the recognition of the IKSP of the ICCs that nurture these important landscapes and ecosystems is inherent to such inscription by the Food and Agricultural Organization (FAO) based on Italy.

Interestingly, another traditional rice-based ecosystem (TRBE) was studied in Poblacion, Bakun, Benguet (Napoleon 2018). It was estimated that there are around 1,429.73 ha devoted to the cultivation of traditional rice varieties (TRVs) and this is around 55% of the total land area devoted to agriculture (Barangay Poblacion

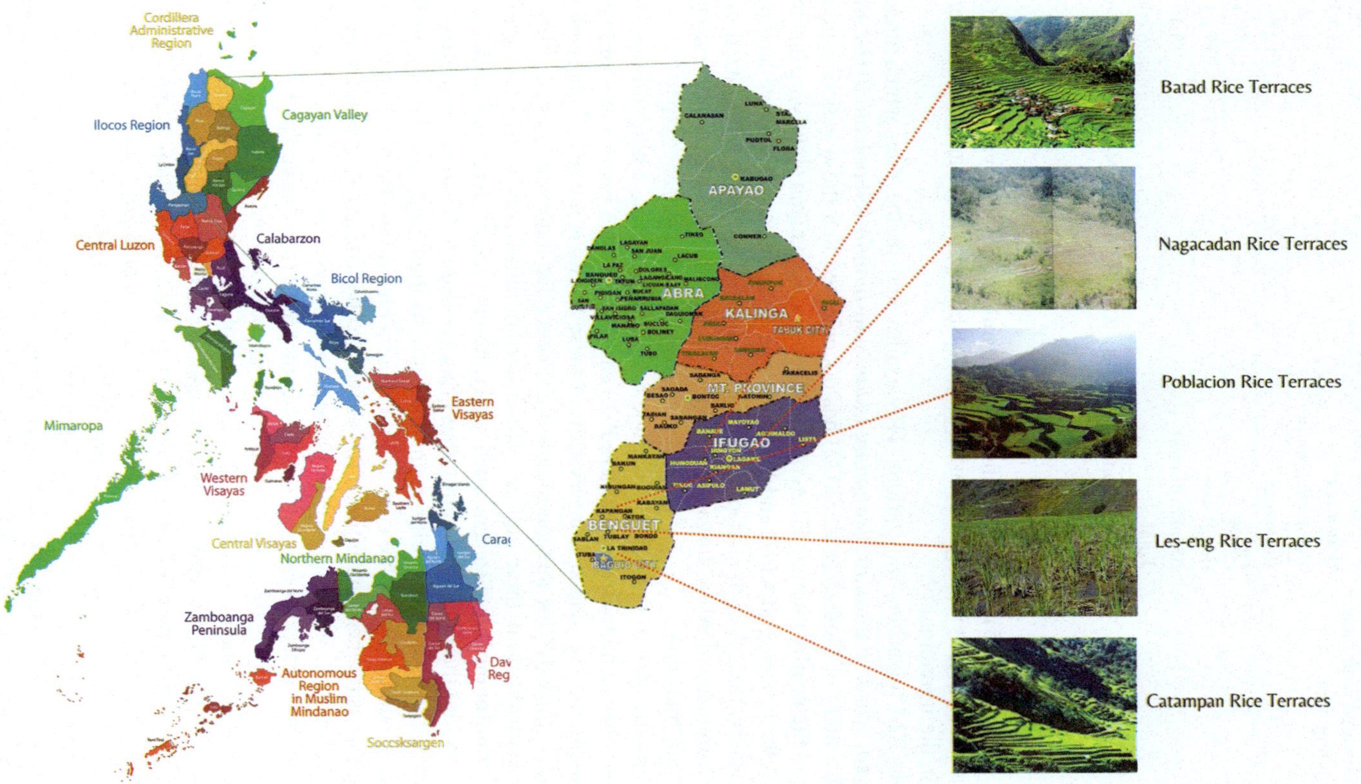

Fig. 22.1 Location of the TRBEs in the Cordillera Administrative Region (Source: CAPS Map; Napoleon 2018; and Gomez and Baniaga 2018)

Table 22.1 The identified TRVs cultivated with their corresponding area

Traditional rice varieties	Area (ha)
Balisanga	11.52
Waray/Balatinaw	14.31
Kintoman	12.56
Oklan	18.54
Bongkitan	15.63
Brando/Blando	13.48
Total	86.04

Source: Napoleon (2018), Gomez (2020), Napoleon and Gomez (2022)

Bakun, Profile 2015) which signifies a high number of cultivations amidst the introduction of high yielding rice varieties and cash crops. There are six documented TRVs in the municipality.

Table 22.1 identifies the various TRVs cultivated in the area which include: *balisanga*, *waray/balitanaw*, *kintoman*, *oklan*, and *blando* (Napoleon 2018). Similarly, the study of Gomez et al. (2022) has documented that these varieties are also present in the nearby municipalities of Kibungan and Kapangan signifying that these varieties have been cultivated and have a very significant cultural value leading to their conservation. As a matter of fact, most of these TRVs are also utilized in the nearby provinces of Abra, Ifugao, and Mountain Province for important festivities, rituals, and traditions.

The unique uses of these TRVs include religious offerings, while portions of the good harvest can be traded or "bartered" to cash. It is also a major raw material for winemaking among others. Conservation of these TRVs is not just anchored to the food that it provides for the community's existence for many decades but also was already embedded in their culture. These communities survive the test of time despite environmental disasters such as typhoons, landslides, earthquakes, and integration into the market economy that bring pressure to these communities.

Local varieties are primarily maintained due to their high adaptability to both predictable and unpredictable climates at different elevations. Being a staple food in subsistence farming communities, TRVs have the characteristics of good aroma and eating quality making them more desirable than commercial rice (Sajise et al. 2012). One of the best practices employed by the community to preserve this variety is the sharing or exchanging of varieties with fellow farmers to have a good acquisition and conservation of good quality seeds. This showed that the diversity of traditional rice varieties in the Cordillera region has been maintained Gomez and Baniaga (2018) in remote and subsistence farming communities, such as Bakun, Kapangan, and Kibungan, despite the influences of social, economic, and technological interventions. Furthermore, data shows that the Cordillera Regions remained to have a high on-farm varietal diversity in terms of TRVs.

Moreover, TRBE site was studied in Palina, Kibungan, where it investigated associated Ecosystem-based Adaptation (EbA) measures (Batani et al. 2019). Palina is considered the 'rice granary' of the municipality of Kibungan and is officially registered as a site for the tourist attraction for rice terraces. Palina is also one site identified by the Department of Agriculture (DA) and the International Rice Research Institute (IRRI), where traditional rice varieties, the so-called 'heirloom' rice still prevails and must be saved and upscaled. In this study, there are more than ten heirloom rice varieties identified. These include *Lasbakan*, *Gayyad*, *Brandon*, *Lablabi*, and *Balatinao* as the most planted varieties. However, research in 2015, listed 22 heirloom rice varieties being propagated in Poblacion, Kibungan, and Palina (Tad-awan et al. 2015).

According to Batani et al. (2019), based on the household survey, many of their respondents indicated that they are still into organic farming of rice varieties and sweet potatoes. They are using sunflower and *Mokusaku* which are organically produced as their fertilizers. In the study of Gomez and Baniaga (2018), they noted that the moisture content of parcels planted with traditional rice and sweet potato is 12.95% and 10.46% which are at satisfactory levels, as shown in Table 22.2. They further observed that the land dedicated to traditional rice cultivation had an organic matter content of 2.08%, the highest among all the cropping systems. The researchers stated that this is consistent with its being an organic production area, since nutrient cycling is expected to be prevalent and is possible because of the material recycling that farmers practice in the area. Organic or natural farming in Palina, Kibungan particularly in camote plots also represents a way of preserving the nutrients of soil mainly because the soil is not disturbed from tilling or plowing, wherein there is less possibility of soil nutrient loss through erosion.

In addition, according to Batani et al. (2019), rice growers particularly local women consider heirloom rice as a niche crop, wherein they prioritize state-sponsored programs on organic heirloom rice production. It is interesting to note that heirloom rice was being appropriated by community members as the reason for advocating organic production to convince vegetable growers not to give up heirloom rice production. Heirloom rice growers, trained under the DA-IRRI heirloom rice program, are reworking the program to favor their advocacy of reverting to pure heirloom rice growing in terraced rice fields converted into vegetable gardens, as shown in Table 22.2.

One of the main things keeping the community in a semi-subsistence mode in Palina is the ongoing cultivation of heirloom rice. One of the important considerations of the community is that the rice fields were inherited from their parents and grandparents, and they intend to continue planting them with rice to preserve their inheritance. Among the Kankanaeys of Palina, planting heirloom rice is 'to continue what the ancestors wanted them to, which is to plant rice for home consumption and food security. It is interesting to note that as part of the preparation for rice planting, *alluyon* or *binnaddang* (Bayanihan in Tagalog, meaning labor exchange and reciprocity) is mobilized for fixing the communal irrigation canal as well as during transplanting and harvesting. Moreover, the researchers observed that water scarcity was a perceived threat to decrease water supply which alerts the community of the need to limit vegetable gardening in favor of heirloom rice production. The traditional practice of *binnadang* (labor reciprocity) was also

Table 22.2 Cropping system and soil properties of Palina, Kibungan

Sampling sites	Cropping system	pH	Soil texture	Moisture content	OM (%)	Nitrogen content (%)	P content (ppm)	K content (ppm)
Purok 1	Rice field	4.4	Silty clay	12.95	2.08	0.104	10.55	62.5
Lingey	Camote farm	4.8	Silty clay	10.46	0.22	0.11	15.50	113.0
Lingey	Rice/vegetable farm	4.8	Silty clay	10.95	1.75	0.088	15.50	133.0
Poen	Vegetable farm	4	Silty clay	11.52	1.75	0.088	15.50	113.0

Source: Gomez and Baniaga (2018)

Table 22.3 Summary of the ethnobotanical survey conducted in Benguet Province

	Local term		Uses		Abundance	
Place	Wild	Cultivated	Wild	Cultivated	Wild	Cultivated
Atok	*Katjan* *Katnay* *Takay-yan*	*Agdey* *Ag-gey*	None	Food	Few	Few
Bakun	*Katnay* *Takay-yan* *Takaynan*	*Aggey* *Agdey*	None	Food	Few	Gone
Bokod	*Katjan* *Takjan*	*Agdey*	Animal fodder, Necklace, Bracelets		Few	Gone
Buguias	*Takay-yan*	*Aggey*	None	Food	Few	Few
Itogon	*Takjan* *Takkay-yan*	*Ag-gey* *Pag-pagey*	None	Food Animal feed	Few	Few
Kabayan	*Takjan* *Takjan katnay* *Takaynan*	*Agdey* *Aggey*	Bracelet Necklace Curtains		Few	Gone
Kapangan	*Takjan*	*Agdey*	Curtains Bracelets Rosaries	Foods Chicken feeds	Few	Few
Kibungan	*Takay-yan* *Takjan*	*Ag-gey*	None	Foods	Few	Few
La Trinidad	*Takjan*	*Aggey*			Gone	Gone
Mankayan	*Takay-yan*	*Ag-gey*	None	Foods	Few	Gone
Sablan	*Takjan* *Katjan*	*Agdey*	None	Foods Animal feeds	Few	Few
Tuba	*Takjan* *Takaynan*	*Agdey*	None	Foods	Few	Gone
Tublay	*Takjan* *Katjan*	*Ag-gey* *Agdey*	None	Foods	Few	Gone

Source: Lirio et al. (2013)

noted which is expressed in terms of community mobilization for repair and maintenance of the communal irrigation system or during rice planting and rice harvesting season, persist because there are heirloom rice terraces to maintain.

Coix and indigenous vegetables: Farmers in Benguet would usually weed out *Coix* plants growing in their rice paddies. Some, however, have the sheer delight of making ornaments out of the *Coix* seeds locally known as *takayan* (Table 22.3). Another interesting observation was finding bundled *Coix* plants with grains being sold during the local Sunday market. The study conducted has identified different varieties of *Coix lacryma-jobi L.* which are edible/cultivated and wild/ornamental. The traditional utilization and conservation of the plant in Benguet was documented, necessary in increasing awareness on the usefulness of this resource for food, medicines, and ornaments which will likewise increase appreciation, conservation, and propagation of *Coix*, thus providing materials for food and livelihood of the rural folks (Lirio et al. 2013).

Moreover, some people, on the other hand, are trying to plant the cultivated type which they consume as food either as cooked porridge, roasted, steamed grains, and as an ingredient in their traditional winemaking called '*tapey*'. Table 22.3 shows the utilization of wild and cultivated *Coix*. Others would also eat raw seeds. Some locals from other municipalities can make crudely made curtains and bracelets, rosaries, earrings, etc. out of the wild *Coix* grains. The crafting of the *Coix* grains into ornaments and similar items is being practiced among the women of Sagpat, Kibungan, and Benguet (Bersamin et al. 2021). It is also good as chicken feed in other areas. The wild types which naturally grow along the river banks of Bokod, Benguet are good animal fodders (Lirio et al. 2013).

Indigenous vegetables play an important role in the Cordilleran diet. Several inputs are necessary for an effort to know and document the indigenous vegetables utilized by old folks and other native respondents. The diversity of the indigenous food crops in the Cordillera documented 49 indigenous vegetables from the 11 municipalities of Benguet and the 4 municipalities in Mountain Province and including Baguio City. These plants belong to 21 families with about 19.15% of the species belonging to *Asteraceae*, followed by 10.64% in *Solanaceae*. Most IVs were observed to be seasonally abundant which provide food and nutritional security to the local community. The rest are distributed among other plant families surveyed by Lirio, Ayyokod and Paing (Lirio et al. 2007).

In terms of frequency, *Rorippa indica*, *Bidens pilosa*, *Solanum nigrum*, *Sonchus arvensis*, and *Cardamine hirsuta* are used as food in most places of the municipalities, where they are present. Similarly, these species have medicinal uses. *B. pilosa*, for example, is popularly eaten, because folks claim that it can prevent goiter. Folks also use the plant in '*tapey*' (rice wine)-making, a tradition they are known for. The value of these representatives and other species not only as food but as medicines as well should be reasons for their wider utilization (Lirio et al. 2006, 2007).

Vitamins serve as chemical partners for the enzymes involved in the body's metabolism, cell production, tissue repair, and other vital processes. Remarkably, *Bidens pilosa* has the highest value of vitamin C (347.5 mg/100 g) and vitamin A (5629.75 μg/100 g), as shown in Table 22.4. Many traditional food crops similarly provide a wide array of health benefits (Lirio et al. 2007).

Moreover, an indigenous plant Panawil (*Leptosolena haenkei C. Presl.*) to the Cordillera Region is an herb locally known to the community. The flowers of *panawil* are gathered and cooked as indigenous vegetables. Its flowers have a sweet-smelling taste, are delicious when combined with fish and meat dishes, and are very acceptable to local taste. Fruits are also gathered for consumption and may serve as food for wild animals. The plant is a good source of phytonutrients, such as terpenoids, flavonoids, tannins, total phenolics, and antioxidants; minerals

Table 22.4 Vitamin contents of IVs

Indigenous vegetable	Vitamin C (mg/100 g)	Vitamin A (μg/100 g)
Amaranthus blitum	44.47	152.98
Bidens pilosa	247.50	5629.75
Cardamine hirsute	30.15	250.81
Cestrum nocturnum	39.25	474.20
Crassocephalum crepidioides	66.25	400.62
Passiflora edulis	17.05	168.30
Pteridium aquilinum	49.53	384.33
Sonchus arvensis	22.42	411.68
Solanum nigrum	70.22	398.70
Brassica chinensis	33.12	396.56

Source: Lirio et al. (2007)

(potassium, iron, and zinc); and vitamins A and C (Paing and Gomez 2021; Paing et al. 2018).

22.3 Forest Ecosystem and the ICC Nexus

The succeeding presentations concerning forest ecosystems as biocultural landscapes are specific field studies that describe the plant diversity and the setting of the locale. Typically, these cases depict studies in areas dominated by ICCs (Kibungan, Benguet), protected areas (Mount Data National Park, MDNP, and Mount Pulag National Park) and/or communal forest (Palina and Sagpat, Kinbungan) (Fig. 22.2).

Protected areas: A floral assessment study conducted at the Mount Data National Park (MDNP) in Sinto, Bauko, Mountain Province revealed that the mossy forest ecosystem was relatively diverse as shown in the Shannon Diversity Index measured ($H' = 2.63$) for the tree species (Gomez 2009, 2010). The more dominant tree species recorded included the following: *Lithocarpus luzoniensis*, *L. jordanae*, *Syzigium santosii*, *Ilex pulogensis*, among others. The species–area curve for the mossy forest ecosystem is depicted in Fig. 22.3. The series of increasing sampling areas corresponding to the number of species encountered are shown in the graph to have a high R^2 value of 0.9079. This is one indication of the relatively high diversity of this type of ecosystem. The species–area curve is a useful concept on the premise that there is a minimal area that can be used for sampling. It should be noted that this graph only shows the tree species; in more correct terms, with the low-lying vegetation present on the forest floor and the others which might be clinging to the tree species, all of these would show the real species richness of the mossy forest ecosystem.

Seemingly, the soils in the area are found toward the more acidic range (Table 22.6). Notably, the organic matter content of the soils is generally high at 19–20% compared to agricultural soils in the area which may just range from 1% to

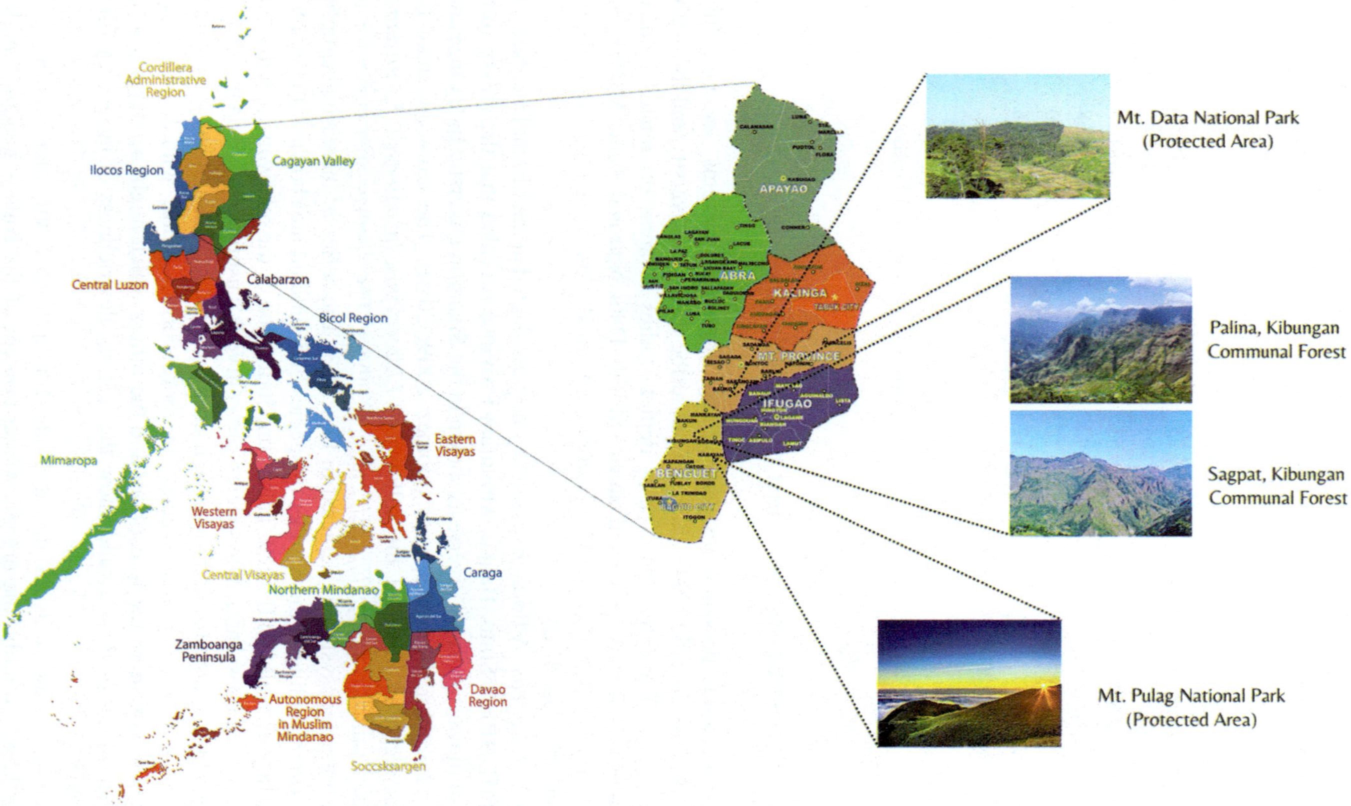

Fig. 22.2 Location of the protected area and communal forests (Source: ACAPS Map)

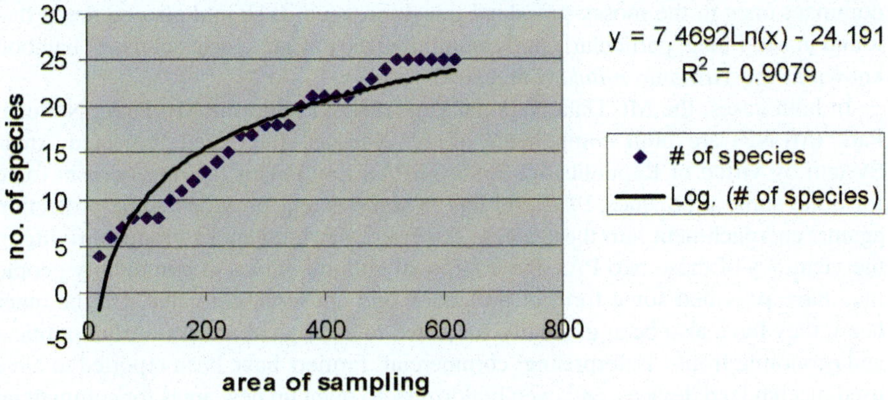

Fig. 22.3 Species–area curve for the mossy forest ecosystem (Source: Gomez 2010)

Table 22.5 Physical properties of the soil in the mossy forest ecosystem

Color	Texture	Soil temperature (°C)	MC(F)%	DM(F)%
Dark, brown–red	Clay loam	14.83	10.19	89.81

Source: Gomez (2010)

Table 22.6 Chemical properties of the soil in the mossy forest

pH		Total N (%)	P (ppm)	K (meq/100 g)	OM %[a]	OM %[b]
Mossy forest	5.00	0.96	4.65	0.80	19.24	35.60

Source: Gomez (2010)
[a] Walkley and Black method
[b] Dry combustion method

2% (based on the result using Walkley and Black method). This implies that the mossy forest soils in the area are expected to be better in terms of fertility, as compared, of course, to the adjacent farm soils, which are severely depleted of organic matter. This is also consistent with the observation that on visual examination, the soil color is of the dark type, an indication that organic matter content is high (Table 22.5).

A floral study was also conducted in the Mount Pulag area specifically the transition zones of the three major vegetation types along its altitude. In terms of pedon, the study showed that the soils in the pine-mossy transition zone had a mean pH value of 5.10, with 31.67% organic matter and 4.67 ppm (Balangen et al. 2000). On the other hand, the mossy-grassland ecotone showed a more acidic pH value of 4.43, with 35% organic matter and 8.67 ppm. These are values that are generally close to those measured in Mt. Data's mossy forest ecosystem. This means that because of the very similar nature of the ecosystem, they oftentimes reveal similarities in other aspects; in this case, the chemical properties of the soil. On plant diversity, it was observed that *Lithocarpus woodii*, *Clethra luzonica*, *Rubus* sp., among others were the most dominant in the pine-mossy transition zone. On the other hand, apart from *L. woodii* and *C. luzonica*, *Gaultheria cumingiana* were the dominant ones in the mossy-grassland transition zone. It should also be noted that found in Mt. Pulag, particularly at the summit is the unique species of dwarf bamboo known as the *Yushania niitakayamensis*.

In both cases, the Mt. Data National Park (MDNP) and the Mt. Pulag National Park (MPNP) are both components of the National Integrated Protected Areas System by virtue of Republic Act 7586 (NIPAS Law). This law has been in force for quite some time since 1992 and has to some extent provided some protection against encroachment into these areas. Together with indigenous groups thriving in the vicinities of these two PAs, the traditional cultural values of community people may have provided some form of protection and conservation value. On the other hand, they have also been gradually subjected to ecological attacks—from political and economic fronts. Enterprising 'commercial' farmers have been reported to have used mechanized devices, and even bulldozers, to open up new areas for commercial vegetable production. Reports in the newspapers have on a number of occasions captured photographs of sporadic land use changes and encroachments into the protected areas.

It is strongly postulated that due to the very favorable physico-chemical properties of the soils in these ecosystems, notably the high organic matter content—and the succeeding nutrient pool emanating from their mineralization—is conducive to supporting agricultural production for highland or semi-temperate crops. This is one of the reasons why farmers tend to open up new areas to support a higher level of crop production. About 48% of the forest cover disturbances are attributed to agricultural expansion (Daipan 2020). This is, of course, a short-term foresight as it addresses only the short-term economic benefits. As stated earlier, this is one of the negative trade-offs of articulating with and being integrated into a bigger scale of the market economy. This is in the context that vegetable production in the highlands is a multimillion industry and a big contributor to local and even national economy, in terms of gross domestic product (GDP) (Aquino 2012). This historical antecedent was also observed with the vegetable-producing areas in Benguet and Mountain Province, where the once mossy forest and pine tree-dominated ecosystems were once robust and abundant and now largely denuded, although these are not within the purview of the NIPAS Law.

Although the MPNP straddles the three provinces, Benguet, Ifugao, and Nueva Vizcaya in its tri-boundary, the management of the MPNP is presently lodged with the Office of the Park Superintendent (PASU), based on the Province of Benguet. The case of the MDNP though is quite different. The LGU in Sinto, Bauko, together with the PAENRO (Provincial Agriculture and Environment Office) of Mountain Province, cooperates to provide the management and protection of such PA. With the present encroachment in MDNP, there have been efforts to advance its disestablishment as a PA (Gomez 2010).

Communal forest of Sagpat, Kibungan, Benguet: Kibungan is in the northwestern part of Benguet province. This is located in the Cordillera region, Northern Luzon, Philippines. The municipality is rich in fauna and flora as well as culture; hence, indigenous knowledge on plant use has been developed. Sagpat, one of the barangay of the municipality, is located 16.67 latitude north, 120.68 longitudes east, with an altitude of 1520 masl or 4986.88 fasl. The communal forest has an area of 1 ha. The *Kankanaey* inhabitants of Sagpat, Kibungan value the resources of their communal forest, since its resources were used to sustain life. Plant resources sustain life as a source of medicine, food, and materials to build shelter and for cooking food for the family. The leaves of medicinal plants are frequently used as decoction and poultice to treat wounds, diarrhea, cough, and skin inflammation. Among these herbal plants, *gipas* (*Sarcandra glabra*) and *gawed* (*Piper betle*) were noted as the common medicinal plants (Bersamin et al. 2021). In addition, identified plants such as *bayabas* (*Psidium guajava*), *laya* (*Zingiber officinale*), and *subusob* (*Blumea balsamifera*) were included in the ten herbal plants identified by the Department of Health (Table 22.7). These indigenous practices of the locals are being supported by the Department of Health for endorsing herbal plants used by indigenous communities as cures for various ailments (Philippine Department of Health 2004).

Concerning edible plants, the community enjoys fruits, root crops, and vegetables. The *pinit* (*Rubus fraxinifolius*), *amti* (*Solanum nigrum*), *bayabas* (*Psidium guajava*), *gatgattang* (*Sonchus arvensis*), *kamote* (*Ipomea batatas*), and *pako* (*Diplazium esculentum*) are among the wildly growing food resources (Bersamin et al. 2021). The term wild refers to non-cultivated plants gathered in the fields (Tardio et al. 2006). The study of Barcelo-Chua (2014) identified that the cultural importance (CI) value of *R. fraxinifolius* is 0.390, while *P. guajava* is 0.061. The value shows that *R. fraxinifoliu* is least used or of minimum importance, while *P. guajava* is a widely used and most important fruit species among the 36 wild fruits in Benguet with 13 municipalities and 140 barangays. The value can be explained from the result shown in Table 22.8 in which *R. fraxinifoliu* utilized plant part is the fruit while for *P. guajava*, the shoot, and fruit.

Other plants such as the *cogon* (*Imperata cylindrica*) and the *buybuy* (tiger grass—*Thysanolaena latifolia*), (Bersamin et al. 2021) study shows that they were utilized for roofing materials and broom making, respectively (Table 22.8). *I. cylindrica* can also be used as an herbal plant for diarrhea, kidney ailments, and colds (Balangcod and Balangcod 2015). *T. latifolia* is identified as native to the Philippines (Co's Digital Flora of the Philippines 2017) which is why it is the most common source of brooms as a cleaning material.

Aside from the abovementioned, plant resources also provide protection especially for children using *dengaw* (*Acorus calamus*) as an amulet, which is believed to ward off evil spirits. A piece of the root can also be pinned on clothes, especially when traveling far distances from the village as protection from being harmed by bad spirits that may be encountered along the way. Interestingly, plants are not only utilized to sustain life but also in use for the dead. The *bayabas* (*Psidium guajava*) and *niyog* (*Cocos nucifera*) plants are used as a substitute for embalming (Table 22.9). Plant resources have diverse uses for the *Kankanaeys* of Sagpat Kibungan, and it would be either for the living or the dead, but the most momentous event is the *cañao*. These refer to social gatherings among local communities in the Cordillera and is usually characterized by the butchering of animals (either pigs or carabaos), local dances with gongs' accompaniment, striking of metals, and amusements. Meat and vegetables are served to the attending participants and guests,

Table 22.7 Medicinal plants used by the Kankanaeys in Sagpat, Kibungan, Benguet, Luzon, Philippines

Local or common name	Scientific name	Plant part used	How plant is used	Ailments/ diseases cured
Atelba	*Viburnum luzonicum* Rolfe	Fruit	Decoction of fruit is given during loose bowel movement (LBM)	LBM
Avocado	*Persea americana* Mill.	Leaves	Decoction of leaves is given during diarrhea and to relieve stomach pain	Diarrhea, stomach pain
Bayabas	*Psidium guajava* L.	Shoots, fruit	Crushed shoots are applied as poultice on wounds; Decoction of shoots is used for allergies and rashes; Fruit and shoot is eaten during diarrhea	Wounds, diarrhea, allergies, rashes
Biday	*Ocimum tenuiflorum* L.	Root	Crushed roots are applied as poultices on wounds	Wounds
Dungawl dengaw	*Acorus calamus* L.	Roots, stem	Crushed roots are applied as poultices on the affected area to relieve muscle pain, and skin allergy and reduce inflammation; Decoction of the stem is given for dysentery	Muscle pain, skin inflammation or allergy, dysentery
Gawed	*Piper betle* L.	Leaves	Leaves are applied directly on the forehead to abate fever	Fever
Gipas	*Sarcandra glabra* (Thunb.)	Leaves, roots	Decoction of leaves and roots is given during colds and urinary tract infections (UTI); It is also used to wash wounds; Leaves are also used as a tea for cleansing	UTI, wounds, colds, anti-oxidant
Lantana	*Lantana camara* L.	Leaves	Decoction of leaves is given during cough	Cough
Laya	*Zingiber officinale* Roscoe	Rhizome	Decoction from the pounded rhizome is given during cough; can also be used as a wash for wounds and after giving birth, and poultice to relieve joint pain; Taken to induce lactation	Cough, wound, Joint Pain, induce lactation
Madre de Cacao	*Gliricidia sepium* (Jacq.)	Leaves	Leaves are applied directly on the forehead to abate fever	Fever
Mahogany	*Swietenia mahagoni* (L.)	Seed	Eaten directly during diarrhea	Diarrhea
Papaitl bibiday	*Ageratin a adenoph ora* (Spreng.)	Leaves	Crushed leaves are applied as a poultice on wounds	Wounds

(continued)

Table 22.7 (continued)

Local or common name	Scientific name	Plant part used	How plant is used	Ailments/ diseases cured
Paragis	*Eleusine indica* (L.) Gaertn.	Leaves and roots	Decoction of leaves is given for cleansing/detoxification and treating dysmenorrhea	Cleansing, hypertension, dysmenorrhea
Pinit	*Rubus fraxinifolius* Poir.	Leaves, trunk	Decoction of leaves and trunk is given during urinary tract infection	UTI
Putputod (horsetail)	*Equisetum ramosissimum*	All parts	Decoction of all parts is given during urinary tract infections and kidney ailments	UTI, kidney ailments
Sipal/sepal	*Drimys piperita* Hook. f.	Fruit	Dried and given during stomach ache; Eaten raw; decoction of the fruit is given during loose bowel movement (LBM)	Stomach ache, LBM
Subusob	*Blumea balsamifera* (L.) DC.	Leaves, roots	Decoction of leaves is given during cough. Roots are boiled and the vapor is inhaled to treat colds	Cough, Colds
Tagumbaw	*Jatropha curcas* L.	Bark, leaves	Crushed leaves and bark are applied as a poultice on fractures	Fracture

Source: Bersamin et al. (2021)

invited, or not invited. Usually, the merriment is spiced by serving rice wine called *tapey* which is made from a special variety of rice. During the gathering, kamote (*Ipomea batatas*) is always served before the sumptuous meal. The food during the gathering is cooked using alnus (*Alnus japonica*) and Benguet pine (*Pinus kesiya*) as firewood (Bersamin et al. 2021). Balangcod and Balangcod (2011) stated that the people of Benguet have developed their ingenuities in harnessing the plant resources around them; therefore, indigenous knowledge on plant utilization in the region is inherent in the local communities.

Communal forest in Palina, Kibungan: In another study, Gomez and Baniaga (2018) estimated the carbon storage of the communal forest in Palina, Kibungan which is a natural pine-dominated forest. The estimated total area of watershed and the communal forest is 315 ha based on the Kibungan Municipal Profile. Based on their estimates taken, around 315 ha of pine forest in Palina has carbon storage of about 93,191 tons or 295.84 tons/ha. This is an enormous amount of storage that exemplifies one of the ecological functions of pine forests. This is also a good indication that the presence of pine trees in Palina plays a significant role in mitigating the atmospheric accumulation of greenhouse gases (GHG), and it also has a role to play in helping farmers adapt to climate change.

Table 22.8 Plants used for food by the Kankanaeys in Sagpat, Kibungan, Benguet, Luzon, Philippines

Local or common name	Scientific name	Plant part used	How the plant is used
Amti (weeds)	*Solanum nigrum* L.	Leaves	Cooked, mixed with other foods
Ayusep	*Vaccinium myrtoides* Miq.	Fruit	Eaten raw
Bayabas	*Psidium guajava* L.	Shoots, Fruit	Cooked or eaten raw
Bilis	*Garcinia vidalii* Merr.		Eaten raw
Binnok	*Medinilla* sp.	Leaves, fruit	Eaten raw
Camote	*Ipomoea batatas* (L.) Lam.	Root	Cooked and mixed with other food
Cassava	*Manihot esculenta* Crantz.	Tuber	Cooked
Climbing beans	*Phaseolus vulgaris* L.	Fruit	Cooked
Gabi	*Colocasia esculenta* (L.) Schott	Leaves, root	Cooked
Gaddang	*Languas haenkei* (*C. Presl*) Merr.	Fruit	Eaten raw
Gagattang (*weeds*)	*Taraxacum officinale* F.H. Wigg.	Leaves	Cooked, mixed with other foods
Kendoy/Kendey	*Rorippa indica* (L.) Hiern.	Leaves	Cooked
Laya	*Zingiber officinale* Roscoe	Rhizome	Cooked, mixed with other foods
Pako	*Diplazium esculentum* (Retz.) Sw.	Leaves, Stem	Cooked
Papaya	*Carica papaya* L.	Fruit	Eaten raw
Pinet, Pinit	*Rubus* sp.	Berries, fruit	Eaten raw
Pomelo	*Citrus maxima* (Burm.) Merr.	Fruit	Eaten raw
Suyok (*Rono*), *bellang*	*Miscanthus sinensis* Anders.	Shoots	Eaten raw
Ul-ek. Utok	*Saurauia elegans* Fern.-Vill.	Fruit	Eaten raw

Source: Bersamin et al. (2021)

The abundance of pine trees and diversity of plant species in the Palina communal forest are still maintained Gomez and Baniaga (2018). According to Batani et al. (2019), a community-like Palina shows the persistence of a consultative tradition. There is consistency in the positions taken by the locals even in terms of managing their watershed and communal forest, wherein the rule is to limit the encroachment of vegetable gardens and strictly no cutting of trees. Another reason why the

Table 22.9 Plants used for ornaments/adornment, rituals/mummification by Kankanaeys in Sagpat, Kibungan, Benguet, Luzon, Philippines

Local or common name	Scientific name	Plant part used	How to plant is used
Bayabas	*Psidium guajava* L.	Leaves, shoots	Mummification, decoction is used to wash the corpse and as a substitute for formalin
Dungaw/ dengaw	*Acorus calamus* L.	Roots	Amulet, attached to clothing to drive away evil spirits
Niyog	*Cocos nucifera* L.	Fruit	The extracted oil is rubbed on the body for the preservation of the dead
Takkayan/ tukkayan	*Coix lacryma-jobi* L.	Fruit	Necklaces, bracelets, earrings, curtains, bags, Christmas tree, basket

Source: Bersamin et al. (2021)

communal forest is protected in Palina is that the residents derive goods and services from the forest. Based on household surveys done by the researchers, residents harvest 80 kg of wild mushrooms annually as part of their home consumption and they have access to many medicinal plants present in the forest. This qualitatively reflected the diversity of plants in the area.

Through the customary rule of *tongtong*, an agreement was made through a written document being kept by the barangay local government unit (barangay officials). Any expansion will have to be stopped via the tongtong mechanism. Aside from community-level customary laws, Barangay Palina follows Municipal Ordinance No. 53-2015 which mandates couples who have applied for marriage licenses with the local civil registry to plant "marriage trees". The *tongtong* and *pan-ngangalatan* are two indigenous mechanisms, where community issues are settled. In the literature, tongtong is a justice system based on agreement (Beta-a, 2002 as cited in Batani et al. 2019) which has the goal of restoring harmonious relationships between and among the community members. *Pan-ngangalatan* is another mechanism of settling disputes, which means to 'talk about.' *Tongtong* is used if 'political issues are to be talked about which involves the wider community and where the dispute is involved, while *pan-ngangalatan* is used for domestic concerns, such as marital issues and everyday concerns.

22.4 Challenges: A Macro-level Perspective

Scaling reality, food regimes, and the Anthropocene: There is a need to observe the biocultural landscape under the lens of a macro-level perspective if we want to understand the nuances of the forces that shape it, particularly during the so-called Anthropocene. Rock and Buchanan (2014) articulate that social problems that impact the situatedness of individuals could be inextricably linked to a global reality. The current socio-economic and ecological conditions that ICCs find themselves in are not isolated from this global reality. Furthermore, ICCs become entangled in the restructuring and rescaling of their social standing. According to Fairclough (2002), restructuring points to "shifts in relations between different domains or fields of social life most obviously between the economic field and other fields, including a 'colonization' of other fields by the economic field." Rescaling on the other hand points to a shift in social life on a local scale, national, regional, to global scale. By analyzing the relational dynamics of ICCs with the impacts of globalization, we can distill a nuanced understanding of the challenges and threats that confront not only the cultural heritage of the indigenous peoples of the Cordillera but also the rich plant biodiversity that supports their very survival.

Globalization and its impacts on ICCs and the biocultural landscape can be analyzed using a Food Regime Theoretical (FRT) perspective postulated by Friedmann and McMichael (1989). By using FRT, we can relate agriculture to the capital accumulation of nation-states and analyze the environmental impacts of the movements and counter-movements that have shaped the neoliberal trajectory of contemporary market economies. In this case, food emerges as the pivotal element upon which environmental and socio-economic policies are developed. Food security has become a global policy dominating the social, economic, and environmental domains of nation-state development programs. FRT provides a theoretical and historical analysis of how politics in food production came to be what it is today and how these policies continue to impact on ICCs and indigenous food systems including the plant diversity of indigenous communities.

Applied to agri-food, a straightforward definition of a food regime is "a rule-governed structure of production and consumption of food on a world scale" (Friedmann 1993; McMichael 2007). Henry Bernstein (2016) enumerates key elements that are identified in different food regimes which he claims "bear on their determinants and drivers; their 'shape', so to speak, and consequences; and their tensions, crises, and transitions, including struggles within and against different food regimes, and responses to their contradictions." These include:

- the international state system;
- international divisions of labor and patterns of trade;
- the 'rules' and discursive (ideological) legitimations of different food regimes;
- relations between agriculture and industry, including technical and environmental change in farming;
- dominant forms of capital and their modalities of accumulation;
- social forces (other than capitals and states);
- the tensions and contradictions of specific food regimes; and,
- transitions between food regimes.

Bernstein uses these elements as a guide in addressing his questions on where, how, and by who is food produced, and consumed and what social and ecological impacts of international relations of food production and consumption have in different food regimes (Bernstein 2016). Similarly, biocultural assets in ICCs, particularly those that are used for food provisioning and other ecosystem services, should be carefully studied based on the aforementioned determinants and drivers.

The corporate food regime: McMichael (2005) notes that the control of finance moved from nation states to international bodies such as the IMF produced conditions that allowed corporations to access lands and resources intensifying profit-making, thereby producing "social catastrophes and environmental degradation." As a result, tension emerges between corporate interest and marginalized people, such as indigenous communities. He further describes the central contradiction in the corporate food regime as that "between a 'world agriculture' (food from nowhere) and a place-based form of agro-ecology (food from somewhere). Friedmann (2009), on the other hand, locates tensions and contradictions by pointing out that there is a potential instability in the aspect of capital transfer. She explains that while power and ownership are crucial aspects in the formation of food regimes, money figures in the "mechanism of dominance—and of the transfer of wealth from a falling hegemon to emerging powers."

The minority: The status of ICCs in the context of nation-states is often the minority. Although the Philippines recognizes its critical role in nation-building and provides them with legal protection, the dominant development framework is characterized by a neoliberal preference for capital. Biological and cultural assets are often collaterals to economic growth and development policies.

Biodiversity in adversity: Given the hegemonic forces of a capitalistic Anthropocene, ICCs and the stewardship of their biocultural domains are spotlighted as crucial agents for ecological sustainability. The preservation of biological assets such as TRVs and other plant species will ensure the viability of incoming generations. Only when the physical, biological, and cultural sovereignties of ICCs are respected and protected from neoliberal forces can we preserve biodiversity. In the Global South, people are increasingly asserting their basic right to healthy and culturally appropriate food produced through indigenous methods that naturally conserve resources and protect biodiversity (Madew and Leung 2019). Local communities the world over are coalescing in a self-mobilized effort to resist the devastation of their self-sustaining culture and system of production.

Simply put, socio-cultural changes operating at the locale have to some extent eroded time-honored cultural values and cosmology about their relations with the bio-physical environment and the resources inherent therein. Socio-economic, political, and technological developments have served to modify these traditional values, albeit as 'double-bladed' instruments. This leads to exacerbated ecological attacks on precious, but scarce land resources brought about by agriculture and other burgeoning industries, as well as built-up areas. Thus, increasing pressure on biological resources, in particular, phytodiversity in this context is a present reality. Hence, the potentials of a number of promising, sustainable, culturally anchored, economically viable and politically acceptable options are essential in the operationalization of biocultural landscapes.

22.5 Summary and Conclusion

Plant biodiversity is an inherent heritage of ICCs that they have protected and valued over generations. For ICCs, humans and their environment operate on a plane of mutuality. Humans are nested in the environment and the environment becomes an available reality for human consciousness. Cultures are developed within these intricate ecological relationships and manifested in the collective wisdom upon which indigenous cultural communities interface with nature. Indigenous communities are the bearers of intergenerational ecological knowledge and the richness of plant biodiversity in ICCs is a testament to the enduring indigenous knowledge systems that are used to sustain the richness and viability of local ecosystems.

Rice as a food/commodity for the ICCs in Benguet conjures a rich discussion for plant biodiversity and biocultural landscape sustainability. The Cordilleras, particularly. Benguet offers rich biodiversity of traditional rice varieties (TRVs). Furthermore, indigenous plants and vegetables are also part of the ICC's heritage.

TRBEs do not only reflect how indigenous communities develop systems of sustenance, but they also give us a glimpse of the strong sense of belongingness and ownership of heritage. Heirloom rice grown by ICCs is a sovereign capital owned by the community. It is an important source of food, and it enriches the cultural identity of provenance. Forest ecosystems provide a sanctuary for biodiversity. Very often, they are protected areas and facilitate the regeneration of endangered species. ICCs play an important role in these protected ecosystems.

However, contemporary models of development threaten the biocultural landscape of indigenous communities. Discourses on global policy cater to development models that subjugate dominance to humans compromising the supposed mutuality between the ecological and anthropological co-existence. Neoliberal frameworks of development continue to reify the capitalistic trajectory of the Anthropocene. A reflexive approach to how we view minorities, knowledge systems, ecological consciousness, and development policies can impact a more sustainable Anthropocene, where biocultural landscapes can flourish and nurture communities who are self-aware of their ecological embeddedness.

Acknowledgements All research participants and the ICCs who shared their time and knowledge in the various researches conducted, Springer Nature, and the Editors (especially Dr. Inocencio E Buot, Jr.) for the opportunity to co-write a book chapter.

References

ACAPS (2022) Maps and Infographics. https://www.acaps.org/

Aquino C (2012) The Philippine vegetable industry. A poverty resource hub of focus on the Global South Philippines. https://www.focusonpoverty.org/the-philippine-vegetable-industry-almost-comatose. 13 Aug 2022

Balangcod TD, Balangcod AKD (2011) Ethnomedical knowledge of plants and healthcare practices among the *Kalanguya* tribe in Tinoc, Ifugao, Luzon, Philippines. Indian Tradit Knowl 10(2): 227–238

Balangcod TD, Balangcod AKD (2015) Ethnomedicinal plants in Bayabas, Sablan, Benguet Province, Luzon, Philippines. Electr J Biol 11(3):63–73

Balangen DF, Latus SC, Sofla RT, Toledo GP, Torres JM (2000) Soil and vegetation characteristics of Mount Pulag National Park. Unpublished BS environmental science thesis. Department of Biology, Benguet State University, La Trinidad, Benguet

Barangay Poblacion Bakun, Benguet profile (2015) Poblacion, Bakun Benguet.

Barcelo-Chua RT (2014) Ethno-botanical survey of edible wild fruits in Benguet, Cordillera Administrative Region, Philippines. Asian Pac J Trop Biomed 4(Suppl 1):S525–S538

Batani RS, Gomez RA Jr, Launio CC, Listino BC (2019) Assessment field, validation and documentation of evidence-based ecosystem adaptation measures with potential for scaling-up in indigenous cultural communities. GIZ project terminal report. Benguet State University, La Trinidad, Benguet

Bernstein H (2016) Agrarian political economy and modern world capitalism: the contributions of food regime analysis. J Peasant Stud 43(3):611

Bersamin AT, Tayaben JL, Balangcod KD, Balangcod AD, Cendana AC, Dom-ogen ET, Licnachan LC, Siadta B, Wong FM, Balangcod TD (2021) Utilization of plant resources among the Kankanaeys in Kibungan, Benguet Province, Philippines. Biodiversitas 2(1): 362–372. ISSN 1412-033X. https://doi.org/10.13057/biodiv/d220144

Buot IE Jr (2008) Sustaining environmental health in the Philippines satoyama landscape. Asian Int J Life Sci 129–138

CO's Digital Flora of the Philippines (2017) Last edited by Pieter B. Pelser, 7 September 2017. https://www.philippineplants.org/Families/Poaceae.html

Daipan BPO (2020) Detection of forest cover disturbances using synthetic aperture radar (SAR) time series data in the Mount Pulag National Park: a key biodiversity area in Luzon, Philippines. Philippine J Sci 150(S1):361–374

Department of Health (2004) Philippine pharmacopeia 1. Department of Health, Bureau of Food and Drugs, Manila

Dublin D, Natori Y, Takahashi Y, Lopez-Casero F (2018) SEPLS: Socio-Ecological Production Landscapes and Seascapes experiences overcoming barriers from around the world. Journal of Conservation International Japan, Institute for Global Environmental Strategies

Fairclough N (2002) Language in the new capitalism. Discourse Soc 13(2):163–166

Friedmann H (1993) The political economy of food: a global crisis. New Left Rev 197:29–57

Friedmann H (2009) Discussion: moving food regimes forward: reflections on symposium essays. Agricult Human V Values 26:335–344

Friedmann H, McMichael P (1989) Agriculture and the state system: the rise and decline of national agricultures, 1870 to the present. Sociol Rural 29(2):93–117

Gomez RA Jr (2013) Food sufficiency in the Terraces: the case of globally-important agricultural heritage systems in Ifugao, Central Cordillera, Philippines. Int J Ecol Conserv 7. https://doi.org/10.7718/ijec.v7i1.723. ISSN 2244-1573

Gomez RA Jr (2020) Spatial and productivity measurements in traditional rice-based ecosystems (TRBEs) in Highland Cordillera. In: Buot E et al (eds) Methodologies supportive of sustainable development in agriculture and natural resource management. SEARCA, Los Banos

Gomez RA Jr (2009) Floral assessment of Mount Data National Park. Philippine Agricultural Resources Research Foundation (PARRFI, Inc.) Professorial Chairholder accomplishment report. PCARRD, Los Banos, Laguna

Gomez RA Jr (2010) Floral assessment of Mount Data National Park. Philippine Inter-university Cooperation Project (PIUC-VLIR) joint project 6 (Business and Environment), Benguet State University and Saint Louis University. Terminal report. PCARRD, Los Banos, Laguna. 14pp

Gomez RA Jr, Baniaga AV (2018) Preliminary report of biophysical assessment in Palina, Kibungan: assessment field, validation and documentation of evidence-based ecosystem adaptation (EbA) measures with potential for scaling-up in indigenous cultural communities. GIZ project. Benguet State University, La Trinidad, Benguet

Gomez RA Jr, Pacardo EP (2005) Landscape change in the Ifugao Rice Terraces with the advent of modern technologies. J Environ Sci Manag 8(1):1–12

Gomez RA Jr, Tad-awan BA, Fagyan FW, Paing JN, Egalla EJD (2022) Native/heirloom rice in the Cordilleras: status, conservation, and utilization. In: Ramamoorthy S, Buot IJ, Chandrasekaran R (eds) Plant genetic resources, inventory, collection and conservation. Springer, Singapore, pp 121–134

Ingosan JB, Labaro JS, Sally JL, Gomez RA Jr. (2008) Agro-ecological appraisal of traditional rice-based ecosystem: the Kapangan experience. BSU Res J 60 & 61

Lirio LG, Paing JN, Ayyokad LO (2006) Indigenous semi-temperate vegetables of Cordillera. Unique Printing Press, Benguet State University La Trinidad, Benguet. ISBN: 971-0330-60-8

Lirio LG, Ayyokad LO, Paing JN (2007) Indigenous semi-temperate vegetables in Cordillera, Philippines. Acta Hortic 752:513–520

Lirio LG, Paing JN, Lan-ew RK (2013) Coix lacryma-jobi Linn. an underutilized grass for food security and economic empowerment of rural communities. Acta Hortic 1:285–291. ISBN 978 90 6605 279 6, Number 979

Madew M, Leung JM (2019) The politics of hunger and the legacy of resistance in postcolonial social work. Crit Rad Social Work 7(3):381–399. https://doi.org/10.1332/204986019X15695497221271

McMichael P (2005) Global development and the corporate food regime. Res Rural Sociol Dev I1: 265–299

McMichael P (2007) Reframing development: global peasant movements and the new agrarian question. Revista Nera 10(10):57–71

Napoleon RD (2018) Watershed functions: irrigation and productivity of Rice Terraces in Poblacion, Bakun, Benguet, Unpublished BS Environmental Science thesis, Department of Biology, Benguet State University, La Trinidad, Benguet

Napoleon RD, Gomez RA Jr (2022) Resiliency of Traditional Rice-based Ecosystems (TRBEs): the Case of Poblacion, Bakun, Benguet, 2022. Paper presented during the 2nd International Conference on Training and Research for Arts, Environment, and Sciences (ICTRAES). Nueva Ecija University of Science and Technology

Paing JN, Gomez RA Jr (2021) Ecological distribution and soil-vegetation analysis of the indigenous vegetable panawil (Leptosolena haenkei C. Presl.) in specific areas of Benguet and Mountain Province, Philippines. Trop Ecol 62:107. https://doi.org/10.1007/s42965-020-00131-4

Paing JN, Pladio LP, Bolayo YD (2018) Environmental factors affecting the phytonutrient contents of indigenous vegetable Panawil (Leptosolena haenkei C. Presl.) in the Cordillera Region, Northern Philippines. J Biodiv Environ Sci 13(3). ISBN No. 220-6663

Rock L, Buchanan C (2014) Social work education in the Caribbean: charting pathways to growth and globalization. In: Noble C, Strauss H, Littlechild B (eds) Global social work: crossing borders, blurring boundaries. Sydney University Press, Sydney

Sajise EE, Borromeo TH, Altoveros NC, Bon SG (2012) Traditional rice in Central Cordillera, Philippines: dynamics of on-farm management of varietal diversity. J Dev Sust Agric 7(1): 75–88

Tad-awan B, Sim J, Batani RS, Tad-awan BS, Laurean CP, Masangcay TD, Sagalla EJ (2015) Heirloom rice production in Kibungan, Benguet. Benguet State University annual report. Benguet State University, La Trinidad, Benguet

Tardio J, Pardo de-Santayana M, Morales R (2006) Ethno-botanical review of wild plants. Bot J Linn Soc 152:27–71

The Stories of Organic Farming Champions in the Philippines and Their Crop Diversity Practices

23

Gloria Luz M. Nelson, Girlie Nora A. Abrigo, and Rowena P. de Guzman

Abstract

Organic farming is sustainable because of its non-use of chemical fertilizers and pesticides. The benefits are a less polluted environment, less soil degradation, and better health for both the farmers and the consumers. Organic farmers from eight provinces in the Philippines were interviewed face to face, and their farms were visited to showcase the various organic farming practices they do to champion organic farming. This, in spite of experiencing low to no income during the transition period. Their stories describe the different crops cultivated, the organic farming technologies used, and the marketing strategies employed. It was observed that the organic farming champions in warm weather conditions planted different crops from those in cool weather conditions. However, regardless of the weather conditions, crop diversification and the use of organic farming technologies are commonly practiced. Furthermore, the stories of each of the 11 organic farmers interviewed highlighted some of their distinct practices. The study concludes that these farmers who champion organic farming are the heroes of tomorrow because of their vocation, advocacy and passion, that contributes to promoting sustainable development and meet the needs of the present and future generations of the growing Filipino population.

Keywords

Organic farming · Crop diversity · Organic farming champions · Philippines

G. L. M. Nelson (✉)
AA Program, Faculty of Education, UP Open University, Laguna, Philippines
e-mail: gmnelson@up.edu.ph

G. N. A. Abrigo · R. P. de Guzman
Department of Social Sciences, College of Arts and Sciences, University of the Philippines Los Baños, Laguna, Philippines

S. Ramamoorthy et al. (eds.), *Plant Diversity in Biocultural Landscapes*,
https://doi.org/10.1007/978-981-19-8649-9_23

23.1 Introduction

The challenge facing agriculture worldwide is the emission of greenhouse gas (GHG) from agriculture, forestry, and fisheries, which have nearly doubled over the past 50 years and are expected to rise by 30% by the year 2050, assuming that the farmers continue their current farming practices (Food Agriculture Organization, 2014, as cited in Asian Development Bank 2017). The Philippines is one of the countries that has the highest risk to different climate hazards (Institute for Economics and Peace 2019). In fact, this will affect 25% of Filipinos who are employed in the agricultural sector (Statista Research Department 2021). The change from conventional to organic farming introduced in 2010 aims to address the United Nation Sustainable Development Goals such as the climate change crisis (SDG 13) and responsible consumption and production (SDG12).

Organic farming is an environmentally sound and sustainable farming system that provides numerous economic, environmental, and social advantages to developing countries (Morshedi et al. 2017). Moreover, organic farming removes exposure to hazardous pesticides both in humans and ecosystems; contributes to better nutrition; ensures a healthy soil and soil microorganisms, which makes nutrients more readily available to plants; reduces genetic changes and insect immunity; costs less with the reduction of various farm inputs; and fosters more ecologically responsible approach (Tal 2018). Organic farming was also recognized as a potential alternative source of income for local farmers (Maohong 2018). In 2016, an international survey on organic agriculture put the Philippines among the highest organic agriculture producers numbering to more than 165,000 in 2016 (Japan International Cooperation Agency 2017).

Some of the important principles of organic agriculture is increasing biodiversity and adapting its practices to local conditions. It encourages the planting of local varieties and species of crops. The practices of crop rotation, multiple cropping, and intercropping contribute to crop diversity and help protect and preserve biodiversity (Gomiero et al. 2011). Since repetitive planting of similar crops on the same plot can deplete soil nutrients, planting a variety of crops through organic farming replenishes the nutrients in the soil as the different crops interact with the soil. This practice has been utilized by many farmers in controlling the nutrients and water intake of the soil, as well as in managing weeds and pests and lessening the disease and risk exposures of their crops (Chongtham et al. 2016). The factors contributing to the impact of organic agriculture on biodiversity have been explained in many studies as well. For one, Niggli (2015) emphasized that most of the studies that they examined showed that the elimination of pesticides, herbicides, and fertilizers has a significant positive impact on species diversity in organic farms. They also confirmed that diversified crop rotation or mechanical weeding contributes to biodiversity in organic farms. In addition, the studies of Tscharntke et al. (2021) and Arlauskienė et al. (2015) showed that mechanical weeding can replace herbicides to further sustain soil fertility and biodiversity. The reason for this is that higher weed coverage can benefit other organisms in the farm. This information related to organic farming practices and their benefits is emphasized in the trainings attended by organic farmers (Nelson et al. 2016, 2019).

Through organic farming, varieties and species of crops that have significantly decreased due to conventional farming can also be restored. Several studies found that organic agriculture contributed to higher levels of biodiversity particularly, on both plants and animals (Bengtsson et al. 2005; Grandi 2008; Gomiero et al. 2011; Underwood et al. 2011; Kavitha and Chandran 2017). The meta-data analysis of research publications on organic farming revealed that more than 80% of existing studies show that organic farming results in greater biodiversity compared to conventional farming systems (Rahmann 2011). For instance, an increase in biodiversity is still evident even in small farms, as farmers can specialize in crop production, horticulture, or growing medicinal herbs (Arlauskienė et al. 2015).

Despite all these environmental and consumer health benefits, organic farming faces a number of challenges. For instance, several studies have revealed how inaccessible organic foods are in terms of the cost, which is brought about by the lack of integrated supply chains and the processing of certification. In a study done in India by Deshmukh and Babar (2015), they also pointed out a lack of market information, difficulty in getting certification, capital-oriented policies of contractors and supply of organic supplements and bio-fertilizers. Dayoub and Korpela (2019) specified that organic farmers must complete specific training to independently engage in processing, packing, storing, organizing, and distributing organic goods, which all incur costs. They also added that in the context of Europe, before a farm may be certified, a 3-year transition phase with lower product pricing must be approved, from which farmers may face some difficulties. Ultimately, fraudulent organic products in the market are extremely destructive to the industry. All these issues pose a threat to the understanding and promotion of OA.

Farmers practicing organic and conventional farming practices have different perspectives and approaches depending on their local conditions (Chongtham 2016). Thus, the various farming methods and strategies have an impact on how farmers manage their farms and how they respond to shifting weather and market situations.

Unlike in the past studies, the present study provides a rich description of various crop diversity practices of champions in their day-to-day farming. Organic farming champions is a coveted title given by the authors because of the following reasons: they adopted organic farming in spite of the fact that it was not popularly accepted among their co-farmers; and some of them started even before the Organic Act Law was passed. It was evident from the stories of the farmers that they were able to influence their co-farmers who were reluctant to adopt organic farming. Thus, they are organic farming champions, because they became influencers in their farming communities. As the organic farming champions contribute to crop diversity, it is important to gain an understanding of the geographical context and their personal attributes. They are champions, since they are providing the best practices in organic farming. Because their stories in the life of being a farmer have been documented, future generations of farmers can adopt these practices to be able to feed the growing

Filipino population, which is at the present time the 13th most populous country in the world (World Population Review 2022).

23.2 Methodology

The study was part of a project funded by the Department of Agriculture–Bureau of Agricultural Research (DA–BAR). The locale of the project study is the nine provinces in the Philippines. However, the 11 organic farming champions interviewed in this study are from eight provinces, as shown in Fig. 23.1.

Organic champions were participants in the focus group discussions (FGDs), a method used to determine the overall knowledge on organic farming technology and organic agriculture programs in their respective provinces. From these participants, 11 organic farming champions were selected based on the following criteria: devoted part or their entire farms to organic farming; planted various crops; influenced co-farmers by conducting training; and their farms are learning centers. Majority of the farmers received certificates of recognition from national and local institutions. Complementing the interviews was the observations from farm visits (see also Fig. 23.1). The farms visited include the farms of the organic champions who were not interviewed before conducting the in-depth interviews, informed consent was obtained and duly signed by each interviewee. The transcriptions served as the data, which was in turn used to narrate the stories of the champions. Therefore, the day-to-day experiences of the organic farming champions were documented to know the different crops and organic farming technologies they use in cultivating their crops. There was no attempt to collect similar data from each of the organic farming champions as what is commonly done in doing case studies (Creswell and Creswell 2019). The write-ups of the interviews were reviewed and validated by the organic farming champions themselves.

23.3 Results and Discussion

23.3.1 Socio-economic Characteristics

The data shown in Table 23.1 are the profile of the organic farming champions and the characteristics, particularly the knowledge, attitudes, and practices of organic farmers, which are necessary to promote organic farming in the Philippines (Nelson et al. 2019). These characteristics of the farmers provide the context for their stories. The organic farming champions are from eight provinces in the Philippines. Two are from each of the provinces of Iloilo, Negros Occidental, and Negros Oriental and one each from Bukidnon, Camarines Sur, Cebu, Davao, and Tarlac. Only one organic farming champion is a female, and she is also the only one who is single. The mean age of the organic farming champions is 60 years. Farmers at least have one child while others have as many as six. Seven of the organic farming champions are college graduates. Everyone is an organic farming practitioner even

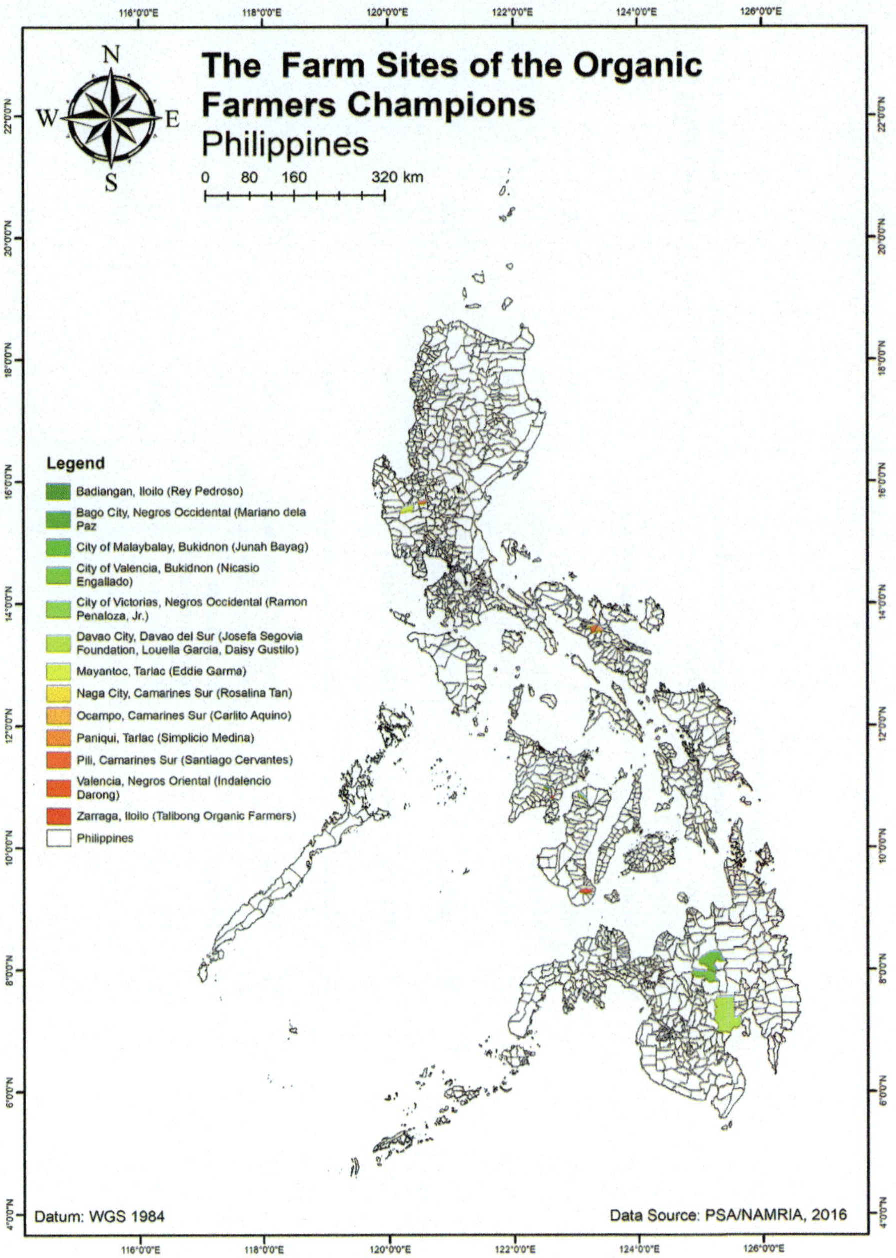

Fig. 23.1 Location of the organic farms

Table 23.1 Socio-demographic characteristics of organic farming champions

Organic Farming Champion (OFC)	Province	Age	Sex	Marital status	Number of children	Highest educational attainment	Year started organic farming	Farm size devoted to organic farming (ha)
1	Tarlac	54	M	Married	2		2009	0.9
2	Camarines Sur		M	Married	1	BA political science	1999	5.7
3	Iloilo		M	Married	2		2006	2.5
4	Iloilo		M	Married		AB political science	2004	1.0
5	Negros Occidental	65	F	Single	0	BS commerce	2006	0.025
6	Negros Occidental	75	M	Married	6	BS pharmacy	1998	20.0
7	Negros Oriental	52	M	Married	3(+)	BS agriculture	2004	1.5
8	Negros Oriental	56	M	Married		High school graduate	2000	2.7
9	Cebu	70	M	Married	3	Diploma in fisheries technology; BS education	2000	2.0
10	Bukidnon	54	M	Married	4	BS agriculture (undergraduate)	1997	3.0
11	Davao	53	M	Married		BS commerce	2004	5.0
Average		*60.3*						*4.03*

before the active promotion of organic farming in the Philippines. The organic farms are as small as 0.025 ha and as large as 20.0 ha. The small size farms are planted with five-leaved chaste tree (*Vitex negundo*) and the bigger farms are planted with more various crops.

There are several reasons why organic farming champions ventured into organic farming. For instance, Farmer 1 started it as a recreation or hobby. However, as time went by, it became a therapy for him and eventually a source of income. According to him, fate brought him into farming "*...naging ito po kasi yung linya natin na kung saan po tayo dadalhin ng ating kapalaran at ito nga po ay ang paghahalaman*". If fate brought Farmer 1 to organic farming, Farmer 2 has been interested in farming, even though he was serving as a soldier. Therefore, when he retired from the service, he went to farming full-time.

On the other hand, Farmers 3, 5, 9, and 11 converted to organic farming primarily for health and food safety reasons. In particular, Farmers 3 and 5 are advocates of the non-use of chemicals. Farmer 3 emphasized that his strategy is to be an organic practitioner first, then as a model to his co-farmers in the community. With this in mind, he closely monitors the farm laborers and trains them well, since organic farming requires intensive labor and technology transfer. Meanwhile, Farmer 5 firmly believes that using chemicals harms people. Because of this, she formulated herbal medicines using organic-based products from their crops. Farmer 11 also shared his realization of wanting to produce safe foods. Moreover, Farmers 3 and 9 cited other reasons why they went into organic farming. These include producing food products with "added value," helping consumers to produce organic products, helping others by providing employment, promoting an environment-friendly practice, generating more savings, increasing income, and minimizing production costs.

The other reasons provided by Farmers 3 and 9 are also similar to why Farmers 4, 6, and 8 started organic farming. Specifically, Farmer 4 ventured into organic farming mainly for economic reasons. He shared that, in his case, organic farming is beyond subsistence, as he can satisfy their family needs from their farm and earns a good income from their organic products. In addition, he can still work and earn from his consultancy on organic farming.

Meanwhile, despite the lack of knowledge in agriculture, Farmer 6 went to organic farming as part of his realization that their previous flower business contributed to the alarming water pollution problem brought about by their intensive use of chemical fertilizers. Therefore, they went into vegetable farming and vermicomposting after he and his wife did much research on organic farming.

Finally, the reason cited by Farmer 8 is related to the production system. He said that he wanted to restore the good quality of the soil. Since his goal is soil rehabilitation, he admitted that his soil required high organic inputs at the start but still lower than conventional agriculture. He also added that since he is not using chemicals, his production cost is lower, and his products are safe to consume and can be sold at a higher price. Similarly, Farmer 10, who has been practicing organic farming since the 1990s, produces more yields and earns more from it as he is not tied to the high cost of chemical inputs.

Regardless of the differences in their attributes and why they started organic farming, these farmers are the champions of organic farming in their provinces. Groups and organizations awarded some, and their peers widely recognized them because of their practice of organic farming.

23.4 Organic Farming Champions and Their Crop Diversity Practices

23.4.1 Vegetable Organic Farmer and His Clients

As chairperson of a producers' farmers organization, he is involved in the teaching of various non-government organizations (NGOs), where farmers gained some expertise. During the foundation day of their municipality, he received an award from the business group under the young men category.

The organic farming practice of this farmer started in 2009. His interest was inspired by his participation in the Organic Congress in their province through the efforts of the Office of the Provincial Agriculturist (OPA). His farm was visited by organic farmers from all over the Philippines as an activity related to the Organic Congress. Since Farmer 1 turned organic, he hardly gets sick. "*Hindi po ako nagkakasakit siguro po dahil sa over exercised na rin ako*" (Farming is a good exercise). "*Yung sa akin malinis kasi bago ako mag-almusal, nililinis ko po yung palayan ko. Kaya malinis yung mga pilapil. Ibig kong sabihin talagang alaga ko ang palay ko*" (Before breakfast, he makes it a habit to clean the *"pilapil"* or farm ridge of the land surrounding his farm).

Almost all the farmers in the organization started as backyard gardeners, whose average size is about 1500 m^2. In the second year, bananas (*Musa*) and mangoes (*Mangifera indica*) were added as additional income. Aside from the additional income from bananas (*Musa*) and mangoes (*Mangifera indica*), banana leaves were used in packaging leafy vegetables to keep their freshness. The farmers in his organization planted mixed vegetables that grow best in high temperatures called *pakbet* which is a popular dish in the northern part of Luzon. In their third year in 2011, they increased their rice farm area. He mentioned that black rice has very low sugar content which is good for those who have diabetes. Since then, Farmer 1 started cultivating black rice which he received from his co-farmer. The size of his black rice farm was 4000 m^2 in 2012.

Organic farming practices are different from conventional farming. Farmer 1 schedules the planting of his various crops in his farm. For example, he knows that it takes them 3 months to harvest Bokchoy/Snow Cabbage (*Brassica rapa, var. pekinensis*). He plants Squash (*Cucurbita maxima*) between October and December. Eggplant (*Solanum melongena*), on the other hand, is planted between February and June. The demand for lettuce (*Lactuca sativa*) is more regular, since it is a basic ingredient for vegetable salad. The decision on what to plant is dependent upon the customers. They are also instrumental in motivating him to go full-time in vegetable farming. He called it "market decision". The kind of vegetables being ordered is rotated from those that are used for salad packs, *pakbet*, and *sinigang*. The salad is a mainstay. The company also prefers parsley (*Petroselinum crispum*) and herbs, such as marjoram (*Origanum majorana*), tarragon (*Artemisia dracunculus*), and basil (*Ocimum basilicum*). The seedlings that they used are being bought in Manila. They are also consulted from time to time as to the best place where to get seedlings.

Farmer 1 shares that he absolutely does not apply chemical fertilizer and only uses vermiculture technique. The waste products are collected from kitchen and farm wastes. African nightcrawlers (*Eudrilus eugeniae*) are used to eating up all the waste products. In turn, he mixed the decomposed matter with microorganisms called vermicompost which is used as fertilizer. When mixed with molasses, the mixture becomes *vermi ethanol* which serves as plant vitamins and insecticide. However, according to Farmer 1, the spraying of microorganisms is as important as good healthy soil. It took him 3 years to rehabilitate his soil. Until the time of the interview, Farmer 1 is not confident that the soil has been fully rehabilitated which means free from toxins or harmful bacteria. To rehabilitate his farm soil, Farmer 1 regularly applies carabao or buffalo (*Bubalus bubalis*) and cow (*Bos taurus*) wastes and also uses carbonized rice hulls (burnt rice hull) mixed with animal waste. Farmer 1 said that he is not necessarily hard working but saw farming as his fate. However, being a migrant to the place, he is thankful that his new community has given their trust and recognized his leadership by electing him Chairperson of the farmers' association, where the members come from the four barangays.

The organic vegetable products that are profitable are lettuce (*Lactuca sativa*). He can still produce his leafy vegetables even in rainy months because he customized a greenhouse built for that purpose. He also said, because of this greenhouse, he can still earn, "*Basta kung panahon ng tag-ulan meron kang greenhouse, malaki talaga ang kita kahit sa mustard lang during rainy season, dahil may greenhouse ako, abot ako sa six thousand.*" Greenhouse is a necessity and can sustain the same income earned during dry months. His harvest from mustard during rainy month's amount to Php 6000.00 (108 USD). The Chinese-like native vegetables such as malabar nightshade (*Basella alba*) and talinum (*alinum triangulare*). He earns Php 4000.00 to Php 8000.00 (72 USD to 144 USD) from his vegetables. His average harvest per week for each commodity is as follows: 50 kg bokchoy (*Brassica rapa, var. pekinensis*), 10 kg lettuce (*Lactuca sativa*) and 20 kg Nalta jute (*Corchorus olitorius*). At present, he cultivates black rice *(Oryza* sativa) and harvests 45 cavans per cropping season. His total weekly income is estimated between Php 1500.00 and Php 2000.00 (27 USD to 26 USD).

Instead of a middle person or trader, an international food company helps them market their products. Walk-in buyers are priests, bankers, and foreigners while occasionally groups of visitors enjoy his produce of sweet potatoes (*Ipomoea batatas*) and corn (*Zea mays*). Farmer 1 credits his accomplishments in meeting his target of producing a variety of crops from the support he receives from various groups. "*Kung hindi ako tinutulungan, hindi ko talaga kakayanin*" (If not for these groups, I will not meet my target yield in farming).

23.4.2 Organic Farming is a Second Calling

It was in 1999 when Farmer 2 was introduced to organic farming, while he was still with the military. He was then farming for pleasure, but when he retired in 2007, he became a full-time farmer and started to cultivate his 5.7 has of farmland. Three hectares of which have been certified by Organic Certification Center of the Philippines (OCCP). He plants organic rice (*Oryza sativa*) of black, white, and red varieties for higher income. Aside from rice, he also produces vegetables, such as lettuce (*Lactuca sativa*), tomatoes (*Solanum lycopersicum* L), bittergourd (*Momordica charantia*), and squash (*Cucurbita maxima*).

Most of his farming practices are from his formal training held in the farmer association training center in August 2012, but he also learned informal organic farming practices since 2010. Unlike conventional farming, organic farming uses organic fertilizer which is made from a mixture of bran and fermented vegetables (fermented solution) and indigenous microorganism (IMO). The use of such beneficial organisms serves not only as fertilizer for the crops, but it also makes the soil fertile. Because he makes his own fertilizer, he does not spend much for farm inputs. Most especially, these farm inputs are found in his own farm, such as making fermented solutions from papaya (*Carica papaya*) and squash (*Cucurbita maxima*) mixed with molasses syrup. He maintains a greenhouse (see Fig. 23.2), where he has some seedlings from seeds from his harvest.

Fig. 23.2 Greenhouse of OFC 2 for seedlings

Farmer 2 is generally satisfied with organic farming, especially because he is proud that his produce is of better quality and is not poisonous (meaning no inorganic fertilizer). In addition, he took it upon himself to introduce the benefits of organic farming to his co-farmers, and when they start farming organically, he becomes happier, "... *mas masaya ako kapag ibang famers ay gumagaya*." (I am happy when other farmers also start practicing organic farming). Part of his advocacy is to encourage, influence and convince his co-farmers to convert to organic farming.

23.4.3 The Story of Conversion to Organic Farming

Farmer 3 has been known as a rice (*Oryza sativa*) and corn (*Zea mays*) farmer since 1998, but it was only in 2006 when he converted to growing organic rice after (*Oryza sativa*) he attended seminars conducted by NGOs and the Department of Agriculture (DA). There was also on-the-job training on appropriate technology. During the first year of conversion, Farmer 3 started with 500 m^2 of vegetables (vegetable gardening). In his second year, he has 0.5 ha rice land and 0.5 ha planted with vegetables. In his third year of conversion, he had 0.75 ha planted with black and red rice (*Oryza sativa*); 0.25 ha planted with vegetables; and 0.5 ha planted with fruit trees with a total of 1.5 ha. At present, he has 2.5 ha: 0.75-ha of rice (*Oryza sativa*) and 1.75 ha are for poultry, fishpond, native chicken (*Gallus domesticus*), and duck (*Anatidae*) raising. His conversion to organic farming took all 3 years. The gradual conversion was needed to neutralize the soil using compost and fermented solution.

There are many lessons that he learned during the process of conversion. For example, during his first year of conversion, he lost about 50% of his original income compared to his earnings in the previous year as a conventional farmer. The temptation to go back to conventional farming was huge. The low production he experienced was due to the acidity of the soil and the lack of farm inputs necessary for production, and the rehabilitation took 3 years and more. Based mostly from his experience, he believes that OA needs a "mind conditioning culture". Today, Farmer 3 is enjoying the fruits of his labor and the operation of his farm is now very economically good, since he utilized organic fertilizers. Farmer 3 now grows vegetables: *pakbet*, *chop suey*, lettuce (*Lactuca sativa*), cabbage (*Brassica rapa*, *var. pekinensis*), cucumber (*Cucumis sativus*), and fruit trees: rambutan (*Nephelium lappaceum*), mangoes (*Mangifera indica*), lanzones (*Lansium domesticum*), and honeybee (*Apis* sp.) production. There are several benefits of organic farming for Farmer 2. First, organic farming eradicates chemical residues, which in turn fight global warming and the effects of climate change. Second, organic farming promotes good health, because it is chemical-free, and third, organic farming afforded him to support the educational needs of his school-aged children.

23.4.4 An Owner of a Self-Sufficient Farm

Farmer 4 attended a variety of trainings and seminars, including organic hog raising, organic nursery establishment, micro-organisms, and organically growing specific crops. After learning about organic farming, he and his wife decided to stop working as employees and focused on organic farming in 2004, "*kasi tumatanda na kami kaya nagdecide na kami ni misis na mag-organic agriculture at hindi na naging empleyado*" (My wife and I are getting older, so we decided to venture into organic agriculture).

Farmer 4 produces the following: taro (*Colocasia esculenta*), ginger (*Zingiber officinale*), squash (*Cucurbita maxima*), silk squash (*Luffa acutangula* Linn.), bitter gourd (*Momordica charantia*), eggplant (*Solanum melongena*), lady's finger (*Abelmoschus esculentus*), beans (*Phaseolus vulgaris*), cabbage (*Brassica rapa*), radish (*Raphanus sativus*), and tomatoes (*Solanum lycopersicum* L) (see Fig. 23.3). In addition, they raised chicken (*Gallus domesticus*), livestock, and tilapia (*Oreochromis niloticus*). Among his produce, taro (*Colocasia esculenta*) is the most profitable during dry months while tomatoes are for rainy months.

According to him, there are economic reasons for converting to organic farming, since they are self-sufficient and derive income from his organic produce and as consultant to organic farmers. Organic farming provided him with a number of opportunities. He says that "*If you are first (to engage in organic farming) in your*

Fig. 23.3 Seedbed of various crops in the organic farm of OFC 4

place, you get funding and linkages." He was able to help mainstream organic farming and became a board member of National Organic Agriculture Board (NOAB). There are health benefits, since "*alam mo kung saan galing ang kinakain mo, walang chemicals, walang pesticides*." (You know that the food you consumed is chemical-free). On the other hand, the social benefits are his involvement in the development of organizations and people's organizations where "*may makakausap ka tungkol sa parehong bagay (OF)*." (where you can discuss common topics). He believes that organic farming mitigates the effects of climate change.

From the outset, he converted the entire area to organic agriculture. "*hindi mo masasabi na organic agriculture kung isa lang. Dapat integrated. May source ka ng carbon, ng nitrogen, at iba pang mga kailangan sa iyong farm. Dapat controlled mo lahat ng inputs and majority ng inputs para ma-guarantee mo na organic lahat*." (You can also say that you have an organic farm if you practice integrated farming. A farmer should have control on almost all of the organic sources of inputs used in the farm). He says that there were no big physical adjustments to their farm, as they were originally using the natural farming method. The adjustments that they went through were more mental or emotional in nature. They had to engage in various farming practices as a part of the conversion.

To start with, there were noticeably less inputs, as they had to do away with chemical fertilizers and pesticides. They began to produce their own seeds. Aside from crops, they raised free-range chickens and hogs which they fed with their own formulation of feeds. One problem they encountered during the formulation of their feeds was that the source of feed components may not be organic. To solve the problem, they ferment the feed components using a trade secret in order to make it organic. They also kept an organic fishpond stocked with tilapia (*Oreochromis niloticus*) and indigenous fish which they fed with water crest (*Ipomoea aquatica*), organic plankton (which they grow by soaking vermicast in the pond).

As a proactive farmer, he simply does not wait for insects and pests to destroy his crops but created a "bio-buffer" zone composed of plants with different repellent properties including forage, Madre cacao (*Gliricidia sepium*), lemon grass (*Cymbopogon citratus*), and other insect repelling plants. If the insect is able to get through the buffer, they use an attractant laced with spices to kill the insects or still resort to manual removal of insects.

23.4.5 From Five-Leaved Chaste Tree Producer to an Herbal Medicine Inventor

The conversion of this farmer started when she noticed that a lot of her farmers and even her co-employer had developed cancer. She became more health conscious after learning from attending trainings and seminars on organic farming about the benefits of non-use of chemicals in farming. It was in 2006 when a small percentage of their land was converted to organic farm. She began growing five-leaved chaste tree (*Vitex negundo*) plant organically: ten during the first year, 20 in the second year, 40 in the third year, and currently 60. The 60 plants are spread in roughly about

250 m^2. In practicing organic farming, she is very conscious not to use any kind of chemical fertilizer, which is substituted using peelings of fruit and cut grass.

Farmer 5 described herself as a pharmacist by profession and she knew the harmful effects of chemical drugs, which is also the reason why she established an herbal company. Her main product is a mixture of five-leaved chaste tree (*Vitex negundo*), lemon grass, and ginger. This Food and Drug Administration (FDA)-registered and Bureau of Food and Drugs (BFAD)-approved herbal medicine is a mucolytic and expectorant used as a remedy for cough and sore throat. She and her partner market their products in Manila, Iloilo, Bacolod, and in a variety of Chinese stores. According to her, some doctors prescribed their herbal medicines to patients. The income from herbal medicine is about Php 30,000.00 (540 USD) in their first month. In 2011, their earnings increased to Php 62,500 (1125 USD) per month. However, this amount was not sustained in the following year, because the approval of the Certified Product Registration (CPR) was delayed.

She cited numerous benefits that one can gain through organic farming. First, organic farming is financially rewarding, as they can mark up the prices of their products. Second, there are health benefits, because there are no chemicals or carcinogens in the food that you are eating. Third, organic farming has social benefits. She met a lot of new people, and her network has become bigger. She now has more possible sources of organic materials and even possible clients for her organic products.

23.4.6 A Farmer Who Values the Common Good over His Personal Interest

A farmer of integrity and a risk taker. He discontinued his cut flower business despite its being financially profitable, because he found out that excessive use of fertilizer is contaminating the water source and has a long-term detrimental effect on the health of the people in his province. "*I will just tell you the effects on our people, farm workers. When we were in the flower business, we were spending 30 thousand pesos a month for hospital bills. We paid for that for the hospitalization of our farm workers. Today, it has not only been reduced but eliminated-entirely (zero). I think that my workers are healthier and happier now because we don't use chemical fertilizers*". His farm workers enjoy better wages, work in a chemical-free environment and are therefore healthier compared to farm workers in conventional farming.

When he converted to organic farming in 1998, he had zero income from his 1-ha vegetable farm which was infested with insects and worms. On his second try the following year, his crops were infested by grubs. In his third year as an organic farmer, the soil in his farm is healthier and considered ecologically balanced through the practice of vermicomposting (see Fig. 23.4). The first 2 years were a learning period. Now, he is a proud owner of a 4-ha organic vegetable farm. Aside from his vegetable farms, he has a 16-ha orchard farm, where he grows bananas (*Musa*), lanzones (*Lansium domesticum*), rambutan (*Nephelium lappaceum*), durian, malay taray (*Artocarpus odoratissimus*), and pomelos (*Citrus maxima*). The total 20-ha

Fig. 23.4 Vermicomposting plot in Negros occidental

farm is now in full productivity. A small portion of his farm is planted with Napier grass (*Cenchrus purpureus*) which they used to feed the African Night Crawlers (*Eudrilus eugeniae*), an important element in vermicomposting technique. Each month he produces 10 tons of vermicompost which he uses to fertilize his farm. Some of the surplus is sold to interested buyers. He produces organically grown free-range chickens. His knowledge on organic farming was enriched due to his attendance to various local and international seminars.

There is plenty of opportunity in organic farming. First, according to him, the demand for organic vegetables has increased now in the market. He sells his organic products in Cebu and Iloilo. During the conduct of the interview, he has a shipment of 50 kg of lettuce in a supermarket in Cebu. Each day, he harvests 25 kg of lettuce. He also harvests carrots and cucumbers. His delivery trucks are not yet refrigerated. The increasing demand for organic products is due to increasing awareness of the benefits of organic products. His products are sold in Market! Market! in Manila and Metro Gaisano. The demand for lettuce is 200 kg a month, but he can only produce 150 kg.

As an organic farmer, Farmer 6 is a household name. He was featured three times in an Agriculture magazine. Every now and then he gets an invitation to appear on TV as a resource person in organic farming.

23.4.7 A Green Thumb-Organic Farmer with a 'Midas Touch'

He was awarded the 2012 Outstanding Farmer for High Value Commercial Crops (HVCC) for Organic Agriculture, where he was featured in the Manila Bulletin last October 11, 12, and 13, 2013. At the age of 52 years, he has been an organic farmer since 2004. He cultivates 1.5 has for organic farming.

He has attended several trainings and seminars related to organic agriculture including seminars sponsored by the Agricultural Training Institute (ATI), such as internal quality system and food handling. He converted to organic farming, because he was planting and selling herbs and was advised to stop using chemical fertilizers. Aside from attending seminars, there were German and Korean visitors who influenced him about the importance of organic farming. His practices include the use of cover crops; composting cacao (*Theobroma cacao*), avocado (*Persea americana*), and devil weed (*Chromolaena odorata*) leaves, animal manure and carbonized rice hull, and molasses; fermentation of seaweeds, madre cacao (*Gliricidia sepium*) and molasses, and animal manure from goats, cows and pigs; and the application of lime after a week. He sells fruits and seedlings of strawberry (*Fragaria* × *ananassa*), a variety of herbs including rosemary (*Salvia rosmarinus*), tarragon (*Artemisia dracunculus*), thyme (*Thymus vulgaris*), Asian and Mexican coriander (*Coriandrum sativum*), sage (*Salvia officinalis*), seven types of basil (*Ocimum basilicum*), Vicks menthol peppermint (*Mentha* × *piperita*), three kinds of green tea (*Camellia sinensis*), lemon balm (*Melissa officinalis*), oregano (*Origanum vulgare*), stevia (*Stevia rebaudiana*), parsley (*Petroselinum crispum*), celery (*Apium graveolens*), Australian leeks (*Allium porrum*), chives (*Allium schoenoprasum*), Chinese onion (*Allium chinense*), dill (*Anethum graveolens*); and vegetables such as asparagus (*Asparagus officinalis*), cucumber (*Cucumis sativus*), broccoli (*Brassica oleracea var. italica*), cauliflower (*Brassica oleracea var. botrytis*), bok choy (*Brassica rapa subsp. chinensis*), mustard (*Sinapis alba*), arugula (*Eruca sativa*), five (5) kinds of spinach (*Spinacia oleracea*), red cabbage (*Brassica oleracea var. capitata f. rubra*), and various kinds of lettuce (*Lactuca sativa*). The abovementioned crops are usually planted in low temperature place such as the farm of the organic farming champion (see Fig. 23.5).

Fig. 23.5 Various crops in the organic farm of OFC 7

When he began practicing organic farming, the income increased by as much as 50% due to higher cost of organic produce as compared to non-organic produce. Moreover, he already has a number of steady clients for his organic produce. He considers everything he plants as profitable, liken to the touch of Midas, especially the herbs which he sells in pots and by the kilogram. For example, thyme (*Thymus vulgaris*) sells for as much as Php 900.00 (16.2 USD) per kg; rosemary (*Salvia rosmarinus*) and tarragon (*Artemisia dracunculus*) at Php 700.00 (12.6 USD) per kg, strawberry (*Fragaria* × *ananassa*), asparagus (*Asparagus officinalis*), parsley (*Petroselinum crispum*), and coriander (*Coriandrum sativum*) at Php 400.00 (7.20 USD) per kg; peppermint (*Mentha* × *piperita*) and basil (*Ocimum basilicum*) at Php 300.00 (5.4 USD) per kg; and celery (*Apium graveolens*) at Php 250.00 (4.5 USD) kg. Aside from organic produce, he also raises Anglo-Nubian goats which he feeds with molasses and corn meal with legumes. The raising of goats is also profitable as they command a high price whether sold alive (Php 5000.00 or 90 USD at 7 months) or sold for their meat (Php 350.00 or 6.3 USD per kg). He supplies organic produce to various hotels, restaurants, and resorts in Dumaguete and Cebu. He also has walk-in customers as he is able to sell some of his produce when people come to visit his farm.

According to him, the practice of organic farming can lead to an abundant supply of toxin-free organic products and employment opportunities. He also cites several benefits that he has gained, since he began organic farming. First, the necessary farm input became less expensive, but the organic products can demand higher prices in the market. The increase in his income has enabled him to send three of his children to college and purchase a vehicle. Second, the practice of organic farming and the use of organic products can lead to improved health. His own family makes use of their organic produce ensuring that they themselves eat healthy food. Third, he meets more people and gains friends and even clients. He became a private consultant on organic farming for foreigners. Engaging in organic agriculture also provided him with opportunities to join various activities and attend trainings.

23.4.8 Organic Rice Producer

Rice (*Oryza sativa*) is the staple food of Filipinos. At the age of 56, he started farming in 1989 and practiced Balanced Fertilization Strategy (BFS) in 2000 and became a System of Rice Intensification (SRI) farmer in 2011. In 2008, he became a local farmer technician and the caretaker of the 2.7-ha OF/SRI demo-farm owned by his hometown. He is a tenant of a 1 ha irrigated land, where he is a rice seed grower of PSB Rc18 variety. He sells rice produced from his farm and from the municipal demo production area. Registered rice seeds from the municipal demo farm are sold at Php 1600.00 (28.8 USD) per 40 kg per bag, while certified seeds of Rc 18 from his farm are sold at Php 1200.00 (21.6 USD) per 40 kg per bag (Php 30.00 or 0.54 USD per kg). At present, there is a pending proposal to increase the price of organically grown rice by Php 5.00 (0.09 USD) per kg. Organic rice is sold to a private marketing company at PhP 34 (0.51 USD) per kg.

He converted to organic farming in 2000 to increase his income, restore the quality of the soil, and produce safe food (without chemicals). Initially, he had to apply high amounts of organic inputs to restore the fertility of his soil, but since these inputs are available on his farm, the cost is still lower as compared to that of chemical farming. Eventually, his income increased because of the lower cost of farm produced inputs and higher production.

To increase his knowledge and skills in organic farming, he attended numerous trainings/seminars on organic fertilizer production, vermicomposting, modified rapid composting, PalayCheck system, and hybrid and commercial rice production. He believes that trainings help farmers in venturing into organic farming, and from these trainings, a farmer can help his fellow farmers.

In terms of the benefits of organic farming, he mentioned that at the start, income will decrease as a result of decrease in yield. He experienced a decrease in income when he first converted to organic farming, but it eventually increased, since organically grown rice has heavier seeds and has higher milling recovery (60–70% as compared to 55–60% for inorganic rice) due to less unfilled grains. Healthwise, organic rice does not result in stomach acidity. Economically, organic products can sustain the needs of his family. At the start, expenses were higher due to the large volume of organic inputs required, while the production was low; thus, income was also low. At present, the expenses in organic farming are lower than conventional, thus income increased.

23.4.9 Organic Farmer Who Practices Intercropping

At the age of 70, he converted his farm into an organic farm mainly for health reasons. He first started as a part-time farmer and became only a full-time farmer in 1990. In 2006, his farm was converted to an organic farm with banana as his main crop, while jackfruit and mango were intercropped with ginger in his 2-ha lowland area and mahogany in another 2-ha upland area. He later eliminated his mango (*Mangifera indica*) and jackfruit (*Artocarpus heterophyllus*) and concentrated on the production of organic banana (*Musa*), since this is the most profitable crop. Aside from lakatan (*Musa acuminata*), he also maintains other varieties of banana such as Cavendish (*Musa acuminata Colla*), senyorita (*Musa acuminata* '*Señorita*'), latundan (*Musa acuminata* × *M. balbisiana*), and also mondo (*Musa* sp.) for home consumption only. He produces vegetables and other crops such as eggplant (*Solanum melongena*), tomato (*Solanum lycopersicum* L), spinach (*Spinacia oleracea*), Malabar nightshade (*Basella alba*), ginger (*Zingiber officinale*), turmeric (*Curcuma longa*), and calamansi (*Citrus* × *microcarpa*), but these are all for home consumption only. He also has root crops planted in sack beds. When his banana harvest is not fit for the market, especially when exposed to drought, these are just boiled or processed into chips for home consumption.

All his organic products are profitable. An organic banana (*Musa*) is being sold at PhP 1.00–3.00 (0.02–0.05 USD) per piece. Bananas are sold to private buyers and the African Nightcrawler (*Eudrilus eugeniae*) and vermicast to private buyers and the Local Government Units (LGUs). The DA is the buyer for banana (*Musa*) suckers. He has a regular direct buyer which he supplies with 5000 pieces of banana per week. The requirement of the buyer is 10,000 pieces or one truckload per week. This is why he is contemplating on expanding his production. He also sells banana suckers at PhP 8.00 (0.14 USD) per piece, vermicast at PhP 5.00 (0.09 USD) per kg and African Night Crawlers (*Eudrilus eugeniae*) at PhP 500.00 (9 USD) per kg.

Organic farming provides additional income and leads to lower expenses, because raw materials for vermicompost such as wastes are readily available on the farm and from the house. Other benefits are longer life and safe food. He also helps his neighbors by giving them work on his farm that helps for the schooling of their children. There is also a higher demand for organic products and people are slowly realizing the benefits of organic farming. To encourage other farmers to go into organic farming, he sells his vermicast/vermicompost at a price which is lower than the prevailing market price.

23.4.10 An Epitome of an Organic Farmer: Entrepreneur

A multi-awarded farmer, he kept his award certificates in albums which he proudly presents to visitors who visit his farm. His most recent award was the 2014 National Organic Achievers Awardee (NOAA) for small individual category. His farm was inherited from his parents. Through time he converted it into an integrated farm. As an organic farmer, he produces his own fertilizers, and his irrigation water is sourced out from the river that flows freely into his rice fields.

His success as an organic farmer can be attributed to many factor: First, the numerous national and international trainings and seminars he attended. Second, he also has formal training in organic farming, since he studied agriculture in college in 1987. While in college, he learned to plant 1000 eggplants (*Solanum melongena*) in a 1200 m^2 lot. He also planted bitter gourd (*Momordica charantia*) in 1989. He, however, did not get a degree in agriculture. It was in 1997 when he became a full-time organic farmer, because according to him, it is good for his health and a good source of income. At present, he has 3 ha of organic rice farm that is family owned. Third, his farm practice is described as a combination of the following practices—natural farming, organic farming, permaculture (permanent culture), and biodynamics. He defined natural farming as farming using fermented juices, while ecological farming is without intervention, where it can be useful for some selected plants. Organic farming, according to him, is the use of organic inputs, such as vermicast, natural compost, liquid fertilizers, and green manuring. Permaculture, on the other hand, is appropriate for fruit trees and it is related to design in the farm that is permanent. He also has a fishpond, where he used to catch and deposit rain water. The pond has a filtration that is used to water his vegetable crops.

The integrated farming or integrated diversified organic farming system (IDOFS) includes organic and biodynamic inputs. He mentioned he uses cow manure as fertilizer, but the manure is not directly applied to his crops but what he does is to bury the manure for 6 months before it can be used as fertilizer. This practice is to prevent the growth of salmonella, commonly found among raw fresh vegetables. In addition, he uses a biodynamics calendar to serve as a guide for the best time to plant. For example, there is a different timing for planting leafy vegetables, fruit bearing plants, root crops and flower bearing plants. He also looks at the position of the moon. Several plants were planted to serve as insect repellant, i.e., cosmos (*Cosmos bipinnatus*), lemon grass (*Cymbopogon citratus*), and fodder cane (*Saccharum spontaneum* Linn.). Bamboo (*Bambusa vulgaris*) with many spikes, and coconut (*Cocos nucifera*) leaves were used as animal traps.

Farmer 10 is known not only as an organic farmer but as an entrepreneur. As a member of organic producers and processors, he was able to market his produce, mainly red and black rice (*Oryza sativa*). Furthermore, he is into "value added products" where he bakes *cookies-otaps*, *barquillos*, piaya, etc using organic rice as the main ingredient. He distributes them to shopping malls, fast-food centers and hospitals, because his products are known to have low glycemic index. He solicited the help of the Department of Trade and Industry (DTI) in packaging his baked products.

Organic farming can only be sustainable if diversification is practiced. He further explained that with only one crop, there will be no substitute when that one and only crop gets denuded with pests and other diseases. The secret of organic farming is a proper mindset reflected in this statement: "farming is for health and not for money". One has to change his/her lifestyle also. Applying chemical according to him is like a sleeping time bomb, when something triggers, it will explode into cancer in the body, and cancer to the pocket. The doctors, according to him, also agree that organic herbal products are good for the body. His last advice is organic farming should involve "*WISE thinking*", meaning W (*W*ater must be conserved), I (*I*nputs must be minimized), S (The practice must be *s*ustainable), and E (Farmers must be an *e*ntrepreneur). Without the latter, according to Farmer 10, organic farming is nothing.

23.4.11 A Proud Son of a Farmer

"*Anak po ako ng magsasaka*". As a son of a farmer, he grew up in the farm and has never stopped doing farm chores. He only started farming after his optional retirement from the government service, where he had worked for 25 years. It was in 1996 when he acquired the rights (no title) to till his own farm. At that time, it was easy to acquire rights which, according to him, was not illegal and at the same time less expensive compared to the titled land. Eventually, he sold his rights to the untitled land which he used to buy titled land. In spite of being a son of a farmer, he did not inherit any farm, but his farmland was procured from his hard-earned savings as a government worker. It was in 2004 when he was introduced to organic farming. He now owns 5-ha land. As an organic farmer, he attended several trainings which as far as he can remember include PalayCheck System, Orchard (on-going since 2008), High Value Commercial Crops: banana (*Musa*), pomelo (*Citrus maxima*), durian (*Durio zibethinus* L), mango (*Mangifera indica*), coconut (*Cocos nucifera*), and palay (*Oryza sativa* Linn)—intercropping (2012). It was way back in 1992 that he started reading about farming. He started making his own organic concoctions, vermicast, vermi tea and Bokashi. One advantage of using organic compost as a fertilizer is that the fruit is usually sweeter. His farm is diversified. He raises pigs (*Sus domesticus*), goats (*Capra aegagrus hircus*) and cows (*Bos taurus*). He sells coconut (*Cocos nucifera*), bananas (*Musa*), and durian (*Durio zibethinus* L) through traders or middlemen. He is still rehabilitating his Pomelo (*Citrus maxima*). His vegetables are sold in the public market.

Sodium chloride is important to all living creatures as long as the amount is right. Salt is inexpensive and readily available. He applies it in his bananas, coconut tree and his other fruits crops which he broadcasts freely at about 1 m from the hill. Most plant disease is caused by fungus, and he has proven in his experiment that salt is an effective way to eliminate fungus from his crops.

When he was asked about the opportunities of organic farming, he mentioned the priority on health and in terms of savings on farm inputs. He said that he did not encounter difficulty in convincing other farmers, because he applied his knowledge. In addition, he said that he is healthier now that he was into organic farming compared to when he was working in a company. He is self-sufficient due to his backyard vegetable gardening.

23.5 Discussion

The distinct characteristic of the organic farming champions is that farming is not only a source of income but more of a vocation and advocacy. Organic farming is a second career for most of them, since more than half have college degrees not related to agriculture. In engaging in organic farming, these organic farming champions are trainers and use their farms as learning centers. Regular attendance to trainings and seminars is one of the motivations of the organic farming champions to continuously engage in organic farming. The non-use of chemicals in nutrient and pest management is another important lesson that they had to learn and to adjust. In addition to undergoing intensive trainings, becoming a member of a farmers' organization gives them the privilege to sell their products in organic trading posts and markets.

In the same light as in the study of Gomiero et al. (2011), organic farming champions in this study learn to cultivate variety of vegetables, staple crops, fruits, root crops and herbs. The type of vegetables that are grown depends upon the characteristics of their farm location. Organic farming champions in lowland areas in warmer weather conditions plant different crops compared to the farmers in the upland with cooler weather conditions. Most of the farms cultivated by the organic farming champions have a combination of the following: annual crops, vegetables and root crops; perennial crops, such as fruit and agroforestry; herbal plants; and fish and farm animals. The integration of crops and livestock raising has also been found among organic farmers outside the Philippines (Arlauskienė et al. 2015; Tscharntke et al. 2021; Gomiero et al. 2011).

In the findings of Arlauskienė et al. (2015), farmers can specialize in crop production and growing medicinal herbs. Meanwhile, the organic farming champions in the study intensify production on specific crops. The importance of rice as a staple crop led an organic farming champion to concentrate on planting rice, including pigmented rice. The medicinal value of five-leaved chaste tree (*Vitex negundo*) and other medicinal plants motivated another organic farming champion to use these crops as ingredients for her organic herbal supplements.

In cultivating various crops, the organic farming champions follow crop management practices consistent with organic farming. In lieu of chemicals, organic farmers make their own fertilizers using a variety of techniques, such as composting, vermicomposting, and making foliar and liquid fertilizers. They, likewise, prepare their own pesticides and insecticides using indigenous herbal extracts. In terms of pest management, they employ a buffer zone of plants with insect repelling properties. The organic farming champions are committed to strictly follow the organic farming practices to achieve a more sustainable livelihood and environment, which is consistent with what Tal (2018) argued that organic farming is an ecological and responsible approach of removing the use of chemical inputs. These practices, in turn, contribute to species diversity in organic farms (Niggli 2015).

Organic farming provides the organic farming champions an opportunity to engage in more diversified market endeavors. The crops that are commercially produced are rice and high value crops. There is a need for organic farming champions to continue developing their diverse products to gain a competitive edge. Some organic farming champions had ventured into value-adding activities, such as producing baked goods, herbal supplements and animal feeds. Other organic farming champions sell their vermicast, vermicompost, compost and their natural fermented solutions. The stories which describe the various farming activities of the organic farming champions can be the model for the next generation of organic farmers.

23.6 Conclusion

The stories of organic farming champions demonstrated that crop diversification and other related practices, such as intercropping, multiple cropping, and non-use of chemical fertilizers and pesticides, are environmentally friendly that helps in mitigating climate change crisis. Organic farming is a lifestyle that resulted in a change in their outlook, where health is more important than wealth. There is no immediate reward in organic farming, but the long-term effect is the achievement of the twin goals of economic growth and sustainable development aspired by an agricultural country, such as the Philippines. At the same time, organic farming fosters the achievement of the 2030 United Nation Sustainable Development Goals of responsible consumption and production (SDG12) and positive action to mitigate climate change (SDG13).

Acknowledgements The authors recognized with deep appreciation the valuable contributions and cooperation of the organic farming champions from the following provinces: Tarlac, Camarines Sur, Iloilo, Negros Occidental, Negros Oriental, Cebu, Bukidnon and Davao.

Special thanks to the Department of Agriculture–Bureau of Agricultural Research (DA–BAR) for funding the project and for the trust, confidence and belief that social scientists can also contribute to the knowledge and understanding of organic agriculture.

References

Arlauskienė A, Kadžiulienė Ž, Deveikytė I, Šarūnaitė L, Semaškienė R, Karbauskienė E et al (2015, June) Crop diversity and productivity under organic farming in Lithuania. In: Proceedings of the Nordic view to sustainable rural development, the 25th NJF Congress, Riga, Latvia. p. 16–18

Asian Development Bank (2017) Achieving sustainable development goals through organic agriculture: empowering poor women to build the future. https://www.adb.org/sites/default/files/publication/384836/swp-15.pdf

Bengtsson J, Ahnström J, Weibull AC (2005) The effects of organic agriculture on biodiversity and abundance: a meta-analysis. J Appl Ecol 42(2):261–269

Chongtham IR (2016) Understanding crop and farm management unpublished doctoral thesis. Swedish University of Agricultural Sciences, Uppsala

Chongtham SK, Singh RP, Singh RK (2016) Effect of crop establishment methods and weed management practices on weed. Indian J Agron 61(1):33–36. https://doi.org/10.5958/2348-7542.2015.00003.0

Creswell JW, Creswell JD (2019) Research design: qualitative, quantitative, and mixed methods approaches. Sage Publications, Thousand Oaks, CA

Dayoub M, Korpela T (2019) Trends and challenges in organic farming in the European Union. Oceania 27:47–43

Deshmukh MS, Babar N (2015) Present status and prospects of organic farming in India. Eur Acad Res 3(4):4271–4287

Gomiero T, Pimentel D, Paoletti MG (2011) Environmental impact of different agricultural management practices: conventional vs. organic agriculture. Crit Rev Plant Sci 30(1–2):95–124

Grandi C (2008) Organic agriculture enhances agrobiodiversity. Biodiversity 9(1–2):33–35

Institute for Economics & Peace (2019) Global Peace Index 2019: measuring peace in a complex world, Sydney, June 2019. https://www.unesco.org/en/world-media-trends/global-peace-index

Japan International Cooperation Agency (JICA) (2017) Local farmers in Cebu go into organic, sustainable farming. https://www.jica.go.jp/philippine/english/office/topics/news/170731.html

Kavitha V, Chandran K (2017) Is organic farming an alternative to Indian intensive farming? Plant Archives 17(2):1447–1452

Maohong B (2018) The rise of organic agriculture in the Philippines and its development. In: Lopez M, Suryomenggolo J (eds) Environmental resources use and challenges in contemporary Southeast Asia, vol 7. Springer, Singapore, pp 189–205. https://doi.org/10.1007/978-981-10-8881-0_9

Morshedi L, Lashgarara F, Hosseini SJF, Najafabadi MO (2017) The role of organic farming for improving food security from the perspective of fars farmers. Sustainability (Switzerland) 9(11). https://doi.org/10.3390/su9112086

Niggli U (2015) Sustainability of organic food production: challenges and innovations. Proc Nutr Soc 74(1):83–88. https://doi.org/10.1017/S0029665114001438

Nelson, GLM, GNA Abrigo, JA O-Dayao, LEP de Gguzman, RP de Guzman (2016) Organic farming: a thriving fashion and passion, selected stories of organic agriculture champions in the Philippines. DA-BAR. [Coffee Table book]

Nelson GLM, Abrigo GNA, de Guzman RP, O-Dayao JA, de Gguzman LEP (2019) Organic farmers in the Philippines: characteristics, knowledge, attitude and practices. J Nat Stud 18(2): 26–43

Rahmann G (2011) Biodiversity and organic farming: what do we know? vTI Agricult Forest Res 3: 189–208

Statista Research Department (2021) Share of people employed in the agriculture industry to the total employment in the Philippines from 2016 to 2020. https://www.statista.com/statistics/713240/philippines-number-of-agriculture-industry-employees/

Tal A (2018) Making conventional agriculture environmentally friendly: moving beyond the glorification of organic agriculture and the demonization of conventional agriculture. Sustainability 10(4):1078. https://doi.org/10.3390/su10041078

Tscharntke T, Grass I, Wanger TC, Westphal C, Batáry P (2021) Beyond organic farming–harnessing biodiversity-friendly landscapes. Trends Ecol Evol 36(10):919–930

Underwood T, McCullum-Gomez C, Harmon A, Roberts S (2011) Organic agriculture supports biodiversity and sustainable food production. J Hunger Environ Nutr 6(4):398–423

World Population Review (2022) World population by country. https://worldpopulationreview.com/. Accessed 13 Aug 2022

Current Status of Seaweed Diversity: Anthropogenic Interventions

24

Chandra Veluchamy, Sonica Divakar, Manoj Sekaran, Akbar John, Inocencio E Buot, Jr, Anantharaman Perumal, Siva Ramamoorthy, Kanagam Nachiappan, Rajasekaran Chandrasekaran, and Kalaivani Thiagarajan

Abstract

Seaweeds are an essential primary producers and are considered the most important life forms in the coastal ecology. They shelter marine and other aquatic organisms by regulating abiotic parameters, offering nutrients in the water column, stabilizing sediments, and providing valuable natural products. Seaweeds are also important to humans in multiple ways, including as a source of medicines, food supplements, industrial chemicals, and a paramount candidate for biofuel research. This puts pressure on natural habitats that hamper the seaweed diversity. The main causes of seaweed diversity are direct and indirect

Kanagam Nachiappan and Rajasekaran Chandrasekaran contributed equally with all other contributors.

C. Veluchamy · S. Divakar · M. Sekaran · S. Ramamoorthy · R. Chandrasekaran · K. Thiagarajan (✉)
Department of Biotechnology, School of Bio Sciences and Technology, Vellore Institute of Technology, Vellore, Tamil Nadu, India

A. John
School of Industrial Technology, Universiti Sains Malaysia (USM), Gelugor, Penang, Malaysia

I. E. Buot, Jr
Institute of Biological Sciences, College of Arts and Sciences, University of the Philippines Los Baños, Laguna, Philippines

A. Perumal
Centre for Advanced Study in Marine Biology, Faculty of Marine Sciences, Annamalai University, Parangipettai, Tamil Nadu, India

K. Nachiappan
Department of Biotechnology, School of Bio Sciences and Technology, Vellore Institute of Technology, Vellore, Tamil Nadu, India

Sri Venkateswara College of Engineering, Chennai, Tamil Nadu, India

S. Ramamoorthy et al. (eds.), *Plant Diversity in Biocultural Landscapes*,
https://doi.org/10.1007/978-981-19-8649-9_24

anthropogenic interventions. Overharvesting of seaweed, recreational, aquaculture operations, discharging pollutants etc., is direct human influences. The indirect effects are reduced light penetration brought on by increased sedimentation from land runoff and eutrophication-induced blooms. Global climate change is influenced by the increase of carbon dioxide levels, UV radiation, global warming, and storm frequency. These consequences of anthropogenic impact on seaweed's diversity have gained global attention. This chapter will be focusing on the anthropogenic intervention on seaweed diversity.

Keywords

Anthropogenic activity · Seaweed · Climate change · Industrial applications and conservation strategies

24.1 Introduction

Oceans cover a significant portion (70%) of our earth surface, whereas less than 3% covers freshwater sources, such as ice, groundwater, freshwater lakes, and rivers. Their physical and chemical properties directly and indirectly affect the marine/freshwater ecosystem (Tom 2008). Algae are a group of photosynthetic organisms in freshwater and marine ecosystems in both unicellular (microalgae) and multicellular (macroalgae) forms. Seaweed, also called marine macroalgae (Kaladharan and Kandan 1997), have a favourable environmental impact on the marine ecology and are crucial in preserving the ecological equilibrium. They are autotrophic, produce a considerable quantity of organic matter through photosynthesis, and are one of the primary sources of oxygen (O_2) generation and CO_2 uptake (Duarte Carlos and Cebrihn 1996). Seaweeds are capable of trapping excess minerals and heavy metals in the marine ecosystem, thereby modulating climatic conditions. 'Ocean afforestation' is applied on seaweed farming, where 53 billion tons of CO_2 is sequestered every year from 9% ocean coverage (Anonymous 2021). Anthropogenic activities in the densely populated and industrialized transition zones between land and ocean ecosystems expose the seaweed to varying degrees of anthropogenic stress. Furthermore, global climate change is influenced by increased UV radiation, CO_2 levels, precipitation, temperature, flow of wind direction, storm frequency, and rising in sea level affects the phenology of the algal realm (Coelho et al. 2000). Modest development and restricted propagule dissemination of seaweed delays response toward human and environmental system. The devastation that humans are causing to the environment is getting worse. Resources are in short supply due to the rapid expansion of the global population, and it will be a challenge on the sustainability of seaweed. In this chapter, we focus on the role of natural macroalgal communities and anthropogenic interventions to comprehend that it is a growing concern to maintain the marine environment's stability.

24.2 Seaweed Diversity

The world's coastline is home to around 8000 marine macroalgae, some of which may reach 270 m deep (Lüning 1991). The diversity of the world's documented seaweed species, mostly found in tropical ocean waters and intertidal zones, accounts for 80% of the diversity of seaweed (Jha and Zi-Rong 2004). The prevalence and diversity of seaweed in a maritime environment are mainly influenced by temperature, tides, depth, and shoreline features (Dhargalkar et al. 2004). A large diversity of marine algae are found along the south–east coast of Tamil Nadu from Mandapam to Kanyakumari, the coastal region of Gujarat, Lakshadweep, and the Andaman–Nicobar Islands (Parthiban and Anantharaman 2018). Oza and Zaidi (2001) reported 844 species of seaweed, including 434 red, 194 brown, and 216 green seaweed in Indian coastline. Among that, 198 species were reported from Gujarat and Tamil Nadu, in which 109 red, 62 green, and 54 brown seaweed (Jha et al. 2009). According to Palanisamy and Yadav (2017), 105 species of marine macroalgae (Seaweed) were reported from Karnataka, with 55 genera belongs to 31 families, and 20 orders. Of these, 40% belong to the Rhodophyceae, 3.2% to the Chlorophyceae, and 28% to the Phaeophyceae. From Okha coast's, Seaweed were reported with a tally of 39 different species, including 10 Phaeophyta, 16 Chlorophyta, and 13 Rhodophyta (Dave et al. 2019). From the Kerala coast, 147 seaweed were reported, which contribute around 17% of total seaweed in India (Palanisamy et al. 2020). The widely distributed species along the Kerala coast were *Centrocersa clavulatum*, *Chaetomorpha* antennina, *Cladophora vagabunda*, *Enteromorpha prolifera*, *Enteromorpha compressa*, *Enteromorpha flexuosa*, *Gelidium micropterum*, *Gelidiopsis variabilis*, *Grateloupita filicina*, *Gracilaria corticata*, *G. lithophila*, *Grateloupita filicina*, *Hypnea musciformis*, *Padina tetrastromatica*, *Sargassum tenerrimum*, *Ulva fasciata,* etc. Seaweeds such as Sargassum and Turbinaria are endemic on Indian coastal region and 16,000 tons are cultivated per annum (Khan and Satam 2003). *Enteromorpha intestinalis*, *Ulva lactuca,* and *Catenella repens* are found to be abundant in the Lower Gangetic delta region (Pramanick et al. 2014). *Ulva* and *Caulerpa* were dominant genera on the Tuticorin coast followed by *Sargassum*, *Padina*, and *Gracilaria* (Canciyal et al. 2014). Roy (2020) reports the new uncommon species *Aghardhiella subulate* and *Fucus vesiculosus* in Olaikuda and Vadakkadu, Rameshwaram. On the tide-dominated Uran coast, Navi Mumbai recorded 19 species of seaweed—16 genera belongs to 15 families and 13 orders (Pawar 2017). Seasonal variations play a substantial part in the occurrence, distribution, and abundance of seaweed diversity. Several workers reported season and geographical specific occurrence of the seaweed. In Hare Island, Gulf of Mannar species, such as *Acanthophora spicifera*, *Corynomorpha prismatica*, *Enteromorpha compressa*, *Hypnea musciformis*, *Kappaphycus alvaerezii*, *Spathoglossum asperum*, *Stoechospermum marginatum*, *Turbinaria ornata*, *T. conoides*, *Ulva reticulata*, and *U. lactuca,* were most abundant in post monsoon season, whereas *Scinaia furcellata*, *Turbinaria ornata*, *T. conoides*, *Rosenvingea intricata*, and *Pocokiella variegate* were completely absent in pre-monsoon season. Pathak et al. (2021) reported that *Ulva lactuca*, *U. fasciata*, *Caulerpa taxifolia*, *C. racemosa*, *Ectocarpus siliculosus*, *Sargassum johnstonii*, and *Padina tetrastromatica* were abundant in the post-monsoon season at Veraval and Sikka coasts of Gujarat. The highest seaweed diversity is seen during the monsoon and post-monsoon seasons on the Karnataka coast (Yadav and Palanisamy 2021). At the Uran coast, maximum species diversity was present during pre-monsoon and post-monsoon than monsoon, whereas higher species composition was recorded during post-monsoon than other seasons (Pawar 2017). Some of the seaweed species were highly adaptable and could sustain all seasons, such as *Gracilaria pygmaea*, *Gracilaria corticata*, *G. verrucosa*, *Padina pavonia*, and *P. tetrastromatica* (Mary Josephine et al. 2013). In Muttom, south–west coast (Arabian Sea) *Ulva lactuca*, *U. fasciata*, *Sargassum duplicate*, *S. wightii*, *Padina tetrastomatica*, *Gracilaria pygmaea*, *and G. corticata* were reported (Domettila et al. 2013) (Table 24.1) represents the distribution of seaweed diversity across the Indian coast.

24.2.1 Emerging Threat to Seaweed Diversity

It is essential to act quickly with a scientific approach to protect seaweed's sustainability in order to maintain its ecological value. Although it is well-recognized that seaweeds are sensitive to physical and chemical changes in the marine environment, little is known about how anthropogenic climate change influences seaweed diversity in the coastal ecosystems. Anthropogenic activities produce a significant amount of pollutants from industrial effluents, domestic waste, agricultural, and urban runoff with varying degrees of impact on the ecosystem (Steneck et al. 2002). The varying pH, density, total dissolved solids, heavy metals, pesticides, fungicides, etc. in the domestic and industrial segments have adversely affected seaweed's growth and development (Araujo et al. 2012).

24.2.1.1 Industrial Pollution

Seaweeds are threatened in developing nations, where they are affected by various human activities. Environmental changes brought on by heavy metals, plastics, other pollutants, and climate change pose a serious threat to the living organisms, especially in maritime ecosystems (Ritter et al. 2008; Walker et al. 2005). Reactive oxygen species (ROS) are produced by heavy metal pollutants, which also cause oxidative damage to seaweed, which can then negatively impact other organisms at various trophic levels. According to Medina et al. (2005), mining-related copper deposition severely impacted algal biodiversity and wiped out all benthic carnivores and several benthic herbivores. CO_2 can impact the physiology of seaweed in the maritime environment (Kroeker et al. 2010), utilizing CO_2 for photosynthesis. However, too much CO_2 emitted from human activities might disturb the availability of carbonate ions and lower the pH of aquatic ecosystem (Orr et al. 2005). Eutrophication may cause some algae species to grow more quickly in the beginning, but sewage discharges hurt sperm motility, hinder fertilization, and may even increase germling mortality. Other eutrophication-related indirect impacts are also projected

Table 24.1 Distribution of seaweed diversity on various Indian coast

Location	Chlorophytes	Phaeophytes	Rhodophytes	References
Okha, Gulf of Kutch	16	10	13	Dave et al. (2019)
Olaikuda and Vadakkadu, Rameshwaram, Tamil Nadu	28	28	28	Roy (2020)
Beyt Dwarka coast, Gulf of Kutch	5	8	26	Kalasariya et al. (2020)
Gulf of Mannar marine biosphere reserve	48	41	48	Veeragurunathan et al. (2022)
St. Mary's island, Karnataka	16	19	17	Yadav and Palanisamy (2021)
Shivrajpur coast, Gujarat	18	22	30	Hakim et al. (2022)
Sindhudurg coastline, Maharashtra	16	10	12	Valanju (2020)
Andaman and Nicobar Islands	79	58	107	Karthick et al. (2021)
Veraval coast, Gujarat	21	14	32	Ishakani et al. (2016)
Tuticorin coast, Tamil Nadu	32	27	27	Canciyal et al. (2014)
Hare Island, Gulf of Mannar	30	28	32	Mary Josephine et al. (2013)
Kudankulam, Gulf of Mannar	15	8	9	Satheesh and Wesley (2012)

to have adverse effects, such as increased sediment cover of substrata, scouring brought on by wind-induced resuspension of sediments, and grazing (Coelho et al. 2000).

24.2.1.2 Mariculture

The seemingly unabated global demand for red seaweed raw materials and k-carrageenan, the commercial production of *Kappaphycus alvarezii*, has expanded in remote tropical regions. However, the nearby marine biota should be negatively impacted by this increasing cultivation of *K. alvarezii* (Veeragurunathan et al. 2021). Due to the fragmentation of a single vegetation type into smaller intact units causing habitat change in a specific area (Eklöf et al. 2006). The major seaweed that encounters their southern range constraints is the common brown seaweed *Ascophyllum nodosum*, which grows on protected intertidal rocky coasts located at the Northern Atlantic (Araujo et al. 2009). The significant genetic variability within populations of the same species of seaweed is due to anthropogenic interventions. To prevent genetic variation, seed stock of seaweed obtained from natural habitat is used for large-scale propagation (Anonymous 2021). Furthermore, there is strong evidence that some seaweed farming practices destroy specific habitats (such as seagrass beds, fish, and coral reefs diversity), but this can be avoided with careful agricultural planning at appropriate geographical location (Kelly et al. 2020).

24.2.1.3 Intervention by Local Inhabitants

The seaweed population gets adversely affected by the anthropogenic activities carried out by the local inhabitants. When seaweed undergoes stress, benthic populations may experience temporary or extended changes in composition, diversity, and richness affecting the ecosystem (Estes et al. 1989; Steneck et al. 2002). Trampling is a frequent anthropogenic activity on rocky intertidal coasts that affects the biomass, variety, cover, and composition of the seaweed (Araujo et al. 2012; Milazzo et al. 2002b). In Europe, *A. nodosum*-dominated intertidal rocky seashore assemblies in Portugal were subjected to experimental trampling, which resulted in a reduction in the native species cover at higher stamping intensities (Araujo et al. 2012). Human activities related to recreation and over harvesting of seaweed significantly disturbed marine fauna either directly or indirectly in the rocky intertidal zones (Milazzo et al. 2002a, 2004; Thompson et al. 2002). Moreover, plastics pose a serious threat to all of the marine life. Microplastics gradually accumulate in the marine environment due to continuous disposal of plastic trash. From the surface to the sediments of the ocean floor, plastics of various sizes are present in marine environments (Taylor and Cole 1994). Creating awareness and educating the local people about the significance of algal diversity and protection are the need of the hours to conserve and sustain the seaweed diversity.

24.2.1.4 Over Harvesting of Indigenous Species

In China, 1700 years ago, seaweed utilization was first documented, even before its usage by humans in the Neolithic era (Yang et al. 2017). Through many years, coastal people have gathered a broad variety of seaweeds with different classifications. Domestic applications of seaweed (for food and animal feed) first took precedence, but subsequently enhanced industrial uses (for cosmetics and fertilisers) led to over-exploitation in the natural habitat (Delaney et al. 2016). Recent global trend in extensive usage of seaweed has driven up the demand as well as hamper the diversity (Rebours et al. 2014; Hafting et al. 2015). The two major countries with the highest cumulative output of seaweed are China and Indonesia, producing nearly 23 million tons in 2014. *Saccharina japonica* and *Undaria pinnatifida* were the two species of red algae (kelp) from the genera *Gracilaria* and *Pyropia*, produced in China for human use (FAO 2016). *Eucheuma/Kappaphycus,* and *Gracilaria, Porphyra, Saccharina, and Undaria* are the top five genera that account for 98% of the worldwide seaweed production (Suo and Wang 1992). The seaweed businesses in Europe, Canada, and Latin America still depend on harvesting natural resources (Rebours et al. 2014). *Lessonia, Macrocystis, Sarcothalia crispata, Gigartina skottsbergii*, and *Chondracanthus chamisoii* are all over-exploited in Chile, whereas *Gracilaria chilensis* is the only species farmed relatively in a smaller quantity (Buschmann et al. 2008). However, over-harvesting of *Gelidium* spp. for microbiology-grade agar production in

Morocco and Japan causes a negative effect on its population dynamics (Fujita et al. 2006; Callaway 2015). Similarly, indiscriminate collection of *Gelidiella acerosa* and *Gracilaria edulis* used for agar production severely affects the seaweed diversity along the Tamil Nadu coast, India (Chennubhotla et al. 1990). Therefore, efforts for large-scale cultivation of wild seaweed using modern scientific interventions are mandatory to supply the industrial requirement and sustain the diversity.

24.2.1.5 Tourism

Marine tourism plays a significant role in the worldwide tourism industry expanding more quickly, especially in tropic regions (Bennett et al. 2003). The two major perspectives on the growing interest in marine tourism are as follows: First, some places have gained recognition for coastline recreation, and next to it, certain marine recreation happenings have gained reputation themselves (Zulaiha et al. 2014). Marine tourism can be considered a sustainable practice, where the marine life forms are reused without harvesting them, but, at the same time, the tourism casts both direct and indirect influences on the marine environment. Seaweeds are regarded as the best bioindicators, since their existence, dominance, and density reveal the status of the environment's natural conditions (Silva et al. 2012). An environment's increased complexity leads to more distinct niches being created, and as a result, more species may coexist (Dean and Connell 1987). The abundance of seaweed is often regulated by environmental conditions or criteria, such as the hard, sturdy substrate that acts as a natural habitat for seaweed (Chasani and Suyono 2020). Numerous recreational activities affect marine life in several ways, such as breakages, stamping, resuspension of sediments, changes in the behavior of marine lifeform as a result of offering food, disturbing animals, debris, and garbage production (Silva et al. 2012). The distribution of several seaweed species has consistent effect of tourism in Maracajaú reefs, Northeast Brazil were examined in two sites, one with tourist activity, and the other with restricted access. The algal strains in the region often frequented by travelers were made up of trivial algae with shorter life series that are readily adaptable to the recurrently disturbing environment. In contrast, the location without tourist presence displayed greater species diversity (Silva et al. 2012). An ecological imbalance in this area encouraged the growth of opportunistic *Caulerpa racemosa* and annual seaweed with short life cycles. Macroalgal diversity is also altered by the disturbances caused by ship's navigational movements, unregulated pollution, construction of roads, resorts, homestays, marinas, jetties, and small villages along the coast to enhance tourism (Davenport et al. 2006; Nyberg 2007).These actions can result in substantial mortality in marine life, with sessile invertebrates and marine algae being extremely vulnerable (Milazzo et al. 2002b). Due to the likelihood that tourism-related activities would damage coral reefs by exploiting seaweed that is typically attached to the substrate, directly disturbing the seaweed habitat (Chasani and Suyono 2020) due to enhanced pollution, the variety of seaweed will diminish or maybe vanished.

24.3 Climate Change Vulnerability

24.3.1 Ocean Acidification

The anthropocene climate has been mainly defined by ocean acidification caused by a substantial rise in atmospheric CO_2. Coral reefs are particularly at risk due to ocean acidification, where low availability of carbonate ions inhibits biogenic calcification and accelerates the breakdown of carbonate substrata. Whereas these patterns eventually affect ecosystem calcification and organisms depending on coral reefs (Doo et al. 2019). Dynamical change in pH and CO_2 level affects the seaweed diversity due to ocean acidification. Seaweed is the major primary producer growing among corals that provide habitat for a variety of herbivorous organisms (Dayton 1985). Ocean acidification is anticipated to have a negative impact on calcifying seaweeds (such as *Halimeda* sp.), because their calcium carbonate skeletons are more difficult to sustain at a lower pH. However, very less attention has been given to how non-calcifying (fleshy) seaweed respond to ocean acidification (Van der Loos et al. 2019). The seaweed may become more vulnerable to physical force due to the subsequent pH decrease, leading to loss of seaweed biomass due to storm frequency brought on by climate change (Kinnby et al. 2021). Increasing CO_2 enhances sporophytic photochemical productivity while inhibiting gametophytic development in kelps, such as *Undaria pinnatifida*. This might have an adverse effect on seedling culture (Gao et al. 2019). Many Rhodophyta, Ochrophyta, and Chlorophyta species can survive in CO_2 conditions. However, some species are not able to complete their life cycles in high CO_2 environments. For example, the filamentous Rhodophyta's reproductive capacity decreased when CO_2 levels increased (Porzio et al. 2011) (Table 24.2) gives important details on the impact of ocean acidification on seaweed.

24.3.2 Sea Level Rise

Accelerated sea level rise (ASLR) is the most certain outcome of global warming. The physical changes caused by the sea level rise can lead to the dislocation of the population, and cause ecological consequences, such as the destruction of reefs, loss of diversity, and altered biophysical properties of coastal areas (Perez et al. 1999). As the sea level rises, seagrass beds and seaweed receives less light across their depth range, which might restrict photosynthesis (Jackson et al. 2001; Short 1980). It reduces distribution, productivity, functional values, and bed structure alteration (Drew 1979; Harley et al. 2012). In Red Sea, seaweeds appeared less during winter and spring, with an increase in sea level covering the reef (Shoubaky and Kaiser 2014). Seaweed, in the presence of other photo-autotrophs, fight for resources, such as light, nutrients, and space for attachment. Their relative effectiveness in obtaining these resources relies on the availability of those resources, as well as environmental stress (Harley et al. 2012). Human activities are modifying the availability of these resources impacting the algal community's growth and its competitiveness (Harley et al. 2012). An upward shift in benthic communities is documented as a result of the

Table 24.2 Impact of ocean-acidification on seaweeds

Model species	Location	Parameters	Response	References
Lobophora rosacea	New Caledonia	Low and variable pH over a long-term timescale and low and constant pH over a short-term timescale	Decrease in some specialized metabolites at low pH	Gaubert et al. (2020)
Rhodophyta, Ochrophyta and Chlorophyta	Ischia Island in the Gulf of Naples, Italy	pH 8.1 to 7.8	Rhodophyta decreased in reproductive capacity as CO_2↑	Porzio et al. (2011)
Rhodophyceae Phaeophyceae and Chlorophyceae	Saint Martin's island, Bangladesh	Diversity and physiochemical analysis	Winter (24.68 °C) and pre-monsoon (29.05 °C) pH↓ in the northwest during the high tide of winter (pH 8.34 ± 0.24)	Hossen et al. (2022)
Calcified coralline algae	Shikine Island, volcanic coasts of Italy and Japan, north–West Pacific,	Molecular tools to analyse the natural inclines under the ocean	↓ Coralline algae diversity with ↑acidification	Peña et al. (2021)
Macrocystis pyrifera	Otago, New Zealand	pH (7.59–8.50) Analyzed Meiospore germination	6–9% reduction in germination success under extreme acidity	Roleda et al. (2012)
Fucus vesiculosus	Southwestern Baltic Sea, Germany	Seasonal reproductive biology	CO_2 affect the temporal development of *Fucus* fertility	Graiff et al. (2017)
Fucus vesiculosus	Island of Helgoland in the German Bight, North Sea	Analyzed growth at different CO_2 concentration	Reduced growth of the algae at high CO_2-concentrations	Gutow et al. (2014)

rise in sea level. However, changes in the relative availability of suitable substratum types and orientations at particular shore levels may also affect changes in relative algal abundance (Vaselli et al. 2008). In addition to sea level rise, water motion also influences seaweed diversity. The water movement in coastal and estuary areas increases as a result of the higher tidal flows caused by the rise in the sea level (Fonseca et al. 1983) with increased current velocity, leaf biomass, breadth, and canopy height of the vegetation (Short 1980). The gametes of seaweed depend on physical factors, notably water movement, for dispersion, since they are non-motile, short-lived, and heavier than water (Lobban and Harrison 1994; Shanks et al. 2003). Thus, water motion has a significant role in determining the fate of seaweed gametes, particularly for settling tetraspores, spermatangia, or carpospores (Lobban and

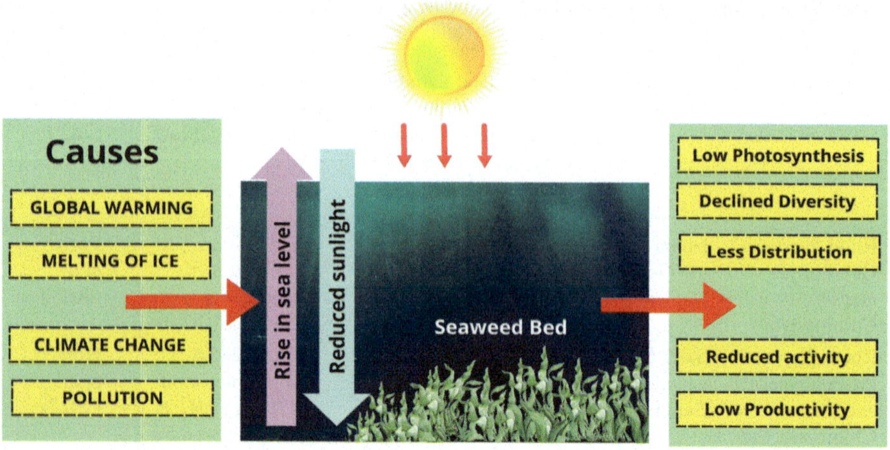

Fig. 24.1 Impact of rising sea level on seaweed bed

Harrison 1994). Figure 24.1 shows the graphical illustration of the causes and consequences of sea level rise on seaweed beds.

24.3.3 Seaweed: Global Economic Scenario

The three major hazards to society that require immediate attention are resource depletion, climate change, and toxicity risk. A bio-based economy and a focus on reducing our dependence on fossil fuels might be beneficial for addressing these issues and achieving sustainable development (Mengal et al. 2018). It can only succeed if renewable resources, particularly biomass, synthesize wide range of products, such as animal feed, food, energy, bio-based materials, and medicines. The scientific community and start-up businesses have recently placed significant emphasis on either developing new materials or displacing the current fossil-based goods (Schütte 2018). According to the Food and Agricultural Organization (FAO), seaweed (30 Mt) were consumed by 2014, mainly obtained from aquaculture, among that 6% were from wild species. Since 1995, improvements in harvesting technologies have increased (176%) seaweed consumption globally (Nakhate and Van Der Meer 2021).

People of China, Japan, and Korea cultivate and utilize seaweed/products in their day-to-day life. These products include processed dairy, meat, and fruit items as well as household goods, including paint, toothpaste, solid air fresheners, cosmetics, and medications (Dhargalkar and Pereira 2005). Seaweed is well-recognized for its richness in minerals, vitamins, polysaccharides, and bioactive compounds with antibacterial, antiviral, and antifungal metabolites (Holdt and Kraan 2011). Among the 35 nations in the Northern and Southern hemispheres engaged in commercial harvesting of seaweed, West Asians are more concerned with the thickeners and gelling qualities of seaweed, and East Asians inculcated in their routine diet. (Pereira and Cotas 2019). Seaweeds are considered a third-generation biofuel feedstock because of its ability to grow without fertile soil, freshwater, or fertilizer and produce more biomass per unit area of cultivation than its terrestrial equivalents. Macroalgae with minimal lignin compared to lignocellulose make it cost-effective to release the sugars. In Scotland, between 2016 and 2019, the Macro Fuel Horizon 2020 project generated 10 tons of seaweed to make 20 L of gasoline supporting 80 km drive in cars (Fasahati et al. 2022). Eliminating CO_2, heavy metal contaminants, and dissolved nutrients control eutrophication, and enhances the sustainability of marine environment.

24.4 Regulation Policies and Conservation Strategies

Seaweed-dominated ecosystems are rapidly changing, shifting to higher latitudes, or even retracting, as a result of anthropogenic factors, such as eutrophication, ocean acidification, global warming, etc. (Wernberg et al. 2016). It is still challenging to monitor changes and biodiversity loss in marine ecosystem, which raises the threat that fading tropical seaweed populations go unnoticed. Apart from these global issues, seaweed cultivation is rapidly affected by unique threats, such as stages of growth, alteration in phenology, etc., at temperate as well as tropical regions (Coleman and Goold 2019).

The genetic diversity of the seaweed has been significantly reduced as a result of excessive exploitation of wild habitat and combined with vegetative farm expansion (Guillemin et al. 2014). Due to the rate of global expansion and interaction with other human activities, a thorough analysis of the potential long-term effects of seaweed production on coastal ecosystems and their potential mitigation is undoubtedly necessary (Brakel et al. 2021). In addition, government support is lacking for seaweed farmers, making it difficult to spread knowledge, provide training, and facilitate the introduction of various agrotechnologies implying seaweed farming and business. As an alternative, encouraging cooperative-led initiatives to create standards of good biosecurity practice might help to remove implementation hurdles and promote initial conversation between farmers and the relevant government agency is essential (Campbell et al. 2022).

24.5 Opportunities by the Government of India

The SOFIA (State of World Fisheries and Aquaculture) in 2020 estimates 32.4 million tons of world's total cultivation of seaweeds, counting both wild and cultured algae. India contributes around 0.02% of global production, which is around 5300 tons of wet seaweed by 2018. With the assistance of funding supplied by the Government of India (GOI), 25,000 tons of total seaweed was harvested in 2020, and it is expected to expand to 1,120,000 tons by 2025. Since 2017, the GOI's department of fisheries has provided funding to a number of research institutions that have collaborated to offer training, support for seaweed production, and established seaweed processing facilities. *Kappaphycus alvarezii* and *Gracilaria* spp. combined to produce 201.72 tons and 255.32 tons of wet weight of biomass between 2017 and 2020. Since 2017, National Fisheries Development has released INR 53.55 million for projects, benefiting 1000 people. Furthermore, the Pradhan Mantri Matsya Sampada Yojana indicates that the Indian government intends to finance around 640 crores over a 5-year period (Singh et al. 2022).

24.6 Conclusion

Seaweed diversity assessment is very essential to infer the changes in marine environment caused by anthropogenic activities, such as over-exploitation, pollution, industrialization, climatic alterations, bio-invasions etc. More subject expert and research exploration is needed to identify and inventory the seaweed biota. Currently, seaweeds are considered an important source for the production of various industrial products, such as alginate, agar, carrageenan, pigments, cosmetics, and therapeutics. Due to this high demand and land squeeze, the demand for seaweed raw material will shoot up shortly, thereby emphasizing the urgent need to strengthen the knowledge of large-scale production mechanisms. It is suggested that more research be done to learn more about how anthropogenic stress impacts the phenology of seaweed and diversity. In addition, a crucial component of seaweed conservation is reducing anthropogenic disruptions, which can be accomplished by raising awareness among the local populace. Moreover, the reduction of anthropogenic disturbances is the major factor, where humans are more concerned and aware of the importance of seaweed conservation. This chapter puts serious efforts to compile information on seaweed diversity to fill the gaps in the scientific community to aid in policy development to improve seaweed farming under rural employment schemes and also frame some conservation strategies. Thus, considering the ecological values of seaweed, it is essential to initiate speedy and scientific action for sustainable management of seaweed to conserve them for posterity.

Acknowledgements The authors acknowlege the Vellore Institute of Technology management and the Dean, School of BioSciences and Technology for providing the infrastructure, constant support and encouragements. Our sincere gratitude to DST—SPLICE—Climate Change Progamme, Govt. of India for providing financial support to Ms. V. Chandra & Mr. S. Manoj—Grant No. DST/CCP/NCC&CV/133/2017 (G).

References

Anonymous (2021). https://thefishsite.com/articles/encouraging-coastal-communities-to-engage-with-seaweed-aquaculture

Araujo R, Vaselli S, Almeida M, Serrão E, Sousa-Pinto I (2009) Effects of disturbance on marginal populations: human trampling on *Ascophyllum nodosum* assemblages at its southern distribution limit. Mar Ecol Prog Ser 378:81–92

Araujo R, Sousa-Pinto I, Serrao EA, Aberg P (2012) Recovery after trampling disturbance in a canopy-forming seaweed population. Mar Biol 159:697–707
Bennett M, Dearden P, Rollins R (2003) The sustainability of dive tourism in Phuket, Thailand. Communities in SE Asia: challenges and responses. p 97–106
Brakel J, Sibonga RC, Dumilag RV, Montalescot V, Campbell I, Cottier-Cook EJ et al (2021) Exploring, harnessing and conserving marine genetic resources towards a sustainable seaweed aquaculture. Plants People Planet 3(4):337–349
Buschmann AH, Hernandez-Gonzalez MDC, Varela D (2008) Seaweed future cultivation in Chile: perspectives and challenges. Int J Environ Pollut 33(4):432–456
Callaway E (2015) Lab staple agar runs low: dwindling seaweed harvest imperils reagent essential for culturing microbes. Nature 528(7581):171–173
Campbell I, Mateo J, Rusekwa SB, Kambey CSB, Hurtado A, Msuya FE, Cottier-Cook EJ (2022) An international evaluation of biosecurity management capacity in the seaweed aquaculture industry. J Environ Manag 304:114112
Canciyal J, Mogalekar H, Jawahar P (2014) Seaweed diversity of Tuticorin coastal waters along south east coast of India. Int J Curr. https://www.academia.edu/13844796/seaweed_diversity_along_tuticorin_coastal_wtaers_along_southeast_coastma
Chasani AR, Suyono EA (2020) Comparison of structure and composition of seaweed population in Porok and Greweng Coasts, Gunungkidul, Indonesia. AIP Conf Proc 2260(1):020011
Chennubhotla VS, Kaliaperumal N, Kalimuthu S, Ramalingam JR, Subbaramaiah K, Rao KR, Rao PV (1990) Seaweed resources of the Tuticorin-Tiruchendur coast, Tamil Nadu, India. J Mar Biol Assoc India 32(1&2):146–149
Coelho SM, Rijstenbil JW, Brown MT (2000) Impacts of anthropogenic stresses on the early development stages of seaweed. J Aquat Ecosyst Stress Recover 7(4):317–333
Coleman MA, Goold HD (2019) Harnessing synthetic biology for kelp forest conservation. J Phycol 55(4):745–751
Dave T, Vaghela D, Chudasama B (2019) Status, distribution, and diversity of some macroalgae along the intertidal coast of Okha, Gulf of Kachchh, Gujarat in India. J Entomol Zool Stud 7(3): 327–331
Davenport J, Estuarine JD, University College Cork (2006) The impact of tourism and personal leisure transport on coastal environments: a review, vol 67. Elsevier, Amsterdam, pp 280–292
Dayton PK (1985) Ecology of kelp communities. Annu Rev Ecol Syst 16:215–245
Dean R, Connell J (1987) Marine invertebrates in an algal succession. I. Variations in abundance and diversity with succession. J Exp Mar Biol Ecol 109(3):195–215
Delaney A, Frangoudes K, Li S-A (2016) Society and seaweed: understanding the past and present. In: Fleurence J, Levine I (eds) Seaweed in health and disease prevention. Elsevier Academic Press, London, pp 7–40
Dhargalkar VK, Pereira N (2005) Seaweed: promising plant of the millennium. Cult Sci J 17(3–4): 60–66
Dhargalkar VK, Kavlekar D, Verlecar XN, Rathod V (2004) Seaweed—a field manual. National Institute of Oceanography, Dona Paula, Goa
Domettila C, Brintha T, Sukumaran S, Jeeva S (2013) Diversity and distribution of seaweed in the Muttom coastal waters, south-west coast of India. Biodivers J 4(1):105–110
Doo SS, Edmunds PJ, Carpenter RC (2019) Ocean acidification effects on in situ coral reef metabolism. Sci Rep 9(1):1–8
Drew EA (1979) Physiological aspects of primary production in seagrasses. Aquat Bot 7 (C):139–150
Duarte Carlos A, Cebrihn J (1996) The fate of marine autotrophic production. Limnol Oceanogr 41(8):1758–1766
Eklöf JS, Henriksson R, Kautsky N (2006) Effects of tropical open-water seaweed farming on seagrass ecosystem structure and function. Mar Ecol Prog Ser 325:73–84
Estes JA, Duggins DO, Rathbun GB (1989) The ecology of extinctions in kelp forest communities. Conserv Biol 3:252–264

FAO (2016) The state of world fisheries and aquaculture 2016 (SOFIA). FAO, Lazio

Fasahati P, Dickson R, Saffron CM, Woo HC, Liu JJ (2022) Seaweed as a sustainable source of bioenergy: techno-economic and life cycle analyses of its biochemical conversion pathways. Renew Sust Energ Rev 157:112011

Fonseca MS, Zieman JC, Thayer GW, Fisher JS (1983) The role of current velocity in structuring eelgrass (Zostera marina L.) meadows. Estuar Coast Shelf Sci 17(4):367–380. https://doi.org/10.1016/0272-7714(83)90123-3

Fujita D, Ishikawa T, Kodama S, Kato Y, Notoya M (2006) Distribution and recent reduction of Gelidium beds in Toyama Bay, Japam, Japan. J Appl Phycol 18(3):591–598

Gao X, Kim JH, Park SK, Yu OH, Kim YS, Choi HG (2019) Diverse responses of sporophytic photochemical efficiency and gametophytic growth for two edible kelps, Saccharina japonica and Undaria pinnatifida, to ocean acidification and warming. Mar Pollut Bull 142:315–320

Gaubert J, Rodolfo-Metalpa R, Greff S, Thomas OP, Payri CE (2020) Impact of ocean acidification on the metabolome of the brown macroalgae Lobophora rosacea from New Caledonia. Algal Res 46:101783

Graiff A, Dankworth M, Wahl M, Karsten U, Bartsch I (2017) Seasonal variations of Fucus vesiculosus fertility under ocean acidification and warming in the western Baltic Sea. Bot Mar 60(3):239–255

Guillemin ML, Valero M, Faugeron S, Nelson W, Destombe C (2014) Tracing the trans-Pacific evolutionary history of a domesticated seaweed (Gracilaria chilensis) with archaeological and genetic data. PLoS One 9(12):e114039

Gutow L, Rahman MM, Bartl K, Saborowski R, Bartsch I, Wiencke C (2014) Ocean acidification affects growth but not nutritional quality of the seaweed Fucus vesiculosus (Phaeophyceae, Fucales). J Exp Mar Biol Ecol 453:84–90

Hafting JT, Craigie JS, Stengel DB, Loureiro RR, Buschmann AH, Yarish C, Edwards MD, Critchley AT (2015) Prospects and challenges for industrial production of seaweed bioactives. J Phycol 51(5):821–837

Hakim MM, Solanki AD, Patel IC (2022) Biological distribution of the rich source of marine algae from the rocky shoreline of the coast of Shivrajpur, Gujarat, India. Egypt J Agricult Res 100(3): 371–380

Harley CDG, Anderson KM, Demes KW, Jorve JP, Kordas RL, Coyle TA, Graham MH (2012) Effects of climate change on global seaweed communities. J Phycol 48(5):1064–1078. https://doi.org/10.1111/j.1529-8817.2012.01224.x

Holdt SL, Kraan S (2011) Bioactive compounds in seaweed: functional food applications and legislation. J Appl Phycol 23(3):543–597

Hossen R, Jaman MH, Nasrin A (2022) Study of seaweed diversity and water. Physico-chemical Properties of Saint Martin's Island, Bangladesh

Ishakani AH, Joshi NH, Ayaz M, Sumara K, Vadher KH (2016) Assessment of seaweed diversity at Veraval coast, Gujarat. J Exp Zool 19(2):863–868

Jackson EL, Rowden A, Attrill M, Bossey SJ, Jones MB (2001) The importance of seagrass beds as a habitat for fishery species. Oceanogr Mar Biol 39:269–304

Jha RK, Zi-Rong X (2004) Biomedical compounds from marine organisms. Mar Drugs 2:123–146

Jha B, Reddy C, Thakur M, Rao M (2009) Seaweed of India: the diversity and distribution of seaweed of Gujarat coast. Springer, New York

Kaladharan P, Kandan S (1997) Primary productivity of seaweed in the lagoon of minicoy atoll of laccadive archipelago. Seaweed Res Utiln 19(1 and 2):25–28

Kalasariya H, Patel V, Patel R, Patel N, Rathwa S (2020) Seaweed diversity study of selected Beytdwarka coast in Gujarat, India. Int J Bot Stud 5(1):141–145

Karthick P, Ramesh C, Mohanraju R (2021) A checklist of seaweed of the Andaman and Nicobar Islands, India: a way forward for seaweed cultivation, food, and drug applications. Environ Monit Assess 193(10):1–17

Kelly EL, Cannon AL, Smith JE (2020) Environmental impacts and implications of tropical carrageenophyte seaweed farming. Conserv Biol 34(2):326–337

Khan S, Satam S (2003) Seaweed mariculture: scope and potential in India. Aquacult Asia 8(4): 26–29

Kinnby A, White JC, Toth GB, Pavia H (2021) Ocean acidification decreases grazing pressure but alters morphological structure in a dominant coastal seaweed. PLoS One 16(1):e0245017

Kroeker KJ, Kordas RL, Crim RN, Singh GG (2010) Meta-analysis reveals negative yet variable effects of ocean acidification on marine organisms. Ecol Lett 13(11):1419–1434

Lobban CS, Harrison PJ (1994) Seaweed ecology and physiology. Cambridge University Press, Cambridge

Lüning K (1991) Seaweed: their environment, biogeography, and ecophysiology. Wiley, New York

Mary Josephine M, Victorial M, Rani S (2013) Current status of seaweed diversity and their seasonal availability at Hare Island, Gulf of Mannar. Sci Res Rep 146–151

Medina M, Andrade S, Faugeron S, Lagos N, Mella D, Correa JA (2005) Biodiversity of rocky intertidal benthic communities associated with copper mine tailing discharges in northern Chile. Mar Pollut Bull 50(4):396–409

Mengal P, Wubbolts M, Zika E, Ruiz A, Brigitta D, Pieniadz A, Black S (2018) Bio-based industries joint undertaking: the catalyst for sustainable bio-based economic growth in Europe. New Biotechnol 40:31–39

Milazzo M, Chemello R, Badalamenti F, Riggio RC, Riggio S (2002a) The impact of human recreational activities in marine protected areas: what lessons should be learnt in the Mediterranean sea? Marine Ecol 23(Suppl. 1):280–290

Milazzo M, Chemello R, Badalamenti F, Riggio S (2002b) Short-term effect of human trampling on the upper infralittoral macroalgae of Ustica Island MPA (western Mediterranean, Italy). J Mar Biol Assoc 82:745–748

Milazzo M, Badalamenti F, Riggio S, Chemello R (2004) Patterns of algal recovery and small-scale effects of canopy removal as a result of human trampling on a Mediterranean rocky shallow community. Biol Conserv 117(2):191–202

Nakhate P, Van Der Meer Y (2021) A systematic review on seaweed functionality: a sustainable bio-based material. Sustainability 13(11):6174

Nyberg CD (2007) Introduced marine macroalgae and habitat modifiers: their ecological role and significant attributes. ISBN: 91-89677-33-1: Department of Marine Ecology, Göteborg University, Sweden

Orr J, Fabry VJ, Aumont O, Bopp L, Doney SC (2005) Anthropogenic ocean acidification over the twenty-first century and its impact on calcifying organisms. Nature 437:681–686

Oza RM, Zaidi SH (2001) A revised checklist of Indian marine algae. CSMCRI, Bhavnagar, p 296

Palanisamy M, Yadav SK, Murthy GSV (2020) Seaweed resources of Kerala Coast and its economic potential. Madras Agric J 107(special):1

Palanisamy M, Yadav SK (2017) Eight new additions to marine macro algal Flora of Karnataka Coast, India. Nelumbo 59(2):187

Parthiban C, Anantharaman P (2018) Diversity and biomass of drift seaweed from the Tuticorin Coast, India. Species 19:72–86

Pathak S, Dubey S, Sharma M, Mishra P, Mahajan SK (2021) Seasonal variation on the density of seaweed species from Veraval and Sikka coast, Gujarat. J Entomol Zool Stud 9(1):1510–1515

Pawar PR (2017) Distribution pattern & species diversity of seaweed at Uran (Navi Mumbai), West Coast of India. Int J Pure Appl Zool 5(1):25–32. https://www.alliedacademies.org/articles/distribution-pattern--species-diversity-of-seaweed-at-uran-navi-mumbai-west-coast-of-india.html

Peña V, Harvey BP, Agostini S, Porzio L, Milazzo M, Horta P et al (2021) Major loss of coralline algal diversity in response to ocean acidification. Glob Chang Biol 27(19):4785–4798

Pereira L, Cotas J (2019) Historical use of seaweed as an agricultural fertilizer in the European Atlantic area. In: Pereira L, Bahcevandziev K, Joshi NH (eds) Seaweed as plant fertilizer, agricultural biostimulants, and animal fodder. CRC Press, Boca Raton, pp 1–22

Perez RT, Amadore LA, Feir RB (1999) Climate change impacts and responses in the Philippines coastal sector. Climate Res 12(2–3 spec. iss. 6):97–107

Porzio L, Buia MC, Hall-Spencer JM (2011) Effects of ocean acidification on macroalgal communities. J Exp Mar Biol Ecol 400(1–2):278–287
Pramanick P, Zaman S, Mitra A (2014) A review on tropical seaweed diversity and benefits. J Chem Biol Phys Sci 4(4):3909–3920
Rebours C, Marinho-Soriano E, Zertuche-González JA, Hayashi L, Vásquez JA, Kradolfer P, Soriano G, Ugarte R, Abreu MH, Bay-Larsen I, Hovelsrud G (2014) Seaweed: an opportunity for wealth and sustainable livelihood for coastal communities. J Appl Phycol 26(5):1939–1951
Ritter A, Goulitquer S, Salaün JP, Tonon T, Correa JA, Potin P (2008) Copper stress induces biosynthesis of octadecanoid and eicosanoid oxygenated derivatives in the brown algal kelp Laminaria digitata. New Phytol 180(4):809–821
Roleda MY, Morris JN, McGraw CM, Hurd CL (2012) Ocean acidification and seaweed reproduction: increased CO2 ameliorates the negative effect of lowered pH on meiospore germination in the giant kelp Macrocystis pyrifera (Laminariales, Phaeophyceae). Glob Chang Biol 18(3): 854–864
Roy S (2020) Seaweed diversity within intertidal zone of Olaikuda and Vadakkadu, Rameshwaram, southeast coast of India. Heliyon 6(7):e04585
Satheesh S, Wesley SG (2012) Diversity and distribution of seaweed in the Kudankulam coastal waters, south-eastern coast of India. Biodiv J 3(1):79–84
Schütte G (2018) What kind of innovation policy does the bioeconomy need? New Biotechnol 40: 82–86
Shanks AL, Grantham BA, Carr MH (2003) Propagule dispersal distance and the size and spacing of marine reserves. Ecol Appl 13(1):159–169. https://doi.org/10.1890/1051-0761
Short F (1980) A simulation model of the seagrass production system. Handbook of seagrass biology: an ecosystem perspective. Garland STPM Press, New York
Shoubaky G, Kaiser M (2014) Monitoring spatial and temporal seaweed variation using remote sensing data in Al-Shoaiba Coast, Red Sea. Br J Environ Climate Change 4(4):409–422
Silva IB, Fujii MT, Marinho-Soriano E (2012) Influence of tourist activity on the diversity of seaweed from reefs in Maracajaú, Atlantic Ocean, Northeast Brazil. SciELO 22(4):889–893
Singh PM, Maity D, Saha S, Dhal NK (2022) Seaweed utilization and its economy in Indian agriculture. Materials Today: Proceedings.
Steneck RS, Graham MH, Bourque BJ, Corbett D, Erlandson JM, Estes JA, Tegner MJ (2002) Kelp forest ecosystems: biodiversity, stability, resilience and future. Environ Conserv 29:436–459
Suo R, Wang Q (1992) Laminaria culture in china. Infofish Int 1:40–42
Taylor RB, Cole RG (1994) Mobile epifauna on subtidal brown seaweeds in northeastern New Zealand. Mar Ecol Prog Ser 115:271–271
Thompson RC, Crowe TP, Hawkins SJ (2002) Rocky intertidal communities: past environmental changes, present status and predictions for the next 25 years. Environ Conserv 29:168–191
Tom S (2008) Garrison-essentials of oceanography, 5th edn. Brooks Cole, England
Valanju N (2020) Biodiversity of seaweed along the coastline of Sindhudurg (Maharashtra). Research Interventions and Advancements in Plant Sciences. 175–182
Van der Loos LM, Schmid M, Leal PP, McGraw CM, Britton D, Revill AT, Hurd CL (2019) Responses of macroalgae to CO2 enrichment cannot be inferred solely from their inorganic carbon uptake strategy. Ecol Evol 9(1):125–140
Vaselli S, Bertocci I, Maggi E, Benedetti-Cecchi L (2008) Assessing the consequences of sea level rise: effects of changes in the slope of the substratum on sessile assemblages of rocky seashores. Mar Ecol Prog Ser 368:9–22
Veeragurunathan V, Mantri VA, Eswaran K (2021) Influence of commercial farming of Kappaphycus alvarezii (Rhodophyta) on native Seaweed of Gulf of Mannar, India: evidence for policy and management recommendation. J Coast Conserv 25(6):1–12
Veeragurunathan V, Mandal SK, Vizhi JM, Grace PG, Gurumoorthy U (2022) Studies on seaweed diversity along the intertidal zone of islands of Gulf of Mannar marine biosphere reserve, India for policy and management recommendation: studies on seaweed diversity along the intertidal

zone of islands of Gulf of Mannar marine biosphere reserve, India for policy and management recommendation. J Coast Conserv 26(4):28

Walker CH, Sibly RM, Peakall DB (2005) Principles of ecotoxicology. CRC Press, Boca Raton

Wernberg T, Bennett S, Babcock RC, De Bettignies T, Cure K, Depczynski M et al (2016) Climate-driven regime shift of a temperate marine ecosystem. Science 353(6295):169–172

Yadav SK, Palanisamy M (2021) Marine macro algal diversity of St. Mary's Island, Karnataka, India. Nelumbo 63(1):235

Yang L-E, Lu Q-Q, Brodie J (2017) A review of the bladed Bangiales (Rhodophyta) in China: history, culture and taxonomy. Eur J Phycol 52:1–13

Zulaiha S, Abidin Z, Mohamed B (2014) A review of SCUBA diving impacts and implication for coral reefs conservation and tourism management. SHS Web Conf 12:01093

Part V

Emerging Frameworks for Conservation

Homegarden Agroforestry for Plant Diversity Conservation in an Urban Landscape: Practices and Prospects

25

Ricardo T. Bagarinao

Abstract

Urbanization is rapidly increasing in recent years. Many habitats and agricultural lands are converted into residential land uses displacing several species as well as reducing food production systems, respectively. However, several households in urban centers are establishing homegarden agroforestry (HGA) within their residences, which can be considered as important ecological systems that could support plant diversity. This paper aims to elucidate the role of homegarden agroforestry for plant diversity conservation especially in urban landscapes, where human activities dominate and in attaining the sustainable development goals. HGA structures and compositions will be discussed. Implications in relation to plant diversity management and sustainability and prospects will be presented.

Keywords

Homegarden agroforestry · Urbanization · Green space · Urban landscape · Plant diversity conservation

25.1 Introduction

Our world is highly and rapidly urbanizing. According to the Population Division of the United Nations (2018a, b), the global urban population has increased from merely 30% in 1950 to 55% in 2018 albeit the rate of increase differs across the

R. T. Bagarinao (✉)
Faculty of Education, University of the Philippines Open University, Laguna, Philippines
e-mail: ricardo.bagarinao@upou.edu.ph

S. Ramamoorthy et al. (eds.), *Plant Diversity in Biocultural Landscapes*,
https://doi.org/10.1007/978-981-19-8649-9_25

regions. The Northern America was found to be the most urbanized region, where 82% of its total population is in urban centers. Asia is about 50% urbanized, while Africa remains rural, since 57% of its total population is still in rural areas.

However, recent estimates (e.g., Ritchie and Roser 2018) indicate that urban shares across all countries will increase in the coming decades, while rural shares are expected to be concentrated in very few countries. In fact, it is projected that by 2050, two-thirds of the world's population will be in urban areas—a significant increase from 54% in 2016 (Ritchie and Roser 2018).

Given this rate of increase, urbanization has become a major driver of ecosystem degradation as well as biocultural landscape conversion. Lyu et al. (2018) had indicated that urbanization has become an inevitable global process that drives land use change, which, in turn, affects ecosystem services and traditional ecological knowledge, and induces severe environmental problems and traditional knowledge erosion. For instance, Seto et al. (2011) reported that increasing urbanization and urban population created a strong pressure on natural resources and high demand for ecological system services. When combined, they can lead to some serious environmental problems, including water crisis, air pollution, plant diversity loss, and alteration of microclimate (Seto et al. 2011). Estoque and Murayama (2013) reported that such scenarios are common among developing countries, where urban centers are largely dependent on the integrity of ecosystems in surrounding rural areas. Moreover, Liang et al. (2008) had indicated that urbanization had resulted in the scarcity of green spaces in urban centers. Loss of plant species is very high and has become an obstacle to the creation and development of ecological-friendly cities. On the other hand, Lasisi and Ekpenyong (2011) had established a link between urbanization and loss of traditional ecological knowledge. They pointed out that urbanization is a major driver for the loss of TEK, including trado-medicine (herbs from leaves, roots, tree barks, etc.), technology, nutrition, and resource conservation. This requires the establishment of an urban development that promotes the conservation of the environment as well as traditional knowledge.

However, common-to-urban centers are green spaces consisting of homegardens. Kumar and Tiwari (2017) describe homegardens as unique land-use systems that consist of multipurpose trees and shrubs in combination with annual, perennial, and seasonal agricultural crops and livestock grown and managed within the compounds of individual homes. With these compositions, several authors (e.g., Paembonan et al. 2018; Eyasu et al. 2020; Kefale 2020; Gifawesen et al. 2020) used homegarden agroforestry (HGA) instead of simply homegardens to refer to such type of food production system. The diversity of its composition is seen to contribute to food security, sustainable development, and biodiversity conservation. In fact, these unique systems have been considered as important ecological as well as agricultural systems in an urban landscape setting that can be tapped to address food security, sustainability, and ecological issues of urbanization especially the plant diversity loss.

This paper aims then to elucidate the role of HGAs in plant diversity conservation especially in urban setting, where human activities dominate and in attaining the sustainable development goals. HGA structures and compositions will be discussed.

Implications in relation to plant diversity management and sustainability and prospects will be presented.

25.2 The Concept of Homegarden Agroforestry

The idea of homegarden agroforestry (HGA) can be traced back to the prehistoric time when hunters and gatherers had accidentally dispersed the seeds of fruit trees near their camp sites (Hutterer 1984); hence, HGAs are considered traditional food production systems. Soemarwoto (1987) reported that since then, HGAs had become associated with elite residences, palaces, temples, and even the homes of the ordinary people. Recently, there is an increasing richness in literature that describes the roles that HGAs are playing in social and ecological systems. For instance, Taylor and Lovell (2021) argued that HGAs offer greater socio-cultural as well as ecological benefits than the annual cropping system. In addition, De Zoysa (2022) reported that homegarden agroforestry plays an important role in increasing species diversity in complex, agro-ecological and sustainable land use systems. Given such a role, they attempted to develop strategies that could promote the use and establishment of sustainable HGAs in a district in Sri Lanka.

Kumar and Nair (2004) pointed out that HGA's definition, structure, and function can vary from place to place according to socio-economic climate and cultural factors, local physical environment, and ecological characteristics. De Zoysa (2022) defined HGAs as the land use systems that involve the deliberate management of multipurpose assembly of trees and shrubs with an intimate association with annual, perennial, and seasonal agricultural crops and in some cases, with livestock within the property of individual homes. Similarly, Kabir and Webb (2008) defined HGAs as a special category of agroforestry that deals with the cultivation of multilayered multipurpose crops and trees in combination with animals around a homestead. On the other hand, Galhena et al. (2013) considered HGAs as important elements of the local food systems that can enhance household food nutrition and food security. They considered such systems as income-generating and livelihood opportunities providers especially for the resource-poor households in the tropical developing countries. Galluzzi et al. (2010) also looked at HGAs as systems that help connect conservation of natural ecological systems and livelihood income by creating a link between the conservation of species and genetic diversity and marketable cultivated species. The plants or species, however, are arranged randomly and conveniently depending on the needs of the farmers, while species density is variable depending on the demand in the market and the needs of the household (Mengistu and Asfaw 2016).

Though scholars vary in their understanding of HGAs, most of them consider HGAs as agroforest systems that are located close to human dwellings (Huai and Hamilton 2009) and integrated into family life (Kefale 2020). As Huai and Hamilton (2009) had emphasized, these production systems tend to be located close to dwellings for reasons of security, special care, and convenience. As such, they are seen to occupy land marginal to field production and receive labor marginal to major household economic activities. In other words, HGAs are marked by low capital input and simple technology, while they consist of two basic components, namely, proximity to human dwellings and diversity of plants that are well-adapted to the local environmental conditions and serving multiple purposes.

25.3 Structures and Composition of HGAs

The structure of HGAs differs from one region to another depending on the garden size and shape, types of crops being cultivated and the intensity of cultivation, and complexity of species diversity (Kefale 2020). However, it can be categorized into vertical and horizontal structures, where the former reflects the degree of complexity of the garden, while the latter reflects the placement of plant species within the garden relative to the owner's house (Megabit et al. 2018). For instance, De Clerck and Negrreos-Castillo (2000) reported that the traditional HGAs in Mexico consist of five vertical layers, namely, tall tree stratum, lower tree stratum, tall shrub stratum, lower shrub stratum, and herbaceous stratum. According to them, the tall tree stratum is the most dominant stratum that stands at 12 m or more. The stratum is dominated by several species of palms, timber trees, and tall fruit-bearing trees, such as avocado and mango trees. On the other hand, the lower tree stratum, which stands at 3–6 m, composed mainly of fruit-bearing trees, such as citrus species. The tall shrub stratum whose height ranges from 1.5 to 3 m makes up about 15% of the total homegarden agroforest species and mainly composed of bananas, papaya, hierba santa, achiote, and chaya. In contrast, low shrub stratum contains 12% of the species in the homegardens and extends from 0.5 to 1.5 m high. It is composed of annual and perennial herbaceous plants including tomatoes, corn, arrowroot, chili, and espelon. The last stratum, i.e., the herbaceous stratum, rose to 0.5 m from the ground. It comprised of herbs and creepers, such as basil, squash, and sweet potatoes.

These layers are different from those located in Thailand, where most of the HGAs have only four vertical layers, namely, emergent, canopy, understory, and ground cover (Gajaseni and Gajaseni 1999). Huai and Hamilton (2009) reported that this layering is like the dry dipterocarp forest ecosystem in the area except that they are lower in height. According to them, the emergent layer rises more than 10 m from the ground and comprises mainly of coconut, areca palm, bamboo, and teak. The canopy layer stands between 5 m and 10 m and composed of mango, tamarind, durian, coral tree, Indian trumpet flower, and plum mango. Standing at less than 5 m, the understory layer is composed of diverse fruit-bearing trees with a height <5 m. The ground cover consists of shrubs and grasses.

In addition, HGAs in arid regions such as Brazil have three strata only. As Albuquerque et al. (2005) had reported, they consist of upper stratum that stands at 7–12 m and composed of fruit and timber trees, the mid stratum rising at 3–7 m and comprising of plants with multiple uses (e.g., *Tabebuia* sp. and *Myracrodruon urundeuva*) and the lower stratum that rises between 1 and 3 m and composed of medicinal plants and forages. In India, Das and Das (2005) reported that HGAs had four-to-five vertical layers, consisting of the emergent, canopy, understory, shrub, and herb layers. According to them, the emergent layer, which had a height of 15 m or more, consists of multipurpose tree species, such as *Artocarpus lakoocha, Bombax ceiba, Tamarindus indica Tetrameles nudiflora,* and *Toona ciliata*. Having a height between 10 and 15 m, the canopy layer is usually composed of *Mangifera indica, Gmelina arborea, Artocarpus chama, Areca catechu, Cassia siamea, Dillenia indica, Lagerstroemia speciosa, Sterculia foetida, Artocarpus heterophyllus,* and Bamboo species. The understory layer is commonly dominated with *Ficus* sp., *Litchi chinensis, Cocos nucifera,* and *Citrus* spp. and stands at a height between 5 and 10 m. The shrub layer is composed of shrubs species such as Hibiscus species and saplings of species forming the emergent and canopy layers. It stands on a height between 1 and 5 m. The herb layer, which consists mainly of vegetables, ornamentals, and medicinal species, is less than 1 m (Das and Das 2005).

While there is a clear layering of the vertical structure of HGAs, the horizontal structure is highly variable and dependent on the size of the available land, educational attainment of the HGA owners, and market demand of crops and other products (Kefale 2020). Paembonan et al. (2018) reported that in South Sulawesi, Indonesia, the proportion of trees to annual crops is dependent on the size of the land owned by the farmers themselves. Small HGAs are dominated by annual crops, while large HGAs have more fruit and timber trees. Patel et al. (2022) reported similar findings in their study on homegardens in India. They observed that the highest density of tree species was observed in large-sized gardens, while the highest density of annual crops is in small-sized gardens. In addition, Asfaw et al. (2015) reported that in North-Western Ethiopia, spaces of HGAs closer to the gardener's house are dominated by herbaceous plants, while those in the middle are planted with coffee under the shade of trees and fruits. Garden borders were planted with spiny plants and trees that are used for pole, fuel, and construction materials.

Being the way plant species are organized in the horizontal plane of the HGAs, Lope-Alzina and Howard (2012) indicated that the horizontal structure is more difficult and complex to characterize compared with the vertical structure. They suggested that this structure should be understood to consist of management zones, where plants are organized according to specific criteria. In a study conducted in North–East India, Das and Das (2005) identified five major management zones in HGAs according to their composition, function, and location: (a) bamboo groves; (b) betel groves; (c) banana groves; (d) vegetable garden; and (e) dense or forest zones. Each zone is dominated by a specific plant species and perform a specific function. For instance, the betel, bamboo, and banana groves are usually dominated by pure stands of betel, bamboo, and banana, which are used for commercial purposes. These zones are located near the owners' house for ease of management and harvest (Das and Das 2005). On the other hand, the dense or forest zones, which are composed of a mixture of trees, fulfil several family needs, such as fodder, fuelwood, timber, etc. Since the trees in these zones include wild or rare tree species, they are also viewed as conservation sites.

On the other hand, Mendez et al. (2001) recorded a total of ten management zones, including the zone of fruit trees and multipurpose trees in their study of Nicaraguan HGAs. Majority of these trees have been used for timber, fuelwood, etc.

In addition, Asfaw et al. (2015) reported at least four management zones that can be observed generally within the horizontal structure of the HGAs they investigated in North-western Ethiopia. These zones include the herbaceous zone, coffee and shade and fruit trees zone, woody zone, and the border zone of spiny stem and tree species. The herbaceous zone is located close to the owner's house and consists of herbaceous plant species, while the coffee and shade and fruit tree zone is in the middle of the HGAs and composed mainly of coffees under the shade and fruit trees. The woody zone is also located in the middle of the HGAs, where trees such as *Eucalyptus* species are used for fuel, pole, and construction materials. The border zone is mainly comprised of spiny stem plant species, such as *Rosa abyssinica*, *Capparis tomentosa*, and *Acacia nilotica*, which are used as live fence of the HGAs. Interestingly, they observed that horizontal structure of the plant species declines as one goes from the zone closer to the owner's residence to out of the fields. Megabit et al. (2018) reported similar observation in their study involving the Bulen District of the same region in Ethiopia, though in this district, the management zones are dominated by perennial and annual crops, such as cucumber, corn, sorghum, common bean, and coffee.

25.4 Homegardens Agroforestry in Urban Settings

HGAs in urban landscapes are looked at as approaches to achieve sustainability specifically to create sustainable livelihoods. As Drescher et al. (2006) had emphasized, diversifying food and income sources can be a good buffer against dwellers' vulnerability to poor economic development in urban environment. HGAs are being seen as part of such diversification. However, in the process of achieving such a goal, other important environmental management aspirations such as biodiversity conservation are being attained. Šiftová (2021) had pointed out that HGAs are now being included in the local as well as regional governments' local biodiversity action plans together with resilience and food or health agendas. Although *spatial densification*, which is the increased demographic pressure on available lands (Drescher et al. 2006), is a common problem in urban areas, HGAs are established through soil-less cultures, such as hydroponics and container gardens. In some cases, rooftop gardening is being done as an alternative to the traditional HGAs especially in densely populated cities. As part of this strategy, HGAs owners utilized fences, bottles, bags, and walls for raising plants (Ranasinghe. 2005). In Cuba, organoponics is a well-established form of HGAs, where efficiency in water conservation is combined with the use of compost and manure for increasing soil fertility in an urban context (Pinderhughes 2004). In an urban area in the Philippines, where spaces are very limited, HGAs were done in either small-raised garden beds or in available open spaces near their houses (Barameda 2017). In other cities, vegetables and ornamentals are grown in containers, which were hang vertically on the walls and fences especially at eye level to free the HGAs owners from bending while tending the plants. In the United States of America, HGAs especially of those Chinese-origin households are exhibiting a unique layered structure, which is supported by trellises that are constructed from found lumber and branches (Taylor and Lovell 2015). Some of them were either located in the backyards of single-family dwellings or in multifamily buildings but with secondary growing areas either in front yards or side yards.

In terms of the kind of plants that are used in urban HGAs, Mosina et al. (2014) reported that HGAs in cities especially in some provinces in South Africa are characterized by higher numbers of plant species that are used for food and medicines. According to them, these species have higher percentages in terms of composition among the HGAs they surveyed, although indigenous plants that show similarities to the natural surrounding vegetation were also observed. In other cities, however, HGAs are characterized by introduced or exotic and ornamental plants. Belonias et al. (2014) reported similar findings in their study in a city in Leyte, Philippines. They reported that more than 70% of the plant species observed in the study site was exotics and used as ornamental plants. The most common plants were erect herbs, followed by trees, shrubs, herbaceous vines, and epiphytes, which were used for ornamentation and provision of fruits, medicines, vegetables, and spices. In contrast, Akinnifesi et al. (2010) had observed more indigenous species than the exotic species in the urban HGAs they investigated in Northeastern Brazil. Of the total 63 fruit tree species observed in the area, almost 60% is indigenous to Brazil (wild and semi-wild domesticates). In Yucatan Peninsula, Mexico, urban HGAs, which are extensions of the rural homegarden practices, are characterized mainly with indigenous plants suitable for food, medicinal, fodder, forage, and fertilizer purposes (Poot-Pool et al. 2018). They contribute significantly toward the conservation of regional and local biodiversity aside from contributing toward local food security and nutrition and strengthening family economy.

25.5 HGAs and Plant Diversity Conservation

Among the environmental resources that are mostly affected by urbanization is plant diversity. Wang et al. (2020) had indicated that urbanization is among the major causes for the loss of plant diversity both at the local and regional scales. The vulnerability of these resources is due to their sedentary lifestyles, which make them highly prone to destruction and extinction (Huot et al. 2013). Urbanization that requires deforestation or vegetation clearing will not only reduce ecosystem services that will benefit both biodiversity and city dwellers but also lead to urban plant diversity loss. Thus, Nguyen et al. (2020) suggested that understanding the spatial and temporal patterns of plant diversity in an urban setting should be an important component of future sustainable urban development. To this effect, analyzing the contribution of homegarden agroforestry to plant diversity conservation in an urban landscape is an important step toward an ecologically well-informed urban development decision-making.

Several studies (e.g., Idohou et al. 2014; Gbedomon et al. 2017; Semu 2018) had indicated the important role of HGAs in in-situ conservation of plant biodiversity. For instance, Eyzaguirre and Watson (2002) had identified HGAs as key places for plant biodiversity conservation, while Eyzaguirre and Linares (2004) considered them as vital refugee camps for several plant species especially those that are not widely grown in the larger agro-ecosystem. Zemede and Woldu (1997) reported that HGAs perform dual function, i.e., for food production as well as for the conservation of agrobiodiversity. Idohou et al. (2014) also reported that HGAs had been the haven for the most ecologically important species. Their findings were consistent with the findings of Maldonado et al. (2013), where ecologically important plant species had dominated in their study sites.

In addition, Zemede (2002) had emphasized that HGAs are important places to arrest the threat in genetic erosion of useful plants through continuous use. For example, Gbedomon et al. (2017) reported in their study involving HGAs in Benin, West Africa that crops and crop wild relatives and wild plant species have been conserved effectively especially in HGAs with multiple functions. Their findings indicate that multifunctional HGAs had higher plant species diversity, where crops and crop wild relatives perform food function, while wild plant species have mainly medicinal function. Furthermore, in a study involving 75 randomly selected HGAs in Holeta Town, Ethiopia, Amberber et al. (2014) reported that about 112 plant species belonging to 93 genera and 43 families were effectively conserved and classified to 14 functional groups. Of the total, 43% were herbs, 29% were trees, 25% were shrubs, and 3% were climbers. They concluded that the HGAs being investigated play a crucial role not only in securing food supply to the household owners but also in in situ conservation of these functional groups of plants. Agustina et al. (2019) also reported a high plant diversity in HGAs in their study involving 90 HGAs in East Java, Indonesia. To cite, about 447 plant species were recorded in the study sites that belonged to 95 families. Of this total, Araceae contains the highest number of species (31), followed by Asteraceae (21), Asparagaceae (20), Euphorbiaceae (17), Myrtaceae, Poaceae Solanaceae (16), Musaceae (14), Lamiaceae (13), Amaryllidaceae and Fabaceae (12), Amaranthaceae (11), and Rutaceae (10).

Tobias et al. (2021) had further expanded the role of homegardens to include their importance in housing endemic and threatened species. In their study involving three homegardens in Luzon, Philippines, they observed that of the 75 recorded species, about 10 were endemic to the Philippines and 18 species were native, while 19 species were listed as threatened in either IUCN Red List (IUCN, 2018-1) or the Philippine Red List (DAO 2017). On the other hand, Caballero-Serrano et al. (2016) looked at the plant diversity in HGAs in relation to the ecosystem services they performed that have either a direct or positive impact on human welfare. They reported that of the 484 plant species they recorded in about 138 HGAs in Sangay Ecuador, about 20 ecosystem services were associated with them, the most important of which were the provision of food, medicine, and ornamentation. Kumari et al. (2009) reported similar findings, where 51%, 36%, and 12% of the 289 plant species that they recorded in 106 randomly selected HGAs in Thimbirigaskatuwa suburban area in western Sri Lanka performed ornamentation, food provision, and medicinal functions, respectively. In addition, Mosina et al. (2014) had also recorded some indigenous species in HGAs in Limpopo Province, South Africa, which includes

Vigna subterranea (L.) Verdc, *Vigna unguiculata* (L.) Walp, and *Erythrina lysistemon* Hutch. (Papilionaceae family) and *Aloe ecklonis* Salm-Dyck and *Aloe* sp. (Xanthorrhoeaceae).

These studies reveal how HGAs owners conserve (either intentionally or unintentionally) important species in their ecosystem (Idohou et al. 2014). These findings support the idea that HGAs can significantly contribute to the conservation of plant species diversity while providing HGAs owners a sustained supply of nutrient-rich food and medicine. As Hemmelgarn and Munsell (2021) had emphasized, HGAs in urban landscapes must be viewed beyond food opportunities, as these dynamic, multi-strata systems perform broader set of functions, including the conservation of plant diversity. Kumar and Nair (2004) had even included environmental, social, medicinal, as well as decorative effects as potential beyond food benefits of HGAs. Akinnifesi et al. (2010) had also indicated that HGAs could offer a unique opportunity for the conservation and sustainable use of plant species that are already endangered in their native habitats. They argued that these species continue to perform a significant role in the ecological as well as socio-economic landscape of the city, making HGAs to be successfully serving as a sustainable, green, environmental model for plant diversity conservation (Poot-Pool et al. 2018).

25.6 Prospects for HGAs

The many studies pointing out to the potential of HGAs for plant diversity conservation can justify its prospective future as an effective approach for managing plant diversity as well as attaining environment-friendly and sustainable urban development. As cited in various studies (e.g., Akinnifesi et al. 2010; Caballero-Serrano et al. 2016; Poot-Pool et al. 2018; Hemmelgarn and Munsell 2021), HGAs can be a multifunctional system that can effectively support plant diversity conservation, ecosystem services restoration and maintenance, food security, and sustainability. Hence, they can be a good prospective system that can be tapped for a deliberate plant diversity management in an urban ecosystem.

HGAs can be an effective repository of traditional knowledge albeit some modifications may be observed under an urban landscape setting. Nonetheless, the idea of conserving plant diversity within a small area can be made possible by repositioning this vast traditional ecological knowledge of the HGAs owners into a human-dominated ecosystems and making it as part of the local community's cultural patterns. For instance, Shimrah et al. (2018) pointed out that the deliberate arrangements (vertical and horizontal) of crops within the HGAs could be based on the traditional knowledge of owners to allow the efficient utilization of nutrients by different plant species, creation of barriers for pests and insects, regulation of soil erosion by larger roots, and minimization of loss in time of natural calamities and physiological stress.

Hemmelgarn and Munsell (2021) had also emphasized that HGAs can serve as models for the potential symbiosis between nature- and human-oriented objectives especially in human-dominated ecosystems, such as urban centers or cities. These objectives are usually in conflict with each other and most of the time are not compatible especially when development is prioritized. The presence, however, of diverse plant species that range from trees to herbs in HGAs can break the "war" between these conflicting objectives. Although Sanderson and Huron (2011) had indicated that these objectives are complex and sustainably harmonizing them are not always easy, the presence of such symbiotic relationship within the HGAs can enhance ecosystem services, including plant conservation and habitat connectivity (Doody et al. 2010). Since cities are composed of heterogeneous mosaics and contain unique interactions between their biophysical and socio-cultural elements, it is important that the management of the emerging urban ecosystem should be deliberate. This is necessary, because though there is a richness of indigenous and non-indigenous species in urban HGAs, their interactions within the anthropogenic communities will not necessarily result in greater stability (Alberti et al. 2003). As Pretty et al. (2009) had pointed out, the resilience of urban ecosystems depends on whether HGA owners designed human-oriented objectives that promote conservation and cultural diversity.

Aside from contributing to the conservation of plant diversity, HGAs are prospective storage of important genetic resources of plants. As Vinceti et al. (2022) had indicated, forest destruction and deforestation are continually becoming a global concern especially with the intensification of climate change and biodiversity loss. Most of the remnant species are now being confined to smaller areas, which may be affected by the changing climate. HGAs, however, can be seen as important repository areas for the genetic resources of important plant species. Not only that they sustain the household food consumption and income generation, HGAs can potentially play a vital role in conserving plant genetic resources, such as fruit trees, nut trees, medicinal plants, and other crop and crop wild relatives. Furthermore, the tree species cultivated in HGAs can sustain genetic diversity in addition to trees found in remnant forests. If several HGA systems are developed in urban areas, it is possible that more and more genetic resources of tree species will be conserved. As discussed above, several HGAs house unique, rare, indigenous, and native plant species. They also harbor a high diversity of the genetic resources of crop plants and their wild relatives. In a study conducted by Heraty and Ellstrand (2016) involving a comparison of the allelic quantity of corn between the commercial populations and garden populations in Los Angeles and Riverside, California, it was observed that the total number of alleles in garden populations were higher than the number of alleles in commercial populations. In addition, all garden populations were observed to be polymorphic, while only 80% in commercial populations were polymorphic. The results indicate a more diverse population in the gardens being investigated. In addition, Mwavu et al. (2016) had reported high measure of diversity indices in the 120 HGAs that they have investigated in Kampala, Uganda. About five (5) families were observed to be species-rich among all the HGAs, namely: Moraceae (11 species), Leguminosae (4), Euphorbiaceae (3), Myrtaceae (3), and Combretaceae (3), while more than 80% of the HGAs had indicated an alpha diversity value of greater than 1. Although there has been no study yet that determines the effectiveness of HGAs in storing plant genetic resources compared to the existing gene banks, the presence of this characteristic of HGAs makes them a suitable “storage area” for such genetic diversity. Given this diversity, they can become refuges for the genetic resources of wild species that are threatened by forest degradation, deforestation, and land conversion due to urbanization.

Aside from directly becoming a repository of these important plant genetic resources, HGAs can also play a vital role in the transitioning from wild to domesticated plant species as Wiersum (2004) had suggested. Since HGAs can be a site for unplanned hybridization among related plant species, they can serve as areas for the “evolution” of plants (Hughes et al. 2007). As indicated in various practices, the germplasm from the wild can be brought under cultivation in HGAs (Vinceti et al. 2022). In the process, plant genetic resources may be further enhanced while at the same time being protected.

HGAs can also play important roles in attaining sustainable development goals. Being one of the earliest surviving traditional food systems, HGAs appear to be a vital approach for accomplishing several SDGs, such as SDG 2 (Zero Hunger), SDG 3 (Good Health and Wellbeing), SDG 8 (Decent Work and Economic Growth), SDG 11 (Sustainable Cities and Communities), SDG 13 (Climate Action) and SDG 15 (Life on Land). With proper planning and taking into consideration how the traditional communities have sustained its practice for so long, HGAs can be viewed as an effective strategy to sustain household food security, healthy living, inclusive, safe, and resilient urban development, and conservation of important ecosystems as well as plant species (Sharma et al. 2022). For instance, Wiersum (2006) reported that HGAs were utilized to reduce the ‘hidden hunger’ and disease caused by micronutrient deficiency among the communities in Java and Sulawesi, Indonesia. HGAs were then utilized as a sustainable source of food albeit in small amount for the households. They were also used as the main source of staple food especially for people in densely populated areas, where the establishment of farms is not feasible (Soemarwoto 1987).

In addition, the ability of HGAs to supply fresh and nutritional food can help improve the health and well-being of the household owners as well as their immediate local communities. In Africa and Asia, HGAs were used as major sources for micronutrients especially for children suffering from vitamins and minerals deficiency. Iannotti et al. (2009) had reported that HGA’s food production programs in countries where the problem is severe were launched to assist households in addressing Vitamin A deficiency. Consequently, the improvement of the quality of diet has facilitated the improvement of health among these children as year-round production of vegetables and fruits were sustained. Selepe and Hendriks (2014) also reported an improvement in the health condition of the children who participated in the homegarden project in Eatonside, South Africa. They observed that with project implementation, dietary diversity of these children had significantly increased with good access to nutrient-rich food produced from the garden. Consequently, the problem of malnutrition, which was alarming at the start of the project, had been addressed with readily available vegetables and other nutritional foods from the HGAs.

Nicholls et al. (2020) had also emphasized that the conventional food production system produces more grains, fats, and sugars than fruits and vegetables, which are easily grown and harvested in HGAs. The HGAs household owners will also tend to consume more nutritious food and eat less processed food, and thereby enhancing their health and wellness. The latter can also result from the increasing good environmental quality as more and more green areas will be established in urban centers and cities. Through this process, HGAs can also contribute to climate action, as they reduce carbon emission by way of sequestering carbon dioxide and other greenhouse gases from the air and reduce the transport of food from rural areas to urban centers. Both mechanisms may mitigate climate change locally. Since the best approach to mitigate climate change is to trap the carbon in the air with the help of vegetation, HGAs with their diversity of plant species can potentially perform this function and convert the gaseous carbon into organic carbon in their biomass and soil. For instance, Kumar (2006) had estimated that carbon concentration above the ground in HGAs can be between 60 and 70 tonnes per hectare, while the organic carbon stock in the soil will come in a range of 80–100 tonnes carbon per hectare in a meter of soil depth. In another study conducted by Lowe et al. (2022) involving 122 HGAs in Sri Lanka, the computed above-ground biomass (AGB), which contains the sequestered carbon in the air, ranges from 0.8 to 139.4 Mg C/ha. The computation was based on the 5140 woody trees, which comes from 100 genera and 45 families recorded in the HGAs.

Although there is no hard data available that shows the effectiveness of HGAs as carbon sink, these studies clearly indicate their potential role in climate change mitigation especially in urban landscapes. If more and more urban HGAs will be established across the region and/or the country, the impact may be synergistic and a considerable reduction in the carbon concentration in the air may be expected.

Meanwhile, HGAs can be viewed as a vital approach to provide decent work and income to households. As suggested by various studies, HGAs provide economic benefits that are beyond the food and nutritional security especially for urban poor families. Calvet-Mir et al. (2012) reported that HGAs had contributed to income generation, economic welfare of households, and improved livelihoods. These are possible, because HGAs products can be sold to augment the household income and/or developed into a small cottage industry. Mitchell and Hanstad (2004) also pointed out that the savings from consuming products from HGAs can contribute to the disposable income of the households. In fact, Galhena et al. (2013) reported that income from HGAs has been utilized by the households to purchase additional food items and spent for education and other domestic purposes. Okigbo (1990) also reported that income from the sale of tree crops and livestock produce in HGAs had been accounted for more than 60% of the total household income in Southeastern Nigeria. Similar findings had been reported by Trinh et al. (2003) in their study involving the HGAs owners in Vietnam. They reported that HGAs in Vietnam were able to provide more than 22% of the cash income of the households, and thereby, improving the financial status of the family.

The prospective use of HGAs to attain SDGs can be possible, because they can manage, use, and conserve natural resources efficiently as compared to other agricultural production systems especially the monocropping and commercial agricultural practices. At the same time, they also provide multiple fundamental ecological functions such as nutrient re-cycling, and photosynthetic route enhancements and ecosystem services such as food, habitat, and recreation (Monroy and García 2013). Garcia-Flores et al. (2016) considered them as the fundamentals for inclusive and sustainable development.

25.7 Summary and Recommendations

The multifunctionality of HGAs can be tapped to address plant diversity loss due to urbanization. This traditional food production system may be harnessed for the conservation of plant species especially in biocultural landscapes affected directly or indirectly by urbanization. The ability of HGAs to allow the co-existence as well as the symbiotic relationship between nature-oriented and human-oriented objectives in human-dominated landscapes can make them a good system for plant diversity conservation and sustainable development. Their ability to provide multiple ecological functions and ecosystem services can also make them a good approach in attaining sustainable development goals in an urban setting. It is, therefore, recommended that the establishment of HGAs in urban centers must be considered as an important part of the development plan in any cities. A policy that requires the inclusion of HGAs in any land development projects must also be formulated to strengthen such a requirement. This is important, because HGAs are unlikely to be preserved with urbanization despite their important contribution to plant diversity conservation and sustained nutrient-rich food and medicinal supply. In addition, research studies that analyze the effectiveness of HGAs as a system for plant genetic resources conservation and carbon sink can be conducted. Finally, the importance and benefits of HGAs (both economic and ecological) must be disseminated to educate and encourage local communities, policymakers, and urban developers to consider them in their decision-making process and development plans. They need to be integrated into any environmental education programs especially targeting the younger generations.

References

Agustina TP, Nisyawati, Walujo EB (2019) Plant diversity and uses of the home garden in Pujon sub-district, Malang Regency, East Java, Indonesia. International Conference on Biology and Applied Science (ICOBAS). Retrieved from https://aip.scitation.org/doi/pdf/10.1063/1.5115659

Akinnifesi FK, Sileshi GW, Ajayi OC, Akinnifesi AI, de Moura EG, Linhares JFP, Rodrigues I (2010) Biodiversity of the urban homegardens of São Luís city, northeastern Brazil. Urban Ecosyst 13:129–146. Retrieved from https://bit.ly/3QXHVAJ

Alberti M, Marzluff JM, Shulenberger E, Bradley G, Ryan C, Zumbrunnen C (2003) Integrating humans into ecology: opportunities and challenges for studying urban ecosystems. Bioscience

53:1169–1179. Retrieved from https://academic.oup.com/bioscience/article/53/12/1169/301939/

Albuquerque UP, Andrade LHC, Caballero J (2005) Structure and floristics of homegardens in northeastern Brazil. J Arid Environ 62:491–506. https://doi.org/10.1016/j.jaridenv.2005.01.003

Amberber M, Argaw M, Zemede A (2014) The role of homegardens for in situ conservation of plant biodiversity in Holeta Town, Oromia National Regional State, Ethiopia. Int J Biodiver Conserv 6(1):8–16. https://doi.org/10.5897/IJBC2013.0583

Asfaw Z, Linger E, Zewudie S (2015) Plant species richness and structure of homegarden agroforestry in Jabithenan District, North-Western Ethiopia. Int J Environ Sci 4(2):52–58. https://bit.ly/3wSAa6y

Barameda TV (2017) Surviving in the city through home gardens: a case study of home gardeners in Barangay UP Campus. Philipp J Soc Develop 9:39–62. Retrieved from https://cswcd.upd.edu.ph/wp-content/uploads/2021/10/PJSD-Vol-9-2017_Barrameda.pdf

Belonias B, Platino CS, Malanguis JM (2014) Agrobiodiversity of home gardens in selected marginal upland villages of Inopacan, Leyte, Philippines. Ann Trop Res 36:48–69. Retrieved from https://bit.ly/3Avt2A3

Caballero-Serrano V, Onaindiab M, Aldayc JG, Caballero D, Carrascod JC, McLarene B, Amigo J (2016) Plant diversity and ecosystem services in Amazonian homegardens of Ecuador. Agric Ecosyst Environ 225:116–125. https://doi.org/10.1016/j.agee.2016.04.005

Calvet-Mir L, Gómez-Bagetthun E, Reyes-García V (2012) Beyond food production: ecosystem services by home gardens. A case study in Vall Fosca, Catalan Pyrenees, northeastern Spain. Ecol Econ 74:153–160. https://doi.org/10.1016/j.ecolecon.2011.12.011

DAO (2017) Establishing the national list of threatened Philippine plants and their categories and the list of other wildlife species. Quezon City. Retrieved from https://www.philippineplants.org/dao-2017-11.pdf

Das T, Das AK (2005) Inventorying plant biodiversity in homegardens: a case study in Barak Valley, Assam, North East India. Curr Sci 89(1):155–163. https://bit.ly/3m1GaF9

De Clerck FAJ, Negrreos-Castillo P (2000) Plant species of traditional Mayan homegardens of Mexico as analogs for multistrata agroforests. Agrofor Syst 48:303–317. https://bit.ly/3wWpkOc

De Zoysa M (2022) Scientific basis of agroforestry homegardens in Matara District, Sri Lanka: present status and improvement needs. Open J Forest 12:60–87. https://doi.org/10.4236/ojf.2022.121004

Doody BJ, Sullivan JJ, Meurk CD, Stewart GH, Perkins HC (2010) Urban realities: the contribution of residential gardens to the conservation of urban forest remnants. Biodivers Conserv 19:1385–1400. https://doi.org/10.1007/s10531-009-9768-2

Drescher AW, Holmer RJ, Iaquinta DL (2006) Urban Homegardens and allotment gardens for sustainable livelihoods: management strategies and institutional environments. In: Nair PK, Kumar B (eds) Tropical homegardens: a time tested example of sustainability. Springer, pp 317–338. https://doi.org/10.1007/978-1-4020-4948-4_18

Estoque RC, Murayama Y (2013) Landscape pattern and ecosystem service value changes: implications for environmental sustainability planning for the rapidly urbanizing summer capital of the Philippines. Landsc Urban Plan 116:60–72. https://www.sciencedirect.com/science/article/abs/pii/S0169204613000777

Eyasu G, Tolera M, Negash M (2020) Woody species composition, structure, and diversity of homegarden agroforestry systems in southern Tigray, Northern Ethiopia. Heliyon 6(12):1–9. https://doi.org/10.1016/j.heliyon.2020.e05500

Eyzaguirre PB, Watson JW (2002) Rome: IPGRI home gardens and in situ conservation of plant genetic resources in farming systems. In: Proceedings of the 2nd International Home Gardens Workshop, 17–19 July 2001, Witzenhausen, Federal Republic of Germany. Retrieved from https://bit.ly/3dvGN8N

Eyzaguirre PB, Linares OF (2004) Home gardens and agrobiodiversity. Smithsonian Books, Washington, 296pp

Gajaseni J, Gajaseni N (1999) Ecological rationalities of the traditional homegarden system in the Chao Phraya Basin, Thailand. Agrofor Syst 46:3–23. https://doi.org/10.1023/A:1006188504677

Galhena DH, Freed R, Maredia KM (2013) Home gardens: a promising approach to enhance household food security and wellbeing. Agric Food Sec 2:8. https://doi.org/10.1186/2048-7010-2-8

Galluzzi G, Eyzaguirre P, Negri V (2010) Home gardens: neglected hotspots of agro-biodiversity and cultural diversity. Biodivers Conserv 19:3635–3654. https://doi.org/10.1007/s10531-010-9919-5

Garcia-Flores JC, Gutierrez-Cedillo JG, Balderas-Plata MÁ, Araujo-Santana MR (2016) Livelihood strategy in the rural area of the Mexican Central Highlands: the family vegetable garden. Agric Soc Desarro [online] 13(4):621–641. Retrieved from https://www.scielo.org.mx/pdf/asd/v13n4/1870-5472-asd-13-04-00621.pdf

Gbedomon RC, Salako VK, Fandohan AB, Idohou AFR, Kakaï RG, Assogbadjo AE (2017) Functional diversity of home gardens and their agrobiodiversity conservation benefits in Benin, West Africa. J Ethnobiol Ethnomed 13(66):2–15. https://doi.org/10.1186/s13002-017-0192-5

Gifawesen ST, Tola FK, Duguma MS (2020) Review on role of home garden agroforestry practices to improve livelihood of small-scale farmers and climate change adaptation and mitigation. J Plant Sci 8(5):134–145. https://doi.org/10.11648/j.jps.20200805.15

Hemmelgarn HL, Munsell JF (2021) Exploring "Beyond-Food" opportunities for biocultural conservation in Urban Forest Gardens. Urban Agric Reg Food Syst 6:e20009. https://doi.org/10.1002/uar2.20009

Heraty JM, Ellstrand NC (2016) Maize germplasm conservation in Southern California's urban gardens: introduced diversity beyond ex situ and in situ management. Econ Bot 70:37–48. https://doi.org/10.1007/s12231-016-9333-3

Huai H, Hamilton A (2009) Characteristics and functions of traditional homegardens: a review. Front Biol China 4(2):151–157. https://doi.org/10.1007/s11515-008-0103-1

Hughes CE, Govindarajulu R, Robertson A, Filer DL, Harris SA, Bailey CD (2007) Serendipitous backyard hybridization and the origin of crops. Proc Natl Acad Sci U S A 104:14389–14394. https://doi.org/10.1073/pnas.0702193104

Huot OB, Nachappa P, Tamborindeguy C (2013) The evolutionary strategies of plant defenses have a dynamic impact on the adaptations and interactions of vectors and pathogens. Insect Sci 20(3): 297–306. https://doi.org/10.1111/1744-7917.12010

Hutterer KL (1984) Ecology and evolution of agriculture in Southeast Asia. In: Rambo TA, Sajise PE (eds) An introduction to human ecology research on agricultural systems in Southeast Asia. University of the Philippines, Los Banos

Iannotti L, Cunningham K, Ruel M (2009) Improving diet quality and micronutrient. Nutrition: homestead food production in Bangladesh. Washington DC

Idohou R, Fandohan B, Salako VK, Kassa B, Gbèdomon RC, Yédomonhan H, Kakaï RLG, Assogbadjo AE (2014) Biodiversity conservation in home gardens: traditional knowledge, use patterns and implications for management. Int J Biodiver Sci Ecosyst Serv Manag 10(2): 89–100. https://doi.org/10.1080/21513732.2014.910554

Kabir ME, Webb E (2008) Can homegardens conserve biodiversity in Bangladesh? Biotropica 40:95–110. https://www.researchgate.net/publication/229726061_Can_Homegardens_Conserve_Biodiversity_in_Bangladesh

Kefale B (2020) Homegarden agroforestry in Ethiopia - a review. Int J Bio-resour Stress Manag 11(4):345–352. https://www.pphouse.org/upload_article/4159914_5_August_2020_Kefale.pdf

Kumar BM (2006) Carbon sequestration potential of tropical homegardens. In: Nair PKR, Kumar BM (eds) Tropical homegardens: a time-tested example of sustainable agroforestry. Springer, pp 185–204

Kumar BM, Nair PKR (2004) The enigma of tropical homegardens. Agrofor Syst 61:135–152. https://doi.org/10.1023/B:AGFO.0000028995.13227.ca

Kumar V, Tiwari A (2017) Importance of tropical homegardens agroforestry system. Int J Curr Microbiol App Sci 6(9):1002–1019. Retrieved from https://www.ijcmas.com/6-9-2017/Vikas%20Kumar%20and%20Anjali%20Tiwari.pdf

Kumari M, Kansuntisukmongkol K, Brockelman WY (2009) Plant diversity in home gardens and its contribution to household economy in suburban areas in Sri Lanka. Environ Nat Resour J 7 (2):12–29. Retrieved from https://ph02.tcithaijo.org/index.php/ennrj/article/view/82565

Lasisi R, Ekpenyong AS (2011) Urbanization and loss of traditional ecological knowledge: lessons from Rumuodomaya Community in Rivers State. Int J Cross-Cult Stud 1(1):54–64. Retrieved from https://bit.ly/3pxT7rM

Liang Y, Li J, Välimäki S (2008) Impact of urbanization on plant diversity: a case study in built-up areas of Beijing. Forest Stud China 10(3):179–188. https://doi.org/10.1007/s11632-008-0036-4

Lope-Alzina DG, Howard PL (2012) The structure, composition, and functions of homegardens: focus on the Yucatán Peninsula. Etnoecológica 9(1):17–41. Retrieved from https://bit.ly/3S62CM0

Lowe WAM, Silva GLLP, Pushpakumara DKNG (2022) Homegardens as a modern carbon storage: assessment of tree diversity and above-ground biomass of homegardens in Matale district, Sri Lanka. Urban For Urban Green 74:127671. https://doi.org/10.1016/j.ufug.2022.127671

Lyu R, Zhang J, Xu M, Li J (2018) Impacts of urbanization on ecosystem services and their temporal relations: a case study in Northern Ningxia, China. Land Use Policy 77:163–173. https://doi.org/10.1016/j.landusepol.2018.05.022

Maldonado B, Caballero J, Delgado-Salinas A, Lira R (2013) Relationship between use value and ecological importance of floristic resources of seasonally dry tropical forest in the Balsas River Basin, México. Econ Bot 67:17–29. https://doi.org/10.1007/s12231-013-9222-y

Megabit B, Muktar M, Lisanework N (2018) Plant species diversity and structure in homegarden agroforestry systems of Bulen District, North-Western Ethiopia. Agric Forest Fisher 7(6): 121–132. https://bit.ly/3PTAgDs

Mendez VE, Lok R, Somarriba E (2001) Interdisciplinary analysis of homegardens in Nicaragua: microzonation, plant use and socioeconomic importance. Agrofor Syst 51:85–96. https://doi.org/10.1023/A:1010622430223

Mengistu B, Asfaw Z (2016) Woody species diversity and structure of agroforestry and adjacent land uses in Dallo Mena District, south-East Ethiopia. Nat Resour 7(10):515–534. https://doi.org/10.4236/nr.2016.710044

Mitchell R, Hanstad T (2004) Small Homegarden Plots and Sustainable Livelihoods for the Poor. LSP Working Paper 11, Rome. Retrieved from https://www.fao.org/3/J2545E/j2545e.htm

Monroy R, García FA (2013) Wild fauna with use value in the traditional fruit orchards of the indigenous community of Xoxocotla, Morelos, Mexico. Etnobiología 11:44–52. Retrieved from https://revistaetnobiologia.mx/index.php/etno/article/view/201/202

Mosina GK, Maroyi A, Potgieter MJ (2014) Comparative analysis of plant use in peri-urban domestic gardens of the Limpopo Province, South Africa. J Ethnobiol Ethnomed 10:35. Retrieved from https://www.ncbi.nlm.nih.gov/pmc/articles/PMC3977957/

Mwavu EN, Ariango E, Ssegawa P, Kalema VN, Bateganya F, Waiswa D, Byakagaba P (2016) Agrobiodiversity of homegardens in a commercial sugarcane cultivation land matrix in Uganda. Int J Biodiver Sci Ecosy Serv Manag 12(3):191–201. https://doi.org/10.1080/21513732.2016.1177595

Nguyen TT, Barber P, Harper R, Linh TVK, Dell B (2020) Vegetation trends associated with urban development: the role of golf courses. PLoS One 15(2):e0228090. https://doi.org/10.1371/journal.pone.0228090

Nicholls E, Ely A, Birkin L, Basu P, Goulson D (2020) The contribution of small-scale food production in urban areas to the sustainable development goals: a review and case study. Sustain Sci 15:1585–1599. https://doi.org/10.1007/s11625-020-00792-z

Okigbo B (1990) Home gardens in tropical Africa. In: Landauer K, Brazil M (eds) Tropical home gardens. United Nations University Press, Tokyo

Paembonan SA, Millang S, Dassir M, Ridwan M (2018) Species variation in home garden agroforestry system in South Sulawesi, Indonesia and its contribution to farmers' income. In: IOP conference series: earth and environmental science, vol 157, pp 1–6. https://iopscience.iop.org/article/10.1088/1755-1315/157/1/012004/pdf

Patel SK, Sharma A, Singh R, Tiwari AK, Singh GS (2022) Diversity and distribution of traditional home gardens along different disturbances in a Dry Tropical Region, India. Front For Global Change 5:822320. https://doi.org/10.3389/ffgc.2022.822320

Pinderhughes R (2004) Alternative urban futures: planning for sustainable development in cities throughout the world. Rowman & Littlefield Publishers, Inc, London

Poot-Pool WS, Cetzal-Ix W, Basu SK, Noguera-Savelli E, Noh-Contreras DG (2018) Urban home gardens: a sustainable conservation model for local plants based on Mexican Urban agri-horticultural practices. In: Nandwani D (ed) Urban horticulture. Sustainable development and biodiversity, vol 18. Springer, Cham. https://doi.org/10.1007/978-3-319-67017-1_4

Pretty J, Adams B, Berkes F, de Athayde SF, Dudley N, Hunn E, Maffi L, Milton K, Rapport D, Robbins P, Sterling E, Stolton S, Tsing A, Vintinner E, Pilgrim S (2009) The intersections of biological diversity and cultural diversity: towards integration. Conserv Soc 7:100–112. Retrieved from http://www.jstor.org/stable/26392968

Ranasinghe (2005) From informal to formal acceptance: leaps and bounds of urban agriculture in the development sphere. Int J Biodiver Sci Manag 1(1):17–24. https://doi.org/10.1080/17451590509618076

Ritchie H, Roser M (2018) Urbanization. OurWorldInData.org. Retrieved from https://ourworldindata.org/urbanization [Online Resource]

Sanderson EW, Huron A (2011) Conservation in the city. Conserv Biol 25:421–423. Retrieved from https://conbio.onlinelibrary.wiley.com/doi/epdf/10.1111/j.1523-1739.2011.01683.x

Selepe MM, Hendriks SL (2014) The impact of home gardens on pre-schoolers nutrition in Eatonside in the Vaal Triangle, South Africa. Retrieved from https://www.ajhtl.com/uploads/7/1/6/3/7163688/article_17_vol.3_2_july_14.pdf

Semu AA (2018) The study of homegarden agrobiodiversity, practices of homegardening and its role for in situ conservation of plant biodiversity in eastern Hararghe, Kombolcha town Oromia regional state Ethiopia. Open J Forest 8:229–246. https://doi.org/10.4236/ojf.2018.82016

Seto KC, Fragkias M, Guneralp B, Reilly MK (2011) A meta-analysis of global urban land expansion. PLoS One 6(8):e23777. Retrieved from https://journals.plos.org/plosone/article?id=10.1371/journal.pone.0023777

Sharma R, Mina U, Kumar BM (2022) Homegarden agroforestry systems in achievement of sustainable development goals. A review. Agron Sustain Dev 42(3):44. https://doi.org/10.1007/s13593-022-00781-9

Shimrah T, Lumleng P, Shimrah C, Khuman YSC, Varah F (2018) Role of traditional homegardens in biodiversity conservation and socioecological significance in Tangkhul community in North-east India. Trop Ecol 59(3):533–539. Retrieved from https://bit.ly/3QC07QD

Šiftová J (2021) Shaping the urban home garden: socio-ecological forces in the management of private green spaces. Land Use Policy 111:105784. https://doi.org/10.1016/j.landusepol.2021.105784

Soemarwoto O (1987) Homegardens: a traditional agroforestry system with a promising future. In: Steppler HA, Nair PKR (eds) Agrofrestry: a decade of development. International Council for Research in Agroforestry, Nairobi. Retrieved from http://apps.worldagroforestry.org/Units/Library/Books/PDFs/07_Agroforestry_a_decade_of_development.pdf?n=28

Taylor JR, Lovell ST (2015) Urban home gardens in the global north: a mixed methods study of ethnic and migrant home gardens in Chicago, IL. Renew Agric Food Syst 30:22–32. Retrieved from https://bit.ly/3QxyAQg

Taylor JR, Lovell ST (2021) Designing multifunctional urban agroforestry with people in mind. Urban Agric Reg Food Syst:1–22. Retrieved from https://digitalcommons.uri.edu/cgi/viewcontent.cgi?article=1047&context=pls_facpubs

Tobias A, Baltazar AM, Taguinod JJ, Buot IEJ (2021) The role of home gardens in conserving threatened plants of the Philippines. Asian J Biodiver 12:87–103. https://doi.org/10.7828/ajob.v12i1.1396

Trinh LN, Watson JW, Hue NN, De NN, Minh NV, Chu P, Sthapit BR, Eyzaguirre PB (2003) Agrobiodiversity conservation and development in Vietnamese home gardens. Agric Ecosyst Environ 97:317–344. https://doi.org/10.1016/S0167-8809(02)00228-1

United Nations (UN) Population Division (2018a). Population facts: the speed of urbanization around the world. United Nations. No 2018/1. Retrieved from https://population.un.org/wup/Publications/Files/WUP2018-PopFacts_2018-1.pdf

United Nations (2018b) World urbanization prospects: online edition. Retrieved from https://esa.un.org/unpd/wup/

Vinceti B, Elias M, Azimov R, Turdieva M, Aaliev S, Bobokalonov F, Butkov E, Kaparova E, Mukhsimov N, Shamuradova S, Turgunbaev K, Azizova N, Loo J (2022) Home gardens of Central Asia: reservoirs of diversity of fruit and nut tree species. PLoS One 17(7):e0271398. https://doi.org/10.1371/journal.pone.0271398

Wang M, Li J, Kuang S, He Y, Chen G, Huang Y, Song C, Anderson P, Lowicki D (2020) Plant diversity along the urban–rural gradient and its relationship with urbanization degree in Shanghai, China. Forests 11(171):2–17. Retrieved from https://bit.ly/3Qzl928

Wiersum KF (2004) Forest gardens as an "intermediate" land-use system in the nature-culture continuum: characteristics and future potential. Agrofor Syst 61–62:123–134. https://doi.org/10.1023/B:AGFO.0000028994.54710.44

Wiersum KF (2006) Diversity and change in homegarden cultivation in Indonesia. In: Nair BMKPKR (ed) Tropical Homegardens: a time-tested example of sustainable agroforestry, vol 3. Springer, Dordrecht. https://doi.org/10.1007/978-1-4020-4948-4_2

Zemede A (2002) Homegardens in Ethiopia: some observations and generalizations. In: Watson JW, Eyzaguirre PB (eds) Homegardens and in situ conservation of plant genetic resources in farming systems. Proceedings of the second international homegardens workshop, July, 2001, Witzenhausen, Germany. Intl Plant Genet Resour Inst, Rome, pp 125–139. Retrieved from https://bit.ly/3JXIGr7

Zemede A, Woldu Z (1997) Crop association of homegardens in Welayta and Gurage in southern Ethiopia. SINET Ethiop J Sci 20(1):73–90. https://doi.org/10.4314/sinet.v20i1.18092

Comprehending the Cultural Landscape of Mount Makiling: A Road to Community Well-Being Necessary to Enhance Ecosystem Integrity

26

Merites M. Buot and Inocencio E Buot, Jr

Abstract

During this fast-paced era, cultural activities in a landscape can be disregarded in favor of monetary gains from socioeconomic activities. However, it is essential to nourish the soul as well; hence, the study aims to understand the cultural landscape of the legendary Mount Makiling, a small mountain mass in Luzon Island, just an hour south of Manila, the Philippine's capital. According to the story, the mountain has a deity and goddess named, *Mariang Makiling*, who has been always there when local people need help. The story of this deity has been closely intertwined with the life of local people even to this day. Interviewing key informants along with review of literature was done to verify the cultural activities in the cultural landscape of Mount Makiling and the corresponding impact on well-being of the community. Results presented emerging themes such as *panata* (religious vows) during local festivities, social gathering, and communing with nature. Respondents expressed their satisfaction with what they are doing as part of their culture, handed to them through oral traditions over generation. During fiestas, they enjoy helping the community and honoring their God, and when they visit Mount Makiling's sacred groves, they enjoy reconnecting with nature and with the mountain deity, Mariang Makiling. Cultural activities are opportunities to socialize and deepen relationships with friends

M. M. Buot (✉)
Department of Human Kinetics, College of Arts and Sciences, University of the Philippines Los Baños, Laguna, Philippines
e-mail: mmbuot@up.edu.ph

I. E. Buot, Jr
Institute of Biological Sciences, College of Arts and Sciences, University of the Philippines Los Baños, Laguna, Philippines
e-mail: iebuot@up.edu.ph

S. Ramamoorthy et al. (eds.), *Plant Diversity in Biocultural Landscapes*, https://doi.org/10.1007/978-981-19-8649-9_26

and families, hence, uplifting personal and community well-being, essential in helping sustain the overall Mount Makiling landscape ecology. A framework is proposed to do more public awareness of the cultural activities that should trigger community participation in activities, supportive of Mount Makiling biodiversity conservation that should enhance landscape integrity for abundant ecosystem services for the local people.

Keywords

Community well-being · Community capital framework · Cultural landscape · Landscape ecology · Mount Makiling · Mariang Makiling

26.1 Introduction

A cultural landscape is a place shared by both culture and nature and has always been associated with activities, events, history, and stories that everyone in the community value (Birnbaum 1994). These are places where local people and the surrounding environment have a long-term harmonious relationship (Bryan 1931; Roundtree and Conkey 1980; Toma and Buisson 2022). Mount Makiling in southern Luzon, Philippines, is a typical example of a cultural landscape. It has served as community space for food production, a source of timber for housing, a place to stay for relaxation and even permanent abode for some. In addition, it is an essential watershed that protects the lower grounds, the urban center and the Laguna Lake, the largest lake of the Philippines and a valuable water resource for many (Castillo et al. 2018; Combalicer et al. 2011; Magcale-Macandog et al. 2021; Papa et al. 2012). Because of these manifold benefits derived from Mount Makiling, many people are concerned about scientific planning for sustainable management to ensure provision of ecosystem services. Besides scientific activities, cultural pursuits are done by the communities primarily to pacify and thank their gods and mountain deities. Mount Makiling has been recognized as a "Vavilov Center," on account of its rich flora and fauna existing for thousands of years (DOST 1991). The mountain is home to rich native biodiversity (Abraham et al. 2010; Lambio and Buot Jr 2011; Magcale-Macandog et al. 2022; Pancho 1973, 1983) forming a unique watershed, an important resource to sustain water requirements of the people and industries around the watershed and the neighboring communities.

Due to increasing numbers of local populations and frequent migration, urbanization has started and integrity of the Mount Makiling landscape is at stake (Magcale-Macandog et al. 2022). Reports from Alcala et al. (2019) and Buot Jr and Osumi (2011) revealed a number of land use types on Mount Makiling that may pose threats on the rich indigenous and endemic species of the mountain. There is a need to take a closer look at the cultural landscape alongside what hard science has been doing. Cultural activities satisfy the soul and local people need to rekindle what they were used to in the past and be one with nature again. Participation in Mount

Makiling landscape protection and conservation might be better enhanced with the full participation of the local people.

The general objective of this chapter is to understand the cultural landscape of Mount Makiling. Specifically, it aims to (a) discuss the legend of Mount Makiling; (b) identify the cultural activities within Mount Makiling landscape; (c) determine the influence of the cultural activities to the well-being of the community; and (d) establish a framework showing that with a good well-being, ecological integrity can be sustained.

26.2 Methodology

A review of the literature in online journals was conducted. Specifically, we limited our searches in Google Scholar, Open Athens, and other refereed online publications. We used the words, such as cultural landscape, Mount Makiling, and ecosystem services in our searches. In addition, key informants were interviewed. The study focused on learning about the perceptions of those who frequently participated in cultural activities around the Mount Makiling cultural landscape for peculiar reasons. To ascertain the pattern of interactions, meanings that they ascribed to their participation, their opinions, and the actions that follow were examined.

26.3 Revisiting the Legend of Mariang Makiling

The Philippines is laden with a rich ensemble of tales and legends. In addition, Mount Makiling has its own tale to tell (Fig. 26.1). It is the fairy, depicted as a typical Filipino woman in the story. It has been told again and again with different variations in different settings (Mojares 1943; Pulhin and Tapia 2005). It was believed that Mariang Makiling was the protector of Mount Makiling (Fig. 26.2). It was always a tale about a kind woman who generously shared the abundance of the mountain biodiversity with the nearby folks and afforded them safety from the danger of the wilderness.

The legend of Mariang Makiling or Maria Makiling has been part of the oral tradition, but was first noticed in the writings of Rizal's La Solidaridad (Rizal 1890). It is the topic of tales told and retold with elementary school kids easily remembering them. In its literary form, Rizal wrote (cited in Mojares 1943, p. 1):

> I thought of Mariang Makiling. I listened amidst the silence of the woods to catch the sound of her melancholic instrument. I slept there at night to see if I could find her perfect figure floating in the air, lighted by the moon that filtered through the thick foliage.
> I saw nothing, and heard nothing.

This illustrates how Mount Makiling and its deity, Mariang Makiling, are closely intertwined with the people's lives. People love the mountains, the trees and above

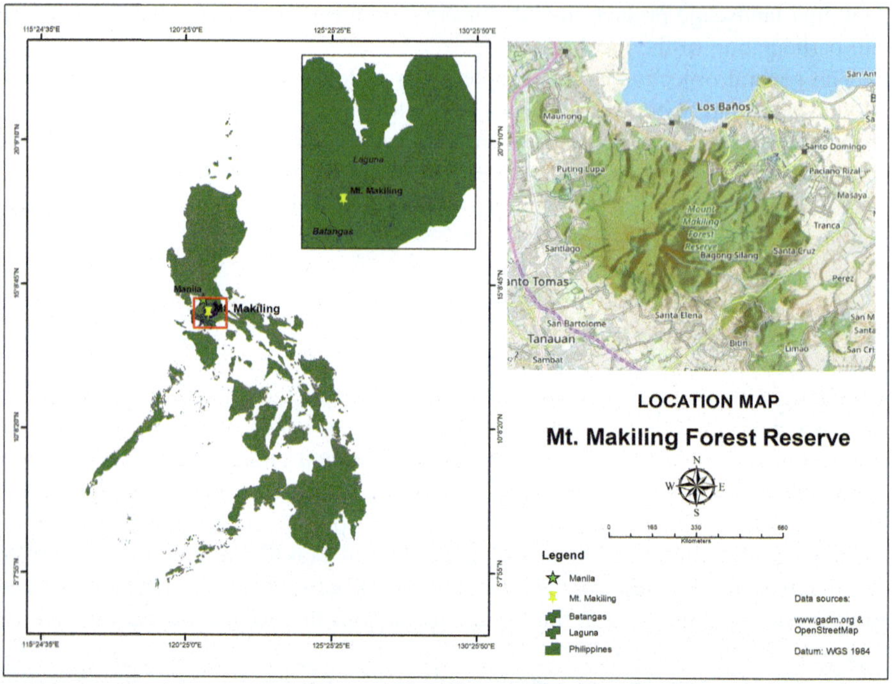

Fig. 26.1 GIS generated map of the Philippines (left) showing the location of Mount Makiling. On the right, is a topographic map of the legendary mountain showing the largest lake of the Philippines, Laguna Lake at its base (prepared by Ms. Marne Origenes, IBS-UPLB)

all, love and longed to see Mariang Makiling as in the writings of Rizal. Mariang Makiling abhors activities leading to environmental destruction. Local people had always been cautious not to earn the ire of the mountain goddess. They support the conservation of Mount Makiling, knowing that misfortune can always happen if they violate the basic tenets of Mariang Makiling (Martin and Proulx 2020; Meadows 1991; Pulhin and Tapia 2005; Taylor 2012).

Mariang Makiling also watched over Laguna de Bay (Fig. 26.3) and influenced the fish catch of the fisherfolk (Mojares 1943; Pulhin and Tapia 2005). The elders' tales made it clear that the local lake and forest were under the influence of mythology. She was also believed to be able to avert natural disasters including earthquakes, floods, and storms to protect the local people.

Mount Makiling appears to have the profile of a sleeping woman when viewed from various angles (Fig. 26.4). People even to this day, imagine this figure to be Mariang Makiling. The present form of this dormant volcanic mountain and watershed remind the local people of the legendary woman who selflessly protects them from whatever harm so long as they protect the mountain as well from overharvesting, shifting cultivation, and deforestation. To the dismay and anguish of the locals, the Mariang Makiling legend came to an end when the deity sadly vanished after purportedly having disagreements with her human love interest.

Fig. 26.2 Icon of Mariang Makiling at the UPLB campus situated at the foot of Mount Makiling (photo by Michiko Karisa Buot)

Fig. 26.3 Legendary Mount Makiling and the Laguna de Bay at the base (photo by Jabez Flores)

Nevertheless, the community around the mountain sustained the cultural activities to appease Mariang Makiling until this day.

Fig. 26.4 Mount Makiling peak has a form of a sleeping woman, thought to be Mariang Makiling (photo by Mayumi Kara Buot)

26.4 Cultural Activities Around Mount Makiling Cultural Landscape

It is difficult to explain, but there are numerous cultural activities that people have historically engaged in, to honor the mountain and perhaps, Mariang Makiling as well. This includes paying homage to the mountain's Peak 2 during Roman Catholic's Holy Week and performing rituals and festivals (Fig. 26.5). Peak 2 is a sacred grove and is one of Mount Makiling's three summits, rising 1100 m above sea level (Alcala et al. 2019; Delos Angeles and Buot Jr 2018; Fernando et al. 2004; Magcale-Macandog et al. 2021, 2022; Pancho 1973, 1983). Going to Mount Makiling's Peak 2 during the Roman Catholic, Christian's Holy Week has become a ritual for the majority of mountaineers and community local people.

Another cultural activity in the communities at the base of Mount Makiling is the Bañamos festival every September 17th to celebrate the hot springs in the locality. The municipality of Los Baños was known as *Mainit* in the past. In Tagalog language, it means warm or hot, which made reference to the numerous hot springs that line the coastline of Laguna de Bay (SEARCA undated). When the local government established this in the final year of the twenty-first century, the curative properties of the hot springs around Los Baños served as the inspiration. It led to the establishment of the Los Baños Therapeutic Massage Center and Health Spa now called "Caesar P. Perez Memorial Therapeutic Massage Center and Health Spa (L.B.M.O. 2021). Nearly 22 years later, the community festival continues to remind the local and surrounding populace of the importance of hot springs to health, affirming the results of the study of Mossabir et al. (2021) on the hot springs around the base of Mount Makiling (Los Baños, Bay, Calamba areas). All of these, all the

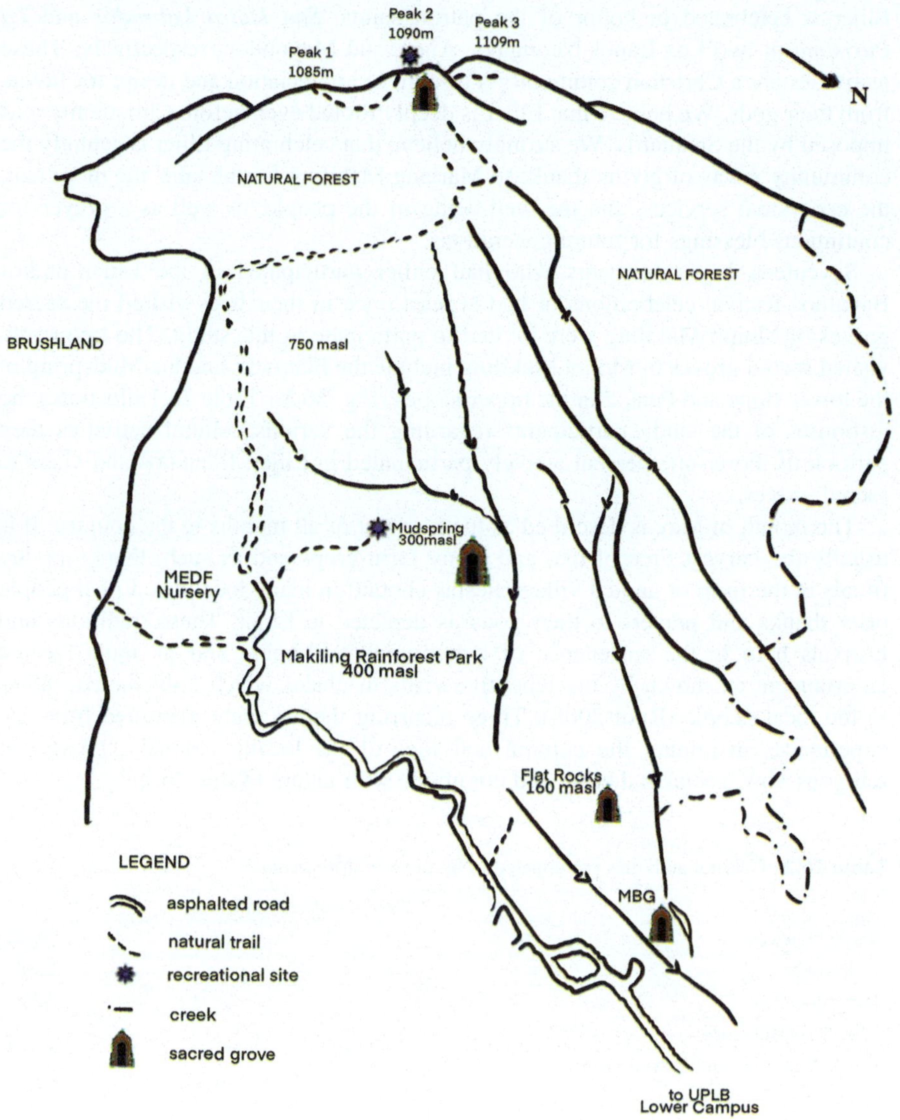

Fig. 26.5 Detailed map of Mount Makiling showing few of the sacred groves (adapted from Lambio and Buot Jr 2011)

more make local people think of and admire the goodness of the mountain deity, Mariang Makiling. The hot springs on Mount Makiling are a popular tourist destination and are thus good for the local economy as well.

One more significant cultural event celebrated every month of May around the base of Mount Makiling and on the coast of Laguna de Bay, is the Elijer (Buot 2008).

Elijer is celebrated in honor of the patron saints, *San Isidro Labrador* and *La Purisima* of two Los Banos barangays, Anos, and Mayondon, respectively. These festivities are a Christian community's gesture of appreciation and desire for favors from their gods. We believe that Elijer is deeply rooted even before Christianity was imposed by the Spaniards. We strongly believe that celebrating Elijer is actually the community's way of giving thanks to Mariang Makiling for the lake, the mountain, the ecosystem services, and the well-being of the people, as well as a prayer for continuous blessings for future generations.

Seventeen key informants who had either participated in the Elijer and/or Bañamos festival celebrations or had at least once in their lives visited the sacred groves of Mount Makiling were invited to participate in this study. The frequently visited sacred groves of Mount Makiling include the Flatrocks and the Mudspring in the lower slope and Peak 2 in the upper slopes (Fig. 26.5). Table 26.1 illustrates the responses of the study participants regarding the various cultural activities they partook of. Forty-one percent actively participated in Elijer, Banamos and visits to sacred groves.

The month of May is described as the merriest of all months in the country. It is usually the harvest time of rice and many farm crops and as such, thanks giving rituals in the form of annual village fiestas abound in many localities. Local people offer thanks and prayers to their gods as depicted in Elijer. These traditions and festivals help in the sustenance of community well-being and its multi-faceted environment as shown by the repetitive visits to places which hold sacred values to the local people (Buot 2008). Three recurring themes were identified from the experiences in joining the cultural activities of the locality, namely, *panata* or religious vow, social gathering and commune with nature (Table 26.2).

Table 26.1 Cultural activities participated in by the key informants

	Number of respondents	%
Sacred Groves	6	35.29
Elijer only	1	5.88
Bañamos only	1	5.88
Elijer and Bañamos	2	11.76
Elijer, Bañamos and Sacred Groves	7	41.18
Total	17	100

Table 26.2 Recurring themes in experiencing the cultural activities of Mount Makiling

Emerging themes	Significant statements
Yearly '*panata*' (religious vow)	"I am pleased with Elijer since this is a way to God and serve the community"
Social gathering	"Togetherness of family and friends" "Chance to talk and enjoy nature"
Commune with nature	"Chance to be with the trees, see things differently." "Learning by seeing, learning by doing"

People who went through difficult situations are obligated to make yearly promises or the so-called, *panata*. Buot (2008) wrote about the family *panata* which had been so, for over generations in barangay Anos and Mayondon. These devotees are unaffected by the oppressive heat of the sun and other challenging circumstances they had to overcome, just to make through and fulfill their *panata*. Starting from the lakeshore, in barangay Mayondon, Elijer devotees walk in a procession and wind up in the village chapel, where people together offer thanks and prayer. Usually, the people pray for blessings of long life, good health, and prosperity (in agriculture and fishery) as in the days of Mariang Makiling.

Other members of the community want to commune with nature (Table 26.2). Regardless of whatever physical circumstances, many local people express their desire to visit the sacred grove in Peak 2. Those who do not make it to the upper sacred grove (Fig. 26.5), just enjoy nature at the sacred groves in the lower slopes, the Mudspring and the Flatrocks (Fig. 26.5). At least, they have paid homage to the legendary mountain and to its fairy/deity who bestows them blessings during the year.

As they do the different cultural activities (Table 26.2), they socialize and catch up with one another's current state and plans, expressing support and strengthening friendships and relationships.

26.5 Influence of the Dynamic Cultural Landscape to Community Well-Being

A dynamic cultural landscape has a positive impact on community well-being. A closer reexamination of Table 26.2 reveals statements of key respondents reflective of satisfaction, happiness, enjoyment, harmony with nature, and overall community well-being. After actively participating in the Elijer celebration, a key informant describes his/her inner joy, at seeing the community and worshipping God, including the deities of Mount Makiling. The social gathering during fiestas, promotes togetherness of families and of their long-time friends as they come back in search of their place identity. Their attachment to each other and of their communities and local stories enhances inner happiness, and these people alone can describe.

People who come back to partake of the cultural activities, clearly demonstrated their attachment to the landscape and the place (Farina 2000; Plieninger et al. 2014; Taylor 2012, 2016; Taylor and Lennon 2012). More and more people believed that a good life must also include social and spiritual well-being, in addition to material and economic abundance (Buot 2017; Buot and Buot Jr 2022; Toma and Buisson 2022; Verschuuren et al. 2014).

Table 26.2 further shows that people coming back to visit the Mount Makiling cultural landscape, want to commune with nature. They want to be with the trees and the sacred mountain, where the trees are protected and productive. Indeed, there is a high sense of community well-being experienced by these people. At this state, brilliant ideas relevant to addressing landscape challenges can come up during group discussions. A high level of community well-being means a faster route to recovery from any manmade or natural problems (Buot and Buot Jr 2022; Buot and Dulce 2019; Lee et al. 2015). It is hypothesized that a high level of community well-being assures ecological integrity (Buot and Buot Jr 2022; Cox et al. 2010; Parrish 2008) which naturally assures ecological integrity.

26.6 Crafting a Framework Showing that Good Cultural Landscape Enhances Community Well-Being and Ecosystem Integrity

A framework (Fig. 26.6) should be established that best charts the need for fully engaging community cultural activities in a cultural landscape. Cultural traditions, practices, rituals, or arts that speak of the community identity and advocacy can be cultivated by the community as a group. This will give the locals that sense of belongingness and pride (Mossabir et al. 2021; Roundtree and Conkey 1980). Every locality normally has this in place, in the form of oral traditions sustained over generations boosting community well-being (Fig. 26.6). Incidentally, holding actual and real cultural activities needs multisectoral community planning to be successful (Buot et al. 2020; Ciftcioglu et al. 2016). The framework for community capitals can also be a very helpful and pertinent guide (Emery and Flora 2006; Flora and Flora 2013). This means that during the planning process, all aspects in the entire locality including the stakeholders' aspirations need to be considered (Atkinson 2014; Buot 2008; Buot Jr 2008a, b; Buot and Buot Jr 2022; Buot and Cardenas 2018; Buot and Dulce 2019). In the context of Mount Makiling cultural landscape, a ridge to rift approach (Global Environment Facility n.d.; Hapinat 2019) as the guide in the planning process would be ideal.

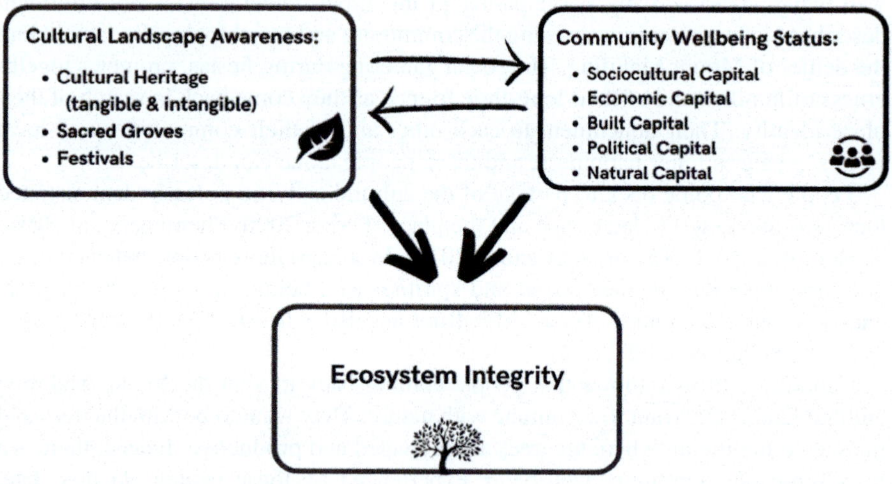

Fig. 26.6 Framework showing that awareness of cultural landscape enhances Community Well-being in order to sustain ecosystem integrity

This means that planning involves the full consideration of Mount Makiling watershed, a biodiversity-rich natural capital, the village, community, and all the government and non-government developmental programs at the lower slopes and base of the legendary mountain (sociocultural capital) and lake shore communities and establishments (economic capital). The cultural landscape of Mount Makiling is quite intricate and planning needs to take into account these networks of complicated details in order to achieve ecosystem integrity (Buot Jr 2008a, b) and, hence, sustain ecosystem services for all Mount Makiling stakeholders—the noble hope of the deity, Mariang Makiling for all the local people.

26.7 Conclusion

The cultural landscape of Mount Makiling, a small mountain mass in southern Luzon, is quite complex. The cultural activities currently practiced (visiting the sacred groves, observing and participating in Elijer and Bañamos festivities) can be traced back to local people's desire to please Mariang Makiling, the deity of the legendary mountain. Key respondents attest that these cultural activities enhance community well-being of the villagers. High degree of community well-being can generate community participants in activities that would spur ecosystem integrity and, hence, sustained ecosystem services of the Mount Makiling watershed.

Acknowledgements The authors are indebted to Ms. Geraldine P. Empleo, Ms. Elvira C. Tamisin, to all the other respondents of this study, and Ms. Therese Anne Virtucio for the art illustration.

References

Abraham ERG, Gonzalez JCT, Castillo ML, Lit IL Jr, Fernando ES (2010) Forest cover and biodiversity profile of the crater area of Mt. Makiling, Luzon, Philippines. Asia Life Sci Suppl 4:48–82

Alcala AA, Delos Angeles M, Buot IE Jr (2019) Fern species diversity across various land use types of mt. Makiling, Luzon Island, Philippines. Biodiversitas 20(9):2437–2445

Atkinson CL (2014) Public policy processes and the environment: implications for a sustainable future. Sustain Account Manag Policy Js 5(4):457–475

Birnbaum CA (ed) (1994) Preservation forum. "Focus on landscape preservation, vol 7, no 3." National Trust for Historic Preservation, Washington DC. Retrieved on August 23, 2022 from https://www.nps.gov/tps/how-to-preserve/briefs/36-cultural-landscapes.htm

Bryan PW (1931) The cultural landscape. Geography 16(4):273–284

Buot MM (2008) Los Baños Folk Festival: its role in sustaining health of the socio-cultural environment. Asia Life Sci (Suppl 2):53–65

Buot MM (2017) Benchmarking concepts of Community Wellbeing in an area vulnerable to disaster due to volcanic eruption. International Journal of Environment and Rural Development 8(1):58–62

Buot MM, Buot IE Jr (2022) Volcanic seascapes: disaster recovery for community wellbeing and habitat protection. In: Pungetti G (ed) Routledge handbook of seascapes, Routledge, New York, pp 385–398

Buot MM, Cardenas VR (2018) Community Wellbeing Index (CWBi) in the area exposed to Mayon Volcanic eruption in Camalig, Albay, Philippines. Sylvatrop 28(2):27–46

Buot MM, Dulce MZ (2019) An index to determine community wellbeing along coastal community in Leyte, Philippines. Environ Asia 12(1):56–67

Buot IE Jr (2008a) A new way of looking at environmental health: focused on man and his environment. Asia Life Sci Suppl 2:1–5

Buot IE Jr (2008b) Sustaining environmental health in Philippine Satoyama landscapes. Asia Life Sci Suppl 2:129–138

Buot IE Jr, Osumi K (2011) Land use type pattern and woody species composition near human disturbed landscapes on Mount Makiling, Luzon Island. Am J Environ Sci 7(40):306–315

Buot MM, Cardenas VR, Dizon JT, Quimbo AT, Nelson GLM (2020) Multisectoral participation in the development of an index for community wellbeing. In: Buot IE (ed) Methodologies supportive of sustainable development in agriculture and natural resources management: selected cases in Southeast Asia. Southeast Asian Regional Center for Graduate Study and Research in Agriculture (SEARCA) and the University of the Philippines Los Baños (UPLB), Laguna, Philippines, pp 243–260

Castillo ML, Castillo LA, Canceran MS, Gonzalvo KJP, Barua LD, Alegre AC, Barredo-Parducho VO, Gestiada EC, Breva RV, Bantayan NC (2018) Distribution, diversity and spatial analysis of tree species in a long-term ecological research plot in Molawin-Dampalit Watershed, Mount Makiling Forest Reserve. Asian J Biodivers 9:12–36

Ciftcioglu GC, Uzun O, Nemutlo FE (2016) Evaluation of biocultural landscapes and associated ecosystem services in the region of Suğla Lake in Turkey. Landsc Res 41(5):538–554. https://doi.org/10.1080/01426397.2016.1173659

Combalicer MS, Kim D, Lee DK, Combalicer EA, Cruz RVO, Im S (2011) Changes in the forest landscape of Mt. Makiling forest reserve, Philippines. For Sci Technol 72:60–67

Cox D, Frere M, West S, Wiseman J (2010) Developing and using local community wellbeing indicators: learning from the experience of community indicators Victoria [online]. The Aust J Soc Issues 45(1):71–88

DOST (1991) A primer. Save Laguna de Bay and Mt. Makiling. 22 p

Delos Angeles M, Buot IE Jr (2018) Altitudinal zonation of ferns along the northeastern slope of Mount Makiling, Philippines. Sylvatrop, Technical J Philip Ecosyst Nat Resour 28(2):47–68

Emery M, Flora CB (2006) Spiraling-up: mapping community transformation with community capitals framework. Commun Develop J Commun Develop Soc 37:19–35

Farina A (2000) The cultural landscape as a model for the integration of ecology and economics. Bioscience 50(4):313–320

Fernando ES, Sun BY, Suy MH, Kony HY, Koh KS (2004) Flowering plants and ferns of Mt. Makiling. ASEAN-Korea Environmental Cooperation Unit (AKECU). 368 p

Flora CB, Flora JL (2013) Rural communities: legacy and change, 4th edn. Westview Press, Boulder, CO

Global Environment Facility (n.d.) From Ridge to Reef Water, environment, and community security GEF action on transboundary water resources. Retrieved on August 23, 2022 from: https://www.thegef.org/sites/default/files/publications/GEF_RidgetoReef2015_r2_Final.pdf

Hapinat HL (2019) From ridge-to-rift: a community based participatory approach to biodiversity conservation of marine resources in Batad, Iloilo, Philippines. FFTC Agricultural Policy Platform (FFTC-AP). Food and Fertilizer Technology Center for the As Ian and Pacific Region. Retrieved on August 23, 2022 from https://ap.fftc.org.tw/article/1407

Lambio IAF, Buot IE Jr (2011) Leaf size zonation pattern of woody species along the northeastern altitudinal gradient of Mount Makiling, Luzon Island, Philippines. Asia Life Sci 20(2):369–383

Lee SJ, Kim Y, Phillips R (2015) Exploring the intersection of community Well-being and community development. In: Lee SJ, Kim Y, Lee RP (eds) Community Well-being and community development conceptions and applications. Springer, pp 1–8

Los Banos Municipality Ordinances (2021) https://losbanos.gov.ph/ordinance

Magcale-Macandog DB, Lapitan FRG, Laruya JM, Valerio JIF, Aguila JCD, Mesina CAL, Santos TMF, Cuevas ANT, Baylon KD, Silapan IM, Dimalobot R, Edrial JD, Larida NJA, Natuel FA, Perez MAGLD, Quinones SGL (2021) Updating of Makiling biodiversity information system (MakiBIS) and analysis of biodiversity data. Biodiversitas 22:211–226

Magcale-Macandog DB, Paller VGV, Torreta NK, Lambio IAF, Hadsall AS, Buot I Jr, delos Angeles MD, Cervancia CR, Quiñones SGL, Laruya JM (2022) Plant diversity of mount Makiling forest reserve: implications to management and conservation. In: Ramamoorthy S et al (eds) Plant genetic resources, inventory, collection and conservation. https://doi.org/10.1007/978-981-16-7699-4_5

Martin CA, Proulx R (2020) Level-2 ecological integrity: assessing ecosystems in a changing world. Perspect Ecol Conserv 18:197–202. (Martin and Proulx (2020) reiterated the challenging process of ecological integrity assessment affected by anthropogenic influences are now doing)

Meadows D (1991) Book review: a sand county almanac by Aldo Leopold. (reissue) New York

Mojares RB (1943) Waiting for Mariang Makiling essays in Philippine cultural history. Ateneo de Manila University Press, pp 1–19

Mossabir R, Milligan C, Froggatt K (2021) Therapeutic landscape experiences of everyday geographies within the wider community: a scoping review. Soc Sci Med 279:113980

Pancho JV (1973) Mt. Makiling as a station for biological research. Philipp Agricult 57:21–33

Pancho JV (1983) Kalikasan, the Philippine journal of biology. Supplement #1 Quezon City Philippines. New Mercury Printing Press, 476 p

Papa RDS, Tordesillas D, Vitug LD, Briones JCA, Zafaralla MT, Kada R (2012) Zooplankton spatial abundance and distribution in the West Bay of Laguna de Bay (Philippines) with notes on potential implications on food and health security. Acta Manilana 60:67–75

Parrish MCKPPB (2008) A wellbeing index based on an enabling environment. Int J Soc Econ 35(3):174–187. https://doi.org/10.1108/03068290810847851

Plieninger T, der Horst D, Schleyer C, Bieling C (2014) Sustaining ecosystem services in cultural landscape. Ecol Soc 19(2):59

Pulhin J, Tapia MA (2005) History of a legend: managing the Makiling forest reserve. In: Search of excellence: exemplary forest management in Asia and the Pacific, 261 p

Rizal J (1890) Mariang Makiling. La Solidaridad. (Arnold H. Warren translation 1968). https://studylib.net/doc/8653687/mariang-makiling-by-jose-rizal%2D%2D1968-arnold-h.-warren-tra?fbclid=IwAR0Hy9efPx-y4tUBv0eNw0u2cR2FK0-AS41GQzjlRsjrEhAR1N_fGuq0N90

Roundtree LB, Conkey MW (1980) Symbolism and the cultural landscape. Ann Assoc Am Geogr 70(4):459–474

Taylor K (2012) Landscape and meaning: context for a global discourse on cultural landscape values. In: Taylor K, Lennon J (eds) Managing cultural landscapes (Key Issues in Cultural Heritage), 2012, eBook Index. Routledge, Abingdon, pp 21–44

Taylor K (2016) The Historic Urban Landscape paradigm and cities as cultural landscapes. Challenging orthodoxy in urban conservation. Landsc Res 41(4):471–480. https://doi.org/10.1080/01426397.2016.1156066

Taylor K, Lennon J (2012) Introduction: leaping the fence. In: Taylor K, Lennon J (eds) Managing cultural landscapes, Routledge key issues in cultural heritage series. Abingdon, Routledge, pp 1–18

Toma TSP, Buisson E (2022) Taking cultural landscapes into account: implications for scaling up ecological restoration. Land Use Policy 120:106233

Verschuuren B, Subramanian SM, Hiemstra W (eds) (2014) Community Well-being in biocultural landscapes: are we living well? Practical Action Publishing Ltd. https://doi.org/10.3362/9781780448374

System Dynamic Modeling of Top Harvested Plant Bioresources in Northern Negros Natural Park

27

Aragon A. Dechimo, Jr, Inocencio E Buot, Jr, Rogelio Andrada, II, Ricardo T. Bagarinao, Maria Celeste N. Banaticla-Hilario, and Marisa J. Sobremisana

A. A. Dechimo, Jr (✉)
Faculty for Teacher Education, Philippine Normal University, Cadiz City, Negros Occidental, Philippines

School of Environmental Science and Management, University of the Philippines, Laguna, Philippines
e-mail: dechimo.aa@pnu.edu.ph; aadechmo@up.edu.ph

I. E. Buot, Jr
Institute of Biological Sciences, College of Arts and Sciences, University of the Philippines Los Baños, Laguna, Philippines

R. Andrada, II
School of Environmental Science and Management, University of the Philippines, Laguna, Philippines

Institute of Renewable Natural Resources, University of the Philippines Los Banos, Laguna, Philippines

R. T. Bagarinao
School of Environmental Science and Management, University of the Philippines, Laguna, Philippines

Faculty of Education (FEd), UP Open University (UPOU), Los Baños, Philippines

M. C. N. Banaticla-Hilario
Institute of Biological Sciences, College of Arts and Sciences, University of the Philippines Los Baños, Los Baños, Laguna, Philippines

M. J. Sobremisana
School of Environmental Science and Management, University of the Philippines, Laguna, Philippines

S. Ramamoorthy et al. (eds.), *Plant Diversity in Biocultural Landscapes*,
https://doi.org/10.1007/978-981-19-8649-9_27

Abstract

The planning for management initiatives of Northern Negros Natural Park (NNNP) as a protected area needs datasets on plant bioresources mostly harvested from various zones in the park. This paper aims to 1) develop a modelthat could predict the forest cover change in NNNP, 2) project scenarios on the impact of harvesting plantbioresources (PBRs) on forest cover change; and 3) analyze the scenario projections on their impact to policy andmanagement practices for PBRs use and forest cover in NNNP. System dynamics modeling methodology wasutilized to develop the logistic growth model utilized for scenario projections. The scenario projections made use ofthe data from the household survey and floral assessment in NNNP. Results showed that the developed projectionmodel entered in STELLA software can project scenarios for top harvested PBRs and forest cover of NNNP. In allprojected scenarios, the collection of top PBRs is still below the species' tipping point of local extinction. Onaverage, the annual harvest is at 0.57% of its estimated stock population in the business as usual scenario,intensified harvest scenario (1.68%), and reduced or ideal harvest rate scenario (0.18%). A low to moderate risk ofextinction was observed in the estimated average species density. The harvest of these species was regulated bythe Expanded National Integrated Protected Area Systems (ENIPAS) Act of 2018. However, the weakimplementation of these regulations contributed to the continuing illegal collection and commercial activities insideNNNP. The results presented in this paper provided a 20-year projection and trends for the PBRs mostly harvestedas important data set needed in planning and management initiatives in the protected area.

Keywords

Projection · System Dynamics Modelling · STELLA · Plant Bioresources

27.1 Introduction

Northern Negros Natural Park as a home to 8714 families (NNNP 2016) of 11 local government units offers ecosystem services to its dependents. Forest resources as one of the commonly extracted resources has provided food, raw materials, and medicines have been subjected to deforestation due to development and increasing human population (Jha and Bawa 2006). Currently, timber and non-timber forest product (NTFP) extraction is considered as important economic activity (do Valle et al. 2007) and in effect has influenced the environmental change (Leaver 2019). It can also yield both positive and negative impacts to the social characteristics of the local community. It benefits several components of the livelihood of people but at the same time increases household inequalities (Kusters et al. 2006). As the forest product extraction contributes to the level and patterns of forest degradation (Alber and Robinson 2013), and its influence reaches the local economy and social structure, it is fitting to project the changes in forest dynamics particularly of the mostly harvested non-timber forest products to obtain relevant information for managing and decision making (Limaei et al. 2017).

Projection model can provide and explore management options and alternatives other than forecasting plant forest plant resources (Limaei et al. 2017). The use of system dynamic modeling promotes the understanding of forest ecosystem's dynamic nature (Bharat et al. 2015) and its utilization will shed light to the conditions of NNNP.

This study will attempt to project scenarios for the mostly harvested non-timber forest products of plant bioresources (PBRs) in Northern Negros Natural Park. Specifically, this study aims to:

1. develop a model that could predict the forest cover change on Northern Negros Natural Park (NNNP),
2. project scenarios (business-as-usual, increased harvest rate, and reduced harvest rate) on the impacts of harvesting top plant bioresources in Northern Negros Natural Park,
3. analyze the scenario projections with respect to their impact to policy interventions and management practices for plant bioresource use in Northern Negros Natural Park.

27.2 Study Area and Methods

Study Area Northern Negros Natural Park (123°07′ to 123°19′E and from 10°35′ to 10°49′N), situated at northern portion of Negros Occidental lying north of Mount Kanlaon and west of Bacolod City the highly urbanized city in the province. It is the largest and the most intact remaining forest stand in the island of Negros covering a span of 80, 454.5 hectare bordering five municipalities (Toboso, Calatrava, E.B. Magalona, Murcia and Don Salvador Benedicto), and six cities (San Carlos, Sagay, Cadiz, Victorias, Silay, and Talisay. NNNP is one of the two sites identified in the NIPAS to support the largest forest areas in Visayas region of the Philippines. It was designated "very high" for the terrestrial and inland water areas for biological importance. The park is composed of three terrestrial forest habitat types, such as the lowland forests, lower montane, and upper montane forests (Fig. 27.1).

Methods *Development of Projection Model:* The list of NTFPs as the mostly extracted plant bioresources from NNNP (Dechimo, A. 2022, "Plant Bioresource Utilization by Local Communities of Northern Negros Natural Park, Negros Island, Philippines" unpublished doctoral thesis, University of the Philippines Los Banos) was used in identifying the plant species to be projected. The study utilized the system dynamics modeling methodology process (Forrester 1961; Sterman 2000) to project the baseline harvest rate or the business-as-usual (BAU) scenario and the two harvest rate scenarios providing the upper and lower extremes of harvest (Gifford et al. 2022), the increased harvest rate (IH) scenario, and the reduced harvest rate

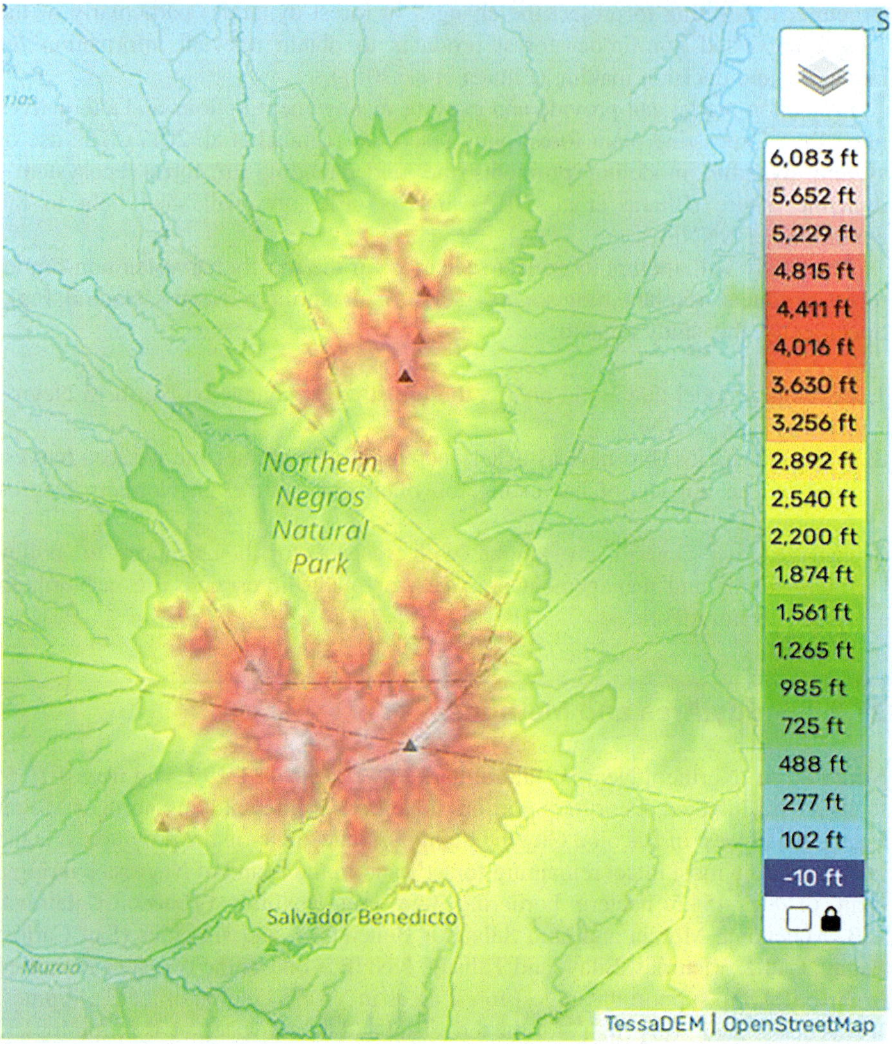

Fig. 27.1 Northern Negros Natural Park, Negros Occidental showing the Elevation Gradient (topographic-map.com, 2022)

(RH) scenario, respectively. To express the harvest rate for both the upper and lower extremes, harvest rates for all species were set to the doubled value for the upper extreme and the baseline rate was halved for the lower extreme harvest. In general, there are five steps in the process, the: (1) problem articulation; (2) development of a dynamic hypothesis; (3) formulation of a simulation model; (4) testing the simulation model; and (5) policy or strategy design, experimentation, and analysis. This study, however, will replicate the method utilized by Purnomo et al. (2012) which

Table 27.1 Plant bioresources mostly harvested in Northern Negros Natural Park, Negros Island, Philippines

Scientific name	Common name	Use(s) in the community
Ludisia discolor	*Jewels Orchid*	Ornamental
Alocasia macrorrhizos	*Badyang*	Ornamental, herbal
Caladium bicolor	*Gabi-Gabi*	Ornamental
Pinanga insignis	*Balasyang*	Food, ornamental
Asplenium nidus	*Manan-Aw*	Ornamental
Alocasia zebrina	*Tigre-tigre*	Ornamental

will include four steps and the fifth step was added for the scenario design analysis. These are: (1) developing of the conceptual model for PBRs in NNNP; (2) specification and execution of the model using STELLA 9.0.2 software; (3) evaluation of the model for the business-as-usual (BAU) scenario; (4) use of the model for intensified extraction rate and reduced extraction rate scenarios; and (5) policy or strategy design, and analysis which will identify the leverage or tipping points through scenario design analysis (Table 27.1).

Scenario Projection for Top Plant Bioresources Harvested: In order to estimate the models for scenario projections, the empirical data where species density, growth rates, and other factors were considered in as explanatory variables. The secondary data on the growth rate or productivity rate of identified top PBRs extracted together with the harvest rates as surfaced by household surveys were vital in the specifications of models in step number 2. The extraction rate, however, was the only variable manipulated for the two scenario modelings in steps number 4 other than the assumptions in climate change impacts. Finally, each identified mostly harvested PBRs underwent the three-scenario design analysis.

Analysis of Scenario Projection in Relation to Policy Intervention and Management Practices: The scenarios projected for every species were put against the existing policies and management practices. This was done to see the interaction between the policies and management practices and the trend of species density surfaced by the scenario models. This is the fifth step of system dynamics modeling according to Purnomo et al. (2012), the policy or strategy design, and analysis that identified the tipping points through scenario design analysis.

27.3 Results and Discussion

Development of the Projection Model The entry details in STELLA 9.0.2 software made use of the separate growth and harvest equations. The generic growth equation (Limaei et al. 2017) that was used in the model is

$$\dot{x} = \frac{dx}{dt} = sx\left(1 - \frac{x}{K}\right) + kA$$

where x is stock level of species X, t is time (year), s is intrinsic growth rate, showing how much a species can grow between successive time periods, K is carrying capacity, indicating the maximum population of a species in an environment over a period without destroying the resource base, k is for other parameters (wind speed, humidity and temperature), and A is altitude (m a.s.l.). It can also be expressed as follows:

$$\dot{x} = \frac{dx}{dt} = sx - \frac{s}{K}x^2 + kA$$

The generic harvest model by Gaoue et al. (2016a, b) which was utilized is

$$a\frac{dx}{dt} = (r_e - \alpha h_n - \beta h_l)x\left(1 - \frac{x}{k}\right) - h_l x$$

where r is the intrinsic growth rate, k is the carrying capacity of the environment, h_l is the lethal harvest rate, h_n is the non-lethal harvest rate, re is the maximum growth rate, α is a parameter to account for the demographic cost of non-lethal harvest of plant bioresource (e.g., harvest of foliage, fruits and bark), and β is a parameter to account for indirect effects of lethal harvest.

There were three scenarios projected for each of the identified PBRs. The first scenario was the business-as-usual (BAU) scenario, where harvest and growth were projected for the next 20 years basing from the current actual values for harvest rates. The second model was the intensified harvest rate or increased harvest rate (IH). It is made by doubling or increasing the harvest rates for each species. The last model was the ideal harvest rate or the reduced harvest rate (RH), where harvest rates were set to or reduced to the most ideal value. Using the STELLA 9.0.2 software, a model was made for every scenario. Figure 27.2 shows the formulated model for the projection of species density which were evaluated for the three scenarios (BAU, IH, and RH) of every species, as PBRs extracted from NNNP.

Scenario Projection for Top Plant Bioresources Harvested *Harvest Quantity of PBRs inside NNNP.* The top PBRs' current harvest rate averaged 15% of their tipping points (Table 27.2) for unstable population in the park. The tipping point value for each species was achieved using optimization (Calkin et al. 2002). The tipping point was identified for each species before it reaches collapse (Boettiger et al. 2015). The baseline average harvest rate can be marked low risk to extinction, since it is still within the sustainable harvest rate for forest products (Gaoue et al. 2016b). This is still observed true even if the harvest rate is intensified (30%) and more so to the reduced harvest rate (8%) (Fig. 27.3).

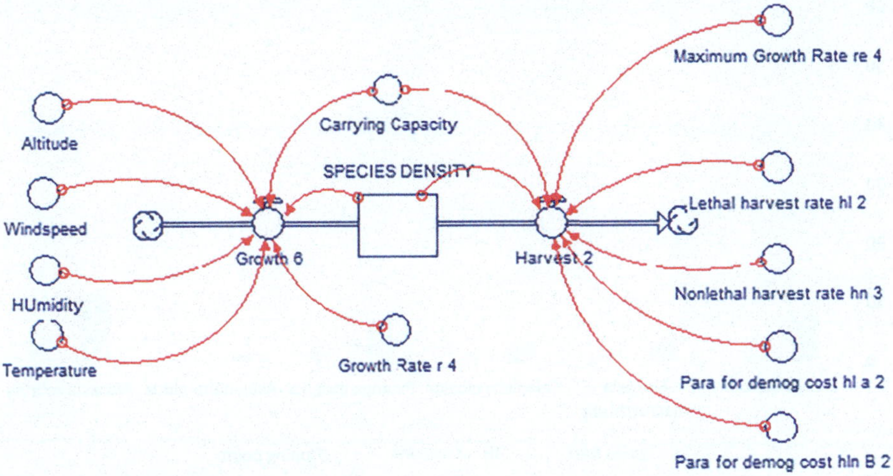

Fig. 27.2 Researcher-made generic model for species density projection for three scenarios of top harvested bioresources in Northern Negros Natural Park

Table 27.2 Proportion of harvest rates to each species' tipping point to unstable population in Northern Negros Natural Park, Negros Island, Philippines

Top PBR species	Vegetative zonation	BAU (%)	IH (%)	RH (%)
Ludisia discolor	Zone II	47	94	24
Alocasia macrorrhizos	Zone IA	6	13	3
Caladium bicolor	Zone IA	21	42	10
Pinanga insignis	Zone IA	18	36	9
Asplenium nidus	Zone IB, IC, Zone II	4	8	2
Alocasia zebrina	Zone IB and IC	29	57	14

Among top PBRs harvested inside the Northern Negros Natural Park, *A. macrorrhizos* and *A. nidus* were the species projected to have lower risk of extinction because of their relatively high projected species density trends. They are commonly found within all the zones (Table 27.1) of the altitudinal gradient. The low species density makes the species vulnerable to extinction when faced with catastrophic events not favorable for its growth, survival, and capacity to recover (Hall and Bawa 1993). This vulnerability and risk of extinction can be true to all the plant bioresources evaluated. The BAU and RH scenarios (Fig. 27.4) show the similar but slightly lower trends for *L. discolor* and *P. insignis*. Both registered a stable species density reaching the highest value at 1134 plants per hectare. At an intensified harvesting (IH) scenario, *L. discolor's* risk of extinction was seen stable. This is beside the fact that the harvest rate at IH scenario is an individual plant lower than its tipping point with the projected highest density of 671 plants per Ha achieved starting the second quarter of the second year. It has the lowest species density of 303 plants per Ha at the start of the projection. All three scenarios presented a stable species density for the next 20 years telling that *L. discolor*

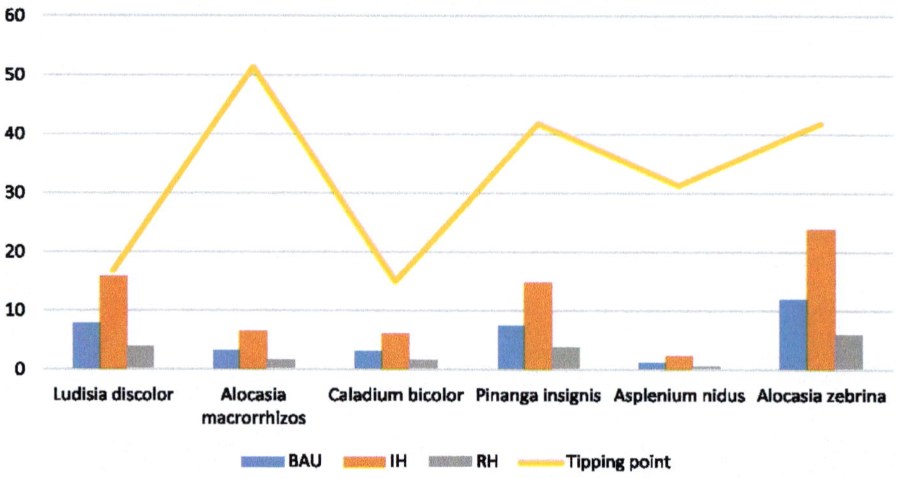

Fig. 27.3 Top PBR harvest rates and tipping points to unstable population in Northern Negros Natural Park, Negros Island, Philippines

could survive given the current harvest rate as baseline for human pressures to the species inside NNNP. Similarly, *P. insignis* was seen stable at 1134 plants per hectare by the first quarter of year 3. It also has registered low stable density at 656 plants per hectare starting the third quarter of year 1 of the projection.

Even though the IH scenario projection for *A. zibrina* (Fig. 27.4) projected lowest in species density trends compared to the BAU and IH scenarios and compared to all the projected species, its projected stable species density of 520 plants per Ha can be considered low risk from extinction if ever catastrophic events happen. BAU scenario stabilized at 752 plants per Ha in as early as year 1, while RH scenario stabilized at the species density of 1587plants per Ha at the quarter three of the third year.

The very low harvest rate of *A. nidus* (Fig. 27.4) in the current (BAU) scenario, the ideal (RH) harvest scenario, and the scenario for intensified harvest (IH) resulted exactly the same high species density for each scenario. Even at the intensified harvest scenario, *A. nidus* collection could still yield its optimum stocks value dictated by the carrying capacity. It has recorded the highest species density for all scenarios at 2510 plants per Ha achieved on the third quarter of the sixth year. These scenarios are ideal and below the sustainable harvest rate for forest products (Gaoue et al. 2016b).

Caladium bicolor (Zone IA) and *A. Zibrina* (Zone IB and IC) were identified to be medium risk to extinction in NNNP because of their current slightly higher harvest rates as the considered human disturbance to the species. *Caladium. bicolor* was projected to have stable species density for all the BAU, IH, and RH scenarios with differences in their values favoring the RH scenario and presenting IH scenario with the lowest stable species density value. Both BAU and RH scenarios for *C. bicolor* started gaining the species density in the first quarter of the initial year

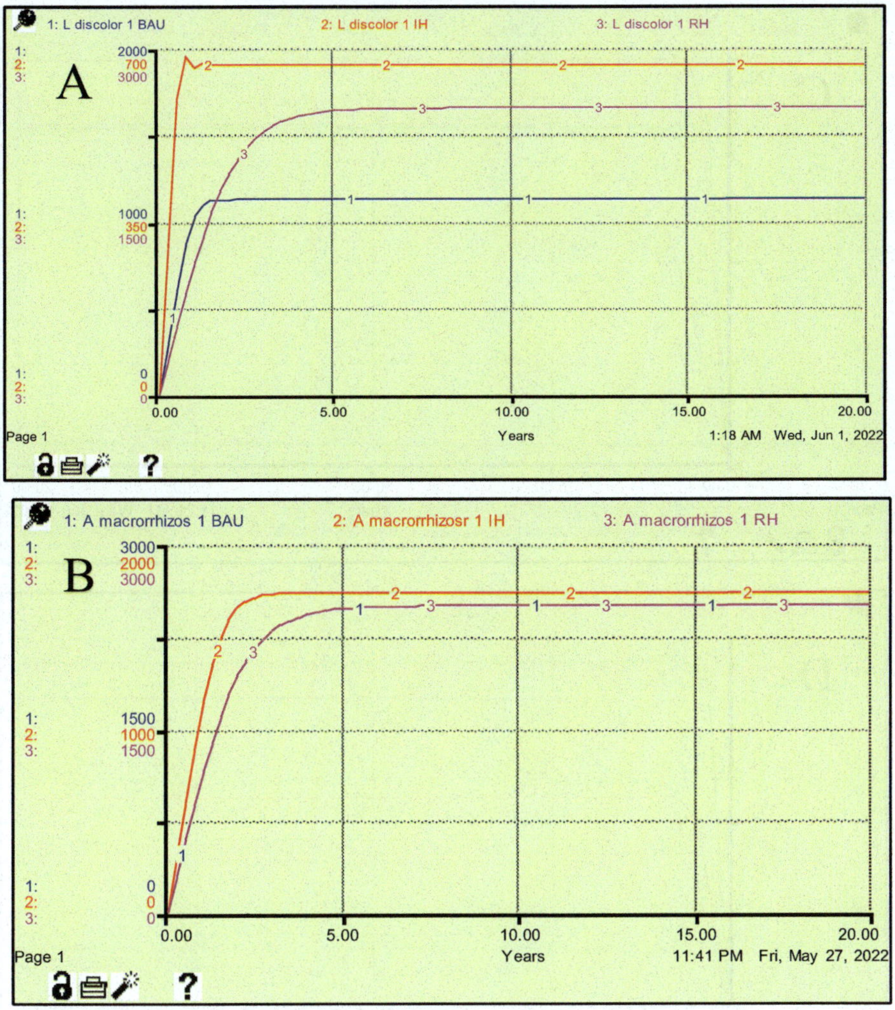

Fig. 27.4 20-year species density projection for three scenarios of (**a**) *L. discolor,* (**b**) *A. macrorrhizos,* (**c**) *C. bicolor*, (**d**) *P. insignis,* (**e**) *A. nidus, and* (**f**) *A. zebrina* inside Northern Negros Natural Park. Red lines represent the business-as-usual scenario for each species, Green for the intensified harvest or increased harvest scenario and Purple for the ideal or reduced harvest scenario

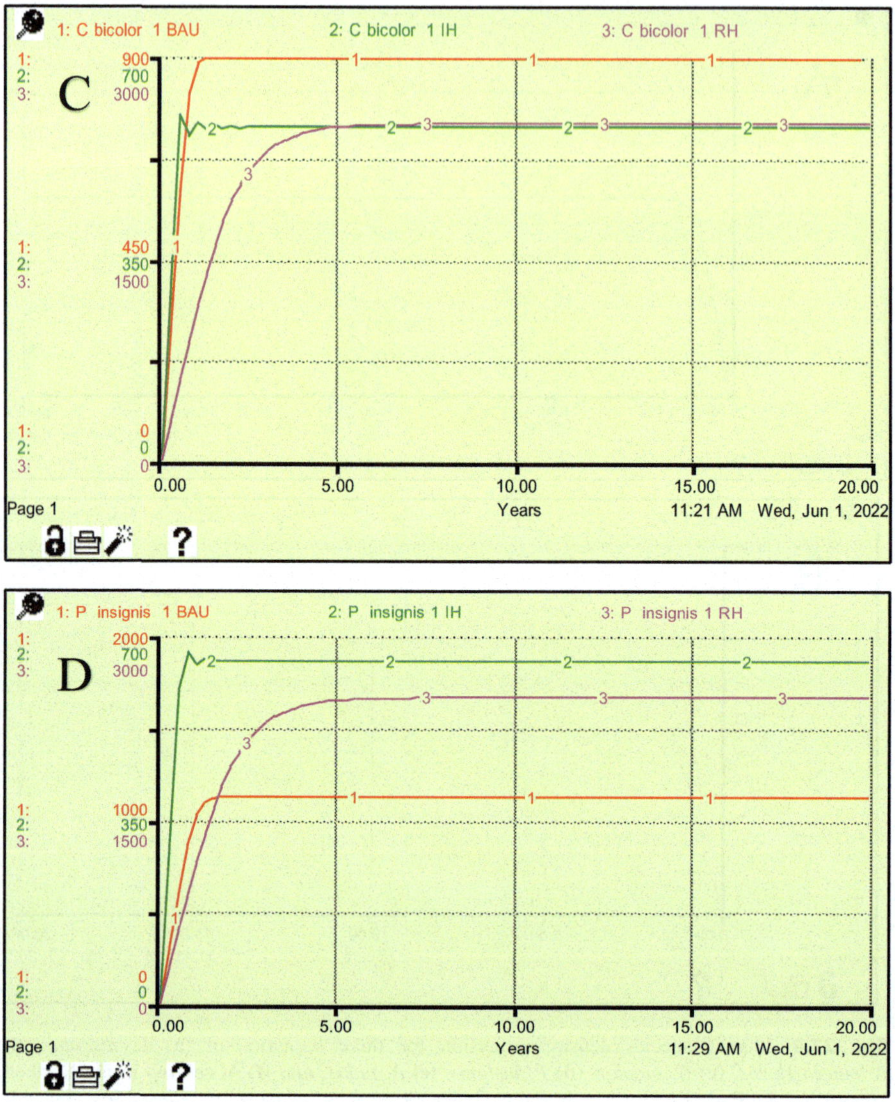

Fig. 27.4 (continued)

which were the lowest value projected for both scenarios. The BAU scenario started stabilizing on the first year until it reaches the highest species density at 896 plants per Ha on the first quarter of the first year. While RH scenario started to reach its stable range of 1148 to 2502 plants per Ha. RH scenario was projected to have the highest species density starting the second quarter of the tenth year at 2502 plants per Ha. The lowest value for species density is vulnerable to the risk of extinction and

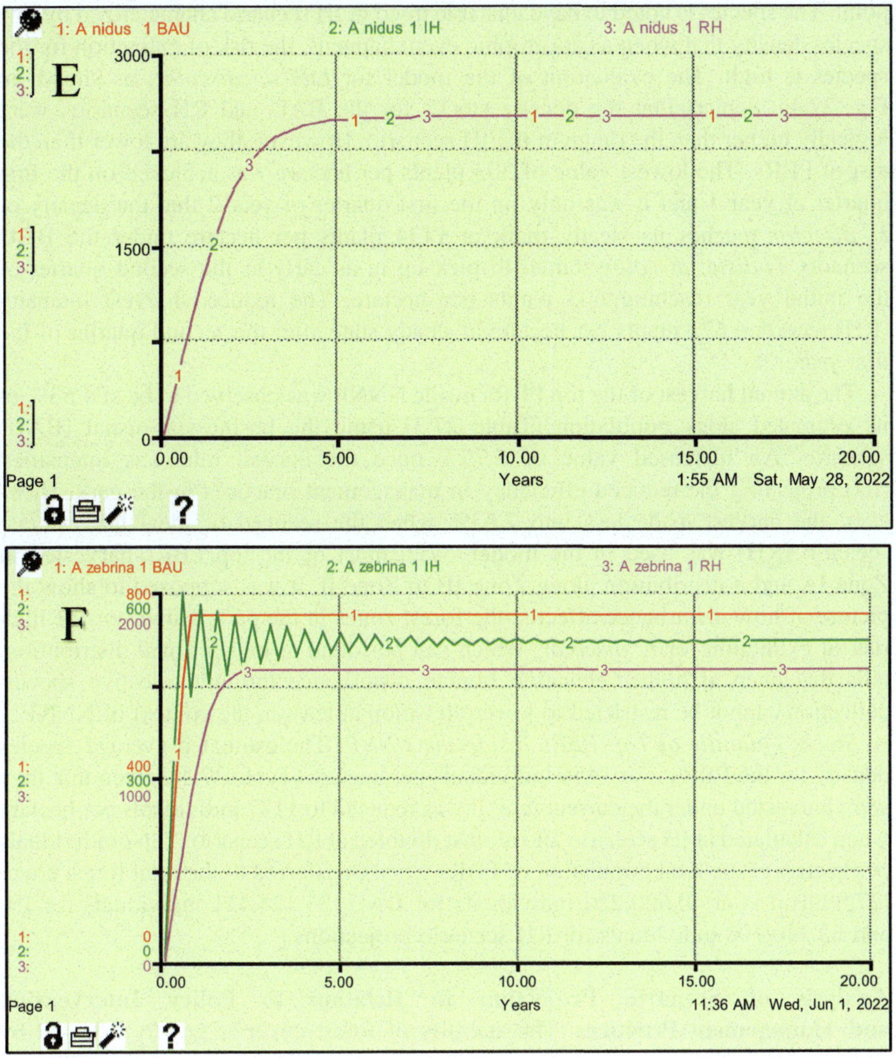

Fig. 27.4 (continued)

the highest projected species density is not projecting a low risk from extinction. These risks were observed to be higher in the lower attitudinal zones and are reduced in the high altitudinal zone particularly in Zone II. Species with large elevation range are low risk to local extinction (Guisan and Theurillat 2000).

The species identified to have higher risk of extinction according to the species density projections is *L. discolor* that is commonly found within Zone II (Table 27.1). It has recorded the highest percentage of BAU scenario to its tipping point. The species is noted to have unstable trend in IH scenario characterized by low species density that when a catastrophic event happens, the risk of extinction for the species is high. The evaluation of the model for *Ludisia discolor,* as shown in Fig. 27.4a, depicts that the density stocks for the BAU and RH scenarios were logically higher than the range of the IH scenario. However, they are lower than the rest of PBRs. The lowest value of 303 plants per hectare was achieved on the first quarter of year 1 and it was only on the first quarter of year 2 that the density of *L. discolor* reaches its steady mark at 1134 plants per hectare under the BAU scenario. *Ludisia. discolor* started to pick up in as early as the second quarter of the initial year reaching 603 plants per hectare. The reduced harvest intensity (RH) recorded 671 plants per hectare at steady state after the second quarter of the first year.

The annual harvest of the top PBRs inside NNNP was observed to be at 4.63% of its estimated stock population (Table 27.3) using the business-as–usual (BAU) scenario. An increased value of 6.75% once the harvest rate was intensified (IH) projecting the reduced efficiency in management practices in the park. However, the harvest made less into 2.63% when the reduced or ideal harvest rate scenario (RH) was used in the model. With most of the top PBRs harvested in Zone IA and a distribution along Zone IB to Zone II, it was supposed to show the picture of how disturbance affected the forest zones in the park. However, the high risk of extinction for *L. discolor,* which can be found at higher zonal distribution, tells that even at higher elevation human disturbance through selective species collection cannot be restricted to lower elevation at least in the context of NNNP.

Stocks Quantity of Top PBRs Inside the NNNP: The estimated average species density for the PBRs is at 1456 individuals per hectare (Table 27.4), given that they were harvested under the current rate. It was reduced to 1187 individuals per hectare when calculated at IH scenario and almost doubled at RH scenario (2267 individuals per hectare). The total population of PBRs when transferred to the total forest cover (27,899Ha) is at 40,620,220 individuals for BAU, 33,124,431 individuals for IH, and 63,240,708 individuals for RH scenario projections.

Analysis of Scenario Projection in Relation to Policy Intervention and Management Practices The stability of forest cover is greatly affected by anthropogenic activities, such as afforestation and harvest activities (Masek et al. 2011). The harvest trend for top PBRs inside the forest cover of NNNP is at most proportion of 1.61% for *A. zebrina* at and at BAU scenario which was obviously

Table 27.3 Percentage of the harvested top PBRs to the stocks quantity in Northern Negros Natural Park, Negros Island, Philippines

TOP PBRs	BAU (%)	IH (%)	RH (%)
Ludisia discolor	1.40	4.02	0.46
Alocasia macrorrhizos	0.14	0.51	0.07
Caladium bicolor	0.36	1.10	0.07
Pinanga insignis	0.68	2.31	0.16
Asplenium nidus	0.05	0.11	0.03
Alocasia zebrina	1.62	4.65	0.39

Table 27.4 Projected average species densities of top PBRs in Northern Negros Natural Park, Negros Island, Philippines

Top PBRs	Average species density (per hectare)			Average species density inside NNNP's forest cover (27,899 hectares)		
	BAU	IH	RH	BAU	IH	RH
Ludisia discolor	1112	666	2348	31,026,317	18,578,464	65,499,365
Alocasia macrorrhizos	2359	1687	2359	65,822,062	47,052,420	65,822,062
Caladium bicolor	887	576	455	24,744,544	16,080,760	12,699,032
Pinanga insignis	1112	651	2359	31,026,317	18,153,555	65,822,062
Asplenium nidus	2359	2359	2359	65,822,062	65,822,062	65,822,062
Alocasia zebrina	745	517	1538	20,784,274	14,434,137	42,912,571

Table 27.5 Harvest rates scenarios and the tipping points to unstable harvesting of the PBRs in Northern Negros Natural Park, Negros Occidental Philippines

Top PBRs	Harvest rates (individual per year per Ha)			
	BAU	IH	RH	Tipping point
Ludisia discolor	8	16	4.0	17
Alocasia zebrina	3	7	1.7	52
Caladium bicolor	3	6	1.6	15
Pinanga insignis	8	15	3.8	42
Alocasia macrorrhizos	1	3	0.6	32
Asplenium nidus	12	24	6.0	42

increased when the harvest scenario was intensified (2.32%) and reduced at ideal rate scenario (0.78%). The stocks density of all the top PBRs is stable at collective average of 1465 individuals per hectare.

Even though the top PBRs harvested are non-timber forest products, the resource extraction activities such as shifting cultivations and reliance to bioresources for food, building materials, and energy are the major causes of forest degradation (Molinario et al. 2020) and thus a threat to forest cover change in the park. The scenarios projected for the PBRs' extraction reflects that the rates are kept below the sustainable harvest rates (Gaoue et al. 2016b) are indicative of either effective management practices for the regulations of PBRs harvest, or the current stocks density level is still high, or both are true.

Table 27.5 shows the harvest rates for every scenario of each PBRs harvested in the natural park. The rightmost column identified the species' risk of extinction based on their species density. The species tend to have higher risk of extinction if they display lower species density (Purvis et al. 2000). The low species density in a landscape like the NNNP in the interest of this paper is brought about by the intensity of harvest rates. The tipping point of the harvest rate is the threshold that a species annual harvest rate can remain stable. The percentage then of the current harvest rate to the tipping point is showing the intensity of the harvest and a determinant to the species stocks density and thus manifest its impact to the forest cover in general. The projections showed that intensified harvest rates resulted low species stocks density.

Ludisia discolor was identified to have the highest risk of extinction relative to the species density projected for the PBRs collected in the Zone II of the park. The current harvest rate of *L. discolor* is eight individuals per year per hectare which is 47% of its tipping point before its stocks density becomes unstable and low. Even considered highest in the risk of extinction, *L. discolor* along with the rest of the PBRs are considered to have low risk of extinction since their percentage to the tipping point harvest rates is slightly higher than the sustainable harvest for non-timber forest products (Gaoue et al. 2016a).

Currently, the Republic Act 11038 or also known as the Expanded National Integrated Protected Area System or ENIPAS and the restrictions imposed by the local village cooperatives or associations are the policies regulating the collection of the identified species moderate in risk of extinction except for the *A. zebrina*. It was listed as species with the status of Vulnerable (Vu) by the Department of Environment and Natural Resources DAO2017-11 an Updated National List of Threatened Philippine Plants and Their Categories. *Alcasia zebrina* was harvested 12 plants per year per hectare at 26% of its tipping point before the harvest becomes unsustainable. Community-based enforcers or commonly known as the BBB or Bantay Bukid Brigade who conducts a regular foot patrol to monitor activities inside the strict protection zone, the regular police checkpoints on the major roads leading out of the protected area together with the very aggressive information and education campaign of the Department of Environment and Natural Resources partnered by the local governments are the contributing factors to the low harvest rate of the vulnerable *A. zebrina* in NNNP, *L. discolor* with 47% BAU harvest percentage of its tipping point, *C. bicolor* with 20%, and *P. insignis* with 21%. *Alocasia. macrorrhizos* (3 plants per hectare per year), *A. nidus* (1 plant per hectare per year), *E. pilosa* (4 plants per hectare per year), and *D. esculentum* (16 plants per hectare per year) are the least risky species from extinction according to their harvest rates in the current harvest (BAU) scenario. They are characterized by having more plants harvested per year in a hectare. *D. esculentum* being listed as "Least Concern" by IUCN Red List also helped in its lower harvest rate.

27.4 Summary and Conclusion

In order to develop scenario projections, the growth model inputs to STELLA software made use of the general harvest and general growth equations that included species density, growth rates, and other explanatory variables. The scenarios projected for the top harvested plant bioresources were shown to be below the sustainable harvest rates, thus indicating a low risk of extinction from the park. The average species densities for each of the top PBRs were also high. In light of the top PBRs gathered in NNNP that were documented, ENIPAS and DENR DAO 2017–11 were successfully implemented. These regulations have proven to be sufficient for the time being to regulate the harvest trends. However, when the current policies are not well-enforced as well as the management practices grow relaxed, threats to biodiversity and forest cover change could increase. This projection study could still be repeated, carried out on a regular basis, and applied to a larger scope to have a clearer modeling and projection of the biodiversity and forest cover change in the park. In the interim, the paper's findings provide important information that should be incorporated into considerations of policy and evaluation and improvement of management practices.

References

Alber H, Robinson EJZ (2013) A review of the spatial economics of non-timber forest product extraction: implications for policy. Ecol Econ 92:87–95. https://doi.org/10.1016/j.ecolecon.2012.01.021

Bharat K, Saini SK, Khosla K, Mehra S, Ravindra N (2015) System dynamics modelling for forest fuelwood management in India. Ambient Sci 2(2):1–7. https://doi.org/10.21276/ambi.2015.02.2.ta01

Boettiger C, Mangel M, Munch S (2015) Avoiding tipping points in fisheries management through gaussian process dynamic programming. Proc R Soc B Biol Sci 282(1801). https://doi.org/10.1098/rspb.2014.1631

Calkin DE, Montgomery CA, Schumaker NH, Polasky S, Arthur JL, Nalle DJ (2002) Developing a production possibility set of wildlife species persistence and timber harvest value. Can J For Res 32(8):1329–1342. https://doi.org/10.1139/x02-056

do Valle DR, Staudhammer CL, Cropper WP (2007) Simulating nontimber forest product management in tropical mixed forests. J For 105(6):301–306. https://doi.org/10.1093/jof/105.6.301

Forrester JW (1961) Industrial dynamics. M.I.T. Press, Cambridge, MA

Gaoue OG, Jiang J, Ding W, Agusto FB, Lenhart S (2016a) Optimal harvesting strategies for timber and non-timber forest products in tropical ecosystems. Theor Ecol 9(3):287–297. https://doi.org/10.1007/s12080-015-0286-4

Gaoue OG, Ngonghala CN, Jiang J, Lelu M (2016b) Towards a mechanistic understanding of the synergistic effects of harvesting timber and non-timber forest products. Methods Ecol Evol 7(4):398–406. https://doi.org/10.1111/2041

Gifford TS, Zobel JM, Shartell LM (2022) Modeling potential changes in rare species habitat from planned timber harvest in Minnesota, USA. Forests 13(2). https://doi.org/10.3390/f13020216

Guisan A, Theurillat J-P (2000) Assessing alpine plant vulnerability to climate change: a modeling perspective. Integr Assess 1(4):307–320. https://doi.org/10.1023/a:1018912114948

Hall P, Bawa K (1993) Methods to assess the impact of extraction of non-timber tropical forest products on plant populations. Econ Bot 47(3):234–247. https://doi.org/10.1007/bf02862289

Jha S, Bawa KS (2006) Population growth, human development, and deforestation in biodiversity hotspots. Conserv Biol 20(3):906–912. https://doi.org/10.1111/j.1523-1739.2006.00398.x

Kusters K, Achdiawan R, Belcher B, Pérez MR (2006) Balancing development and conservation? An assessment of livelihood and environmental outcomes of nontimber forest product trade in Asia, Africa, and Latin America. Ecol Soc 11(2):20

Leaver JM (2019) Response of bird functional diversity to forest product harvesting in the Eastern Cape, South Africa. For Ecol Manag 445:82–95

Limaei SM, Lohmander P, Olsson L (2017) Dynamic growth models for continuous cover multi-species forestry in Iranian Caspian forests. J For Sci 63(11):519–529. https://doi.org/10.17221/32/2017-JFS

Masek JG, Cohen WB, Leckie D, Wulder MA, Vargas R, de Jong B, Smith WB (2011) Recent rates of forest harvest and conversion in North America. J Geophys Res 116:G00K03. https://doi.org/10.1029/2010jg001471
Molinario G, Hansen M, Potapov P, Tyukavina A, Stehman, a. S. (2020) Contextualizing landscape-scale forest cover loss in the Democratic Republic of Congo (DRC) between 2000 and 2015. Land 9(1):23. https://doi.org/10.3390/land9010023
NNNP (2016) Survey and Registration of Protected Area Occupants (SRPAO) report. Northern Negros Natural Park Protected Area Superintendent Office, PENRO Negros Occidental, Department of Environment and Natural Resources, Bacolod City
Purnomo PK, Widianingsih LP (2012) The influence of environmental performance on financial performance with corporate social responsibility (CSR) disclosure as a moderating variable: evidence from listed companies in Indonesia. Rev Integr Bus Econ Res 1(1):57
Purvis A, Gittleman JL, Cowlishaw G, Mace GM (2000) Predicting extinction risk in declining species. Proc R Soc B Biol Sci 267(1456):1947–1952. https://doi.org/10.1098/rspb.2000.1234
Sterman JD (2000) Business dynamics: systems thinking and modeling for a complex world. Irwin McGraw-Hill, Boston, MA

Heritage Negotiations in Manila and Pampanga: The Use of Toponyms in the Study of Two Biocultural Landscapes in the Philippines

28

Honey Libertine Achanzar-Labor

Abstract

Toponyms or place names depict biocultural landscapes, where human activities and interactions are frequent. These may point to certain species native to a geographical area, or to a result of an exchange of goods, trade, or migration. Regardless of the basis, people tend to protect a species for its cultural or medical value, albeit at times, they over-harvest certain species for economic reasons and could be agents to landscape degradation. Changes in biocultural landscape may be observed then through time, while the toponym of the area is retained. Far from an arbitrary pairing of form and meaning, toponyms indeed reveal that names reflect the historical, cultural, or socio-economic experience of the people who use them (Radding and Western, Geogr Rev 100:394–412, 2010).

The scenarios of the old cities of Manila and Pampanga—two of the first royal cities established during Spanish occupation in the sixteenth century (Blair and Robertson, The Philippine Islands, 1493-1898, vol. 3. Ohio: Arthur H. Clark Company, 1911)—will be presented using archival and historical sources which point to the floral landscape associated with big water bodies and their corresponding geocultural value. Regional toponyms and their relations with local flora will be investigated, to study whether toponyms can be used as source of information for relationships between heritage studies and local flora. Archival data will be triangulated with data coming from local and international studies and with current reports of policies and/or projects from the private sector as well as local and national government.

The destruction of homelands, fragmentation of habitats, and post-capitalist conditions of modernity is countered by thoughtful remembrance of tradition.

H. L. Achanzar-Labor (✉)
University of the Philippines Manila, Manila, Philippines
e-mail: hrachanzarlabor@up.edu.ph

S. Ramamoorthy et al. (eds.), *Plant Diversity in Biocultural Landscapes*,
https://doi.org/10.1007/978-981-19-8649-9_28

The appropriateness of conservation measures carried out on varying aspects of the bicultural landscape, some of which are directed toward particular sustainable development goals, is also analyzed.

The study concludes showing how toponyms can be of use in current studies of biocultural landscape, as these may indicate how change can be brought about by (1) non-spatial narratives of vulnerability and historical injustice, (2) the agency of the vulnerable (Brauch, Paul J. Crutzen and the anthropocene: a new epoch in earth's history. The Anthropocene: politik—economics— society—science, vol. 1. Springer, Cham, 2021), and (3) the amplification of conservation issues in the Anthropocene that address particular sustainable development goals. It is in amplifying these narratives that heritage studies can broaden and diversify the discourse on the Anthropocene and, therefore, affirm its relevance in this new geological age.

Keywords

Anthropocene · Heritage · Pampanga · Manila · Toponym · Conservation · Flora · Nila · Indigo

28.1 Introduction

Toponyms or place names, especially those derived from a topographical feature, depict biocultural landscapes, where human activities and interactions are frequent. These may point to certain species native to a geographical area, or to a result of an exchange of goods, trade, or migration. Regardless of the basis, people tend to protect a species for its cultural or medical value, albeit at times, they over-harvest certain species for economic reasons and could be agents to landscape degradation. Changes in biocultural landscape may be observed then through time, while the toponym of the area is retained. Far from an arbitrary pairing of form and meaning, toponyms indeed reveal that names reflect the historical, cultural, or socio-economic experience of the people who use them (Radding and Western 2010).

The chapter is an inquiry on the manifest cultural heritage of a place and its relation to its biocultural origin. It considers examples of cultural heritage practice in the old cities of Manila and Pampanga—two of the first royal cities established during Spanish occupation (Blair and Robertson 1911)—and addresses how plants/flora factor in its development in areas associated with big water bodies. The destruction of homelands, fragmentation of habitats, and post-capitalist conditions of modernity is countered by thoughtful remembrance of tradition. The appropriateness of conservation measures carried out on varying aspects of the bicultural landscape, some of which are directed toward particular sustainable development goals, is also analyzed.

The scenarios of Manila and Pampanga will be presented using archival and historical sources which point to their floral landscape and its geocultural value. Regional toponyms and their relations with local flora will be investigated, to study whether toponyms can be used as source of information for relationships between heritage studies and local flora. Archival data will be triangulated with data coming from local and international studies and with current reports of policies and/or projects from the private sector as well as local and national government.

When the Spanish conquistador Miguel Lopez de Legazpi and his men sailed to Manila Bay in May of 1571, they found a sheltered natural harbor on the eastern side of a bay and at the mouth of the Pasig River. They found about 40 settlements clustered around the wide bay and along the banks of the river, and the largest of the settlements was called Tondo. The political and military center of the region, however, was **Maynila**, a Tagalog polity along the Pasig River, of less than a thousand people, which was guarded by a fort boasting of *lantakas* (culverins) and smaller cannons peeking out from behind its palisade of trunks of trees and coconut palms.

With an area of 1994 km^2, and a coastline of 190 km, Manila Bay is situated in the western part of Luzon and is bounded by Cavite and Metro Manila on the east, Bulacan, and **Pampanga** on the north, and Bataan on the west and northwest. The bay drains approximately 17,000 km^2 of watershed area, with the Pampanga River contributing about 49% of the freshwater influx. Coastal and marine habitats in the area include upland forests, mangrove, mudflats, sandy beaches, sea grass, and coral reefs.

28.2 Manila

The Muslims and native converts of the old settlement of Maynila were bound by blood, trade, and friendship to the sultan of Brunei (Scott 2021), and they nurtured connections of commerce and culture with neighboring Muslim centers spanning the Jolo archipelago in southern Philippines; Borneo; Aceh, at the northern tip of Sumatra; the Moluccas, within the Indonesian archipelago; Malacca, in the Malay Peninsula; and Pattani, in the southern region of Thailand. At various times of the year, Chinese merchants also arrive in trading junks laden with wares to do business with the coastal Malays, who favored their jars and dishes (Reyes 2017).

With its safe harbor, a strategic position that allowed for intraregional trade, and established Sino-Filipino-Muslim trade networks, Manila presented an ideal location for socio-political–economic interactions (Fig. 28.1). Legazpi liked what he saw. As he had done a few years earlier in the Visayas, in central Philippines, he expended great effort in putting down local resistance and set about establishing a permanent colonial settlement. He deposed its chieftain, Rajah Sulayman, negotiated a series of pacts and alliances, and instituted Manila as the seat of power of the Spanish government and one of the royal Spanish cities in the Philippines in the Spanish

Fig. 28.1 Peter van den Keere, Amsterdam, 1598. *Insulae Philippinae*. This small map is historic for being the first map of the Philippines already within its established historical borders. Source: Three Hundred Years of Philippine Maps. Metropolitan Museum of the Philippines, 2012

East Indies,[1] along with Cebu, Naga, and Iloilo. The Spaniards ruled the land for more than three hundred years, until 1898 (Karnow 1989).

Manila was reorganized into a Spanish *ciudad* by Miguel Lopez de Legazpi on 24 June 1571 (Fig. 28.2). This fortified Spanish city, hence Intramuros, was built directly on top of Maynilà, the native settlement of Rajah Matanda and Rajah Sulayman, which Legazpi took possession of on May 19, 1571 (Blair and Robertson 1911, pp. 173–174).

28.2.1 Manila as Toponym

Manila, as the place is known today, is derived from the Sanskrit word *nila* which literally means "dark blue". This shade of blue used to be solely associated with India, its country of origin, hence, is more popularly known as indigo (Jones 2008).

The name *Maynilà*, the shortened form of the "may(roon) nila," points to the presence of these indigo-yielding plants growing in the polity, where Tagalog is

[1] The overseas territories of the Spanish Empire in Asia and Oceania from 1565 to 1901, governed from Mexico City and Madrid through the captaincy general in Manila.

Fig. 28.2 Francois Valentijn, Dordrecht, 1724. *De Stad Manilha* (The City of Manila). Source: Three Hundred Years of Philippine Maps. Metropolitan Museum of the Philippines, 2012

spoken (Ocampo 2022), rather than as a settlement that trades in indigo dye, since this was founded several hundred years before indigo dye extraction became an important economic activity in the area in the 18th century (Baumgartner 1975).

The native Tagalog name for the indigo plant, *tayum* (Roces 1991), also finds use in the toponym of a street in Manila: Tayuman. A busy street in the heart of Manila now known as a hub for religious-supplies (Christy 2012), its name indicates that it may have been a place where fabrics were dyed with an indigo pigment. It may also mean a place, where *tayom* plants used to grow (Limos 2020).

Scott (2021) in his book Barangay: Sixteenth-Century Philippine Culture and Society refers to Pedro San Buenaventura (1613) *Vocabulario de la lengua Tagala* as a rich source of information on the local ethnography during the age of Spanish contact, among these, on the Tagalog practice of weaving and dyeing:

Ordinary Tagalog blankets were called *kumot*; large red ones, *minalot*; *tinalaban* was a striped cloth in different colors, which was called *sabasabat* if these were in black, blue, and white.

Galungang and *labayan* were reels for winding thread, *tugas* was to wash thread in preparation for dyeing, and *sapad* was to soak it in blue dye.

Dyestuffs included *suga* (saffron or pomegranate), *tayom* (indigo), and red aguish or talk roots and bangkuro bark, while cotton husks themselves made a bright red dye.

Fig. 28.3 *Indigofera suffroticosa mill*

Indigofera is a large genus of over 750 species of flowering plants belonging to the pea family Fabaceae. They are widely distributed throughout the tropical and subtropical regions of the world (Gao and Schrire 2022). Several species, especially *Indigofera tinctoria* and *Indigofera suffruticosa*, are used to produce the dye indigo.

Dye is obtained from the plant's leaves: these are soaked in water and fermented in order to convert the glycoside indican naturally present in the plant to the blue dye indigotin. The precipitate from the fermented leaf solution is mixed with a strong base, such as lye.

Today, most dye is synthetic, but natural dye from *Indigofera tinctoria* is still available, marketed as natural coloring, where it is known as *tarum* in Indonesia and *nila* in Malaysia.

A number of indigenous communities the Philippines still persist as well in naturally dyeing their fabrics to this day. The Blaan and T'boli still dye their *abaca* using local plants grown in their area. Leaves from the *knalum* (*Diospyrus sp.)* tree give them black and the roots of the *lagu* tree (*Morinda citrifolia*) give them red. Ifugao weavers in northern Luzon use plant dyes from *Ficus septica*, or *hawili*, for black, and they also dunk their yarns in the mud of the rice terraces. The Itnegs of Abra, on the other hand, continue to use *tayum* (*Indigofera tinctorial), sapang* (*Caeselpinia sappan*), and other plant dyes in their textiles.

Marian Pastor Roces (1991) in her book *Sinaunang Habi* documents three species of genus *Indigofera* that have been used in the Philippines before synthetics arrived, all three indicate their use or presence in the Tagalog-speaking communities:

> *Indigofera suffroticosa mill* [*bila-bila* in Ivatan, *piyasak* in Maranaw, *tina-tinaan* in Tagalog, *tayum* in Panay Bisaya] (Fig. 28.3), *Indigofera hirsuta* ([*kisul*, or *lipang-aso*, in Tagalog; *tagom a bombolen* in Maranaw; *uyun-uyun bantud* in Hanunuo Mangyan] (Fig. 28.4), *Indigofera tinctoria* [*dagum* in Subanun, *tayum-tayuman* in Tagalog, *tayum* in Bisaya] (Fig. 28.5).

Fig. 28.4 *Indigofera hirsuta*

Fig. 28.5 *Indigofera tinctoria*

In Indonesia, the Sundanese use *Indigofera tinctoria* (known locally as *tarum* or *nila)* as a dye for batik. The plant is called *tarum* as distinct from *nila*, the color indigo, in Malaysia and Brunei.

Despite the said blood and trade relationship of Maynila with Brunei, the latter's political influence over Maynila is not considered to have extended to military or political rule.

Intermarriage was a common strategy for large thalassocratic states, such as Brunei, to extend their influence, and for local rulers such as those from Maynila to help strengthen their family claims to nobility (Junker 1998). As the representation of social status in Brunei is objectified through the act of consumption of traditional woven textiles (Wahsalfelah 2010), the natives of Maynila, whose link to nobility was premium, would have also placed importance to *nila* as a source of natural textile dye.

The employment of traditional woven textiles as a ceremonial dress and part of regalia in the royal court in Brunei plays an important role in emphasizing the tangible expression of status and power. At all court events, the color of the dress is specified according to the status of the wearer, as well as the time when the function is held. Traditional or royal dignitaries wear different types, pattern, designs, motifs, and colors of traditional woven sinjang, arat (belt), and dastar, according to rank. Each male dignitary carries a kris. The traditional textiles for the Cheteria (common nobles), for instance, have the same pattern and design for all levels of wearers, but distinguished by way of the color differences. The color for the chief Cheteria is dark purple. Light purple is for Cheteria 4. Cheteria 16 would wear blue, while Cheteria 32 wears magenta (Wahsalfelah 2010).

The emergence of cheaper and more readily available synthetic dyes practically killed the use of indigo as well as indigo trade in the Philippines. However, its use has been kept alive in indigenous communities that continue to use the natural dye for their weaves.

The Integs of Abra continue to use *tayum* (*Indigofera tinctoria*), *sapang* (*Caeselpinia sappan*), and other plant dyes in their textiles. *Indigofera tinctoria* is also still used by the Subanun who refer to it as dagum. *Indigofera suffroticosa mill* is still used by the Ivantan, the Maranaw, and the Panay Bisaya. *Indigofera hirsuta* is still likewise grown by the Maranaw and the Hanunuo Mangyan (Palasi 2022).

Efforts have also been made it revive the use of indigo, among this, through the Non-Timber Forest Products Exchange Program Philippines, which started working on the Indigo Project to promote the use of non-timber forest products—such as indigo—to give alternative livelihood to forest dwellers, such as indigenous peoples.

The Indigo Project now works with indigenous populations in upland and rural communities in the Philippines and Indonesia, including members of Mangyan Alangan, Mangyan Hanunoo and Mangyan Iraya of Mindoro; Higaonon of Bukidnon, Palaw'anon and Tagbanua of Palawan, T'boli of South Cotabato, Maguindanao weavers and those of Negros (Arceo-Dumlao 2016).

28.3 Pampanga

The province of Pampanga lies north of Manila Bay. The name *La Pampanga* was given by the Spaniards, who encountered locales living along the banks (*pampáng*) of the Pampanga River, the second largest river on the island of Luzon and the country's fifth longest river. The establishment of Pampanga in 1571 makes it the first Spanish province in Luzon (Fig. 28.6).

Sixteenth-century Pampanga was simply the basin of the great river system of that name from the Candaba Swamp to the Manila Bay delta. The rest of the province and "Upper Pampanga" (Nueva Ecija) was the forest land with mountains on the west, north, and east inhabited by the Alaguetes, Ilongots, Negritos, and Zambals (Scott 2021).

Just like Manila, Pampanga has retained its toponym, the importance of the mighty Pampanga River preceded whatever change may have occurred in the place. Likewise, tree-based names of the old towns of Pampanga have remained unchanged pointing to the vast forests that once characterized the province.

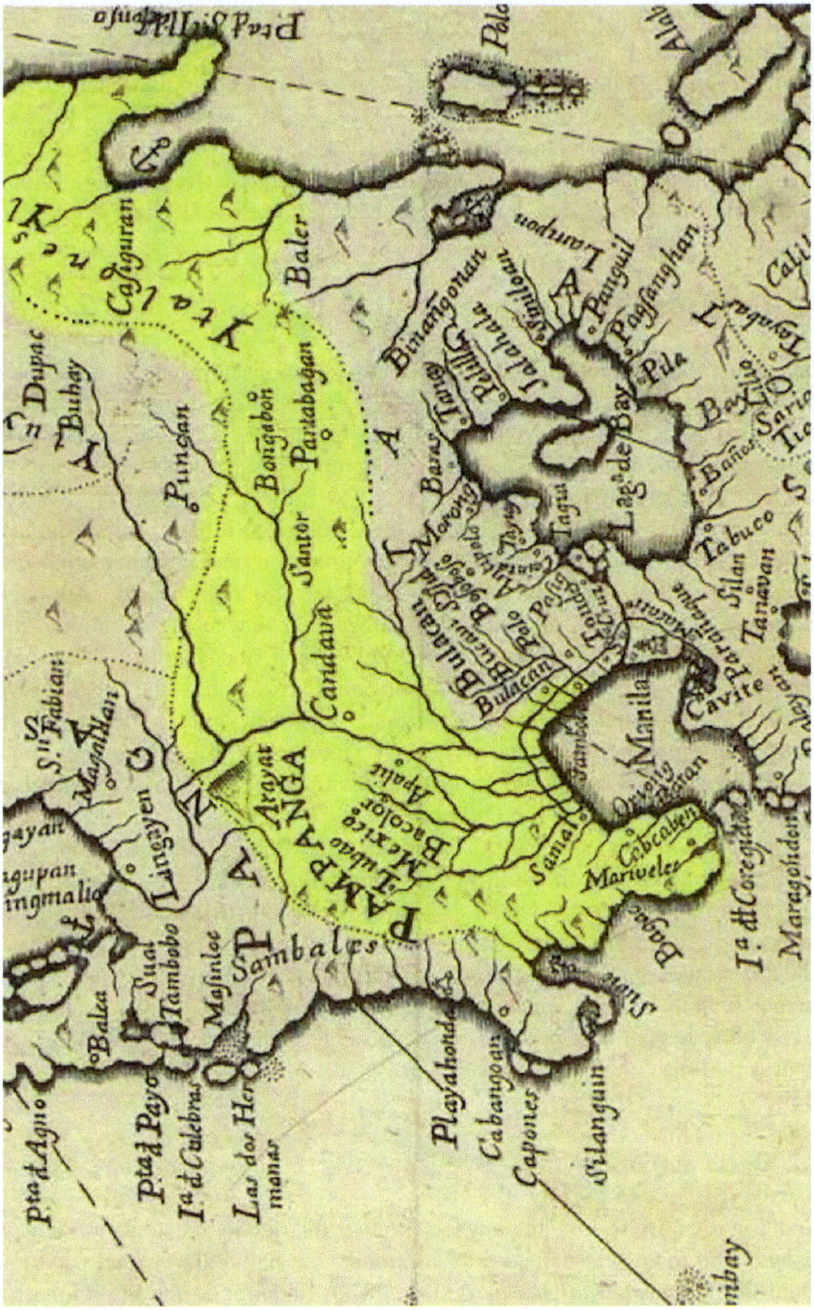

Fig. 28.6 Map shows territories under the province of La Pampanga in the year 1625. File: Pampanga before Partitions, 1625.jpg (2021, July 15). *Wikimedia Commons, the free media repository*. Retrieved from https://commons.wikimedia.org/w/index.php?title=File:Pampanga_before_Partitions,_1625.jpg&oldid=575284218

In the 1980s, Pampanga's land use has been cut down to 40% Forest, 33% Grassland, 24% Agricultural land, and 3% Wetlands (National Water Resources Board 1983). Philippine forests have long been logged for timber products. Forest cover has continuously declined from about 68% in 1876 to a mere 18% in 2001. Forest destruction and degradation and conversion of unique habitats have threatened many plant species.

The continuing disappearance of such vital habitats for many plant species sets one of the greatest challenges for all Filipinos: to halt the destruction of Philippine plant diversity that is so essential to meet their own present and future needs.

28.3.1 Candaba as Toponym

Among the Philippine encomiendas or private estates established by the Spanish colonial government in the sixteenth century (Bataan, Betis y Lubao, Macabebe, Candaba, Apalit, Calumpit, Malolos, Binto, Guiguinto, Caluya, Bulacan, and Meycauayan), four have remained part of Pampanga up until today: Candaba, Betis, Macabebe, and Apalit.

Candába may have been the first and oldest settlement in Pampanga. Linguistic evidence indicates that during prehispanic times, terrestrial navigation only recognized two directions—paraláya or "going to Bunduk Aláya" and paráuâ (Parma) or "going to dáuâ or Kandáuâ"—suggesting that in the beginning, there was only two directional references: Bunduk Alava and Candába. The oldest archeological artifact in Pampanga was also found in Candaba—the Candaba Neolithic Adze—a 5000-year-old stone tool, made of basalt with the length of 36.4 cm. width of 9.8 cm and thickness of 4.53 cm (Beyer 1947). The Adze could have been used then for cutting down trees and making boats; it is now considered a National Cultural Treasure in the Philippines.

Boat production was necessary in Candaba, the lowest point in Central Luzon. The Candaba Swamp encompasses about 32,000 ha, made of freshwater ponds, swamps, and marshes surrounded by seasonally flooded grasslands. The entire area becomes submerged underwater during the wet season.

Candaba now is basically a fishing and farming community. It is noted for its communal fishing grounds encompassing some 430 km^2 of highly arable land. Produced here are the best produce of the province: watermelon and muskmelon (Melindres 2004).

28.3.2 Betis as Toponym

The old town of Betis in Pampanga is now a district in the town of Guagua, Pampanga. The town was named after a lumber tree namesake, which used to be abundant in the area—*Madhuca betis* (Fig. 28.7). Endemic to the Philippines, the tree reaches a height of 30 m and grows in primary forests at low altitudes. Its

Fig. 28.7 *Madhuca betis*

two-toned colored leaves are attractive, but the tree is known to emit an undesirable smell during flowering season.

More importantly, the pale redwood is hard and heavy, and is considered excellent for purposes requiring great strength and durability, like heavy construction and ship building. Its seeds contain oil for illumination, while its barks and leaves have reported medicinal values.

Depleted due to indiscriminate logging and *kaingin*, the tree is no longer found in Betis, but its legacy remains in old furnitures once carved by the local craftsmen. The town Betis remains known for its quality wood products, primarily, intricately carved furniture, as well as religious statues and relief sculptures, offshoot of an early Augustinian mission in the sixteenth century. Augustinian historian Fr. Isacio Rogriguez wrote in his book, “Al Servicio del Evangelio,” that the Augustinians in 1591 already had a mission in Betis and Lubao consisting of 20,000 persons and four convents (Darang 2013); they taught the people of Betis how to carve wood into religious images.

One of its most famous residents, Maestro Apung Juan Flores, created the elaborately carved chandelier and rosette ceiling decoration in the Malacañang Palace. Tondo artist Antonio Galleron and National Artist Vicente Manansala trained under him (Mawis-Aliston 2020).

28.3.3 Macabebe as Toponym

Another ancient town in Pampanga is Macabebe, so-called as it is situated along the shores or banks (Kapangpangan: *bebe*) of Río Grande de Pampanga. The river's routes and its northern tributaries provided the pathways to the early major settlements in Pampanga, and the English translation of its toponym—'*surrounded by rivers*'—describes the historical heritage of the ancient town.

This grand access to the riverine water system facilitated transport of logs to and from Macabebe, and encouraged the extensive practice of wood carving and boat building. "Macabebe was able to attack Legazpi's fleet with forty *karakoas* only two weeks after he landed in Manila." (Scott 2021).

Karakoas were large outrigger warships made and used by native Filipinos, notably the Kapampangans and the Visayans, during seasonal sea raids. The *karakoa* was distinct from other traditional Philippine sailing vessels in that they were equipped with platforms for transporting warriors and for fighting at sea. As boats were only for long-distance transportation in the Philippines then, all commercial and political contacts depended upon them (Clark 2020).

This suggests that Kapampangans were among the navigators called "Luzones" who traded with Borneo, Malacca, Sumatra, and the Moluccas in pre-Hispanic times (Scott 2021).

The need to address the demand of Spanish galleon crew factored to the end of the era of boat building in Pampanga. A Manila galleon customarily carried about 130 sailors: officers and skilled crew were usually Spanish or Mexican, but the other members of the crew were largely Filipino, with a few Mexicans. A common ratio of Filipinos to Spaniards on a Manila galleon was five-to-one (La Follette et al. 2018).

The making of *karakoas* is of a bygone era in Pampanga, but Macabebe is still known for its sculptural works. Religious sculptures, known as "santos," usually with the image of Jesus Christ, The Blessed Virgin Mary or patron saints, are found in almost every traditional Catholic home and church. Regardless of the subject or object carved, may this be a karacoa of pre-colonial times or a Catholic saint, what is paramount is the access to wood ideal for carving in Pampanga.

Pablo "Ambo" Bautista, a businessman based on Macabebe, is credited to have popularized the craft of carving Santos. He trained his sons to also become craftsmen in their own rights, contributing to the development of santo-making. Eventually, more artists and craftsmen had put up their shops in Macabebe, thus cementing the town's reputation as the "Home of the Santeros." Wood, ivory, cement and/or fiber glass are used to produce ecclesiastical art pieces in Macabebe (Mawis-Aliston 2020).

In line with the celebration of Macabebe's 443rd founding anniversary, the local government launched on 16 January 2018 the first Santero Festival to celebrate its flourishing statue-making industry. The festival is dedicated to the "Santeros" instrumental in conserving this tradition in Macabebe (Pavia 2018).

28.3.4 Apalit as Toponym

Another old Kapampangan town named after a tree is Apalit, which got its name from the country's national tree, the narra or *Pterocarpus indicus* (Fig. 28.8). Apalit is old Kapampangan for the narra tree.

The narra tree is a large deciduous reddish hardwood that is native to most Asian countries.

Widely introduced due to its many uses, the narra was declared the national tree of the Philippines in 1934 by Governor-General Frank Murphy of the Insular Government of the Philippine Islands.

Narra or apalit is acknowledged as an excellent timber in southern Asia and is categorized as one of the most beneficial timbers in the Philippines. Its wood is used for a lot of products, such as cabinetry, cart wheels, carving, construction, furniture, and even musical instruments.

Apalit is known for its excellent blacksmiths, as chisels are a demand in a region known for wood work; it also used to be known for lush forests of Apalit trees that served as landmark for traders and visitors who regularly visited the settlement along the banks of Rio Grande de la Pampanga.

The town is now famous for its *Apung Iru* Fluvial Procession, one of the most important water-based intangible cultural heritage of the Philippines. Celebrated every June 27–29 of each year, the feast of St. Peter (A*pung Iru*) is observed with a fluvial parade along the Pampanga River.

Fig. 28.8 *Pterocarpus indicus*

Anticipation mounts as *Apung Iru* is transferred from a wooden boat to a pagoda decorated with multi-colored flowers. From the banks of the river, throngs would acknowledge the passing *Apung Iru* by waving leafy branches and fronds or by making the sign of the cross. With excitement reaching fever pitch, brave souls would dunk themselves in the waters of the river, unmindful of the danger as they swim alongside the pagoda (Tiatco 2010).

The image, with its ivory face and hands, dates from the last quarter of the 1700s. It is said that Don Pedro Armayan-Espíritu y Macam (d. 1904) or his parents acquired the image from an aunt, Doña Máxima Santos, in exchange for a considerable parcel of agricultural land in Apalit. Its safekeeping was passed on from one generation to the next. Last 1975, the Saint Peter's Mission, Inc. was established by the clan with the stipulation that *Apung Iru* and his feast be maintained by the generations to come.

28.4 Summary and Conclusion

Toponyms in Manila and Pampanga refer to local flora significant to the development of biocultural heritage; they bear the name of historical nomenclatures that have undergone more than three centuries of colonial rule. Despite changes in the biocultural landscape, some of these place names remain unaltered and, in fact, serve as biohistorical and geohistorical markers.

The name Manila has remained unaltered due to its established name in the trade network. The presence of the plant *nila* in the area is proof of its having participated in the regional network of exchange, a source of pride for any polity. The choice of Manila as the colonial administrative center is, in fact, also a confirmation of the importance of the polity the locales with their datus have previously established.

The armed struggle of the colonized natives in Manila toward their deemed displacement bespeaks their pride of place, of being part of a free/independent polity that has engaged in trade with from different parts of the region. The Spaniards observed this when they reached Manila; and they conquered and made it their administrative center, to show their king that they have indeed conquered a land worth governing even from so far a distance.

The retainment of toponyms in Pampanga is a confirmation of the importance of the great river to the people, not only as a source of water for washing and cleansing, as a means of communication, but also as a superhighway where people, logs (as medium/material for shelter, boats, and other carved implements) as well as other goods can be transported.

The toponyms in Pampanga also affirm the region's master wood carvers, its gentry who is known to be engaged in work-working for centuries. Their main occupation may no longer be boat building—of *karakoas*, riverine boats, and later, the Spanish galleons—but they continue to be acclaimed as master carvers, this time of furniture, churches, and its furnishings.

To end, the study shows how toponyms may be of use in current studies of bicultural landscape, as these may indicate how change can be brought about by

(1) non-spatial narratives of vulnerability and historical injustice, (2) the agency of the vulnerable (Brauch 2021), and (3) the amplification of conservation issues in the Anthropocene that address particular sustainable development goals. It is in amplifying these narratives that heritage studies can broaden and diversify the discourse on the Anthropocene and, therefore, affirm its relevance in this new geological age.

References

Arceo-Dumlao T (2016) Indigo trade getting rid of the blues. Philippine Daily Inquirer. Retrieved from https://business.inquirer.net/208627/indigo-trade-getting-rid-of-the-blues

Baumgartner J (1975) Manila - Maynilad or Maynila? Philipp Q Cult Soc 3(1):52–54. Retrieved from http://www.jstor.org/stable/29791188

Beyer HO (1947) Outline review of Philippine Archaeology by Islands and Provinces. Philipp J Sci 77(3–4):205–399. Retrieved from https://philjournalsci.dost.gov.ph/images/pdf_upload/pjs1947/PJS_Vol_77_No3_4_Jul_Aug_1947.pdf

Blair EH, Robertson JA (1911) Relation of the conquest of the island of Luzon. In: The Philippine Islands, 1493-1898, vol 3. Arthur H. Clark Company, Ohio. Retrieved from https://quod.lib.umich.edu/cgi/t/text/pageviewer-idx?c=philamer;cc=philamer;q1=blair;op2=and;op3=and;rgn=works;rgn1=author;rgn2=title;rgn3=title;idno=AFK2830.0001.003;didno=AFK2830.0001.003;view=image;seq=00000171

Brauch HG (2021) The Anthropocene concept in the natural and social sciences, the humanities and law – a bibliometric analysis and a qualitative interpretation (2000–2020). In: Benner S, Lax G, Crutzen PJ, Pöschl U, Lelieveld J, Brauch HG (eds) Paul J. Crutzen and the anthropocene: a new epoch in earth's history. The Anthropocene: politik—economics— society—science, vol 1. Springer, Cham. https://doi.org/10.1007/978-3-030-82202-6_22

Christy B (2012) Ivory worship. National Geographic Magazine. Retrieved from http://nordmansustainability.com/wp-content/uploads/2012/09/Ivory_worship.pdf

Clark J (2020) Karako: Warships from Philippine history. Retrieved from https://www.aswangproject.com/karakoa-warship-philippines/

Darang J (2013) Betis, Pampanga woodcarving crafts: an Augustinian legacy. Philippine Daily Inquirer. Retrieved from https://lifestyle.inquirer.net/141625/betis-pampanga-woodcarving-crafts-an-augustinian-legacy/#ixzz7bvQCEK00

Gao X, Schrire BD (2022) Indigofera L. Flora of China. eFlora. Accessed 21 August 2022. http://www.efloras.org/florataxon.aspx?flora_id=2&taxon_id=116399

Jones R (2008) Loan-words in Indonesian and Malay (dalam bahasa Inggeris). Yayasan Obor Indonesia, Jakarta

Junker LL (1998) Integrating history and archaeology in the study of contact period Philippine chiefdoms. Int J Hist Archaeol 2(4):291–320

Karnow SA (1989) Miguel López de Legazpi. In our image: America's empire in the Philippines. Random House, New York

La Follette C, Deur D, González E (2018) The galleon's final journey: accounts of ship, crew, and passengers in the colonial archives. Or Hist Q 119(2):210–249. Retrieved from https://pdxscholar.library.pdx.edu/cgi/viewcontent.cgi?article=1180&context=anth_fac

Limos MA (2020) Here's what the names Balintawak, Tayuman, and Antipolo mean. Esquire Magazine. Retrieved from https://www.esquiremag.ph/long-reads/features/lost-meanings-of-places-names-philippines-a00293-20201228-lfrm

Mawis-Aliston V (2020) Pampanga's role in Philippine architecture. Philippine Daily Inquirer. Retrieved from https://business.inquirer.net/304854/pampangas-role-in-philippine-architecture#ixzz7aWyS9Zmo

Melindres R (2004) The utilization of Candaba swamp from prehistoric to present time: evidences from archaeology, history and ethnography. Bhatter Coll J Multidiscip Stud 4:81–93
National Water Resources Board (1983) Pampanga river basins, Report No. 24-3A, Framework Plan
Ocampo A (2022) Araw ng Maynila, araw ng Kastila. Philippine Daily Inquirer. Retrieved from https://opinion.inquirer.net/154337/araw-ng-maynila-araw-ng-kastila-2
Palasi K (2022) Philippine natural dyes: a short overview. Philippine Textile Council, Habi. Retrieved from https://www.habiphilippinetextilecouncil.com/blogs/what-we-do/philippine-natural-dyes-a-short-overview
Pavia J (2018) Santero festival redefines Macabebe. iOrbit New online. Retrieved from https://iorbitnews.com/santero-festival-redefines-macabebe/
Radding L, Western J (2010) What's in a name? Linguistics, geography, and toponyms. Geogr Rev 100:394–412. https://doi.org/10.1111/j.1931-0846.2010.00043.x
Reyes RA (2017) Flaunting it: how the galleon trade made Manila, circa 1571–1800. Early Am Stud 15(4):683–713. Retrieved from https://eprints.soas.ac.uk/24768/1/Reyes_Flaunting%20It.pdf
Roces MP (1991) Sinaunang habi: Philippine ancestral weaves. Nikki Books, Philippines
San Buenaventura P (1613) Vocabulario de la lengua Tagala, Pila
Scott WH (2021) Barangay: Sixteenth-century Philippine culture and society. Ateneo de Manila University Press, Quezon City
Tiatco AP (2010) Libad nang Apung Iru and pamamaku king krus: performances of ambivalence in Kapampangan cultural spectacles. Drama Rev 54(2):91–102
Wahsalfelah NH (2010) Transformation in mode of clothing in Brunei Darussalam and its impact on identities. Borneo Res J 4:193–208. Retrieved from https://www.researchgate.net/265852235_TRANSFORMATION_IN_MODE_OF_CLOTHING_IN_BRUNEI_DARUSSALAM_AND_ITS_IMPAC T_ON_IDENTITIES

Liaforestry-Sustainable Forest Management Model Using Multi-purpose Tree Species for Reforestation and Climate Change Mitigation, Lesson Learned from Meru Betiri National Park

29

Hari Sulistiyowati, Arif Mohammad Siddiq, and Maman Surahman

Abstract

The need to share information related to plant diversity of the biocultural landscape during the Anthropocene epoch in Indonesia. Rehabilitation areas of Wonoasri resort are selected to expose the plant diversity of tropical forest succession in Meru Betiri National Park (MBNP) because of human intervention. The study has been subjected to review reforestation projects and their success in the park, and to analyze lessons learned from liaforestry-sustainable forest management model using multi-purpose plant species (MPPS) for reforestation and climate change mitigation. The pilot case representative is demonstration plots located at the rehabilitation areas within MBNP. Literature reviews and plot samplings are used to collect data. Quantitative data are analyzed using ecological approaches, while qualitative data are analyzed using comprehensive descriptive analysis. Without intervention of human activity, plant diversity of tropical forests in MBNP showed high H′ Shannon–Wiener Index which is about $H > 2.5$ on average. The implementation of liaforestry-sustainable forest management model using multi-purpose plants impacts on increase plant species composition and diversity that impact on bird diversity, increase people (rehabilitation farmer) income, improve structure and function of secondary tropical forest and increase carbon stocks that result in sustainable climate change mitigation. Intensive communication individually or in group to build their trust and commitment for partnership are implemented for this community-based management.

H. Sulistiyowati (✉) · A. M. Siddiq
The University of Jember, Jember, Indonesia
e-mail: sulistiyowati.fmipa@unej.ac.id

M. Surahman
The Meru Betiri National Park, Jember, Indonesia

S. Ramamoorthy et al. (eds.), *Plant Diversity in Biocultural Landscapes*,
https://doi.org/10.1007/978-981-19-8649-9_29

Keywords

Liaforestry · Forest management · Mitigation · Meru Betiri National Park

29.1 Introduction

Indonesia is a country that has extensive tropical rain and species diversity, the fourth largest plant in the world. These plant species are depicted in forests that are spread throughout Indonesia including protected forest (Indrawan et al. 2007; Sari et al. 2020). The protected forests are forest areas that have the main function as system protection life support to regulate water management, prevent flooding, control erosion, prevent seawater intrusion, and maintain soil fertility. Protected forests in Indonesia have an important function in maintaining ecosystems and world biodiversity. As part of protected forests, the primary tropical forests of Meru Betiri National Park (MBNP) are forest that naturally regenerates from native species and have clear indications that within there are no significant human activities that affect the original ecological changes.

The phenomenon of forest degradation in Indonesian tropical forest has a significant increase (Sunderline and Resosudarmo 1997; Tsujino et al. 2016). Over-human population is an indirect factor that has an impact on the high demand for land and forest conversion (Zhao et al. 2017; Vadrevu et al. 2019). Cases of forest degradation have occurred on several islands in Indonesia, such as Sumatra (Basyuni et al. 2018), Kalimantan (Susanto et al. 2018), Sulawesi (Supriatna et al. 2020), Papua (Koalisi Indonesia Memantau 2021), and Java (Prasetyo et al. 2009; Rudiarto et al. 2020). The degradation forest is evenly found in lowland forest areas, highlands, and as dramatically even in conservation areas. Therefore, conscious efforts are needed to recover the degraded forest condition.

Plant diversity as a community property of the MBNP primary tropical forests is very dynamic and is of great interest. These plants provide food, medicinal, building materials, culture, or other resources for the people, specifically those who live close to Protected Forests. However, during the Anthropocene epoch, there were threats and disturbance resulting in several protected forest areas In Indonesia, especially in a biocultural landscape where human activities and interactions are frequent. The plant diversity of the primary tropical forest at MBNP areas has decreased. Recover degraded forest through reforestation is urgent need by tree planting.

Tropical reforestation, especially within deforested or degraded protected areas, is critical to prevent global biodiversity loss and climate change. According to (Budiharta et al. 2014), reforestation could potentially mitigate the impacts of deforestation, yet knowledge on how to efficiently allocate funding for tree plantations and reforestation is still in its infancy. One of the approaches that can be made to restore degraded tropical forests is by involving the community as implemented at the rehabilitation zone of Meru Betiri National Park (MBNP). Communities who manage and conserve forests are parties influencing as well as affected by deforestation and forest degradation (Muttaqin et al. 2019; Setiajiati et al. 2019).

To avoid wood harvesting illegally, the planted tree species have to consider any tree species that is categorized as non-wood forest products. The enrichment planting with multi-purpose tree species (MPTS) is carried out in an attempt to accelerate the complexity of forest structure and functioning. The MPTS which are mostly non-timber forest product plants such as medicinal, fruit, or leaf-producing plants are selected to avoid wood harvesting illegally by people. For the protection of the existing tree species growing at the degraded forest to be cut by the people, liana species is planted. Some lianas such as Piperaceae family which are medicinal plants have high economic value in the market and are also endemic to tropical forests. The lianas need tree species to support them to climb, so this climbing mechanism can be used as strategy to cover and save the existing tree species at the degraded area for being cut.

The combination of MPTS and Liana planting for reforestation model is successfully implemented in rehabilitation area of MBNP as sustainable forest management model. It is because the target of increasing plant density and diversity in rehabilitation areas is achieved. The successful implementation of this model can be used as a lesson learned. This lesson learned is useful as a basis for designing future scenarios from past and current experiences gained from implementing various forest rehabilitation projects. It can be of use to the wider conservation community, but for this to happen, lessons need to be systematically collected and disseminated in a coherent manner to the broader conservation and development communities and, importantly, to policy makers. To prevent forest damage caused by community activities around the area, it is necessary to restore forest areas by involving the community. Therefore, this study has been subjected to review reforestation projects and their success in the park, and to analyze lessons learned from liaforestry-sustainable forest management model using multi-purpose plants for reforestation and climate change mitigation.

29.2 Methodology

The pilot case representative is demonstration plots located at the rehabilitation area of Meru Betiri National Park (MBNP). Literature reviews are used to get information related to plant diversity and community in the MBNP area. Plot samplings are used to collect data of plant composition, dominance, and diversity in the two-reference primary tropical ecosystems and demonstration plots of rehabilitation areas (Donglo, Curah Malang, and Pletes Sites) in Wonoasri Resort, MBNP (Fig. 29.1).

Bird observation using point count (Bibby et al. 2000) with a 40×40 m^2 scale in the rehabilitation area. These observations were made with a duration of 30 min on every point count (PC). Observations include the bird's species, number of individuals, activities, and the position of the strata occupied. Observation of the morphology and behavior of birds using binoculars (Nikon Aculon 10×50), camera (Canon EOS 60D DSLR), telephoto lens (Tamron 75–300 mm), and field stationery. Furthermore, the bird identification process uses the reference of (MacKinnon et al. 2010).

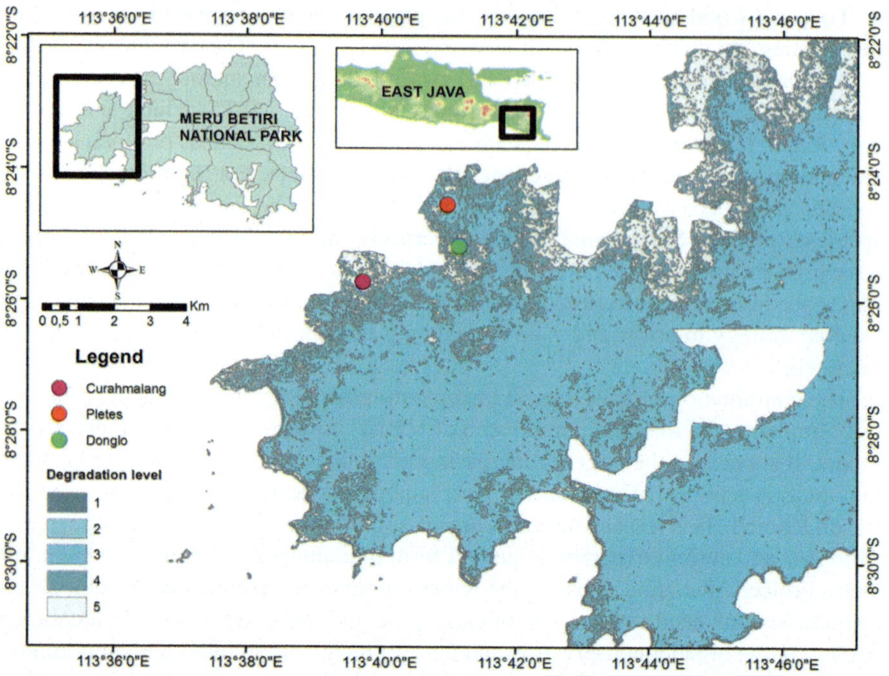

Fig. 29.1 Area of demonstration plot in Wonoasri Resort at MBNP (Source: Google earth 2022)

Quantitative data are analyzed using ecological approaches, while qualitative data are analyzed using comprehensive descriptive analysis. Index diversity of Shannon Wiener is used for calculating Plant diversity (Magurran 1988). Furthermore, the existence of specific birds is used as species indicators to determine the change of forest diversity at rehabilitation areas. Important value (IV) is used to analyze the dominance and codominant species of plant communities (Barbour et al. 1987). Construction of forest vertical strata occupied by birds using spatially explicit individual-based forest simulator (SexI-FS) software version 2.1.0. As for the process of illustrating the distribution of birds based on the vegetation strata, using a graphic design application.

29.3 Liaforestry-Sustainable Forest Management Model-Lesson Learned from Meru Betiri National Park (MBNP)

29.3.1 Plant Diversity in Tropical Forest of MBNP

Tropical forests are the most diverse ecosystems on land, because they hold the vast majority of the world's terrestrial species (Hill and Hill 2001; Turner 2001; Adler 2013; Pillay et al. 2022). Biodiversity would be the only hope to help people to improve the carrying capacity of the earth planet to support life systems and ensure sustainable development. Among these, the tropical forest of Meru Betiri National Park (MBNP) has natural resource potential. The plant diversity of MBNP provides functioning ecosystems that supply oxygen, clean air and water, pollination of plants, food and protection for fauna, and many ecosystem services.

Plant diversity of ecosystem references in the primary tropical forest of MBNP is presented in Table 29.1 and Fig. 29.2. A total of 37 tree species are found at the first reference ecosystem (primary tropical forest at Donglo Site) with index Shannon–Wiener (H′) of 3.106. On the other hand, 28 tree species are found at the second reference ecosystem (primary tropical forest at Curah Malang Site) with index

Shannon–Wiener (H′) of 2.856. Based on these results, without any intervention of people, plant diversity of the forest ecosystem is high ($H' > 2.5$). Plant diversity has a close relationship with the composition, structure, function, and stability of plant communities. Thus, high plant diversity represents that most of the species composition is dominant indicating a stable ecosystem. According to (Allan et al. 2011) that more diverse communities have been shown to have higher and more temporally stable ecosystem functioning than less diverse ones, suggesting that they should also have a consistently higher level of functioning over time.

Furthermore (Wang et al. 2017), said that plant diversity has important effects on community stability. The diversity of communities can promote ecosystem stability and increase nutrient supply and productivity (Chaturvedi et al. 2021). Forests are dynamic living systems and they are continually changing through time controlled by environmental factors, such as soil type, water regimes, vegetation history, climate, and invasive species.

Nine tree species including Artocarpus elasticus Reinw. ex Bl., *Ficus benjamina* L., *Harrisonia perforata* (Blanco) Merr, *Lagerstroemia speciosa* (L) Pers, *Protium javanicum* Burm. F., *Pterospermum acerifolium* Linn. Wild., *Pterospermum javanicum* Jungh, *Schoutenia ovata* Korth., and *Streblus spinosus* (Blume) Corner are found in both representative sites of primary tropical forest. While in Curah Malang site, the tropical forest is dominated by *Pterospermum acerifolium* Linn. Wild.–*Grewia acuminata* Juss (IV = 34.124 and 29.373), the tropical forest in Donglo Site (Fig. 29.3), is dominated by *Diospyros campanulata* Bakh–*Tetrameles nudiflora* R.Br (IV = 29.629 and 28.924). All the tree species found in these sites are typical plant species composition of the tropical forest of MBNP. These species are also found among 57 tree species in Tropical Evergreen Aglaia–Streblus Forest of MBNP (Sulistiyowati and Buot Jr 2020).

Plant diversity also greatly affects the survival of animals, one of which is the bird community. The birds are very dependent on the vertical strata of the forest (stratification). Stratification provides a protection space or cover type which is a limiting factor (welfare factor) for birds. Several studies have shown that birds will use nagging branches from trees in the forest for nesting, perching, and resting while stalking prey (Partasasmita et al. 2021). Therefore, degradation of the structure and composition of vegetation will have an effect on distribution and birds (Thongsoulin et al. 2019).

Table 29.1 Plant diversity of MBNP at Curah Malang and Donglo Sites as reference ecosystem of primary tropical forest (no human intervention)

	Curah Malang site		Donglo site	
No.	Scientific name	H'	Scientific name	H′
1	*Aglaea argentea*	0.073	*Aleurites moluccana* L. Willd	0.104
2	*Alstonia scholaris*	0.015	*Artocarpus elasticus* Reinw. ex Bl.	0.036
3	*Artocarpus elasticus*	0.015	*Barringtonia racemosa* L. Spreng	0.058
4	*Bauhinia purpurea*	0.073	*Callicarpa candicans* var. candicans	0.008
5	*Bombax ceiba* L.	0.015	*Calophyllum inophyllum* L.	0.062
6	*Caesalpnia sappan*	0.027	*Cannarium hirsutum* Wild	0.130
7	*Cassia siamea* Lmk.	0.047	*Diospyros cauliflora* De Wild	0.248
8	*Diospyros campanulata*	0.251	*Dracontomelon dao* (Blanco) Merr Rolfe	0.081
9	*Ficus benjamina* L	0.065	*Dysoxylum amoroides*	0.031
10	*Ficus glomerata*	0.065	*Ficus benjamina* L.	0.107
11	*Ficus hispida*	0.126	*Ficus septica* Burm.f	0.101
12	*Flacourtia rukam*	0.027	*Ficus variegata* Blume	0.036
13	*Garuga floribunda*	0.015	*Grewia acuminata* Juss	0.186
14	*Harrisonia perforata* (Blanco) Merr	0.065	*Harrisonia perforata* (Blanco) Merr	0.186
15	*Klenhovia hospital*	0.042	*Lagerstroemia speciosa* (L) Pers	0.101
16	*Lagerstroemia speciosa* Pers	0.027	*Leea indica* (Burm.f.) Merr	0.070
17	*Melia azedarach*	0.065	*Mallotus barbatus* (Wal) Muell	0.098
18	*Mischocarpus sundaicus*	0.027	*Mangifera indica* L.	0.233
19	*Muntingia calabura*	0.101	*Pangium edule* Reinw	0.031
20	*Persea grastissima*	0.108	*Protium javanicum* Burm. F.	0.021
21	*Pipturus incanus*	0.027	*Pterospermum acerifolium* Linn. Wild.	0.256
22	*Protium javanicum* Burm. F.	0.080	*Pterospermum javanicum* Jungh	0.122
23	*Pterospermum acerifolium* Linn. Wild.	0.203	*Schoutenia ovata* Korth.	0.036
24	*Pterospermum javanicum* Jungh	0.038	*Sterculia foetida* L.	0.058
25	*Schizostachyum silicatum*	0.218	*Streblus asper* Lour.	0.088
26	*Schleichera oleosa*	0.114	*Streblus spinosus* (Blume) corner	0.062
27	*Schoutenia ovata* Korth.	0.120	*Streblus taxoides* (Roth) Kurz	0.273
28	*Spathodea campanulata*	0.047	*Syzygium aqueum* (Burm. f.)	0.026
29	*Streblus spinosus* (Blume) corner	0.108		
30	*Sterculia campanulata*	0.126		
31	*Streblus asper*	0.218		
32	*Syzygium densiflora*	0.120		
33	*Tectona grandis*	0.177		

(continued)

Table 29.1 (continued)

	Curah Malang site		Donglo site	
No.	Scientific name	H'	Scientific name	H′
34	*Terminalia bellirica* Roxb.	0.015		
35	*Tetrameles nudiflora*	0.027		
36	*Toona sureni* Merr.	0.056		
37	*Vitex heterophylla*	0.173		
		3.106		2.85125

High species composition and plant diversity of the reference ecosystem at both sites are potential value for other animals, specifically insects, primate (such as *Trachypithecus auratus, Macaca fascicularis*), deers, Bos javanicus, birds, or others. The tree species provide a variety of food resources, such as seeds, fruits, leaves, and nectars. According to (Nevo et al. 2019), the plant species with fleshy fruits offer animals rewards, such as sugar, protein, and fat, to feed on their fruits and disperse their seeds. The animals can capture visual and olfactory signals indicating the presence and ripeness of the food resources of tree species.

The forest canopy creates microclimates and also buffers the effects of precipitation that facilitates the growth of epiphytes. Both structure and function of canopy trees and epiphytes control variation in forest microclimates (Nakamura et al. 2017). The tree composition and diversity also provide a complexity of layering canopies that facilitate for mating, nesting, resting, and socialization of other animals. Insects including bees, ants, wasps, and butterflies get benefits of goods from the tree species but also support as pollinators. The primates move through the canopy foliage, using their long, prehensile tails to free their hands to pick fruit, leaves, and other edibles. Some birds are hanging by one foot as it uses its other to hold a piece of fruit that it was gnawing, and others are calling of their flocks hopping around in the branches.

In this case, birds have an important role in utilizing vegetation as a source of food. Simberloff and Dayan (2021) stated that birds can be grouped based on their behavior and type of food, such as granivores, insectivores, etc. In the Donglo-Pletes rehabilitation forest ecosystem, birds with these five categories were found, i.e., Insectivora (40.63%), Granivore (21.88%), Carnivora (18.75%), Frugivora (15.63%), and Nectovora (3.13 %), respectively (Fig. 29.4). Insectivorous birds in this area have the highest number of species. This shows that the abundance and diversity of insects, especially tree fauna, are higher. Likewise, the presence of granivore, frugivore, and nectivorous birds in this area indicates that the diversity of flowering plants as a source of bird food is available. Furthermore, the presence of carnivores and raptors also indicates that there is the presence of prey, such as snakes, lizards, and other small vertebrates.

The existence of the five categories above are indicators for the complex ecosystem of primary tropical forest (reference ecosystems). The availability of food sources and the opportunity to build nests in the strata are the main factors. This can also be an answer in looking at the ecosystem recovery process that is focused on this area, so that in terms of bioindicators in forest monitoring, the presence of birds that occupy each vertical strata can be used.

Fig. 29.2 Ecosystem references of Curah Malang site (above) and Donglo Site (below) in the primary tropical forest of MBNP

29.3.2 Human Activity Impact on Forest Degradation

Forest ecosystems such as tropical rain forests have an important role as a carbon pool (carbon pool) for a certain period of time and hold CO_2 emissions into the atmosphere (Wood et al. 2012; Kothandaraman et al. 2020). Deforestation that occurs in various world ecosystems that have an impact on increasing global

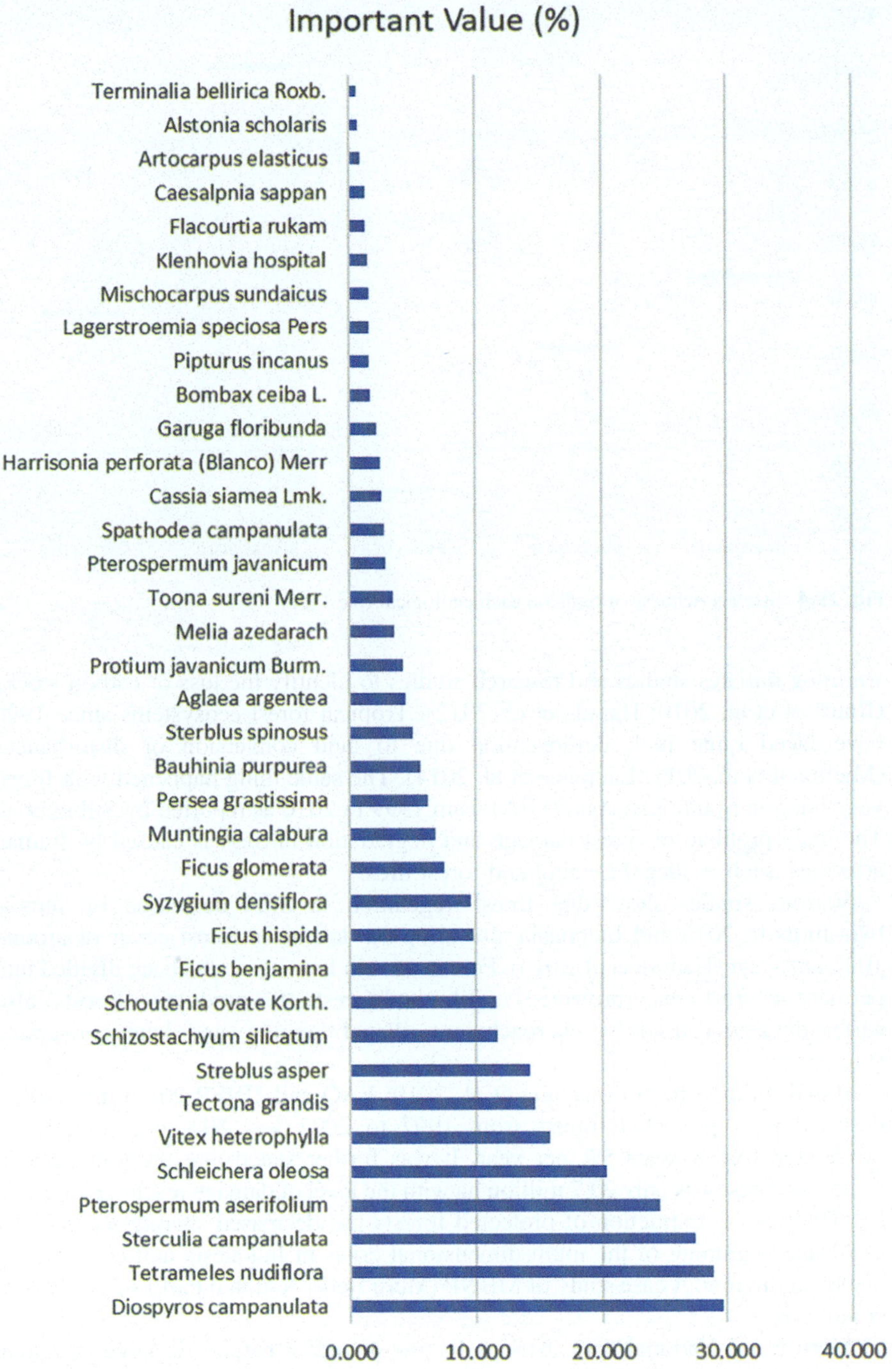

Fig. 29.3 Important value (IV) of *Diospyros campanulata* Bakh–*Tetrameles nudiflora* R.Br tropical forest as reference ecosystem in Donglo site

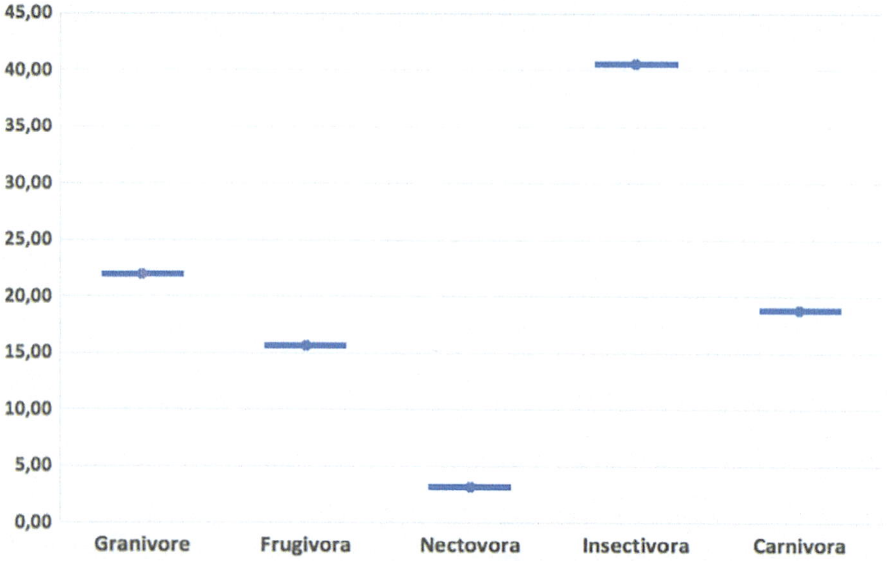

Fig. 29.4 Species richness of birds on each ecological role

warming initiates studies and research studies to identify the loss of carbon stocks (Butchart et al. 2010; Harada et al. 2012). Tropical forest ecosystems since 1990 have faced quite high deforestation due to land conversion or disturbances (Miettinen et al. 2011; Laurance et al. 2014). The same thing happened with forest vegetation in South East Asia (SEA) from 1999 to 2010 as reported by Stibig et al. The main problem of forest damage and degradation in SEA is caused by human activities, such as illegal logging and forest fires.

Various studies show that forest vegetation in SEA decreased by almost two-thirds in 2010 and Indonesia also shows a decline in forest cover of around 1021 km^2/year (Hansen et al. 2013). Forest areas in Indonesia which are divided into permanent forest areas (protected) and limited forests (for production forests) also suffer damage caused by encroachment, illegal logging, and land conversion (Fig. 29.5).

Based on data from (Tacconia et al. 2019; FAO and UNEP 2020), the rate of destruction of protected forests from 1997 to 2002 was 10% per year, while production forests were 5% per year. It was further stated that the total area of protected forest was only 20.7 million ha with the level of damage reaching 4 million ha. The rate of destruction of protected forests has decreased sharply since 1997/ 1998, the beginning of the multi-dimensional crisis in Indonesia and continues to decline until now. A case study of MBNP (Meru Betiri National Park) shows similar conditions.

Meru Betiri National Park, besides having abundant natural resource potential, also holds the potential for latent conflict in a small part of the ten buffer villages. Problems, conflicts, and potential conflicts with the community are like "coals in the

Fig. 29.5 Forest degradation at Meru Betiri National Park

husk" which at any time will smolder, dim, and die, and it is just a matter of how to deal with solving these problems. From the incident, conflicts are always triggered by the arrest of people who commit violations in the area in the form of illegal logging, hunting, and other violations. This condition shows that there is no evaluation or study of the motive or mode of the violation. There were ten big violations happening from 1997 to 2020, such as looting ex-production forest areas, burning and vandalism of resort offices, Police officer hostage, destruction of the generator house, Solarcell and speed boat theft, warehouse destruction, damage to patrol motorbikes in the area, and destruction of office, Rajegwesi Resort Post, Facilities Resort Sukamade, and officer abuse.

Most of the incidents were triggered by the settlement of illegal logging cases, as an illustration in the last 10 years from 2011 to 2020 illegal logging cases reached 66% then hunting by 14% (Surahman 2021). There are three groups of people who commit illegal logging violations in the TN area. Meru Betiri, namely, (a) paymaster; (b) labor; (c) own/partial needs; and (d) power/person (backing), with motives, such as encouragement of family economic needs, greed, self-enrichment, and extravagance/consumerism. Meanwhile, the mode used is in the form of: (a) destruction of state assets (office buildings, and vehicles); (b) mass mobilization/demo; and (c) threatening/holding officer's hostage. As an illustration that the paymaster or the dealer is often referred to as the outsider, the person who finances the operations in the field including buying chainsaws and buying stolen goods from the workers and ensuring that the actions of the perpetrators in the field are safe from legal entanglement (involvement of unscrupulous persons); from the data collection, there are as many as 5 paymaster each has 7–8 workers.

Meanwhile, laborers or members of the community are groups of people who carry out direct looting in the forest and the proceeds are sold to paymaster; usually, this group of workers consists of 5–15 people with the division of tasks, among others, as loggers/cutters, transporters, and guards. Most of the labor groups are poor people, because there are no other alternative businesses to meet the needs of their family life. In carrying out the action, there is usually a person on duty to monitor the movement of officers, and if caught, they are prepared for mass mobilization actions to terror and asset destruction.

If the handling of illegal logging cases is resolved through legal channels, it will be difficult to resolve, this in addition to requiring high costs, time and energy, social relations with the community and terror against officers will continue to occur, so it will reduce feel safe and comfortable at work. There needs to be a new paradigm shift in work, especially in efforts to solve the problem of community pressure on the area. In the legislation, it is very clear that all community problems in the area can be resolved through conservation partnerships.

In addition, alternative solutions for the community, especially illegal logging actors through increasing productive economic efforts, need to be carried out and for paymaster/investors of course resolved through legal channels (Surahman 2021). Thus, the problem of illegal logging will be resolved without causing any impact. Conservation partnerships create togetherness and a sense of kinship; gain communities as partners in conservation efforts; reduce destructive community pressure on the area; and improve prosperous society, so the forest will sustain. Furthermore, increased knowledge and competence of human resources are through community empowerment in developing knowledge and potential-based creative economy clusters so as to reduce the level of dependence on the TNMB rehabilitation area.

There is no single solution in solving every problem. Social, economic, and cultural conditions of the community are important factors to be considered in formulating patterns or concepts of problem solving. Likewise, how to build communication and determine key factors is an integral part of problem-solving efforts.

29.3.3 Liaforestry Model

29.3.3.1 Destruction Identification

Protected forest areas, especially those experiencing deforestation, are widely used by the community as agricultural land which has an impact not only on decreasing biodiversity and reducing carbon stocks but also initiating the distribution and colonization of exotic species (Fig. 29.6). More human presence in the landscape increases the risk of exotic invasive species spread such as *Chromolaena odorata*, *Lantana camara*, *Synedrella nodiflora*, and *Sida acuta* (Setyawati 2013; Sulistiyowati et al. 2021; Liana et al. 2021; Rizkiah et al. 2021) become invasive and a barrier to the regrowth of native plants in the area (Fig. 29.5).

The desire to rehabilitate or restore degraded areas to forest again has been carried out for years since 1997, resulting in the emergence of rehabilitation zones in several protected areas. Various efforts to restore fertility and ecological functions of the area have been carried out by carrying out various programs that are still in favor of the community, such as the multi-purpose tree species (MPTS) plant seed assistance program, land management training, or forming community conservation institutions. However, this policy still faces many obstacles, because there are no significant changes and because community-based secondary forests have not been

Fig. 29.6 Some exotic species found at the rehabilitation areas of MBNP

realized. This is possible due to the lack of supervision from managers and intense coordination regarding program development.

The need for reforestation of complex forest structures in rehabilitated or degraded areas requires a high commitment from both parties. According to (Wang et al. 2020), proper management of activities will be able to restore biodiversity and ecological functions of the area. Tree planting for reforestation is commonly done to regain ecosystem integrity and enhance human well-being in deforested and degraded areas. For this reason, a recovery model for degraded areas is needed by maximizing the growth of endemic plant diversity, and maximizing will be able to increase productivity capacity, ecological function, and biodiversity value. Therefore, various methods of reforestation through planting seeds of endemic plants have been developed in degraded areas (Lamb et al. 2006; Toledo et al. 2018). The agro-ecosystem or agroforestry approach is the most appropriate method to meet the needs of the community around the area and to suppress the growth of exotic plants in degraded forest ecosystems. The combination of planting agricultural crops under MPTS endemic tree stands can provide both economic and ecological benefits.

Case study of the TNMB rehabilitation area in Wonoasri village covers an area of 290 ha, but the rehabilitation program (as shown in Fig. 29.7) that has been implemented has not succeeded in returning it to secondary forest, because the density of tree stands is $<16\%$ or 234 stems/ha and the species diversity is very low ($H < 1$). In addition, the stratification of the area for the last 15 years is homogeneous, because it only consists of two strata, namely, tree stands and agricultural or horticultural crops. The similarity of plant species planted is the main cause of the uniformity of the plant strata in question. In addition, the low diversity of plants also has an impact on the low diversity of insect or birds' communities as bioindicators and soil fertility in the Wonoasri rehabilitation area, the results of the study show that the average H of insects is 0.33, while the soil organic matter content is very low, which varies between 0.796 and 0.964% or

Fig. 29.7 Forest degradation of Wonoasri areas in MBNP (representative of rehabilitation area)

<<1%. Based on these characteristics, the rehabilitation area was categorized into severe damage.

Compared to the reference ecosystems, the forest density is 1443 stems/ha and the plant diversity index was high ($H > 2.5$). Furthermore, the stand stratification is more complex, because it consists of various strata, namely, herbs/seedlings, shrubs, and trees from various types of plants. Therefore, reforestation has been done to increase both plant density and diversity in the rehabilitation areas of Wonoasri Resort-MBNP. Increasing plant density has an important role as carbon sinks and stores and prevents carbon dioxide emissions into the atmosphere (Houghton 2007). On the other hand, the decrease in vegetation cover contributes to the increase in global warming (Miettinen et al. 2011; Laurance et al. 2014; Hansen et al. 2013), because the sequestration rate of carbon is low. Therefore, new community-based management has been done to solve the problems without problem by implementation of sustainable **liaforestry** model management.

29.3.3.2 Community-Based Management

Community-based management of liaforestry model was introduced using simple language that can be understood for the people living around the MBNP area. The stages of the program need are clearly stated to increase confidence in the community that the implemented reforestation model is not just a one-time project. The reforestation model implemented is a sustainable program that is continuously monitored to achieve targets over a period of 5–10 years. Targets of community-based management are local or district governments, village governments, community leaders, leaders of farmer groups, conservation cadres, and farmers who manage rehabilitation areas in Resort Wonoasri of MBNP (Fig. 29.8).

The program objectives are, namely, to increase the density and diversity of endemic plants, to decrease carbon emission through land-based climate change mitigation, and to increase income of the farmer. Specific program targets were informed, such as: (1) increasing the productivity of the rehabilitation area on a 255 ha rehabilitation area and developing demonstration plots as a concrete example of ecosystem restoration in the rehabilitation area, (2) implementation of the memorandum of understanding: the rehabilitation area agreement which is a cooperation agreement between MBNP and the rehabilitation land farming community, (3) increasing soil buffering against damage and soil function in the TNMB rehabilitation area, (4) increasing knowledge and competence of human resources through community empowerment in developing knowledge and potential-based creative economy clusters so as to reduce the level of dependence on the TNMB rehabilitation area, and (5) establishment of forest alleys and yards in an effort to increase carbon sequestration and community welfare.

Individual and community approaches were used to gain trust and commitment of farmers who manage the rehabilitation areas of MBNP. Individual approaches from community leaders, farmer group leaders, and local government heads were taken to obtain suggestions for the success of the program. In groups, focus group discussions can be conducted to explain the details of the program and to identify the weaknesses of the programs that have been carried out, so that the same mistakes will not be repeated. Concept of "planting, picking, processing and selling" through the liaforestry model is also introduced to the community to gain people's trust in program implementation.

29.3.3.3 Demonstration Plot Development

Ecosystem restoration can be briefly defined as "returning the original forest ecosystem that existed before deforestation occurred". However, it should be understood

Fig. 29.8 Introducing the liaforestry model through focus group discussion

that ecosystem restoration cannot restore all types of flora and fauna that once lived in the original forest before deforestation in one stage. The main objective of forest restoration is to restore the structure and function of the original ecosystem, by planting key tree species that play an important role in the ecology of natural forests. The success of restoration activities can be measured by the return of the terraced canopy structure, the increase in the number of returning species (especially rare species or key species), improvement of soil conditions, and the recovery of populations of certain types of flora or fauna (Briones 2014; Paul et al. 2010).

The difficulty in forest restoration the tree species planted by the farmers who utilized the rehabilitation area. They commonly only select and plant tree species they like. Therefore, reforestation of fast-growing tree species is implemented as a model to increase the density and diversity of plant species in the rehabilitation area. In addition (Liu et al. 2021), said that reforestation can recover soil water content and increase freshwater supply to human society. The implementation of this ecosystem recovery model is meaningless without the involvement of managers, communities, and stakeholders.

A comprehensive recovery model is needed by involving all parties with the types of plants that benefit all parties. The types of plants above and below the canopy that are productive are absolutely necessary to provide economic benefits for the cultivators. Selection of plant species is important, because community retention requires intense communication with the community about the potential of the plant in question. Changing the community's perspective on the use of protected area land requires an intensive approach, individually and in groups, and takes a long time to adapt.

So far, the reforestation model using the Agroforestry technique is one model that is considered successful, because it combines forest plants with agriculture that is able to prevent overgrazing, bush fires, logging, or harvesting firewood as well as to prevent erosion and flooding. The type of plants to be planted must be endemic to preserve the original germplasm. On this occasion, the ecosystem recovery model introduced is the DEMONSTRATION PLOT LIAFORESTRY, which provides examples of good practice in managing degraded forest ecosystems using **LIa**na plants that are endemic to areas of high economic value as plants under the canopy of MPTS or multi-purpose tree species (FORESTRY) that can also be planted with intercropping plants such as stink lily (*Amorphophallus muelleri*), root crops or the Zingiberaceae group which are endemic to the region. Both plants provided by ICCTF (International Climate Change Trust Fund)–BAPPENAS (National Planning Institution) were distributed to 525 rehabilitation farmers to be planted at the rehabilitation areas of 255 ha in 2017.

A survey to the locations of degraded forest ecosystems as demonstration plots is required to observe field conditions. Information on field conditions is absolutely necessary to determine which areas need to be treated specifically as demonstration plots. Field conditions that need to be observed are the type and condition of the existing canopy stands and the types of agricultural crops planted, the density of the stands, the maintenance of the land, and the commitment of the farmers. The total number of MPTS and liana seedlings are determined by the density of tree stands of the managed rehabilitation area per farmer to achieve density standard (500 stems/ha).

The condition of the canopy that is still maintained is very important as a host that shelters the liana plants that climb/wrap around the stems of these plants. A fairly high density of stand plants is needed as a host for endemic lianas to be planted, the more stands the more lianas that can be climbed/wrapped on the stems (maximum two lianas per plant stand). Farmer commitment is absolutely necessary to ensure plant growth. Based on these three things, it is possible to select which land can be used as demonstration plots with special treatment, so that it can become good practice for smallholders in the vicinity.

The types of plants offered for planting are endemic MPTS stand types (target of 500 stem/ha), so in addition to restoring ecosystem diversity and ecological functions, it is also to continue to provide economic value for smallholders from the fruit or seeds that it produces. Examples of MPTS that can be offered include *Artocarpus* sp (breadfruit, jackfruit, and breadnut), *Mangifera indica* (mango), *Manilkara* sp (sapodilla), *Durio zibethinus* (durian), *Garcinia mangostana* (mangosteen), *Nephelium lappaceum* (rambutan), *Anacardium occidentale* (cashew), *Persea americana* (avocado), *Psidium guajava* (guava), *Salacca edulis* (salak); shade-tolerant species such as *Areca catechu* (betel palm), *Annona muricata* (soursop), *Cinnamomum* sp (cinnamon), *Lansium domesticum* (langsep), *Myristyca fragrans* (nutmeg), *Arenga saccharifera* (Aren) or other species endemic to the area.

The types of lianas that can be planted are climbing or twining types that have high economic value, such as javanease pepper, pepper, or other types of lianas (Piperaceae) that are endemic to the area (Fig. 29.9). The advantages of planting this high economic value liana plant are that it is easy to maintain, the selling price of the fruit/seed is expensive, and its propagation is easy, because it is only done vegetatively. In addition, these species are not favored by monkeys, or other wild herbivores because of their bioactive content, so that production yields are not reduced at harvest.

Organic cultivators are expected to use ZPT (Plant Growing Substances), flower growing substances (ZPB), biological pesticides, and organic fertilizers, so that maximum growth is free of chemicals, so that they have high economic value. The onion skin, banana stem, coconut coir (coco fiber), jackfruit or mango leaves are biological resources used for ZPT and ZPB. Recently, the farmer used variety of biological resources, such as egg, Mojo fruit (*Aegle marmelos*), Bintaro fruit (*Cerbera manghas*), and others. This is because the average consumer wants organic plants that are free of chemical pesticides. This model of cultivation also reduces pollution to the ecosystem.

29.3.3.4 Postimplementation of Liaforestry Model

Plant Diversity as Indicator Species for Secondary Forest Succession

Using functional groups of vegetation cover to assess the recovery of ecosystem functions in regenerating tropical forests of rehabilitation areas is an approach to identify the success of the program. The vegetation cover used is categorized into herbs, shrubs, and tree species.

Fig. 29.9 Planting the liana (Javanease pepper) and MPTS used for liaforestry model

The plant diversity of herbs, shrubs, and tree species increases from <<1 to >1.5 in average of rehabilitation areas after the implementation of liaforestry model. Based on the data sampling, the index diversity of Shannon Wiener of the herb community is 2.412, shrub community is 1.6, and tree community is 2.5. The results indicate that some species dominates the area, specifically for herb and tree communities. However, the most tree species composition of the rehabilitation area (Fig. 29.8) is dominated by the MPTS compared to that of primary tropical forest.

The 36 species composition of the herb community is dominated by *Pueraria javanica–Synedrella nodiflora* (Important Value/IV is 34.592% and 25.714%). On the other hand, 13 species of shrub community are dominated by *Chromolaena odorata* L.–*Mimosa pudica* L. (IV is 87.933% and 38.354%). Meanwhile, 23 species composition of the tree community is dominated by *Mangifera indica* L.–*Parkia roxburghii* Merr. (INP 89.028% and 57.022%). Successional age was the key factor in controlling species richness and diversity, while local environmental properties emerged as the main factors shaping community composition. Increasing the plant diversity after the implemented model is an indicator of the success of the reforestation (Fig. 29.10) as stated by (Hilmers et al. 2018) that the number of species of producer and consumer groups generally followed a U-shaped pattern in response to forest succession.

Fig. 29.10 Increasing plant diversity post-implementation of liaforestry model at Pletes site (representative of rehabilitation areas)

Increased density and diversity of plant species will directly have an impact on increasing soil fertility, arthropod and other fauna diversity as well as increasing forest capacity as carbon pools or stored carbon reserves while reducing CO_2 gas emissions. The tree density of the 255 ha Wonoasri rehabilitation area in its existing condition can absorb 3204 tCO_2 of carbon and produce 2318 tO_2.

However, after the implementation of the liaforestry model, this rehabilitation area management activity is able to increase plant density by ±82,745 stems/ha on an area of 255 ha and ±2000 stem of Javanese pepper and pepper plants in a demonstration plot area of 10 ha. It gives an impact on an increase in carbon uptake to 4596 tCO_2 and production of 3325 tO_2. Increased oxygen production by 28% from 679 tO_2/day to 870.26 tO_2/day Javanese pepper contributed 3.27 tC and 11.99 tCO_2. If natural succession succeeds in triggering the density of tree stands by 80–100%, then carbon uptake and oxygen production will also increase in the next >5 years.

Based on the increase in the level of carbon sequestration and oxygen production mentioned above, the success of the program has an impact on reducing 1403 tCO_2. Greenhouse gas (GHG) emission mitigation will continue to increase when the density and diversity of plants increase in line with the natural succession of the rehabilitation ecosystem. This is also a form of support for the East Java GHG Regional Action Plan, especially the forest rehabilitation program in protected forest areas with a target for emission reductions in the East Java Province in the Forestry sector in 2013–2020 of 6,221,572 tCO_2eq based on East Java Governor Policy No 67 in 2012.

In addition to the ecological functions obtained by implementing the liaforestry ecosystem recovery model, namely, the reduction of carbon dioxide emissions, and the contribution of oxygen of the rehabilitation land ecosystem area. Another benefit obtained from this program is the additional economic value from planting Javanese chili or pepper which can be gained after the age of the plant is 1 year to 1 and a half years.

Fig. 29.11 Recently condition of degraded forest at Donglo Site as representative of rehabilitation areas

At the age of 1 year, Java chili plants produce fruit as much as 3 ounces (wet weight)/stem at a price of 10,000 IDR (benchmark minimum price). Therefore, if they have ten stems, on average they will get an additional Rp. 100,000/month. When the Javanese pepper is 3 years, there will be an average production increase of 2 kg wet weight/stem/month at a price of 120,000 IDR, so that with ten stems of Javanese pepper, it will get a minimum production of 20 kg wet weight with a result of Rp. 600,000 per month. In 2022, as shown in Fig. 29.11, the community harvests 5–7 kg/week with the price of 65,000 IDR/kg dry weight or 1,690,000 IDR/month on average from Javanese pepper production as additional income. Other economic benefits from the plant diversity at the rehabilitation areas are fruit production from mango, jackfruits, breadfruits, or other MPTS planted by the farmer. All these incomes achieve the minimum income standard of Jember County which is about 2,355,662.91 IDR/month.

Bird Community as Indicator Species for Secondary Forest Succession

The forest succession after liaforestry model implementation at the rehabilitation area gives impact on the diversity of birds (Siddig et al. 2016; Rolo et al. 2017). The factors responsible for shaping the bird composition are intrinsic to the habitat (e.g., provide certain resources to birds or have a particular structural or floristic composition in itself). As stated by (Ramírez-Soto et al. 2018), each disturbance type has bird

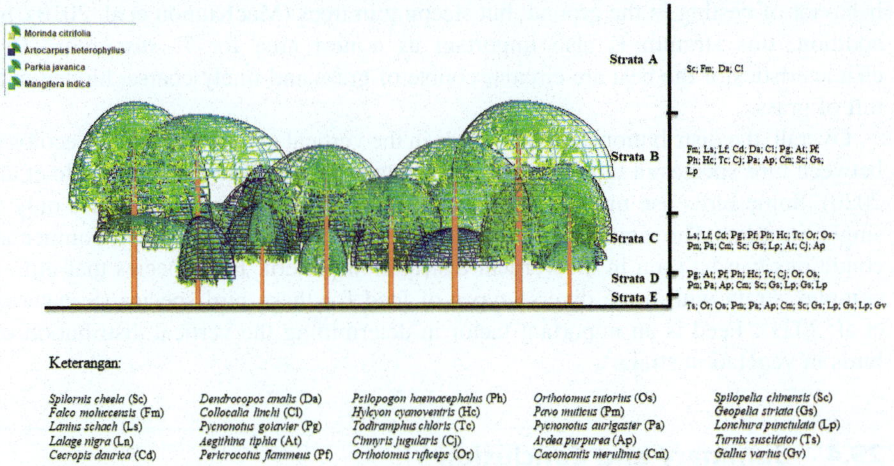

Fig. 29.12 Distribution of birds on vertical stratification

species composition that reflects its ecological characteristics, such as availability of food resources and certain microhabitat features.

The composition of birds that occupy the rehabilitation area (Pletes and Donglo Sites as sampling demonstration plots) after the implementation model is 25 species from 19 families. Bird groups occupy five strata (A–E) (Fig. 29.12). Strata A is occupied by raptors, swifts, and swallows. The upper canopy is used by raptors to rest or monitor prey, while areas that tend to be open are used by raptors as a foraging area. This is also stated by (Thirgood et al. 2003) that raptors have a foraging habit in open areas. Strata A characteristics in this area only reach 16–17 m, so it has not become a nest of raptor options. Raptors from Accipitridae generally prefer tree heights above 40 m (Lohmus 2006; Misna et al. 2006). On the other hand, the existence of the swift and swallow here is only found flying around with the occasional dip to catch prey in the form of small insects.

Strata B–C is the layer with the highest bird species richness (19 species). Both of these indicate the abundant availability of food for the bird community in this area. Some of them are common species, such as *P. aurigaster*, *P. goiavier*, *G. striata*, *T. chloris*, and *A. tiphia*. The abundance of *Pycnonotus aurigaster* is high (4–5 individuals)/group. This species has an active habit in groups and prefers forest edge habitats or secondary plants (MacKinnon et al. 2010; Ponpituk et al. 2020; Cholifatullah et al. 2020) stated that the Pycnonotidae group utilizes the vertical strata of forest ecosystem vegetation to find food, socialize, mate, build nests, and parental care. In this case, the rehabilitation area is dominated by secondary and fruiting plants, such as *M. citrifolia* and *Ma. indica*, causing *A. aurigaster* species to be found in high abundance.

Strata D–E (shrubs, herbs, and forest floor) are used by several species of birds in their ecological activities. One of them is the family Phasianidae which has a unique behavior of nesting in the ground, but sleeping in trees (MacKinnon et al. 2010). In addition, this stratum is also important as a nest area for *T. suscitator*. The characteristics of the nest are circular course of grass and finely coarse, hidden in a tuft of grass.

Overall, the distribution pattern of birds in the vertical strata related to the ecology between bird species in utilizing different strata to support life (Rajaonarivelo et al. 2020). Some birds use more than one stratum, while other bird species use only a single stratum. The ability of a bird species to survive against environmental conditions can be seen in the vertical distribution pattern. Bird species that utilize complex strata with more diverse types of feed for these bird species (Simamora et al. 2021). Feed is an important factor in determining the vertical distribution of birds in vegetation strata.

29.4 Summary and Conclusion

Based on the lessons learned from liaforestry-sustainable forest management model using multi-purpose plant species (MPPS) for reforestation and climate change mitigation, we conclude that

1. Community-based management for reforestation can be done by intensive communication individually or in group to build their trust and commitment for partnership
2. During the Anthropocene, liaforestry model of sustainable reforestation can:

a. increase plant species composition and diversity from $H < 1$ to $H > 2.5$ that impact on bird diversity
b. giving additional income of 1,690,000 IDR/month on average to the farmer
c. improve complexity structure and function of secondary tropical forest
d. increase 22% carbon stocks that result in sustainable climate change mitigation

Acknowledgements This study was made possible through the financial and logistic support of the Research Group of Natural Resources Ecosystem Conservation (KeRis_Dimas NREC)–Sentra Tropical Natural Resources Conservation (Sentra T-NRC)–CDAST, University of Jember, Meru Betiri National Park, National Planning Institution (BAPPENAS), and International Climate Change Trust Fund (ICCTF). Thank you to the Springer Nature Editorial office for giving the opportunity to share lessons learned from liaforestry-sustainable forest management model using multi-purpose plants for reforestation and climate change mitigation.

References

Adler GH (2013) Rainforest ecosystems, animal diversity. Encyclopedia of biodiversity, 2nd edn. Elsevier, Amsterdam, pp 304–312

Allan E, Weisser W, Weigelt A, Roscher C, Fischer M, Hillebrand H (2011) More diverse plant communities have higher functioning over time due to turnover in complementary dominant species. PNAS 108(41):17034–17039

Barbour MG, Burk JH, Pitts WD (1987) Terrestrial plant ecology. Benjamin/Cummings Publishing Company, San Francisco, p 604

Basyuni M, Sulistiyono N, Wati R, Hayati R (2018) Deforestation trend in North Sumatra over 1990-2015. IOP Conf Ser Earth Environ Sci 122:1–7

Bibby C, Burgess N, Hill D, Mustoe S (2000) Bird census techniques. Academic, London

Briones MJI (2014) Soil fauna and soil functions: a jigsaw puzzle. Front Environ Sci 2:22

Budiharta S, Meijaard E, Erskine PD, Rondinini C, Pacifici M, Wilson KA (2014) Restoring degraded tropical forests for carbon and biodiversity. Environ Res Lett 9:114020

Butchart SBM, Walpole M, Collen B, Van Strien A, Scharlemann JPW, Almond REA, Jonathan EM, Baillie B, Bomhard C, Brown J, Bruno KE, Carpenter GM, Carr J, Chanson AM, Chenery J, Csirke NC, Davidson F, Dentener M, Foster A, Galli JN, Galloway P, Genovesi RD, Gregory M, Hockings V, Kapos J, Lamarque F, Leverington J, Loh MA, Mcgeoch L, Mcrae A, Minasyan MH, Morcillo TEE, Oldfield D, Pauly S, Quader C, Revenga JR, Sauer B, Skolnik D, Spear D, Stanwell-Smith SN, Stuart A, Symes M, Tierney TD, Tyrrell A, Vié J-C, Watson R (2010) Global biodiversity: indicators of recent declines. Science 328:1164–1168

Chaturvedi RK, Anshuman T, Raghubanshi AS, Singh JS (2021) Functional traits indicate a continuum of tree drought strategies across a soil water availability gradient in a tropical dry forest. For Ecol Manag 482:118740

Cholifatullah F, Ramadani A, Nurhasanah AH, Asmara ATW, Setia TM (2020) Utilization of tree canopies by avifauna in two types of habitat in Telaga Warna Nature Tourism Park, Bogor, West Java. IJOBB 4(2):62–70

FAO and UNEP (2020) The state of the world's forests 2020. Forests, biodiversity and people. FAO, Rome. https://doi.org/10.4060/ca8642en

Hansen M. C., D. Thau, P. V. Potapov, S. V. Stehman, C. O. Justice, R. Moore, S. J. Goetz, J. R. G. Townshend, M. Hancher, T. R. Loveland, S. A. Turubanova, A. Kommareddy, A. Tyukavina, A. Egorov, And L. Chini. 2013. High-resolution global maps of 21st-century forest cover change. Science 342: 846-850.

Harada K, Prabowo D, Aliadi A, Ichihara J, Ma H-O (2012) How can social safeguards of REDD+ function effectively conserve forests and improve local livelihoods? A case from Meru Betiri National Park, East Java, Indonesia

Hill J, Hill R (2001) Why are tropical rain forests so species rich? Classifying, reviewing and evaluating theories. Prog Phys Geogr 25(3):326–354

Hilmers T, Friess N, Bässler C, Heurich M, Brandl R, Pretzsch H, Müller RS (2018) Biodiversity along temperate forest succession. J Appl Ecol 55(6):2756–2766

Houghton RA (2007) Balancing the global carbon budget. Annu Rev Earth Planet Sci 35(1): 313–347

Indrawan M, Primack RB, Supriatna J (2007) Biologi konservasi. Yayasan Obor Indonesia, Jakarta

Koalisi Indonesia Memantau (2021) Planned deforestation: forest policy in Papua. Greenpeace, Jakarta

Kothandaraman S, Dar JA, Sundarapandian S, Dayanandan S, Khan ML (2020) Ecosystem-level carbon storage and its links to diversity, structural and environmental drivers in tropical forests of Western Ghats, India. Sci Rep 10:13444

Lamb D, Erskine PD, Parrotta JA (2006) Restoration of degraded tropical forest landscapes. Science 310(5754):1628–1632

Laurance WF, Sayer J, Cassman KG (2014) Agricultural expansion and its impacts on tropical nature. Trends Ecol Evol 29:107–116

Liana Y, Sulistiyowati H, Siddiq AM (2021) Distribution pattern of the exotic plant (Synedrella Nodiflora (L) Gaertn) in the rehabilitation zone, resort of Wonoasri, Meru Betiri National Park. BioLink 8:63–71

Liu T, Jiang K, Tan Z, He Q, Zhang H, Wang C (2021) A method for performing reforestation to effectively recover soil water content in extremely degraded tropical rain forests. Front Ecol Evol 9:643994. https://doi.org/10.3389/fevo.2021.643994

Lohmus A (2006) Nest-tree and nest-stand characteristics of forest-dwelling raptors in east-central Estonia: implications for forest management and conservation. Proc Estonian Acad Sci Biol Ecol 55(1):31–50

MacKinnon J, Phillipps K, Bv B (2010) Seri Panduan Lapangan pengenalan Burung-burung di Sumatera, Jawa, Bali, dan Kalimantan. LIPI-Birdlife International, Bogor

Magurran AE (1988) Ecological diversity and its measurement. Princeton University Press, Princeton

Miettinen J, Shi C, Liew SC (2011) Deforestation rates in insular Southeast Asia between 2000 and 2010. Glob Chang Biol 17:2261–2270

Misna A, Gunawan H, dan Yoza D. (2006) Karakteristik Pohon Bersarang Elang Brontok (Nisaetus cirrhatus) di Siak Riau. J Riau Biol 1(2):187–193

Muttaqin MZ, Alviya I, Lugina M, Hamdani FAU, Indartik. (2019) Developing community-based forest ecosystem service management to reduce emissions from deforestation and forest degradation. Forest Policy Econ 108:101938

Nakamura A, Kitching RL, Cao M, Creedy TJ, Fayle TM, Freiberg M, Hewitt CN, Itioka T, Koh LP, Ma K, Malhi Y, Mitchell A, Novotny V, Ozanne CMP, Song L, Wang H, Ashton LA (2017) Forests and their canopies: achievements and horizons in canopy science. Trends Ecol Evol 32(6):438–451

Nevo O, Razafimandimby D, Valenta K, Jeffrey JAJ, Reisdorff J, Chapman CA, Ganzhorn JU, Ayasse M (2019) Signal and reward in wild fleshy fruits: does fruit scent predict nutrient content? Ecol Evol 9(18):10534–10543

Partasasmita R, Atsaury ZIA, Husodo T (2021) The use of forest canopy by various bird species in tropical forest montana zone, the Nature Reserve of Mount Tilu, West Java, Indonesia. Biodiversitas 18(2):453–457

Paul M, Catterall CP, Pollard PC, Kanowski J (2010) Recovery of soil properties and functions in different rainforest restoration pathways. For Ecol Manag 259(10):2083–2092

Pillay R, Venter M, Aragon-Osejo J, González-del-Pliego P, Hansen AJ, Watson JEM, Venter O (2022) Tropical forests are home to over half of the world's vertebrate species. Front Ecol Environ 20(1):10–15

Ponpituk Y, Siri S, Safoowong M, Suksavate W, Marod D, Duengkae P (2020) Temporal variation in the population of bulbuls (Family Pycnonotidae) in lower montane forest, Northern Thailand. Biodiversitas 21(8):3644–3649

Prasetyo LB, Kartodihardjo H, Adiwibowo S, Okarda B, Setiawan Y (2009) Spatial model approach on deforestation of Java island Indonesia. JIFS 6:37–44

Rajaonarivelo JA, Andrianarimisa A, Raherilalo MJ, Goodman M (2020) Vertical distribution and daily patterns of birds in the dry deciduous forests of central western Madagascar. Trop Zool 33(1):36–52

Ramírez-Soto R, Rodríguez-Mesa R, Villa-Bonilla B, Hernández IS, Inzunza ER (2018) Using birds to assess and track forest restoration. Trop Conserv Sci 11:1–6

Rizkiah DW, Sulistiyowati H, Siddiq AM (2021) Distribution patterns of exotic plant chromolaena odorata, in rehabilitation Zone at Donglo block, resort of Wonoasri, Meru Betiri National Park. JRBA 3(1):1–6

Rolo V, Olivier PI, van Aarde R (2017) Tree and bird functional groups as indicators of recovery of regenerating subtropical coastal dune forests. Restor Ecol 25(5):788–797

Rudiarto I, Rahmawati I, Sejati AW (2020) Land degradation and community resilience in rural mountain area of Java Indonesia. Springer, Cham, pp 449–460

Sari R, Fauziah AIP, Susandarini R, Makmur I (2020) The plant species diversity of Lasitae protected nature forest and nearby area, district of Barru, South Sulawesi. J Trop Biodiv Biotechnol 5(2):89–99

Setiajiati F, Karyaatmadja B, Sutedja I, Kuswondho H, Satria P, Sejati PS, Maharani RS (2019) Lesson learned from social forestry practice in a forest and climate change project in Kalimantan, Indonesia. IOP Conf Ser Earth Environ Sci 363(1):012001

Setyawati T (2013) The list of twenty selected invasive plant species in Indonesia. In: Soesilo HD (ed) Draft report on Presence of IAS in some conservation areas in Indonesia. GEF/UNEP Project: removing barriers to invasive species management in production and protection forest of South East Asia. UNEP, Bogor

Siddig AAH, Ellison AM, Ochs A, Villar-Leemand C, Laub MK (2016) How do ecologists select and use indicator species to monitor ecological change? Ecol Indic 60:223–230

Simamora TI, Purbowo SD, Laumonier Y (2021) Looking for indicator bird species in the context of forest fragmentation and isolation in West Kalimantan, Indonesia. Glob Ecol Conserv 2:1–20

Simberloff D, Dayan T (2021) The guild concept and the structure of ecological communities. Annu Rev Ecol Syst 22:115–143

Sulistiyowati H, Buot IE Jr (2020) Chapter 8 ecological valuation of the structure and dynamics of the forest biomass at the tropical evergreen Aglaia-Streblus Forest of Meru Betiri National Park, Indonesia. In: Buot A, Inocencio E (eds) Methodologies supportive of sustainable development in agriculture and natural resources management: selected cases in southeast Asia. University of the Philippines Los Baños, Laguna

Sulistiyowati H, Rahmawati E, Wimbaningrum R (2021) Spatial distribution patterns of Lantana camara L. population as invasive alien species in Pringtali Savana Bandealit resort Meru Betiri National Park. Jurnal Ilmu Dasar 22(1):2442

Sunderline WD, Resosudarmo IAP (1997) Laju dan Penyebab Deforestasi di Indonesia: Penelaahan Kerancuan dan Penyelesaiannya. Occas Paper Cifor 9(1):1–25

Supriatna J, Shekelle M, Fuad HAH, Winarni WL, Dwiyahreni AA, Farid M, Mariati S, Margules C (2020) Deforestation on the Indonesian island of Sulawesi and the loss of Primate Habitat. Gecco 24:1–14

Surahman M (2021) Menyelesaikan masalah tanpa masalah Mandala dan Jawara alternatif solusi. Balai Taman Nasional Meru Betiri, Jember

Susanto E, Lestari N, Hapsari M, Krisdyatmoko. (2018) Driving factors of deforestation in Indonesia: a case of Central Kalimantan. J Studi Pemerintahan 9(4):511–532

Tacconia L, Rodrigues RJ, Maryudi A (2019) Law enforcement and deforestation: lessons for Indonesia from Brazil. Forest Policy Econ 108:101943

Thirgood SJ, Sm R, Graham IM (2003) What determines the foraging distribution of raptor on heather moorland? Oikos 100(1):15–24

Thongsoulin P, Phanthuamath V, Islam MS, Thongmanivong S, Kang Y (2019) The relationship between birds diversity and habitats in Phou Khao Khouay (PKK) National Park of Lao PDR. Open J For 9(2):143–158

Toledo RM, Santosa R, Verheyen K, Perring MP (2018) Ecological restoration efforts in tropical rural landscapes: challenges and policy implications in a highly degraded region. Land Use Policy 75:486–493

Tsujino R, Yumoto T, Kitamura S, Djamaluddin I, Darnaedi D (2016) History of forest loss and degradation in Indonesia. Land Use Policy 57:335–347

Turner IM (2001) Rainforest ecosystems, plant diversity in encyclopedia of biodiversity, 2nd edn. Elsevier, Amsterdam, pp 313–321

Vadrevu K, Heinimann A, Gutman G, Justice C (2019) Remote sensing of land use/cover changes in South and Southeast Asian Countries. Int J Digital Earth 12(10):1099–1102

Wang H, Lü G, Zhou Y, Cao J (2017) Effects of functional diversity and functional redundancy on the stability of desert plant communities under different water and salt gradients. J Ecol 37(23): 171–180

Wang W, Feng C, Liu F, Li J (2020) Biodiversity conservation in China: a review of recent studies and practices. Environ Sci Ecotechnol 2:100025

Wood TE, Cavaleri MA, Reed SC (2012) Tropical forest carbon balance in a warmer world: a critical review spanning microbial- to ecosystem-scale processes. Biol Rev 87:912–927

Zhao J, Yang Y, Zhao Q, Zhao Z (2017) Effects of ecological restoration projects on changes in land cover: a case study on the Loess Plateau in China. Sci Rep 7:1–12

Conservation of IUCN Threatened Zingiberaceae Species in Tropical Asia: Challenges, Gaps and Opportunities

30

Maria Celeste N. Banaticla-Hilario and Ma. Regina B. Altamirano

Abstract

We gathered information from the IUCN Red List, databases, and literature to review the conservation status, nature of threats, and the challenges, gaps, and opportunities in conserving Zingiberaceae species in tropical Asia that are threatened with extinction. Only 45% of the 1449 species recorded have been assessed, of which 297 species (21% of all evaluated species) are threatened. Southeast Asia is a conservation hotspot in terms of species diversity and endemism, concentration of threatened species, and high deforestation rates. Most of the activities that directly threaten species cause deforestation and forest degradation, with most of the threatened species not being intentionally collected. About 75% of the species are conserved either through in situ or ex situ methods, or both. We identified information gaps, and recommended a pragmatic conservation approach that promotes and enhances the threatened ginger species' use and values, which may improve their chances of surviving the Anthropocene extinction bottleneck.

Keywords

Gingers · Southeast Asia · In situ and ex situ conservation · Protected areas · Anthropocene

M. C. N. Banaticla-Hilario (✉)
Institute of Biological Sciences, College of Arts and Sciences, University of the Philippines Los Baños, Los Baños, Laguna, Philippines
e-mail: mbhilario1@up.edu.ph

M. R. B. Altamirano
Division of Biological Sciences, College of Arts and Sciences, University of the Philippines Visayas, Miagao, Iloilo, Philippines

S. Ramamoorthy et al. (eds.), *Plant Diversity in Biocultural Landscapes*,
https://doi.org/10.1007/978-981-19-8649-9_30

30.1 Introduction

The ginger family or Zingiberaceae is composed of around 1900 species (WCVP 2022) of aromatic, perennial herbs, usually with creeping rhizomes, and with oil cells in their vegetative parts (except in *Costus*) (Lock 1985). There are 50–60 genera spread across the tropics and subtropics, where Southeast Asia is the center of diversity (Mohamad and Kalu 2019; Saensouk et al. 2016; Sakai et al. 2013; Setiawan et al. 2022). Throughout the world, various species of gingers are used locally and commercially as medicines, foods, spices, food additives, ornamental plants, fragrances, cosmetics or in folk rituals and religious ceremonies (Ibrahim et al. 2007; Larsen et al. 1999; Yeats 2013). "Ginger" refers to *Zingiber officinale* Roscc., which has been cultivated in China and India since ancient times, and spread to most tropical and subtropical countries to become one of the world's most important and most widely used spices (Ravindran et al. 2007). Other members of the family (hereby generally referred to as "gingers"), such as species of *Alpinia* Roxb., *Curcuma* L., and *Etlingera* Giseke have been explored for their medicinal properties and potential for domestication (Itokawa et al. 2008; Ghosh and Rangan 2013; Afzal et al. 2013; Ud-Daula and Basher 2019). Such widespread and domesticated members of Zingiberaceae are likely to continue to benefit from human activities, and survive the Anthropocene extinction bottleneck (Malhi et al. 2014; Kress and Krupnick 2022). This paper focuses on threatened ginger species; these include species that are currently considered non-useful to humans, and wild species that are over-exploited and face extreme human-induced stressors. For these threatened species, it is predicted that climate change impacts and further human encroachment into protected areas will likely result in them being truly endangered and at strong risk of extinction (Marsh et al. 2022; Kress and Krupnick 2022).

This chapter aims to provide an overview of the threatened gingers of Tropical Asia and their conservation. It presents the taxonomic and geographic distribution of the IUCN-threatened species and investigates existing conservation-related information, such as the nature of threats and conservation initiatives that aim to address these threats. It also assesses the current in situ and ex situ conservation status of the threatened species, and identifies conservation priorities and gaps that must be addressed, challenges that must be overcome, and opportunities that can be explored in the future. Finally, it discusses the potential of biocultural and other new approaches to ginger conservation in Asia.

30.2 Methods

We obtained the checklist of Tropical Asian species from the Zingiberaceae Resource Center (ZRC) website (Newman et al. 2005). The ThreatSearch database (BGCI 2022a) was also used to look for reported threatened species that are not in the ZRC checklist. The scientific names were counterchecked using the Plants of the World Online (POWO 2022) and World Checklist of Vascular Plants (WCVP 2022) databases. Only species with accepted names are included in this research.

The IUCN conservation statuses of the accepted species were then determined (IUCN 2022). We adopted IUCN's classification of threatened species as those that are assigned to any of these three categories: (1) vulnerable (high risk of extinction in the wild); (2) endangered (higher risk of extinction in the wild); and (3) critically endangered (extremely high risk of extinction in the wild) (IUCN 2022). We limit our study to IUCN Red-listed species, since they were subjected to the same standards of assessment. We recognize that by relying on IUCN, we are restricting our scope to well-studied taxa and areas, and species with global assessments only.

Information on the threatened species' distribution range, threats, uses, and conservation actions were sourced from the IUCN Red List data accessed from June 13 to June 30, 2022 (IUCN 2022). The Global Biodiversity Information Facility (GBIF.org 2022) and ZRC (Newman et al. 2005) databases were used to validate IUCN geographic distribution data as well as species presence in protected areas and in ex situ collections. PlantSearch (BGCI 2022b) was also utilized in identifying species under ex situ conservation. A literature search covering publications from the past 10 years was conducted to gather data on the latest developments in Zingiberaceae research and conservation.

30.3 Results and Discussion

30.3.1 Threatened Ginger Species of Tropical Asia

Table 30.1 shows the Zingiberaceae genera that are present in Tropical Asia, and their corresponding species count according to IUCN conservation status (IUCN 2022). A total of 1449 accepted species under 56 genera were recorded in the region. Only 45% of the species in this list have been assessed by the IUCN. Among the species evaluated, 95 are Data Deficient, (i.e., lacking information for a complete assessment), 211 are of Least Concern, and 45 were Near Threatened (Table 30.1).

Twenty-one percent of Tropical Asian gingers species are threatened, consisting of 297 species distributed across 33 genera (Table 30.1). Of these, 114 are Vulnerable, 129 Endangered, and 53 Critically Endangered (IUCN 2022). These threatened species, their distribution, reported threats, and presence in in situ and ex situ sites are all listed in Appendix Table 30.5. One species, *Amomum sumatranum* (Valeton) Škorničk. & Hlavatá is already extinct in the wild, but will still be considered as a threatened species in this chapter. Among the large genera (with > 100 species), *Etlingera* has the greatest number of threatened species (52), followed by *Zingiber* Mill. (43), *Alpinia* (21), and *Globba* L. (21) (Table 30.1; Appendix Table 30.5).

Several small genera (<10 species) contain very high proportions of threatened species (Table 30.1; Appendix Table 30.5). The lone members of the monotypic genera *Cyphostigma* Benth., *Nanochilus* K. Schum., *Parakaempferia* A.S. Rao & D. M. Verma and *Siamanthus* K. Larsen & J. Mood are all threatened, as well 100% of the species in *Siliquamomum* Baill. and *Newmania* N.S.Lý & Škorničk. Other highly threatened genera are *Adelmeria* Ridl. (6 out of 9 species) which is endemic to the Philippines, *Geostachys* (Baker) Ridl. (16 out of 24) and Borneo-endemics

Table 30.1 Species count of Tropical Asian Zingiberaceae genera, based on IUCN conservation status (IUCN 2022)

Genus	Number of species per category							NE	Total number
	EW[a]	CR[a]	EN[a]	VU[a]	NT	LC	DD		
Adelmeria		1	3	2	1	1		1	9
Alpinia		1	8	12	2	16	10	138	187
Amomum	1	1	4	4	1	10	7	71	99
Boesenbergia		2	1	7		7	7	63	87
Burbidgea			2	2		1			5
Camptandra						3			3
Caulokaempferia								11	11
Cautleya								2	2
Conamomum			3	2		4		2	11
Cornukaempferia								5	5
Curcuma		6	7	4	2	12	18	67	116
Cyphostigma[b]			1						1
Distichochlamys								4	4
Elettaria								1	1
Epiamomum			4	1		1			6
Etlingera		12	20	20	21	33	9	29	144
Gagnepainia						1		1	2
Geocharis			3				2	1	6
Geostachys		6	4	6	1	2	5		24
Globba		3	10	8	4	29	7	45	106
Haniffia			1	1		2			4
Hedychium		2	3	4	1	1	3	48	62
Hellenia				1		1		1	3
Hemiorchis								3	3
Hornstedtia		1	2	3	1	3	1	25	36
Johoralia[b]								1	1
Kaempferia						1	2	40	43
Kedhalia[b]								1	1
Lanxangia					1	1			2
Larsenianthus		1						3	4
Leptosolena[b]					1				1
Meistera			3	6	3	8	2	18	40
Myxochlamys			1		1				2
Nanochilus[b]		1							1
Newmania		1	4						5
Paracostus						1			1
Parakaempferia[b]		1							1
Perakalia[b]								1	1
Plagiostachys			4	2		10	2	9	27
Pleuranthodium			3	1		6	4	9	23
Pommereschea						1			1

(continued)

Table 30.1 (continued)

Genus	Number of species per category							NE	Total number
	EW[a]	CR[a]	EN[a]	VU[a]	NT	LC	DD		
Rhynchanthus								2	2
Riedelia								77	77
Roscoea								15	15
Scaphochlamys		2	9	6	1	14	1	9	42
Siamanthus[b]				1					1
Siliquamomum			3	1					4
Stachyphrynium								1	1
Stadiochilus[b]								1	1
Sulettaria		1		2				12	15
Sundamomum			7			6		1	14
Tamijia								2	2
Tapeinochilos								1	1
Vanoverberghia				2		1		1	4
Wurfbainia			2	1		13	1	6	23
Zingiber		11	17	15	5	22	14	72	156
Total/category	1	53	129	114	46	211	95	800	1449

The categories are: *EW* extinct in the wild, *CR* critically endangered, *EN* endangered, *VU* vulnerable, *NT* neat threatened, *LC* least concern, *DD* data deficient, *NE* not evaluated

[a] Considered as threatened

[b] Monotypic genus

Epiamomum A. D. Poulsen & Škorničk (5 out of 6) and *Burbidgea* Hook. f. (4 out of 5).

30.3.2 Potential Extinction Hotspots?

The distribution of threatened species in Tropical Asia is displayed in Fig. 30.1. Based on biogeographic region, high concentrations of threatened taxa and in particular, threatened endemic species are observed in Southeast Asia, specifically in the Sundaland biogeographic region (Peninsular Malaysia, Borneo and Sumatra). Borneo, politically divided among Brunei, Malaysia and Indonesia, is the most diverse hotspot for threatened gingers, sheltering 96 species under 20 genera. Three genera (*Burbidgea, Epiamomum* and *Myxochlamys* A. Takano and Nagam.) and 87 species are endemic to the island. The meta-analysis study of De Bruyn et al. (2014) identified Borneo and Indochina as the main areas of plant and animal species diversification in the Southeast Asian region. Peninsular Malaysia, and the islands of Sumatra and Sulawesi also exhibit high levels of endemism and richness of threatened species (Fig. 30.1). Compared to other tropical regions in the world, Southeast Asia has a large number of globally threatened species included in the IUCN Red List (Sodhi et al. 2010). Unless the existing threats to plant species are eliminated or effectively reduced, it is entirely conceivable that Southeast Asia,

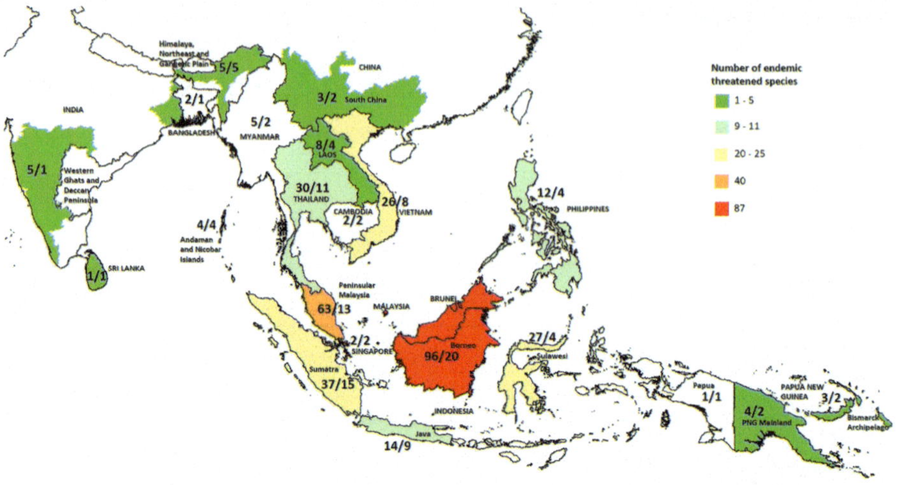

Fig. 30.1 Geographic distribution of IUCN threatened Zingiberaceae species in Tropical Asia. The numbers of threatened taxa (number of species/number of genera) are indicated. Geographic areas are colored according to the number of endemic species. The figures are based on IUCN distribution data (IUCN 2022) validated with occurrence data from the Global Biodiversity Information Facility (GBIF.org 2022) and Zingiberaceae Resource Center (Newman et al. 2005–present) databases

especially Sundaland, will transform from a center of diversity, to an extinction hotspot of ginger species.

The distribution of threatened species by country is shown in Table 30.2. Malaysia has the largest number of threatened ginger taxa (150 species/21 genera), followed by Indonesia (113 species/24 genera), Thailand (30 species/11 genera), and Vietnam (26 species/8 genera). Malaysia and Indonesia also have the highest number of endemic species (86 and 66, respectively). The number and names of endemic genera are also provided in Table 30.2. Two endemic genera have been recorded in Indonesia, while one endemic genus has been recorded each in India, Sri Lanka, Thailand, Vietnam, and Philippines. So far, we did not find any information about the threatened ginger species in Timor Leste.

30.4 Nature of Threats

30.4.1 Threats from Anthropogenic Activities

Figure 30.2 shows the contribution of the different types of direct threats to the threatened gingers, based on the total number of species affected. The classification of threats to the species follows IUCN (2022) threats classification scheme (version 3.2).The three major threat types are: (1) agriculture and aquaculture, where planting of annual and perennial non timber crops (no. 2a) and establishment of wood and pulp plantations (no. 2b) were recorded in 165 and 101 species, respectively;

Table 30.2 Country-level species and genus counts of threatened Zingiberaceae species in Tropical Asia

Region/country	Number of species						Number of genera		Endemic genus/genera
	VU	EN	CR	EW	Total	Endemic	Total	Endemic	
Region									
India	5	6	6		17	13	10	1	*Parakaempferia*
Sri Lanka	1				1	1	1	1	*Cyphostigma*
Bangladesh	1	1			2		1		
East Asia (Southern part)									
China	3				3	1	2		
Southeast Asia (mainland)									
Myanmar	3	2			5		2		
Thailand	15	13	2		***30***	10	11	1	*Siamanthus*
Cambodia	2				2		2		
Laos	4	4			8	5	4		
Vietnam	6	16	4		***26***	20	8	1	*Newmania*
Southeast Asia (maritime)									
Malaysia	***65***	***63***	***22***		***150***	***86***	***21***		
Singapore	1		1		2	1	2		
Brunei	11	4			15	1	9		
Philippines	1	5	6		12	11	4	1	*Adelmeria*
Indonesia	**41**	**50**	**21**	1	***113***	***66***	***24***	2	*Myxochlamys; Nanochilus*
Australasia									
Papua New Guinea	3	4			7	5	2		

The names of endemic genera are also provided. The highest counts are italicized

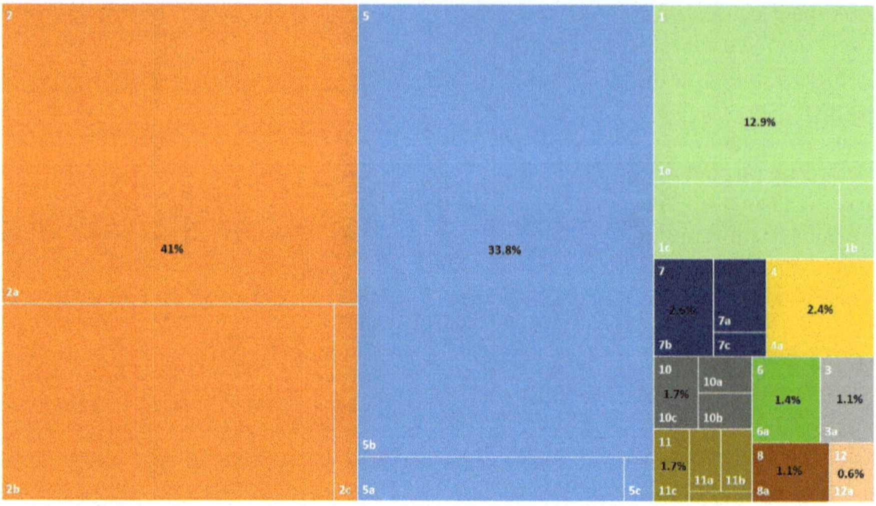

Code	Threat	Code	Threat
1	Residential and commercial development	7	Natural systems modification
1a	Housing and urban areas	7a	Fire and fire suppression
1b	Commercial and industrial areas	7b	Dams and water management/use
1c	Tourism and recreation areas	7c	Other ecosystem modifications
2	Agriculture and aquaculture	8	Invasive and other problematic species, genes, diseases
2a	Annual and perennial non-timber crops	8a	Invasive non-native/alien species/disease
2b	Wood and pulp plantations	10	Geological events
2c	Livestock farming and ranching	10a	Volcanoes
3	Energy production and mining	10b	Earthquake/tsunamis
3a	Mining and quarrying	10c	Avalanches/landslides
4	Transportation and service corridors	11	Climate change and severe weather
4a	Roads and railroads	11a	Habitat shifting and alteration
5	Biological resource use	11b	Droughts
5a	Gathering terrestrial plants	11c	Temperature extremes
5b	Logging and wood harvesting	11d	Other impacts
5c	Fishing and harvesting aquatic resources	12	Other options
6	Human intrusions and disturbance	12a	Other threats
6a	Recreational activities		

Fig. 30.2 Threats to tropical Asian Zingiberaceae. Data was obtained from the IUCN Red List records of all threatened species (IUCN 2022). The relative frequency values of each threat type are provided

(2) biological resource use, where logging and wood harvesting (no. 5b) was recorded in 205 species; and (3) residential and commercial development, where development of housing and urban areas (no. 1a) was recorded in 60 species (Fig. 30.2; Appendix Table 30.5). Other reported threats on ginger species include mining and quarrying (no. 3a), building of roads and railroads (no. 4a), fire and fire suppression (no. 7a), and dams and water management/use (no. 7b) (Fig. 30.2; Appendix Table 30.5). Most of the direct threats reported broadly correspond to

proximate causes of deforestation (Geist and Lambin 2002), or deforestation and forest degradation (Hosonuma et al. 2012).

Most Zingiberaceae species are found in lowland primary and secondary forests occurring in stream banks, plateau areas, ravines, roadsides, karsts, and disturbed areas (Mohamad and Kalu 2019; Saensouk et al. 2016; Setiawan et al. 2022). Deforestation and forest degradation harm the Zingiberaceae species in a lot of ways, causing stresses at the ecosystem and species levels. Removing trees from the ecosystem removes the canopy cover and alters the microclimate above-ground and in the soil (Yeats 2013). Such eco-climatic changes threaten the survival of ginger species, as they generally prefer shaded and humid habitats (Yeats 2013; Setiawan et al. 2022). Habitat fragmentation reduces the presence of pollinators and seed dispersers, limiting the populations' ability to reproduce and disperse (Yeats 2013). Immense loss and fragmentation of ginger habitats may isolate subpopulations and individuals geographically and reproductively, increasing their risk of extinction in the future (Niissalo et al. 2017). The lone population of the vulnerable species *Amomum petaloideum* (S.Q. Tong) T.L. Wu in Laos has become locally extinct, as their habitat was displaced by a rubber tree plantation (Leong-Skornickova et al. 2019a). The Kalimantan-endemic *Etlingera balikpapanensis* A.D. Poulsen already lost ~49% of its populations within the last three generations, due to deforestation and habitat degradation, which continue to heighten its risk of extinction (Robiansyah et al. 2021). Gingers that occur at higher altitudes in montane forests, such as certain species of *Alpinia, Amomum* L., *Hornstedtia* Retz. and *Vanoverberghia* Merr. (Docot et al. 2019; Mohamad and Kalu 2019) are not spared from the threats of deforestation as Zeng et al. (2018) discovered that agricultural expansions in Southeast Asia have encroached a large portion of the highlands. Lau (2014) observed that the disturbances in the high-altitude habitats of *Geostachys* are comparable to those in the lowland habitats.

Unfortunately, more than 50% of the forests in Southeast Asia has been denuded (Lai 2022), and the region is still undergoing deforestation at an alarming rate (Koh and Sodhi 2010; Zeng et al. 2018). Moreso, Southeast Asian lowland forests have become extremely imperiled habitats, as their accessibility makes them an easy target for logging, agriculture, and other deforestation activities (Lau 2014; WWF 2017; Zeng et al. 2018). Sundaland has lost 70% of its original lowland forest cover (Wilcove et al. 2013). Expansion of palm oil plantations contributes significantly to the decline of forest ecosystems and biodiversity loss in Indonesia, Malaysia, and Thailand, which happens to be the largest producers of palm oil in the world (Dhandapani 2015; Cazzolla Gatti et al. 2019; Jaroenkietkajorn et al. 2021; Robiansyah et al. 2021). In addition, pulp wood and timber plantations as well as logging activities continue to play a major role in the deforestation of the region (WWF 2017; Lai 2022). There is an urgent need to prioritize conservation in Southeast Asia in order to prevent massive biodiversity loss (Sodhi et al. 2010; Duckworth et al. 2012). It is projected that the region is at risk of losing 13–42% of its local populations by the onset of the next century (Koh and Sodhi 2010).

Roads and railroads (no. 4a), recreational activities (no. 6a), and presence of invasive non-native species (no. 8a) were reported as threats to some ginger species

(Fig. 30.2; Appendix Table 30.5). Creation of roads and railroads not only results in habitat fragmentation but also paves the way for the entry of agricultural crops, invasive species, tourists, and poachers. Ecotourism activities can also cause habitat disturbances. Species that grow along trails, such as *Geostachys secunda* (Baker) Ridl. in Gunung Batu Puteh, Malaysia, are in danger of being trampled on (Lau 2014). Placement of dumpsites located near forest borders has reduced the quality of the ecosystem, reducing the populations of *Geostachys rupestris* Ridl. in Gunung Jerai, Malaysia (Lau 2014). It is also possible that tourists can unintentionally introduce invasive species that may disrupt the balance in the ecosystem and displace the populations of the threatened species. A few invasive species have been reported as threats, namely: *Lantana camara* L. and *Hedychium coronarium* J. Koenig in the Philippines (Docot 2020a, b); *Mimosa pigra* L. in Thailand (Romand-Monnier and Contu 2013a); and *Piper aduncum* L. in Papua New Guinea (Romand-Monnier and Contu 2013b).

Of the 297 threatened species, only 18 species were recorded to be gathered as target species (no. 5a, Fig. 30.2; Appendix Table 30.5). These species are used as food, medicine and as ornamental plants. At present, there are no existing studies on the harvest utilization rates of these species. For the majority of the threatened species (94%), anthropogenic threats do not directly use the species of concern, but unintentionally increase the risk of their extinction.

30.4.2 Other Threats

A small proportion of threatened ginger species are endangered by non-anthropogenic factors. Geological events such as volcanic eruptions, tsunamis, and landslides were reported for species from the Philippines, Papua New Guinea, and Nicobar Islands (no. 10 in Fig. 30.2; Appendix Table 30.5). These natural calamities can cause habitat loss and eventual extinction of narrowly endemic species. Climate change and severe weather conditions (e.g., habitat shifting and alteration, droughts, and temperature extremes) create habitat disturbances that may cause mortality of individuals and populations. This threat type (no. 11) has been reported in a few species from Borneo, but it is highly possible that it will affect more species globally in the future. Having a very restricted geographic range is considered a threat, since the entire species can be wiped out by a single unforeseen event, such as a volcanic eruption or a forest fire. This threat (no. 12a in Fig. 30.2) was only recorded in four species from Malaysia, in spite of the fact that 67 threatened ginger species in Tropical Asia were reported to occur in one locality (Appendix Table 30.5).

No threats were reported for 13 species despite their threatened conservation status (Appendix Table 30.5). Their populations should be occasionally observed to check for impacts of climate change and other newly emerging threats.

30.5 In Situ Protection and Future Challenges

In situ or on-site conservation aims to protect ecosystems, so that the species can continue to thrive in their natural habitats. It involves population monitoring and management of target species to ensure that they continue to reproduce and adapt successfully to the changing eco-climatic conditions in their natural environment (Heywood and Dulloo 2005). In situ conservation of threatened species is built on the premise that the threats that put species in danger of extinction will be eliminated or managed when the species are harbored in PAs. However, results of assessments on the effectiveness of PAs in conserving habitats and species populations have been mixed (Geldmann et al. 2013; Heywood 2017). Deforestation in many protected areas is on-going, and in protected landscapes in tropical Asia, is expected to continue or increase in magnitude in the future (Jayathilake et al. 2021). The protection provided by PAs is also compromised by inadequate resources, management planning and implementation (Leverington et al. 2010; Heywood 2017). Despite these limitations, PAs are key components of biodiversity conservation efforts, and also play important socio-economic and cultural roles, such as serving as heritage sites, contributing to local livelihoods and national economies and climate change mitigation and adaptation, among others (Heywood 2017; Watson et al. 2014).

Table 30.3 presents the extent of in situ conservation of threatened ginger species in Tropical Asia, in terms of number of in situ sites, as well as the number of conserved species. Malaysia has the largest number of protected areas (PAs) and the largest number of in situ-conserved species, accounting for 68% of its threatened ginger species. Malaysia also has the PAs that harbor the highest number of species, namely: Gunung Mulu National Park (18 species from 11 genera); Taman Negara National Park (14 species), Kinabalu Park (10 species), and Lambir Hill National Park (10 species). These species-rich PAs can be designated as centers of in situ protection, where the ginger habitats and populations can be regularly monitored. In terms of species coverage, Sri Lanka and the Philippines have the highest percentage of threatened species with in situ protection (100% and 92%, respectively). There is a pressing need to locate more Zingiberaceae species and survey more PAs in Indonesia, given that it has the second largest number of threatened species, yet there is only 33% species coverage. To date, we could not find any accessible records of in situ-conserved threatened ginger species in Southern China, Cambodia, Myanmar and Papua New Guinea.

Field studies on ginger biodiversity and species biology are essential in estimating the number of species within the PA, pinpointing their populations and understanding their interactions with their environment. Field surveys of diversity, habitats, phenology and ethnobotanical uses of Zingiberaceae species in some PAs have been documented in Thailand (Nontasit et al. 2015; Saensouk et al. 2016; Saensouk and Saensouk 2021; Ragsasilp et al. 2022), the Philippines (Naïve 2017; Acero et al. 2019; Acma et al. 2020) and Indonesia (Pitopang et al. 2019). Data produced from such studies are essential in both in situ and ex situ conservation planning. Once located, populations of the target species must be monitored and

Table 30.3 In situ conservation of threatened Zingiberaceae in Tropical Asia

Country	Number of protected areas	Number of species conserved	Species coverage (%)
Bangladesh	2	1	50
Brunei	4	7	47
India	9	10	53
Indonesia	*31*	*37*	33
Laos	6	6	75
Malaysia	*54*	*102*	68
Philippines	8	11	*92*
Singapore	1	1	50
Sri Lanka	3	1	*100*
Thailand	15	18	60
Vietnam	16	17	65

The number of protected areas sheltering threatened Zingiberaceae species is provided. Species coverage refers to the proportion of species that are conserved in situ. The largest numbers are italicized

evaluated on a regular basis. These activities are crucial in detecting dwindling populations that must be rescued.

There are very few accounts of ginger population assessments in protected areas, however. Robiansyah et al. (2021) re-assessed *Etlingera balikpapanensis* populations in two protected areas in Indonesia and classified the species as endangered based on IUCN standards. Similarly, Kumar and Singh (2018) assessed the populations of *Globba spathulata* Roxb. and *Hemiorchis pantlingii* in Dampa Tiger Reserve in India and classified both species as "vulnerable".

PAs are not spared from the threats of climate change. Extreme climate conditions and changing weather patterns can cause habitat disturbances that can negatively affect the gingers and other forest species. Temperature projections reveal that low lying PAs in Borneo are susceptible to increasing temperatures brought by climate change (Scriven et al. 2015). This would be detrimental to the Zingiberaceae, since its members are mainly found in lowland forests. Scriven et al. (2015) recommended increasing the connectivity of PAs along the altitudinal gradient to allow the species to migrate to cooler habitats. The reduction of pollinators in isolated PAs is detrimental to ginger populations, as pollinators are specific to species groups within the family (Sakai et al. 2013). In their study of the critically endangered species *Zingiber singapurense* Škorničk., Niissalo et al. (2017) concluded that the risk for extinction increases, as the habitat becomes smaller and more isolated. Expanding the geographic scope of PAs and creating forest corridors between PAs can alleviate the impacts of climate change and habitat fragmentation to species (Scriven et al. 2015).

Given the concerns about the effectiveness of in situ conservation, narrow-endemics, and species that are present in one or two PAs need urgent conservation action. Island endemics, which constitute 55% of the threatened ginger species in Tropical Asia (Appendix Table 30.5; IUCN 2022) are vulnerable to extinction due to their limited gene pool (Mathew 2017). Among the threatened species, 194 (65%)

are under in situ protection (Appendix Table 30.5) but 180 of these are recorded in just one or two PAs. As mentioned earlier, species that are restricted to one locality have the potential to be eliminated by a single stochastic event. For rare, highly threatened species, we recommend a combination of in situ and ex situ conservation measures. The former will allow the species to undergo natural evolutionary processes in its natural habitat, while the latter will ensure the survival of individuals and populations and possibly provide a reservoir of plant material for forest regeneration in the future.

30.6 Ex Situ Conservation Efforts and Limitations

Ex situ conservation is the maintenance and propagation of species in facilities outside their natural habitats (e.g., botanical gardens, genebanks). It eliminates the threats to species by removing them from their ecosystem. Although conserving ex situ heavily limits sexual reproduction and natural evolution, it secures the survival of populations of species that are on the brink of extinction. Of the 104 threatened ginger species that are conserved ex situ, 76 are found in just one or two facilities/sites (Appendix Table 30.5). The survival of these plant species with very limited ex situ coverage will depend on the quality of the population or accession (in terms of reproductive ability and adaptability to ex situ conditions), and the success of maintenance, storage propagation and distribution in the facility, and also of natural habitat restoration. Conserved accessions or populations should be as genetically diverse as possible to provide a variety of traits that can be tapped by plant breeders and other scientists, and to help the populations adapt to natural conditions once they are restored back in the wild. For the critically endangered *Zingiber singapurense*, it was recommended that all its distinct genotypes be represented in ex situ collections in order to increase its chance of survival (Niissalo et al. 2018).

Ex situ collections provide a repository of materials (e.g., seeds, rhizomes, and living plants) that can be used for research, propagation, ecological restoration, and other purposes. This way, natural populations are spared from the pressures of plant collection by researchers and other users. The most widely conserved threatened species are *Burbidgea schizocheila* Hackett (31 sites), *Hedychium muluense* R.M. Sm. (24 sites) and *Hedychium hasseltii* Blume (20 sites), which are all cultivated as ornamental plants (Appendix Table 30.5). Based on the records, The Royal Botanic Garden Edinburgh (RBGE) maintains the largest number (43) of threatened Tropical Asian ginger species (GBIF.org 2022; IUCN 2022; BGCI 2022b). The Zingiberaceae Living Collection of RBGE was collected, classified, and established by taxonomists and other experts in the field, and it continues to contribute to the understanding and conservation of the ginger family (Yeats 2013). In Singapore, samples of populations of all the threatened Zingiberaceae species have been secured and maintained in nurseries, where they will be mass-propagated and re-introduced to new habitats (Niissalo et al. 2017). The RBGE and Singapore Botanic Gardens can serve as common repositories of threatened ginger species until the national genebanks are fully capable of conserving their plant resources ex situ.

However, not all species are suitable for ex situ conservation. Souvannakhoummane (2014) reported the inability of some ginger species to survive the conditions in Pha Tad Ke Botanical Garden in Laos. The critically endangered *Boesenbergia stenophylla* R.M. Sm. does not respond well to in vitro propagation and is recommended to be primarily conserved in situ (Toyat et al. 2015). The growth and reproduction of species with a narrow ecological niche (like *B. stenophylla*) require specific environmental conditions that may be difficult and/or too expensive to simulate outside their natural habitats.

30.7 Conservation Coverage of Threatened Species

Figure 30.3 shows the extent of conservation done in our set of threatened Zingiberaceae species. We refer to species that are under both in situ and ex situ conservation as "fully covered". Overall, 70 of threatened ginger species (23.6%) are fully covered; 29 species (9.8%) are only conserved ex situ, 124 species (41.9%) are only protected in situ, and 73 species (24.6%) are not conserved at all (Appendix Table 30.5). The vulnerable and endangered species groups exhibit similar trends, where majority of the species are under in situ protection only (43.6% and 46.5%, respectively) (Fig. 30.3). The proportions of unconserved species are also comparable in these groups (18.4% for vulnerable and 18.6% for endangered) (Fig. 30.3). Alarmingly, more than half (52.8%) of the critically endangered species are

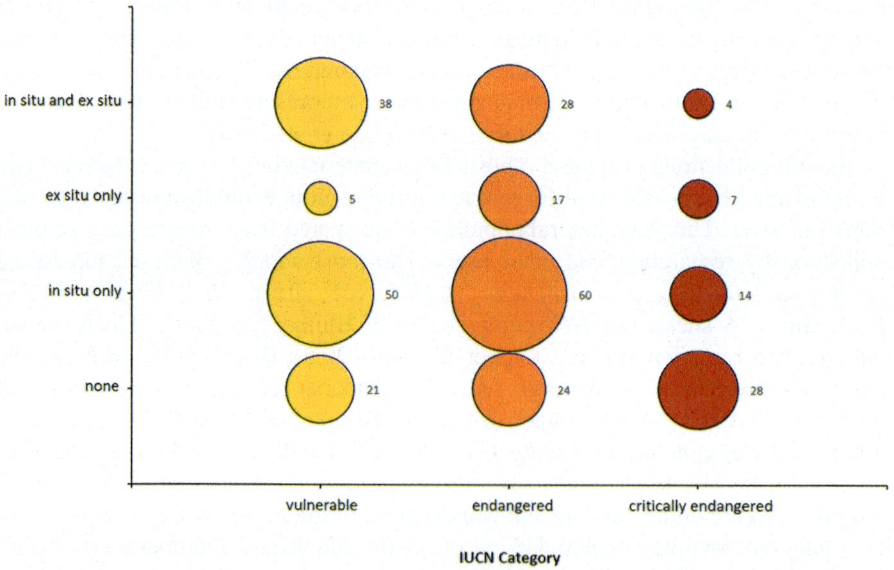

Fig. 30.3 Conservation coverage of threatened Zingiberaceae species in Tropical Asia. Each color represents an IUCN category, and the circles indicate the number of threatened Tropical Asian ginger species under a conservation coverage

unconserved and only 7.5% are fully covered (Fig. 30.3). *Amomum sumatranum*, which is categorized as extinct in the wild, does not have any records of both in situ and ex situ conservation. In 2019, BGCI reported it to be conserved in one ex situ locality (Olander 2020a). However, our recent BGCI search revealed that it was no longer present in any ex situ site (BGCI 2022a, b). The IUCN status of this species has not been changed to "extinct" but we fear that *Amomum sumatranum* has succumbed to extinction. There is a very high probability that the critically endangered and unconserved species will be extinct in the near future, unless drastic conservation measures are taken.

Ideally, all threatened ginger species, regardless of their IUCN category (i.e., vulnerable, endangered, critically endangered), should be prioritized and covered by both in situ and ex situ conservation. Based on our data synthesis (Fig. 30.3), this goal is still far from being achieved. It may even be too late to rescue the rare, critically endangered species that are not yet conserved, just like the case of *Amomum sumatranum*. Some species in PAs remain susceptible to threats (Kumar and Singh 2018; Robiansyah et al. 2021), while some species cannot be rescued by ex situ measures (Souvannakhoummane 2014; Toyat et al. 2015). In addition, the tropical Asian countries do not have the same level of commitment and resources for biodiversity conservation. Hence, it is imperative to devise strategic conservation plans that can save as many species in the region as possible.

30.8 Conservation and Information Gaps

Designing effective and efficient conservation plans requires a complete picture of the actual diversity in the region, and adequate information about the species' biology and ecology. Filling these gaps will render a solid foundation for developing both in situ and ex situ conservation methods.

More than half (55%) of the tropical Asian ginger species are not yet evaluated by IUCN (Table 30.1). There are 800 species across 40 genera that have to be assessed. Large numbers of unassessed species belong to those of the genera *Alpinia* (138), *Zingiber* (72), *Amomum* (71), *Curcuma* (67) and *Boesenbergia* (63). In addition, there are 95 data deficient species that are still awaiting complete assessment (Table 30.1). We recommend prioritizing the assessment of narrow-endemic and monotypic genera *Johoralia* C.K. Lim., *Kedhalia* C. K. Lim, *Perakalia* C.K. Lim and *Stadiochilus* R.M. Sm. and of the genera that are entirely unassessed, namely: *Caulokaempferia* K. Larsen, *Cautleya* (Royle ex Benth.) Hook.f., *Cornukaempferia* Mood and K. Larsen, *Distichochlamys* M.F. Newman, *Hemiorchis* Kurz, *Rhynchanthus* Hook.f., *Riedelia* Oliv., *Roscoea* Sm. and *Tamijia* S. Sakai & Nagam. Assessments of the species' extinction risk need to be expedited, as these will help prevent extinctions amidst increasing environmental pressures (Lughadha et al. 2020).

A small number of threatened species was reported in Bangladesh, Myanmar, Cambodia, Papua New Guinea and the island groups in eastern Indonesia (i.e., Lesser Sunda Islands, Malaku Islands, and Western New Guinea). Along with

Timor Leste, these areas are poorly explored yet potentially rich in Zingiberaceae species. These areas should be prioritized for species surveys, field studies, and capacity building for plant research and conservation.

Studies on threatened ginger species under ex situ and in situ conservation are very limited. Yeats (2013) documented the cultivation of *Etlingera* species at RBGE, in which growing conditions, propagation techniques and pest control were thoroughly discussed. Niissalo et al. (2017) conducted a comprehensive analysis of all the ginger species in Singapore and concluded that more than half of the species are at risk of extinction due to fragmented habitats. More research is needed to optimize the protection and handling of the various threatened species. Population structure, seed storage behavior, suitable propagation techniques and species-specific requirements for growth and reproduction are important aspects of conservation that must be ascertained.

30.9 Biocultural and Other Conservation Approaches

The Anthropocene is marked by conditions that drive the extinction and speciation of biological organisms, indicating that human activities are intertwined with non-human environmental factors in affecting the functions, variation patterns, biogeography, ecology and survival of plant populations, species and communities (Bliege Bird and Nimmo 2018; Thomas 2020). Majority of the IUCN-listed, threatened species of tropical Asian Zingiberaceae are not useful to humans, hence falling in the category of "losers of plant biodiversity in the Anthropocene" as classified by Kress and Krupnick (2022). These "loser species" are subjected to extreme anthropogenic threats and will likely not survive the Anthropocene (Kress and Krupnick 2022). While some conservationists postulate the importance of saving biodiversity in its entirety, a new pragmatic approach to conservation has emerged, focusing mainly on specific target species that are worthy of saving (Heywood 2017). We recommend the proceeding strategies to emphasize the value and promote the conservation of threatened ginger species in Tropical Asia.

30.9.1 Local Community Involvement

Local communities near PAs can actively participate in conserving threatened ginger species. Education and awareness (e.g., about threatened species in the area, sustainable use of plant resources) are vital in protecting endangered plant populations. Establishment of home gardens or community nurseries should be encouraged, so that the locals will not have to collect plants from the wild. Home gardens are considered as a form of "quasi in situ" conservation, where plants extracted from their natural habitats are grown and maintained under natural or semi-natural conditions (Volis and Blecher 2010). Such new man-made habitats contain fewer species but may support rare plants, contributing to the species richness and diversity in the area (Thomas 2013). For instance, in Bueng Kan province, Thailand, 23.88% of the Zingiberaceae species are represented in home gardens (Ragsasilp et al. 2022).

Biodiversity-rich religious sites can be suitable for in situ conservation (Singh and Garg 2014). The Kurdi Angod sacred site in the Western Ghats region of India shelters a plethora of medicinal, endemic, rare and threatened species, including the vulnerable *Curcuma pseudomontana* J. Graham (Singh and Garg 2014). Such places are considered sacred and protected by the natives.

Local communities residing within or near specific ecosystems possess traditional ecological knowledge (TEK) that can significantly contribute to the understanding and conservation of the area. TEK refers to the people's knowledge, beliefs and practices with regard to the species and ecosystems that surround them (NPS 2022). TEK is accumulated over hundreds or thousands of years of interacting with nature, and is passed down from generation to generation (Susanti and Zuhud 2019; NPS 2022). In Thailand where Zingiberaceae diversity is high, a relatively large number of species is used as food and spices (20–32 species), medicine (33), ornamental plants (21–34) and ritual plants (18–19) (Saensouk and Saensouk 2021; Ragsasilp et al. 2022). Certain species are also used as cosmetics and dyes (Ragsasilp et al. 2022). In species-rich Borneo and Sulawesi, *Etlingera* species are considered as local food (Poulsen 2006, 2012). In Peninsular Malaysia, ginger species are frequently used in traditional medicine (Ibrahim et al. 2007). TEK should be documented and preserved, as it provides a different insight on the diversity and utility of Zingiberaceae species that can be integrated to PA management plans. Sadly, the loss of biodiversity due to deforestation is accompanied by diminishing TEK. The exodus of younger generations (for education and job opportunities) inhibits the progress and preservation of traditional knowledge (Pradheeps and Poyyamoli 2013; Susanti and Zuhud 2019). The decline of TEK can in turn contribute to biodiversity loss as lack of awareness and concern for forest species makes them more vulnerable to threats (Cahyaningsih et al. 2021b).

30.9.2 Promoting and Enhancing Species Use and Value

Kress and Krupnick (2022) implied the two fates of useful species: (1) widely cultivated, domesticated crops are bound to thrive in the Anthropocene, while (2) useful, wild species will eventually decline due to overharvesting and exploitation. In our set of threatened gingers, 73 species have recorded uses for different purposes, wherein 13 species are reportedly threatened by collecting/gathering (Table 30.4). Although it may seem counterintuitive, promoting the use of threatened species, when done strategically, can lessen their risk of extinction. At the local level, home gardens and community nurseries should be activated and tapped as the primary source of plant materials. Promoting the use of wild gingers will hopefully, increase the demand for them, and induce their cultivation. It will also benefit the people by broadening their options for food and medicine and by providing a potential source of income (in the case of ornamental plants). On the other hand, genebank accessions and botanical garden collections can be used for

Table 30.4 Uses of threatened Zingiberaceae species in Tropical Asia

Scientific name	Status	Distribution	Part/s used	Purpose	Remarks	Reference
Alpinia beamanii R.M.Sm.	Endangered	Indonesia (Kalimantan); Malaysia (Sabah, Sarawak)	Fruits	Food		Lamb et al. (2013)
Alpinia congesta Elmer[a]	Endangered	Philippines	Fruits	Food		Docot (2019a)
Alpinia diversifolia (Elmer) Elmer	Vulnerable	Philippines	Stem base	Food		Elmer (1915)
Alpinia manii Baker	Vulnerable	India (Andaman Islands)	Sap of stems and leaves	Bee repellant	Potentially ornamental	Waman et al. (2018)
Alpinia romblonensis Elmer	Vulnerable	Philippines	Fruits	Food		Crane (2000)
Amomum sumatranum (Valeton) Škorničk. & Hlavatá	Extinct in the wild	Indonesia	Sap	Medicine		Heyne (1987), The National Development Planning Agency (2003) as cited by Cahyaningsih (2021a)
Boesenbergia siphonantha (King ex Baker) M.Sabu, Prasanthk. & Škorničk.	Vulnerable	India (Andaman Islands); Thailand; Vietnam	Rhizomes	Medicine	For diarrhea and other digestive disorders; potentially ornamental	Tran et al. (2019)
Boesenbergia stenophylla R.M. Sm.[a]	Critically endangered	Indonesia (Kalimantan); Malaysia (Sabah, Sarawak)	Rhizomes Stem	Medicine	For abdominal pains, alcohol intoxication and food poisoning For infections and digestive disorders	Toyat et al. (2015), Susanti and Zuhud (2019)

Curcuma candida (Wall.) Techapr. & Škorničk.[a]	Vulnerable	Myanmar; Thailand	Young inflorescences	Food		Leong-Skornickova et al. (2019b)
Curcuma caulina J. Graham[a]	Endangered	India (Maharashtra)	Tubers	Food		Romand-Monnier and Contu (2013c)
			Leaves	Fiber	For making paper	
Curcuma colorata Valeton	Endangered	Indonesia (Java)	Leaves and rhizomes	Medicine	For diarrhea and malaria	Heyne (1987), Eisai (1995) as cited by Cahyaningsih (2021a)
Curcuma coriacea Mangaly & M. Sabu[a]	Endangered	India (Kerala)		Ornamental		West Demmy and Burch (1998)
Curcuma pseudomontana J. Graham[a]	Vulnerable	India (Kerala, Karnataka, Andhra Pradesh)	Tubers	Food Medicine	Used to treat dysentery, leprosy, heart disease, jaundice and body swellings and as a galactagogue	Sasikumar (2005), Ravindran et al. (2007), Tyagi (2005), Jagtap et al. (2006), Rama Rao (2006)
Curcuma rhabdota Sirirugsa & M.F. Newman[a]	Vulnerable	Cambodia; Laos; Vietnam		Ornamental		Leong-Skornickova et al. (2019c)
Curcuma sahuynhensis Škorničk. & N.S. Lý	Endangered	Vietnam	Young inflorescences and young shoots	Food		Tran and Leong-Škorničková (2019)
Curcuma sumatrana Miq.[a]	Vulnerable	Indonesia (Sumatra)	Leaves	Food Medicine	For skin disorders	Nurainas and Ardiyani (2019)
Curcuma supraneeana (W.J. Kress & K.Larsen) Škorničk.[a]	Critically endangered	Thailand		Ornamental		Leong-Skornickova et al. (2019d)

(continued)

Table 30.4 (continued)

Scientific name	Status	Distribution	Part/s used	Purpose	Remarks	Reference
Curcuma vitellina Škorničk. & H.Đ. Trần	Endangered	Vietnam	Young inflorescences	Food		Leong-Škorničková et al. (2019)
Etlingera aulocheilos A.D. Poulsen	Endangered	Indonesia (Sulawesi)	Aril, leafy shoots Young shoots Leaves	Food Medicine Construction material	For fevers, liver diseases, sore knees and scabies As thatching material	Poulsen (2012)
Etlingera biloba A. D.Poulsen	Endangered					
Etlingera borealis A.D.Poulsen	Endangered					
Etlingera caudata A.D.Poulsen	Critically endangered					
Etlingera chlorodonta A.D. Poulsen	Critically endangered					
Etlingera cylindrica A.D. Poulsen	Endangered					
Etlingera doliiformis A.D. Poulsen	Critically endangered					
Etlingera eburnea A.D.Poulsen	Endangered					
Etlingera echinulata A.D. Poulsen	Endangered					

Etlingera flavovirens A.D. Poulsen	Critically endangered					
Etlingera heliconiifolia (K. Schum.) A.D. Poulsen	Vulnerable					
Etlingera hyalina A.D.Poulsen	Endangered					
Etlingera mucida A.D.Poulsen	Critically endangered					
Etlingera mucronata A.D. Poulsen	Endangered					
Etlingera orophila A.D.Poulsen	Endangered					
Etlingera penicillata (K. Schum.) A.D. Poulsen	Endangered					
Etlingera polycarpa ssp. *ligulata* A.D. Poulsen	Vulnerable					
Etlingera serrata A.D.Poulsen	Critically endangered					
Etlingera spinulosa A.D.Poulsen	Endangered					

(continued)

Table 30.4 (continued)

Scientific name	Status	Distribution	Part/s used	Purpose	Remarks	Reference
Etlingera sublimata A.D. Poulsen	Endangered					
Etlingera tubilabrum A.D. Poulsen	Vulnerable					
Etlingera urophylla A.D. Poulsen	Critically endangered					
Etlingera xanthantha A.D. Poulsen	Critically endangered					
Etlingera yessiae A.D.Poulsen	Vulnerable					
Etlingera muriformis A.D. Poulsen	Vulnerable	Indonesia (Kalimantan)	Fruits	Food		Poulsen and Olander (2019a)
Etlingera aurantia A.D.Poulsen	Vulnerable	Brunei; Indonesia (Kalimantan); Malaysia (Sabah, Sarawak)	Fruits	Food		Poulsen (2006)
Etlingera baculutea A.D. Poulsen & Ibrahim	Vulnerable	Indonesia (Kalimantan); Malaysia (Sabah, Sarawak)	Core of shoots	Food		Poulsen (2006)
Etlingera baramensis S.Sakai & Nagam.	Vulnerable	Indonesia (Kalimantan); Malaysia (Sarawak)	Fruits	Food		Poulsen (2006)
			Roots	Cosmetic	As perfume	

Etlingera barioensis A.D. Poulsen	Endangered	Malaysia (Sarawak)	Shoots Leaves	Food Food packaging	For wrapping rice	Poulsen (2006)
Etlingera burttii A. D.Poulsen	Endangered	Malaysia (Sabah, Sarawak)	Shoots and fruits	Food		Poulsen (2006)
Etlingera corneri Mood & Ibrahim	Vulnerable	Malaysia (Peninsular Malaysia); Thailand		Ornamental		Saw (2019)
Etlingera fenzlii (Kurz) Škorničk. & M.Sabu	vulnerable	India (Nicobar Islands)	Aril Rhizomes	Food Medicine Insect repellant	For coughs, colds, skin diseases	Leong-Škorničková et al. (2009), Mathew (2017)
Etlingera kenyalang A.D. Poulsen & H.Chr.	Endangered	Malaysia (Sarawak)	Shoots and fruits Leaves	Food Religious	As part of an ornament in Gawai Kenyalang Festival	Poulsen and Olander (2019a)
Etlingera loerzingii (Valeton) R.M.Sm.	Vulnerable	Indonesia (Sumatra)	Shoots and inflorescences	Food		Poulsen et al. (2009)
Etlingera longipetiolata (B.L.Burtt & R.M. Sm.) R.M.Sm.	vulnerable	Malaysia (Sabah, Sarawak)	Shoots and fruits	Food		Poulsen and Olander (2019b)
Etlingera peekelii (Valeton) R.M.Sm.	Vulnerable	Papua New Guinea (Bismarck Archipelago)	Leaves Fruits	Food packaging Cultural practice	As substitute for betel nut	Poulsen (2020)
Etlingera venusta (Ridl.) R.M.Sm.[a]	Vulnerable			Ornamental		Saw (2019)

(continued)

Table 30.4 (continued)

Scientific name	Status	Distribution	Part/s used	Purpose	Remarks	Reference
		Malaysia (Peninsular Malaysia)				
Globba laeta K. Larsen[a]	Endangered	Thailand	Rhizomesinflorescences	Medicine Religious Ornamental	For stomachache and flatulence	Ragsasilp et al. (2022), Saensouk et al. (2016)
Larsenianthus arunachalensis M. Sabu, Sanoj & Rajesh Kumar	Critically endangered	India (Arunachal Pradesh)			Potentially ornamental	Sabu (2019)
Plagiostachys crocydocalyx (K. Schum.) B.L.Burtt & R.M.Sm.	Endangered	Brunei; Indonesia (Kalimantan); Malaysia (Sabah, Sarawak)	Fruits	Food		Lamb et al. (2013), Poulsen (2006)
Pleuranthodium peekelii (Valeton) R.M.Sm.	Endangered	Papua New Guinea (North Solomons, Bismarck Archipelago); Solomon Islands	Young inflorescences	Food		Lofthus (2019)
Siliquamomum tonkinense Baill.[a]	Vulnerable	China (Yunnan); Vietnam	Rhizomes	Medicine		Leong-Skornickova et al. (2019e)
Vanoverberghia sepulchrei Merr.	Vulnerable	Philippines	Fruits	Food		Docot et al. (2016)
Vanoverberghia vanoverberghii (Merr.) Funak. & Docot	Vulnerable	Philippines	Fruits	Food		Docot (2019b)
Zingiber album Nurainas	Endangered	Indonesia (Sumatra)	Young inflorescences	Food		Nurainas (2019)

Zingiber chrysostachys Ridl.	Endangered	Malaysia (Peninsular Malaysia); Thailand	Young inflorescences Rhizomes	Food Medicine Ornamental	As tonic and laxative	Ragsasilp et al. (2022)
Zingiber collinsii Mood & Theilade	Vulnerable	Cambodia; Vietnam		Ornamental		Leong-Skornickova et al. (2019f)
Zingiber jiewhoei Škorničk.	Vulnerable	Laos	Young inflorescences and young shoots	Food		Leong-Škorničková and Souvannakhoummane (2019)
Zingiber mawangense Noor Ain & Meekiong	Critically endangered	Malaysia (Sarawak)			Potentially ornamental	Olander (2019)
Zingiber monophyllum Gagnep.[a]	Endangered	Vietnam		Ornamental		Leong-Skornickova et al. (2019g)
Zingiber niveum Mood & Theilade[a]	Endangered	Laos	Young inflorescences	Food Ornamental		Souvannakhoummane (2019)
Zingiber pachysiphon B.L. Burtt & R.M.Sm.	Vulnerable	Indonesia (Kalimantan); Malaysia (Sabah, Sarawak)		Ornamental		Lamb et al. (2013)
Zingiber raja C.K. Lim & Kharuk.	Endangered	Malaysia (Peninsular Malaysia); Thailand		Ornamental		Olander (2020b)
Zingiber singapurense Škorničk.	Critically endangered	Singapore		Ornamental		Leong-Škorničková (2019)

(continued)

Table 30.4 (continued)

Scientific name	Status	Distribution	Part/s used	Purpose	Remarks	Reference
Zingiber sulphureum Burkill ex Theilade	Endangered	Malaysia (Peninsular Malaysia)			Potentially ornamental	Larsen and Saksuwan Larsen (2006)

[a] Threatened by collecting/gathering, based on IUCN records (IUCN 2022)

research and commercial purposes. This strategy can only be applied to species that are adequately represented in ex situ and quasi in situ collections and to those that are easy to propagate.

Species with medicinal properties have high conservation value, as they are widely used in both traditional and contemporary medicine. In Indonesia, *Amomum sumatranum* and the endangered *Curcuma colarata* are included in the list of medicinal plants that are prioritized for conservation (Cahyaningsih et al. 2021a). In India, the vulnerable species *Curcuma pseudomontana* has been reported as one of the medicinal plants that require immediate in situ protection (Pradheeps and Poyyamoli 2013; Wagh and Jain 2013). Chemical profiling of non-useful, threatened gingers with sufficient ex situ plants can reveal the species' beneficial properties and increase their potential for commercial use and cultivation. Members of the genus *Alpinia* were found to have anticancer, anti-inflammatory, and antimicrobial properties (Ghosh and Rangan 2013), while antioxidant activity was observed in *Cornukaempferia* species (Chumroenphat et al. 2021). The endangered species *Newmania sontraensis* possess rhizomes and leaves that are rich in essential oils (Tuan et al. 2021), while the leaves of the vulnerable species *Etlingera fenzlii* exhibit antifungal properties (Aravind and Nasreen 2022).

Some threatened ginger species can be promoted as ornamental plants (Table 30.4). However, this can only be done successfully once the most suitable propagation techniques and protocols for each species have been determined. Niissalo et al. (2018) emphasized the importance of mass propagation in providing for the horticulture market and in preventing poaching. In Singapore, several ginger species have been successfully planted in parks and street sides, and have become a part of the local landscape (Niissalo et al. 2017).

Unutilized wild ginger species still merit conservation, since these crop wild relatives could possess nutritional, medicinal and economic potentials that are yet to be realized. Wild Zingiberaceae species could also exhibit some useful traits (e.g., stress tolerance) that can be tapped by plant breeders in improving cultivated ginger varieties. In Andaman and Nicobar Islands, *Alpinia manii*, a vulnerable species, is one the crop wild relatives that urgently need conservation (Waman et al. 2018). In the same region, it was observed that the critically endangered *Zingiber odoriferum* displays disease- and pest resistance under natural conditions, making it a good candidate for conservation and further evaluation (Sharma et al. 2018; Niissalo et al. 2018).

30.9.3 Genetic Studies to Aid Conservation Planning

Population genetic studies are vital in conservation planning, as they determine the extent of diversity and its distribution within and/or between the species of concern. Using next-generation sequencing, Niissalo et al. (2018) analyzed the genetic diversity of all the known populations of *Zingiber singapurense* and detected very low population differentiation and only 13 genotypes within the species, indicating lack of sexual reproduction. Based on this variation pattern, they recommended specific conservation measures, including: (1) conservation of pollinator, if needed; (2) removal of excess hybrid individuals from the populations; (3) ex situ maintenance of all existing clones; (4) reintroduction and monitoring of propagated plants in humid secondary forest habitats; (5) increasing the population size to a minimum of 250 individuals via controlled breeding; (6) documented cross pollinating of plants; and (7) ensuring that seedlings that will be reintroduced in a particular site will be as unrelated as possible (Niissalo et al. 2018). Simple sequence repeats (SSR) and random amplified polymorphic DNA markers (RAPD) revealed a similar variation pattern in *Boesenbergia stenophylla*, where no genetic variation was observed among three subpopulations in Borneo (Toyat et al. 2015). This implies that sampling just one of the three subpopulations is enough to represent the genetic diversity in the locality. Depending on the goal and the capacity of the ex situ facility, genetic data can help in assessing collection gaps, pinpointing highly diverse and genetically distinct populations, and in identifying duplicates or genetically identical populations or accessions.

30.10 Summary and Conclusion

Our review of the IUCN-listed threatened ginger species in tropical Asia highlights the importance of Southeast Asia, particularly the Sundaland region, in terms of species diversity and endemism, concentration of threatened species, and high rates of deforestation. There are countries and island groups in tropical Asia that are poorly explored- and need to be assessed for threatened gingers that may also be endemics and are more vulnerable to threats. Assessment of the remaining (55%) of the tropical Asian ginger species that are still not evaluated by IUCN needs to be expedited, and narrow-endemics, monotypic genera, and genera that have not been evaluated before should be prioritized.

Most of the IUCN-listed species are directly threatened by human activities and actions that cause deforestation and forest degradation, and are not intentionally collected. Currently, there is documented evidence of natural disturbances, climate change, and restricted geographic range as threats to populations; the impacts of these and other non-anthropogenic factors need to be further studied.

About 75% of the threatened ginger species are conserved either through in situ or ex situ methods, or both, while the rest are not placed under any conservation scheme. The non-covered species include more than half of all critically endangered species. Species-rich PAs can be designated as centers of in situ protection, where the ginger habitats and populations can be regularly monitored. Most field studies that have been conducted on threatened tropical Asian gingers in PAs are on aspects of diversity, habitats, phenology and ethnobotany, while long-term studies involving population assessment and monitoring, which are essential for conservation planning and management, have been very few.

In the face of issues and concerns surrounding the effectiveness of PAs to conserve habitats and species, and the anticipated increase in the threats from both anthropogenic factors and climate change, we recommend a pragmatic conservation approach that employs a combination of both in situ and ex situ processes and methods, to promote and enhance the threatened ginger species' use and values to man; done strategically, this may improve their chances of surviving the Anthropocene extinction bottleneck.

Local communities near PAs should be engaged to actively participate in information, education and awareness, and conservation activities. TEK that promote sustainable use and management of gingers should be incorporated into PA management plans. For threatened ginger species that are adequately represented in quasi in situ and ex situ collections, promoting their use can induce their cultivation as this broadens options for food and medicine and potential income sources for the local people. Threatened ginger species with medicinal properties have high potential for commercial use and cultivation, while those with ornamental value can be mass produced, once the most suitable propagation techniques and protocols for each species have been determined. Threatened ginger species with currently no known use should be treated as crop wild relatives that could possess nutritional, medicinal, and economic potentials that are yet to be realized, and could also exhibit some useful traits that can be tapped for improving cultivated ginger varieties, especially in the context of a changing climate. Genetic studies will help pinpoint highly diverse and genetically distinct populations that should be prioritized for ex situ conservation.

Appendix (Table 30.5)

Table 30.5 IUCN threatened Zingiberaceae species in Tropical Asia, their threats, and presence in in situ and ex situ conservation sites (IUCN 2022)

Scientific name	IUCN category	Country[a]	IUCN threat types[b]	Presence[c] in PA/s	Number of ex situ sites[d]
Adelmeria dicranochila Docot & Banag[e]	Endangered	Philippines	2a, 8a	Yes	
Adelmeria isarogensis Docot & Banag[e]	Endangered	Philippines	2c, 8a, 10a	Yes	1
Adelmeria leonardoi Docot & Banag[ef]	Critically endangered	Philippines	2a, 8a	Yes	
Adelmeria paradoxa (Ridl.) Merr.[e]	Vulnerable	Philippines	1a, 1c, 2a	Yes	
Adelmeria pinetorum (Ridl.) Ridl.[e]	Vulnerable	Philippines	1a, 2a, 3a, 4a, 5b	Yes	
Adelmeria undulata Docot and Banag[e]	Endangered	Philippines	2a, 8a	Yes	

(continued)

Table 30.5 (continued)

Scientific name	IUCN category	Country[a]	IUCN threat types[b]	Presence[c] in PA/s	Number of ex situ sites[d]
Alpinia amentacea R.M. Sm.[e]	Vulnerable	Indonesia; Malaysia	2b, 5b		
Alpinia argentea (B.L. Burtt & R.M.Sm.) R.M. Sm.[e]	Vulnerable	Malaysia	2b, 5b	Yes	1
Alpinia beamanii R.M. Sm.[e]	Endangered	Indonesia; Malaysia	2b, 5b	Yes	
Alpinia congesta Elmer[ef]	Endangered	Philippines	5a, 10a	Yes	
Alpinia corneri (Holttum) R.M.Sm.	Critically endangered	Malaysia	2a, 2b, 5b		
Alpinia diversifolia (Elmer) Elmer[e]	Vulnerable	Philippines	1a, 1c, 2a, 3a, 5b	Yes	
Alpinia hansenii R.M. Sm.[e]	Endangered	Malaysia	2b, 5b	Yes	1
Alpinia havilandii K. Schum.[e]	Vulnerable	Brunei; Malaysia	2b, 5b	Yes	1
Alpinia ligulata K. Schum.[e]	Vulnerable	Brunei; Indonesia; Malaysia	2b, 5b	Yes	1
Alpinia macrostephana (Baker) Ridl.[f]	Vulnerable	Malaysia	12a	Yes	
Alpinia manii Baker[e]	Vulnerable	India	5b	Yes	1
Alpinia microlophon Ridl.[e]	Endangered	Indonesia; Malaysia	1a, 2a, 2b, 5b	Yes	
Alpinia mollissima Ridl.[f]	Vulnerable	Malaysia	12a	Yes	
Alpinia ptychanthera K. Schum.[e]	Vulnerable	Malaysia	1a, 5b	Yes	
Alpinia rafflesiana Wall. ex Baker	Vulnerable	Indonesia; Malaysia; Thailand	2a, 2b, 5b	Yes	7
Alpinia romblonensis Elmer[e]	Vulnerable	Philippines	1a, 1c	Yes	
Alpinia scabra (Blume) Náves	Endangered	Indonesia; Malaysia; Thailand	2a, 2b, 5b	Yes	1
Alpinia seimundii Ridl.[f]	Vulnerable	Malaysia	12a	Yes	
Alpinia tamacuensis R.M. Sm.	Endangered	Indonesia; Malaysia	5b	Yes	
Alpinia velutina Ridl.	Endangered	Vietnam	5b, 7b	Yes	
Alpinia vitellina (Lindl.) Ridl.	Endangered	Malaysia	1a, 2b, 5b		1

(continued)

Table 30.5 (continued)

Scientific name	IUCN category	Country[a]	IUCN threat types[b]	Presence[c] in PA/s	Number of ex situ sites[d]
Amomum billburttii Škornička. & Hlavatá	Endangered	Malaysia	1a, 2b, 5b		
Amomum curtisii (Baker) Škornička. & Hlavatá	Endangered	Malaysia; Thailand	2a, 2b, 5b		1
Amomum exertum (Scort.) Škornička. & Hlavatá	Endangered	Indonesia; Malaysia	1a, 2b, 5b	Yes	
Amomum odontocarpum D.Fang	Vulnerable	China; Laos; Vietnam	2a, 5b	Yes	
Amomum petaloideum (S.Q.Tong) T.L.Wu	Vulnerable	China	2a		4
Amomum puberulum (Ridl.) Škornička. & Hlavatá[ef]	Vulnerable	Indonesia			
Amomum riwatchii M. Sabu & Hareesh[f]	Critically endangered	India	4a, 10c	Yes	
Amomum smithiae (Y.K. Kam) Škornička. & Hlavatá	Vulnerable	Malaysia; Thailand	1a, 2b, 5b		
Amomum stenosiphon K. Schum.[e]	Endangered	Malaysia	1a, 2a, 2b, 5b		
Amomum sumatranum (Valeton) Škornička. & Hlavatá[ef]	Extinct in the wild	Indonesia	1a, 1b		
Boesenbergia aurantiaca R.M.Sm.[e]	Endangered	Malaysia	5b, 11a, 11b, 11c	Yes	6
Boesenbergia flavoalba R.M.Sm.[e]	Vulnerable	Brunei; Malaysia	2a, 11a, 11c	Yes	
Boesenbergia flavorubra R.M.Sm.[e]	Vulnerable	Brunei; Malaysia	11b, 11c	Yes	
Boesenbergia grandis R. M.Sm.[e]	Vulnerable	Malaysia; Indonesia	11d	Yes	1
Boesenbergia loerzingii (Valeton) K.Larsen ex M. F.Newman, Lhuillier & A.D.Poulsen[ef]	Critically endangered	Indonesia	1a, 2a, 5b		
Boesenbergia orbiculata R.M.Sm.[e]	Vulnerable	Brunei; Malaysia	2a, 11a	Yes	1
Boesenbergia roseopunctata (Ridl.) I. M.Turner[ef]	Vulnerable	Indonesia	5b		
Boesenbergia siphonantha (King ex	Vulnerable		1a	Yes	1

(continued)

Table 30.5 (continued)

Scientific name	IUCN category	Country[a]	IUCN threat types[b]	Presence[c] in PA/s	Number of ex situ sites[d]
Baker) M.Sabu, Prasanthk. & Škorničk.		India; Thailand; Vietnam			
Boesenbergia stenophylla R.M.Sm.[e]	Critically endangered	Indonesia; Malaysia	2a, 5a	Yes	1
Boesenbergia urceoligena A.D. Poulsen[e]	Vulnerable	Brunei	11c	Yes	
Burbidgea longilora (Ridl.) R.M.Sm.[e]	Vulnerable	Malaysia	2a, 2b, 5b		
Burbidgea pauciflora Valeton[e]	Endangered	Brunei; Indonesia; Malaysia	2a, 2b, 5b	Yes	1
Burbidgea schizocheila Hackett[e]	Vulnerable	Malaysia	2b, 5b	Yes	31
Burbidgea stenantha Ridl.[e]	Endangered	Indonesia; Malaysia	2b, 5b	Yes	9
Conamomum citrinum Ridl.	Endangered	Indonesia; Malaysia	2a, 2b, 5b	Yes	
Conamomum cylindraceum (Ridl.) Škornička. & A.D.Poulsen	Vulnerable	Indonesia; Malaysia	2a, 2b, 5b	Yes	
Conamomum cylindrostachys (K. Schum.) Škornička. & A. D.Poulsen	Vulnerable	Indonesia; Malaysia	2b, 5b	Yes	
Conamomum spiceum (Ridl.) Škornička. & A.D. Poulsen	Endangered	Malaysia	5b		
Conamomum squarrosum (Ridl.) Škornička. & A.D. Poulsen	Endangered	Malaysia	1a, 2a, 5b	Yes	
Curcuma bhatii (R.M. Sm.) Škornička. & M. Sabu[e]	Critically endangered	India	2c		1
Curcuma candida (Wall.) Techapr. & Škornička.	Vulnerable	Myanmar; Thailand	2a, 5a, 5b	Yes	2
Curcuma caulina J. Graham[e]	Endangered	India	1a, 1c, 2a, 2c, 3a, 5a, 7b		1
Curcuma colorata Valeton[e]	Endangered	Indonesia	5b		6

(continued)

Table 30.5 (continued)

Scientific name	IUCN category	Country[a]	IUCN threat types[b]	Presence[c] in PA/s	Number of ex situ sites[d]
Curcuma coriacea Mangaly & M.Sabu[e]	Endangered	India	1a, 1c, 2a, 2b, 2c, 3a, 5a, 5b, 7b	Yes	
Curcuma corniculata Škorničk.	Endangered	Laos	2a		
Curcuma leonidii Škorničk. & Luu[f]	Critically endangered	Vietnam	1a, 2a	Yes	
Curcuma newmanii Škorničk.[f]	Critically endangered	Vietnam	1c, 2a		1
Curcuma prasina Škorničk.	Endangered	Thailand	6a	Yes	1
Curcuma pseudomontana J.Graham	Vulnerable	India	2a, 3a, 5a, 5b, 7a, 8a		
Curcuma pygmaea Škorničk. & Šída f.[f]	Critically endangered	Vietnam	2b		
Curcuma rhabdota Sirirugsa & M.F.Newman	Vulnerable	Cambodia; Laos; Vietnam	2a, 5a	Yes	4
Curcuma sahuynhensis Škorničk. & N.S.Lý	Endangered	Vietnam	1a, 5b	Yes	
Curcuma sumatrana Miq.[e]	Vulnerable	Indonesia	1c, 4a, 5a	Yes	2
Curcuma supraneeana (W.J.Kress & K.Larsen) Škorničk.[f]	Critically endangered	Thailand	2a, 3a, 5a		1
Curcuma vamana M. Sabu & Mangaly[e]	Critically endangered	India	1a, 2c, 4a		1
Curcuma vitellina Škorničk. & H.Đ.Trần	Endangered	Vietnam	2a,6a, 7b	Yes	1
Cyphostigma pulchellum (Thwaites) Benth.[e]	Endangered	Sri Lanka	1a, 2a, 5b	Yes	
Epiamomum angustipetalum (S.Sakai & Nagam.) A.D.Poulsen & Škorničk.[e]	Endangered	Brunei; Malaysia	2b, 5b	Yes	
Epiamomum borneense (K.Schum.) A.D.Poulsen & Škorničk.[e]	Endangered	Brunei; Indonesia; Malaysia	2b, 5b	Yes	
Epiamomum epiphyticum (R.M.Sm.) A.D.Poulsen & Škorničk.[e]	Vulnerable	Malaysia	2a, 2b, 5b	Yes	

(continued)

Table 30.5 (continued)

Scientific name	IUCN category	Country[a]	IUCN threat types[b]	Presence[c] in PA/s	Number of ex situ sites[d]
Epiamomum hansenii (R.M.Sm.) A.D.Poulsen & Škorničk.[e]	Endangered	Malaysia	1a, 2a, 2b, 5b	Yes	
Epiamomum pungens (R.M.Sm.) A.D.Poulsen & Škorničk.[e]	Endangered	Indonesia; Malaysia	1a, 2a, 2b, 5c	Yes	
Etlingera amomoides A. D.Poulsen & Mood[e]	Endangered	Malaysia	2b, 5b		
Etlingera aulocheilos A. D.Poulsen[e]	Endangered	Indonesia	2a, 5b		1
Etlingera aurantia A.D. Poulsen[e]	Vulnerable	Brunei; Indonesia; Malaysia	2b, 5b	Yes	
Etlingera baculutea A.D. Poulsen & Ibrahim[e]	Vulnerable	Indonesia; Malaysia	2b, 5b	Yes	
Etlingera balikpapanensis A.D. Poulsen[e]	Endangered	Indonesia	2b, 5b, 7a	Yes	
Etlingera baramensis S. Sakai & Nagam.[e]	Vulnerable	Indonesia; Malaysia	2b, 5b	Yes	
Etlingera barioensis A.D. Poulsen[e]	Endangered	Malaysia	2b, 5b	Yes	
Etlingera biloba A.D. Poulsen[e]	Endangered	Indonesia	2a, 5b	Yes	
Etlingera borealis A.D. Poulsen[e]	Endangered	Indonesia	2a, 5b		
Etlingera burttii A.D. Poulsen[e]	Endangered	Malaysia	2b, 5b		3
Etlingera caudata A.D. Poulsen[ef]	Critically endangered	Indonesia	2b, 5b	Yes	
Etlingera chlorodonta A. D.Poulsen[ef]	Critically endangered	Indonesia	2a, 5b		
Etlingera corneri Mood & Ibrahim	Vulnerable	Malaysia; Thailand	2a, 5b	Yes	6
Etlingera corrugata A.D. Poulsen & Mood[e]	Endangered	Malaysia	2b, 5b	Yes	1
Etlingera cylindrica A.D. Poulsen[e]	Endangered	Indonesia	2a, 5b	Yes	
Etlingera dictyota A.D. Poulsen & A.L.Lamb[e]	Vulnerable	Indonesia; Malaysia	2a, 5b		
Etlingera doliiformis A. D.Poulsen[ef]	Critically endangered	Indonesia	2a, 5b, 7a		
Etlingera eburnea A.D. Poulsen[e]	Endangered	Indonesia	2a, 5b		

(continued)

Table 30.5 (continued)

Scientific name	IUCN category	Country[a]	IUCN threat types[b]	Presence[c] in PA/s	Number of ex situ sites[d]
Etlingera echinulata A.D. Poulsen[e]	Endangered	Indonesia	2a, 5b		
Etlingera facifera A.D. Poulsen[e]	Vulnerable	Indonesia; Malaysia	5b	Yes	
Etlingera fenzlii (Kurz) Škorničk. & M.Sabu[e]	Vulnerable	India	10b	Yes	1
Etlingera flavovirens A. D.Poulsen[ef]	Critically endangered	Indonesia	2a, 5b		1
Etlingera gracilis (Valeton) R.M.Sm.[ef]	Critically endangered	Indonesia	1a, 2a		
Etlingera grandiflora (Valeton) R.M.Sm.[e]	Endangered	Papua New Guinea	2a, 2b, 5b		1
Etlingera grandiligulata (K.Schum.) R.M.Sm.[e]	Vulnerable	Indonesia		Yes	
Etlingera heliconiifolia (K.Schum.) A.D.Poulsen[e]	Vulnerable	Indonesia	2a, 5b	Yes	1
Etlingera hyalina A.D. Poulsen[e]	Endangered	Indonesia	2a, 5b		1
Etlingera insolita A.D. Poulsen[ef]	Critically endangered	Indonesia	2a, 5b		
Etlingera kenyalang A.D. Poulsen & H.Chr.[e]	Endangered	Malaysia	5b	Yes	
Etlingera lagarophylla A. D.Poulsen[ef]	Vulnerable	Malaysia	2a, 4a, 12a		
Etlingera loerzingii (Valeton) R.M.Sm.[e]	Vulnerable	Indonesia	1a, 2a, 5b	Yes	1
Etlingera longipetiolata (B.L.Burtt & R.M.Sm.) R.M.Sm.[e]	Vulnerable	Malaysia	2a, 5b		
Etlingera mucida A.D. Poulsen[ef]	Critically endangered	Indonesia	2a, 5b	Yes	
Etlingera mucronata A. D.Poulsen[e]	Endangered	Indonesia	2a, 5b		
Etlingera muriformis A. D.Poulsen[ef]	Vulnerable	Indonesia	5b		
Etlingera newmanii S. Sakai & Nagam.[ef]	Critically endangered	Malaysia	2a		
Etlingera orophila A.D. Poulsen[e]	Endangered	Indonesia		Yes	
Etlingera palangkensis A. Takano & Nagam.[ef]	Critically endangered	Indonesia	2a, 5b		
Etlingera peekelii (Valeton) R.M.Sm.[e]	Vulnerable	Papua New Guinea	2a, 5b		

(continued)

Table 30.5 (continued)

Scientific name	IUCN category	Country[a]	IUCN threat types[b]	Presence[c] in PA/s	Number of ex situ sites[d]
Etlingera penicillata (K. Schum.) A.D.Poulsen[e]	Endangered	Indonesia	2a, 5b		
Etlingera polycarpa ssp. ligulata A.D.Poulsen[e]	Vulnerable	Indonesia	2a, 5b	Yes	1
Etlingera serrata A.D. Poulsen[ef]	Critically endangered	Indonesia	2a, 5b		
Etlingera solaris (Blume) R.M.Sm.	Vulnerable	Indonesia	2a, 5b	Yes	1
Etlingera spinulosa A.D. Poulsen[e]	Endangered	Indonesia	2a, 5b	Yes	
Etlingera sublimata A.D. Poulsen[e]	Endangered	Indonesia	2a, 5b, 5c		
Etlingera subulicalyx (Valeton) R.M.Sm.[e]	Endangered	Indonesia	1a, 2a, 5b	Yes	1
Etlingera tubilabrum A. D.Poulsen[e]	Vulnerable	Indonesia	2a, 5b	Yes	1
Etlingera urophylla A.D. Poulsen[ef]	Critically endangered	Indonesia	2a, 5b		
Etlingera venusta (Ridl.) R.M.Sm.	Vulnerable	Malaysia	5a	Yes	10
Etlingera vestita (Valeton) R.M.Sm.[e]	Vulnerable	Papua New Guinea	2a, 5b		
Etlingera xanthantha A. D.Poulsen[ef]	Critically endangered	Indonesia	2a, 5b		
Etlingera yessiae A.D. Poulsen[ef]	Vulnerable	Indonesia	2a, 5b		
Geocharis fusiformis (Ridl.) R.M.Sm.	Endangered	Malaysia; Philippines	2a, 2b, 5b		1
Geocharis radicalis (Valeton) B.L.Burtt & R. M.Sm.[e]	Endangered	Indonesia	1a, 2a, 5b	Yes	
Geocharis rubra Ridl.	Endangered	Indonesia; Malaysia	1a, 2a, 2b, 5b		
Geostachys belumensis C. K.Lim & K.H.Lau	Vulnerable	Malaysia	5b	Yes	
Geostachys chayanii Mayoe	Endangered	Thailand	5b	Yes	
Geostachys decurvata (Baker) Ridl.	Critically endangered	Malaysia; Thailand	1c, 5b	Yes	2
Geostachys erectifrons K. H.Lau, C.K.Lim & Mat-Salleh[f]	Vulnerable	Malaysia	6a	Yes	

(continued)

Table 30.5 (continued)

Scientific name	IUCN category	Country[a]	IUCN threat types[b]	Presence[c] in PA/s	Number of ex situ sites[d]
Geostachys leucantha B. C.Stone[f]	Critically endangered	Malaysia	5b		
Geostachys maliauensis C.K.Lim & K.H.Lau[e]	Endangered	Malaysia	5b	Yes	
Geostachys megaphylla Holttum[f]	Critically endangered	Malaysia	2a,6a		
Geostachys penangensis Ridl.	Endangered	Malaysia; Indonesia; Thailand	1c	Yes	
Geostachys primulina Ridl.	Critically endangered	Malaysia	2c, 5b		
Geostachys rupestris Ridl.[f]	Critically endangered	Malaysia	1c	Yes	
Geostachys secunda (Baker) Ridl.[f]	Vulnerable	Malaysia	1c		
Geostachys sericea (Ridl.) Holttum[f]	Vulnerable	Malaysia	1c, 5a, 6a		
Geostachys smitinandii K.Larsen	Vulnerable	Thailand	1c, 5a, 5b	Yes	
Geostachys sumatrana Valeton[e]	Endangered	Indonesia	5b	Yes	
Geostachys tahanensis Holttum[f]	Vulnerable	Malaysia	1c	Yes	
Geostachys taipingensis Holttum[f]	Critically endangered	Malaysia	1c		
Globba acehensis A. Takano & H.Okada[e]	Endangered	Indonesia	1a, 1c, 5b	Yes	
Globba albobracteata N. E.Br.[e]	Vulnerable	Indonesia	1a, 1c, 2a	Yes	2
Globba andersonii C.B. Clarke ex Baker[e]	Endangered	India; Bhutan	4a, 5b, 10c	Yes	
Globba bracteolata Wall. ex Baker	Endangered	Myanmar; Thailand; India	1c		
Globba colpicola K. Schum.	Endangered	Thailand	1a		1
Globba flagellaris K. Larsen	Vulnerable	Thailand	2a	Yes	1
Globba glandulosa Ridl.[ef]	Vulnerable	Malaysia	2a, 5b	Yes	
Globba holttumii S.N. Lim	Endangered	Malaysia	1a, 2b, 5b	Yes	

(continued)

Table 30.5 (continued)

Scientific name	IUCN category	Country[a]	IUCN threat types[b]	Presence[c] in PA/s	Number of ex situ sites[d]
Globba laeta K.Larsen	Endangered	Thailand	2a, 4a, 5a, 5b, 7a, 8a	Yes	2
Globba maculata Blume	Critically endangered	Indonesia; Malaysia	1a, 2a, 5b	Yes	
Globba magnibracteata Y.Y.Sam[f]	Critically endangered	Malaysia	4a, 5b	Yes	
Globba multifolia A. Takano & H.Okada[e]	Endangered	Indonesia	1a, 2a, 5b	Yes	
Globba praecox Chokthaweep., K.J. Williams & Paisooks.	Vulnerable	Thailand; Myanmar	3a	Yes	
Globba pumila Ridl.[e]	Endangered	Malaysia; Indonesia	2a, 2b, 5b		
Globba radicalis Roxb.	Endangered	Bangladesh; India; Myanmar	2a, 2c, 5b		2
Globba ranongensis Picheans. & Tiyaw.[f]	Endangered	Thailand			
Globba spathulata Roxb.	Vulnerable	Bangladesh; India; Myanmar	2a, 4a, 5b, 7a, 7c	Yes	
Globba talangensis A. Takano & H.Okada[e]	Vulnerable	Indonesia	2a	Yes	
Globba tembatensis Y.Y. Sam & Julius[f]	Critically endangered	Malaysia	5b, 7b	Yes	
Globba unifolia Ridl.	Vulnerable	Malaysia; Thailand	1a, 2a, 2b, 5b	Yes	5
Globba variabilis Ridl.	Vulnerable	Indonesia; Malaysia; Singapore; Thailand	2a, 2b, 5b	Yes	4
Haniffia albiflora K. Larsen & Mood	Vulnerable	Thailand		Yes	5
Haniffia cyanescens (Ridl.) Holttum	Endangered	Malaysia	2a, 2b, 5b	Yes	
Hedychium collinum Ridl.	Vulnerable	Malaysia	5b		1
Hedychium hasseltii Blume[ef]	Critically endangered	Indonesia	2a, 5b		20
Hedychium hirsutissimum Holttum	Vulnerable	Malaysia		Yes	
Hedychium horsfieldii Wall[e]	Vulnerable	Indonesia	1a, 2a, 5b		2

(continued)

Table 30.5 (continued)

Scientific name	IUCN category	Country[a]	IUCN threat types[b]	Presence[c] in PA/s	Number of ex situ sites[d]
Hedychium lineare R.M. Sm.[e]	Endangered	Indonesia; Malaysia	2a, 2b, 5b	Yes	
Hedychium macrorrhizum Ridl.[f]	Vulnerable	Malaysia			
Hedychium muluense R. M.Sm.[e]	Endangered	Indonesia; Malaysia	2a, 2b, 5b	Yes	24
Hedychium roxburghii Blume[e]	Endangered	Indonesia	2a, 5b	Yes	9
Hedychium simile Blume[ef]	Critically endangered	Indonesia	2a, 5b		
Hellenia borneensis (A.D.Poulsen) Govaerts[e]	Vulnerable	Indonesia; Malaysia	2a, 2b, 5b, 7b	Yes	1
Hornstedtia deliana Valeton[e]	Vulnerable	Indonesia	2a, 5b		
Hornstedtia mollis (Blume) Valeton[e]	Endangered	Indonesia	5b	Yes	1
Hornstedtia pusilla Ridl.[f]	Critically endangered	Malaysia	2b		
Hornstedtia reticosa Valeton[e]	Vulnerable	Indonesia	5b	Yes	
Hornstedtia rubra (Blume) Valeton	Endangered	Indonesia	2a, 2b, 5b	Yes	1
Hornstedtia striolata Ridl.	Vulnerable	Malaysia	5b		
Larsenianthus arunachalensis M.Sabu, Sanoj & Rajesh Kumar[ef]	Critically endangered	India	10c	Yes	
Meistera calcarata (Lamxay & M.F. Newman) Škorničk. & M. F.Newman	Vulnerable	Laos		Yes	1
Meistera celsa (Lamxay & M.F.Newman) Škorničk. & M.F. Newman	Endangered	Laos; Vietnam	2a, 4a, 5b	Yes	
Meistera cerasina (Ridl.) Škorničk. & M.F. Newman[e]	Vulnerable	Brunei; Indonesia; Malaysia	2a, 2b, 5b	Yes	1
Meistera gyrolophos (R.M.Sm.) Škorničk. & M.F.Newman	Vulnerable	Indonesia; Malaysia	2a, 2b, 5b	Yes	
Meistera lappacea (Ridl.) Škorničk. & M.F. Newman	Vulnerable	Indonesia; Malaysia; Thailand	2a, 2b, 5b	Yes	1

(continued)

Table 30.5 (continued)

Scientific name	IUCN category	Country[a]	IUCN threat types[b]	Presence[c] in PA/s	Number of ex situ sites[d]
Meistera ochrea (Ridl.) Škorničk. & M.F. Newman	Endangered	Indonesia; Malaysia	2a, 2b, 5b	Yes	1
Meistera oligantha (K. Schum.) Škorničk. & M. F.Newman[e]	Vulnerable	Malaysia	2a, 2b, 5b	Yes	
Meistera sceletescens (R.M.Sm.) Škorničk. & M.F.Newman[e]	Vulnerable	Brunei; Malaysia	1a, 2a, 2b, 5b	Yes	
Meistera stephanocolea (Lamxay & M.F. Newman) Škorničk. & M. F.Newman	Endangered	Laos		Yes	1
Myxochlamys amphiloxa R.J.Searle[e]	Endangered	Indonesia	5b		
Nanochilus palembanicus (Miq.) K.Schum.[ef]	Critically endangered	Indonesia	2b, 5b		
Newmania gracilis Škorničk., Q.B.Nguyen & H.Đ.Trần	Endangered	Vietnam	2a, 4a, 5b, 7b		
Newmania orthostachys N.S.Lý & Škorničk	Endangered	Vietnam	2a, 2b, 5b	Yes	
Newmania serpens N.S. Lý & Škorničk.[f]	Critically endangered	Vietnam	2a, 2b, 5a		
Newmania sessilanthera Luu & Škorničk.	Endangered	Vietnam	2b, 5b		
Newmania sontraensis H.Đ.Trần, Luu & Škorničk.	Endangered	Vietnam	1c, 2a	Yes	
Parakaempferia synantha A.S.Rao & D.M.Verma[eff]	Critically endangered	India	5b	Yes	
Plagiostachys bracteolata R.M.Sm.[e]	Vulnerable	Malaysia	2a, 2b, 5b	Yes	
Plagiostachys brevicalcarata Julius & A.Takano[e]	Endangered	Malaysia	2a, 2b, 5b	Yes	
Plagiostachys crocydocalyx (K.Schum.) B.L.Burtt & R.M.Sm.[e]	Endangered	Brunei; Indonesia; Malaysia	2b, 5b	Yes	
Plagiostachys lasiophylla Gobilik & A.L.Lamb[e]	Vulnerable	Malaysia	2a, 2b, 5b	Yes	
Plagiostachys megacarpa Julius & A.Takano[e]	Endangered	Malaysia	2a, 2b, 5b	Yes	1

(continued)

Table 30.5 (continued)

Scientific name	IUCN category	Country[a]	IUCN threat types[b]	Presence[c] in PA/s	Number of ex situ sites[d]
Plagiostachys nicobarica M.Sabu, Sanoj & Prasanthk.[ef]	Endangered	India	10b	Yes	1
Pleuranthodium comptum (K.Schum.) R.M.Sm.[e]	Endangered	Papua New Guinea	5b		
Pleuranthodium papilionaceum (K. Schum.) R.M.Sm.[e]	Vulnerable	Papua New Guinea	2a, 8a		
Pleuranthodium pedicellatum (Valeton) R. M.Sm.	Endangered	Indonesia; Papua New Guinea	5b		
Pleuranthodium peekelii (Valeton) R.M.Sm.	Endangered	Papua New Guinea; Solomon Islands	5b, 10a, 10b, 10c		1
Scaphochlamys abdullahii Y.Y.Sam & Saw	Endangered	Malaysia	5b	Yes	
Scaphochlamys argentea R.M.Sm.[e]	Endangered	Malaysia	1a, 2a, 2b, 5b	Yes	
Scaphochlamys atroviridis Holttum[f]	Critically endangered	Malaysia	1a, 2a, 2b, 5b		
Scaphochlamys baukensis Y.Y.Sam	Endangered	Malaysia	1a, 4a, 6a	Yes	
Scaphochlamys biru Meekiong[ef]	Vulnerable	Malaysia	2a, 2b, 5b	Yes	
Scaphochlamys burkillii Holttum	Endangered	Malaysia; Indonesia	1a, 2a, 2b, 5b	Yes	
Scaphochlamys calcicola A.D.Poulsen & R.J. Searle[e]	Endangered	Malaysia	2a, 2b, 5b		
Scaphochlamys concinna (Baker) Holttum	Endangered	Malaysia	1a, 1b, 2b	Yes	
Scaphochlamys cordata Y.Y.Sam & Saw[f]	Critically endangered	Malaysia	5b	Yes	
Scaphochlamys gracilipes (K.Schum.) S. Sakai & Nagam.[e]	Vulnerable	Malaysia	1c	Yes	
Scaphochlamys grandis Holttum	Vulnerable	Malaysia	2a, 5b	Yes	
Scaphochlamys lanceolata (Ridl.) Holttum	Endangered	Malaysia	2a, 2b	Yes	2

(continued)

Table 30.5 (continued)

Scientific name	IUCN category	Country[a]	IUCN threat types[b]	Presence[c] in PA/s	Number of ex situ sites[d]
Scaphochlamys laxa Y.Y. Sam & Saw	Vulnerable	Malaysia	5b	Yes	
Scaphochlamys longifolia (Ridl.) Holttum	Vulnerable	Malaysia	4a, 5b	Yes	
Scaphochlamys perakensis (Ridl.) Holttum	Endangered	Indonesia; Malaysia	2a	Yes	1
Scaphochlamys polyphylla (K.Schum.) B. L.Burtt & R.M.Sm.[e]	Vulnerable	Indonesia; Malaysia	1a, 2a, 2b, 5b	Yes	
Scaphochlamys tenuis Holttum	Endangered	Malaysia	5b	Yes	
Siamanthus siliquosus K. Larsen & Mood[f]	Vulnerable	Thailand	5b	Yes	5
Siliquamomum alcicorne Škorničk. & H.Đ.Trần	Endangered	Vietnam	2a, 5b, 7b		
Siliquamomum oreodoxa N.S.Lý & Škorničk.	Endangered	Vietnam	2a, 4a, 5b	Yes	1
Siliquamomum phamhoangii Luu & H.Đ. Trần	Endangered	Vietnam		Yes	
Siliquamomum tonkinense Baill.	Vulnerable	China; Vietnam	2a, 5a, 5b	Yes	3
Sulettaria lambirensis (R.M.Sm.) A.D.Poulsen & M.F.Newman[e]	Vulnerable	Malaysia	2a, 5a, 5b, 7a, 11b	Yes	
Sulettaria longituba (Ridl.) A.D.Poulsen & Mathisen	Vulnerable	Brunei; Indonesia; Malaysia	2b, 5b	Yes	
Sulettaria multiflora (Ridl.) A.D.Poulsen & Ardiyani	Critically endangered	Indonesia; Malaysia	2a, 2b	Yes	
Sundamomum borealiborneense (I.M. Turner) A.D.Poulsen & M.F.Newman[e]	Endangered	Indonesia; Malaysia	2a, 2b, 5b	Yes	
Sundamomum flavoalbum (R.M.Sm.) A.D.Poulsen & M.F.Newman[e]	Endangered	Indonesia; Malaysia	5b	Yes	
Sundamomum longipedunculatum (R.M.	Endangered	Malaysia	2a, 2b, 5b	Yes	

(continued)

Table 30.5 (continued)

Scientific name	IUCN category	Country[a]	IUCN threat types[b]	Presence[c] in PA/s	Number of ex situ sites[d]
Sm.) A.D.Poulsen & M.F. Newman[e]					
Sundamomum luteum (R.M.Sm.) A.D.Poulsen & M.F.Newman[e]	Endangered	Malaysia	5b	Yes	
Sundamomum oligophyllum (A.J.Droop) A.D.Poulsen & M.F. Newman[ef]	Endangered	Indonesia	2a		
Sundamomum pseudofoetens (Valeton) A.D.Poulsen & M.F. Newman	Endangered	Indonesia	5b	Yes	
Sundamomum somniculosum (S.Sakai & Nagam.) A.D.Poulsen & M.F.Newman[e]	Endangered	Malaysia	2a, 2b, 5b	Yes	
Vanoverberghia sepulchrei Merr.[e]	Vulnerable	Philippines	1a, 2a	Yes	4
Vanoverberghia vanoverberghii (Merr.) Funak. & Docot[e]	Vulnerable	Philippines	2a	Yes	
Wurfbainia bicorniculata (K.Schum.) Škorničk. & A.D.Poulsen[ef]	Vulnerable	Indonesia	2a, 2b, 5b		
Wurfbainia blumeana (Valeton) Škorničk. & A. D.Poulsen[e]	Endangered	Indonesia	2a, 5b		
Wurfbainia staminidiva (Gobilik, A.L.Lamb & A. D.Poulsen) Škorničk. & A.D.Poulsen[e]	Endangered	Indonesia; Malaysia	1a, 2a, 2b, 5b		
Zingiber albiflorum R.M. Sm.[e]	Critically endangered	Malaysia	2a, 2b, 5b	Yes	1
Zingiber album Nurainas[e]	Endangered	Indonesia	1c, 2a	Yes	
Zingiber argenteum Mood & Theilade[ef]	Critically endangered	Malaysia	2a, 2b, 5b		
Zingiber arunachalensis A.Joe, T.Jayakr., Hareesh & M.Sabu[ef]	Critically endangered	India	4a, 10c		
Zingiber atroporphyreum Škorničk. & Q.B.Nguyen	Endangered	Vietnam		Yes	1

(continued)

Table 30.5 (continued)

Scientific name	IUCN category	Country[a]	IUCN threat types[b]	Presence[c] in PA/s	Number of ex situ sites[d]
Zingiber aurantiacum (Holttum) Theilade	Vulnerable	Malaysia	1a, 2a, 2b, 5b	Yes	1
Zingiber cardiocheilum Škorničk. & Q.B.Nguyen	Vulnerable	Vietnam		Yes	1
Zingiber chlorobracteatum Mood & Theilade[e]	Vulnerable	Malaysia	1a, 2a, 2b, 5b		
Zingiber chrysostachys Ridl.	Endangered	Malaysia; Thailand	1a, 2a, 2b, 5b	Yes	
Zingiber collinsii Mood & Theilade	Vulnerable	Cambodia; Vietnam	2a, 5b, 6a, 7b	Yes	8
Zingiber eborinum Mood & Theilade[ef]	Critically endangered	Malaysia	2a, 2b, 5b		2
Zingiber engganoense Ardiyani[ef]	Critically endangered	Indonesia	5b		
Zingiber flagelliforme Mood & Theilade[e]	Endangered	Malaysia	2a, 2b, 5b	Yes	3
Zingiber flammeum Theilade & Mood[e]	Endangered	Malaysia	2a, 2b, 5b		2
Zingiber fraseri Theilade	Endangered	Malaysia; Thailand	1a, 2a, 2b, 5b	Yes	1
Zingiber incomptum B.L. Burtt & R.M.Sm.[e]	Vulnerable	Malaysia	1a, 2a, 2b, 5b	Yes	1
Zingiber jiewhoei Škorničk.[f]	Vulnerable	Laos	6a		2
Zingiber lambii Mood & Theilade[ef]	Critically endangered	Malaysia	5b	Yes	
Zingiber lecongkietii Škorničk., H.Đ.Tran & Sída f.	Endangered	Vietnam	2a, 5b		1
Zingiber leptostachyum Valeton[e]	Vulnerable	Indonesia; Malaysia	2a, 2b, 5b	Yes	1
Zingiber loerzingii Valeton[ef]	Vulnerable	Indonesia	1c	Yes	
Zingiber longibracteatum Theilade[e]	Vulnerable	Malaysia; Thailand	1a, 2a, 5b	Yes	4
Zingiber macradenia K. Schum.[e]	Vulnerable	Indonesia	1a, 2a		6
Zingiber macrocephalum (Zoll.) K.Schum.[ef]	Critically endangered	Indonesia	1a, 2a, 5b		
Zingiber martini R.M. Sm.[e]	Endangered	Malaysia	2a, 2b, 5b		2

(continued)

Table 30.5 (continued)

Scientific name	IUCN category	Country[a]	IUCN threat types[b]	Presence[c] in PA/s	Number of ex situ sites[d]
Zingiber mawangense Noor Ain & Meekiong[ef]	Critically endangered	Malaysia	1a, 2a, 5b		
Zingiber mellis Škorničk., H.Đ.Tran & Sída f.	Endangered	Vietnam	2a, 5b, 6a		1
Zingiber microcheilum Škorničk., H.Đ.Tran & Sída f.	Endangered	Vietnam	7c	Yes	1
Zingiber monophyllum Gagnep.	Endangered	Vietnam	2a, 5a	Yes	
Zingiber niveum Mood & Theilade	Endangered	Laos	5a	Yes	5
Zingiber odoriferum Blume[e]	Critically endangered	Indonesia	5b		3
Zingiber pachysiphon B. L.Burtt & R.M.Sm.[e]	Vulnerable	Indonesia; Malaysia	2a, 2b, 5b	Yes	
Zingiber petiolatum (Holttum) Theilade	Vulnerable	Malaysia; Thailand	1a, 2a, 2b, 5b	Yes	
Zingiber phillippsiae Mood & Theilade[e]	Vulnerable	Malaysia	4a	Yes	1
Zingiber porphyrosphaerum K. Schum.[e]	Endangered	Malaysia	1a, 2b, 5b	Yes	
Zingiber pseudopungens R.M.Sm.[e]	Vulnerable	Brunei; Indonesia; Malaysia	2a, 2b, 5b	Yes	3
Zingiber raja C.K.Lim & Kharuk.	Endangered	Malaysia; Thailand	1a, 2a, 5b	Yes	
Zingiber singapurense Škorničk.[e]	Critically endangered	Singapore	1a	Yes	1
Zingiber sulphureum Burkill ex Theilade	Endangered	Malaysia	2b, 5b	Yes	3
Zingiber ultralimitale Ardiyani & A.D. Poulsen[ef]	Vulnerable	Indonesia		Yes	1
Zingiber velutinum Mood & Theilade[ef]	Critically endangered	Malaysia	2b		
Zingiber vinosum Mood & Theilade[e]	Endangered	Malaysia	1a, 2a, 2b	Yes	1

(continued)

Table 30.5 (continued)

Scientific name	IUCN category	Country[a]	IUCN threat types[b]	Presence[c] in PA/s	Number of ex situ sites[d]
Zingiber wrayi Ridl.	Endangered	Malaysia; Thailand	1a, 2a, 5b	Yes	6

[a] Cross-referenced with data from GBIF (GBIF.org 2022) and ZRC (Newman et al. 2005–present)
[b] Threat types—(1a) housing and urban areas, (1b) commercial and industrial areas, (1c) tourism and recreation areas; (2a) annual and perennial non-timber crops, (2c) livestock farming and ranching, (3a) mining and quarrying, (4a) roads and railroads, (5a) gathering terrestrial plants, (5b) logging and wood harvesting, (5c) fishing and harvesting aquatic resources, (6a) recreational activities, (7a) fire and fire suppression, (7b) dams and water management/use, (7c) other ecosystem modifications, (8a) invasive non-native/alien species/disease, (10a) volcanoes, (10b) earthquake/tsunamis, (10c) avalanches/landslides, (11a) habitat shifting and alteration, (11b) droughts, (11c) temperature extremes, (11d) other impacts, (12a) other threats (IUCN 2022)
[c] With additional data from ZRC (Newman et al. 2005–present)
[d] Updated data from PlantSearch (BGCI 2022b) and GBIF (GBIF.org 2022)
[e] Endemic to one island
[f] Recorded in a single locality

References

Acero KML, Amoroso VB, Lumista HP, Mendez NP, Acma FM (2019) Species composition and distribution of Zingiberaceae in Mt. Hamiguitan Expansion Site, Davao Oriental, Philippines. J Trop Biol Conserv 16(October):121–136

Acma FM, Mendez NP, Amoroso VB (2020) Zingiberaceae of Marilog Forest Reserve, Southern Philippines: its richness and endemism. Asian J Biodiv 11(1):73–87. https://doi.org/10.7828/ajob.v11i1.1388

Afzal A, Oriqat G, Akram Khan M, Jose J, Afzal M (2013) Chemistry and biochemistry of terpenoids from curcuma and related species. Nature 3(1):1–55. https://doi.org/10.1080/22311866.2013.782757

Aravind S, Nasreen K (2022) Antifungal activity of Etlingera fenzlii against its Rhizospheric fungal isolates. January, 47–52. https://doi.org/10.35629/7781-07014852

BGCI (2022a) ThreatSearch online database. Botanic Gardens Conservation International, Richmond. Available at www.bgci.org/threat_search.php. Accessed 28 July 2022

BGCI (2022b) PlantSearch online database. Botanic Gardens Conservation International, Richmond. Available at https://tools.bgci.org/plant_search.php. Accessed on 15 July 2022

Bliege Bird R, Nimmo D (2018) Restore the lost ecological functions of people. Nat Ecol Evol 2(7): 1050–1052. https://doi.org/10.1038/s41559-018-0576-5

Cahyaningsih R, Magos Brehm J, Maxted N (2021a) Setting the priority medicinal plants for conservation in Indonesia. In: Genetic resources and crop evolution, vol 68. Springer, Cham. https://doi.org/10.1007/s10722-021-01115-6

Cahyaningsih R, Magos Brehm J, Maxted N (2021b) Gap analysis of Indonesian priority medicinal plant species as part of their conservation planning. Glob Ecol Conserv 26:e01459. https://doi.org/10.1016/j.gecco.2021.e01459

Cazzolla Gatti R, Liang J, Velichevskaya A, Zhou M (2019) Sustainable palm oil may not be so sustainable. Sci Total Environ 652:48–51. https://doi.org/10.1016/j.scitotenv.2018.10.222

Chumroenphat T, Saensouk S, Saensouk P (2021) Chemical composition and antioxidant activity of three species of cornukaempferia in Thailand. Biodiversitas 22(9):4036–4044. https://doi.org/10.13057/biodiv/d220952

Crane E (2000) The world history of beekeeping and honey hunting. Taylor & Francis, London

De Bruyn M, Stelbrink B, Morley RJ, Hall R, Carvalho GR, Cannon CH, Van Den Bergh G, Meijaard E, Metcalfe I, Boitani L, Maiorano L, Shoup R, Von Rintelen T (2014) Borneo and Indochina are major evolutionary hotspots for Southeast Asian biodiversity. Syst Biol 63(6): 879–901. https://doi.org/10.1093/sysbio/syu047

Dhandapani S (2015) Biodiversity loss associated with oil palm plantations in Malaysia: Serving the need versus Saving the nature.

Docot RVA (2019a) Alpinia congesta. The IUCN red list of threatened species 2019: e. T117263632A124279648. https://doi.org/10.2305/IUCN.UK.2019-2.RLTS.T117263632A124279648.en. Accessed on 15 June 2022

Docot RVA (2019b) Alpinia vanoverberghii. The IUCN red list of threatened species 2019: e. T117271173A124280275. https://doi.org/10.2305/IUCN.UK.2019-2.RLTS.T117271173A124280275.en. Accessed 25 June 2022

Docot RVA (2020a) Adelmeria dicranochila. The IUCN red list of threatened species 2020: e. T132925112A132925153. https://doi.org/10.2305/IUCN.UK.2020-1.RLTS.T132925112A132925153.en. Accessed 15 June 2022

Docot RVA (2020b) Adelmeria isarogensis. The IUCN red list of threatened species 2020: e. T132925190A132925256. https://doi.org/10.2305/IUCN.UK.2020-1.RLTS.T132925190A132925256.en. Accessed 13 June 2022

Docot RVA, Gomez MGM, Aquino RBC, Barretto III AB, Orihara SM, Mintu CB (2016) Morphoanatomy and vegetative propagation of the Philippine endemic Leptosolena haenkei C.Presl and Vanoverberghia sepulchrei Merr. (Zingiberaceae) by rhizome cutting using NAA and Kinetin. Thai J Botany 8(1):111–126

Docot RVA, Banag CI, Tandang DN, Funakoshi H, Poulsen AD (2019) Recircumscription and revision of the genus Vanoverberghia (Zingiberaceae). Blumea 64(2):140–157. https://doi.org/10.3767/blumea.2019.64.02.05

Duckworth JW, Batters G, Belant JL, Bennett EL, Brunner J, Burton J, Challender DWS, Cowling V, Duplaix N, Harris JD, Hedges S, Long B, Mahood SP, McGowan PJK, McShea WJ, Oliver WLR, Perkin S, Rawson BM, Shepherd CR, Wirth R (2012) Why South-east Asia should be the world's priority for averting imminent species extinctions, and a call to join a developing cross-institutional programme to tackle this urgent issue. Dens Sapiens 5(2):1327

Eisai (1995) Medicinal herb index in Indonesia, 2nd edn. PT, Eisai

Elmer ADE (1915) Notes and description of Zingiberaceae. Leaflets Philipp Bot 8(116):2885–2920

GBIF.org (2022) GBIF occurrence download. https://doi.org/10.15468/dl.z2frfb. Accessed 17 August 2022

Geist HJ, Lambin EF (2002) Proximate causes and underlying driving forces of tropical deforestation. Bioscience 52(2):143. https://doi.org/10.1641/0006-3568(2002)052[0143:PCAUDF]2.0.CO;2

Geldmann J, Barnes M, Coad L, Craigie ID, Hockings M, Burgess ND (2013) Effectiveness of terrestrial protected areas in reducing habitat loss and population declines. Biol Conserv 161: 230–238. https://doi.org/10.1016/j.biocon.2013.02.018

Ghosh S, Rangan L (2013) Alpinia: the gold mine of future therapeutics. 3 Biotech 3(3):173–185. https://doi.org/10.1007/s13205-012-0089-x

Heyne K (1987) Tumbuhan Berguna Indonesia Jilid 1–3 (the useful plants of Indonesia volume 1–3). Yayasan Sarana Wana Jaya. Badan Litbang Kehutanan, Jakarta

Heywood VH (2017) Plant conservation in the Anthropocene – challenges and future prospects. Plant Divers 39(6):314–330. https://doi.org/10.1016/j.pld.2017.10.004

Heywood VH, Dulloo ME (2005) *In situ* conservation of wild plant species: a critical global review of best practices. In: IPGRI technical bulletin 11. IPGRI, Rome

Hosonuma N, Herold M, De Sy V, De Fries RS, Brockhaus M, Verchot L, Romijn E (2012) An assessment of deforestation and forest degradation drivers in developing countries. Environ Res Lett 7(4):044009. https://doi.org/10.1088/1748-9326/7/4/044009

Ibrahim H, Khalid N, Hussin AK (2007) Cultivated gingers of peninsular Malaysia gardens. Bull Singapore 59(2):71–88

Itokawa H, Shi Q, Akiyama T, Morris-Natschke SL, Lee K-H (2008) Recent advances in the investigation of curcuminoids. Chin Med 3(1):11. https://doi.org/10.1186/1749-8546-3-11

IUCN (2022) The IUCN red list of threatened species. Version 2021-3. https://www.iucnredlist.org. Accessed from 13 June 2022 to 30 June 2022

Jagtap SD, Deokule SS, Bhosle SV (2006) Some unique ethnomedicinal uses of plants used by the Korku tribe of Amravati district of Maharashtra, India. J Ethnopharmacol 107(3):463–469

Jaroenkietkajorn U, Gheewala SH, Scherer L (2021) Species loss from land use of oil palm plantations in Thailand. Ecol Indic 133(December):108444. https://doi.org/10.1016/j.ecolind.2021.108444

Jayathilake HM, Prescott GW, Carrasco LR, Rao M, Symes WS (2021) Drivers of deforestation and degradation for 28 tropical conservation landscapes. Ambio 50(1):215–228. https://doi.org/10.1007/s13280-020-01325-9

Koh L, Sodhi N (2010) Conserving Southeast Asia's imperiled biodiversity: Scientific, management, and policy challenges. Biodivers Conserv 19:913–917. https://doi.org/10.1007/s10531-010-9818-9

Kress WJ, Krupnick GA (2022) Lords of the biosphere: plant winners and losers in the Anthropocene. Plants People Planet 2021:350–366. https://doi.org/10.1002/ppp3.10252

Kumar P, Singh P (2018) Conservation assessment of two rare gingers (Zingiberaceae) from Dampa Tiger Reserve, Mizoram, India. J Threat Taxa 10(5):11599–11605. https://doi.org/10.11609/jott.3797.10.5.11599-11605

Lai O (2022) Deforestation in Southeast Asia. Earth.Org. March 7, 2022. https://earth.org/deforestation-in-southeast-asia/

Lamb A, Gobilik J, Ardiyani M, Poulsen AD (2013) A guide to gingers of Borneo. Natural History Publications (Borneo), Kota Kinabalu

Larsen K, Saksuwan Larsen S (2006) Gingers of Thailand. The Botanical Garden Organization, Queen Sirikit Botanic Garden, Chiang Mai

Larsen K, Ibrahim H, Khaw SH, Saw LG (1999) Gingers of Peninsular Malaysia and Singapore. Natural History Publications (Borneo), Kota Kinabalu, p 135

Lau K (2014) The conservation of Peninsular Malaysian Geostachys (Zingiberaceae). Gardens' Bull Singapore 66(1):3–14

Leong-Škorničková J (2019) Zingiber singapurense. The IUCN red list of threatened species 2019: e.T132717094A132717102. https://doi.org/10.2305/IUCN.UK.2019-3.RLTS.T132717094A132717102.en. Accessed 27 June 2022

Leong-Škorničková J, Souvannakhoummane K (2019) Zingiber jiewhoei. The IUCN red list of threatened species 2019: e.T131722834A131722848. https://doi.org/10.2305/IUCN.UK.2019-3.RLTS.T131722834A131722848.en. Accessed 26 June 2022

Leong-Škorničková J, Sabu M, Prasanthkumar MG (2009) Transfer of Amomum fenzlii, a Nicobar Islands endemic, to Etlingera (Zingiberaceae). Gardens' Bull Singapore 60(2):381–387

Leong-Škorničková J, Tran HD, Newman MF, Dinh Quang D (2019) Curcuma vitellina. The IUCN red list of threatened species 2019: e.T201889A132688601. https://doi.org/10.2305/IUCN.UK.2019-3.RLTS.T201889A132688601.en

Leong-Skornickova J, Tran HD, Newman M, Lamxay V, Bouamanivong S (2019a) Amomum petaloideum. The IUCN red list of threatened species 2019: e.T202217A132694706. https://doi.org/10.2305/IUCN.UK.2019-3.RLTS.T202217A132694706.en. Accessed 17 June 2022

Leong-Skornickova J, Tran HD, Newman M, Lamxay V, Bouamanivong S (2019b) Curcuma candida. The IUCN red list of threatened species 2019: e.T201926A132692308. https://doi.org/10.2305/IUCN.UK.2019-3.RLTS.T201926A132692308.en. Accessed 18 June 2022

Leong-Skornickova J, Tran HD, Newman M, Lamxay V, Bouamanivong S (2019c) Curcuma rhabdota. The IUCN red list of threatened species 2019: e.T201886A132688107. https://doi.org/10.2305/IUCN.UK.2019-3.RLTS.T201886A132688107.en. Accessed 18 June 2022.

Leong-Skornickova J, Tran HD, Newman M, Lamxay V, Bouamanivong S (2019d) Curcuma supraneeana. The IUCN red list of threatened species 2019: e.T201917A132691416. https://doi.org/10.2305/IUCN.UK.2019-3.RLTS.T201917A132691416.en. Accessed 18 June 2022

Leong-Skornickova J, Tran HD, Newman M, Lamxay V, Bouamanivong S (2019e) Siliquamomum tonkinense. The IUCN red list of threatened species 2019: e.T201918A132691577. https://doi.org/10.2305/IUCN.UK.2019-3.RLTS.T201918A132691577.en. Accessed 25 June 2022

Leong-Skornickova J, Tran HD, Newman M, Lamxay V, Bouamanivong S (2019f) Zingiber collinsii. The IUCN red list of threatened species 2019: e.T201915A132691250. https://doi.org/10.2305/IUCN.UK.2019-3.RLTS.T201915A132691250.en. Accessed 26 June 2022

Leong-Skornickova J, Tran HD, Newman M, Lamxay V, Bouamanivong S (2019g) Zingiber monophyllum. The IUCN red list of threatened species 2019: e.T201913A132690971. https://doi.org/10.2305/IUCN.UK.2019-3.RLTS.T201913A132690971.en. Accessed 27 June 2022

Leverington F, Costa KL, Pavese H, Lisle A, Hockings M (2010) A global analysis of protected area management effectiveness. Environ Manag 46(5):685–698. https://doi.org/10.1007/s00267-010-9564-5

Lock M (1985) Flora of tropical East Africa-Zingiberaceae (1985) (No. 31). CRC Press, Boca Raton

Lofthus Ø (2019) Pleuranthodium peekelii. The IUCN red list of threatened species 2019: e.T117417763A124283887. https://doi.org/10.2305/IUCN.UK.2019-3.RLTS.T117417763A124283887.en. Accessed 24 June 2022

Lughadha NE, Bachman SP, Leão TCC, Forest F, Halley JM, Moat J, Acedo C, Bacon KL, Brewer RFA, Gâteblé G, Gonçalves SC, Govaerts R, Hollingsworth PM, Krisai-Greilhuber I, de Lirio EJ, Moore PGP, Negrão R, Onana JM, Rajaovelona LR, Walker BE (2020) Extinction risk and threats to plants and fungi. Plants People Planet 2(5):389–408. https://doi.org/10.1002/ppp3.10146

Malhi Y, Gardner TA, Goldsmith GR, Silman MR, Zelazowski P (2014) Tropical forests in the Anthropocene. Annu Rev Environ Resour 39:125–159. https://doi.org/10.1146/annurev-environ-030713-155141

Marsh SME, Hoffmann M, Burgess ND, Brooks TM, Challender DWS, Cremona PJ, Hilton-Taylor C, de Micheaux FL, Lichtenstein G, Roe D, Böhm M (2022) Prevalence of sustainable and unsustainable use of wild species inferred from the IUCN red list of threatened species. Conserv Biol 36(2):1–14. https://doi.org/10.1111/cobi.13844

Mathew SP (2017) On floristic diversity of Andaman-Nicobar Islands with special reference to insular germplasm conservation outside the islands. Abrahamia 3(August):104–130

Mohamad S, Kalu M (2019) Assessment of Zingiberaceae (Tribe Alpinieae) from North East Sarawak, Malaysia. IOP Conf Ser Earth Environ Sci 269(1). https://doi.org/10.1088/1755-1315/269/1/012032

Naïve MAK (2017) Zingiberaceae of Kalatungan Mountain Range, Bukidnon, Philippines. Biosci Discov 8(3):311–319

Newman M, Sangvirotjanapat S, Pullan M (2005) Zingiberaceae resource centre. Online database available from https://padme.rbge.org.uk/ZRC/

Niissalo MA, Leong-Škorničková J, Khew GS, Webb EL (2017) Very small relict populations suggest high extinction debt of gingers in primary forest fragments of a tropical city. Am J Bot 104(1):182–189. https://doi.org/10.3732/ajb.1600313

Niissalo MA, Leong-Škornicková J, Webb EL, Khew GS (2018) Pedigree analyses and next-generation sequencing reveal critically low regeneration in extremely threatened Zingiber singapurense (Zingiberaceae). Bot J Linn Soc 187(2):346–361. https://doi.org/10.1093/botlinnean/boy018

Nontasit N, Kanlayanapaphon C, Mekanawakul M, Nualmangsar O (2015) Taxonomic studies and traditional uses of zingiberaceae in khao luang national park, nakhon si thammarat province, Thailand. Walailak J Sci Technol 12(8):643–658

NPS (2022) National Park Service. Traditional Ecological Knowledge. https://www.nps.gov/subjects/tek/index.htm. Accessed 8 August 2022

Nurainas (2019) Zingiber album. The IUCN Red List of Threatened Species 2019: e.T132721576A132721585. https://doi.org/10.2305/IUCN.UK.2019-2.RLTS.T132721576A132721585.en. Accessed 26 June 2022

Nurainas, Ardiyani M (2019) Curcuma sumatrana. The IUCN red list of threatened species 2019: e.T117310829A124281750. https://doi.org/10.2305/IUCN.UK.2019-2.RLTS.T117310829A124281750.en. Accessed 18 June 2022

Olander SB (2019) Zingiber pachysiphon. The IUCN red list of threatened species 2019: e.T117468438A124284942. https://doi.org/10.2305/IUCN.UK.2019-2.RLTS.T117468438A124284942.en. Accessed 27 June 2022

Olander SB (2020a) Amomum sumatranum. The IUCN red list of threatened species 2020: e.T152357598A154696297. https://doi.org/10.2305/IUCN.UK.2020-2.RLTS.T152357598A154696297.en. Accessed 17 June 2022

Olander SB (2020b) Zingiber raja. The IUCN red list of threatened species 2020: e.T117468914A124285027. https://doi.org/10.2305/IUCN.UK.2020-2.RLTS.T117468914A124285027.en. Accessed 27 June 2022

Pitopang R, Damry R, Hamzah B, Zubair MS, Amar AL, Fathurahman F, Basri Z, Poulsen AD (2019) Diversity of Zingiberaceae and traditional uses by three indigenous groups at Lore Lindu National Park, Central Sulawesi, Indonesia. J Phys Conf Ser 1242(1):012039. https://doi.org/10.1088/1742-6596/1242/1/012039

Poulsen AD (2006) Etlingera of Borneo. Natural History Publications (Borneo) in association with Royal Botanic Garden Edinburgh, Kota Kinabalu

Poulsen AD (2012) Etlingera of Sulawesi. Natural History Publications (Borneo) Sdn. Bhd, in association with Royal Botanic Garden Edinburgh and University of Oslo Natural History Museum, Kota Kinabalu (Sabah, Malaysia), Edinburgh (United Kingdom) and Oslo (Norway)

Poulsen AD (2020) Etlingera peekelii. The IUCN red list of threatened species 2020: e.T117322216A124282292. https://doi.org/10.2305/IUCN.UK.2020-2.RLTS.T117322216A124282292.en. Accessed 21 June 2022

Poulsen AD, Olander SB (2019a) Etlingera muriformis. The IUCN red list of threatened species 2019: e.T117321804A124282252. https://doi.org/10.2305/IUCN.UK.2019-2.RLTS.T117321804A124282252.en. Accessed 21 June 2022

Poulsen AD, Olander SB (2019b) Etlingera longipetiolata. The IUCN red list of threatened species 2019: e.T117321427A124282212. https://doi.org/10.2305/IUCN.UK.2019-2.RLTS.T117321427A124282212.en. Accessed 21 June 2022

Poulsen AD, Ardiyani M, Kasuma YWC (2009) Etlingera loerzingii (Zingiberaceae) - a gorgeous torch ginger from sumatra grown in botanic gardens. Bulletin Kebun Raya Indonesia 12(1)

POWO (2022) Plants of the world online. Facilitated by the Royal Botanic Gardens, Kew. Published on the Internet. http://www.plantsoftheworldonline.org/. Retrieved 9 June 2022

Pradheeps M, Poyyamoli G (2013) Ethnobotany and utilization of plant resources in Irula villages (Sigur plateau, Nilgiri Biosphere Reserve, India). J Med Plant Res 7(6):267–276

Ragsasilp A, Saensouk P, Saensouk S (2022) Ginger family from Bueng Kan Province, Thailand: diversity, conservation status, and traditional uses. Biodiv J Biol Divers 23(5):2739–2752. https://doi.org/10.13057/biodiv/d230556

Ram Rao N (2006) Wild plant genetic resources of Godavari valley of eastern Ghats in Andhra Pradesh. Eastern Ghats Eptris-Envis Newsl 12(2):3–6

Ravindran PN, Nirmal Babu K, Sivaraman K (2007) Turmeric: the Genus Curcuma. CRC Press, Boca Raton

Robiansyah I, Pribadi DO, Helmanto H, Yuswandi AY, Wihermanto. (2021) Population and conservation status reassessment of Etlingera balikpapanensis A.D. Poulsen, an endemic ginger

in East Kalimantan. IOP Conf Ser Earth Environ Sci 762(1):012008. https://doi.org/10.1088/1755-1315/762/1/012008
Romand-Monnier F, Contu S (2013a) Globba laeta. The IUCN red list of threatened species 2013: e.T44393199A44442842. https://doi.org/10.2305/IUCN.UK.2013-1.RLTS.T44393199A44442842.en. Accessed 22 June 2022
Romand-Monnier F, Contu S (2013b) Pleuranthodium papilionaceum. The IUCN red list of threatened species 2013: e.T44392442A44509845. https://doi.org/10.2305/IUCN.UK.2013-1.RLTS.T44392442A44509845.en. Accessed 24 June 2022
Romand-Monnier F, Contu S (2013c) Curcuma caulina. The IUCN red list of threatened species 2013: e.T44392545A44510728. https://doi.org/10.2305/IUCN.UK.2013-1.RLTS.T44392545A44510728.en. Accessed 17 June 2022
Sabu M (2019) Larsenianthus arunachalensis. The IUCN red list of threatened species 2019: e.T117416792A124283657. https://doi.org/10.2305/IUCN.UK.2019-3.RLTS.T117416792A124283657.en. Accessed 23 June 2022
Saensouk P, Saensouk S (2021) Diversity, traditional uses and conservation status of zingiberaceae in Udorn Thani Province, Thailand. Biodiversitas 22(8):3083–3097. https://doi.org/10.13057/biodiv/d220801
Saensouk S, Saensouk P, Pasorn P, Chantaranothai P (2016) Diversity and uses of Zingiberaceae in Nam Nao National Park, Chaiyaphum and Phetchabun provinces, Thailand, with a new record for Thailand. Agric Nat Resour 50(6):445–453. https://doi.org/10.1016/j.anres.2016.08.002
Sakai S, Kawakita A, Ooi K, Inoue T (2013) Variation in the strength of association among pollination systems and floral traits: evolutionary changes in the floral traits of Bornean gingers (Zingiberaceae). Am J Bot 100(3):546–555. https://doi.org/10.3732/ajb.1200359
Sasikumar B (2005) Genetic resources of curcuma: diversity, characterization and utilization. Plant Genet Resour 3:230–251
Saw LG (2019) Etlingera corneri. The IUCN red list of threatened species 2019: e.T117318138A124282052. https://doi.org/10.2305/IUCN.UK.2019-3.RLTS.T117318138A124282052.en. Accessed 19 June 2022
Scriven SA, Hodgson JA, McClean CJ, Hill JK (2015) Protected areas in Borneo may fail to conserve tropical forest biodiversity under climate change. Biol Conserv 184:414–423. https://doi.org/10.1016/j.biocon.2015.02.018
Setiawan E, Ardiyani M, Miftahudin, Poulsen AD, Chikmawati T (2022) The diversity and distribution of Alpinia zerumbet clade in West Malesia. Biodiversitas 23(4):1734–1744. https://doi.org/10.13057/biodiv/d230403
Sharma TVRS, Singh S, Singh PK (2018) Crop wild relatives in Andaman and Nicobar Islands, India. Genet Resour Crop Evol 65(8):2303–2317. https://doi.org/10.1007/s10722-018-0680-z
Singh RKR, Garg A (2014) The translocated Kurdi Angod sacred site - A conservatory of RET plants of the western ghats. Curr Sci 106(1):11–12
Sodhi NS, Posa MRC, Lee TM, Bickford D, Koh LP, Brook BW (2010) The state and conservation of Southeast Asian biodiversity. Biodivers Conserv 19(2):317–328. https://doi.org/10.1007/s10531-009-9607-5
Souvannakhoummane K (2014) The conservation of Zingiberaceae in Lao PDR. In: The 3rd Xishuangbanna International Symposium Botanical Gardens and Climate, 3. https://www.researchgate.net/publication/275346612_The_Conservation_of_Zingiberaceae_in_Lao_PDR
Souvannakhoummane K (2019) Zingiber niveum. The IUCN red list of threatened species 2019: e.T117468246A124284922. https://doi.org/10.2305/IUCN.UK.2019-3.RLTS.T117468246A124284922.en. Accessed 27 June 2022
Susanti R, Zuhud EAM (2019) Traditional ecological knowledge and biodiversity conservation: the medicinal plants of the Dayak Krayan people in Kayan Mentarang National Park, Indonesia. Biodiversitas 20(9):2764–2779. https://doi.org/10.13057/biodiv/d200943
The National Development Planning Agency (2003) Indonesian biodiversity strategy and action plan national document. The National Development Planning Agency

Thomas CD (2013) The Anthropocene could raise biological diversity. Nature 502(7469):7. https://doi.org/10.1038/502007a

Thomas CD (2020) The development of Anthropocene biotas. Philos Trans R Soc B Biol Sci 375: 1794. https://doi.org/10.1098/rstb.2019.0113

Toyat JA, Abdullah AN, Go R, Abdullah TL, Saleh G, Jiwan M, Kundat FR, Magiman MM (2015) Conservation strategies for jerangau merah (Boesenbergia stenophylla) using DNA profiling and micropropagation. In: Proceedings - Kuala Lumpur International Agriculture, Forestry and Plantation. Hotel Putra, Kuala Lumpur

Tran HD, Leong-Škorničková J (2019) Curcuma sahuynhensis. The IUCN red list of threatened species 2019: e.T131724352A131724370. https://doi.org/10.2305/IUCN.UK.2019-3.RLTS.T131724352A131724370.en. Accessed 18 June 2022

Tran HD, Sabu M, Sangvirotjanapat S (2019) Boesenbergia siphonantha. The IUCN red list of threatened species 2019: e.T117304453A124281205. https://doi.org/10.2305/IUCN.UK.2019-3.RLTS.T117304453A124281205.en. Accessed 17 June 2022

Tuan DQ, Dien D, Pham TV, Le TQ, Ho DV, Nhan LT, Anh LT, Huong PT, Hoai NT (2021) Chemical composition of essential oil from the rhizomes and leaves of Newmania Sontraensis H.Đ.Trân, Luu & Škornick (Zingiberaceae) from Vietnam. J Essent Oil Bearing Plants 24(6): 1260–1268. https://doi.org/10.1080/0972060X.2022.2026253

Tyagi DK (2005) Pharma forestry: field guide to medicinal plants. Atlantic, New Delhi

Ud-Daula AFM, Basher MA (2019) Genus Etlingera - a review on chemical composition and antimicrobial activity of essential oils. J Med Plant Res 13(7):135–156. https://doi.org/10.5897/jmpr2019.6740

Volis S, Blecher M (2010) *Quasi in situ*: a bridge between *ex situ* and *in situ* conservation of plants. Biodivers Conserv 19(9):2441–2454. https://doi.org/10.1007/s10531-010-9849-2

Wagh VV, Jain AK (2013) Status of threatened medicinal plants of Jhabua district, Madhya Pradesh. Ann Plant Sci 2(10):395–400

Waman AA, Bohra P, Mane S (2018) Conservation and utilization of wild relatives of important spices and plantation crops in Andaman and Nicobar Islands, India- an overview. Curr Agric Res J 6(3):320–327. https://doi.org/10.12944/carj.6.3.10

Watson JE, Dudley N, Segan DB, Hockings M (2014) The performance and potential of protected areas. Nature 515(7525):67–73. https://doi.org/10.1038/nature13947

WCVP (2022) World checklist of vascular plants, version 2.0. Facilitated by the Royal Botanic Gardens, Kew. http://wcvp.science.kew.org/. Retrieved 15 August 2022

West Demmy E, Burch D (1998) Gingers - a rising Star in Florida gardens. Proc Florida State Hortic Soc 111:190–191

Wilcove DS, Giam X, Edwards DP, Fisher B, Koh LP (2013) Navjot's nightmare revisited: logging, agriculture, and biodiversity in Southeast Asia. Trends Ecol Evol 28(9):531–540. https://doi.org/10.1016/j.tree.2013.04.005

WWF (2017) The environmental status of Borneo 2016 report. 16.http://wwf.panda.org/downloads/wwf_hob_measures_report_2012_final_for_web.pdf

Yeats H (2013) The history and cultivation of Etlingera - the torch gingers - at the Royal Botanic Garden Edinburgh. Sibbaldia 11:71–86

Zeng Z, Estes L, Ziegler AD, Chen A, Searchinger T, Hua F, Guan K, Jintrawet A, Wood F, E. (2018) Highland cropland expansion and forest loss in Southeast Asia in the twenty-first century. Nat Geosci 11(8):556–562. https://doi.org/10.1038/s41561-018-0166-9

The manufacturer's authorised representative in the EU is Springer Nature Customer Service Centre GmbH, Europaplatz 3, 69115 Heidelberg, Germany. If you have any concerns regarding our products, please contact ProductSafety@springernature.com

Printed and bound by CPI Group (UK) Ltd, Croydon, CR0 4YY
01/10/2026
02225868-0019